MANAGING QUALITY:
Integrating the Supply Chain

THIRD EDITION

MANAGING QUALITY:

Integrating the Supply Chain

S. Thomas Foster

Brigham Young University

UPPER SADDLE RIVER, NEW JERSEY 07458

Library of Congress Cataloging-in-Publication Data

Foster, S. Thomas.
Managing quality : integrating the supply chain / S. Thomas Foster.—3rd ed.
p. cm.
Includes bibliographical references and index.
ISBN 0-13-220644-7
1. Quality of products. 2. Quality control. I. Title.
HF5415.157.F67 2007
658.4′013—dc22

2006009189

AVP/Executive Editor: Mark Pfaltzgraff
VP/Editorial Director: Jeff Shelstad
Product Development Manager: Pamela Hersperger
Editorial Assistant: Barbara Witmer
Product Development Manager, Media: Nancy Welcher
AVP/Executive Marketing Manager: Debbie Clare
Associate Director, Production Editorial: Judy Leale
Senior Managing Editor, Production: Cynthia Regan
Production Editor: Denise Culhane
Permissions Coordinator: Charles Morris
Associate Director, Manufacturing: Vinnie Scelta
Manufacturing Buyer: Michelle Klein
Design/Composition Manager: Christy Mahon
Cover Design: Kiwi Design
Cover Illustration/Photo: Steven Puetzer / Photonica / Getty Images, Inc.
Director, Image Resource Center: Melinda Reo
Manager, Rights and Permissions: Zina Arabia
Manager, Visual Research: Beth Brenzel
Manager, Cover Visual Research & Permissions: Karen Sanatar
Image Permission Coordinator: Nancy Seise
Composition/Full-Service Project Management: GGS Book Services
Printer/Binder: Hamilton Printing
Typeface: 10/12 Times Ten Roman

Credits and acknowledgments borrowed from other sources and reproduced, with permission, in this textbook appear on appropriate page within the text.

Pearson Education LTD.
Pearson Education Singapore, Pte. Ltd
Pearson Education, Canada, Ltd
Pearson Education–Japan
Pearson Education Australia PTY, Limited
Pearson Education North Asia Ltd
Pearson Educación de Mexico, S.A. de C.V.
Pearson Education Malaysia, Pte. Ltd

10 9 8 7 6 5 4 3 2
ISBN 0-13-220644-7

To my wife,
Casie (MEC); my children, Kimberlee, Amie, Stephen, Daniel,
Matthew, Josh, and Rocky; my father, Steve; and my grandchildren, Kiley and Riley
In Loving Memory of
Jeanne M. Foster

Brief Contents

Contents

Preface

Welcome to the third edition *of Managing Quality: Integrating the Supply Chain.* You will notice that there has been a small but important change to this title. We are using the theme of supply chain management as a unifying theme for quality improvement. Previous adopters of *Managing Quality* will note that the coverage of quality topics is just as comprehensive as ever. We simply adopt the unifying theme of the supply chain to enhance our emphasis on the integration of systems with customers, suppliers, technology, and people. This is in response to changes in the marketplace and our customers—you! We think that you will find that your customers—the students—will find this change makes your quality management course ever more relevant and interesting. Of course, the new edition of the text has been updated with many changes to keep our coverage of quality topics on the cutting edge. In addition, chapter highlights have been increased by 50% and end-of-chapter problems by 100%. Again, our adopters asked for these changes and we have responded. Following are the core values that permeate this text:

MAJOR THEMES

Supply Chain as a Unifying Theme Today's firms are evermore focused on improving supply chain performance. Key to this improvement is quality management. As we look upstream, we need to develop our suppliers. Downstream, we focus on customer service and after-sales service. Implicit in this process is service design. In your classes, you can drive these concepts home by emphasizing the systems view implicit in supply chain management. This unifying theme provides a linkage between the roots of quality management (Shewhart and Deming) with new developments such as Six Sigma and service quality. *For clarification, this is not a supply chain management text. This is a quality management text that utilizes supply chain management as a unifying theme.*

Integrative Approach Workers and managers in organizations are somewhat limited by their particular functional preparation and specialization (going back to their educational training). This narrow presentation filters how they analyze and cognitively interpret information. However, quality management has emerged as a discipline that is not owned by any of the functional areas such as operations management, supply chain management, human resources, or marketing. We all have to work together to satisfy customers.

Contingency Approach This is a concept we have emphasized for a long time that is beginning to get traction in the research and practitioner literature. We passionately believe that the future of quality management will involve learning the contingencies associated with managing quality. There is no "one way" or "magic pill" that companies can implement to improve quality. Therefore, the contingency approach is used to instruct students how to assess the current position of the firm and identify an effective strategy for improvement based on a profound understanding of their company, market, customers, etc. Thus, improvement is based on the contingent variables that are operative in the firm as it exists. This contingency approach is introduced in Chapter 1 and permeates the rest of the text.

The author and over 300 universities around the world have successfully taught quality management using this contingency approach. The contingency approach coupled with the unifying theme of the supply chain will make this pedagogically even more powerful. To effectively manage quality, a few conditions must be present—students must understand their businesses, understand the quality body of knowledge, understand the available tools, and have a method for planning quality based on this knowledge. This text provides a basis for accomplishing this—when combined with an instructor's insight.

NEW FEATURES

- New title: Supply chain management is used as a unifying theme to help students understand the systems view needed to improve quality.
- Increased coverage of customer relationship management systems (CRM).
- Added coverage of Covey's eight habits. So many companies have adopted this approach that we felt that students should be familiar with Covey.
- More emphasis on performance metrics and measurement—how to do it right and how not to do it wrong.
- Section on supply chain strategy.
- New discussion of quality management in China—in addition to the United States, Europe, and Japan.
- The Baldrige discussion has been updated to the 2006 standard.
- Cost analysis of warranties—from both the producer's and consumer's perspectives.
- Service Transaction Analysis topic coverage.
- New tools such as rainbow charts, spider charts, force-field analysis, and dashboarding.
- Increased discussion of process mapping and extended processing mapping.
- Replaced QS9000 coverage with an in-depth discussion of ISO/TS 16949 as a model for assessing supplier quality.
- More discussion of form versus substance in quality implementation. This is needed in a time of quality hucksterism.
- 40% more Quality Highlights and A Closer Look at Quality boxes.
- 100% more end-of-chapter problems.
- This is the most up-to-date, cutting-edge textbook on the market.
- Many other changes.

CHANGES TO THIS EDITION

Active Models There are interactive Excel spreadsheets in the accompanying CD-ROM that correspond to examples in Chapters 12 and 13 and that allow the student to explore and better understand important quantitative concepts. Students or instructors can adjust inputs to the model and, in effect, can answer a whole series of "what if" questions that are provided (e.g., What if variation in the process changes? What if the process indicates changes are needed? What if we change the sample size?). These Active Models are great for classroom presentation and/or homework.

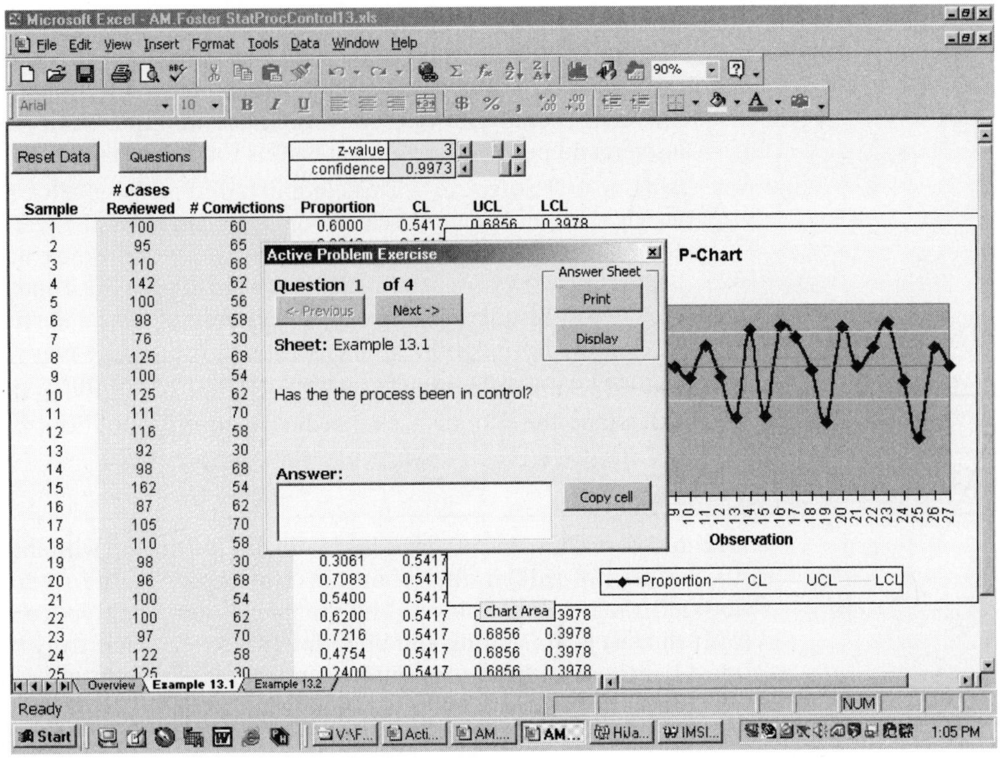

FOR THE STUDENT

Companion Web Site By logging on to *www.prenhall.com/foster*, students will be able to find the following resources by chapter: online quizzes, direct links to company web sites, quality tools, and knowledgebase topics.

Student CD-ROM A CD-ROM, with the following, is in the back of each copy of the book.

- **Excel Files**—for examples in Chapters 12, 13, and 14.
- **Active Models**—for selected textbook examples.
- **Video Clips**—selected video clips (one to two minutes in length) that illustrate chapter-related topics.
- **Excel Quality**—plug-in files for selected examples in the text.

FOR THE INSTRUCTOR

Besides the changes and additions to the text, we've made substantial revisions to the support materials for this book.

Instructor's Resource Center A password protected site at www.prenhall.com/foster gives instructors access to the following resources:

- **Instructor's Manual**— This manual, created by Howard Flomberg of Metropolitan State College of Denver, includes solutions to practice problems, case study questions, sample syllabi, and teaching tips.

- **PowerPoint Presentations—** A set of PowerPoint presentations, also created by Howard Flomberg, is available for each chapter.
- **Test Item File—** The Test Item File, updated by Bill Roach of Washburn University, contains a variety of true/false, multiple-choice, fill-in-the-blank, short answer, and problem-solving questions for each chapter.
- **TestGen-EQ Software—** This computerized package allows instructors to custom design, save, and generate classroom tests based on the test bank questions provided in the Test Item File. The test program permits instructors to edit, add, or delete questions from test banks; edit existing graphics and create new graphics; analyze test results; and organize a database of tests and student results. This software allows for flexibility and ease of use. It provides many options for organizing and displaying tests, along with a search and sort feature.

ACKNOWLEDGMENTS

The author wishes to first thank his family for putting up with the fences left unmended, the horses not ridden, the mountains not explored, the rivers not rafted, the snow unskied, the music not played, and the many hours spent in front of a computer screen over the last years writing and updating this book. It has truly been a labor of love for me. I believe that these concepts are important for the future of the nations.

I wish to thank my parents, who always emphasized the importance of education as a means of achieving a happy life. I thank Everett E. Adam, Jr., for mentoring me. I wish to thank my colleagues at Boise State University and my new colleagues at Brigham Young University for providing encouragement for this project. Pat Shannon has been a wonderful friend and mentor, I have learned a great deal from him. Dave Groebner, Phil Fry, Robert Minch, and Gary Green have been especially helpful. Thanks to all of my students and those individuals who have hired me as a consultant. These people have helped me to pursue lifelong learning.

I also wish to thank the following reviewers for their thoughtful comments: Scott A. Dellana, East Carolina University; Dana Johnson, Michigan Technological University; David Lewis, University of Massachusetts at Lowell; Dave Magee, Lexington Community College; Patricia Nemetz Mills, Eastern Washington University; and Rathel R. Smith, Southwest Missouri State University.

Alana Bradley, my editor at Prentice Hall, deserves recognition for the great encouragement she has given to me. Finally, I am thankful for my faith that keeps me progressing eternally.

About the Author

Dr. Tom Foster is a professor, researcher, and consultant in the field of quality management. Among Dr. Foster's areas of expertise are strategic quality planning, service quality, Six Sigma, government quality, and the role of technology in improving quality. Tom is a professor of quality and supply chain management at Brigham Young University and has also taught at Pennsylvania State University and Boise State University. He received his PhD from the University of Missouri-Columbia.

Dr. Foster has professional experience in manufacturing, financial services operations, and international oil exploration. He has consulted for over 20 companies, including Trus Joist MacMillan, the United States Department of Energy, Hewlett-Packard, Heinz Frozen Food, and Cutler Hammer/Eaton Corporation. Dr. Foster served on the 1996 and 1997 boards of examiners for the Malcolm Baldrige National Quality Award and has served as a judge for state awards.

Tom is on the editorial boards of the *Journal of Operations Management, the Quality Management Journal, Benchmarking: An International Journal*, and the *Quality Observer*. He has published over 40 quality-related research articles in journals such as *Decision Sciences*, the *International Journal of Production Research*, the *Quality Management Journal*, and *Quality Progress*. He is listed in *Who's Who in America* and *Who's Who in the World*. Dr. Foster is founder of *www.freequality.org*, was awarded the ASBSU Outstanding Faculty Award, and served as guest editor for the *Journal of Operations Management* special issue on Supply Chain Management. In addition, he was winner of the 2002 Decision Sciences Institute Innovative Education Award.

Tom has five childern, two grandchildren, and is married to the former Casie Pratt. In his spare time, he skis, enjoys the Rocky Mountains, and plays his Fender Stratocaster.

PART ONE

Understanding Quality Concepts

To understand quality in the supply chain, we need a common language. In the general public, the language of quality is imprecise and inconsistent. The language of quality professionals is much more precise and consistent.

To understand the advanced concepts in the later chapters, in Chapters 1 through 3 we build a conceptual foundation of quality theory. This forms the basis of the contingency approach. To apply quality improvement on a contingent basis, one needs to understand the foundation that has been laid by leaders in the quality movement such as W. Edwards Deming, Joseph Juran, Philip Crosby, Kaoru Ishikawa, and others. These people have made huge contributions to the well-being of the world, and a knowledge of their teachings and ideas is necessary for quality application.

In Chapter 3 we consider important frameworks, such as ISO 9000:2000, the Deming Prize, and the Baldrige criteria. These provide models for improvement that are being used in many countries around the world.

CHAPTER 1

Differing Perspectives on Quality

You know a dream is like a river
Ever changing as it flows.
—GARTH BROOKS

RECOGNIZING DIFFERENT PERSPECTIVES ON QUALITY

Quality management involves flows. There are process flows, information flows, material flows, and flows of funds. Each of these flows has to operate effectively, efficiently, and with outstanding quality. Like a river, we refer to upstream flows and downstream flows. The sums of these flows construe the supply chain.

Considering the **supply chain** causes us to think about quality differently. One of the problems with quality efforts has been that they tend to be too internally oriented. The supply chain causes us to expand our vision as we *internalize* processes that had previously been *externalized*. These include **upstream** processes relating to our dealing with suppliers—negotiating, selecting, and improving supplier performance—and **downstream** processes—delivering products and services and serving customers.

The supply chain encompasses many differing functions and processes. It includes all of the core activities from the raw materials stage to after-sales service. To execute all of these processes correctly involves integrating differing functions, expertise, and dimensions of quality. This need for integration increases the need for flexible, cross-functional, problem-solving and employees who can adapt to rapidly changing markets.

There are many different definitions and dimensions of quality in the supply chain. We will present several of these definitions and dimensions in this chapter. For the present, you should view quality as a measure of goodness that is inherent to a product or service. Employees working for the same firm often view quality differently. Think of the different functions that are involved in creating products and services. These include design engineering, marketing, operations, cost accounting, financial management, and others throughout the supply chain. A product design engineer might feel that customer satisfaction is mostly influenced by product design and product attributes and might take great pains to design a product that satisfies the customer. However, the product also needs to satisfy marketing's need for quick design cycle times and accounting's need for low-cost products. So perceptions differ on a variety of levels, including what our goals for the product or service are. A Closer Look at Quality 1-1 illustrates this point by comparing people's perceptions of differing musical technologies. The discussion shows just how perceptions of quality can vary. Although CDs have many technological advantages, including improved durability and less surface noise, many consumers and performers long for the intangible "warmth" of LPs.

A CLOSER LOOK AT QUALITY 1-1

WHICH ARE BETTER, CDS OR LPS?

In the early 1980s, the introduction of the digital compact disc (CD) revolutionized the music world. The predecessor of the CD, the long-playing record (LP), was made of pressed vinyl that was subject to wear after repeated use. CDs were much more *durable* than LPs. LPs had a much greater tendency for surface noise, and many closets were filled with old, scratchy LP records that owners were loath to throw away. LP records also had limited capacity. For example, if too many songs were pressed on a record, the record would be more likely to skip, because the groove in the record was too narrow for the stylus (record player needle). CDs were capable of holding much more music for the same price.

In 1982, Sony and Philips launched the CD as the "perfect sound forever." However, music lovers complained that CDs sounded too sterile and lacked the "warmth" of the LPs. Jazz pianist Keith Jarrett complained "CDs lose the subtlety where expression lies." An early adopter of digital technology, musician Neil Young, lamented "digital was a disaster." He stated that digital music was "an insult to the brain and heart and feelings."

These feelings fueled a move away from digital CDs and back to LPs among some customers and performers. For example, the popular rock group Pearl Jam from Seattle, Washington, released a recording in vinyl LP format only. Several reissues of recordings by classic blues, jazz, rock, and rap artists such as Muddy Waters, Albert Collins, Johnny Copeland, Koffee Brown, The Manhattan Transfer, Aerosmith, and the Buzzcocks have been marketed in vinyl.

CD technology is being supplanted by digital video disc (DVD) technology with up to 17 times the storage capacity of the CD, improved video capability, and better-quality audio as well as MP3 and other digital technologies. What is clear is that the meaning of quality varies drastically from person to person. It is left to each of us to decide. Which do we like best?

Perceptions affect every aspect of our world—including the business world. In order to communicate effectively about quality, managers need to recognize that differences in perceptions of quality exist. Although this observation may not seem too startling, many managers have strong opinions about what quality is. Sometimes these opinions can be at variance with the beliefs of the majority of their customers. This may hurt the competitiveness of a firm. For that reason, in this chapter we study quality from a variety of perspectives. Later we provide a means for recognizing and resolving differences in perception. Finally, we introduce contingency view of quality management that we emphasize throughout this book.

WHAT IS QUALITY?

If you ask ten people to define quality, you probably will get ten definitions.

Product Quality Dimensions

There are several definitions of quality, or **quality dimensions**. One of the most respected collections of quality dimensions was compiled by David Garvin[1] of the Harvard Business School. Garvin found that most definitions of quality were either **transcendent, product-based, user-based, manufacturing-based**, or **value-based.** What does each of these terms mean?

[1]Garvin, D., "What Does 'Product Quality' Really Mean?" *Sloan Management Review* (Fall 1984):25–43.

Transcendent: Quality is something that is intuitively understood but nearly impossible to communicate, such as beauty or love.

Product-based: Quality is found in the components and attributes of a product.

User-based: If the customer is satisfied, the product has good quality.

Manufacturing-based: If the product conforms to design specifications, it has good quality.

Value-based: If the product is perceived as providing good value for the price, it has good quality.

Using these five definitions of quality, Garvin developed a list of eight quality dimensions (see Table 1-1). These dimensions describe product quality specifically in the following paragraphs.

Performance refers to the efficiency with which a product achieves its intended purpose. This might be the return on a mutual fund investment, the fuel efficiency of an automobile, or the acoustic range of a pair of stereo speakers. Generally, better performance is synonymous with better quality.

Features are attributes of a product that supplement the product's basic performance. These include many of the "bells and whistles" contained in products. A visit to any television or computer retail store will reveal that features, such as surround sound, HDTV capability, plasma, and size, are powerful marketing tools for which customers will pay a premium. A full-line television retail store may carry televisions priced from $200 to $12,000. This range represents a 6,000% price premium for additional features!

Reliability refers to the propensity for a product to perform consistently over its useful design life. A subfield in quality management has emerged, called *reliability management*, based on the application of probability theory to quality. A product is considered reliable if the chance that it will fail during its designed life is very low. If a refrigerator has a 2% chance of failure in a useful life of 10 years, we say that it is 98% reliable.

Conformance is perhaps the most traditional definition of quality. When a product is designed, certain numeric dimensions for the product's performance will be established, such as capacity, speed, size, durability, or the like. These numeric product dimensions are referred to as *specifications*. The number of ounces of pulp allowed in a half-gallon container of "pulp-free" orange juice is one example. Specifications typically are allowed to vary a small amount called a *tolerance*. If a particular dimension of a product is within the allowable range of tolerance of the specification, it conforms.

The advantage of the conformance definition of quality for products is that it is easily quantified. However, it is often difficult for a service to conform to numeric specifications. For example, imagine trying to measure the quality of a counselor's work

TABLE 1-1 Garvin's Product Quality Dimensions

Garvin's Product Quality Dimensions
Performance
Features
Reliability
Conformance
Durability
Serviceability
Aesthetics
Perceived quality

SOURCE: Reprinted from "What Does Product Quality Really Mean?" by D. Garvin, *Sloan Management Review*, Fall 1984, by permission of publisher. Copyright 1984 by Sloan Management Review Association.

versus that of a carmaker. Because counseling is intangible, it is almost impossible to measure.

Durability is the degree to which a product tolerates stress or trauma without failing. An example of a product that is not very durable is a light bulb. Light bulbs are damaged easily and cannot be repaired. In contrast, a trash can is a very durable product that can be subjected to much wear and tear.

Serviceability is the ease of repair for a product. A product is very serviceable if it can be repaired easily and cheaply. Many products require service by a technician, such as the technician who repairs your personal computer. If this service is rapid, courteous, easy to acquire, and competent, then the product generally is considered to have good serviceability. Note that different dimensions of quality are not mutually exclusive.

Aesthetics are subjective sensory characteristics such as taste, feel, sound, look, and smell. Although vinyl interiors in automobiles require less maintenance, are less expensive, and are more durable, leather interiors generally are considered more aesthetically pleasing. In terms of aesthetics, we measure quality as the degree to which product attributes are matched to consumer preferences.

Perceived quality is based on customer opinion. As we said in the beginning of this chapter, quality is as the customer perceives it. Customers imbue products and services with their understanding of their goodness. This is perceived quality. We can witness an example of the effect of perceived quality every year in college football polls that rank teams. In many cases, the rankings are based on past records, team recognition, tradition of the university, and other factors that are generally poor indicators of team quality on a given Saturday. In the same way that these factors affect sportswriters' perceptions, factors such as brand image, brand recognition, amount of advertising, and word of mouth can affect consumers' perceptions of quality.

The Garvin list of quality dimensions, although it is the most widely cited and used, is not exhaustive. Other authors have proposed lists of additional quality measures, such as safety. Carol King[2] identified dimensions of service quality such as *responsiveness, competence, access, courtesy, communication, credibility, security*, and *understanding*. Allowed time, you probably could think of additional dimensions as well.

Service Quality Dimensions

Service quality is even more difficult to define than product quality. Although services and production share many attributes, services have more diverse quality attributes than products. This often results from wide variation created by high customer involvement. For example, the consumer of a fountain pen probably will not care that the factory worker producing the pen was in a foul mood (as long as the quality of the pen is good). However, excellent food served in a restaurant generally will not suffice if the server is in a foul mood. In addition, a consumer probably will not consider a pen poor quality if he or she is in a bad mood when using the pen. However, food and service in a restaurant could be excellent and still be perceived poorly if the patron is feeling badly.

Parasuraman, Zeithamel, and Berry, three marketing professors from Texas A&M University, published a widely recognized set of service quality dimensions. These dimensions have been used in many service firms to measure quality performance. The Parasuraman, Zeithamel, and Berry dimensions are defined here (see Table 1-2).

[2]King, C., "A Framework for a Service Quality Assurance System," *Quality Progress* 20, 9 (1987):27–32.

The organic view of the organization sees the whole as the sum of different parts uniting to achieve an end. The heart and the liver do not perform the same function in a body, but they each perform processes that are necessary for survival of the whole. Just as the body is subject to breakdown when different parts do not perform properly, so are organizations. Unfortunately, firms do not have the magnificent communication network (i.e., the central nervous system) to coordinate activities that human bodies have. For this reason, firms must constantly improve their communication. Recognizing fundamental differences between how different functions view quality is an important first step in understanding and resolving problems associated with mismatches of quality perceptions within organizations.

As organizational processes become more cross-functional, many of these communications issues will find resolution. However, experience with cross-functional teams has been difficult for many firms because of poor communication skills among team members. Therefore, it is expected that cognitive differences between different functions will continue to be a major problem that firms must overcome.

This section of the chapter views quality management from the perspectives of several different functions. Many of the topics discussed in this chapter are presented in concept only. More in-depth discussions of these topics appear in later chapters. This chapter is designed to lay out the field of quality management from an interdisciplinary, integrative perspective. The functions that are discussed here include supply chain management, engineering, operations, strategic management, marketing, finance/accounting, human resources, and management information systems.

A Supply Chain Perspective

Supply chain management grew out of the concept of the value chain. The value chain includes **inbound logistics**, **core processes**, and **outbound logistics**. Other functions such as human resources, information systems, and purchasing support these core processes. Operations, logistics, and marketing are the primary participants in the supply chain. In recent years, supply chain management has moved to the forefront in importance. This is largely due to the opportunity for cost savings along with quality and service improvement. There are many important quality-related activities that are part of supply chain management. We will discuss these separately as upstream activities, core processes, and downstream activities.

Upstream activities include all of those activities involving interaction with suppliers. **Supplier qualification** involves evaluating supplier performance to determine whether or not they are worthy providers. This often requires grading suppliers using established criteria such as conformance rates, cost levels, and delivery reliability. Many times, **supplier filters** are used such as **ISO 9000:2000**, an international standard. This means that you can filter suppliers based on whether or not they are ISO 9000:2000 registered. Related standards include **QS9000** and **ISO/TS 16949** (an automotive standard). **Supplier development** activities include evaluating, training, and implementing systems with suppliers. This often includes the use of **electronic data interchange (EDI)** to link customer purchasing systems to supplier enterprise resource planning systems. Where needed, **acceptance sampling** may be needed to determine whether supplier products meet requirements. **International sourcing** is an important supply chain issue with many companies—especially in China. This will be discussed in more depth in Chapter 3.

Core process activities include traditional process improvement as well as **value stream mapping**. This requires flowcharting processes to determine where customer value is created as well as identifying non-value-added process steps. Value stream mapping also involves analyzing processes from a systems perspective such that upstream and downstream effects of core process changes can be evaluated. **Six sigma** is a procedure for implementing quality improvement analysis to reduce costs and improve product, service, and process design. Six sigma black belts become supply chain quality consultants who can lead value-adding improvements. The steps in six sigma include **define, measure, analyze, improve, and control (DMAIC)**-related activities. A major tool used in six sigma is the **design of experiments (DOE)**.

Downstream activities include shipping and logistics, customer support, and focusing on delivery reliability. Supply chain management has also focused more attention on **after-sale service**.

An Engineering Perspective

Engineering is an applied science. As such, engineers are interested in applying mathematical problem-solving skills and models to the problems of business and industry. One outgrowth of this approach is the field of operations research. In the early twentieth century, Sir R. A. Fisher and other researchers in England expanded the field of mathematical statistics to problems related to variation experienced in the production area.

Engineers typically are employed in manufacturing production environments. However, more and more engineers are being hired into services firms requiring a strong technical component. For example, American Airlines employs a large operations research engineering staff that optimizes flight schedules.

Two of the major emphases in engineering are the areas of product design and process design. **Product design engineering** involves all those activities associated with developing a product from concept development to final design and implementation. Figure 1-1 demonstrates the six steps in the engineering life cycle for the design of products. The product design process results in a final design, possibly generated by a computer-aided design (CAD) system. Product design is the key because quality is assured at the design stage.

Product and process design are fields of engineering that have experienced major changes in recent years. Whereas traditionally they have been considered separate and in most cases sequential activities, **concurrent engineering** has resulted in the simultaneous performance of these activities. Typically, concurrent engineering involves the formation of cross-functional teams. This allows engineers and managers of differing disciplines to work together simultaneously in developing product and process designs. The result of concurrent design has been improved quality and faster speed to market for new products.

Engineers also have applied statistical thinking to the problem of *reliability*. As already discussed, reliability management is concerned with assessing and reducing the propensity of a product to fail. Reliability engineers use probability theory to determine the rate of failure a product will experience over its useful life. **Life testing** is a facet of reliability engineering that concerns itself with determining whether a product will fail under controlled conditions during a specified life. Also, reliability engineers are interested in knowing if failure of certain product components will result in failure of the overall product. If a component has a relatively high probability for failure that will affect the overall function of a product, then **redundancy** is applied so that a backup system can take over for the failed primary system. Many redundant systems are used on the NASA space shuttle in the case of primary system breakdown. After all, if a hard drive crashes in space, it is not easy to find a replacement close by.

FIGURE 1-1 Design Life Cycle

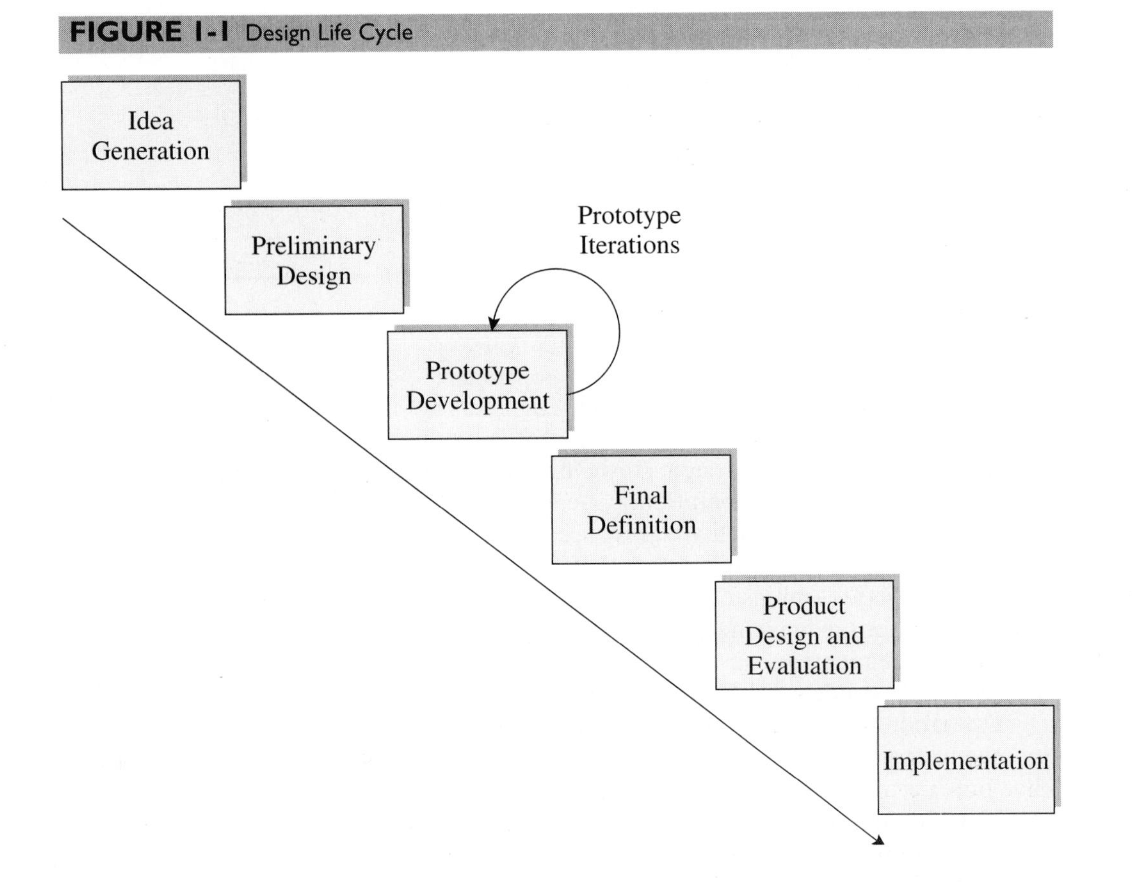

Another engineering-related contribution to quality management is the field of **statistical process control (SPC).** Statistical process control is concerned with monitoring process capability and process stability. If a process is capable, it will consistently produce products that meet specification. If a process is stable, it will only exhibit random or common variation. This type of variation is often acceptable, if kept within limits. The control process as specified by Shewhart in his book *Statistical Method from the Viewpoint of Quality Control* (1939) is shown in Figure 1-2. This is the process underlying SPC. A hypothesis is specified that the process meets a given specification, data about the process are gathered, and a hypothesis test is performed to see if the process is stable. SPC is discussed in greater depth in Chapters 12, 13, and 14.

In summary, the engineering view of quality is technically oriented, focusing on statistics and technical specification that are needed to produce high-quality products. Only recently have engineers begun to interact with customers in meaningful ways.

An Operations Perspective

The operations management view of quality is rooted in the engineering approach. However, operations management has grown beyond the technical engineering perspective. In many ways, because of the close interplay between operations management and engineering, engineering also has extended its view of quality management.

Operations was the first functional field of management to adopt quality as its own. This effort began in the mid- to late 1970s and soon was integrated into other functional areas of management. Like engineers, operations managers are concerned

FIGURE 1-2 Shewhart's Control Process

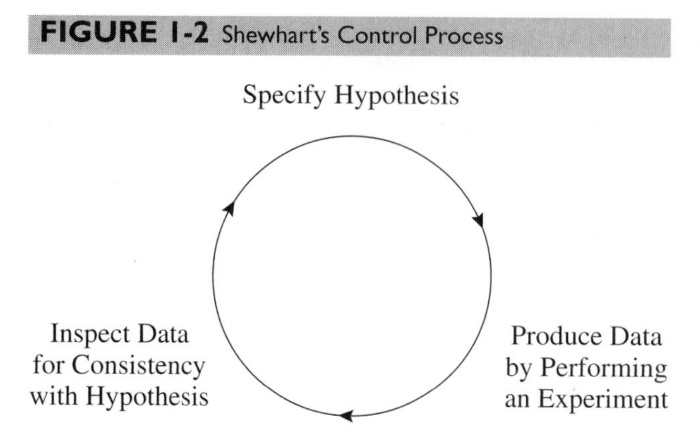

about product and process design. However, rather than focusing on only the technical aspects of these activities, operations concentrates on the management of these activities. Initially, operations quality was focused almost entirely on SPC. Beginning in the early 1980s, statistical quality control (SQC) courses became more managerial in nature, including teachings by W. E. Deming, an important quality expert, and others. Today, operations management has developed into an integrative field combining concepts from engineering, operations research, organizational theory, organizational behavior, and strategic management to address quality problems.

Operations management (OM) uses the **systems view** that underlies modern quality management thinking (see Fig. 1-3). The systems view involves the understanding that product quality is the result of the interactions of several variables, such as machines, labor, procedures, planning, and management. OM focuses on the management and continual improvement of conversion processes. This systems view focuses on interactions between the various components (i.e., people, policies, machines, processes, and products) that combine to produce a product or service. The systems view also focuses management on the *system as the cause of quality problems*.

In recent years, a major advance in operations management has been the improved understanding of the operations–marketing interface. This interface has resulted in an increased focus on the customer. For example, many firms, such as Harley-Davidson and John Deere, have involved their customers in product design by including the customers in product design teams. This has helped operations managers externalize their views to the customer as well by making the customers part of the design process. This outsider, or externalized, view is important because operations managers in firms still tend to be focused heavily on meeting production schedules, sometimes at the expense of good quality. (It is a sign of maturity in operations managers when they are willing to miss a shipment deadline because quality isn't right. Too often manufacturing managers are still willing to say "ship it" when they know that the product they are shipping is defective.)

Operations management has migrated toward a more strategic view. Ferdows and Demeyer[3] linked this strategic view of OM to quality management by proposing a model (see Fig. 1-4) in which quality was identified as the base on which lasting improvement in other competitive dimensions (dependability, speed of delivery of

[3]Ferdows, K., and Demeyer, A., "Lasting Improvement in Manufacturing Performance: In Search of a New Theory," *Journal of Operations Management* 9, 2 (1990):168–184.

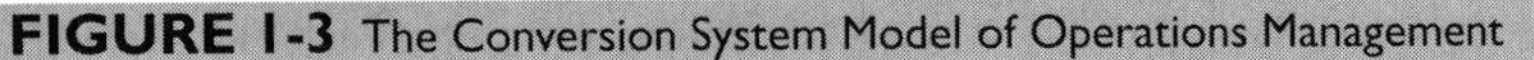

FIGURE 1-3 The Conversion System Model of Operations Management

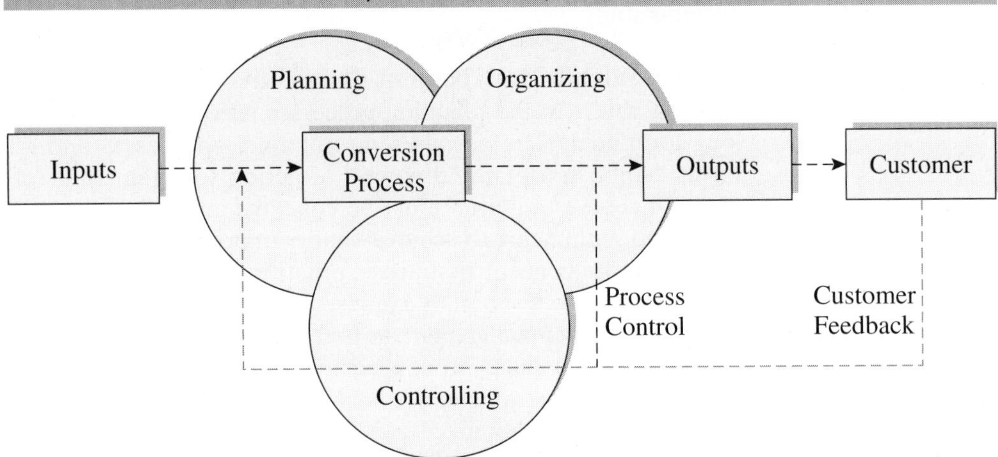

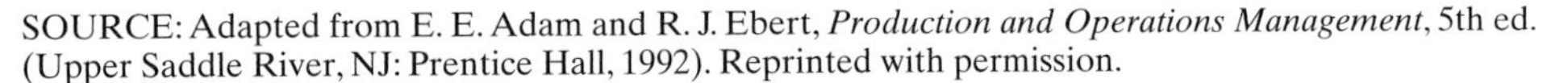

SOURCE: Adapted from E. E. Adam and R. J. Ebert, *Production and Operations Management*, 5th ed. (Upper Saddle River, NJ: Prentice Hall, 1992). Reprinted with permission.

concept to market, and cost efficiency) were established. For example, service and product quality are important foundations for firms, such as Wal-Mart, that desire to be low-cost competitors. This strategic view also has led to a better understanding of the relationship between quality and other competitive variables such as profitability, cost leadership, and operational success. Quality is a significant variable for predicting a firm's success.

In summary, the historically, internalized view of operations managers has become externalized. Still, the customer needs to become more central in the thinking of many operations managers who still tend to be too product focused. This will occur as operations management becomes more service focused. One common complaint among critics of operations management is that too much credence is given to fads of the day rather than honestly improving the fundamentals of the business. To the field of operations

FIGURE 1-4 An Operations Management Competence Model

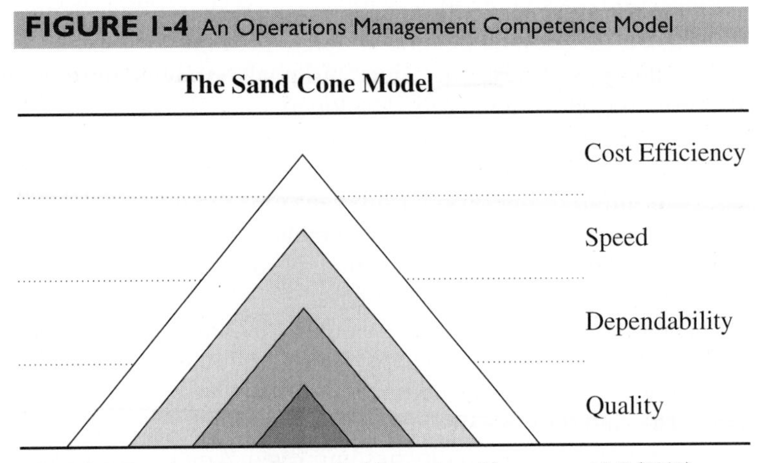

SOURCE: Reprinted from *Journal of Operations Management* 9, 2 (1990): 168–184, "Lasting Improvements" by K. Ferdows and A. Demeyer, with permission from Elsevier Science.

management's benefit, it is OM that has elevated quality management as a key area of business study.

A Strategic Management Perspective

Strategy refers to the planning processes used by an organization to achieve a set of long-term goals. The keys here are planning processes and a long-term orientation. Firms establish a planned course of action to attain their objectives. Further, this planned course of action must be cohesive and coherent in terms of goals, policies, plans, and sequencing to achieve quality improvement.

When the concept first arose, practitioners treated quality-related strategic planning as if it were a separate exercise from strategic planning. However, we soon realized that quality management, to become pervasive in a firm, needed to be included in all the firm's business processes, including strategic planning. Thus quality-related goals, tactics, and strategies are becoming more a part of the strategic planning process instead of a separate entity.

Company strategies are rooted in the building blocks of mission and core values. An organization's *mission* states why the organization exists. For example, the mission of the Saturn Motor Company states that the company exists "to market vehicles developed and manufactured in the United States that are world leaders in quality, cost, and consumer satisfaction through the integration of people, technology, and business systems and to transfer knowledge technology and experience throughout General Motors." The *core values* of an organization refer to guiding operating principles that simplify decision making in that organization. Therefore, if a company states "environmental consciousness" as a core value, it will institute policies leading to practices that favor a clean environment. Companies go to great extents to establish, communicate, and reinforce a sense of mission and values in their organizations because mission and values strongly influence organizational culture. Organizational culture is often seen as a major determinant (and sometimes roadblock) to the successful implementation of quality improvement.

The quality movement has greatly influenced strategy process in recent years. *Strategy process* refers to the steps an organization uses in the development of its strategic plans. Although we discuss this in greater depth in Chapter 4, strategic planning processes often involve a comprehensive environmental analysis that includes the remote, operating, and external environment that influence quality performance.

Figure 1-5 shows a generic strategic planning process and its components. Based on the analysis of mission, vision, and goals, strategic options, and business-level and functional-level strategies are developed.

Development of functional-level strategies helps improve the coherence and alignment of strategic plans within an organization. *Alignment* refers to consistency between different operational subplans and the overall strategic plan. For example, if an organization pursues a quality emphasis from a strategic standpoint, then the company should pursue supply chain, human resources, budgetary, or marketing courses of action that support a quality emphasis.

The ultimate goal of strategic quality planning is to aid an organization to achieve sustainable competitive advantage. In many markets, such as the auto industry, it is becoming difficult to sustain a competitive advantage based on quality alone. By the late 1970s and into the 1980s, Japanese cars were perceived as superior. Now, because American automotive quality has improved, American, Japanese, and European autos are all perceived as having relatively high quality. Therefore, Japanese automakers are having a difficult time differentiating themselves based on quality alone. This is also

FIGURE 1-5 A Generic Strategic Planning Process

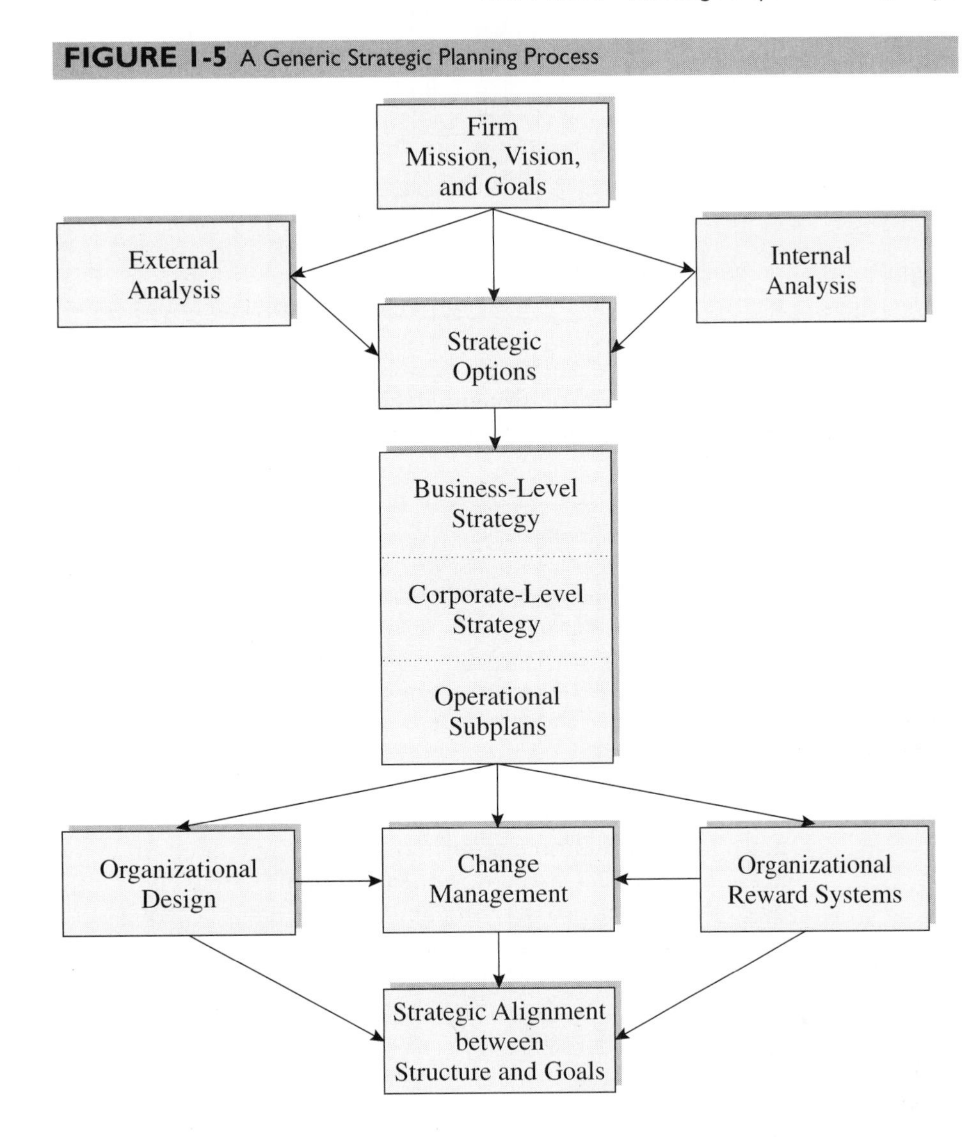

true in many other markets, where quality is an order qualifier. In some markets, such as semiconductors, where yield rates are low, competitive advantage in cost can be gained by a quality conformance strategy. Thereby, the quality–cost combination can be used as an order winner.

Madu and Kuei propose a strategy process based on plan–do–check–act (see Fig. 1-6). In this approach, *plan* represents strategy formulation, *do* refers to implementing strategy, *check* relates to evaluation and control, and *act* results in full-scale strategy implementation.

As quality has become integral to competitiveness, strategic planning for quality has become more important. Research shows that quality is still the major competitive concern of CEOs. Quality Highlight 1-1 shows how General Electric has made quality a key strategic imperative. This Highlight also demonstrates the interrelatedness among strategy, finance, and operations in achieving strategic objectives.

FIGURE 1-6 A Plan–Do–Check–Act Approach to Strategic Quality Planning

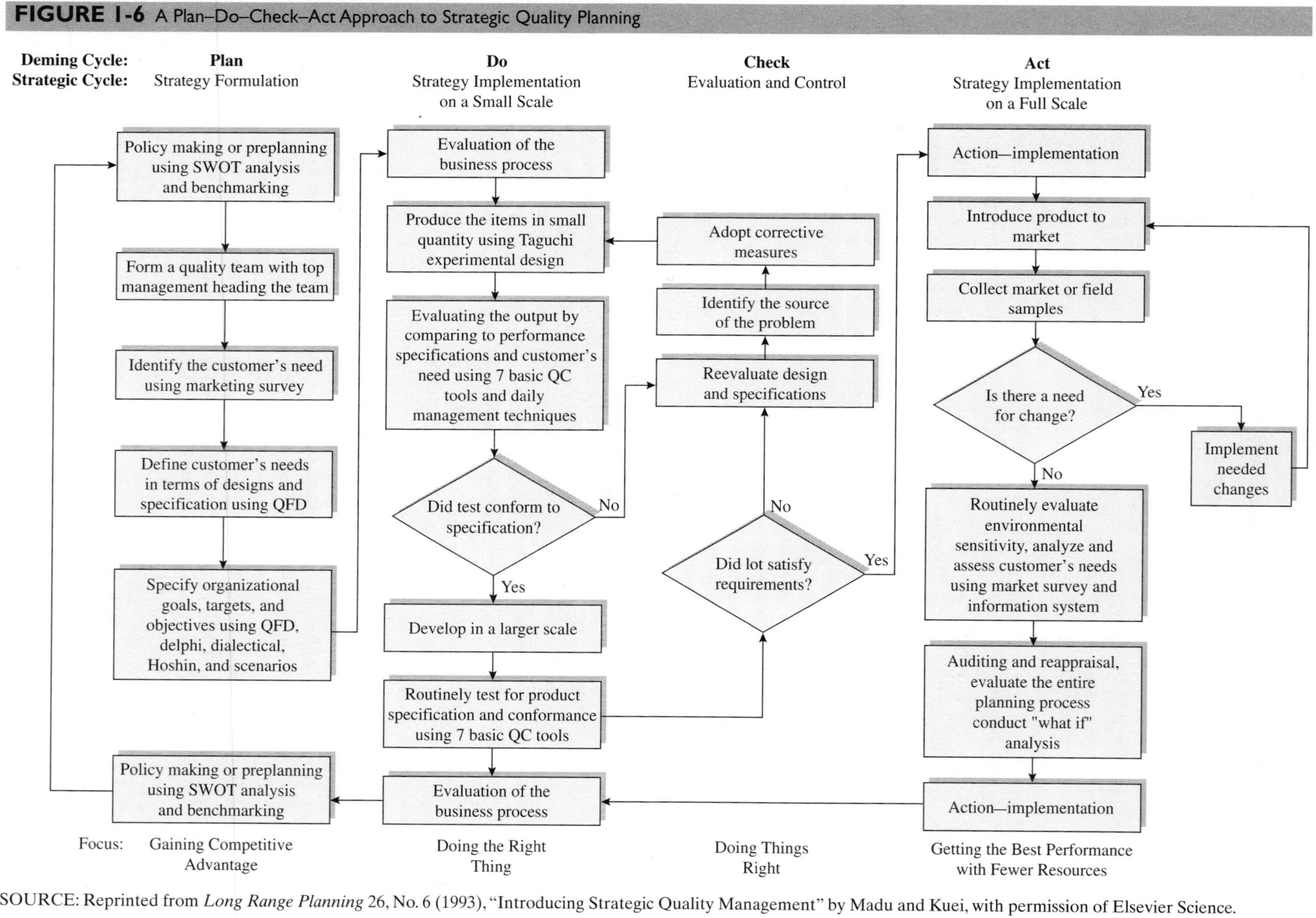

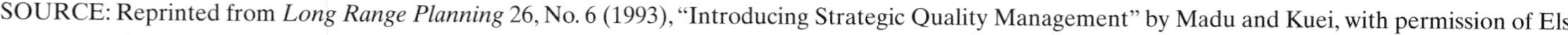
SOURCE: Reprinted from *Long Range Planning* 26, No. 6 (1993), "Introducing Strategic Quality Management" by Madu and Kuei, with permission of Elsevier Science.

QUALITY HIGHLIGHT 1–1

QUALITY STRATEGY AT GE

www.GE.com

For years, the exuberant management style of Jack F. Welch,[a] chair of General Electric Co., helped drive GE ahead. Now, the company has reached an especially notable milestone: It is the most profitable company in the United States.

Although GE is one of the world's most profitable companies, its management nevertheless faces a vexing problem: What can they possibly do for an encore after the retirement of Jack Welch? Many other executives are worrying about that, too. Investors are also wondering; they fear a slowdown in earnings growth.

To keep GE ahead, managers have devised an array of corporate strategies. They put exceptionally heavy reliance on the quality control program that far outstrips run-of-the-mill efforts, shifting GE's sales emphasis from manufacturing products to supplying services, pushing profitable niche acquisitions, and rapidly expanding abroad.

The quality control program was "a mammoth undertaking; I mean, I can't even begin to describe the size of this undertaking," stated Welch. A sure sign of top management's determination was that 40% of GE executive bonuses, which ran as high as $1 million, depended on implementation of the program. Previously, bonuses were only based on profit and cash flow.

It could significantly bolster profits some security analysts say, if only because it may eliminate costly, embarrassing blunders. One of GE's main mantras for growth has been new-product development. But some GE products ran into spectacular design and manufacturing snafus.

- GE's locomotive unit in Erie, Pennsylvania, built motors for new rail cars that were put into service on a major transit line in Montreal. But in heavy snows the electric motors shorted out and broke down, stranding 8,000 commuters. After GE discovered it had insulated the motors inadequately and had misdesigned auxiliary power systems, the entire transit line was shut down for 19 days while the company fixed the problem.
- A GE gas turbine sold to utility power plants around the world began cracking because of faulty design. GE's cost of fixing the huge turbines climbed from $150 million to $200 million, the company says.
- Just four months before a new GE jet engine was to power Boeing 777s for British Airways, the engine failed in a test and had to be redesigned. The plane's delivery was seven weeks late. In addition, a Federal Aviation Administration report cited sloppy GE manufacturing as a reason for delaying permission for extended flights by the plane over water. Permission was granted a year later than expected.

GE denied that such problems forced GE to adopt its new quality program. "We are not in trouble," they said, citing the robust profits. But they conceded that "the time wasted, the money wasted, in field fixes, in quality problems, in working things out, across corporate America, across the world, is enormous."

GE's quality program, which was borrowed from Motorola, Inc., involves training "Black Belts" for four months in statistical and other quality-enhancing measures. The Black Belts then spend full time roaming GE plants and setting up quality improvement projects. The program is producing a variety of benefits. "Your customers are happy with you, you are not firefighting, you are not running in a reactive mode." GE hopes the program, by preventing costly snafus, will save $7 billion to $10 billion over the next decade and thus bolster profits.

[a]Carley, W., "To Keep GE's Profits Rising, Welch Pushes Quality Control Plan," *Wall Street Journal* (January 13, 1997):A1.

A Marketing Perspective

Traditionally, the term *marketing* has referred to activities involved with directing the flows of products and services from the producer to the consumer. More recently, in a trend known as **relationship management,** marketing has directed its attention toward satisfying the customer and delivering value to the customer.

More and more companies are basing sales commissions on perceptual measures of customer satisfaction rather than merely volume of sales. The reasons for this are obvious. Studies show that the value of the loyal customer is much greater than an individual transaction. For example, the profit on a single pizza might be $5, whereas the same customer, if he or she orders one pizza per month over 10 years, is worth $600. This figure becomes $3,600 if the satisfied customer influences five friends to buy one pizza a month over 10 years. If all customers are satisfied, sales increase exponentially! Therefore, more firms are focusing on relationship management. This increases the importance of high levels of customer service and after-sales support.

The marketer focuses on the perceived quality of products and services. As opposed to the engineering-based conformance definition of quality, perceived quality means that quality is as the customer views it. Therefore, marketing efforts are often focused on managing quality perceptions.

The primary marketing tools for influencing customer perceptions of quality are price and advertising. However, these are imperfect mechanisms for influencing perceptions of quality. Toothpaste selling for two dollars is not necessarily better than toothpaste costing one dollar. The link between price and quality could be significant if all products were priced based on cost of materials and production only. However, not all products are priced this way.

Advertising and quality levels might be related. However, the relationship is not as straightforward as one would hope. Tellis and Fornell[4] proposed a contingency theory of advertising and quality saying that this relationship is "more likely when product quality is produced at lower cost and consumers rely less on advertising for their information." Their research showed that the positive effects of quality also were more pronounced later in the life cycle of a product. In these later stages, products are more likely to be standardized. This would give consumers more time to become informed about products and give firms more time to standardize and control costs.

Marketing is also concerned about systems. The marketing system involves the interactions between the producing organization, the intermediary, and the final consumer (see Fig. 1-7). Because of this relationship, it is often very difficult for firms and organizations to agree on who the customer is.

For example, identification of the customer is particularly difficult for service and governmental agencies. The Department of Energy (DOE) facility in Idaho Falls, Idaho, in preparing for the Energy Quality Award competition, spent many days defining their customer. One group believed that the parent body, the DOE in Washington, DC, was the primary customer because they were the funding body for the Idaho Falls office. Others developed much more inclusive lists such as regulatory agencies—the Nuclear Regulatory Commission, the Environmental Protection Agency—the taxpayers, the state of Idaho, and others. Although it might always seem obvious who the customer is to the casual observer, it is not always clear to those who are involved with the business.

[4]Tellis, G., and Fornell, C., "The Relationship between Advertising and Product Quality over the Product Life Cycle: A Contingency Theory," *Journal of Marketing Research* 25, (February 1988):64–71.

FIGURE 1-7 A Marketing System

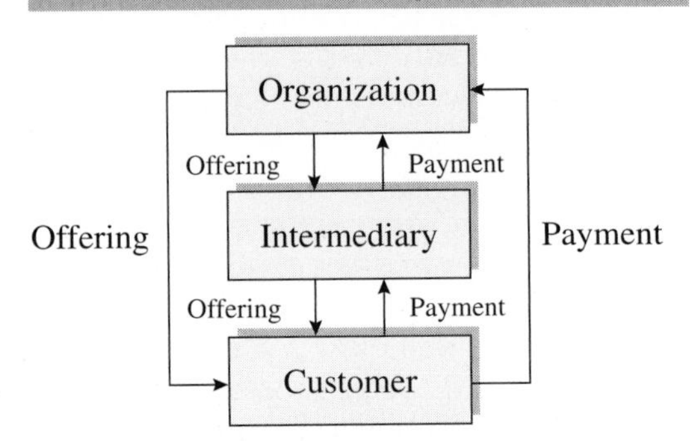

SOURCE: Adapted from R. Bagozzi, *Principles of Marketing Management* (Chicago: SRA, 1986).

Another important contribution of the marketing perspective has been the focus on service at the time of the transaction and after-sales support.

One of the ways marketing has helped to improve product and service quality has been to interact closely with engineering and operations in product design. The role of marketing in design has been to bring the voice of the customer[5] into the design process.

Customer service surveys are important tools for assessing the multiple dimensions of quality. Surveys provide a means for developing multidimensional perceptual measures of quality.

In short, the marketing perspective on quality is unique because the customer is the focus of marketing-related quality improvement. In trying to satisfy customer needs, marketing often wants to develop specialized products for different customers to perfectly satisfy customer needs. This can make life more difficult for producers because operations wants to standardize products to reduce processing complexity. Often, quality strategies result in a compromise between these two polar positions.

A Financial Perspective

One of the most commonly asked questions about quality management is "will it pay us financial benefits?" The answer to this question is an unqualified "maybe." As we read about General Electric in Quality Highlight 1-1, management was pursuing quality improvement as a means of reducing waste and increasing profitability. Implemented correctly, improved quality reduces waste and can lead to reduced cost and improved profitability. However, these returns tend to be long term rather than short term. W. E. Deming, the influential quality expert, made the first theoretical attempt to link quality improvement to financial results through the "Deming value chain." In his value chain (see Fig. 1-8), Deming linked quality improvement to reduction of defects and improved organizational performance. He also stressed quality as a way to increase employment.

The finance function is primarily interested in the relationships between the risks of investments and the potential rewards resulting from those investments. The goal of finance is to maximize return for a given level of risk. A comptroller for a large U.S. corporation stated this as "helping the customer to decide where to buy assets." In this

[5]Griffin, A., and Hauser, J., "The Voice of the Customer," *Marketing Science* 12, 1 (1993):1–26.

FIGURE 1-8 The Deming Value Chain

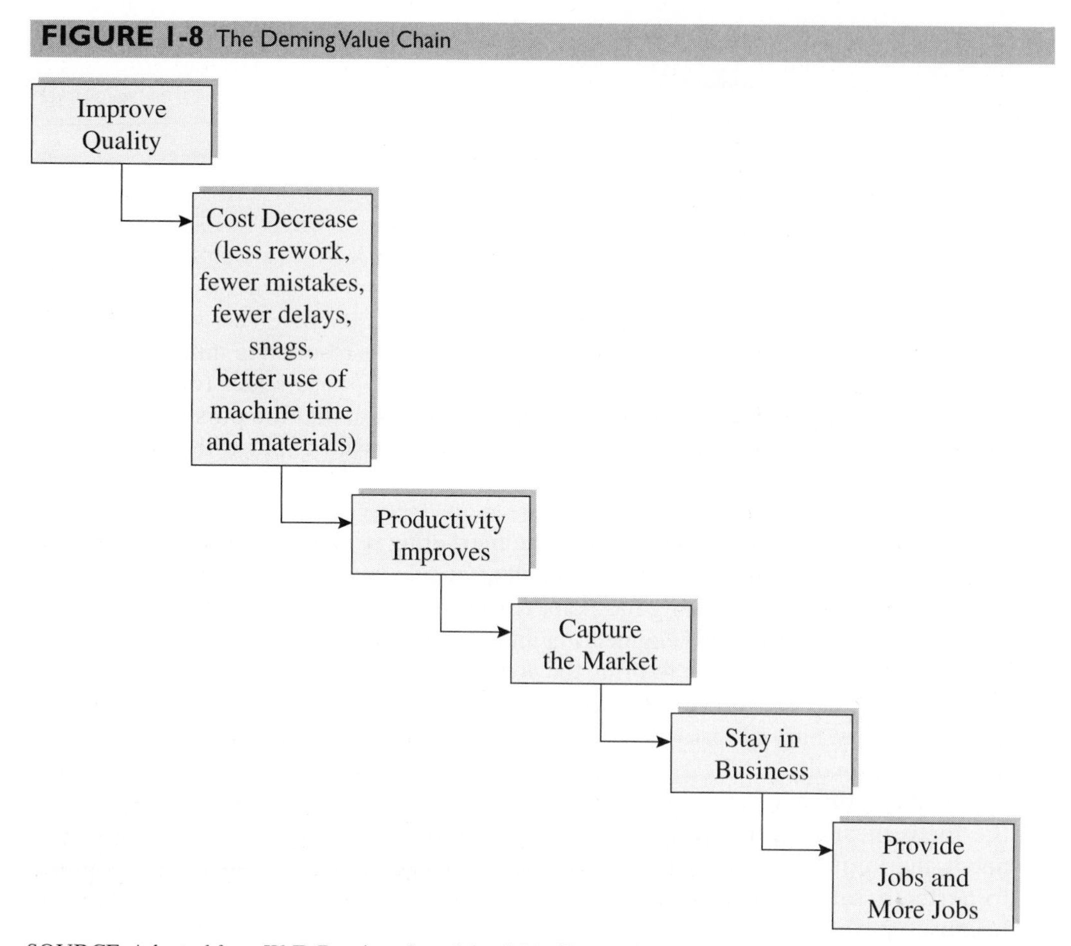

SOURCE: Adapted from W. E. Deming, *Out of the Crisis* (Boston: MIT/CAES, 1986).

sense, this comptroller viewed his primary customers as the stockholders (i.e., owners) of the corporation.

Communication relating to quality might be made more difficult for comptrollers and treasurers because accounting is the primary language of the financial function. Joseph Juran, another influential quality expert, referred to this communication problem when he stated that "*the language of management is money*." One way to translate quality concerns is to identify and measure the costs of quality. These quality-related costs can be in lost sales because of a poor reputation for reliability. Also, training and inspection cost money. Therefore, tradeoff and break-even analyses can be performed using the various costs of quality. Often what is discovered is that although improving quality seems expensive, the savings from reducing scrap, defects, and rework results in favorable returns on investment. This is why companies such as Motorola, Westinghouse, and General Electric are willing to pay millions of dollars to pursue quality.

However, the relationship between quality improvement and financial success is confounded by several intervening variables. For example, a firm investing a great deal of effort and money to establish a quality program for a product in the decline stage of the product life cycle may not receive the expected benefits. Top management involvement that is limited to lip service often results in great expenditure, and great effort, but eventually failure. *The pursuit of quality does not safeguard a company against bad management.*

Another concept that affects financial officers' perceptions of quality improvement is the **law of diminishing marginal returns**. According to this law, there is a point at which investment in quality improvement will become uneconomical. Figure 1-9 shows a quadratic economic quality level model. According to this model, the pursuit of higher levels of quality will result in higher expenditures. Hence, to invest beyond the minimum cost level will result in noneconomic decisions. This view is at odds with the ethic of continual improvement. Such debate has resulted in much controversy in the quality field. Although we will save resolution of these issues for Chapter 4, it should be emphasized that quality evangelizers who claim that the pursuit of quality is eternal and that any investment in quality is justified will be met with skepticism by financial officers trained in economics.

In summary, the financial perspective on quality relies more on quantified, measurable, results-oriented thinking. This has influenced quality thinking as quality professionals have had to seek approval for funding quality improvement efforts. If the objective of a firm is to return value to its shareholders, then the financial view toward quality must be well understood and used.

The Human Resources Perspective

Human resources (HR) managers are involved in enabling the workforce to develop and use its full potential to meet the company's objectives. Understanding the HR perspective on quality is essential because it is impossible to implement quality without the commitment and action of employees. Although leadership is an important antecedent to successful quality efforts, the involvement and participation of employees is just as key. After all, it is the rank and file that implements quality throughout the organization.

Of particular interest to HR managers is **employee empowerment**. Empowering employees involves moving decision making to the lowest level possible in the organization. For example, empowerment can involve something fairly minor, such as allowing

FIGURE 1-9 Basic Economic Quality Level Model

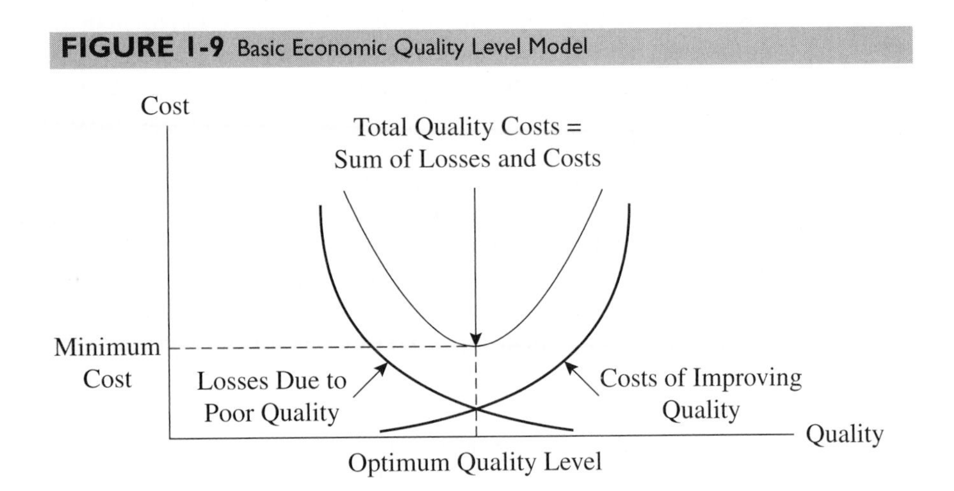

employees to replace broken or worn-out tools without management approval. In more spectacular instances, empowerment has resulted in the elimination of management as employees do their own scheduling, design, and performance of work.

The topic of empowerment is closely related to **organizational design**. HR managers are involved in many aspects of organizational design, such as the design of reward systems, pay systems, organizational structure, compensation, training mechanisms, and employee grievance arbitration.

HR balances the needs of the employee and the organization by advocating the employee to management and advocating the company's needs to employees. *Quality management flourishes where the workers' and the company's needs are closely aligned.* When needs are aligned, actions that are good for the company are also good for the employee.

Job analysis is a major function of HR.[6] Job analysis involves collecting detailed information about a particular job. This information includes tasks, skills, abilities, and knowledge requirements that relate to certain jobs. The information is used to define a job description that is used in setting pay levels. Job analysis sometimes has limited the ability of organizations to adapt to the flexibility needed for quality management. Important HR functions are recruitment and hiring of employees. A process called **selection** is employed. Traditionally, selection involved finding workers who have the technical preparation to perform the tasks associated with a job. Fast learners are becoming more valued by today's organizations. Of course, this does not mean that nuclear engineers will be replaced by former romance languages majors. However, it could mean that groups of engineers might report directly to a romance languages expert.

Training allows firms to standardize the approaches their employees use in solving unstructured problems. For example, when examiners for the prestigious Malcolm Baldrige National Quality Award perform site visits for applicants, they expect that top managers and low-ranking employees will use similar processes for solving problems. This is called *vertical deployment of quality management*. Similarly, examiners will expect that different departments and units within a firm will use similar approaches to solve problems. This is called *horizontal deployment*. It is unlikely that firms can achieve such standardization without effective training.

HR departments typically administer and oversee performance appraisal and evaluation. Traditionally, these evaluations involved face-to-face reporting sessions with employees and supervisors. Although some critics, such as Deming, have found this system ineffective, many companies believe performance evaluations are a key method for motivating employees. One quality-related approach to improving the process of performance evaluation is the **360-degree evaluation**, in which an employee's peers, supervisors, and subordinates are involved in evaluating the worker's performance. The College of Business at the University of North Carolina at Greensboro performs 360-degree teaching evaluations for its faculty. Each semester, students fill out teaching evaluations. Fellow faculty members attend each other's classes and read each other's student evaluations every semester. Then, a meeting is held at which the faculty evaluates and critiques each other's teaching performance. In this setting, the focus is on constructive criticism. Finally, the professor's supervisor uses the student and peer input to complete an annual evaluation. This approach seems to be effective in improving teaching performance.

---

[6]Costigan, R., "Adaptation of Traditional Human Resources Processes for Total Quality Environments," *Quality Management Journal* 2, 3(1997):7–24.

TABLE 1-3 HRM versus TQHRM

	Traditional HRM	*TQHRM*
Process characteristics	Unilateral role	Consulting role
	Centralization	Decentralization
	Pull	Release
	Administrative	Developmental
Content characteristics	Nomothetic	Pluralistic
	Compartmentalized	Holistic
	Worker-oriented	System-oriented
	Performance measures	Satisfaction measures
	Job-based	Person-based

SOURCE: Adapted from R. Cardy and G. H. Dobbins, "Human Resources Management in a Total Quality Environment," *Journal of Quality Management* 1, 1 (1996):3.

Although research has been performed to study HR variables that influence customer satisfaction, the results are as yet somewhat inconclusive.[7] However, quality management concepts recently have been effective in influencing HR practice. This has resulted in the growth of a field called **total quality human resources management (TQHRM)**.[8] Table 1-3 shows differences between traditional HR management and TQHRM. TQHRM involves many of the concepts of quality management to provide a supportive and empowered environment.

In summary, the focus of quality management is to manage properly the interactions among people, technology, inputs, processes, and systems to provide outstanding products and services to customers. HR managers have been very active in advocating quality approaches to improve organizational performance. Therefore, an HR focus on human performance provides important insights to quality thinking.

Is Quality Management Its Own Functional Discipline?

A quick read of the *Wall Street Journal* in any given week will reveal job openings for quality managers and engineers. The companies that run these ads seek people who, like you, are committed to and interested in quality management. However, the roles of these departments and specialists are changing in the new century of quality.

Historically, the quality management department performed a policing function in the firm. Quality managers were responsible for quality conformance and spent their time ferreting out causes of defects. However, in the late 1950s, Armand Feigenbaum, a well-known quality consultant who is discussed in Chapter 2, and others showed the limitations of this approach. Thus the movement began toward the total involvement of employees spawning total quality management (TQM). TQM was the 1980s term to describe quality management programs.

With total involvement, the role of the quality department moved from a technical, inspection role to a supportive training and coaching role. As a manager or a quality specialist, you will be asked to either arrange or perform quality-related training. Thus the abilities to conduct effective training and to facilitate teams are important tools for the quality professional.

---

[7]Walman, D., and Gopalakrishnan, M., "Operational, Organizational, and Human Resource Factors Predictive of Perceptions of Service Quality," *Journal of Quality Management* 1, 1 (1996):91–108.
[8]Cardy, R., and Dobbins, G. H., "Human Resources Management in a Total Quality Environment: Shifting from a Traditional to a TQHRM Approach," *Journal of Quality Management* 1, 1 (1996):5–20.

Is quality management its own discipline? Yes and no. Consultants, quality engineers, six sigma black belts, trainers, coaches, and managers are still needed. Therefore, the demand for quality specialists persists. However, because the eventual goal is to completely immerse the organization in quality thinking and commitment, the need for the specialist decreases with time. Therefore, a strong knowledge of quality is best coupled with technical expertise in other areas such as materials management, supply chain management, finance, accounting, operations management, HR management, strategy, industrial engineering, or myriad other disciplines. Indeed, the eventual goal of many companies is to completely distribute the quality management function throughout the firm.

In this section of the chapter we have discussed many different functional perspectives toward quality. Skilled management must recognize that these functional differences exist and provide communication in a way that completely addresses the different perspectives. Recognition that these multiple dimensions exist improves understanding among these disparate coalitions and helps all members of an organization to work toward a common goal.

THE THREE SPHERES OF QUALITY

One way to conceptualize the field of quality management is known as the **three spheres of quality**. These spheres are quality control, quality assurance, and quality management, and their functions overlap as seen in Figure 1-10.

The first sphere is **quality control**. The control process is based on the scientific method, which includes the phases of analysis, relation, and generalization. In the *analysis* phase, a process is broken into its fundamental pieces. *Relation* involves understanding the relationships between the parts. Finally, *generalization* involves perceiving how interrelationships apply to the larger phenomenon of quality being studied. Activities relating to quality control include the following:

- Monitoring process capability and stability
- Measuring process performance

FIGURE 1-10 Three Spheres of Quality

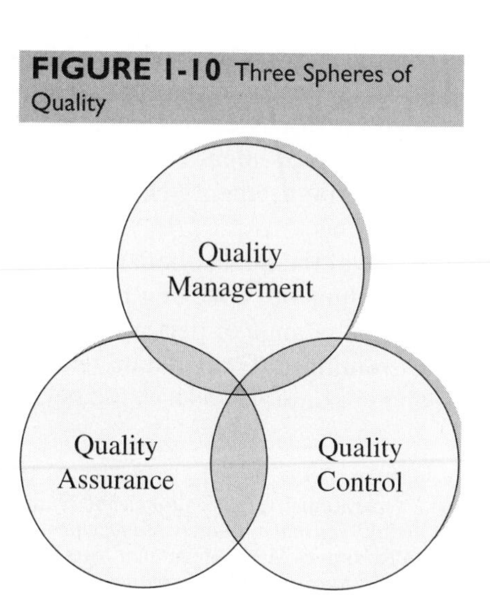

- Reducing process variability
- Optimizing processes to nominal measures
- Performing acceptance sampling
- Developing and maintaining control charts

Quality assurance refers to activities associated with guaranteeing the quality of a product or service. Often, these activities are design-related. This view of quality states that quality control is reactive rather than proactive by detecting quality problems after they occur. Given this, the best way to ensure quality is in the design of products, services, and processes. Quality assurance activities include tasks such as

- Failure mode and effects analysis
- Concurrent engineering
- Experimental design
- Process improvement
- Design team formation and management
- Off-line experimentation
- Reliability/durability product testing

The management processes that overarch and tie together the control and assurance activities make up **quality management**. See Quality Highlight 1-2 for an example of a company with effective quality management. The integrative view of quality management supports the idea that quality is the responsibility of all management, not just quality managers. For this reason, a number of managers, supervisors, and employees are involved in quality management activities such as

- Planning for quality improvement
- Creating a quality organizational culture
- Providing leadership and support
- Providing training and retraining
- Designing an organizational system that reinforces quality ideals
- Providing employee recognition
- Facilitating organizational communication

Many quality-related activities can occur simultaneously within the framework of the three spheres of quality. Because these activities overlap, communication between the protagonists performing the different activities becomes key.

OTHER PERSPECTIVES ON QUALITY

Although we have discussed a variety of perspectives on quality, there are a few others.

The Value-Added Perspective on Quality

A customer-based perspective on quality that is used by services, manufacturing, and public sector organizations involves the concept of value. A **value-added** perspective on quality involves a subjective assessment of the efficacy of every step of the process

QUALITY HIGHLIGHT 1–2

FEDERAL EXPRESS CORPORATION

www.fedex.com

FEDERAL EXPRESS AT A GLANCE

Conceived by Chairman and Chief Executive Officer Frederick W. Smith, Federal Express began operations in 1973. At that time, a fleet of eight small aircraft was sufficient to handle demand. The firm's cargo fleet is now the world's largest at more than 652.

FedEx's "People–Service–Profit" philosophy guides management policies and actions. The company has a well-developed and thoroughly deployed management evaluation system called SFA (survey/feedback/action) that involves a survey of employees, analysis of each group's results by the work group's manager, and a discussion between the manager and the work group to develop written action plans for the manager to improve and become more effective. Data from the SFA process are aggregated at all levels in the organization for use in policy making.

Training of frontline personnel is a responsibility of managers, and "recurrency training" is a widely used instrument for improvement. Consistently included in listings of the best U.S. companies to work for, FedEx has a "no layoff" philosophy, and its "guaranteed fair treatment procedure" for handling employee grievances is used as a model by firms in many industries. Frontline employees can participate in a program to qualify themselves for management positions. In addition, Federal Express has a well-developed recognition program for team and individual contributions to company performance.

SERVICE QUALITY INDICATORS

To spur progress toward its ultimate goal of 100% customer satisfaction, FedEx replaced its old measure of quality performance, percent of on-time deliveries, with a 12-component index that comprehensively describes how its performance is viewed by its customers. Each item in the service quality indicator (SQI) is weighted to reflect how significantly it affects overall customer satisfaction.

To reach its aggressive quality goals, the company has set up one cross-functional team for each service component of the SQI. A senior executive heads each team for each service component of the SQI and ensures the involvement of frontline employees, support personnel, and managers from all parts of the corporation when needed. Two of these corporate-wide teams have a network of more than 1,000 people working on improvements.

Employees are encouraged to be innovative and to make decisions that advance quality goals. FedEx provides employees with the information and technology they need to continuously improve their performance. An example is the Digitally Assisted Dispatch System (DADS), which communicates to some 30,000 couriers through screens in their vans. The system enables quick response to pickup and delivery dispatches and allows couriers to manage their time and routes with great efficiency.[b]

[b]This information is adapted from the Malcolm Baldrige National Quality Award "Profiles of Malcolm Baldrige Award Winners," National Institute for Standards and Technology, Gaithersburg, MD, 2005.

for the customer.[9] A value-added activity can be pinpointed by asking, "Would this activity matter to the customer?" In other words, in most cases, a value-added activity will have economic value to the customer.

[9]Shannon, P., "The Value Added Ratio," *Quality Progress* 30, 3 (March 1997):94–97.

Cultural Perspectives on Quality

International marketers have long noted that there are differences in tastes and preferences between cultures and nations. For example, Mexican food sold in the United States by the major restaurant chains is very different from the food sold in Mexico. This is so because American restaurant chains have found that customers prefer the Americanized versions of these foods.

Although it is somewhat obvious that differences in tastes and preferences exist between different cultures, it is not so obvious that approaches to quality improvement may differ according to culture. However, differences do exist. For example, Japanese companies tend to stress conformity and uniformity in producing quality. This results in high consistency among Japanese products. American firms are more likely to favor empowerment approaches that reward creativity and individualism. Certainly, cultures that are more class conscious or command-and-control oriented might have trouble delegating decision making to lower levels of employees.

ARRIVING AT A COMMON UNDERSTANDING OF QUALITY USING A CONTINGENCY PERSPECTIVE OF QUALITY

Businesses differ in key areas such as mission, core competence, customer attributes, target markets, technology deployment, employee knowledge, management style, culture, and a myriad of other environmental variables. **Contingency theory** presupposes that there is no theory or method for operating a business that can be applied in all instances. A coherent quality strategy will need to address these key environmental variables. For example, a company that defines part of its mission as "valuing and satisfying our customers through personalized service" likely will pursue a different technological approach toward its customers than a company with the mission of "applying technology to solving customer problems." One approach implies personalized service in interacting with customers, whereas the second company focuses on electronic data interchange interfaces with the customer. Both companies are focused on satisfying the customer. The difference is that they pursue different paths and strategies to achieve customer service.

The contingency approach to quality also helps to settle the different perceptions concerning the definition of quality. By adopting a contingency philosophy, we find that the definitions and dimensions of quality applied within organizations will, and should, vary. The definitions of quality used by the Department of Agriculture, Ford Motor Company, and the University of Missouri at Columbia will not be the same. Different definitions of quality also might exist within an organization, even though their quality definitions should be consistent. In an organization that adopts the contingency approach, the dimensions of quality will depend on the environment in which the company operates. This approach provides useful flexibility to managers in pursuing quality.

SUMMARY

There are many different perspectives on quality management. We found that customers and producers viewed quality differently. A focus on production and a focus on services provide two very different perspectives on quality.

There is even a great deal of disagreement about an appropriate definition of quality. A contingency perspective on quality shows that different definitions of quality are appropriate for different organizations.

The functional perspectives on quality vary greatly. As we understand these different functional perspectives, we form the basis for alignment in strategies and improvement in quality communication.

The fundamental areas of quality control, quality assurance, and quality management within the field of quality focus us on different aspects of quality. By designing plans and systems in each of these areas simultaneously, we develop a robust system of quality improvement that will set the stage for improved competitiveness.

Something else that becomes apparent is that quality improvement, despite the simplicity of many principles, requires a complex mix of systems design, organizational design, rewards design, and process design. As discussed throughout this text, these issues can be tied together in a strategic framework.

Key Terms

- Acceptance sampling
- Aesthetics
- After-sale service
- Assurance
- Concurrent engineering
- Conformance
- Contingency theory
- Core processes
- Define, measure, analyze, improve, and control (DMAIC)
- Design of experiments (DOE)
- Downstream
- Durability
- Electronic data interchange (EDI)
- Empathy
- Employee empowerment
- Features
- Inbound logistics
- International sourcing
- ISO 9000:2000
- ISO/TS 16949
- Job analysis
- Law of diminishing marginal returns
- Life testing
- Manufacturing-based
- Organizational design
- Outbound logistics
- Perceived quality
- Performance
- Product-based
- Product design engineering
- QS9000
- Quality assurance
- Quality control
- Quality dimensions
- Quality management
- Redundancy
- Relationship management
- Reliability
- Responsiveness
- Selection
- Serviceability
- Service reliability
- Six sigma
- Statistical process control (SPC)
- Strategy
- Supply chain
- Supplier development
- Supplier filters
- Supplier qualification
- Systems view
- Tangibles
- 360-degree evaluation
- Three spheres of quality
- Total quality human resources management (TQHRM)
- Transcendent
- Upstream
- User-based
- Value-added
- Value-based
- Value stream mapping

Discussion Questions

1. Why is *quality* a difficult term to define? How can we improve our understanding of quality?
2. Briefly discuss Garvin's eight dimensions of quality. Is Garvin's multidimensional approach a step forward in improving our understanding of quality? Why or why not?
3. Is there a difference between service quality and product quality? If so, what are the implications of these differences for a manager of a service business, such as a restaurant or a retail store?
4. Define the concept of *empathy*. Provide an example of empathy as a dimension of service quality.
5. Why is communication within an organization an important part of the quality improvement process?

6. Compare and contrast the engineering perspective and the marketing perspective of quality. How could an overemphasis on the engineering perspective work to the disadvantage of a business organization?
7. Describe the "systems view" that underlies modern quality management thinking. Which of the perspectives of quality discussed in Chapter 1 is most closely aligned with the systems view?
8. Why is planning an important part of the quality management process? How could a firm's quality management initiatives be adversely affected if planning was not a part of the process?
9. Research has shown that quality is still a major competitive concern of CEOs in American corporations. Is this level of concern about quality warranted? Please explain your answer.
10. Summarize Jack Welch's perspective on the importance of quality. Do Welch's perspectives strengthen or weaken your confidence in the future of the General Electric Corporation? Why?
11. What is meant by the phrase *cost of quality*? How can this phrase help a firm address its quality concerns?
12. What are the major differences between traditional human resource management and total quality human resource management? How does total quality human resource management transcend traditional human resource management in regard to providing an environment that is supportive of quality concerns?
13. Describe the three spheres of quality. How do these spheres provide another way to place the field of quality in perspective?
14. Discuss the value-added perspective on quality. What are the implications of this perspective for the manager of a business organization?
15. Discuss the application of contingency theory to quality management? Does contingency theory make sense to you? Why or why not?
16. Should a firm consider the law of diminishing marginal returns when striving to improve quality? Why or why not?
17. Are the perspectives of quality independent of one another? If not, describe ways in which they are interrelated.
18. How can an understanding of the multiple dimensions of quality lead to improved product and service designs?
19. What is your concept of quality? Is it multidimensional, or does it focus on a single dimension such as features, reliability, or conformance? Explain your answer.
20. Describe an instance in which you and a coworker (or superior) perceived the needs of a customer very differently. How did your differences in perception influence how each of you wanted to meet your customer's needs?

Cases

Case 1-1 FedEx: Managing Quality Day and Night

FedEx Homepage: *www.fedex.com*

As darkness falls across America and most businesses are locking up for the evening, one company is gearing up for a long night's work. FedEx, the world leader in the overnight package delivery market, delivers more than 5.3 million packages per business day. Most of us know FedEx as the overnight delivery company with white delivery vans, courteous drivers, and the distinctive purple-and-orange FedEx logo. But behind what the casual observer sees is a very complex company with the capacity to deliver millions of packages to millions of addresses around the globe overnight. Throughout the course of virtually every day and night, FedEx mobilizes its army of 218,000 employees, 69,000 vans and trucks, and 652 planes to get the job done.

For FedEx, getting the job done means managing quality 24 hours a day, with a watchful eye on customer expectations. The company's goals are simple: 100% customer satisfaction, 100% on-time deliveries, and 100% accurate information available on every shipment to every location around the world. Although these sound like far-fetched goals, the company goes to great lengths to try to make them a reality. One of the principal weapons that FedEx uses in pursuit of its goals is its total commitment to quality management.

> Quality management at FedEx encompasses all of its operations. Although the company is the acknowledged leader in the air freight industry, a formal Quality Improvement Process (QIP) plays an integral role in all of the company's activities.[10]

At the heart of the QIP program is the philosophy that quality must be a part of the way that FedEx does business, not part of the time, but all of the time. As a result, themes such as "Do it right the first time," "Make the first time you do it the only time anyone has to," and "Q = P" (quality = productivity) are important parts of the FedEx culture. To reinforce these themes, the company teaches its employees the 1–10–100 rule. According to the rule, if a problem is caught and fixed as soon as it occurs, it costs a certain amount of time and money to correct. If a mistake is caught later in a different department or location, it may cost 10 times that much to repair. And if a mistake is caught by a customer, it may cost 100 times as much to fix.

A number of substantive strategies have been implemented by FedEx to support its quality efforts. Quality action teams (QATs) design work processes to support new product and service offerings. A set of service quality indicators (SQI) has been established to determine the main areas of customers' perception of service. Through careful tracking of these indicators, the company generates a weekly summary of how well it is meeting its customer satisfaction targets. An SQI team works through problems revealed by the indicators. For example, if problems were being created by confusion in FedEx labeling instructions, the team would work on improving the clarity of the instructions. Some of the company's tactics to ensure total quality are extraordinary. For example, every night FedEx launches an empty airliner from Portland, Oregon, bound for Memphis. The jet follows a course that brings it close to several FedEx terminal airports. The purpose of the jet is to swoop down and pick up FedEx packages if any of the company's regularly scheduled airplanes is experiencing mechanical difficulty.

Along with a focus on its external customers, FedEx's approach to quality also involves strengthening the bonds between its internal customers, or employees. To reinforce this notion, the company asks all its employees to ask the following three questions when they interface with a coworker:

1. What do you need from me?
2. What do you do with what I give you?
3. Are there any gaps between what I give you and what you need?

The company also reaches out to its employees in a number of substantive ways. In 1973, the company adopted its People–Service–Profit (PSP) philosophy, which articulates the view that when people are placed first, service and profit follow. An aggressive training program, competitive wages and benefits, profit sharing, bonuses, and a state-of-the-art employee grievance process are all elements of the PSP philosophy. Employee recognition also plays an important role in the company's quality pursuits. For example, each quarter FedEx divisions select their best-quality success story, which is entered in a company-wide competition. Presentations are made by the finalists before the company's CEO, executive vice president, and other top managers. The award for being a finalist is a gold quality pin for each member of the team and the opportunity to be interviewed on the company's internal television network.

The quality efforts practiced by FedEx have paid off. The company has achieved a remarkable 99.7% on-time delivery level. The list of awards the company has won are too numerous to publish. The most impressive are the Malcolm Baldrige National Quality Award, the AT&T Top Performer Award, the Quality Carrier of the Year Award presented by Merck

---

[10] *www.fedex.com.*

Pharmaceuticals, and the Company of the Year Distinguished Service Award presented by the National Alliance of Businesses. Will FedEx's pursuit of quality end here? Asked if winning the Malcolm Baldrige National Quality Award signifies that FedEx has achieved the ultimate level of quality, CEO Fred Smith said, "receipt of the award is simply our license to practice." Apparently, the quest for improved quality at FedEx will continue, day and night. ■

DISCUSSION QUESTIONS

1. What is FedEx's "common language" of quality? Is it important for a company to establish a "common language" of quality? If so, why?
2. There are several different perspectives of quality, including the operations perspective, the strategic perspective, the marketing perspective, the financial perspective, the HR perspective, and the systems perspective. Which of these perspectives are being emphasized by FedEx? Why?
3. Is FedEx's level of emphasis on quality appropriate? Why or why not?

Case 1-2 Granite Rock Company: Achieving Quality through Employees

Granite Rock Homepage: ***www.graniterock.com***

Granite Rock, a California-based mining and construction company, was founded in 1900, but its journey toward improved quality did not start until the mid-1980s. The managers at Granite Rock knew that a resulting decline in customer satisfaction was inevitable and responded to this self-assessment by deciding it needed to become more customer focused. At Granite Rock, this meant not only learning more about the customer but also providing its employees the training and skills necessary to properly implement the company's new philosophy.

At first this focus on HR management was emphasized through employee training. As explained by CEO Bruce Woolpert, you can't have employees out telling customers "yes" unless everyone else in the company knows how to follow up on "yes." As a result, an aggressive training program was implemented involving on-the-job training and classroom time for the majority of the firm's employees. A program called IPDP (Individual Personal Development Plan) was created to help each employee take responsibility for his or her own training program. As the program gained momentum, remarkable things started to happen. One employee indicated that he wanted to learn to read better. As a result of that request, the company initiated a reading program that has helped a number of Granite Rock's employees improve their reading skills. As part of the company's effort to reduce process variability and increase product reliability, many employees were trained in statistical process control, root-cause analysis, and other quality-minded competencies. Today, Granite Rock spends up to $1,600 per year for training of each employee, which is more than 10 times the average for the mining and construction industries.

An equally important step in equipping its employees to contribute to the firm's efforts was establishing an atmosphere of trust between management and the rank-and-file employees. That effort has been pursued in a number of ways, and the trust developed throughout the company has resulted in the effective implementation of employee empowerment and teamwork. In addition, the company has done away with its conventional performance appraisal system (feeling that it was ineffective) and has incorporated performance appraisal into each employee's IPDP. As part of this plan, at least once a year, each employee meets with his or her supervisor to discuss job responsibilities, review accomplishments, assess skills, and set skill and career development goals. Although the IPDP is not mandatory for the company's unionized workers, overall participation stands at

83%. Because the IPDP approach to training and appraisal prepares employees to assume greater responsibilities, Granite Rock promotes heavily from within.

Today, Granite Rock, having won the Malcolm Baldrige National Quality Award, is aggressively pursuing its quality initiatives, with the cooperation of its 400 employees. ■

DISCUSSION QUESTIONS

1. Rather than focusing on human resource management (HRM) as a means of supporting its quality initiatives, Granite Rock could have chosen another area as its focal point (i.e., marketing, operations, information systems, and so on). How does a focus on HRM support a company's quality initiatives?
2. Discuss the different components of Granite Rock's HRM initiatives. How can each of these components support the company's quality efforts?
3. Discuss CEO Woolpert's feelings about communication with the customers (paragraph 2). What happens when others in the company don't know what has been promised to the customer? How can quality management help to overcome this situation?

CHAPTER 2

Quality Theory

> Experience alone, without theory, teaches
> management nothing about what to do to improve quality
> and competitive position, nor how to do it.
> —W. EDWARDS DEMING[1]

INTRODUCTION

For some reason, the word *theory* conjures negative thoughts among many students and businesspeople. There is a supposition that an emphasis on theory will somehow distance us from the real world and what "really works." However, theories form the basis for much of what happens around us, and most of us use them every day without knowing it. For instance, mechanical theories are at work when you drive your car. When you turn on a light, theories are at work harnessing the power of electricity. The airplanes overhead benefit by Bernoulli's theory. The organizations we work for are based on theories proposed by generations of organizational theorists. The money we spend is managed by macroeconomic theory. Because there can be no understanding of the phenomena that surround us without theory, shouldn't there be a general theory of quality management? In other words, is there a model that explains how organizations such as Federal Express, Motorola, and Dell have achieved such high levels of quality within the past generation?

There are several theories on quality improvement in practice currently. In this chapter we learn about the experts in this field, their theories, and how these theories have been used by various organizations.

WHAT IS THEORY?

The term *theory* is often misused. Generally, theory is a "coherent group of general propositions used as principles of explanation for a class of phenomena."[2] For example, it might have been observed that many companies that have implemented quality improvement have experienced improved worker morale. Therefore, a theoretical model of quality and worker morale might be developed as shown in Figure 2-1.

The model in Figure 2-1 has two variables: quality improvement and worker morale. The arrow implies causality. Because the head of the arrow points to worker morale, this is the *dependent variable*, with quality improvement being the *independent variable*. Thus we can see that our small theoretical model is testable. To test our model, we formulate a hypothesis, gather data under controlled conditions, and analyze the data. There is no way to prove the theory. The results of our statistical research will only support the theory or fail to support the theory.

---

[1]Deming, W. E., *Out of the Crisis* (Boston: MIT CAES, 1986), p. 19.
[2]*Random House Webster's College Dictionary* (1996), p. 1,384.

FIGURE 2-1 A Theoretical Model Relating Quality Improvement to Worker Morale

Quality Improvement ——+——→ Worker Morale

For a theory to be complete, it must have four elements[3]: what, how, why, and who-where-when. The *what* of the theory involves which variables or factors are included in the model. The variables in Figure 2-1 are the *what* of that model.

The *how* of a theoretical model involves the nature, direction, and extent of the relationship among the variables. For example, in our simple quality/morale model, we posited that quality improvement positively influences worker morale. This is represented by the plus sign in the model.

The *why* of the theory is the theoretical glue that holds the model together. For example, what are the psychological dynamics that could cause quality improvement to increase worker morale? Suppose we theorize that organizations using quality improvement are more organic in nature, stressing worker empowerment. Suppose further that worker empowerment lifts worker self-esteem by stressing increased decision making and control over his or her own job. Finally, we pose that this increase in self-esteem results in improved worker morale.

The *who-where-when* aspects place contextual bounds on the theory. For example, we might add a condition to our model stating that morale is improved only in companies that receive strong leadership by top management to implement quality improvement.

As is shown in Figure 2-2, theories are established in one of two ways. A theory that is generated by observation and description is said to have been developed by the process of **induction.** The process of induction is useful but is also subject to observer bias and misperception. Suppose we developed our quality/morale theory after visiting a series of Chicago, Illinois, firms that had implemented quality improvement. During our site visits, we found that employee morale was high in all the sites we visited. Our inductive observations of quality influencing morale might be correct. However, as committed researchers, we are so enthralled by our discovery that we fail to notice that the Cubs (the Chicago baseball franchise) just won the World Series of baseball.

FIGURE 2-2 Inductive versus Deductive Reasoning

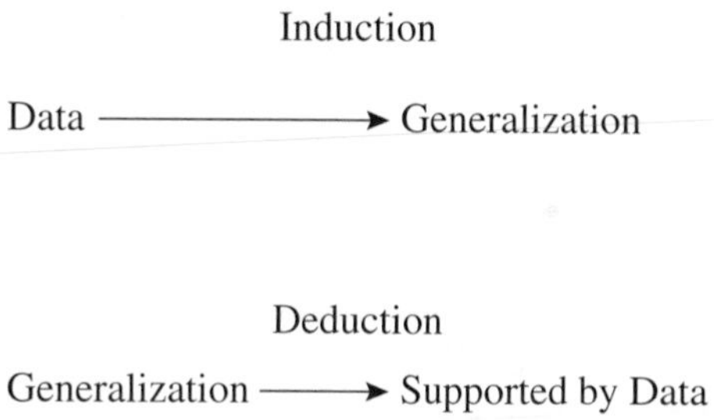

Induction

Data ——→ Generalization

Deduction

Generalization ——→ Supported by Data

[3]Dubin, R., *Theory Development* (New York: Free Press, 1978).

Therefore, our inductively produced model of quality/morale might have been confounded by an intervening variable (the glorious win by the Cubs).

The more common theoretical approach used by researchers is **deduction.** Using deduction, researchers propose a model based on prior research and design an experiment to test the theoretical model. These studies follow the scientific method discussed in Chapter 1.

Many of the concepts and models proposed in this chapter are developed by induction. Therefore, when considering concepts put forth by experts such as Deming, Juran, and Crosby, you should note that their principles are based on years of experience with a wide variety of firms that have improved quality. Their models and principles are also laden with personal biases, judgments, and values. This is fine, as long as you recognize these limitations to their models.

Is There a Theory of Quality Management?

As yet, there is not a unified theory explaining quality improvement in the supply chain that is widely accepted by the quality community. In fact, as we saw in Chapter 1, the literature concerning quality is contradictory and somewhat confusing. Different theories have been proposed by practitioners and researchers. Some of these theories have been drawn from organizational theory, behavioral theory, and statistical theory.

The differing approaches to quality improvement represent competing philosophies that have sought their places in the marketplace of ideas. Practicing quality managers must become familiar with these philosophies and apply those that are appropriate to their particular situations.

The diversity of approaches to quality contributes to variability in the approaches used by companies and increases the chance for failure in organizations. Some of these approaches are proven and others aren't. This has spawned myriad consulting firms and consultants. As shown in A Closer Look at Quality 2-1, some of these are legitimate and some are in it for profit only. From the 1980s to the present, there has been an explosion of consulting firms around the world that taught a variety of means for achieving quality. In the following sections we discuss the problem of the fragmentation of the quality message from a managerial point of view.

A CLOSER LOOK AT QUALITY 2-1

THE PRODUCT THAT IS QUALITY[a]

The message has gotten out. In recent years, hundreds of thousands of employees have attended quality-related training. Welcome to Quality, Inc. It is an amorphous industry of trainers and consultants, selling not only advice but also courses, lectures, workbooks, videos, and more than a little folderol. "There are lots of seminar serums, training transfusions, program prescriptions, and video vaccinations," states Stanley Cheransky, of Gunneson Group International, Inc. The Juran Institute sells a $15,000 do-it-yourself kit complete with 16 videotapes, 10 workbooks, a leader's manual, overhead transparencies, and a five-day course to teach someone how to run the tapes.

Firms that once did little more than human resource training now fly the quality banner. "They've moved to where the fish are biting," says George Labovitz, president of ODI, a consultancy

[a]Reprinted from the 1991 Bonus Issue of *Business Week* by special permission of The McGraw-Hill Companies.

Continued

group. Even some quality associations, such as the American Society for Quality (ASQ), are raking in big bucks from selling books, training, seminars, magazines, and conferences.

In this pell-mell rush, some of the advice dispensed isn't of the highest quality, says G. Howland Blackiston, president of the Juran Institute. And, he implies, some of it is deliberately obscure. "It's become an opportunity to make money. If people were offering a product senior managers could understand, they would get laughed out of the boardroom." Adds Dean Silverman, executive vice president of Temple, Barker, and Sloane, Inc., "For some companies, the investment has paid off. For many others, quality programs have turned out to be just another drag on the bottom line, costing the buck without producing the bang."

For companies besieged by swarms of slick brochures and salespeople, this experience can be bewildering. "There must be 20 pieces of mail a day from firms selling quality, and a lot of it is junk," states Mary Dolan, director of quality implementation at Campbell Soup Co. "There are few consultants who can offer something new. Most of them are repackaging and selling the same ideas and concepts."

Of course, many companies sign up with new products in an effort to compete with others in the world economy. Many firms also are pressuring their suppliers to sign up. "This is about far more than the quality you build into a product," explains David Garvin, a professor at the Harvard Business School. "It's a different set of ideas about management. There are not that many initiatives where the CEO can stand up and say, 'this is mine and I can put my imprint on the company with it.' Quality fits the bill."

The government's Baldrige award, too, has fueled the boom. Former Baldrige judges now staff several consulting firms, such as the Juran Institute. These consultant groups help counsel clients about how to win the award. The award itself has inspired many products, including books, seminars, and videos.

Whatever the product, it's likely to be sold with the fervor of a religious cult. Each quality guru boasts his or her own set of commandments, rituals, and disciples. Within each approach, corporate managers are confronted by a numbing maze of acronyms and buzzwords. There is *TQC, TQM, fishbone diagramming, cause and effect, poka yoke, big Q*, and *little q*. "It's a lot of alphabet soup," complained Deming.

Another shortcoming of consultant-driven approaches to improving quality is the overemphasis on training seminars. "I have not met a client who said that training was enough to get things done," stated George Stalk, a well-known management consultant. "Transferring awareness into bottom-line results is hard. Top management doesn't understand the limitations of much of the quality advice."

Some say the same is true for other kinds of consulting. "A firm could send all of its employees to a Tom Peters seminar, and they would all come back all fired up; but they would be ill-equipped to make any changes in the organization," says David Quady, a consultant with Chevron's in-house quality improvement group. "You need awareness, but then you need to provide education to provide it."

The most successful companies have put their own stamp on quality campaigns, mounting their own massive training efforts internally. "You can't buy these quality systems off the shelf," advises Harvard's Garvin. "The most successful efforts tailor ideas to the organization. All the Baldrige winners have developed their own house brands of quality."

The bottom line is that firms must be wary of the advice they are purchasing. The flashy, feel-good messages they are hearing may not make them feel good in the long run. Alternatively, the fact that a flawed medium is used does not render the message incorrect. There is much that is known about quality management. As well, there is much still to be learned.

LEADING CONTRIBUTORS TO QUALITY THEORY: W. EDWARDS DEMING

W. Edwards Deming (see Fig. 2-3) was widely accepted as the world's preeminent authority on quality management prior to his death on December 24, 1993. Deming gained credibility because of his influence on Japanese and American industry. In the

FIGURE 2-3 W. Edwards Deming

© Catherine Karnow/Corbis.

late 1970s, when it became apparent that many Japanese products were of better quality than U.S. products, U.S. managers were surprised to learn that the Japanese had learned quality management from W. E. Deming, an American. In fact, the Japanese still use the original lectures given by Deming to train new generations of businesspeople.

Although Deming is best known for his emphasis on the management of a system for improving quality, his thinking was based on the use of statistics for continual improvement. In the 1920s, Deming worked in the Western Electric Hawthorne plant. Trained in engineering and mathematical physics at the University of Wyoming and Yale University, he came to know Walter Shewhart, who influenced his thinking about improving quality through the use of statistics.

After working at the Hawthorne plant, Deming worked in government jobs with the U.S. Department of Agriculture and the Bureau of the Census, where he helped develop statistical sampling techniques. During World War II he worked with U.S. defense contractors to use statistics to identify systematic quality problems occurring within defense-related products.

After the war, Deming was sent to Japan by the U.S. Secretary of War to work on a population census. During this time, the Japanese Union of Scientists and Engineers asked him to provide lectures on statistical quality control applications. While in Japan, Deming became impressed by the precision and single-mindedness with which the Japanese pursued quality improvement. Late in his life, Deming commented that he had consulted around the world and had found that Japan's commitment to quality was unparalleled. In his mind, this unwavering pursuit of quality improvement was the genius of the Japanese people. When the United States discovered that it was lagging the Japanese in quality, large corporations such as General Motors and Ford hired Deming to help them develop quality management programs.

Toward the end of his career, Deming gave seminars, wrote books, taught classes, and published articles to explain his approach to quality management. This led to wide dissemination of the Deming approach to quality. In part, because of the lack of focus in America, the results have been somewhat mixed.

Deming stressed that consumers were well served by insisting that service and product providers deliver high quality. He believed that the more consumers demanded high-quality products and services, the more firms would continually aspire to higher levels of performance. In fact, this has happened in the United States. As opposed to 20 years ago, consumers now expect high-quality products at a reasonable cost.

Deming's mantra was "continual neverending improvement." The goal of higher levels of quality would perhaps never be completely met, but firms would continually exercise themselves to get better and better. This is why quality improvement is often referred to as a journey where the elusive destination is never reached.

Deming's 14 Points for Management

Although Deming espoused the belief that theory was important to the understanding of quality improvement, the closest he ever came to expounding a theory was in his 14 points for management. The foundation for the 14 points was Deming's belief that the historic approach to quality used by American management was wrong in one fundamental aspect: Poor quality was not the fault of labor; it resulted from poor management of the system for continual improvement. Although this might now seem obvious, at the time Deming taught this, it was a revelation to managers. Taken as a whole, the 14 points for management (see Table 2-1) represent many of the key principles that provide the basis for quality management in many organizations.[4]

1. *Create constancy of purpose toward improvement of product and service with the aim to become competitive, stay in business, and provide jobs*. Constancy of purpose means that management commits resources—over the long haul—to see that the quality job is completed. This is in contrast to managers who wish to achieve quick returns and get bottom-line results after embarking on quality "programs." In recent years, more people have come to realize that U.S. management is too short-term oriented in its thinking. Unfortunately, quality improvement requires time to be effective. The Japanese experience is instructive. Deming helped to begin the Japanese quality revolution in the early 1950s. As was already stated, the Japanese displayed remarkable commitment to and focus on quality improvement. Even with all this effort, the Japanese were not really recognized as world leaders in quality until the late 1970s. When this long effort is contrasted with the average term of service of the American CEO, one begins to understand the daunting task facing U.S. firms. Fortunately, it has been suggested that although it took Japanese firms 25 years to achieve this level of

TABLE 2-1 Deming's 14 Points

1. Create constancy of purpose.
2. Adopt a new philosophy.
3. Cease mass inspection.
4. End awarding business on the basis of price tag.
5. Constantly improve the system.
6. Institute training on the job.
7. Improve leadership.
8. Drive out fear.
9. Break down barriers between departments.
10. Eliminate slogans.
11. Eliminate work standards.
12. Remove barriers to pride.
13. Institute education and self-improvement.
14. Put everybody to work.

SOURCE: Adapted from W. E. Deming, *Out of the Crisis* (Boston: MIT/CAES, 1986), pp. 18–96.

[4]Deming, W. E., *Out of the Crisis* (Boston: MIT/CAES, 1986).

attainment, it might take American firms only 10 years because they can learn from the Japanese and others who have achieved high quality. Less fortunately, because of the American emphasis on quick returns, 10 years might be an eon to many U.S. firms.

2. *Adopt a new philosophy. We are in a new economic age.* Western management must awaken to the challenge, must learn its responsibilities, and must take on leadership of change. Planned obsolescence was the order of the day when Deming first discussed the new economic age. During this time, automobiles were designed to last 60,000 to 80,000 miles and then be replaced, and Deming referred to an age in which Americans would no longer accept defective products. Now that many firms have excellent quality at a reasonable cost, they are turning to service quality to make the next big advances. More and more, goals for reduction of defects are being replaced by goals for improvement in customer satisfaction. Similarly, specification measurements are being replaced by customer service metrics as the important measures of quality.

3. *Cease dependence on mass inspection to improve quality.* Eliminate the need for inspection on a mass basis by building quality into the product in the first place. In many companies still, the quality assurance (note the misuse of the word *assurance* here) department performs in-process and final inspection of a product. In this scenario, responsibility for quality lies with the quality department. However, by the time the quality department inspects the product, either the quality is built in or it is not built in. At this point, it is too late to add quality.

Deming's alternative is **quality at the source**. This means that all workers are responsible for their own work and perform needed inspections at each stage of the process to maintain process control. Of course, this is possible only if management trusts and trains its workers properly.

4. *End the practice of awarding business on the basis of price tag alone.* Instead, minimize total cost. Move toward a single supplier for any one item, based on a long-term relationship of loyalty and trust. Traditionally, U.S. firms maintained many suppliers. The theory behind this approach was that competition among suppliers would improve quality and decrease cost. In reality, however, the existence of many suppliers caused an overemphasis on cost and an increase in variability. For example, if a metal fabrication company has multiple suppliers, the result is great variability in the makeup and consistency of the incoming stock. The alternative supply chain approach used by many firms is **just-in-time (JIT) purchasing**. As shown in Table 2-2, this approach minimizes the number of suppliers used, resulting in decreased variability. Also long-term contracts are used that result in the ability to develop and certify suppliers. Often these certifications are based on quality standards such as the Malcolm Baldrige National Quality Award Criteria or the ISO 9000:2000 international standard for quality systems. In other cases, supplier certification is based on an internally developed standard.

5. *Improve constantly and forever the system of production and service, to improve quality and productivity, and thus constantly decrease cost.* This point focuses on the management of the system of production.

The system of production includes product design, process design, training, tools, machines, process flows, and a myriad of other variables that affect the system of production and service. In the final analysis, management is responsible for most of the system design elements because it is management that has the authority and the budget to implement systems. Therefore, the workers can be held responsible only for their inputs to the system. Mediocre or poor performance of a system is most often the result of the poor performance of management.

TABLE 2-2 Japanese Just-in-Time Purchasing versus U.S. Purchasing Practices

Purchasing Activity	*JIT Purchasing*	*Traditional Purchasing*
Purchase lot size	Small lots with frequent deliveries	Large lots with less frequent deliveries
Selecting supplier	Single source of supply for a given part in nearby geographic area with a long-term contract	Multiple sources of supply for a given part and short-term contracts
Evaluating supplier	Emphasis on product quality, delivery performance, and price, but *no* percentage of reject from supplier is acceptable	Emphasis on product quality, delivery performance, and price, but about 2% reject from supplier is acceptable
Receiving inspection	Counting and receiving inspection of incoming parts is reduced and eventually eliminated	Buyer is responsible for receiving, counting, and inspecting all incoming parts
Negotiating and bidding process	Primary objective is to achieve product quality through a long-term contract and fair price	Primary objective is to get the lowest possible price
Determining mode of transportation	Concern for both inbound and outbound freight and on-time delivery; delivery schedule left to the buyer	Concern for outbound freight and lower outbound costs; delivery schedule left to the supplier
Product specification	"Loose" specifications; buyer relies more on performance specifications than on product design, and the supplier is encouraged to be innovative	"Rigid" specifications; buyer relies more on design specifications than on product performance, and suppliers have less freedom in design specifications
Paperwork	Less formal paperwork; delivery time and quantity level can be changed by telephone calls	Requires great deal of time and formal paperwork; changes in delivery date and quantity require purchase orders
Packaging	Small standard containers used to hold exact quantity and to specify precise specifications	Regular packaging for every part type and part number with no clear specifications on product content

SOURCE: Adapted from S. Lee and A. Ansari, "Comparative Analysis of Japanese Just-in-Time Purchasing and Traditional U.S. Purchasing Systems," *International Journal of Operations and Production Management* 5, 4 (1986): 5–14.

6. *Institute training on the job.* People must have the necessary training and knowledge to perform their work. Many companies employing laborers have found they must design job-related training. It should be noted that training, although a necessary condition for improvement, is not sufficient to guarantee successful implementation of quality management. The design of effective training is important to quality improvement.

7. *Improve leadership.* The aim of supervision should be to help people, machines, and gadgets to do a better job. Supervision of management is in need of overhaul as well as supervision of production workers. All quality experts agree that leadership is key to improving quality. If assembly line employees become enthused about quality management, they will be able to make some small improvements to their organization. However, this improvement can occur only within the realm of influence of the employee. For wide-ranging improvements to occur, upper management must be involved. It is upper management that has the monetary and organizational authority to oversee the implementation of quality improvement. Without management support and leadership, quality improvement efforts will fail.

8. *Drive out fear so that everyone may work effectively for the company.* For the most part, Deming was referring to those situations in which employees were fearful to change or even admit that problems existed. Many of these problems still exist. At

times, employees who surface problems and seek to create change are considered troublemakers or dissatisfied. It might be true that such employees are dissatisfied. However, does an organization want employees who are satisfied with the status quo? Often employees who seek to create change should be most prized. Some fear comes from making recommendations for improvement and having those recommendations ignored.

Another type of fear should be recognized by top managers who desire to improve quality. Many employees view process-improvement efforts as disguised excuses for major layoffs. Recent efforts at reengineering corporations have been synonymous with layoffs. Often, after pursuing the easy solution of downsizing, the same cultural and organizational barriers that impeded improvement are still there. However, the company has lost the ability to be creative and really improve its ability to increase value to the customer. One solution was offered by a major midwestern defense contractor. It developed a written policy stating that management reserved the right to reduce staffing levels as a result of economic downturns. However, the written policy also stated, "no layoffs will result from productivity or quality improvement projects or efforts." Many Japanese firms overcome this same fear issue by offering lifetime employment.

9. *Break down barriers between departments*. People in research, design, sales, and production must work as a team to foresee problems of production and use that may be encountered with the product or service. In many companies, the time it takes to get design and marketing concepts to market is extremely long. Ingersoll-Rand produces a hand grinder that once required four years to develop a new generation of the product. One employee was quoted as saying, "It took us longer to develop a new generation of the product than it took the Allies to win World War II." In the new competitive environment, such delays in design can jeopardize a company's ability to compete. Honda nearly bankrupted Yamaha in the 1980s because of Honda's ability to introduce new designs to market rapidly. One reason for slow design cycles was the **sequential or departmental approach to design**. This approach requires product designers, marketers, process designers, and production managers to work through organizational lines of authority to perform work. The alternative is **parallel processing in focused teams** who work simultaneously on designs.

10. *Eliminate slogans, exhortations, and targets for the workforce that ask for zero defects and new levels of productivity*. Such exhortations only create adversarial relationships because the bulk of causes of low quality and low productivity belong to the system and thus lie beyond the power of the workforce. In Deming's view, exhortations to "get it right the first time" and "zero defects forever" can have the opposite of the intended effect. By pressuring employees to higher levels of productivity and quality, managers place the onus for improvement on the employees. If systems or means for achieving these higher levels of performance are not provided, workers can become jaded and discouraged. Examples of providing systems to employees might be to provide better training, to empower employees to make process decisions, and to provide a strategic structure that ensures alignment of key strategic goals and operational subgoals.

11. *Eliminate work standards on the factory floor*. Eliminate management by objective. Eliminate management by numbers and numeric goals. Substitute leadership. Deming was very much opposed to work measurement standards on the shop floor. It should be noted that work standards are used worldwide, and companies such as Lincoln Electric have been very successful with the skilled use of such standards.

However, often work standards are implemented improperly. It is obvious that if quantity becomes the overriding concern, then quality suffers. More subtle, if work standards are in place, employees who perform at high levels might lose the impetus to continually improve because they already will have satisfied standards.

Management by objective refers to a process of setting annual goals, typically during a performance appraisal, that are binding on the employee. Although goals are set for employees, systems often are not provided by management to attain these goals. For this reason, Deming disdained performance appraisals.

12. *Remove barriers that rob workers of their right to pride in the quality of their work.* The responsibility of supervisors must be changed from sheer numbers to quality. Too often, hourly laborers are hired to perform only the physical tasks assigned by management. Such workers often suffer from low morale and low commitment to the organization. Unskilled managers often add to this problem by reinforcing the fact that employees cannot be trusted with decisions and self-determination. Once, while discussing self-directed work teams, a manager commented that he feared he was "turning the asylum over to the inmates." Such attitudes were obvious on the shop floor, and employees were very dissatisfied with their state. The upside is that after seeing the results of self-directed teams, this same manager became one of the biggest allies of the process of employee empowerment.

13. *Institute a vigorous program of education and self-improvement.* Point 6 referred to training on the job. Point 13 relates to more generalized education. Many quality experts have argued that firms must exhibit the ability to increase and "freeze" learning. Learning in an organization is a function of the creativity of employees and the ability of the organization to institutionalize the lessons learned over time. This is difficult in firms that have high employee turnover. One of the benefits of the ISO 9000:2000 international quality standard is the requirement that firms document their processes and improvements to the processes. Procedure manuals can help make learning permanent. However, this not enough. Organizational learning requires a structure that reinforces and rewards learning. Such an organization is difficult to create in a command-and-control environment because command-oriented managers will not understand what it takes to allow employees to achieve their best.

14. *Put everybody in the company to work to accomplish the transformation.* The transformation is everybody's job. Point 14 reinforces the fact that everyone in the organization is responsible for improving quality. This again reinforces the fact that a total system for improving quality is needed that includes all the people in the organization.

The Deadly Diseases

Deming outlined deadly diseases that he felt would keep the United States or any other country from achieving top quality and competitiveness in a world market. These deadly diseases are shown in Table 2-3.

A Theory Underlying the Deming Method

It has been argued that the 14 points do not represent a theory but do represent the artifacts of a theory. Anderson, Rungtusanatham, and Schroeder[5] propose a theoretical causal model underlying the Deming management method. Using a Delphi-based

[5]Anderson, J., Rungtusanatham, M., and Schroeder, R., "A Theory of Quality Management Underlying the Deming Management Method," *Academy of Management Review* 19, 3 (1994):472–509.

TABLE 2-3 Deming's Deadly Diseases

1. Lack of constancy of purpose
2. Emphasis on short-term profits
3. Evaluation of performance, merit rating, or annual review
4. Mobility of management
5. Running a company on visible figures alone
6. Excessive medical costs for employee health care
7. Excessive costs of warrantees

SOURCE: Adapted from W. E. Deming, *Out of the Crisis* (Boston: MIT/CAES, 1986), pp. 97–148.

process, these University of Minnesota researchers developed the model in Figure 2-4. This model shows that visionary leadership is an antecedent to an organizational system supporting learning. Organizational learning leads to process management, which provides process outcomes or the desired objectives resulting from the process such as high quality or low cost. This focus on process improvement results in higher employee satisfaction and customer satisfaction.

The importance of this theoretical model is that it can provide a basis for researchers to better understand quality improvement. As trained observers, by studying the relationships among variables, researchers can help managers understand what is necessary for quality improvement.

LEADING CONTRIBUTORS TO QUALITY THEORY: JOSEPH M. JURAN

Along with Deming, Joseph Juran (see Fig. 2-5), born in Romania in 1904, was responsible for the growth of quality in the past half-century. He also visited the Japanese Union of Scientists and Engineers to teach quality concepts. Juran took a more strategic and planning-based approach to improvement than does Deming. Juran promotes

FIGURE 2-4 Theoretical Model Underlying the Deming Method

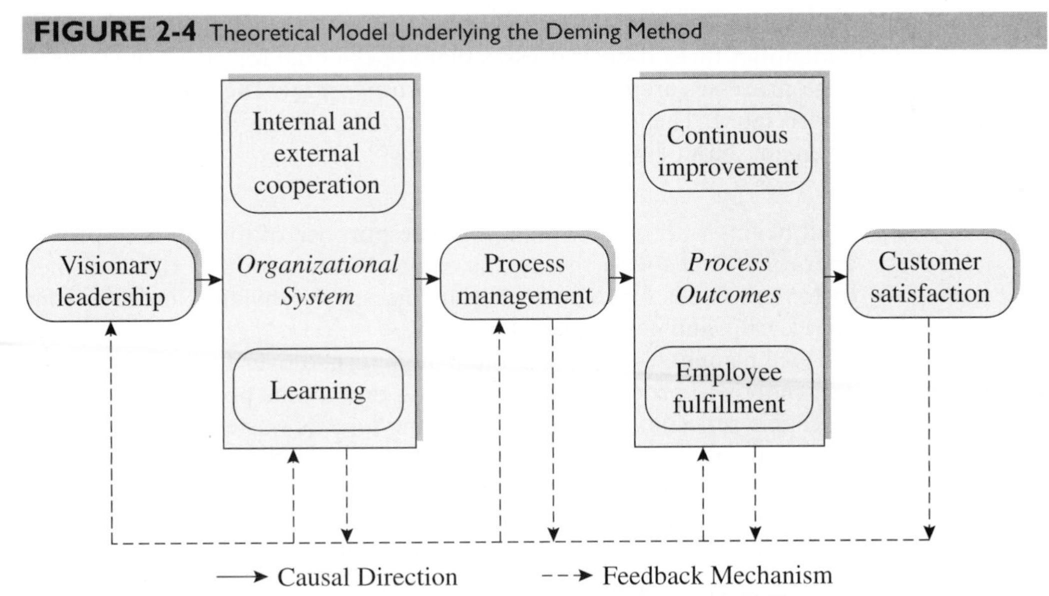

SOURCE: Adapted from J. Anderson, M. Rungtusanatham, and R. Schroeder, "A Theory of Quality Management Underlying the Deming Management Method," *Academy of Management Review* 19, 3 (1994): 481.

FIGURE 2-5 Joseph M. Juran

Juran Institute, Inc.

the view that organizational quality problems are largely the result of insufficient and ineffective planning for quality. He argues that companies must revise strategic planning processes and achieve mastery over these processes. The means proposed by Juran establishes specific goals to be reached and plans for reaching those goals. In addition, Juran's process assigns clear responsibility for meeting the goals and bases rewards on results achieved.

The Juran Trilogy

Juran identifies three basic processes that are essential for managing to improve quality. These processes are referred to as the *Juran trilogy*. The points of the Juran trilogy are interrelated. The three aspects of Juran's trilogy are planning, control, and improvement. Juran discusses these as follows[6]:

> It all begins with quality planning. The purpose of quality is to provide the operating forces with the means of producing products that can meet the customer's needs, products such as invoices, polyethylene film, sales contracts, service calls, and new designs for goods.
>
> Once planning is complete, the plan is turned over to the operating forces. Their job is to produce the product. As operations proceed, we see that the process is deficient: 20% of the operating force is wasted, as the work must be redone due to quality deficiencies. The waste then becomes chronic because of quality deficiencies. The waste becomes chronic because it was planned that way.
>
> Under conventional responsibility patterns the operating forces are unable to get rid of that planned chronic waste. What they do instead is carry out quality

[6]Juran, J. M., *Juran on Planning for Quality* (Boston: Free Press, 1988).

control to prevent things from getting worse. Control includes putting out the fires, such as sporadic spikes.

Control versus Breakthrough

Another important Juran concept is control versus breakthrough. According to Juran, control is a process-related activity that ensures that processes are stable and provides a relatively consistent outcome. Control involves gathering data about a process to ensure that the process is consistent. Control is discussed in more depth in Chapters 12 and 13.

Breakthrough improvement implies that the process has been studied and that some major improvement has resulted in large, nonrandom improvement to the process. The difference between control and breakthrough can be understood when considering a disease such as polio. Control activities involved improving health by quarantining people who had the disease. Breakthrough improvement occurred with the development of the polio vaccine that eradicated the disease.

It is important to understand that control and breakthrough-related activities should occur simultaneously. This distinction clarifies a false dichotomy sometimes associated with continuous improvement versus reengineering. Some managers believe that continuous improvement prevents companies from pursuing large improvements because the focus on detail may lead to neglect of larger needed changes. However, there is nothing about continuous improvement that precludes large improvements. The optimal set of improvement activities probably involves some mix of continuous improvement and breakthrough improvement activities.

Project-by-Project Improvement

Juran teaches that improvement in organizations is accomplished on a project-by-project basis "and in no other way." The project-by-project approach advocated by Juran is a planning-based approach to quality improvement.

In planning for quality improvements, Juran states that managers must prioritize which projects will be undertaken first. Organizations involve a hierarchy of languages (see Fig. 2-6). In this hierarchy, the work that is done operationally at the lowest level is performed by analysts who speak in the language of things. The language of things is typically technical, including jargon and engineering terminology. In order to help determine what projects should be undertaken, these technical people must use the language

FIGURE 2-6 The Hierarchy of Languages

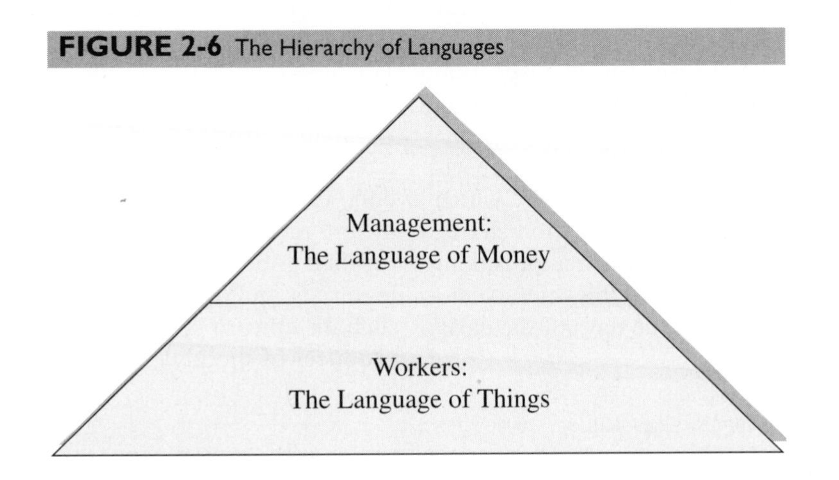

of management, that is, money. Therefore, projects that are identified for possible adoption are prioritized based on financial return. One of the problems many companies experience when implementing quality improvement is showing bottom-line results.

Juran has had a very profound impact on the practice of quality management worldwide. A Closer Look at Quality 2-2 discusses him further.

A CLOSER LOOK AT QUALITY 2-2

JURAN ON THE PAST CENTURY OF QUALITY[b]

Well into his nineties, Juran wrote about the history of quality. Following is a synopsis of some of his recollections.

THE TAYLOR REVOLUTION

The driving force of the Century of Productivity was the movement known as *scientific management*. It was launched by the American engineer and manager Frederick W. Taylor. It made a basic change in managerial practice—the separation of planning from execution. The premise behind the change was that the workers and supervisors of that era lacked the education base needed to do planning. Hence Taylor gave the planning function to managers and engineers. He limited the supervisors and workers to the function of executing the plans.

RESPONSE TO THE TAYLOR REVOLUTION

The upper managers responded by revising the organization. They moved the inspectors out of the production departments and into a central inspection department headed by a chief inspector. To provide added independence, the chief inspector reported to the plant manager or to the vice president for manufacturing. Those central inspection departments became the quality workhorses during the first half of the twentieth century.

In due course, the central inspection departments grew into the quality departments, which today are a feature of so many organization charts. Often enough, they are headed by a vice president for quality who reports directly to the chief executive officer. It is a far cry from the days of the early 1900s.

Creation of the central quality departments also led to two developments which have done a lot of damage: First, many upper managers concluded that quality was the responsibility of the quality department. This belief then made it easier for departments such as production to give top priority to other parameters. Second, upper managers became detached from the quality function. Many concluded that by delegating quality to the quality manager, they could devote their own time to other matters. As they did so, they became progressively less and less informed about quality. Then when the crisis came, they lacked the knowledge needed to choose a proper course of action.

In retrospect, the use of inspection to attain quality involved inherent weaknesses such as high costs and shaky habits. Nevertheless, it made companies competitive in quality on condition that their competitors used the same strategy. That condition was largely met until the Japanese quality revolution came over the horizon.

THE JAPANESE QUALITY REVOLUTION

We now turn to the events that followed World War II. By far, the most important of these was the Japanese quality revolution, which opened the way for Japan to become an economic superpower.

Japan's efforts to achieve greatness through military conquest had failed. Now it would have to be done through trade. Lacking natural resources, this meant importing materials, processing them into finished goods, selling those goods, importing more materials, and so on. The major obstacle to creating such an upward spiral was Japan's reputation as a producer of shoddy goods.

[b]Juran, J. M., *Architect of Quality*, McGraw-Hill, NY, 2004.

To improve that reputation required some fundamental changes in habit patterns. The Japanese CEOs were prepared to make such changes—the shock of losing the war had opened their minds. So they set out to improve their quality reputation. Through the Keidanren (the Federation of Economic Organizations) and the JUSE (Japanese Union of Scientists and Engineers) the companies acted collectively:

1. They sent teams abroad to learn how foreign countries achieved quality.
2. They translated foreign literature into Japanese.
3. They invited two American experts—Deming and myself—to give lectures.

Deming's lectures were on statistical methods, especially the Shewhart control chart. My lectures were on managing for quality, especially on the concept and methodology of annual quality improvement. Let me here deal with a widespread misconception.

Some people believe that had these two Americans not given their lectures, the Japanese quality revolution would not have happened. In my view, this belief has no relation to reality. Had Deming and I never gone, the Japanese quality revolution would have taken place without us.

Each of us did bring to Japan a structured training package the Japanese had not yet developed. In that sense, each of us gave the Japanese a degree of a jump start. But we also did the same for many other countries, none of which succeeded in building such a revolution. That is why I tell my audiences that the unsung heroes of the Japanese quality revolution were the Japanese managers.

Pareto Analysis

Joseph Juran identified an economic concept that he applied to quality problems. This economic concept is called **Pareto's law** or **the 80/20 rule**. This law is named after Vilfredo Pareto (1848–1923), an Italian economist, who modeled income distributions in Milan, Italy. Pareto found that 80% of the wealth in Milan was held by about 20% of the population. Applying Pareto's law to quality problems, imagine a grocer who decides to survey his customers to investigate where quality improvement is needed in his store. He offers a free pound of hamburger to customers who agree to fill out the survey. One of the open-ended questions in the survey asks, "Where do we most need to improve?" Do you suppose that the responses will be uniformly distributed among several problems? No. The majority of the respondents will identify a certain aspect of the grocer's business as the major quality issue. For example, out of 100 respondents, 80 answer, "the meat department needs to be improved." Therefore, in this case, 80% of the dissatisfaction with quality is related to a single area: the meat department.

Using Pareto's law, we see that the majority of quality problems are the result of relatively few causes. Juran dichotomizes the population of causes of quality problems as the "vital few" and the "trivial, but useful, many." Pareto analysis is used as one of the basic seven tools of quality.

LEADING CONTRIBUTORS TO QUALITY THEORY: KAORU ISHIKAWA

The founder of the Japanese Union of Scientists and Engineers (JUSE) was the distinguished business leader Ichiro Ishikawa. His son, Kaoru Ishikawa, went on to lead JUSE during its growth years and became the foremost Japanese leader in the Japanese quality movement.[7] Kaoru Ishikawa provided tools that worked well within the Deming and Juran frameworks.

[7] Many of the Ishikawa concepts discussed here are derived from his book, *Guide to Quality Control* (White Plains, NY: Quality Resources, 1968).

The Basic Tools of Quality

Ishikawa was a great believer in training. In fact, under his guidance, training was a key component of the mission of JUSE, including statistical quality control training. Perhaps Ishikawa's greatest achievement was the development and dissemination of the basic seven tools of quality (B7). As the developer of these tools, Ishikawa is credited with *democratizing statistics*. Although statistical quality control hitherto had been the domain of specialized statisticians, Ishikawa felt that to be successful, firms must make everyone responsible for statistical analysis and interpretation.

Although not strictly a theoretical contribution, the tools are presented in Chapter 10 because they are used for continuous improvement. The major theoretical contribution of Ishikawa is his emphasis on total involvement of the operating employees in improving quality. Ishikawa is credited for coining the term *company-wide quality control* in Japan.

Ishikawa's Quality Philosophy

Ishikawa spent his life working to improve quality in Japan. His ideas were synthesized into 11 points that made up his quality philosophy (see Table 2-4). Ishikawa is often overlooked in the United States; however, every firm that pursues quality improvement uses his tools. By democratizing statistics, he allowed for the complete involvement of the workforce in improving quality and performance.

LEADING CONTRIBUTORS TO QUALITY THEORY: ARMAND FEIGENBAUM

During the years when quality was overlooked as a major competitive factor in the United States, two books were used by most every quality professional. These books were *Statistical Quality Control*[8] by Eugene Grant and Richard Leavenworth and *Total Quality Control*[9] by Armand Feigenbaum. Whereas the approach of the former was statistically oriented, Feigenbaum's book studied quality in the context of the business organization. Feigenbaum's primary contribution to quality thinking in America was his assertion that the entire organization should be involved in improving quality. He was the first in the United States to move quality from the offices of the specialist back to the operating workers. This occurred in the 1950s.

TABLE 2-4 Ishikawa's Eleven Points

1. Quality begins with education and ends with education.
2. The first step in quality is to know the requirements of the customer.
3. The ideal state of quality control is when inspection is no longer necessary.
4. Remove the root causes, not the symptoms.
5. Quality control is the responsibility of all workers and all divisions.
6. Do not confuse the means with the objectives.
7. Put quality first and set your sights on long-term objectives.
8. Marketing is the entrance and exit of quality.
9. Top management must not show anger when facts are presented to subordinates.
10. Ninety-five percent of the problems in a company can be solved by the seven tools of quality control.
11. Data without dispersion information are false data.

SOURCE: Adapted from K. Ishikawa, *Guide to Quality Control* (White Plains, NY: Quality Resources, 1968).

[8]Grant, E., and Leavenworth, R., *Statistical Quality Control* (New York: McGraw-Hill, 1988, original 1946).
[9]Feigenbaum, A., *Total Quality Control* (New York: McGraw-Hill, 1983, original 1951).

TABLE 2-5 Feigenbaum's 19 Steps

1. Total quality control is defined as a system of improvement.
2. Big Q quality (company-wide commitment to TQC) is more important than little q quality (improvements on the production line).
3. Control is a management tool with four steps.
4. Quality control requires integration of uncoordinated activities.
5. Quality increases profits.
6. Quality is expected, not desired.
7. Humans affect quality.
8. TQC applies to all products and services.
9. Quality is a total life-cycle consideration.
10. Control the process.
11. A total quality system involves the entire company-wide operating work structure.
12. There are many operating and financial benefits of quality.
13. The costs of quality are a means for measuring quality control activities.
14. Organize for quality control.
15. Managers are quality facilitators, not quality cops.
16. Strive for continuous commitment.
17. Use statistical tools.
18. Automation is not a panacea.
19. Control quality at the source.

SOURCE: Adapted from A. Feigenbaum, *Total Quality Control* (New York: McGraw-Hill, 1991, original 1951).

Feigenbaum proposes a three-step process to improving quality. These steps involve *quality leadership, quality technology*, and *organizational commitment*. Leadership is the motivating force for quality improvement. Quality technology includes statistics and machinery that can be used to improve technology. Organizational commitment includes everyone in the quality struggle.

Major impediments to improving quality included the four deadly sins of hothouse quality, wishful thinking, producing overseas, and confining quality to the factory. *Hothouse quality* refers to quality programs that receive a lot of hoopla and no follow-through. This is a failing of many firms that do not commit resources *over time*. *Wishful thinking* occurs with those who would pursue protectionism to keep American firms from having to compete on quality. *Producing overseas* is a panacea sometimes undertaken by managers who wish that out of sight, out of mind could solve quality-related problems. Even well-run companies can fall prey to this thinking. The disk-memory division of Hewlett-Packard off-shored its production from the United States to Malaysia in a move to overcome its noncompetitiveness in the market. The move to Malaysia did nothing to solve its problems related to design and process. HP's problems have not gotten better since the move—they have gotten worse. *Confining quality to the factory* means that quality historically has been viewed as simply a shop-floor concern. In actuality, it is everyone's responsibility.

The 19 Steps of TQC

Armand Feigenbaum proposed 19 steps for improving quality (see Table 2-5). These 19 steps outline his approach to the total quality control (TQC) system, which emphasize organizational involvement in improving quality.

LEADING CONTRIBUTORS TO QUALITY THEORY: PHILIP CROSBY

Philip Crosby has been the most successful in marketing his quality expertise of all the leading quality authors and thinkers. Although he began his career as a podiatrist (which he disliked), Crosby pursued a career as a reliability engineer with Crosley Corporation in Indiana. Later, he worked for Martin Corporation as a quality manager, and then he served as the director for quality at International Telephone and Telegraph. In 1979 he founded Crosby and Associates of Winter Park, Florida—the world's largest and most successful quality consulting company. In 1991 he sold Crosby and Associates and founded Career IV to help train executives. Prior to his death in August 2001, he repurchased Crosby and Associates to "fix their image."

Crosby became very well known for his authorship of the book *Quality Is Free*.[10] The primary thesis of this book is that quality, as a managed process, can be a source of profit for an organization. Crosby specified a quality improvement program consisting of 14 steps (see Table 2-6). These steps underlie the Crosby zero-defects approach to quality improvement. His approach also emphasized the behavioral and motivational aspects of quality improvement rather than statistical approaches. In his 14 steps, Crosby prescribed actions for management and workers within the context of his program.

Crosby identified a period of enlightenment during which management becomes attuned to the importance of quality. This enlightenment can be achieved through exposure to videos, books, and seminars and by a sense of needing to respond to competitive challenges. A quality improvement team is then established. This team consists of a member from each department within the organization. Organizational quality measures are established and continually reviewed by the team.

Next, quality-related costs are evaluated. This effort is to be coordinated by the controller's office by establishing the location of areas where corrective action would be most profitable. Quality awareness is emphasized, which addresses the philosophical underpinnings of the Crosby approach. This step addresses the importance of the worker in assuring quality and demonstrating the need to make the employee aware of the necessity for quality improvement.

The final steps are corrective action, establishing an ad hoc committee for the zero-defects program, supervisory training, the establishment of a zero-defects day, error-cause removal, employee recognition, and the establishment of quality councils. These steps then become ingrained by beginning the process over again.

TABLE 2-6 Crosby's 14 Steps

1. Make it clear that management is committed to quality.
2. Form quality improvement teams with representatives from each department.
3. Determine how to measure where current and potential quality problems lie.
4. Evaluate the cost of quality and explain its use as a management tool.
5. Raise the quality awareness and personal concern of all employees.
6. Take formal actions to correct problems identified through previous steps.
7. Establish a committee for the zero-defects program.
8. Train all employees to actively carry out their part of the quality improvement program.
9. Hold a zero-defects day to let all employees realize that there has been a change.
10. Encourage individuals to establish improvement goals for themselves and their groups.
11. Encourage employees to communicate to management the obstacles they face in attaining their improvement goals.
12. Recognize and appreciate those who participate.
13. Establish quality councils to communicate on a regular basis.
14. Do it all over again.

SOURCE: P. Crosby, *Quality Is Free: The Art of Making Quality Certain* (New York: Mentor Executive Library, 1979). Reproduced with permission of The McGraw-Hill Companies.

[10]Crosby, P., *Quality Is Free: The Art of Making Quality Certain* (New York: Mentor Executive Library, 1979).

Although he prescribes quality teams consisting of department heads, Crosby did not promote the same kind of strategic planning proposed by Deming and Juran. Crosby adopted a human resources approach similar to Deming's in that worker input is valued and is encouraged as central to the quality improvement program.

LEADING CONTRIBUTORS TO QUALITY THEORY: GENICHI TAGUCHI

The Taguchi method was first introduced by Dr. Genichi Taguchi to AT&T Bell Laboratories in the United States in 1980. Because of its increased acceptance and utilization, the Taguchi method for improving quality is now believed to be comparable in importance to the Deming approach and to the Ishikawa concept of total quality control. Taguchi's method is a continuation of the work in quality improvement that began with Shewhart's work in statistical quality control and Deming's work in improving quality. Objectives of the Taguchi method are synopsized in Table 2-7.

Among the unique aspects of the Taguchi method are the Taguchi definition of quality, the quality loss function (QLF), and the concept of robust design.

Definition of Quality

A number of definitions of quality have been identified in the literature. The traditional definition of quality was conformance to specifications. However, as discussed in Chapter 1, in recent years this narrow definition of quality has been extended.

Taguchi also diverges from the traditional definition of quality. In Taguchi's terms, *ideal quality* refers to a reference point for determining the quality level of a product or service. This reference point is expressed as a target value. Ideal quality is delivered if a product (or a service tangible) performs its intended function throughout its projected life under reasonable operating conditions without harmful side effects. In services, because production and consumption of the service often occur simultaneously, ideal quality is a function of customer perceptions and satisfaction. Taguchi measures service quality in terms of loss to society if the service is not performed as expected.

Quality Loss Function

Taguchi doesn't agree with traditional quality thought as it relates to specification. Normally, when specifications are set, a target is specified with some allowance for variation. Taguchi doesn't agree with the notion of allowable variation. He states that any deviation from target specs results in loss to society. Alternatively, traditional thinking implies that there is no loss to society if a measurement is near to being out of specification but, nevertheless, remains within the established specification limits. For example, if the quality standard for a restaurant is that 6 (±1) ounces of french fries be included in each order of french fries, an average of 6.90 ounces of french fries per order over long periods of time would increase costs by 15%, which could translate into hundreds or thousands of dollars in lost profit, depending on the size of the firm.

TABLE 2-7 The Taguchi Method

The Taguchi method provides

1. A basis for determining the functional relationship between controllable product or service design factors and the outcomes of a process.
2. A method for adjusting the mean of a process by optimizing controllable variables.
3. A procedure for examining the relationship between random noise in the process and product or service variability.

The 6.90 value is closer to being out of specification than within specification, but traditional thinking asserts that no loss is incurred by the company because the variation is within specification limits (i.e., because 6.9 < 7, the process is in spec). Again, Taguchi would assert that processes should result in a fill of 6 ounces.

Robust Design

The Taguchi concept of robust design states that products and services should be designed so that they are inherently defect-free and of high quality. Taguchi devised a three-stage process that achieves robust design through what he terms *concept design, parameter design*, and *tolerance design*.

This section has emphasized the conceptual foundation to the Taguchi method of quality improvement. Chapter 14 further explores this approach to quality improvement from a quantitative perspective.

LEADING CONTRIBUTORS TO QUALITY THEORY: THE REST OF THE PACK

The contributors mentioned to this point have stood the test of time, but it is important to remember that as practitioners, they suffered from the biases associated with inductive reasoning. However, as we see in this section, inductive reasoning might be preferred to the deductive development of untested theories. In the following paragraphs, several well-known quality authors are considered. Some of these individuals, such as Robert Camp and Stephen Covey, have made important contributions to our understanding of quality. Others have been proven wrong. Underlined in this section is the risk associated with following unsound quality approaches.

Robert C. Camp

Robert C. Camp is the principal pioneer of benchmarking. **Benchmarking** is the sharing of information between companies so that both can improve. This was thought impossible just a few years ago, but his efforts within Xerox Corporation have proved otherwise. As a result of the work of Xerox and other corporations, benchmarking is now a very important, proven practice used worldwide. Camp's best-selling book, *Benchmarking: The Search for Industry Best Practices That Lead to Superior Performance*, is an outstanding handbook. Benchmarking is discussed in greater depth in Chapter 6.

Stephen R. Covey's "8" Habits

Stephen Covey is a management consultant from Utah who leads FranklinCovey, one of the most successful management consulting companies in the world. Dr. Covey, a former professor at Brigham Young University, is best known for his book, *The 7 Habits of Highly Effective People*.[11] His approach to management is value-based in that he proposes that people in management live a life that balances professional with personal and spiritual growth.

According to Dr. Covey, our beliefs affect how we interact with others, which in turn affects how they interact with us. As a result, we need to focus on how we approach our lives rather than focusing on external factors that affect our lives. Implicit in many of Covey's basic teachings are many quality management principles from people such as Deming. These principles are woven together with a values-based approach to life.

---

[11]Covey, S., *The 7 Habits of Highly Effective People*, Free Press, NY, 2004.

His seven habits include:

1. *Be proactive.* This is the ability to control one's environment, rather than have it control you, as is so often the case. Managers need to control their own environment, using self-determination and to demonstrate the power to respond to various circumstances.
2. *Begin with the end in mind.* This means that the manager needs to be able to see the desired outcome and concentrate on activities that help in achieving that end.
3. *Put first things first.* Managers need to personally manage themselves and implement activities that aim to achieve the second habit—looking to the desired outcome. Covey states that habit two is the first, or mental, creation; habit three is the second, or physical, creation.
4. *Think win-win.* This is the most important aspect of interpersonal leadership because most achievements are based on cooperative effort. Therefore, the aim needs to be win-win solutions for all.
5. *Seek first to understand and then to be understood.* By developing and maintaining positive relationships through good communications, the manager can be understood and can understand subordinates.
6. *Synergize.* This is the habit of creative cooperation—the principle that collaboration often achieves more than could be achieved by individuals working independently.
7. *Sharpen the saw.* This involves learning from previous experience and encouraging others to do the same. Covey sees development as one of the most important aspects in being able to cope with challenges and aspire to higher levels of ability.

Covey recently published a book with an eighth habit:

8. *Find your voice, and inspire others to find theirs.* This invites the merging of talent, passion, and conscience to achieve and to help others achieve. This is implicit in a life of service.

Covey is discussed here because wherever you work, around the world, people will be discussing Covey's habits. You need to be familiar with them.

Tom Peters

Tom Peters is a noted author, consultant, and speaker who is widely recognized. The Stanford-trained coauthor of the book *In Search of Excellence*, Peters gives very popular seminars. His approach to studying quality in excellent companies was empirically based. The research for his book involved a case study of several firms and resulted in eight basic practices found in the excellent firms. These eight practices include a bias for action, getting close to the customer, promoting entrepreneurship, productivity through people, value-driven management, sticking to the core competencies, lean staff, and implementing appropriate amounts of supervision and empowerment. *In Search of Excellence* is a thought-provoking, though methodologically loose, approach to assessing business practice.

Peters's recent work has been less rigorous, as evidenced by the title of his recent book, *In Pursuit of Wow!* This latest work is more in the realm of entertainment coupled with serious management thinking.

Michael Hammer and James Champy

Michael Hammer and James Champy urged a form of deductive reasoning combined with entertainment that has resulted in unfortunate consequences for many people and companies. The product of this collaboration is termed **reengineering**. The underlying precept of reengineering is sound: Firms can become inflexible and resistant to change and must be able to change in order to become competitive. The problem is in the process they promoted in the book *Reengineering the Corporation*. This process involves the CEO of the corporation developing a business case (the Harvard University B-School method) followed by a set of recommendations. He or she then charges others with rapidly implementing the recommendations without further study or analysis.

Hammer and Champy have been surprisingly candid about the failings of reengineering, admitting to a 70% or higher failure rate. (See A Closer Look at Quality 2-3.)

By ignoring the necessity for attention to detail and analysis that has characterized many world-class firms, Hammer and Champy led many firms to make radical changes that have led to major failures. If there is a lesson to be learned from the reengineering failures, it is this: *Some quality and performance improvement approaches are brainchildren. Others have been observed to work in a number of organizations, in a variety of cultures, and in a number of economic sectors. Avoid the former until they become the latter*. This is common sense that has been validated by unfortunate experience. The risk of failure must be considered in decision making. Analysis and attention to detail cannot be overlooked. Fortunately, there are well-founded approaches to organizational redesign that can be applied and have been applied for decades. The backlash against reengineering is likely to make these efforts less positively viewed.

Granted that, to an extent, the message Hammer developed was lost by managers who misapplied his ideas, the human costs have been great. As discussed in A Closer Look at Quality 2-3, the objective should not be to adopt whatever tool is currently "hot." This search for the next hot concept causes firms and organizations to focus fundamentally on the wrong questions.

VIEWING QUALITY THEORY FROM A CONTINGENCY PERSPECTIVE

By now you might be asking, With all this disagreement about how to approach quality improvement, how should I proceed? Perhaps you have gained more empathy for CEOs and business leaders who are distrustful of quality management as a field. There is a great deal of contradictory information about how firms should improve quality.

This mass of contradictory information is not unique to quality. A similar state exists in finance, where much divergent advice is given on how best to invest funds. In marketing, new approaches are constantly emerging. In fact, much of leadership and management has to do with being able to sort through and make sense of conflicting information.

Because a variety of approaches can work to improve quality, it is best to focus on fundamental questions such as: What are our strengths? Where are our competencies? In what areas do we need to improve? What are our competitors doing to improve? What is our organizational structure? These are fundamental questions that are asked during the self-assessment phase of strategic planning.

Once these questions are answered, a profound understanding of the business emerges. Based on this knowledge of the business, an understanding of the major approaches to quality improvement will provide a means for selecting those points, philosophies, concepts, and tools that will form the basis for improvement.

As has already been stated, Baldrige winners and other firms well known for quality do not adopt only one quality philosophy. For example, firms that are successful in quality do not adopt a blanket "Deming approach to quality." The successful firms

A CLOSER LOOK AT QUALITY 2-3

HAMMER RECANTS (SORT OF), OR IN SEARCH OF THE LOST PRODUCT TO SELL[c]

This boxed excerpt is a continuation of the theme in A Closer Look at Quality 2-1. It explains how Michael Hammer and others market the consulting products that have made them very wealthy.

Michael Hammer, the management guru whose ideas launched tens of thousands of pink slips, wants to drive home a new message that some of his followers have missed. At a conference in Boston, the cofounder of the reengineering movement appeared before 437 managers at GE Fanuc Automation North America. This organization embraced Dr. Hammer's idea that hierarchies should be smashed and replaced with streamlined "process" teams made up of marketing, manufacturing, sales, and service people who use computers to combine tasks and who work without a lot of supervisors.

But Mr. Borwhat (the Fanuc manager) told the gathering that, unlike a lot of companies, GE was not going to lay off a lot of workers. In fact, the company was to increase workers and revenues.

"What can I say?" Dr. Hammer asked, turning toward the managers whose companies paid $2,200 a head for them to attend his three-day meeting. "It's right. The real point of this is longer-term growth on the revenue side. It's not so much getting rid of people. It's getting more out of people."

Three years after Dr. Hammer and consultant James Champy launched the hottest management fad of the 1990s with their best-selling book *Reengineering the Corporation*, Dr. Hammer pointed out a flaw: He and other leaders of the $4.7 billion reengineering industry forgot about people. "I wasn't smart enough about that," he says. "I was reflecting my engineering background and was insufficiently appreciative of the human dimension. I've learned that's critical."

So have a lot of businesses, disillusioned by the backlash against layoffs, overwork, and constant upheaval stemming from their efforts to adopt Dr. Hammer's model. Companies learned that simply cutting staff, rather than reorganizing the way people in different functions work, won't yield the "quantum leaps" in performance Hammer and Champy heralded in their book.

HOT NEW PRODUCTS

As a result, Dr. Hammer and other consulting giants are scrambling to remodel their next strategies. They may even choose to turn them in for new models as they look for the next big thing. Hot candidates include knowledge management and growth strategy. The search for something better has caught fire as the appeal of reengineering has waned. When a Boston consulting group asked executives at 1,000 companies to rate various management tools, reengineering didn't score above average on any of several matrixes.

James Champy, the consultant who left the firm he founded, CSC for Perot Systems, is broadening his approach. "There's a need to figure out how to change culture and behavior more quickly." He uses "business transformation" as the working title for his new blend of reengineering, strategy, and cultural change. But he adds, "I'd like to find a better label."

The old model of reengineering certainly has caused trouble. Levi Strauss and Co., which worked with Dr. Hammer, put the brakes on its $850 million reengineering effort after management created turmoil by demanding that 4,000 white-collar workers reapply for their jobs as part of reorganization into "process" groups.

But even as skepticism over reengineering mounts, Dr. Hammer remains an oracle. He has a sense of mission. "After I got tenured at MIT, I got bored and quit. My wife thought I was crazy."

[c]White, J., "Re-engineering Gurus Take Steps to Remodel Their Stalling Vehicles," *Wall Street Journal* 228, 105 (November 26, 1996):A1.

adopt aspects of each of the various approaches that help them improve. This is called the *contingency perspective*.

The keys to the contingency approach are an understanding of quality approaches, an understanding of the business, and the creative application of these approaches to the business. Thus the optimal strategy will apply quality philosophies and approaches to business on a contingency basis. From your own perspective, you need to make correct quality-related decisions. In doing this, you should consider the different quality experts in this chapter and choose those concepts and approaches that make sense for you.

RESOLVING THE DIFFERENCES IN QUALITY APPROACHES: AN INTEGRATIVE VIEW

There are many differences among the approaches to quality management along the supply chain espoused by the experts mentioned in this chapter. However, rather than focusing on differences, it is instructional to review the literature to identify common themes and messages.

Table 2-8 provides a list of variables that are addressed by Deming, Juran, Crosby, Taguchi, Ishikawa, and Feigenbaum. Also included is the services approach to quality by Parasuraman, Zeithamel, and Berry. (cited in Chapters 1 and 6)

By studying the common themes of the various experts, it is clear that some occur often, whereas others are more idiosyncratic. The target in Figure 2-7 shows the variables that are at the core of quality management and those variables that, although important, are less widely supported. The following list of variables is not intended to be all encompassing. At the same time, these are variables that firms should address when seeking to improve performance. The core variables include the following:

Leadership

The role of the leader in being the champion and major force behind quality improvement is critical. The implication is clear—companies having weak leadership will not achieve a market advantage in quality. The task for leaders is to become conversant with quality management approaches. They must then be willing to lead by example, not just by words.

TABLE 2-8 Quality Improvement Content Variables

Variables	*Deming*	*Juran*	*Crosby*	*Taguchi*	*Ishikawa*	*Feigenbaum*	*PZB*
Leadership	y	y	y		y	y	
Information analysis	y	y			y	y	
Strategic planning		y	y			y	y
Employee improvement	y	y	y		y	y	y
Quality assurance of products and services	y	y		y	y	y	y
Customer role in quality	y	y				y	y
Role of quality department	y	y				y	
Environmental characteristics and constraints	y					y	
Philosophy driven	y	y	y	y	y	y	
Quality breakthrough		y					
Project/team-based improvement		y	y	y	y		

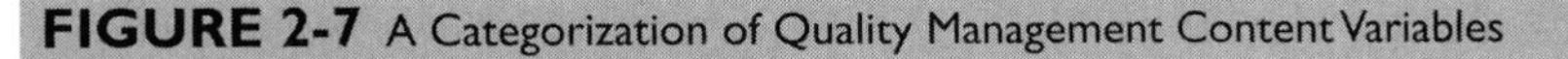
FIGURE 2-7 A Categorization of Quality Management Content Variables

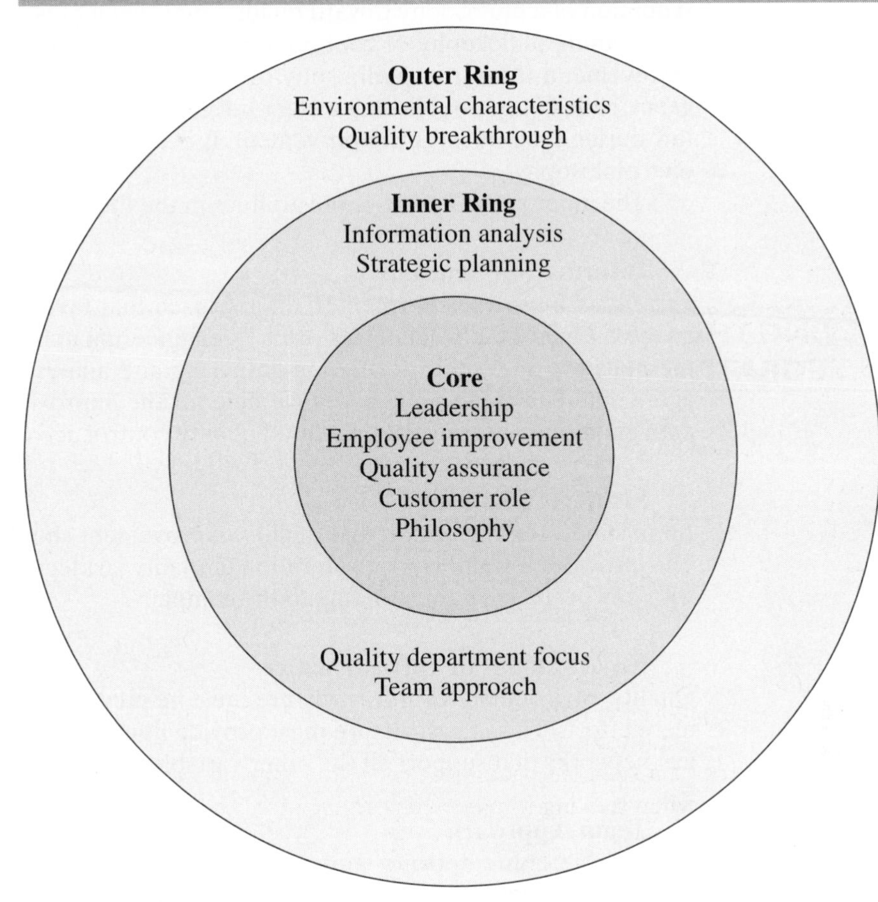

Employee Improvement

Once the leadership is enlightened and motivated to go forward in the quality effort, employees must be trained and developed. This training is a long-term undertaking that requires firms to invest in their employees. When budgeting for training, the direct training delivery costs are not the only costs that should be budgeted. There are also indirect costs associated with temporary lost productivity and time spent in training.

Quality Assurance

Quality experts agree that quality can be assured only during the design phase. Although statistical inspection is an important approach to improving quality, it is inherently reactive. Therefore, efforts must be invested in designing products, services, and processes so that they are consistently of high quality. McDonalds Corporation has built a huge competitive advantage by investing in processes and standardizing those processes so that they always produce the same products regardless of location.

Customer Focus

An understanding of the customer is key to quality management efforts. Unless firms are gathering data about customers and analyzing these data, they are poorly informed about customer needs and wants. The mantra of many businesses today involves customer satisfaction. This is closely related to customer service as a central role for quality.

Quality Philosophy

Adoption of a philosophy toward quality improvement is also important. Whether it is the Deming philosophy of continual improvement, the Ishikawa philosophy of total involvement, the Juran philosophy of project-by-project improvement, or a contingency philosophy, establishing a clear message provides a company with a map to follow during their quest for improvement. It is up to each organization to determine its own philosophy.

The inner and outer rings of variables in the literature include the following:

Information Analysis

Fact-based improvement refers to an approach that favors information gathering and analysis. One of the weaknesses of the reengineering approach was that it overlooked the need for in-depth information gathering and analysis. Most of the experts cited agree that data gathering is a key variable for the improvement of quality. Included in data gathering are statistically related quality control activities.

Strategic Planning

Juran supported the notion that quality improvement should be strategically planned. This provides a framework for a rational quality strategy that will provide alignment with key business factors relating to the company.

Environment or Infrastructure

Quality environment or infrastructure must be created that supports quality management efforts. This infrastructure must provide human resource systems and technological networks that support all the other variables mentioned here.

Team Approach

One of the contemporary approaches to quality management learned from the Japanese is teamwork. More and more, companies are forming cross-functional teams to achieve process improvement. Many firms also have formed teams to manage key processes. In some organizations, employees' jobs are defined by the teams in which they are involved. In such organizations, the firm is a collection of loosely related teams performing the work of the firm.

Two outer-ring variables are major themes emerging from several of the experts and have been adopted widely.

Focus of the Quality Department

As a result of the dispersion of responsibility for quality, the role of the quality department has changed significantly. Rather than performing the policing function, these departments are filling more of a coaching role. Also, the knowledge these quality specialists have is useful for training and in-house consulting. Many firms have now turned their quality departments into profit centers by "selling" their services, on a consulting basis, to other firms.

Breakthrough

The need to make large improvements is not precluded by continuous improvement. Firms must find ways to achieve radical improvements. The process used to achieve this often involves technology or organizational redesign. Analysis and data are necessary for successful breakthrough implementation.

THEORETICAL FRAMEWORK FOR QUALITY MANAGEMENT

As we have discussed in this chapter, many variables build the framework for a quality management theory (see Fig. 2-8). Quality management begins with leadership. The organizational leaders have the authority and monetary ability to drive quality assurance, employee improvement, and the permeation of a corporate philosophy.

The quality philosophy influences and guides the organizational leader in making decisions concerning quality strategy. The philosophy also helps guide decisions concerning quality assurance and employee improvement.

Leadership, quality assurance, philosophy, and employees are encompassed by a focus on the customer. From this perspective, the customer is the focus of all the activities of the firm.

The outer boxes in Figure 2-8 refer to activities and processes that help to improve the core systems relating to people. The breakthrough improvement, team building, data gathering, strategic planning, and quality department coaching are some of the major activities forming the interrelated set of activities making up the quality system.

FIGURE 2-8 A Theoretical Framework for Quality Management

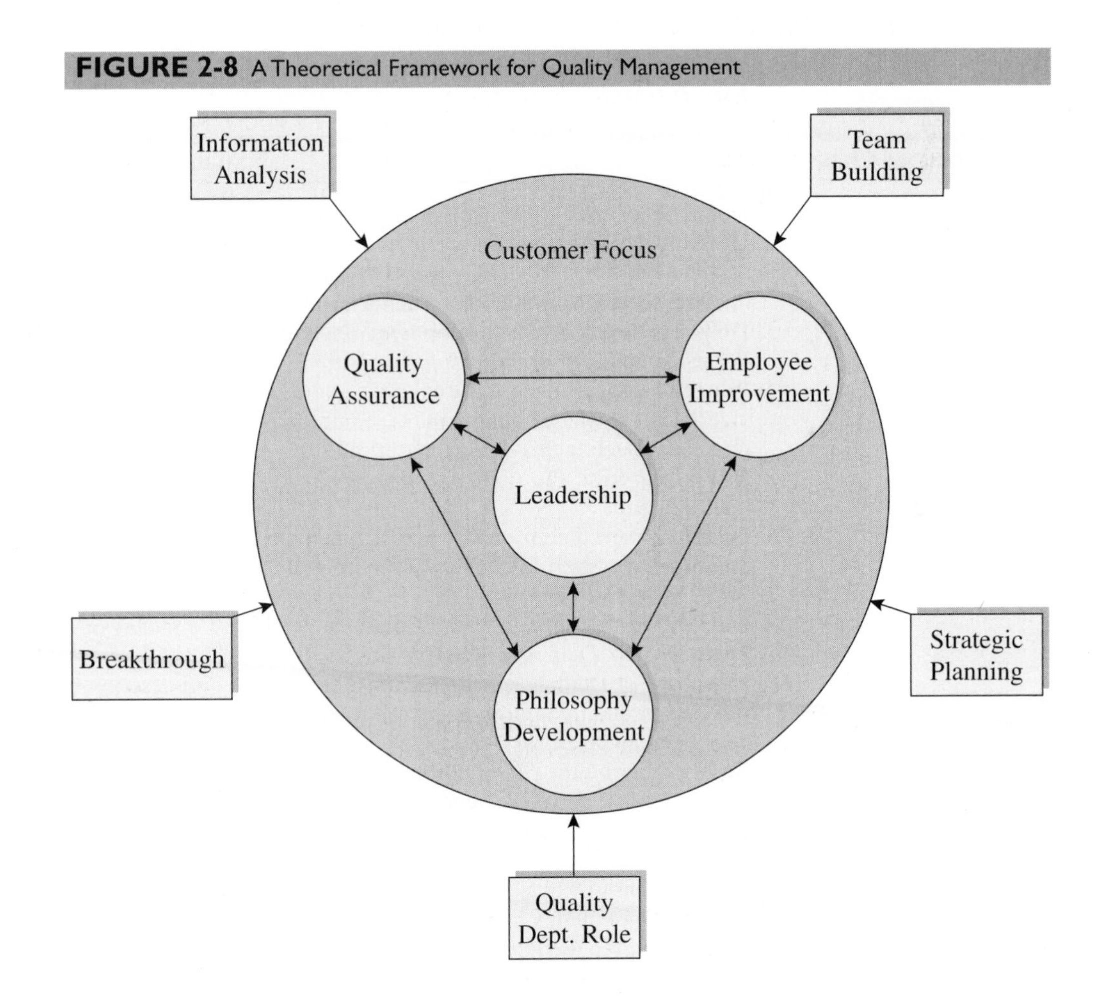

SUMMARY

This chapter briefly introduced the major voices in the growth of the quality movement. Although there is no single theory that is widely adopted to explain the quality phenomenon, the basis for a theory exists. The variables that are proposed in Figure 2-8 form the basis of a model that is testable. We propose that the interaction of these variables forms the basis for a quality system that will result in improved quality.

We introduced a variety of models. The models discussed have been applied in many different organizations and settings all over the world. Therefore, these approaches have worked in a variety of contexts. These are not philosophies that someone just "thought up." They should be pondered and, when appropriate, implemented to achieve organizational improvement.

KEY TERMS

- Benchmarking
- Deduction
- Induction
- Just-in-time (JIT) purchasing
- Parallel processing in focused teams
- Pareto's law (the 80/20 rule)
- Quality at the source
- Reengineering
- Sequential or departmental approach to design

DISCUSSION QUESTIONS

1. Define theory. Why are theories important for managing quality in the supply chain?
2. Describe the differences between induction and deduction. If you developed a theory based solely on your experiences of quality practices in business organizations, would you be basing your theory on induction or deduction? Why?
3. Do you believe that the development of a unified theory of quality management is possible? What is a unified theory?
4. Why do managers need to be cautious about purchasing material (e.g., courses, workbooks, videos, and so on) on quality management from trainers and consultants? How would you go about selecting this type of material?
5. Briefly describe the contributions W. Edwards Deming made to the field of quality management.
6. Deming believed poor quality was not the fault of workers but resulted from poor management of the system for quality improvement. Do you agree with Deming's stand on this issue? Why or why not?
7. Deming was not an advocate of mass inspection as a means of ensuring product quality. Please explain Deming's beliefs in this area.
8. Select one of Deming's 14 points for management and describe how this point could have resulted in quality improvements in a business or volunteer organization with which you have been involved.
9. Briefly describe the contributions that Joseph M. Juran made to the field of quality management. What do you believe was Juran's most significant contribution?
10. Is the concept of scientific management compatible with employee empowerment? Why or why not?
11. Does the phrase "quality is the responsibility of the quality department" reflect a healthy perspective of quality management? Please explain your answer.
12. Briefly describe the Japanese quality revolution following World War II. What can modern-day managers learn from studying the history of this era?

13. What was Joseph Juran's primary contribution to quality thinking in America? Discuss Juran's three-step process to improving quality.
14. Hothouse quality refers to those quality programs that receive a lot of hoopla and no follow-through. Provide several examples of management practices that can lead to hothouse quality. How can hothouse quality be avoided?
15. Compare and contrast Deming's, Juran's, and Crosby's perspectives of quality management. What are the major similarities and differences between their perspectives?
16. Describe Taguchi's perspective of ideal quality. Does this perspective have practical applications? If you were a manager, would you consider using the Taguchi method? Why?
17. According to Hammer and Champy, reengineering programs fail 70% of the time. Why do you think that reengineering programs have such a high failure rate? Can you think of ways to improve the success rate of reengineering programs?
18. Describe how the contingency perspective helps us understand why a single approach to quality management may never emerge.
19. How can a philosophy of quality improvement help a firm in its overall efforts of improving the quality of its products and services?
20. Do you believe that CEOs and business managers should be skeptical about the quality movement, or should they embrace the quality movement and try to involve their firms in as many quality initiatives as possible? Please explain your answer.

Cases

Case 2-1 Rheaco, Inc.: Making a Quality Turnabout by Asking for Advice

Rheaco Homepage: ***www.rheaco.com***

Rheaco, a Grand Prairie, Texas, company that manufactures high-precision parts for the aerospace and defense industries, was in trouble. The company was behind on 50% of its deliveries, was receiving complaints from its customers about product quality, and was experiencing internal scheduling and capacity problems. To make matters worse, its customer base was shrinking as a result of cutbacks in defense spending, and many of its customers were reducing the number of suppliers that they maintained in an effort to improve their own product quality.

Rather than giving up, the top managers at Rheaco sought help. Rather than hire a costly consultant, Rheaco asked the Automation and Robotics Research Institute (ARRI) at the University of Texas for assistance. The ARRI is a university-based institute that works with private manufacturing firms in an effort to disseminate advanced manufacturing concepts and philosophies. Initially, ARRI personnel assisted Rheaco's top management team in articulating a vision statement, examining the company's strengths and weaknesses, and pinpointing areas of concern. A number of concerns were identified, which were contributing both directly and indirectly to Rheaco's problems. Surprisingly, a strength identified by ARRI's analysis was that Rheaco's employees, despite the company's difficulties, had an overall positive attitude. This quality no doubt contributed to Rheaco's ability to eventually work through its problems and return to profitability.

A team of Rheaco and ARRI personnel tackled the company's problems ranging from poorly structured manufacturing processes to low cooperation among departments. Rather than telling Rheaco what to do, the ARRI team worked with Rheaco's management and frontline employees to further define problem areas and develop solutions. An Enterprise Excellence Plan was developed, which acted as a road map for Rheaco's improvement efforts. Consistent with this plan, the following initiatives were implemented, all of which were new to Rheaco:

- Cellular manufacturing
- Just-in-time inventory control
- Total quality management
- Employee empowerment

Each of these initiatives was implemented with a clear rationale and with the support of Rheaco's management and employees. For example, Rheaco had a problem in the area of product flow. Cellular manufacturing is a technique designed to improve the product flow rate by placing all the parts associated with a given product area close to one another. This technique reduces travel time, improves communication, and facilitates continuous flow of the product. The implementation of cellular manufacturing improved the efficiency of Rheaco's operations to the extent that one Rheaco employee remarked, "Material flowed like water, and people moved to the work instead of work to the people." As a result of this initiative, in one particular instance, the time required to produce a component decreased from 40 to 2 hours.

Other improvements were made, particularly in the areas of shipping and receiving, inventory control, and human resource management. After ARRI had been working with Rheaco for a period of time, the company started identifying and correcting problems on its own, which is exactly what is supposed to happen. The mission of ARRI is to transfer advanced manufacturing concepts and philosophies to a private firm and then to withdraw. ARRI also helps the firms that it works with establish relationships with other ARRI-assisted companies, which was important to Rheaco.

Rheaco got back on its feet, largely as a result of its willingness to ask for help. Its flow rate dramatically improved, its manufacturing capacity increased by 300%, and the company solidified customer relationships. The Rheaco story is a reminder that companies cannot always go it alone in terms of achieving higher quality and improving manufacturing effectiveness. There are many organizations at the federal, state, and community levels that are equipped to provide business organizations assistance at little or no cost. ■

DISCUSSION QUESTIONS

1. Many companies fail in their efforts to improve quality without ever having asked for advice. In your opinion, what are some of the reasons that inhibit firms from asking for timely advice? If you were a manager at Rheaco, would you have sought out an agency like the ARRI?
2. Discuss ARRI's recommendations to Rheaco. How did these recommendations help Rheaco improve its product quality?
3. ARRI's initial evaluation of Rheaco indicated that Rheaco's employees, despite the company's difficulties, had an overall positive attitude. Do you believe that this factor contributed to ARRI's ability to provide Rheaco advice? Why or why not?

Case 2-2 Has Disney Developed a Theory of Quality Guest Services Management?

Disney Homepage: ***www.disney.go.com***
Walt Disney World: ***www.disneyworld.disney.go.com/***

As you approach the Magic Kingdom at Walt Disney World in Orlando, Florida, the recorded voice on the monorail announces that you are about to arrive at a "magical place" that appeals to both the young and the young at heart. As you enter the park, the surroundings are truly magical. The flower beds are beautiful, the grounds are clean, the buildings are spotless, and if you're lucky, you might even catch a glimpse of Mickey, Goofy, Tigger, or Winnie the Pooh. After a while, it is easy to start believing in the myth—that this is indeed a magical place.

But what is the Magic Kingdom—really? At its essence, it is a carefully conceived, masterfully executed theatrical production. The attention to detail is remarkable. Everything that happens in the park is carefully scripted, from the way the Disney characters interact with children to the way guests are moved through the park. For instance, if you stand at the base of Cinderella's Castle and look back toward the entrance to the park, you'll notice a subtle difference between the sidewalks that lead guests either to the right or to the left as they exit Main Street and enter the attractions area. The sidewalk to the right is wider than the sidewalk to the left. Why? Because through years of experience Disney has learned that people have a natural tendency to turn to the right rather than to the left. By building bigger sidewalks on the right, Disney is better able to handle the early morning crowds. Similarly, when you stand in line waiting to go on a ride, you'll notice that the line is designed to snake back and forth rather than extend in a straight line. Disney also has learned over the years that lines appear to be "shorter" if they snake back and forth rather than extend in a straight line. By reducing the perceived length of its lines, Disney is able to increase customer satisfaction.

Through this and similar examples, what Disney has done is develop a "theory" (or theories) of how guests will behave in the Magic Kingdom and its other parks. As a result, the company is able to "predict" how its guests will move through its park, how they will jockey for position in line, and how they will react to a variety of circumstances. This knowledge enhances Disney's ability to deliver a high level of customer service. The lines move smoothly, the rides are easy to board, the directions are simple and clear, and the souvenir shops are right where they need to be, not by accident, but because Disney knows what works. Far from being magic, it is carefully executed guest service management based on Disney's "predictions" of how its guests will behave. At times, Disney's understanding of its guests seems almost fanatical. For instance, at the company's theme parks, most of the drinking fountains come in pairs, one high and one low, to accommodate a parent and a child. The drinking spouts are directed toward one another, so if a parent and child drink at the same time, the parent can watch the child, rather than being turned in the opposite direction. That way, the parent and the child both feel secure and can share a drink and a smile.

In addition to anticipating how its guests will behave, Disney also enhances the quality of its guest services by "setting the stage" for good quality experiences. For instance, at the Polynesian Hotel directly across the Seven Seas Lagoon from the Magic Kingdom, you can hear Hawaiian music playing underneath the water in the hotel's main pool. At the Wilderness Lodge, you'll notice that pine needles cover the grounds. The funny thing is, there are no pine trees nearby. The pine needles are periodically brought to the property by Disney employees and spread out over the grounds. Other areas of customer service and guest relations are equally as surprising.

Can a company's approach to quality be based on "theories"? At the Magic Kingdom, it appears to be. By developing theories of their customers and other relevant activities, companies like Disney are able to enhance the quality of their products and services. ■

DISCUSSION QUESTIONS

1. Is Disney's level of emphasis on anticipating the behavior of its guests appropriate, or does the company expend too much effort in this area? Explain your answer.
2. Is it appropriate to think in terms of developing a "theory" of how guests will behave in a theme park or any other setting? If so, why?
3. Think about the last time that you visited a theme park. Were your expectations met? Did you have a sense that the operator of the park attempts to "anticipate" the behavior of the guests? If so, provide some specific examples.

CHAPTER 3

Global Supply Chain Quality and International Quality Standards

Global competition is played out by different rules
and for different stakes at each level.
—C. K. PRAHALAD and GARY HAMEL

INTRODUCTION

International trade is not a new phenomenon. The Roman, Greek, Egyptian, Chinese, Prussian, and other great empires were built on international trade. Columbus encountered the Americas for Queen Isabella of Spain when he was trying to establish a trade route to the East Indies across the Atlantic Ocean from Europe.

Although international trade has existed for a long time, the volume of international trade exploded after World War II and has continued to reach tremendous levels. This international diversity can be seen all around us. Probably, the watch you wear, the computer you use, the car you drive, or the frying pan you use to prepare breakfast are not produced in the country where you live. The nationalities of products are even obscured as companies become more internationally dispersed. The most famous electric guitar in the world is the Fender Stratocaster. If you go to your local music shop, you will find that Stratocasters vary in cost from $500 to around $3,000. Some of the variation in cost is because of different features and model names. However, closer inspection to the headstock will show that much of the price variation has to do with where the guitar is produced. Fender Stratocasters can be produced in Japan, Mexico, Korea, and the United States. You might wonder which nationality of Fender Stratocaster garners the highest price on the free market. The result might surprise you. American-made Stratocasters are the highest priced and the most coveted by guitar players. The Japanese models are judged by most guitarists to be of inferior quality. The Mexican and Korean guitars generally are considered beginner's guitars—not for the serious musician. However, the perceived quality might not match the reality. Certainly, for many products, Japanese, European, Chinese, or other nations may be the preferred source.

The task of managing quality is affected by this increased globalization. This chapter discusses the opportunities and obstacles created by globalization. The differences between regions of the world also include discussions of various quality approaches that have been developed in those regions and the awards that act as quality barometers within each.

MANAGING QUALITY FOR THE MULTINATIONAL FIRM (MNF)

Firms must cope with more diversity now than in the past. One of the causes of this diversity challenge has been the increased emphasis on international trade over the past

FIGURE 3-1 U.S. Trade 1960–2005

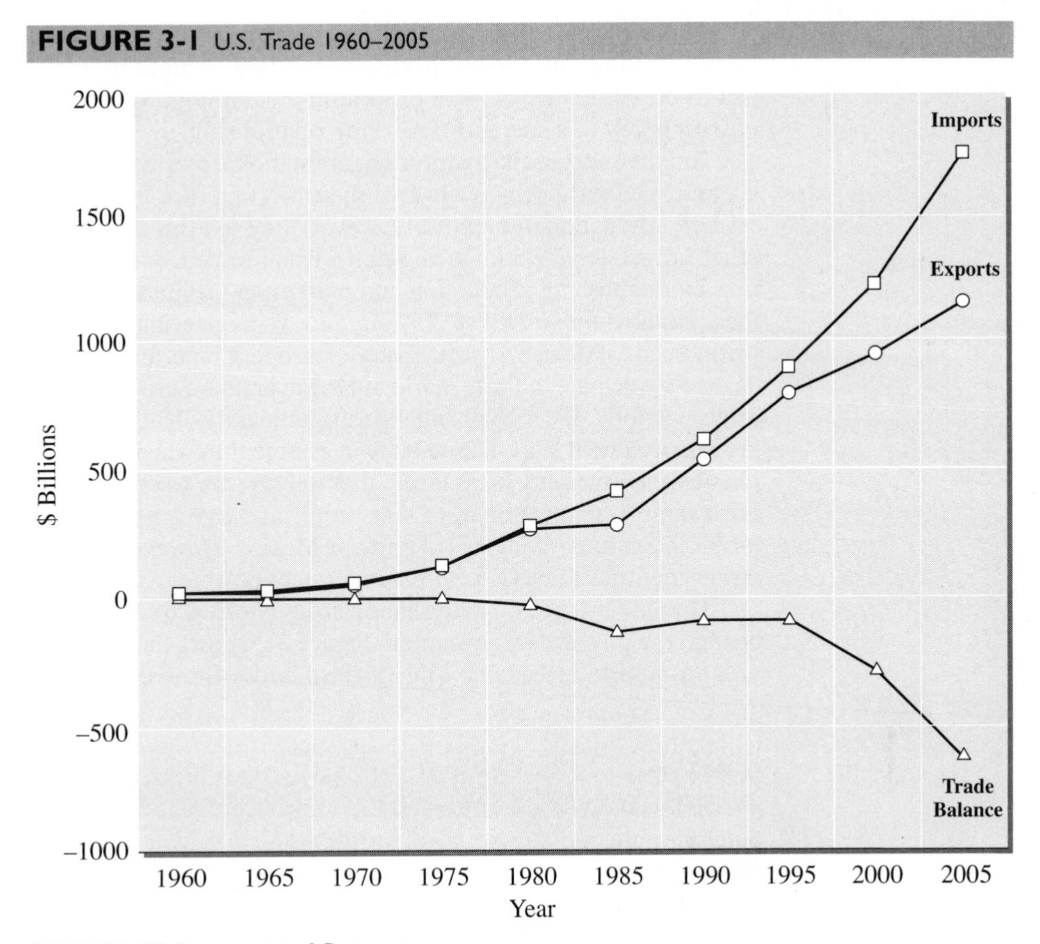

SOURCE: U.S. Department of Commerce.

four decades. This growth in international trade has occurred as companies have sought new markets. Figure 3-1 shows that although the trade deficit has remained relatively constant over time, both imports and exports of products have been increasing steadily.

There are a variety of mechanisms that firms use in globalizing. The first is **licensing**. By licensing, a U.S. corporation can allow foreign firms to sell in restricted markets while using the design of the original designer. Licensing often involves the sale of the same product with another trademark in different countries. Through licensing, firms are able to reach international markets without having to establish international supply chains or marketing arms.

Firms also seek international markets through joint ventures or **partnering**. This agreement is often reached when two firms have technology, products, or access to markets that each other wants. For example, in the 1980s when the Syrian National Petroleum Company desired to exploit a large oil discovery within its own borders, it needed technology to extract the oil. As a result, the Syrian Petroleum Company entered into a partnership with the international firms Deminex, Shell Oil Company (Netherlands), and Pecten International Company (Shell U.S.A.) to profitably explore, extract, and pipe the oil to export to foreign markets. In spite of the fact that Syria represented a serious political risk, the foreign partners did not fear nationalization and were willing to enter into the agreement because Syria needed technology, capital, and knowledge that the foreign partners provided. With both licensing and partnering, the risk to the company is loss of proprietary technology. In this case, the Deming attribute of trust is key to a successful working relationship.

Another approach to capturing international markets is **globalization**. The benefits of licensing and partnering were that the exporting firm did not have to globalize to make sales in international markets. However, they did this at the cost of sharing profits with other firms. Globalization means that a firm fundamentally changes the nature of its business by establishing production and marketing facilities in foreign countries. We refer to these firms as *multinational corporations*. With growing economies in many parts of the world, such as Mexico, Brazil, Eastern Europe, China, and Russia, firms will need to globalize to participate in these markets. However, there are effects of globalization that firms often overlook. By globalizing, firms significantly change the physical environment, the task environment, and the societal environment in which they operate. By changing their **physical environment**, firms locate themselves near to or far away from natural resources. For example, semiconductor firms requiring large amounts of water probably will not locate in Saudi Arabia or arid parts of Mexico. However, they may locate in one of the Asian countries to be close to ready supplies of water as well as expanding markets.

The advantage of saving labor costs is often overemphasized when deciding to change the physical environment. Figure 3-2 shows the fiscal contrast between making an automobile in Mexico and in the United States. It shows that overall, because

FIGURE 3-2 Wages and Costs in Mexico (Average Hourly Compensation for Manufacturing Workers in U.S. Dollars)

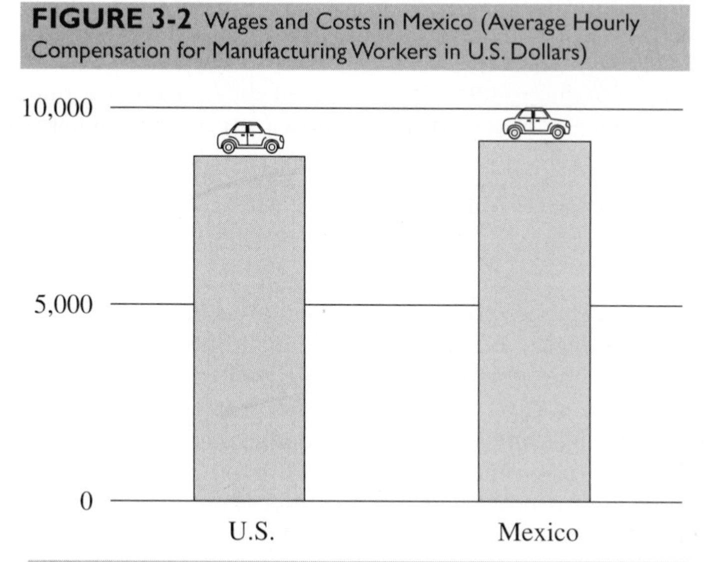

The cost of making a car

	U.S.	Mexico
Parts, components	$7,750	$8,000
Labor	700	140
Supply chain costs	300	1,000
Inventory	20	40
Total	**$8,770**	**$9,180**

SOURCE: Bureau of Labor Statistics, Office of Technology Assessment (1998).

component costs and supply chain costs are higher in Mexico, the wage savings alone is not sufficient to make automobile production less expensive in Mexico.

The **task environment** of the firm has to do with the operating structure that the firm encounters when globalizing. The economic structures, skills of the employees, compensation structure, technologies, and government agencies all vary when globalizing. The regulatory structures that firms encounter when globalizing require an understanding of international law. Although many view the U.S. government as somewhat regulatory, many times firms find themselves having to deal with very complex regulatory structures when establishing operations in countries such as Germany, Japan, or Spain. Technological choices vary as firms globalize. What works at home may not serve customers adequately abroad. For example, tobacco producers who globalize find that mass-production technologies used in the United States are not flexible enough for the European Economic Community, where regulations concerning tobacco products vary a great deal from country to country.

The **social environment** facing globalizing corporations refers to cultural factors such as language, business customs, customer preferences, and patterns of communication. The cultural factors facing globalizing firms are often the most complex and difficult issues they will encounter. For example, the American businessperson who likes to have breakfast meetings will be frustrated in Spain, where people do not go to work until later in the day. In Brazil, a potential client telling you that he or she will meet you "unless it rains" means the client does not intend to show up at the meeting. To Brazilians, the "unless it rains (*se não chouver*)" qualifier is a polite way for them to explain that they will not be there. From a quality point of view, the efficiency-minded American way of moving customers through a system may not translate well to South American customers who are used to more personalized service. For example, in a Brazilian pharmacy, the customer goes to one counter to select the product of choice, then to a window to pay for the product, and then to another window to receive the product in pink wrapping paper. Brazilians want their products wrapped so that they do not display their personal goods in public.

As shown in Figure 3-3, physical, task, and social environments have implications for the choices made in improving quality, particularly in the area of quality management. Differences in the physical, task, and social variables all add to the complexity and

FIGURE 3-3 Global Factors That Affect Quality-Related Decisions.

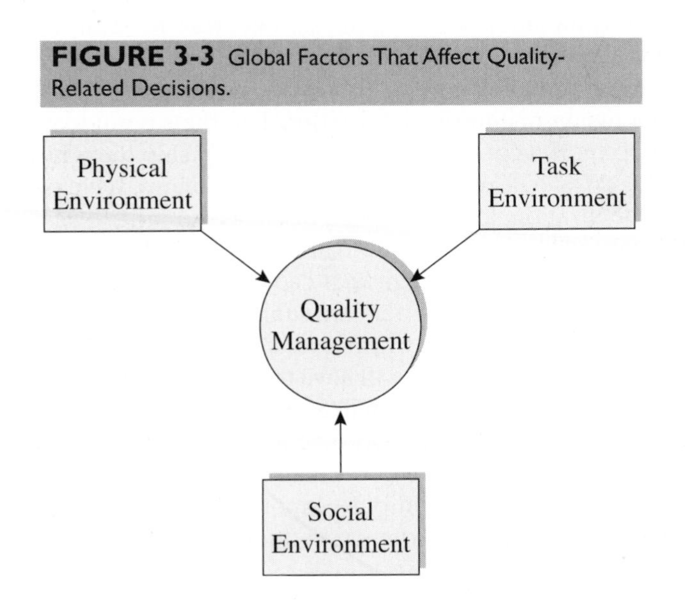

TABLE 3-1 U.S. Patent Applications (2003)

Rank	*Company (Country)*	*No. of Patents*
1	IBM (U.S.A.)	3,415
2	Canon (Japan)	1,992
3	Hitachi (Japan)	1,893
4	Matsushita (Japan)	1,786
5	Hewlett Packard (U.S.A.)	1,759
6	Micron (U.S.A.)	1,707
7	Intel (U.S.A.)	1,592
8	Philips (Netherlands)	1,353
9	Samsung (Korea)	1,313
10	Sony (Japan)	1,311
11	Fujitsu (Japan)	1,302
12	Mitsubishi (Japan)	1,243
13	Toshiba (Japan)	1,184
14	NEC (Japan)	1,181
15	GE (U.S.A.)	1,139
16	AMD (U.S.A.)	905
17	Fuji (Japan)	804

SOURCE: U.S. Patent and Trademark Office, 2004.

variability that firms experience. Therefore, there is significant tension between a need for standardization/high central control and loss of control because of decentralization.

Market diversity drives the need for culture-specific research and development (R&D). Although this can be true within a nation that is culturally diverse, such as the United States, it is especially true on the global front. There are few products, such as Coca-Cola, that translate well across different cultural horizons. Many companies must adapt their products to the preferences of the markets they are serving. This greatly increases the complexity of international marketing and R&D. For example, in some European countries, McDonald's sells wine with their meals because it is the European habit to drink wine with meals. Often globalized firms also must apply for patents in different places in the world to protect their domestic patents. Patent applications are an important indicator of R&D productivity. As shown in Table 3-1, only 4 of the top 12 applicants for patents in the United States in 2003 were American companies. Six of the other companies were Japanese, one was Korean, and one was from the Netherlands. Quality Highlight 3-1 shows how one company has adapted its supply chain practice to provide high-quality service on a worldwide basis.

Another means of entering international markets is not to globalize but to become an **exporter**. Exporters produce their products and ship them internationally, incurring high shipping costs but avoiding many of the problems, such as loss of control associated with globalization, that have been discussed so far. However, success on a multinational scale may be more difficult to attain for exporters because they never develop the marketing expertise and logistical capabilities associated with entering foreign markets. Many times pure exporters are subject to limitations that resident companies do not have in terms of import tariffs and import restrictions. For example, Kodak U.S.A. exporting film into Brazil will have to pay a tariff equal to three times its market value. However, Kodak making film in Sao Paulo, Brazil, will pay no import tariffs and enjoy a cost advantage in the market over a company that chooses to export to Brazil.

Companies wishing to export to Japan have found it easier to establish partnerships in Japan than to engage in pure exportation. It is very difficult to navigate the complex Japanese marketing structure without partnerships and long experience. This

QUALITY HIGHLIGHT 3-1

SUPPLY CHAIN QUALITY IN THE GLOBAL CONTEXT

www.national.com

National Semiconductor (NSC), of Santa Clara, California, has sought advantage by sourcing its output all over the world. These sourcing firms represent partnerships that NSC has developed around the globe. However, dealing with a global network of partners can be difficult. Because of the global nature of its distribution, 40% of NSC's order-to-delivery cycle was taken up with distribution and supply chain processes.

Today, customers need much faster response times. So NSC resorted to globalized logistics. However, the company's logistical system was poorly suited to the demands of globalized logistics. The company's network was a tangle of unnecessary interchanges, propped up by 44 different international freight forwarders and 18 different air carriers. What NSC realized was that its primary competence was making semiconductors. It was not adept in the logistics of transporting them.

To compensate for the poorly developed logistics systems, NSC maintained "just-in-case" inventory centers around the world. In the existing system, delivery could vary from 5 to 18 days. This was too much variation and made NSC unresponsive to customer requirements because of the poor response times.

After analyzing the supply chain marketplace, NSC decided to partner with Business Logistics Service, a subsidiary of Federal Express. Because logistics was not the primary competence of NSC, Federal Express could provide the support needed for NSC to improve its customer service. Therefore, the partnership lowered costs for NSC and provided income for Federal Express. This was important because with semiconductors the time from production to purchase can be three times faster than for other products. Therefore, a repeatable, reliable distribution system was needed to meet customer requirements for dependable delivery.

often involves ceding partial ownership of the firm to Japanese partners to obtain access to complicated distribution networks.[1]

Another issue with exporting is that the United States developed a negative quality image during the 1970s and 1980s. This image improved greatly in the 1990s as America reached quality levels similar to Japan's in many markets. Research by Macy, Foster, and Barringer[2] shows that quality still is a significant factor in helping U.S. exporters achieve success. Figure 3-4 shows an empirically derived model for quality-based success for exporting companies. In this model, it is shown that firms that are characteristically entrepreneurial in nature and that plan their exports effectively tend to have higher quality. For exporting firms, quality leads to lower price and greater export success.

Although there are a variety of means for entering the global market, one thing is certain: The global market is a reality that must be addressed. Gibson Guitar in Bozeman, Montana, does not operate in a vacuum. Besides exporting guitars overseas, they must produce products that meet international standards of quality or they will not be able to sell their guitars—even in Bozeman. Through local vendors or by Internet order, Bozeman guitarists can buy many quality guitars that are made by Takamine, Yamaha, Ibanez, or other high-quality foreign companies that are competing in the local market.

[1]Abegglen, J., and Stalk, G., *Kaisha, The Japanese Corporation* (New York: Basic Books, 1985).
[2]Barringer, B., Foster, S., and Macy, G., "The Role of Quality in Determining Export Success," *Quality Management Journal* 6, 4 (1999):55–70.

FIGURE 3-4 Export Quality Model

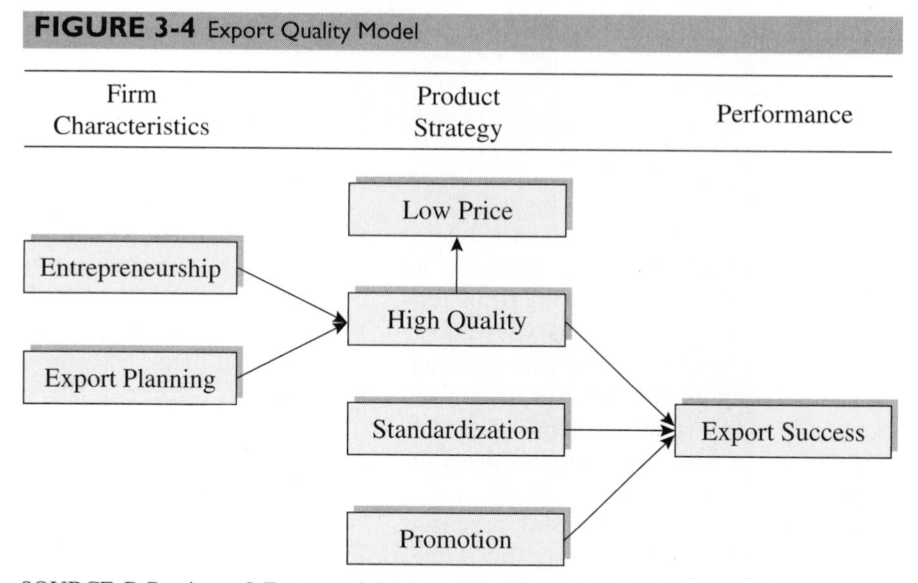

SOURCE: B. Barringer, S. Foster, and G. Macy, "The Role of Quality in Determining Export Success," *Quality Mangement Journal* 6, 4 (1999):64.

Next we continue the theme of integrative quality by exploring quality improvement within U.S., Japanese, and European contexts.

QUALITY IMPROVEMENT: THE AMERICAN WAY

America has been called the birthplace of modern quality management because it is home to Shewhart, Deming, Juran, and others. The U.S. military also was an early adopter of many quality techniques. Originally, the main interest in quality in the United States was in the application of statistics to solve quality problems. In recent years, the approach has become much more behavioral as teams and other approaches have been applied. We begin by discussing a very important model for quality management in the United States, the Malcolm Baldrige National Quality Award.

THE MALCOLM BALDRIGE NATIONAL QUALITY AWARD

At the end of Chapter 2 we discussed several operational variables that should be addressed in assessing our future directions for quality improvement. The power of these variables is that they focus management on systemic issues rather than the tactical, day-to-day problems. However, management needs a means for assessing the approaches it has employed to improve operating performance. One of the most powerful self-assessment mechanisms is the **Malcolm Baldrige National Quality Award (MBNQA)**(Figure 3-5). The success of the Baldrige award in the United States has influenced international practice and has formed the basis for several international awards.

FIGURE 3-5 Baldrige Award

The MBNQA process is open to small (less than 500 employees) and large firms (more than 500 employees) in the manufacturing, health care, education, and service sectors. The MBNQA is not open to public-sector and not-for-profit organizations. However, there are other quality-related options available to these organizations, such as the Senate Quality Recognition, the President's Award, and various state quality awards that are based on the Baldrige. Thus, although the award itself is not available to all organizations, the criteria are available for assessment under a variety of different mechanisms. There can be only two winners, by category, for a given year. This limits the number of potential winners to six per year in the business category. However, the number of winners always has been less than six.

Some of the key characteristics of the quality award include the following attributes:

- The criteria focus on business results. Companies must show outstanding results in areas such as financial performance, customer satisfaction, customer retention, product performance, service performance, productivity, supplier performance, and public citizenship. In order to win the award, applicants must show that they consistently have performance levels at best-in-class and best-of-the-best. Business results for the Baldrige winner must be exemplary and set the standard for the rest of the country. The importance of business results is reflected in the scoring of the Baldrige. The weight for results has ranged from 25% to 45% of the total score. The reason for the emphasis on results has to do with two related factors. It is quite embarrassing if a Baldrige winner has financial trouble, and the Baldrige winner must be able to serve as a role model to other firms. The ability to serve as a role model is enhanced by strong business results.

- The Baldrige criteria are nonprescriptive and adaptable. Although the focus of the Baldrige is on results, the means for obtaining these results are not prescribed. Therefore, the criteria are not tactically prescriptive. For example, the criteria do not specify which tools, techniques, or organization a company should use to improve. However, at the strategic level, the Baldrige criteria can be viewed as prescriptive. In explanation, a company that does not gather data from its customers, or a company that does not have an effective benchmarking program, will not be Baldrige-qualified. Therefore, the core values and key characteristics of the Baldrige criteria can be viewed as strategically prescriptive.
- The criteria support company-wide alignment of goals and processes. Once organizational strategy is formulated, connecting and reinforcing measures are developed that support and monitor strategic outcomes. Measures aid in alignment and ensure consistency of purpose while supporting innovation. Alignment between strategic goals and operational subplans helps foster a learning-based system.
- The criteria permit goal-based diagnosis. The criteria and scoring guidelines provide assessment dimensions. These are approach, deployment, and results. The approach defines the method or system for addressing a particular performance objective. Deployment implies that the approach has been implemented in the organization. Results show the outcomes of the approach and deployment. By assessing approach, deployment, and results in several areas, firms are able to assess their current strengths and areas for improvement. Once areas for improvement are identified, these can be prioritized and tackled one by one.

The model for the MBNQA consists of seven interrelated categories that compose the organizational system for performance. As shown in Figure 3-6, the seven categories are leadership; strategic planning; customer and market knowledge; measurement, analysis, and knowledge management; human resource focus; process management; and business results.

Notice that the basis of the Baldrige model is information and analysis. This confirms the core value of management by fact. Business results are also highlighted, reinforcing the results orientation of the Baldrige framework and criteria. Each of these seven major categories is divided into *items* and *areas to address*. The items are denoted by a decimal number, such as 1.1, 2.2, and so on. The areas are given a letter, such as 1.1.a, 2.2.b, and so forth. The number of items and areas varies from year to year because the award criteria are constantly updated by the Baldrige staff, examiners, and judges.[3]

Category 1 (Table 3-2) shows the award criteria for leadership. This category is used to evaluate the extent to which top management is personally involved in creating and reinforcing goals, values, directions, customer involvement, and a variety of other issues. Within category 1, the applicant outlines what the firm is doing to fulfill its

---

[3]Tables 3-2 through 3-9 contain Baldrige verbiage. The complete Baldrige criteria can be downloaded for free from *www.NIST.gov*.

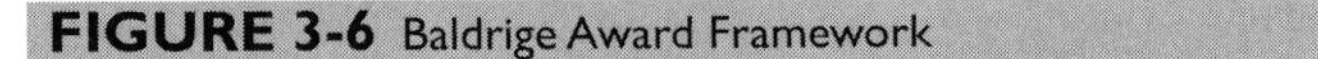

FIGURE 3-6 Baldrige Award Framework

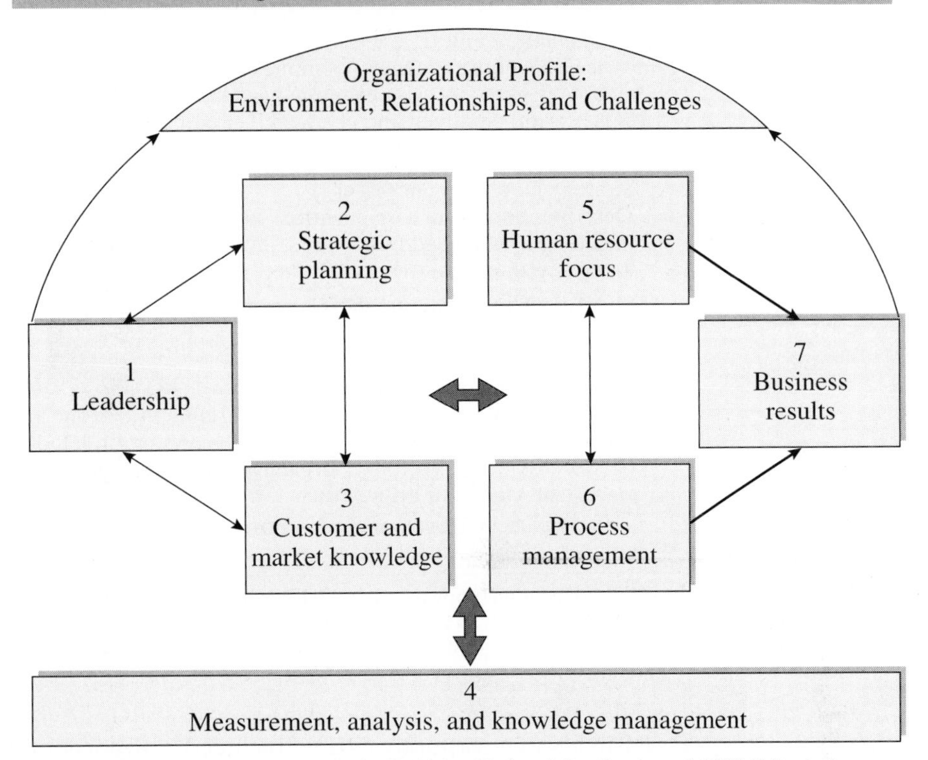

SOURCE: Foundation for the Malcolm Baldrige National Quality Award, 2005 Criteria for Performance Excellence, 2005

TABLE 3-2 Baldrige Category I

Leadership (120 pts.)
The *Leadership* Category examines HOW your organization's SENIOR LEADERS guide and sustain your organization. Also examined are your organization's GOVERNANCE and HOW your organization addresses its ethical, legal, and community responsibilities.

1.1 Senior Leadership (70 pts.)
Describe HOW SENIOR LEADERS guide and sustain your organization. Describe HOW SENIOR LEADERS communicate with employees and encourage high PERFORMANCE.

a. VISION and VALUES
b. Communication and Organizational PERFORMANCE

1.2 Governance and Social Responsibilities (50 pts.)
Describe your organization's GOVERNANCE system. Describe HOW your organization addresses its responsibilities to the public, ensures ETHICAL BEHAVIOR, and practices good citizenship.

a. Organizational GOVERNANCE
b. Legal and ETHICAL BEHAVIOR
c. Support of KEY Communities

SOURCE: Foundation for the Malcolm Baldrige National Quality Award, 2006 Criteria for Performance Excellence, 2006.

TABLE 3-3 Baldrige Category 2

Strategic Planning (85 pts.)
The *Strategic Planning* Category examines HOW your organization develops STRATEGIC OBJECTIVES and ACTION PLANS. Also examined are HOW your chosen STRATEGIC OBJECTIVES and ACTION PLANS are deployed and changed if circumstances require, and HOW progress is measured.

2.1 Strategy Development (40 pts.)
Describe HOW your organization establishes its strategy and STRATEGIC OBJECTIVES including HOW you address your STRATEGIC CHALLENGES. Summarize your organization's KEY STRATEGIC OBJECTIVES and their related GOALS.

a. Strategy Development PROCESS
b. STRATEGIC OBJECTIVES

2.2 Strategy Deployment (45 pts.)
Describe HOW your organization converts its STRATEGIC OBJECTIVES into ACTION PLANS. Summarize your organization's ACTION PLANS and related KEY PERFORMANCE MEASURES or INDICATORS. Project your organization's future PERFORMANCE on these KEY PERFORMANCE MEASURES or INDICATORS.

a. ACTION PLAN Development and DEPLOYMENT
b. PERFORMANCE PROJECTION

SOURCE: Foundation for the Malcolm Baldrige National Quality Award, 2006 Criteria for Performance Excellence, 2006.

responsibility as a corporate citizen. Included with corporate citizenship is a documentation of measures and facts relating to how the company responds to its regulatory environment.

Category 2 (Table 3-3) focuses on how the company establishes strategic directions and how it sets its tactical action plans to implement the strategic plans. In addressing *how* the company establishes strategic direction, the applicant should outline methods, measures, deployment, and evaluation/improvement factors relating to establishing strategic plans.

Item 2.1 focuses on strategy development, taking into account customers, competition, risks, capabilities, and suppliers. Item 2.2 asks for the applicant to identify the company's strategy and action plans and how they are deployed. Item 2.2 addresses how strategy is formulated relating to employees and human resources.

Category 3 (Table 3-4) addresses the customer and market knowledge. To be successful in serving the customer, firms must understand the product and service attributes that are important to the customer. This is documented as well as how the firm assesses the relative importance of product or service features. The processes for listening to and learning from customers and markets also must be evaluated, improved, and kept current with changing business needs.

If customers are satisfied, they often become loyal customers. Think of where you shop for groceries. You are not likely to shop at different stores each time. If you find a grocer with a convenient location, good stock, reasonable prices, and good service, you become a loyal customer. To engender this loyalty, firms must demonstrate that they have systems in place for determining customer contact requirements and for communicating these requirements to employees so that they can provide satisfactory service. Companies also must demonstrate processes that are effective in capturing and managing customer complaints. Complaints must be resolved promptly according to

TABLE 3-4 Baldrige Category 3

Customer and Market Focus (85 pts.)
The *CUSTOMER and Market Focus* Category examines HOW your organization determines the requirements, expectations, and preferences of CUSTOMERS and markets. Also examined is HOW your organization builds relationships with CUSTOMERS and determines the KEY factors that lead to CUSTOMER acquisition, satisfaction, loyalty and retention, and to business expansion and SUSTAINABILITY.

3.1 CUSTOMER and Market Knowledge (40 pts.)
Describe HOW your organization determines requirements, expectations, and preferences of CUSTOMERS and markets to ensure the continuing relevance of your products and services and to develop new opportunities.
a. CUSTOMER and Market Knowledge

3.2 Customer Relationships and Satisfaction (45 pts.)
Describe HOW your organization builds relationships to acquire, satisfy, and retain CUSTOMERS; to increase CUSTOMER loyalty; and to develop new opportunities. Describe also HOW your organization determines CUSTOMER satisfaction.
a. CUSTOMER Relationship Building
b. CUSTOMER Satisfaction Determination

SOURCE: Foundation for the Malcolm Baldrige National Quality Award, 2006 Criteria for Performance Excellence, 2006.

best-in-industry standards. Complaint data, once captured, are aggregated and analyzed for use in improving customer service processes. Notice only approach and deployment are documented in Category 3.

In improving customer satisfaction, the firm must have a system for gathering and analyzing customer satisfaction data. In order to score high, the company will have continually improved the processes for gathering, analyzing, and deploying customer service information. Additionally, this information will be used for process and service improvement.

Category 4 (Table 3-5), measurement analysis, and knowledge management, relates to the firm's selection, management, and use of information to support company processes and to improve firm performance. These data include both financial information, such as sales, assets, and liabilities, and nonfinancial data, such as operating measures of quality, productivity, and speed of response to customer requests. These measures are only introduced at this point; no results are provided. The applicant then describes how these measures are integrated into an information system that can be used to track and improve company performance. Once data and measures are tracked in an information system, people must deploy the data to ensure that company goals and objectives are being achieved. Robust processes for improving the data should be described.

Category 5 (Table 3-6) deals with the human resource focus. The workforce is to be enabled to develop and use its full potential, aligned with company objectives. This involves developing an internal environment conducive to full participation and personal growth, including human resources development. This initiative is directed toward process and performance improvement. Employees must be empowered to understand and respond to changes in customer needs and requirements. To facilitate this learning, skill sharing and open communication with employees are necessary to provide a cohesive work system. Finally, the applicant outlines the systems in place that provide compensation and recognition.

TABLE 3-5 Baldrige Category 4

Measurement, Analysis, and Knowledge Management (90 pts.)
The *Measurement, Analysis, and Knowledge Management* Category examines HOW your organization selects, gathers, analyzes, manages, and improves its data, information, and KNOWLEDGE ASSETS. Also examined is HOW your organization reviews its performance.

4.1 Measurement, Analysis, and Review of Organizational Performance (45 pts.)
Describe HOW your organization measures, analyzes, aligns, reviews, and improves its PERFORMANCE at all LEVELS and in all parts of your organization.

a. PERFORMANCE Measurement
b. PERFORMANCE ANALYSIS and Review

4.2 Information and Knowledge Management (45 pts.)
Describe HOW your organization ensures the quality and availability of needed data and information for employees, suppliers and PARTNERS, and CUSTOMERS. Describe HOW your organization builds and manages its KNOWLEDGE ASSETS.

a. Data and Information Availability
b. Organizational Knowledge Management
c. Data, Information, and Knowledge Quality

SOURCE: Foundation for the Malcolm Baldrige National Quality Award, 2006 Criteria for Performance Excellence, 2006.

TABLE 3-6 Baldrige Category 5

Human Resource Focus (85 pts.)
The *Human Resource Focus* Category examines HOW your organization's WORK SYSTEMS and your employee LEARNING and motivation enable employees to develop and utilize their full potential in ALIGNMENT with your organization's overall objectives, strategy, and ACTION PLANS. Also examined are your organization's efforts to build and maintain a work environment and employee support climate conducive to PERFORMANCE EXCELLENCE and to personal and organizational growth.

5.1 Work Systems (35 pts.)
Describe HOW your organization's work and jobs enable employees and the organization to achieve HIGH PERFORMANCE. Describe HOW compensation, career progression, and related workforce practices enable employees and the organization to achieve HIGH PERFORMANCE.

a. Organization and Management of Work
b. Employee PERFORMANCE Management System
c. Hiring and Career Progression

5.2 Employee Learning and Motivation (25 pts.)
Describe HOW your organization's employee education, training, and career development support the achievement of your overall objectives and contribute to HIGH PERFORMANCE. Describe HOW your organization's education, training, and career development build employee knowledge, skills, and capabilities.

a. Employee Education, Training, and Development
b. Motivation and Career Development

5.3 Employee Well-Being and Satisfaction (25 pts.)
Describe HOW your organization maintains a work environment and an employee support climate that contribute to the well-being, satisfaction, and motivation of all employees.

a. Work Environment
b. Employee Support and Satisfaction

SOURCE: Foundation for the Malcolm Baldrige National Quality Award, 2006 Criteria for Performance Excellence, 2006.

The Baldrige criteria promote the ethic of developing employees through learning. The training must support key performance plans and needs by developing employees for the long term. Systems for designing education and training must be continually improved. Finally, assessment of the value of training versus the cost is an important attribute to be included in the application.

Employee well-being and satisfaction are important attributes of a high-performance work environment. The process for developing and maintaining a healthy and safe work environment is documented in area 5.3.a. Key factors and measures must be in place to monitor the efficacy of these processes. Employee support services and understanding of the key factors that influence employee satisfaction are included in the item.

Category 6 (Table 3-7) examines the key aspects of process management. These aspects include customer focus in design, delivery process design for services and products, support processes, and processes relating to partners. The design of products must be changed and upgraded to reflect changes in customer requirements and technology. Systems are outlined that measure and assess these changes in requirements. Once products are designed, production and logistical processes also must be designed to meet quality and operational objectives. The mark of a successful design process is the trouble-free introduction of products and services. This is a result of careful attention to design and experience with product and service introduction.

Once products and processes are designed, there needs to be a means for gathering data and evaluating whether the products and services meet the customer's expectations. This information can be used to achieve better performance, and it can be fed back into the design process to improve existing products and services.

Item 6.2 relates to the management of support processes. Performance feedback is given to support personnel to aid them in improving processes.

Category 7 (Table 3-8) documents the results of the other six categories and requires a series of tables and graphs that demonstrate the operational and business results of the firm. Although all the information provided by the applicant is considered by Baldrige examiners to be highly sensitive, Category 7 is often considered the

TABLE 3-7 Baldrige Category 6

Process Management (85 pts.)

The *Process Management* Category examines the KEY aspects of your organization's PROCESS management, KEY product service, and business PROCESSES for creating CUSTOMER and organizational VALUE and KEY support PROCESSES. This Category encompasses all KEY PROCESSES and all work units.

6.1 Value Creation Processes (45 pts.)
Describe HOW your organization identifies and manages its KEY PROCESSES for creating CUSTOMER VALUE and achieving business success and growth.

a. VALUE CREATION PROCESSES

6.2 Support Processes and Operational Planning (40 pts.)
Describe HOW your organization manages its KEY PROCESSES that support your VALUE CREATION PROCESSES. Describe your PROCESSES for financial management and continuity of operations in an emergency.

a. Support PROCESSES
b. Operational Planning

SOURCE: Foundation for the Malcolm Baldrige National Quality Award, 2006 Criteria for Performance Excellence, 2006.

TABLE 3-8 Baldrige Category 7

Business Results (450 pts.)
The *Business Results* Category examines your organization's PERFORMANCE and improvement in KEY business areas—product and service outcomes, CUSTOMER satisfaction, financial and marketplace PERFORMANCE, human resource RESULTS, operational PERFORMANCE and leadership and social responsibility. PERFORMANCE LEVELS are examined relative to those of competitors.

7.1 Product and Service Outcomes (100 pts.)
Summarize your organization's KEY product and service PERFORMANCE RESULTS. SEGMENT your RESULTS by product and service types and groups, CUSTOMER groups, and market SEGMENTS, as appropriate. Include appropriate comparative data.
a. Product and Service Results

7.2 Customer-Focused Results (70 pts.)
Summarize your organization's KEY CUSTOMER-focused RESULTS, including CUSTOMER satisfaction and CUSTOMER-perceived VALUE. SEGMENT your RESULTS by product and service types and groups, CUSTOMER groups, and market SEGMENTS as appropriate. Include appropriate comparative data.
a. CUSTOMER-Focused Results

7.3 Financial and Market Results (70 pts.)
Summarize your organization's KEY financial and marketplace PERFORMANCE RESULTS by CUSTOMER or market SEGMENTS, as appropriate. Include appropriate comparative data.
a. Financial and Market Results

7.4 Human Resource Results (70 pts.)
Summarize your organization's KEY human resource RESULTS, including WORK SYSTEM PERFORMANCE and employee LEARNING, development, well-being, and satisfaction. SEGMENT your RESULTS to address the DIVERSITY of your workforce and the different types and categories of employees, as appropriate. Include appropriate comparative data.
a. Human Resource Results

7.5 Organizational Effectiveness Results (70 pts.)
Summarize your organization's KEY operational PERFORMANCE RESULTS that contribute to the improvement of organizational effectiveness. SEGMENT your RESULTS by product and service types and groups and by market SEGMENTS, as appropriate. Include appropriate comparative data.
a. Organizational Effectiveness Results

7.6 Leadership and Social Responsibility Results (70 pts.)
Summarize your organization's KEY GOVERNANCE, SENIOR LEADERSHIP, and social responsibility RESULTS, including evidence of ETHICAL BEHAVIOR, fiscal accountability, legal compliance, and organizational citizenship. SEGMENT your RESULTS by business units, as appropriate. Include appropriate comparative data.
a. Leadership and Social Responsibility Results

SOURCE: Foundation for the Malcolm Baldrige National Quality Award, 2006 Criteria for Performance Excellence, 2006.

most sensitive by the applicant. For example, because reporting requirements are less exacting for privately held firms, many of the results reported are proprietary. In spite of this, if a firm wishes to qualify for a site visit, the requested information must be made available to the examiners.

Since we have discussed the MBNQA, you might be interested to know who Malcolm Baldrige was. A Closer Look at Quality 3-1 gives some details about his life.

The Baldrige Process

For the firm applying for the Baldrige award, the first step is eligibility determination. Because the Baldrige pertains only to for-profit firms chartered in the United States, eligibility must first be determined by mailing an eligibility determination form to the National Institute of Standards and Technology (NIST). Once eligibility is established, the applicant sends the completed application to NIST. The application is then

A CLOSER LOOK AT QUALITY 3-1

WHO WAS MALCOLM BALDRIGE?

Howard Malcolm Baldrige was the U.S. secretary of commerce from 1980 to 1987. He died while in office, doing what he loved best. He was riding his horse, practicing for the calf-roping event in a California rodeo, when the horse fell and fatally injured its rider. Secretary Baldrige was 64.

The competitive spirit that spurred Secretary Baldrige to win rodeo events characterized his life. A graduate of Yale University, he entered the army just prior to World War II and quickly rose through the ranks. After the war, he took a job as a supervisor in a small Connecticut foundry. His business career was marked by much success. Toward the end of his career, he led a turnaround effort at Scovill Corporation in Waterbury, Connecticut, based on a steadfast commitment to quality and customer service.

Baldrige believed in good government and was active in politics. His political activity, along with his passion for quality and reputation as a quality leader, led to his appointment as the secretary of commerce in 1980. As secretary of commerce, his top priority was strengthening American industry's ability to compete successfully in international markets. He believed that the role of the U.S. government was to establish fair trade practices and that the role of private industry was to improve its technology and gain a global reputation for quality products and services.

The Baldrige award was a great tribute to Malcolm Baldrige. Some of the participants in the Baldrige application process have been passionate about the value of the program. Following are two quotations that illustrate this point:

> In my opinion, win or lose, the greatest value in applying for the Baldrige award is the feedback report compiled by the examiners. This objective evaluation prepared by a team of well-trained, hard working experts provided the information and focus necessary for us to cause positive changes in our organization.
>
> —Henry A. Bradshaw
> Armstrong Building Products

> Participating in the Baldrige process energized improvement efforts. The energy resulted from the team motivation that occurs when pursuing a common goal. That trend has continued. We have reduced the number of in-process defects to only one-tenth what they were at the time we won the Baldrige.
>
> —Phil Roether
> Texas Instruments

The legacy of Malcolm Baldrige lives on in the benefits derived from the recipients and observers of the Malcolm Baldrige National Quality Award competition. Much can be learned about quality by becoming a keen observer of the companies that win the coveted award.

subjected to first-round review by Baldrige examiners. During this review, examiners read and score the applications. Judges then review the scoring to determine which applicants will continue to *consensus*. During the consensus phase, between five and eight examiners who have scored the application participate in a conference call to determine a consensus score for each of the scoring items.

Once consensus is reached, judges receive a consensus report from the senior examiner leading the examiner team. Judges then make a site-visit determination. At this point, applicants scoring sufficiently high are granted a site visit. In the past, simply the granting of a Baldrige site visit has been cited as evidence of high-quality processes. These firms sometimes refer to themselves as "**Baldrige qualified**."

The site visit consists of a team of four to six examiners visiting a company over a period not to exceed one week. Typically, the site visit consists of two to three days at the company site and another two days in the hotel to prepare the site-visit reports for the Baldrige judges. These reports show the results of the site visit. The purpose of the site visit is to *verify and clarify* those portions of the Baldrige application having the greatest impact on the judges' scores. During the first round of scoring, examiners identify strengths, areas for improvement, and site-visit issues. Prior to the site visit, the site-visit team reviews and prioritizes the site-visit issues. Members of the site-visit team then assign responsibilities to team members to complete during the site-visit. After these assignments are complete, the examiners finish their reports while on the site visit. The results are transmitted for use by the judges in final determination of a winner.

One of the most important outcomes of the Baldrige process is examiner feedback to applicant companies. As shown in Figure 3-7, feedback reports are provided to the applicants as part of the assessment process. Many firms have found these comments helpful in identifying gaps in deployment in their improvement processes. As the Baldrige process has matured and become more focused on overall company performance, the feedback reports have become more useful for improving overall management of processes and systems. The feedback report is one of the major benefits of the Baldrige process. The feedback report includes

- The *scoring summary*, which is a synthesis of the most important strengths and areas for improvement for each of the seven Baldrige categories.
- The *individual scoring range*, which provides a 20-point scoring range (e.g., item 2.1: 40%–60%). This gives insight concerning the relative areas of strength and areas that need improvement.
- The *scoring distribution*, which provides the percentage of applicants for a particular year that scored in each of the eight scoring bands.
- The *examiner comments*, which give feedback concerning the organization. These are created from actual examiner comments and are the meat of the feedback report.

Baldrige Scoring

The Baldrige applications are scored with supporting written guidelines. These guidelines are provided in Table 3-9.

One of the unique attributes of Baldrige scoring is the 50% anchor. For an approach and deployment (A&D) item to score 50%, there must be a sound, systematic approach that is responsive to the overall purposes of the item. There also must be fact-based improvement processes in place in key areas, with emphasis on improvement rather than reaction to problems. Also, the approach is well deployed, though some areas or work units may be in very early stages of deployment. Translated, this means that a firm must have an approach that addresses all the aspects of the item, and the approaches must be deployed. This means that if you are doing everything according to the Baldrige criteria, you will score 50%. To score in the higher scoring bands, you must demonstrate in the application not only that you have all the major aspects of a quality system, but the quality system also must have been refined and improved over an extended period of time. Quality Highlight 3-2 shows how Clarke American Checks, Inc., won the Baldrige award by improving and refining its processes.

Being a Baldrige Examiner

Appointment to the board of examiners for the Malcolm Baldrige National Quality Award is a very prestigious designation. In recent years, the board of examiners has consisted of between 200 and 300 members from across the United States. Examiners

FIGURE 3-7 Baldrige Process

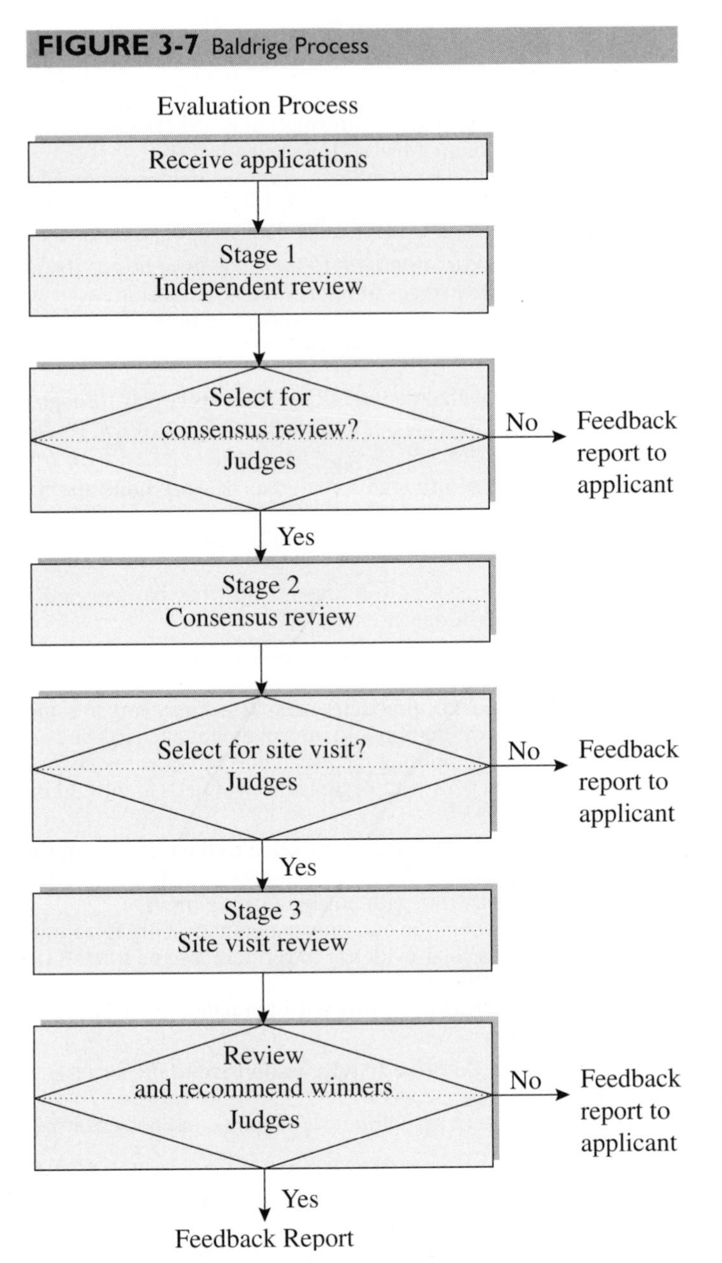

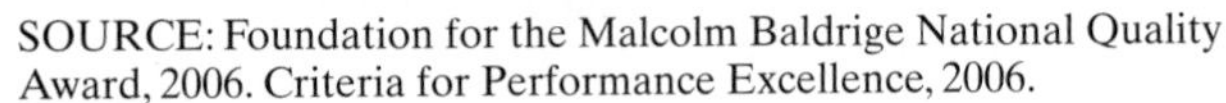

SOURCE: Foundation for the Malcolm Baldrige National Quality Award, 2006. Criteria for Performance Excellence, 2006.

for the board of examiners come from a variety of fields. There are CEOs, academics, physicians, quality specialists, consultants, and retirees. However, all these people have one thing in common: They are committed to the core values of the Baldrige award. They demonstrate this commitment by being willing to give up approximately 10% of their year to serve on the board with no compensation. Besides scoring applications, examiners are asked to write training cases, update the criteria, and help to improve the Baldrige process. The examiners are expected to exhibit the highest professionalism,

TABLE 3-9 Baldrige's Scoring Guidelines

SCORE	*PROCESS*
0% or 5%	• No SYSTEMATIC APPROACH is evident; information is ANECDOTAL (A) • Little or no DEPLOYMENT of an APPROACH is evident. (D) • An improvement orientation is not evident; improvement is achieved through reacting to problems. (L) • No organizational ALIGNMENT is evident; individual areas or work units operate independently. (I)
10%,15%, 20%, or 25%	• The beginning of a SYSTEMATIC APPROACH to the BASIC REQUIREMENTS of the Item is evident. (A) • The APPROACH is in the early stages of DEPLOYMENT in most areas or work units, inhibiting progress in achieving the BASIC REQUIREMENTS of the Item. (D) • Early stages of a transition from reacting to problems to a general improvement orientation are evident. (L) • The APPROACH is ALIGNED with other areas or work units largely through joint problem solving. (I)
30%, 35%, 40%, or 45%	• An EFFECTIVE, SYSTEMATIC APPROACH, responsive to the BASIC REQUIREMENTS of the Item, is evident. (A) • The APPROACH is DEPLOYED, although some areas or work units are in early stages of DEPLOYMENT. (D) • The beginning of a SYSTEMATIC APPROACH to evaluation and improvement of KEY PROCESSES, is evident. (L) • The APPROACH is in early stages of ALIGNMENT with your basic organizational needs identified in response to the other Criteria Categories. (I)
50%, 55%, 60%, or 65%	• An EFFECTIVE, SYSTEMATIC APPROACH, responsive to the OVERALL REQUIREMENTS of the Item, is evident. (A) • The APPROACH is well DEPLOYED, although DEPLOYMENT may vary in some areas or work units. (D) • A fact-based, SYSTEMATIC evaluation and improvement PROCESS and some organizational LEARNING are in place for improving the efficiency and effectiveness of KEY PROCESSES. (L) • The APPROACH is ALIGNED with your organizational needs identified in response to the other Criteria Categories. (I)
70%, 75%, 80%, or 85%	• An EFFECTIVE, SYSTEMATIC APPROACH, responsive to the MULTIPLE REQUIREMENTS of the Item, is evident. (A) • The APPROACH is well DEPLOYED, with no significant gaps. (D) • Fact-based, SYSTEMATIC evaluation and improvement and organizational LEARNING are KEY management tools; there is clear evidence of refinement and INNOVATION as a result of organizational-level ANALYSIS and sharing. (L) • The APPROACH is well INTEGRATED with your organizational needs identified in reponse to the other Criteria Items. (I)
90%, 95%, or 100%	• An EFFECTIVE, SYSTEMATIC APPROACH, fully responsive to the MULTIPLE REQUIREMENTS of the Item, is evident. (A) • The APPROACH is fully DEPLOYED without significant weaknesses or gaps in any areas or work units. (D) • Fact-based, SYSTEMATIC evaluation and improvement and organizational LEARNING are KEY organization-wide tools; refinement and INNOVATION, backed by ANALYSIS and sharing, are evident throughout the organization. (L) • The APPROACH is well INTEGRATED with your organizational needs identified in response to the other Criteria Items. (I).

SOURCE: Foundation for the Malcolm Baldrige National Quality Award, 2006 Criteria for Performance Excellence, 2006.

maintain absolute confidentiality, and be prompt and organized in responding to the requirements of the position.

The strictest standards to avoid conflict of interests are maintained by examiners. Examiners must divulge all their investments to NIST. They also must document for whom they have worked or consulted and who the major competitors, suppliers, and customers are for their present and former employers. This conflict-of-interest information is used by NIST to assign examiners to applicants.

QUALITY HIGHLIGHT 3-2

CLARKE AMERICAN CHECKS, INC.

www.clarkeamerican.com

An example of a firm that won the Baldrige award is Clarke American Checks of San Antonio, Texas. Clarke American supplies personalized checks, checking-account and bill-paying accessories, financial forms, and a growing portfolio of services to more than 4,000 financial institutions in the United States. Founded in 1874, the company employs over 3,000 people in 15 states.

In addition to filling more than 50 million personalized check and deposit orders every year, Clarke American provides 24-hour service and handles more than 11 million calls annually.

Clarke American competes in an industry that has undergone massive consolidation. Three major competitors have 95% market share and vie for the $1.8 billion U.S. market for check-printing services supplied to financial institutions.

The company is organized into a customer-focused matrix of 3 divisions and 11 processes. It is in a nearly continual state of organizational redesign, reflecting ever more refined segmentation of its partners and efforts to better align with these customers' requirements and future needs.

In the early 1990s, when an excess manufacturing capacity in check printing triggered aggressive price competition, Clarke American elected to distinguish itself through service. Company leaders made an all-out commitment to ramp up the firm's "First in Service" (FIS) approach to business excellence. Comprehensive in scope and systematic in execution, the FIS approach defines how Clarke American conducts business and how all company associates are expected to act to fulfill the company's commitment to superior service and quality performance.

FIS is the foundation and driving force behind the company's continuous improvement initiatives. It aligns Clarke American's goals and actions with the goals of its partners and the customers of these financial institutions. The company uses this single-minded organizational focus to accomplish strategic goals and objectives.

The company's key leadership team (KLT)—consisting of top executives, general managers of business divisions, and vice presidents of processes—establishes, communicates, and deploys values, direction, and performance expectations. Leadership responsibilities go beyond task performance. KLT members are expected to be role models who demonstrate commitment and passion for performance excellence. Each year, associates evaluate the competencies of executives and general managers in 10 important leadership areas. Senior leaders also keep a scorecard to track their progress in implementing company strategies and in demonstrating key behaviors, a tool adopted from a previous Baldrige award winner.

Goals, plans, processes, measures, and other vital elements of performance improvement are clearly documented and accessible to all. However, the company also places a premium on two-way, face-to-face communication. For example, KLT members and other senior local management lead monthly FIS meetings in their respective divisions or processes. These 2- to 3-hour meetings are held at all Clarke American facilities, and all associates are expected to attend. Agenda items include competitive updates, reviews of company goals and direction, key-project progress reports, associate and team recognition, and question and answer periods.

Strategic and annual planning is tightly integrated through the company's goal deployment process. Ambitious long-term (3 to 5 years) business goals help to guide the development of short-term (1 to 3 years) objectives and the selection of priority projects necessary to accomplish them.

From orientation and onward, associates are steeped in the company's culture and values: customer first, integrity and mutual respect, knowledge sharing, measurement, quality workplace, recognition, responsiveness, and teamwork. And they are schooled regularly in the application of standardized quality tools, performance measurement, use of new technology, team disciplines, and specialized skills.

Work teams and improvement teams carry out efforts to attain operational improvements spelled out in "run the business" goals. Cross-functional

Continued

project teams attend to "change the business" initiatives. Sharing of knowledge across teams, a company value, is facilitated through systematic processes.

In the last decade, Clarke American has invested substantially in new technology, using it to improve performance and to deepen relationships with partners through its offerings of customer management solutions. For example, with a major information technology supplier, the company developed digital printing capability that enables it to provide faster and more customized products and services to financial institutions and their customers. New technology has led to major reductions in cycle time and errors, nearly complete elimination of hazardous materials, less waste, and dramatically improved quality.

Clarke American, its partners, and its associates are reaping the benefits of these and other improvements. Company-conducted telephone surveys of partner organizations consistently show a 96% satisfaction rate. Partner loyalty ratings have increased from 41% to 54%. In independent surveys commissioned by the company every 18 months, Clarke American's customer-satisfaction scores for all three partner segments are trending upward and top those of its major competitors.

Among Clarke American associates, overall satisfaction has improved from 72% to 84%, when survey participation reached 98%, comparable to the world-class benchmark. Rising associate satisfaction correlates with the 84% increase in revenue earned per associate in the last decade. Annual growth in company revenues has increased from a rate of 4.2% to 16%, compared with the industry's average annual growth rate of less than 1% over the last decade.

SOURCE: NIST, *Profiles of Baldrige Winners* (Gaithersburg, MD: NIST, 2006).

State Awards

In recent years, the number of applicants for the Malcolm Baldrige National Quality Award has decreased. There are numerous reasons for this. A major reason is that applications declined in number once firms realized how difficult it would be to actually win the Baldrige award. Another reason is the existence of state awards. In 2005, more than 1,000 firms in the United States applied for state quality awards. The state awards are important because they give firms a basis to become familiar with the Baldrige process. Many recent winners of the Baldrige award have previously won state awards.

A review of the different state award programs reveals three categories of approaches to state awards. The first approach is the **full-Baldrige approach**. In these states, the Baldrige criteria have been adopted, and firms apply using the Baldrige criteria. In these cases, the criteria are used, but the scores required to win the state awards are lower than those for the national awards. This approach occurs often in more populated states (e.g., Missouri, New York, and Florida) that have well-funded award programs.

An approach that some other states have taken is the **Baldrige-lite** approach. This approach uses the Baldrige criteria but with a simplified process and/or application. This occurs in states such as Massachusetts and California.

The third approach is the **multilevel approach**. Using the multilevel approach, often the top level includes the full-Baldrige criteria. At the second level, a Baldrige-lite approach is used. Then, in lower levels, recognition is provided for firms that are putting forth significant effort toward improving performance.

The state award programs are becoming important vehicles for promoting involvement in the Baldrige process. Top management might become interested in applying for a state quality award where the competition is less demanding than the Baldrige award. State awards also provide important education for companies through conferences and the literature they provide. Examiner training given through state award programs helps to develop expertise among employees of local companies. These examiners can then disseminate this training within the firms where they work. For more information on the Baldrige award, go to *www.nist.gov*.

QUALITY IMPROVEMENT: THE JAPANESE WAY

The Japanese must be credited with raising worldwide quality to a new level of competitiveness. In the late 1970s, they created competition through quality as their automobiles and electronic products were exported to the nations of the world in huge numbers. Using quality as a competitive weapon to win orders in the marketplace, the Japanese provided an example for the rest of the world that has benefited producers and consumers all over the world.

Deming Prize

Before discussing Japanese quality approaches, we will first touch on the Deming Prize. The **Deming Prize** for quality was established in 1951 by the Japanese Union of Scientists and Engineers (JUSE). The award was funded by the proceeds from Deming's book on statistical process control that resulted from his teachings in Japan. The award now commemorates the distinguished service to Japan by W. E. Deming. The Deming Prize is awarded to individuals and groups who have contributed to the field of quality control. The examination and award process is performed under the direction of the JUSE Deming Award Committee. The Deming Prize process is open to non-Japanese firms. For example, Florida Power and Light and AT&T have won the Deming Prize.

The Deming Prize is awarded in three categories: Deming Application Prize for Division, Deming Application Prize for Small Business, and Quality Control Award for Factory. Unlike the Baldrige, there is no limit on the number of companies that can receive the award in a given year. The Deming Prize award ceremony is nationally televised in Japan.

The Deming Prize is much more focused on processes than is the Baldrige. This is reflected in the categories and items contained in the Deming Prize. Table 3-10 compares categories of the Deming, Baldrige, and European Quality Award competitions. A review of these categories shows the Baldrige is more general and managerial.

TABLE 3-10 Comparison of the Baldrige Award, Deming Prize, and European Quality Award

Baldrige Award	*Deming Prize*	*European Quality Award*
Leadership	Policy	Leadership
Strategic planning	Organization and operations	Policy and strategy
Customer and market focus	Collecting and using information	People management
Information and analysis	Analysis	Resources
Human resource focus	Planning for the future	Processes
Process management	Education and training	Customer satisfaction
Business results	Quality assurance	People satisfaction
	Quality effects	Impact on society
	Standardization	Business results
	Control	

At the same time, the Deming Prize is more prescriptive. We briefly discuss each of the Deming categories in the next paragraphs.

Policy

Japanese firms are well known for policy formation and deployment. To the Japanese, policy has essentially the same meaning strategy has for Americans. The first category of the Deming Prize addresses the areas of management of quality policy, policy formation, policy correctness and consistency, use of statistical methods, policy communication, checking policy, and consistency between short- and long-term policies.

Organization and Operations

The organization and operations portion of the Deming application requires the applicant to document processes for clarifying authority and responsibility, processes for delegating authority, cooperation among divisions, committees and their activities, use of staff, use of quality control activities, and quality control audits. The use of statistical approaches is included in policy formation and deployment.

Collecting and Using Information

The information that is collected for this item includes information that is gathered on the outside and inside. Applicants document how they pass these data through divisions and how rapidly it is transmitted. Finally, firms document how they process the data and analyze them statistically.

Analysis

The analysis category of the Deming Prize covers the areas of selection of priority problems and themes, correct use of analytical methods, and use of statistical methods. Technology is also addressed in this category. Processes are expected to be analyzed for quality with constructive use of analysis. Also, firms must document what they are doing with employee suggestion systems.

Planning for the Future

To plan effectively for the future, firms must understand their present condition well, which is the documentation required for this category. Policies must be reinforced for corrective action. Plans must exist and be documented for promoting total quality control (TQC), and these plans must be included in the long-term planning processes.

Education and Training

Applicants document plans and accomplishments relating to education and training. The quality and control foci of training are documented, along with statistical thinking. Essentially, this category examines the firm's practices relating to education and training—as they relate to quality.

Quality Assurance

As discussed in Chapter 1, quality assurance has to do with the design of products and the new product development process. Safety and liability issues are addressed in the application. Also, areas of process inspection, design, capability, quality assurance, and use of statistical methods are documented by the applicant.

Quality Effects

Quality effects relate to the documentation of benefits, outcomes, and results of quality improvement. These benefits include tangible and intangible benefits. Applicants also document the expected versus realized benefits of quality management.

Standardization

The Deming Prize rewards standardization. The systems of setting standards, monitoring performance against the standards, and revision of standards are documented in the Deming application. Again, statistical methods are required for monitoring progress against the standards.

Control

Finally, the last category of the Deming Prize is control. Essentially, the applicants describe how they are using statistical quality control (SQC), control points, control items, statistical thinking, and the current state of control in the company.

As can be seen, the focus of the Deming Prize is quite different from that of the Baldrige. Although the Baldrige has become very managerial in nature, the Deming focuses more on the nuts and bolts of quality improvement. Each approach has its strengths and weaknesses. Although the Deming is focused on the statistical methods, a complete picture of the management system may not emerge. At the same time, because the Baldrige is so diffuse in its focus, some of the fundamentals of quality improvement may be overlooked by examiners. To counter this problem in design with the Deming Prize, Japan has developed another quality award. This second award is called the Japan Quality Award and is closer in spirit to the Baldrige.

Other Japanese Contributions to Quality Thought

As Juran stated, the genius of the Japanese was in their ability to maintain a focus on the minutia and detail associated with process improvement. They also set the world standard for efficient, clean, and waste-free processes. In this section we discuss the Japanese approach to production and service. The attempt will be to identify those contributions that are uniquely Japanese.

Lean Production

Two views emerge in the literature that pertain to **lean** manufacturing. The first view of lean is a philosophical view of waste reduction. This view asserts that anything in the process that does not add value for the customer should be eliminated. Given this view, quality problems cause scrap and rework and are wasteful. The second view of lean is a systems view stating that lean is a group of techniques or systems focused on optimizing quality processes. An example of this view is the lean production system refined by the Toyota Motor Company and spread to the rest of the world. For our purposes, we will combine the philosophical and systems views to define lean as *a productive system whose focus is on optimizing processes through the philosophy of continual improvement.*

Lean as a Philosophy

As we showed in Chapter 2, philosophy is an important element in improving quality. Perhaps it is because of the difficulty associated with communicating quality that philosophies are so important. Words and definitions help us to communicate on a cerebral level. Philosophies, once internalized, help individuals and organizations to communicate on a feeling-based level. For Toyota Motor Company, the focus was on the continual reduction of waste. Shigeo Shingo, the industrial engineer who was fundamental in helping Toyota to reduce waste, identified a group of seven wastes that workers could address in improvement processes (Table 3-11).

TABLE 3-11 Shingo's Seven Wastes

1. *Waste of overproduction.*
2. *Waste of waiting.*
3. *Waste of transportation.*
4. *Waste of processing itself.*
5. *Waste of stocks.*
6. *Waste of motion.*
7. *Waste of making defective products.*

Japanese Total Quality Control (TQC)

Just as the Japanese lean approach requires attention to detail in every aspect of the process, so does the TQC approach. This attention to detail runs deep in the Japanese culture. About 1,200 years ago, during the eighth century A.D., Japanese swordsmiths hammered 10,000 microlayers of steel into the world's finest blades.

Of course, the Japanese philosophy of continual improvement is reflected in Deming's 14 points for management outlined in Chapter 2. However, beyond Deming's 14 points, there are several Japanese contributions to quality thought and practice that we outline here.

Visibility

An important aspect of the Japanese approach to quality is visibility. Often, when problems exist in business, the first reflex is to hide the problems as though they don't exist. The Japanese approach does the opposite. In the Japanese approach to quality, problems must be made visible before they can be addressed. Among the approaches to improving quality is inventory reduction. Excess work-in-process inventory has the effect of hiding problems. Therefore, it is eliminated.

Another visibility technique the Japanese use is called **andon**, or warning lights. Whenever a defect occurs on the line, the line is stopped. This halts production in several workstations, not just one workstation. As a result, workers from the production line all converge on the process where the warning light went off. Teams are used to identify and eliminate the fundamental causes of the defect. Once the cause is discovered and fixed, work resumes as normal. The lean process adds to visibility by stopping all the steps in the process when one step has a problem. The approval to stop the line whenever there is a problem is called **line-stop authority**.

In-Process Inspection

Another contribution of the Japanese was to teach the rest of the world about **in-process inspection**. With in-process inspection, all work is inspected at each stage of the process, and the workers inspect their own work. This approach gives workers the authority to make quality-related decisions. The Japanese approach to inspection was in direct contrast to the American approach of inspecting quality at the final stage of production through a quality department specialist.

$N = 2$ Technique

The **$N = 2$ technique** is an alternative to acceptance sampling. In traditional acceptance sampling (discussed in Chapter 9), when a company receives a shipment from its suppliers, the shipment is sampled and a determination is made as to whether the shipment should be accepted or rejected. Usually, an acceptance sampling plan involves rules such as

If 2 or fewer defects, accept the lot.

If more than 2 defects, reject the lot.

The $N = 2$ technique involves developing and maintaining a close relationship with suppliers so that it is known if the supplier's processes are in statistical control. If the supplier's processes are in control and capable, and if the first and last pieces in the lot meet specification, then it is concluded that the entire lot of materials will meet specification. Therefore, only a sample size of 2 (the first and last pieces) is needed for acceptance inspection.

Total Involvement of Workforce

As you may recall, Deming's 14th point stated that all the operating forces should be put to work to improve quality. The Japanese are masters at gaining organizational commitment to quality. By deploying quality improvement throughout the organization, we all become responsible for the aspects of quality we influence in a day's work. This includes vertical deployment and horizontal deployment of quality management. **Horizontal deployment** means that all departments are involved in quality. **Vertical deployment** means that all levels of management and workers are actively involved in quality.

The Five *S*'s

Many Japanese firms have adopted the five *S*'s in an effort to improve operations. The five *S*'s are a sequential process that companies follow to literally "clean up their acts." The *S*'s are

1. *Seiri:* Organizing by getting rid of the unnecessary. This may include old files, forms, tools, or other materials that have not been used within the past 2 or 3 years.
2. *Seiton:* Neatness that is achieved by straightening offices and work areas.
3. *Seiso:* Cleaning plant and equipment to eliminate dirtiness that can hide or obscure problems.
4. *Seiketsu:* Standardizing locations for tools, files, equipment, and all other materials. This often involves color coding and labeling areas so that materials are always found in a standard location.
5. *Shetsuke:* Discipline in maintaining the prior four *S*'s.

Quality circles are natural work teams made up of workers who are empowered to improve work processes and are used by Japanese companies to involve employees in improving processes and process capability. Using quality circles, Japanese employees brainstorm quality improvement methods and identify causes of quality problems using quality tools.

Preventive Maintenance (PM)

Japanese manufacturers are known for their approach to maintenance of equipment and machines. The maintenance technique taught by the Japanese is **preventive maintenance**. The idea behind preventive maintenance is that the worst condition a machine should ever be in is on the day you purchase the machine. By maintaining scheduled maintenance and improvement to equipment, machinery actually can improve with age. In the 1980s, the Toyota Kamigo Plant 9 in Japan won the Deming Prize with aged equipment. The key was that the equipment was maintained very well. With preventive maintenance, heavy, unscheduled maintenance is still performed by shop engineers and maintenance specialists. However, regular cleaning, fluid changing, and light maintenance are handled on a regularly scheduled basis by the people who operate the machinery.

QUALITY IMPROVEMENT: THE EUROPEAN WAY

In the late 1970s and early 1980s, Europe found itself in a position similar to that of the United States. European producers of products were finding the Japanese to be formidable competitors and realized they needed to change.

It is difficult to tell where Europe now stands concerning quality management. Because of radical differences in infrastructure, politics, and business practices, it is easy to overgeneralize. **ISO 9000:2000** is the European standard for quality that has been expanded worldwide. At the same time, there has been a perception in the quality community that Europe is behind the Japanese and the United States in improving quality. For example, French and German customer service has remained low by American and Japanese standards and has not shown great improvement. However, companies such as BMW, Mercedes, and Seimens produce products that set the standard for quality.

Culture plays a greater role in European quality practices than it does in the United States. For instance, many foreigners view German service as poor. There is a historical basis for this perception. Since medieval times, shopkeepers were considered part of the land-owning, privileged class in Germanic regions. As such, they enjoyed an elevated status in society. Therefore, a patron who entered an establishment would often look up to the service provider as someone of a higher social class. This created a situation in which the patron actually would be in a position of thanking the service provider for providing service.

Waiters in fine Parisian restaurants are viewed by many visitors from other countries as rude. However, in France, the culinary experience is considered high art. Much as an art expert would be aghast if someone without a trained eye belittled great art, the French waiter might be displeased when an untrained customer ordered the wrong wine. Those are only two examples of why, if service is to be improved, not only business practices will have to change in parts of Europe but also ingrained culture must change.

European businesses are on the horns of a particularly difficult dilemma. Europe is a loose federation of sovereign nations. Each country is trying to protect its own culture while, at the same time, trying to cooperate with the other countries to introduce unified standards. Two types of quality recognition are widely used in Europe. These are the European Quality Award (EQA) and ISO 9000:2000 certification.

European Quality Award

In 1988, a group of 14 large European companies created the European Foundation for Quality Management in reaction to increased competition from overseas, the quick success of the Baldrige award in the United States, and the recognition that changes were needed if Europe were to compete in the world market. The European Foundation for Quality Management administers the European Quality Award (EQA). The EQA has two levels. The highest level is the EQA for the most accomplished applicant in a given year. The second level given is the European Quality Prize for other firms that meet the award criteria.

The model for the EQA is shown in Figure 3-8. Because the EQA is similar to the Baldrige in tone and process, we will emphasize some of the differences between the

FIGURE 3-8 European Quality Award Model

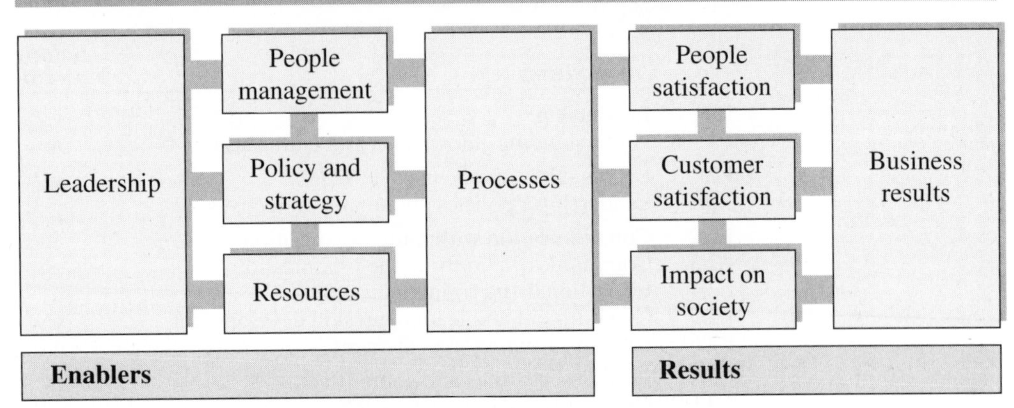

SOURCE: European Foundation for Quality Management, Brussels, Belgium, 2006.

two. The differences are found primarily in the categories of people satisfaction, impact on society, and business results. People satisfaction addresses the perceptions of employees concerning their employer. Items in this category include working environment, perception of management style, career planning and development, and job security. Whereas the Baldrige criterion of human resources and development focuses more on those things that lead to customer service and improved products, the EQA focuses more on employee satisfaction as an outcome of the quality system. From this standpoint, employee satisfaction becomes an indicator of satisfactory management.[4]

The EQA criterion of impact on society asks the applicant to document how the company is viewed by the society it affects. This includes the company's approach to quality of life, the environment, and the preservation of global resources. Therefore, charitable activities, leisure-related activities, and employment stability are all important aspects of the quality system for the Europeans.

A comparison of the Baldrige, Deming, and European awards is presented in Table 3-12. Like the Baldrige, the category of business results focuses on operational and financial results. However, the EQA focuses also on corporate social responsibility in the corporate results category. This includes an assessment of results relating to social and ecological factors.

ISO 9000:2000

On a worldwide basis, ISO 9000:2000 has had a much more significant impact than any of the quality standards or recognitions in terms of the number of companies that have implemented the approach. The focus of ISO 9000:2000 is for companies to document their quality systems in a series of manuals to facilitate trade through supplier conformance. Once the quality system is documented, ISO 9000:2000 registration states that there is a quality system in place and the quality system is being adhered to.

[4]Nakhai, B., and Neves, J., "The Deming, Baldrige, and European Quality Awards," *Quality Progress* (April 1994):33–37.

TABLE 3-12 Comparison of Elements in Three Awards

No.	*Criteria Items*	*Baldrige Award*	*Deming Prize*	*European Quality Award*
1.0	**Leadership**	y	y	y
1.1	Understanding, knowledge, and enthusiasm	n	y	n
1.2	Leadership's involvement in TQM	y	y	y
1.2.1	Leadership's policies and quality objectives	y	y	y
1.2.2	Communication within the organization	y	y	y
1.2.3	Communication outside the organization	y	y	y
1.2.4	Educational and training programs	n	y	n
1.2.5	Commitment and leadership skill development	y	n	y
1.2.6	Human resources management	n	y	n
1.3	Leadership's roles and commitment	y	y	y
1.3.1	Leader's roles and responsibilities	y	y	y
1.3.2	Leader's commitment performance	y	y	y
1.4	Impact on society	y	y	y
2.0	**Resources Management**	y	y	y
2.1	Information management, utilization, and analysis	y	y	y
2.1.1	Information and data collection	y	y	y
2.1.2	Information and data management	y	y	y
2.1.3	Information and data analysis and utilization	y	y	y
2.2	Material resources management	n	n	y
2.2.1	Suppliers selection and communication	n	n	n
2.2.2	Material resources management	n	n	y
2.3	Technology resources management	n	n	y
2.3.1	Technology selection	n	n	y
2.3.2	Technology exploitation	n	n	y
2.3.3	Human skill development	n	n	y
2.4	Financial resources management	n	n	y
2.4.1	Financial strategy	n	n	y
2.4.2	Financial decision making	n	n	y
2.4.3	Financial statement management	n	n	y
2.4.4	Management of shareholder	n	n	y

SOURCE: G. A. Beharis, "A Comparative Assessment of Some Major Quality Awards," *International Journal of Quality and Reliability Management* 12, 9 (1996):34.

ISO is the *Organization for International Standards* of Geneva, Switzerland (the Greek word *isos* means "equal"). The ISO standard was developed so that an international standard for documentation of quality systems could be applied in many different cultures. The ISO standards are very broad and non-specific, so they can be adapted to many different industries.

ISO 9000:2000 Basics

In this section we will discuss ISO 9000:2000 and its requirements. ISO 9000:2000 is not a prescription for running a business or firm. However, its requirements provide a recognized international quality standard that businesses can follow. It is interesting to note that more than 400,000 companies have registered using the ISO standard. To effectively use the ISO standard, you need to plan your processes, follow those processes, ensure that those processes are effective, correct deficiencies in your current processes, and continually improve your processes.

The original version of ISO 9000 was implemented internationally in 1994 after the International Organization for Standardization (ISO) Technical Committee (TC) 176 worked for eight years to develop the standard. The ISO 9000:1994 standard closely mirrored the prior British Standard 5750 in form and substance. By 1997, the ISO 9000 family of documents had become quite cumbersome, with 20 required elements and 12 standards. ISO 9000:2000 is a much simplified document with only six requirements and three standards. The three standards are

1. ISO 9000:2000—Quality management systems: Fundamentals and vocabulary.
2. ISO 9001:2000—Quality management systems: Requirements. This specifies the requirements of a quality management system. These requirements are used for internal implementation, contractual purposes, or third-party registrations.
3. ISO 9004:2000—Quality management: Guidelines for Performance Improvement. This broader document provides guidelines for objectives that are not included in ISO 9001:2000. These include continual improvement and enhancing overall performance.

ISO 9001:2000 consists of five clauses. These include

Clause 4: Quality Management System
Clause 5: Management System
Clause 6: Resource Management
Clause 7: Product Realization
Clause 8: Measurement, Analysis, and Improvement

As shown in Table 3-13, the quality management system is documented using a variety of requirements. These include how you develop, design, implement, and maintain your quality management documents. Also, you must demonstrate how you use quality-related documentation to manage your quality system.

As shown in Table 3-14, the ISO 9001:2000 standard emphasizes top management's role in the quality management system. Its requirements outline management's responsibilities in developing and maintaining a quality management system. Again, procedures are documented for each of these processes, and audits are used to ensure that these procedures are followed.

As shown in Table 3-15, resource management requirements are outlined. These include providing needed resources, personnel, facilities, and the environment necessary to get the work done. Emphasis is placed on training and developing employees.

Table 3-16 shows requirements for product realization. These are all the requirements—including processes, documents, customer requirements, specifications, designs, and quality processes—needed to produce a product. Part of this requirement

TABLE 3-13 ISO 9001:2000 Quality Management System Requirements

4.0 **Quality Management System**

4.1 **General Requirements**
The organization shall establish, document, implement, and maintain a quality management system and continually improve its effectiveness in accordance with the requirements of the international standard.

4.2 **Documentation Requirements**
Quality management system documentation will include a quality policy and quality objectives; a quality manual; documented procedures; documents to ensure effective planning, operation, and control of processes; and records required by the international standard.

SOURCE: International Standards Organization, Geneva, Switzerland, 2003.

TABLE 3-14 ISO 9001:2000 Management Requirements

5.0 **Management System**
5.1 **Management Commitment**
a. Communication of meeting customer, statutory, and regulatory requirements
b. Establishing a quality policy
c. Establishing quality objectives
d. Conducting management reviews
e. Ensuring that resources are available
5.2 Top management shall ensure that customer requirements are determined and are met with the aim of enhancing customer satisfaction.
5.3 Management shall establish a quality policy.
5.4 Management shall ensure that quality objectives shall be established. Management shall ensure that planning occurs for the quality management system.
5.5 Management shall ensure that responsibilities and authorities are defined and communicated.
5.6 Management shall review the quality management system at regular intervals.

SOURCE: International Standards Organization, Geneva, Switzerland, 2003.

TABLE 3-15 ISO 9001:2000 Resource Management Requirements

6.0 **Resource Management**
6.1 The organization shall determine and provide needed resources.
6.2 Workers will be provided necessary education, training, skills, and experience.
6.3 The organization shall determine, provide, and maintain the infrastructure needed to achieve conformity to product requirements.
6.4 The organization shall determine and manage the work environment needed to achieve conformity to product requirements.

SOURCE: International Standards Organization, Geneva, Switzerland, 2003.

TABLE 3-16 ISO 9001:2000 Product Realization Requirements

7.0 **Product Realization**
7.1 The organization shall plan and develop processes needed for product realization.
7.2 The organization shall determine requirements as specified by customers.
7.3 The organization shall plan and control the design and development for its products.
7.4 The organization shall ensure that purchased product conforms to specified purchase requirements.
7.5 The organization shall plan and carry out production and service under controlled conditions.
7.6 The organization shall determine the monitoring and measurements to be undertaken and the monitoring and measuring devices needed to provide evidence of conformity of product to determined requirements.

SOURCE: International Standards Organization, Geneva, Switzerland, 2003.

deals with selecting and developing suppliers in a way that ensures that purchased components satisfy requirements.

Table 3-17 lists the requirements for measurement, analysis, and improvement. Clause 8 has to do with analyzing process data and using the data to improve operations and service to the customer. This includes performing internal audits and monitoring and measuring processes and products. Also included in Clause 8.5 are corrective and preventive action for improvement.

TABLE 3-17 ISO 9001:2000 Measurement, Analysis, and Improvement Standards

8.0 **Measurement, Analysis, and Improvement**
8.1 The organization shall plan and implement the monitoring, measurement, analysis, and improvement process for continual improvement and conformity to requirements.
8.2 The organization shall monitor information relating to customer perceptions.
8.3 The organization shall ensure that product that does not conform to requirements is identified and controlled to prevent its unintended use or delivery.
8.4 The organization shall determine, collect, and analyze data to demonstrate the suitability and effectiveness of the quality management system, including
   a. Customer satisfaction
   b. Conformance data
   c. Trend data
   d. Supplier data
8.5 The organization shall continually improve the effectiveness of the quality management system.

SOURCE: International Standards Organization, Geneva, Switzerland, 2003.

Quality Management Principles Underlying ISO 9000:2000

The following eight principles provide the foundation for ISO 9000:2000. These are mentioned in ISO 9000:2000 and 9004:2000 but not in ISO 9001:2000. They are

1. Customer focus
2. Leadership
3. Involvement of people
4. The process approach
5. A systems approach to management
6. Continual improvement
7. Factual approach to decision making
8. Mutually beneficial supplier relationship

Most of these quality management principles are self-explanatory and are discussed throughout this book. TC 176 has done an excellent job of responding to the critics of the original standard by focusing on these quality management principles. Note the similarities between these principles and the Baldrige categories.

Selecting a Registrar

The ISO 9000:2000 process is very different from any of the awards processes discussed previously. To many who have implemented ISO 9000:2000, the selection of the registrar is the most important step in the ISO 9000:2000 process. This is also the messiest step of the ISO process. There is no centralized authority that qualifies ISO registrars. Although no European country requires ISO 9000:2000 registration, many firms require it. In fact, many countries officially encourage their firms to use only ISO-registered suppliers. However, certain ISO registrars have memoranda of agreement with the departments of commerce of individual countries. This means that out of the hundreds of registrars that exist in the world, only a relative few may be recognized in any particular country. Also, many registrars are not recognized officially anywhere. Therefore, when selecting a registrar, firms must be careful. They should check with customers and departments of commerce and customers within the countries to which they plan to export. This could save severe problems and a waste of money in the long run.

The ISO 9000:2000 Process

While many companies have now transitioned from ISO 9001 (1994) to ISO 9000:2000, we will identify a process for the first-time registering firm. The registration process for ISO 9000:2000 (including the new ISO 9001:2000) typically takes several months from initial meeting to final registration audit. This time frame differs from client to client, but each process usually follows the steps outlined in Figure 3-9.

Step 1 is inquiry, where the client contacts registrars to investigate the terms for registration. The prospective client then makes a final selection of a registrar with whom he or she is comfortable.

In Step 2, the client contracts with the registrar. In this process, registration steps are determined, and a price is negotiated. A client-signed quotation or purchase order leads to the first stage of the certification process. Some clients may wish to have a pre-assessment or gap-analysis audit.

Step 3 often involves a phase 1 audit. At this stage, the registrar performs an onsite audit of the documented quality system against the applicable standard.

Step 4 is the certification audit. Every element of the ISO 9000:2000 standard is audited several times during the registration process. A representative sample of an organization's business processes are chosen for any audit. During each three-year period, 100% of the organization is audited. The audit program is a valuable tool that provides a clearly and mutually defined process and snapshot of auditing—past, present, and future.

FIGURE 3-9 An Example ISO Registration Process

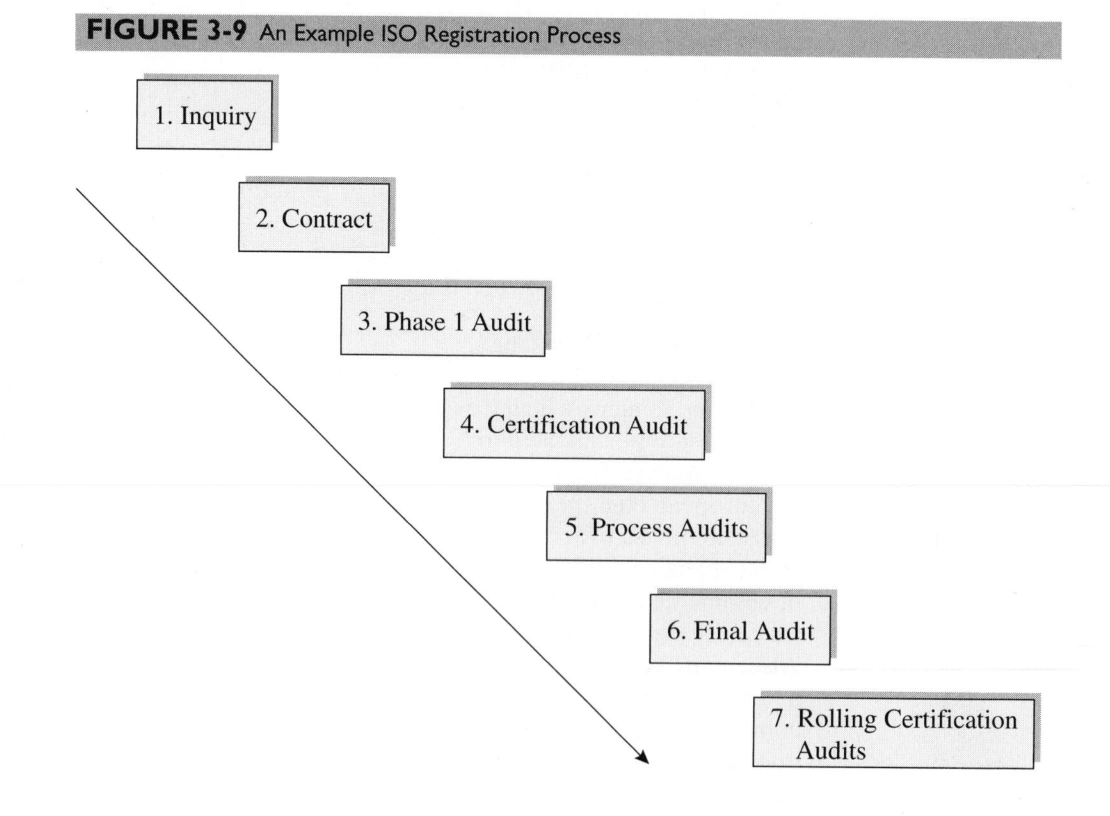

Step 5 may involve process audits (optional). The client may choose business processes for auditing to the applicable standard, allowing the client to learn and experience the registrar's auditing methods and style.

Step 6 involves the final certification audit. Once the client's documented quality system has met the applicable standard, the registrar will conduct an audit to determine the system's effective implementation. This may involve interviewing the process owners and responsible personnel as designated in the documented quality system for processes chosen from the audit program.

After certification, Step 7 involves rolling certification audits. These are sometimes referred to as *surveillance audits*, where the registrar returns on either 6-month or annual cycles.

ISO 14000

Given the success of ISO 9000:2000, ISO embarked on developing an international standard for environmental compliance called ISO 14000. ISO 14000 is a series of standards that provide guidelines and a compliance standard. The compliance standard is ISO 14001, Environmental Management Systems.

ISO 14001 uses the same basic approach as ISO 9000:2000 with documentation control, management system auditing, operational control, control of records, management policies, audits, training, statistical techniques, and corrective and preventive action. In addition to these controls, ISO 14001 includes quantified targets, established objectives, emergency and disaster preparedness, and disclosure of environmental policy. Such a system may provide the basis for developing a comprehensive environmental management system. Table 3-18 presents the ISO 14001 elements. The process for documenting these elements and seeking registration mirrors the ISO 9000:2000 process. Again, a key process has to do with selecting the appropriate registrar.

Outside of Europe, such as in the United States, firms have approached ISO 14000 carefully, with many firms deciding not to adopt the standard.[5] Is this because American firms are less committed to environmental quality than are European firms? The answer is probably "no." ISO 14000 is very risky for U.S. firms. As a result of the self-study process incorporated in the ISO standard, firms can possibly discover violations regarding some environmental topic, such as hazardous waste. It would seem that these firms should be able to then clean up this waste. However, it is not so simple. Once these firms discover variances, they are required to report these variances to the U.S. Environmental Protection Agency (EPA). Even if the firm discovers and cleans up environmental problems, the EPA has made it clear the firm will be subject to fines and penalties, which could include shutting down the business! As a result, firms are reticent to begin the process of self-discovery until the EPA changes its policy. Therefore, environmentally conscious firms that may want to improve their environmental management systems through ISO 14000 are potentially being dissuaded from doing so by the EPA. It will take several years for the courts to settle many of these issues. Until these issues are settled, adoption of ISO 14000 may be slow in many countries. Despite these problems,

[5]Hale, G., and Hemenway, C., "ISO 14001 Will Likely Join the Regulatory Framework," *Quality Digest* 16, 2 (1998):29–34.

TABLE 3-18 ISO 14001 Elements

ISO14001:2004	
4.1	General Requirements
4.3.1	Environmental Aspects
4.2	Environmental Policy
4.4.6	Operational Control
4.4.7	Emergency Preparedness
4.3.2	Legal and Other Requirements
4.3.3	Objectives, Targets and Programme(s)
4.4.1	Resources, Roles, Responsibility and Authority
4.4.2	Competence, Training and Awareness
4.4.3	Communication
4.4.4	Documentation
4.4.5	Control of Documents
4.5.1	Monitoring and Measurement
4.5.2	Evaluation of Compliance
4.5.3	Nonconformity, Corrective Action and Preventive Action
4.5.4	Control of Records
4.5.5	Internal Audit
4.6	Management Review

SOURCE: International Standards Organization, Geneva, Switzerland, 2006. Used with permission of Capaccio.

some American firms, such as Micron Technologies of Boise, Idaho, have adopted ISO 14000. It is expected that firms with little environmental exposure will adopt ISO 14000 first.

QUALITY IMPROVEMENT: THE CHINESE WAY

As can be seen in Figure 3-10, total trade between the United States and China has doubled in the years from 2000 to 2005. In that same time, Chinese imports to the United States have increased from $100 billion to $200 billion. Among the important reasons for this growth in trade have been the opening of the Chinese markets to foreign trade, the constant drumbeat of U.S. and European offshoring to China, and the low cost of doing business in China. Given the growing importance of China in international trade, the question then arises, what about Chinese quality?

Is There Such a Thing As Chinese Quality Management?

What led up to this internationalization from the historically isolationist China? With the death of Chairman Mao Tse-tung in 1976, Deng Xiaoping took the reigns of economic reform leading to an "open-door" policy for trade and economic transfer. This resulted in the creation of a socialist market economy.

FIGURE 3-10 Total Volume of Trade Between the United States and China

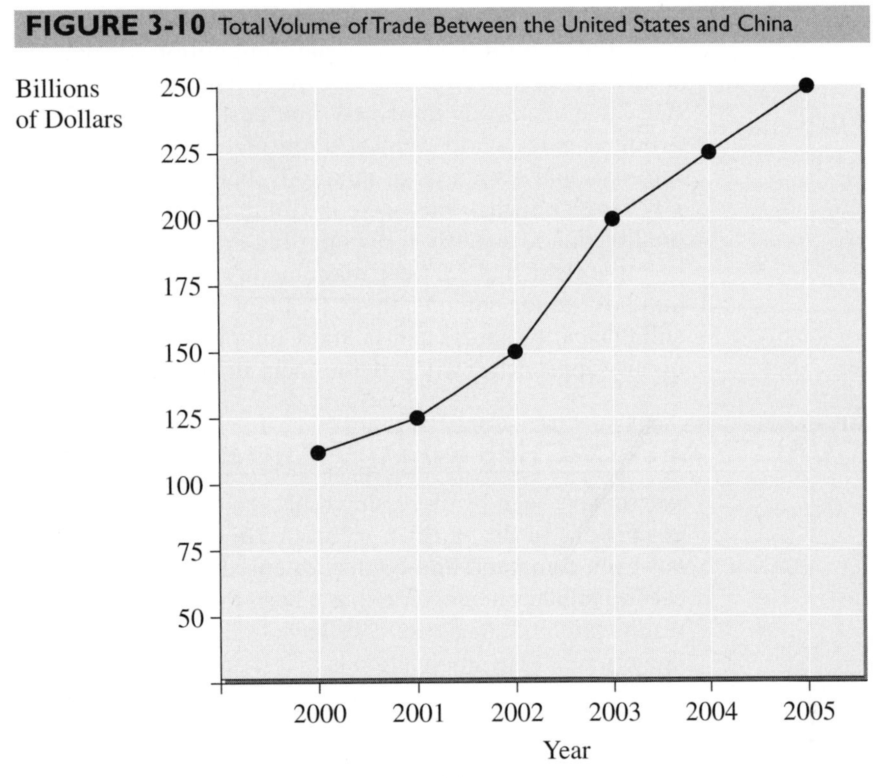

SOURCE: U.S. Department of Commerce, 2005.

Historically, China has been known for fine porcelain, silks, bronze, and construction. Trade between China and the West has existed for many centuries as is evidenced by the ancient Silk Road. Chinese products were held in such high regard that traders were willing to brave the foreboding Taklimakan desert, or "road of death," to bring Chinese products to European markets. These goods were of exceeding high quality and value. However, it is not clear that China has continued this legacy of quality in the modern era.

Some factors have led to low quality in China.[6] Researchers state that these factors include the low level of education of some Chinese workers and lack of experience among agricultural workers who have moved into the industrial sector. These same workers are often unfamiliar with the consumer goods they are making and may not have the commitment necessary to provide long-term resources to quality improvement as they work for a short period of time and then return to their homes to farm.

Another factor omnipresent in Chinese business is *guanxi*, or "influence." *Guanxi* can range from personal relationships to bribery. Here is an example of how *guanxi* can influence perceptions of quality in China:

> We wanted to give some government officials some biscuits as presents (customary in China). The day that they came they said, "By the way, we want double." If they wanted to penalize the company they could go to the media and say that Freshbake Biscuits' products were contaminated.

[6]Glover, L., and Siu, N., "The Human Resource Barriers to Managing Quality in China," *The International Journal of Human Resource Management,* 11, 5(2000); 867–882.

In the last quarter of a century, the United States and Europe have had to improve quality as a result of international competition. As time passes, China will be subject to similar competitive pressures. Both Japan and Mexico have seen increases in salaries that have negatively impacted their cost advantages. While it is not clear that Chinese products meet world standards for quality, over time, Chinese firms will need to meet international standards of quality if their standard of living is to increase proportionally. Foreign firms who locate in China to take advantage of low cost and forget about quality will do so at their own peril in the long term.

It is clear that Chinese manufacturers are aware of quality management practices and have made some changes to adapt certain practices.[7] What form that will take is still unclear. What is clear is that China is an economic giant. The adoption of modern quality approaches will provide them further competitive advantage in the future.

ARE QUALITY APPROACHES INFLUENCED BY CULTURE?

As we have seen by discussing tools, techniques, and approaches adopted within different regions of the world, a picture of diversity emerges. The U.S. approach historically has been command-and-control oriented. This might be the result of a history of political and military management as a basis for business management. Americans tend to be results oriented. As a result, we have seen that the focus in America is also very bottom-line oriented. The implication is that if quality does not provide bottom-line results, it is often not something to be pursued.

The Japanese approach is based on an ethic of consistency and emphasis on reduction of waste. There are cultural underpinnings for this approach. The Japanese land mass is smaller in size than California. Given that many people are contained in a small land mass, it is no surprise that a premium would be placed on reducing waste. Also, consider the fact that the Japanese have almost no natural resources. For this reason, almost all raw materials must be imported at great cost. Again, this causes a cultural ethic of reduced waste. Finally, it is fairly obvious in observing Japanese business practices that they prize conformance and consistency.

The Europeans have adopted broad standards that can be adapted to the diverse nation-states in the Economic Union. There is also an apparent concern among the Europeans to satisfy employees, focus on work as art, and care for the environment. There are strong unions in Europe that cause the need for sensitivity toward employees. The work-as-art ethic has roots in medieval times when the artisan ethic was developed. As an example, in Germany, an aspiring shipbuilder would find employment as an apprentice with a master shipbuilder. After years as an apprentice, the student would become a journeyman shipbuilder, until one day, possibly, becoming a master shipbuilder. Again, the cultural roots of management and quality improvement run deep in different cultures.

Are other cultures developing quality methods that will be instructive for world competitors? The Russians have long used what is now called **group technology** to produce products and to improve production. Several Asian Tiger nations, such as Malaysia, South Korea, and Singapore, have begun to make inroads in the field of quality and will become formidable competitors in the future. Chinese quality management is a work in process. We will keep watching.

---

[7]Pun, Kit-Fai, "Cultural Influences on Total Quality Management Adoption in Chinese Enterprises: An Empirical Study," *Total Quality Management*, 12, 3 (2001): 323–342.

Summary

In an increasingly globalizing economy, it is important to understand the approaches that various nations use to improve quality. It is clear that the trend is toward greater participation in a global economy. As a result, the worker of the future will need to be able to adapt to approaches having roots in other cultures.

In this chapter we discussed the global economy and the role played by quality in that world economy. We considered different quality models from different regions such as the Malcolm Baldrige Award, the Deming Prize, the European Quality Award, and ISO 9000:2000.

From an integrative perspective, it is reasonable to borrow from all these models if that will help your firm perform better. The underlying theme in this chapter is the importance of learning from other cultures to compete effectively.

Key Terms

- Andon
- Baldrige-lite
- Baldrige qualified
- Deming Prize
- Exporter
- Full-Baldrige approach
- Globalization
- Group technology
- Horizontal deployment
- In-process inspection
- ISO 9000:2000
- Lean
- Licensing
- Line-stop authority
- Malcolm Baldrige National Quality Award (MBNQA)
- Multilevel approach
- $N = 2$ technique
- Partnering
- Physical environment
- Preventive maintenance
- Quality circles
- Social environment
- Task environment
- Vertical deployment

Discussion Questions

1. What are the advantages or disadvantages of licensing as a means of gaining access to foreign markets?
2. What are the advantages and disadvantages of globalization? Provide an example of a firm that has engaged in globalization. What are some of the potential advantages and disadvantages of globalization for this particular organization?
3. What motivates U.S. firms to compete for the Malcolm Baldrige National Quality Award (MBNQA)? How could a firm benefit from participating in the MBNQA competition, even if it did not apply for the award?
4. Category 3 of the MBNQA criteria focuses on the importance of the customer in assessing the quality of the products and services that a firm sells. Why do you think the authors of the Baldrige criteria included this category? How is the customer important in assessing the quality of the products and services that a company sells?
5. Category 5 of the MBNQA criteria focuses on human resources and development. Why do you think the authors of the Baldrige criteria included this category? Why is human resource management an important consideration in quality planning and management?
6. Do you believe the MBNQA should include a category for not-for-profit organizations? If so, what adjustments should be made in the Baldrige criteria to make it applicable for this category?
7. If you were the CEO of a manufacturing firm, would you encourage your firm to apply for the MBNQA? Why or why not?

8. If you were presented the opportunity to be a Baldrige examiner, would you accept it? Why or why not? Make your answer as substantive as possible.
9. How can firms use lean production to improve quality? Is lean a useful concept for both service and manufacturing organizations?
10. Describe the concept of "visibility" in the context of the Japanese total quality approach. How does the concept of visibility help a firm identify problems in its production processes?
11. In what sense does excess inventory act as a "security blanket" for manufacturing firms?
12. Describe the concept of line-stop authority. If you were an operator in a production facility, would you want to have line-stop authority? Why or why not?
13. Why is it a good idea for workers to inspect their own work?
14. In what ways are the Malcolm Baldrige National Quality Award and the Deming Prize similar, and in what ways are they different?
15. Is it appropriate to use the criteria for quality awards as a framework for organizational improvement and change? Why or why not?
16. What are the major substantive differences between the quality awards discussed in the chapter and ISO 9000:2000? Are they intended for similar or entirely different purposes?
17. Describe the purpose and the intent of the ISO 9000:2000 program. What are the advantages of becoming an ISO 9000:2000-certified company? Are there any disadvantages?
18. Describe the purpose and the intent of the ISO 14000 program.
19. What are the pros and cons for an American firm that is considering pursuing ISO 14000 certification? If you were the CEO of a manufacturing firm, would you pursue ISO 14000? Why or why not?
20. Do you believe quality approaches are influenced by culture? How?

Cases

Case 3-1 University of Wolverhampton: Becoming an ISO 9000 University[8]

Related Web page: *www.wlv.ac.uk*

August 1, 1994, was a red-letter day most University of Wolverhampton employees will never forget. That day, the university became the "first university in the United Kingdom, and perhaps the world, to gain ISO 9000 registration for the entire organization." The core business of the university is "the design and delivery of learning experiences with provision for research and consultancy services."

The British Standards Institute spent four days at the university in June 1994 to assess Wolverhampton's systems after which the university was recommended for registration. This occurred after three days of corporate effort involving the university's quality assurance personnel and several staff. This effort was undertaken in addition to the normal external accreditation programs the university usually pursues. The requirement was a great deal of additional work. However, it was believed that if the internal systems were well documented, this could actually facilitate working with the external accrediting bodies. Thus ISO registration, in the long run, could simplify the lives of Wolverhampton staff. However, this wasn't the main objective for pursuing ISO 9000. There existed a strong desire to provide a "rational and documented system base for the pursuit of total quality in a large—by UK standards—and complex organization.

[8]Storey, S., "Passion and Persistence: Becoming an ISO 9000 University," *TQM in Higher Education* (November 1994):1–2.

The system reflects the activities of the university. Among the diverse activities and topics that had to be documented were the articles of government, the financial regulations, the health and safety requirements, the university charter, and many other items. Ultimately, these items were incorporated into the quality system. For the university, "the quality system that works not only defines what you control absolutely, but how you deal with what controls you." This includes all of the procedures, rules, laws, regulations, permissions, and vetoes that exist within a university system.

Used correctly, ISO 9000 improves systems. However, this does not mean that quality is improved, per se. Wolverhampton employees believe that the quality of administrative services have improved in serving students and the general public. On the teaching and research side, there is no evidence that quality has improved for this organization.

The information gathering process was rather expensive. It required commitment from the top. Wolverhampton had this commitment. Another important attribute was simply hardheaded commitment to a goal by those involved in the process. The ISO 9000 system is now open and well known. According to Wolverhampton employees, it was well worth the trouble. ■

DISCUSSION QUESTIONS

1. The University of Wolverhampton cites benefits relating to ISO 9000. What do you view as the benefits of ISO 9000 adoption and implementation?
2. More generally, what are the strengths and weaknesses of ISO adoption in academic institutions.
3. Would ISO registration be a useful tool for your college or university? Is there anything unique about the University of Wolverhampton that would make ISO 9000 more useful there?

Case 3-2 Wainwright Industries: An Entirely New Philosophy of Business Based on Customer Satisfaction and Quality

Related Web page: *wainwrightindustries.com*

Baldrige Award and Wainwright Industries:
www.quality.nist.gov/wainwright_Industries.htm

In the early 1980s, Wainwright Industries, a manufacturer of stamped and machine parts, was facing nothing less than a crisis. Increased competition, along with intensified customer scrutiny, was forcing Wainwright to either improve quality or lose its competitive stature. In the face of this challenge, the employees of the company, led by CEO Arthur D. Wainwright, decided to make radical changes. It was clear that business as usual with a few minor improvements would not save the company. What Wainwright needed was an entire new philosophy of doing business based on quality and total customer satisfaction.

To determine how to achieve this objective, Wainwright used the criteria for the Malcolm Baldrige National Quality Award as a roadmap. Drawing input from all levels of the company, the top management team led the process by setting goals, developing implementation strategies, and establishing key quality standards. Initially, the company emphasized three principles: employee empowerment, customer satisfaction, and continuous improvement. As a creative way of demonstrating the importance of working together, the company adopted the duck as its mascot, based on the fact that ducks fly in formation as a means of supporting one another in flight. In addition, whenever a duck falls out of formation, it suddenly feels the drag and resistance of trying to fly alone and quickly returns to the flock. Wainwright used this analogy to support the concepts of teamwork and employee empowerment, which were integral parts of the company's quality improvement efforts.

Along its journey toward improved quality, a number of specific initiatives were implemented. Lean manufacturing, statistical process control, computer-aided design, cross-training, profit-sharing, and quality-minded manufacturing initiatives were put in place. Special emphasis was placed on training and benchmarking. Since it initiated its quality program, the company has spent up to 7% of its annual payroll on training. To demonstrate its resolve in this area, the company has made training an important criterion for employee advancement. Wainwright has benchmarked against a number of companies, including firms in the textiles, chemical, and electronics industries. For instance, after studying Milliken & Company, a previous Baldrige award winner, Wainwright implemented an employee suggestion program that has been very effective.

Along with the changes mentioned previously, Wainwright also has changed its culture to make it more egalitarian and quality minded. The employees at Wainwright (including the CEO) now all wear the same uniform, eat in the same cafeteria, and park in the same parking lot. Office walls have literally been torn down and replaced with glass, based on the premise that if the managers can watch the frontline employees work, the frontline employees should be able to watch the managers work too. As a result of these changes, the managers of the company have become coaches and facilitators rather than supervisors and disciplinarians. This important change has helped facilitate the teamwork atmosphere that is supportive of high quality and total customer satisfaction.

The results of the company's continuous improvement efforts are linked to five strategic indicators, including safety, internal customer satisfaction, external customer satisfaction, design quality, and business performance. The status of each of these criteria is tracked by "mission control," a room set aside to document the company's efforts. In mission control, each customer's satisfaction is documented with a plaque, a current monthly satisfaction rating, and a red or green flag indicating the customer's status relative to objectives.

As a result of these initiatives, Wainwright has met the challenge. It has not only survived, but has emerged as an industry leader. The company has earned the status of preferred supplier to a growing number of quality-conscious customers and has received special recognition from General Motors, Ford, and IBM-Rochester. The goal of six-sigma quality is being pursued. Perhaps most importantly, in the last decade, overall customer satisfaction has increased from 84% to 95%, and the company's market share, revenues, and profits are at record levels. Ironically, the company was one of the recipients of the Malcolm Baldrige National Quality Award, the very award against which the company benchmarked in its early days of quality improvement. ■

DISCUSSION QUESTIONS

1. In its pursuit of improved quality, Wainwright emphasized two sets of initiatives: one based on improvements in its manufacturing operations (i.e., just-in-time manufacturing, computer-aided design) and the other based on human resource management (i.e., employee empowerment, profit sharing). Why was it necessary for Wainwright to emphasize both of these sets of initiatives? How are they related?
2. What is an egalitarian culture? How does the development of an egalitarian culture help a company like Wainwright Industries become more quality minded?
3. Although quality is important for every product or service, it may be particularly important for the precision auto parts industry. Do you agree with this statement? Why or why not?

PART TWO

Designing and Assuring Quality

Much of the traditional approach to quality was reactive and after the fact. We have learned through experience that you can assure quality only through the proper design of products and services. The basis for assuring quality is strategic planning that prioritizes and plans quality improvement. The question is: "If we want to achieve our goals, how are we going to get there?"

The voice of the customer is key in designing products. We must learn how to gather information about the customer. In Chapter 5 you will learn the tools used by firms who know their customers.

By benchmarking with other firms and competitors, we learn the voice of the market. To stay ahead, a firm must know what is happening around it. This provides a logical progression from average, to market leader, to world-class firm.

We then focus on designing products and services. Once we know what the customers want, how do we design goods and processes that will satisfy these wants?

Part of designing and assuring quality is the ability to deliver products to the customer. Thus supply chain management has become a key link in providing top-quality service.

CHAPTER 4

Strategic Quality Planning

Quality is not just a control system;
quality is a management function.
—DAVID GARVIN
Harvard Business School[1]

Quality is strategic. This may seem somewhat obvious, but the actions of companies implementing quality measures often obscure this fact. This is especially true when a company is in a reactive mode and does not use effective planning. In this chapter we discuss important aspects of strategic quality planning. Strategic planning has two important dimensions. These are *content* and *process*. Strategy content answers the question of what is to be contained in the strategic plan. Strategy process consists of the steps used to develop the strategy.

In this chapter we first discuss content and then process. Finally, we will look at quality results and whether quality has been shown to yield bottom-line results along the supply chain.

STRATEGY CONTENT

Why is quality planning important? As we have discussed in previous chapters, quality improvement is a planned managerial activity. As will be shown in this chapter, quality improvement involves identifying potential improvements, prioritizing potential areas for improvement, and planning the implementation of projects and improvements.

What are the content variables that should be included in strategic quality planning? Among the variables we discuss are time, leadership, quality costs, generic strategies (cost, differentiation, and focus), order winners, and quality as a core competency. These content variables outline key considerations when developing a strategic plan. These considerations are either explicitly or implicitly addressed in the strategic planning processes discussed later in the chapter.

THE IMPORTANCE OF TIME IN QUALITY IMPROVEMENT

Video Clip: Mission at the Ritz

We discuss two aspects of time: the time it takes to achieve business goals as a result of quality and the speed at which companies improve. Although several studies address important antecedents to quality improvement, many allude to, but do not specifically address, the quality-related variables of time and speed of quality improvement. Real-life experience shows that time is a key variable in improving quality. A major study of best quality-related practices undertaken by Ernst and Young[2] was critical of total quality

[1]Garvin, D., *Operations Strategy* (Englewood Cliffs, NJ: Prentice Hall, 1992).
[2]Ibid.

management (TQM) programs for not providing bottom-line results. At the same time, the Ernst and Young study advocated the *gradual* implementation of TQM. A comprehensive study by the U.S. General Accounting Office[3] stated that, on average, 3.5 years were required for companies to begin to see significant results from quality improvement programs. In a study of the U.S. auto industry, Narasimhan, Ghosh, and Mendez[4] found a 2.26-year lag between quality improvement and customer recognition of quality improvement. Shigeo Shingo[5] stated that 25 years were required for Toyota Motor Company to achieve significant improvement and that this time could be reduced to 10 years for competitors. W. Edwards Deming[6] consistently stated that continuous quality improvement was a slow process that required commitment of resources and time. A review of these studies and writings suggests that time is an important variable to consider when managing successful quality improvement programs. Time is also an important component of strategy. Strategic planning implies planning for the long term. Thus strategic planning is important for continuous quality improvement.

Firms will seek after and attempt to attain rapid quality improvement in order to obtain the benefits associated with improved quality, such as greater market share and increased sales. However, setting short-term goals for higher quality levels and managing toward those goals actually may prove detrimental to the firm. Managerial action that will lead to an optimal rate of quality improvement requires an understanding of the effects of rapid quality improvement. A study by Foster and Adam[7] focused on this issue of speed of quality improvement. Although their findings were preliminary, in a case study of five plants in one company, they found those plants that improved quality conformance more quickly did not see costs improve as much as plants that improved more slowly. How could this be? Let's examine this issue more closely.

We will call one of the approaches that some managers use the "management by dictate" model. Using management by dictate, we set numeric goals for the coming year. For example, top management might say: "I want to see a 50% reduction in defects during the coming year." Thus a numerical goal has been put in place for lower-level managers to meet. This is analogous to what Deming (discussed in Chapter 2) referred to as creating goals and not providing systems to achieve the goals. According to Donald Wheeler, a professor at the University of Tennessee, when goals such as these are set, one of three things will occur:

1. People will achieve the goals and incur positive results.
2. People will distort the data.
3. People will distort the system.

Achieving the goals is what management hopes will occur. Management truly would like to think that a goal can be attained without providing systems. Distorting the data may range from creative "cooking of the data" to finding honest data that shed the best light on the system in question. Distorting the system also can occur. For example,

---

[3]United States General Accounting Office, "Management Practices: U.S. Companies Improve Performance through Quality Efforts" (Washington, DC: GAO Report to the Honorable Don Ritter, House of Representatives, 1991).
[4]Narasimhan, R., Ghosh, S., and Mendez, D., "A Dynamic Model of Product Quality and Pricing Decision on Sales Response," *Decision Sciences* 24, 5 (1993):893–908.
[5]Shingo, S., Foreword in *Study of Toyota Production System from Industrial Engineering Viewpoint* (Tokyo: Japan Management Association, 1981).
[6]Deming, W. E., *Out of the Crisis* (Boston: MIT/CAES, 1986).
[7]Foster, S. T., and Adam, E. E., "Examining the Impact of Speed of Quality Improvement on Quality-Related Costs," *Decision Sciences* 27, 4 (1996):623–646.

if we set a goal for reduction of defects, we can define defects as those occurring in final inspection. We can then implement more rigorous in-process inspection that will eliminate defects before they arrive at final inspection. Voila! A simultaneous reduction in defects and increase in costs. This is an example of distorting the system. A Closer Look at Quality 4-1 discusses an example of how measurement systems can be distorted.

The key, then, is for firms to put in place a process that will allow learning to occur. The plan–do–check–act cycle discussed in Chapter 1 allows for organizational learning and freezing of learning to take place. Management that is looking for a quick fix to long-term problems probably will not be too satisfied with this fact.

We still do not know the optimal time for learning to take place. Indeed, there are stories of companies that have had the capacity to achieve rapid improvement. However, one suspects that each of these companies also provided enormous amounts of support for employees to achieve the goals. Providing such an infrastructure is the work of management because they have the budget and authority to do it.

A CLOSER LOOK AT QUALITY 4-1

PROBLEMS WITH MEASURING EDUCATIONAL PERFORMANCE[a]

When measurement systems are tied to performance ratings and money, people begin to distort the results and the systems. Such is the case with the No Child Left Behind Act passed by Congress. At the core of the No Child Act is testing designed to evaluate the performance of schools and teachers.

At Sunset Ridge School in Northfield, Illinois, students complained that someone had changed answers to math questions on their state achievement tests. An inquiry found altered multiple-choice questions on 90% of the tests taken by the school's eighth graders. School officials say an eighth-grade teacher promptly resigned as a result. It appears that even among rich, top-performing schools, teachers who are under pressure to show stellar results on state-mandated standardized tests are cheating.

An unfortunate by-product of the No Child law has been an unanticipated increase of cheating by teachers and administrators. From Pittsburgh to Milwaukee to Worcester, Massachusetts, and Spokane, Washington, hundreds of teachers, principals, and administrators have been accused of doing dishonest things to boost their schools' test scores. Among the transgressions committed by these role models include changing test answers, handing out tests and answers in advance, blocking weak students from taking exams, and giving students extra time.

The rash of test rigging is driven largely by the goal of the No Child law, which is to assess school and not student performance. If a group fails for two consecutive years, parents can move their students to another school—and districts can lose funding. Eventually, failing schools can be closed. In addition, some states have tied teacher bonuses to student scores. According to Steven D. Levitt, a University of Chicago economics professor, "Teacher cheating is rising because the incentives to do so are increasing."

Here are some examples of the cheating:

- Changing answers: Teachers in California, Illinois, and Michigan have been found changing answers.
- Coaching during exams: Instructors have helped students with their tests in Massachusetts, Nevada, and Washington.
- Handing out tests in advance: This has occurred in Pennsylvania and Wisconsin.
- Not counting weaker students: Among the states of dubious distinction are New York, Tennessee, and Texas.

[a]Adapted from B. Grow, "A Spate of Cheating—by Teachers," *Business Week*, July 5, 2004, pp. 94–95.

LEADERSHIP FOR QUALITY

Leadership is a key strategic variable for quality management. A leader organizes, plans, controls, communicates, teaches, advises, and delegates. The existence of a leader implies the existence of a follower. Therefore, the **leading** involves a power-sharing relationship between two or more individuals where the power is distributed unevenly. For example, the leader will have more monetary and organizational authority than does the follower. Thus the leader will need to share authority for the follower to complete work assignments.

Leadership is the process by which a leader influences a group to move toward the attainment of **superordinate goals**. Superordinate goals are those goals that pertain to achieving a higher end that benefits not just the individual but the group. In some organizations, leaders emerge because they are the most powerful or the dominant voice. Some leaders are selected because they have the highest intellect. This occurs in many educational organizations where the leaders are chosen on the basis of the number of articles written or status within an academic field. In still other organizations, leaders are appointed. This can be by either popular voice, by a board of directors, or by upper management.

For followers to have power, leadership must share its power. As a result, leadership is about the sharing of power. This power takes many forms:

Power of expertise. Sometimes a leader has special knowledge (or is perceived to have special knowledge). Professors are leaders in the classroom because they have knowledge that they are sharing with the students. This type of power tends to have very narrow parameters in that the followers will follow only within the confines of the leader's expertise.

Reward power. If a leader has rewards that he or she can bestow on subordinates in return for some desirable action, the leader has reward power. This is often the case in the granting of raises, promotions, rewards, recognition, or a variety of other incentives. Prior to the fall of communism in Russia, leaders often used this mechanism to maintain their authority. Perks and privileges came to those who favored the party bosses.

Coercive power. If the leader has power to punish the follower for not following rules or guidelines, the leader has coercive power. Such power often results in unintended responses, such as the follower giving up or circumventing the leader's rule surreptitiously. Many times the follower will rebel and attempt to even the power relationship. In the 1920s, in response to unions, managers used police powers to squash unions. This resulted in violent responses from the unions.

Referent power. If a leader is charismatic or charming and is followed because he or she is liked, then the leader has referent power. A case of referent power is the mentor who is admired by his or her protegees who want to be like the mentor. Often people will follow referent leaders on the basis of reputation alone, imbuing the referent leader with qualities the leader may or may not possess. John Kennedy was well liked by those around him. He also was viewed as charismatic and intelligent. This made people want to follow him.

Legitimate power. As a result of the positions that different people hold within an organization, the manager has the obligation to request things of subordinates, and the subordinates have the duty to comply with the request.

This is the exercise of legitimate power. Legitimate power comes with the position. It has certain responsibilities and authorities. A newly appointed leader may have to rely on this positional authority in the early part of his or her tenure as a leader.

Leadership Dimensions

We have discussed leadership types. To better understand leaders, we now discuss leadership dimensions. These leadership dimensions help to define the effectiveness of leaders. One dimension of leadership is the **trait dimension**. In his landmark research, Stogdill[8] found that leaders in organizations tended to be taller than the average. They also tended to be more intelligent than the average—although not too intelligent. Evidently, people are intimidated by leaders who are too bright. Leaders are also people who tend to be very productive and reach high levels of performance.

Another dimension of leadership research has to do with **leader skills.** For example, the grid in Table 4-1 shows a set of skills that effective leaders exhibit. Four important skills for leaders are knowledge, communication, planning, and vision. The different quadrants identify skill sets that are needed by managers.

In quadrant 1, knowledge helps the leader accept risk and moderate the stress associated with the risk by using coping mechanisms or healthy outlets. In quadrant 2, the leader must be able to communicate with other leaders and subordinates. In quadrant 3, the leader must be able to plan and make decisions. Finally, in quadrant 4, the leader must be able to formulate a coherent vision of the future toward which to plan.

Leaders are often categorized by **leader behavior**. The leader behavior approach discusses how leaders behave to identify specific leadership styles and to study the effect of leadership style on subordinate satisfaction and performance.

These approaches to categorizing managers are useful in that they can help us understand the dynamics of leadership. From the quality literature, a few dominant themes emerge concerning leadership. First, leaders who are successful in leading quality-related efforts must be worthy of trust. Why is trust such an important variable? One reason trust is important is because quality improvement efforts are often associated with productivity and efficiency improvement efforts that have resulted in reductions in the workforce. Therefore, it is important that leaders are consistent over time in their

TABLE 4-1 Leadership Skills

Quadrant 1: Knowledge	*Quadrant 2: Communication*	*Quadrant 3: Planning*	*Quadrant 4: Vision*
Acceptance of diversity	Assertiveness	Structuring (for task accomplishment)	Assessing the climate (internal and external)
Developing competence	Conflict management	Decision making	Identifying opportunities
Health/wellness	Team building	Evaluation skills	
Learning style	Trust building	Task and time management	
Time management	Motivating others		
Ethics	Recruiting others		
Risk taking	Effective speaking		
Coping skills	Effective writing		
	Effective listening		
	Image building		

[8]Stogdill, R., "Personal Factors Associated with Leadership," *Journal of Applied Psychology* 25 (1948):35–71.

actions so that employees will trust the leader enough to pursue quality. One major U.S. corporation overcomes this trust issue with a written policy stating: "The company reserves the right to adjust work forces in accordance with market and economic changes. However, the company will never have a reduction in force as a result of quality and process improvement efforts." Leaders have pointed to this policy as a key ingredient in the organization's successful quality improvement efforts. If they do eliminate or simplify processes, employees are given at least a lateral move into another job. Downsizing related to quality improvement is always handled through **attrition** (not hiring new employees as older employees either quit or retire) in this company.

Trust is especially important in union environments where a strong adversarial relationship between management and labor may exist. Clauses such as the previous statement can be negotiated with unions to get their approval for forming teams and adopting flexible work rules. Leadership must be absolutely consistent and fair in gaining and earning trust or else future potential for quality improvement will be reduced.

Employees also must trust that if they make recommendations for improvement, the recommendations will be taken seriously and considered for implementation by management. *Nothing can damage a quality improvement effort faster than management's failure to consider implementing changes that employees recommend.* Employees may begin to think "nothing will really change." Quality specialists who find themselves in companies not implementing a majority of employee suggestions need to work with management to increase the percentage.

Another important attribute of quality managers is commitment over the long term. Too many leaders enter quality improvement efforts with the expectation that things will improve overnight. As we have discussed, quality management is hard, painstaking, slow work. Often, victories in quality are rare and should be celebrated—especially at first. Consumer Research Incorporated (CRI), the small service company that won the Malcolm Baldrige National Quality Award, has weekly employee celebrations where the employees share good news and give each other pats on the back. Management of CRI admits that sometimes reasons for the pats on the back are difficult to find. However, the positive reinforcement has been central to their improvement of business results and customer service.

Commitment to quality means that leaders provide funding, slack time, and resources for quality improvement efforts to be successful. This commitment is measured in decades, not quarters or budget cycles. One company embarked on an expensive training process only to throw all its past efforts away by deciding to cut training in a budget cycle two years later. At last glance, this company's quality and process improvement efforts had all but disappeared.

As we see in the case of Solectron (Quality Highlight 4-1), the leadership of its CEO is key to its success. In order to foster a well-run workplace, leaders must be able to resolve conflict effectively.

QUALITY AND ETHICS

Quality appears to be good business. Quality is also good ethics. It is unethical to ship defective products knowingly to a customer. Reliable products and low defect rates reflect an ethical approach of management's care for its customers. This ethic is stated in the well-known mission statement of a New Bedford, Massachusetts, shipbuilder; "We build good ships. At a profit if we can, at a loss if we must. But, we build good ships."

QUALITY HIGHLIGHT 4-1

SOLECTRON CORPORATION

www.solectron.com

Solectron Corporation is an independent producer of high-tech manufacturing services. This manufacturing includes the assembly of printed circuit boards and subsystems for computer makers and electronics product producers. In addition, Solectron provides system-level assembly services, such as assembly of PCs and mainframe computers. Activities performed by Solectron include design, production, assembly, consultation, and testing. Solectron has achieved outstanding results because of its strategic planning system and the personal leadership provided by its management.

By focusing on customer satisfaction, exploiting advanced manufacturing technology, and stressing continuous improvement in operations and service, the company has reached high levels of quality and efficiency, making it best-in-class and a world leader in production. Solectron is an American company that has competed successfully in international markets. Many competitors of Solectron are now customers because they found it was better to outsource to Solectron than to produce many products in-house. In addition, about 90% of all new work comes from returning satisfied customers.

Assessing Customer Needs

Solectron focused its planning processes on the customer. Solectron does not compete with its customers in designing and marketing products. Although it offers an original equipment manufacturers (OEM) design service, usually the company produces to its customers' specifications and designs. As a result of understanding its customers, the company develops strategies to meet its customers' requirements in the areas of service, quality, and cost. Solectron continually monitors customer satisfaction levels and conducts exhaustive research on competitors and markets.

Surveys of customers are conducted on a weekly basis. The results of these surveys go directly to the CEO, who reviews the information with top management in one of three weekly meetings on quality-related issues. The survey information is used to grade the performance of each of Solectron's nine divisions.

Culture of Continuous Improvement

Solectron has developed a culture that reinforces continuous improvement. Developing this culture has required arduous strategic planning. A top management team is involved in a crusade to revitalize American manufacturing through quality. This team sets corporate targets and then works with teams to set supporting goals in functional areas. The company has pursued several strategies to achieve a high-energy, customer-focused workforce. These strategies include a strong family orientation, an effective communication system, and an innovative reward and recognition program. These strategies have helped the company to weather the rapid growth it has experienced. Management is participative with a focus on coaching and a high degree of autonomy for workers. The team-focused approach to employee involvement relies on training and mentorship to overcome barriers to a multilingual workforce with more than 20 ethnic backgrounds.

Each Solectron customer is served by two teams. These teams ensure quality performance and on-time delivery. A project planning team is involved in planning, scheduling, and defining customer requirements. A quality control team meets weekly to monitor and evaluate production with the aim of preventing potential problems before they occur.

The Best and Getting Better

Solectron uses a comprehensive information system, organized in a customized relational database that allows constant monitoring of internal quality

Continued

performance and process control indicators. Key performance data are charted in all departments. Employees are trained and empowered to take corrective action.

Solectron works with suppliers to improve the quality and reliability of incoming components and subassemblies. Statistical process control (SPC) charts track and review results daily. These results are recorded on an automated SPC database. Division managers and the corporate quality director track these results on a daily basis. The company is partnering with its suppliers to improve design quality as a future strategic emphasis.

Solectron performs strategic planning in selecting its future technological choices. Investments in advanced technologies are guided by an evaluation of the customer's future requirements and top management's emphasis on enhancing manufacturing capability. All this effort is paying off for Solectron. As indicated by its chief quality measure, customer satisfaction, the company has won scores of superior performance awards in recent years. After a quality audit, a major customer rated Solectron as the "best contract manufacturer of electronic assemblies in the U.S." Solectron's strategic focus is now expanding as it adds production facilities around the world.

Companies focusing on their customers often develop a set of ethics that includes valuing employees. This is reflected in education, training, health, wellness, and compensation programs that show empathy for the employees. Increasingly, environmental friendliness is seen as an ethical concern. As a result, more companies are implementing recycling programs and making efforts to improve environmental practices.

With the Enron debacle, the problems of the accounting firm Arthur Andersen, the Internet bust, and other highly publicized fiascos, there is one overlooked variable that must resurface—*integrity* in doing business. Floyd Harmston, a University of Missouri-Columbia emeritus economist, once stated that the entire U.S. economy was based on one simple principle: "When I give you a check, there are funds sufficient in my account to cover the check. When this fails to happen, the monetary system is damaged."

Harmston's words seem wiser now than ever. In recent years, with the focus on growth in companies, downsizing, and discussion of "new business models," we need to make sure that we can look at ourselves in our mirrors before we go to bed at night.

Good quality management is good ethics. Would we want to knowingly provide poor service or ship bad product? J. R. Simplot made a promise to Ray Kroc at McDonalds that McDonalds would never run out of French fries. With a handshake, they sealed their deal, and Simplot has kept his promise.

Integrity gets down to honesty. Are we honest to our customers, employees, colleagues, family members, and ourselves? This must be the basis for business. There is not a new business model that obviates the need for integrity.

QUALITY AS A STRATEGY

In Chapter 1 we raised the question of whether quality is sufficient to win orders in the marketplace. As you will recall, we stated that although quality can still win orders in some markets, in many markets quality has become an order qualifier. This means that high-quality production is an essential ingredient to participation in the market. In Chapter 3 we discussed research showing that quality is still an effective tool in successfully exporting in the international market.

We now discuss quality as a strategy from the perspective of generic strategies. These generic strategies are *cost, differentiation*, and *focus*.

Costs of Quality

One of the generic means of competing is cost. Traditionally, this meant the lowest-priced items in the industry. Many companies compete on cost. For example, Kmart competes on cost. New definitions of cost are expansive, considering the summation of costs over the life of a product. This includes service, maintenance, and operating costs for the product. For example, it is recognized that ink-jet color computer printers are relatively inexpensive to purchase. However, owners of these machines have found that the color cartridges dry up rapidly and that the replacement costs for the cartridges are quite high. As we discuss later, the life-cycle costs for many products may be staggering when environmental costs are considered. In this section we discuss the cost of quality and how quality can help decrease the cost of doing business.

There are two broad categories of cost: costs due to poor quality and costs associated with improving quality. At a minimum, management must understand these costs in order to formulate policy concerning quality improvement. In addition, Taguchi and others have provided insights into the issue of quality costs. As an example, the title of the classic book by Crosby, *Quality Is Free*, reveals an interest in the costs of quality.

PAF Paradigm[9]

The PAF paradigm translates quality costs into three broad categories, which are then subdivided into other categories. The three categories are prevention, appraisal, and failure costs (hence the acronym PAF).

Prevention costs are those costs associated with preventing defects and imperfections from occurring. Prevention costs are the most subjective of the three categories of costs. Prevention costs include costs such as training, quality planning, process engineering, and other costs associated with assuring quality beforehand (see Table 4-2).

Two caveats associated with the collection of quality-related costs include (1) there may be some debate as to whether these costs are all related to quality and (2) persons who work in prevention often do not keep accurate records of all costs.

Appraisal costs are associated with the direct costs of measuring quality. These can include a variety of activities such as lab testing, inspection, test equipment and materials,

TABLE 4-2 Prevention Costs

The cost of setting up, planning, and maintaining a documented quality system
Quality planning: establishing production process conformance to design specification procedures, and designing of test procedures and test equipment
Quality and process engineering (including preventive maintenance)
Calibration of quality-related production equipment
Supplier quality assurance
Supplier assessment
All training
Robust design
Defect data analysis for corrective action purposes
Time spent on quality system audits

[9]For a more in-depth discussion of quality-related costs, see Plunkett, J. J., and Dale, B., "A Review of the Literature on Quality-Related Costs," *International Journal of Quality and Reliability Management* 4 (1988):247–257; Carson, J. K., "Quality Costing: A Practical Approach," *International Journal of Quality and Reliability Management* 3, 1 (1986):54–65; and Foster, S., "An Examination of the Relationship between Conformance and Quality Related Costs," *International Journal of Quality and Reliability Management* 13, 4 (1996):50–63.

TABLE 4-3 Appraisal Costs

Laboratory acceptance testing
Inspection and tests by inspectors
Inspection and tests by noninspectors
Setup for inspection and test
Inspection and test materials
Product quality audits
Review of test and inspection data
On-site performance tests
Internal test and release
Evaluation of materials and spares
Supplier monitoring
ISO 9000:2000 qualification activities
Quality award assessments

losses because of destructive tests, and costs associated with assessments for ISO 9000:2000 or other awards (see Table 4-3).

These costs have undergone a fundamental change as U.S. companies have progressed in quality. For example, these costs were traditionally uncomplicated to assess because appraisal was performed by a centralized quality control function. The concept of in-process inspection has made it difficult to measure appraisal costs accurately. In addition, appraisal and auditing costs have been impacted by assessment activities associated with ISO 9000:2000 and the Malcolm Baldrige award as companies have undertaken these assessment programs.

Failure costs are roughly categorized into two areas of costs: internal failure costs and external failure costs. **Internal failure costs** are those associated with on-line failure, whereas **external failure costs** are associated with product failure after the production process. This includes failure after the customer takes possession of the product (see Table 4-4).

Accounting for Quality-Related Costs

One of the impediments to the collection of quality cost data has been the lack of acceptable accounting standards for these costs. For example, the standard accounting definition for quality is "meeting specifications." This narrow definition limits organizations desiring to quantify and measure customer requirements as a means of improving service to the customer. A reason for this is that accounting rules require definitions that are not open-ended or open to alternative interpretations.

TABLE 4-4 Failure Costs

Cost of troubleshooting	Lost production because of manpower availability problems (this refers to idle time brought about by failure to plan manpower efficiently)
Reinspection of stocks after defect detection	Lost production caused by system problems (i.e., material or instructions not available; cost of idle time only)
Disruption of production schedules	Concessions (design and engineering time)
Complaint handling and replacements plus extra time with customers	Process waste (including the waste commonly regarded as unavoidable)
Warranty (taking care not to duplicate previous item)	Cost of product scrapped at product audit
Cost of holding higher levels of stock as a buffer against quality failure	Cost associated with disposition of all scrap
Cost of corrective maintenance to plant	
Cost of corrective action to product (redesign, repair)	

Example 4-1: Quality Costs in Action

Problem: Macaluso's manufacturing company has gathered the following quality-related costs. You are hired as a consultant to evaluate these costs and to make recommendations to management. Compute the ratio of prevention and appraisal costs to failure costs.

Annual Quality Costs		
Failure costs		
Defective products	$	5,276
Engineered scrap		17,265
Nonengineered scrap		125,274
Consumer adjustments		623,980
Downgrading products		1,430,678
Lost goodwill		Not evaluated
Customer policy changes		Not evaluated
TOTAL		2,202,473
Appraisal costs		
Receiving inspection	$	35,765
Line 1 inspection		42,234
Line 2 inspection		53,567
Spot checking		63,766
TOTAL		195,332
Prevention costs		
Quality training	$	14,500
Process engineering		
Corporate		125,678
Plant		39,124
Product redesign		16,422
TOTAL		195,724
Grand total		2,593,529

Solution: Ratio of appraisal to failure costs:

$$195{,}332/2{,}202{,}473 = .0887$$

Ratio of prevention to failure costs:

$$195{,}724/2{,}202{,}473 = .0889$$

Ratio of prevention and appraisal to failure costs:

$$(195{,}332 + 195{,}724)/2{,}202{,}473 = .1776$$

Proportion of total quality costs:

Prevention: $195{,}724/2{,}593{,}529 = .0755$

Appraisal:

$$195{,}332/2{,}593{,}529 = .0753$$

Failure:

$$2{,}202{,}473/2{,}593{,}529 = .8492$$

This analysis shows that failure costs are very high compared with the prevention and appraisal costs. Increasing prevention and appraisal activities (and costs) could result in a significant decrease in failure costs.

Lundvall-Juran Quality Cost Model

The PAF categorization of costs is a useful way of understanding costs. Using the law of diminishing marginal returns, quality costs can be modeled to show the tradeoffs between these costs. This tradeoff model, called the *Lundvall-Juran model*, is shown in Figure 4-1.

The Lundvall-Juran model is a simple economic model. It states that as expenditures in prevention and appraisal activities increase, quality conformance should increase. For example, the more we spend on training and developing our employees, the more benefit we should get. As conformance improves, failure costs will lessen as well. This is an interesting case because if these statements are true, there should be an economic quality level that minimizes quality-related costs, and this flies in the face of the idea of continuous improvement proposed by Deming and others.

Differentiation through Quality

Think of a product you have been desiring for some time that is priced significantly above the average market price for such a product. It might be a very expensive, brand-name stereo receiver or TV that you want because of its special appeal. Chances are that

FIGURE 4-1 Lundvall-Juran Model

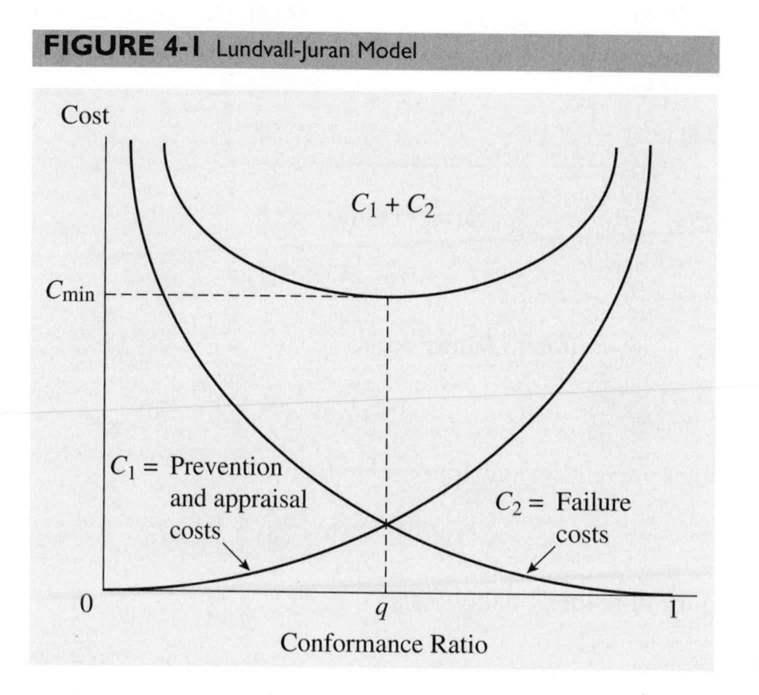

such a product benefits from differentiation. A Harley-Davidson is an example of a product that is priced far above the average price for a motorcycle, yet dealers seem to have trouble keeping enough of them in stock. Differentiation is achieved by a competitor if the consumer *perceives* the product or service to be unique in an important way. Neiman-Marcus, the retailer, charges many times the prices charged by its competitors for some products. A Rolex watch or Bang & Olufson stereo invests the owner with a certain status that other products do not. These are examples of differentiated products.

In the early 1980s, Japanese automakers differentiated their products based on mileage and quality. The quality aspect allowed Japanese automakers to gain significant market share in the United States. There is evidence that by the early 1990s, U.S. automakers had closed the quality gap sufficiently so that the Japanese could no longer differentiate on quality. This is the case in many markets. It is increasingly difficult to differentiate products based on quality alone. However, there is still much room to differentiate based on service to the customer in many markets.

Focus through Quality

The third generic strategy is to focus the product. For example, think of a product that is particularly regional or is marketed to a particular segment of the population. This limited customer group or segment of the market is the object of the focus strategy. Chrysler is building high-powered specialty cars, such as the Prowler, that are marketed to a small, affluent segment of the market. As the baby-boom generation ages, many more companies are segmenting products that can be marketed to this age group. For example, more fitness clubs for the elderly are springing up around the country. Such a focus strategy can be very profitable. Consumer Research Incorporated reports that they reduced the number of customers they served by 40% while at the same time doubling profits. They found that by focusing on only their very large clients for advertising services, they were able to get more business with these clients and simultaneously achieve higher service ratings.

The three generic strategies of cost, differentiation, and focus have been identified as important strategic decisions. A company that emphasizes cost will use different approaches to producing quality products than will a company that emphasizes differentiation or focus.

Order Winners

Terry Hill[10] of the London Business School defined a process for setting strategy that is centered on the identification of the order-winning criterion (OWC). Although the OWC is generally associated with manufacturing strategy, the same concept can be applied to service strategy.

Table 4-5 provides an overview of the planning framework defined by Terry Hill. This framework addresses several of the problems occurring in manufacturing. At times, there is a mismatch or misalignment between corporate objectives and decisions and operational subplans. There is a close relationship between the Hill model and generic strategies that we have discussed. First, the organization determines its competitive priorities and defines how it wins orders in the marketplace. For example, if the company wins orders based on focusing on small niche markets, marketing strategies will be developed to market a wide variety of specialized products. At the same time, this agreement on order winners allows the manufacturing people to make process

[10]Hill, T., *Manufacturing Strategy* (Homewood, IL: Irwin, 2000).

TABLE 4-5 Hill's Strategy Framework

Corporate Objectives	*Marketing Strategy*	*How Do Products Win Orders in the Marketplace?*	*Manufacturing Strategy* — *Process Choice*	*Manufacturing Strategy* — *Infrastructure*
Growth Survival Profit Return on investment Other financial measures	Product markets and segments Range Mix Volumes Standardization versus customization Level of innovation Leader versus follower alternatives	Price Quality Delivery speed reliability Demand increases Color range Product range Design leadership Technical support being supplied	Choice of alternative processes Tradeoffs embodied in the process choice Process positioning Capacity size timing location Role of inventory in the process configuration	Function support Manufacturing planning and control systems Quality assurance and control Manufacturing systems engineering Clerical procedures Payment systems Work structuring Organizational structure

SOURCE: T. Hill, *Manufacturing Strategy* (Homewood, IL: Irwin, 2000). Reproduced with permission of The McGraw-Hill Companies.

choices and infrastructural decisions that support wide variety. The process choice then might be a flexible manufacturing system using short setup and change-over times.

The key to the Hill model is reaching consensus on the OWC. The process for doing this involves segmenting the business into smaller markets that can each be identified with an order-winning criterion. This provides an understanding of the markets the company is serving. Products are chosen for each market, and marketing provides sales forecasts for the identified markets. Strategic debate then occurs, resulting in the selection of an OWC. From this, manufacturing strategy is formulated.

Quality as a Core Competency

Prahalad and Hamel[11] have identified the strategic concept of core competence. They describe core competence as consisting of

> communication, involvement, and a deep commitment to working across organizational boundaries. It involves many levels of people and all functions. World class research in, for example, lasers or ceramics can take place in corporate laboratories without having an impact on any of the businesses of the company. The skills that together constitute core competence coalesce around individuals whose efforts are not so narrowly focused that they cannot recognize the opportunities for blending their functional expertise with those of others in new and interesting ways.
>
> Core competencies do not diminish with use. Unlike physical assets, which do deteriorate over time, competencies are enhanced as they are applied and shared. But competencies still need to be nurtured and protected; knowledge fades if it is not used. Competencies are the glue that binds existing businesses. They are also the engine for new business development. Patterns of diversification and market entry may be guided by them, not just by the attractiveness of markets. (p. 79)

[11] Prahalad, C., and Hamel, G., "The Core Competence of the Corporation," *Harvard Business Review* (May–June, 1990):79–91.

Using the Prahalad and Hamel definition of competency, quality—in and of itself—probably is not a core competency. However, for firms operating in rapidly evolving markets or industries, the ability to change can be more important than the actual changing technology of the moment. Hence organizations producing outstanding products or services with a good understanding of processes are better positioned to operate in the changing market because they can introduce new products rapidly with fewer quality-related holdups. Therefore, core competency is built on the foundation of a long-term commitment to quality and continual process improvement.

QUALITY STRATEGY PROCESS

For the most part we have been discussing strategy content so far in this chapter. *Content* refers to the variables, definitions, components, and concepts that are included in the strategy. With the discussion of the Hill model and core competency, we are now ready to discuss strategy process. *Process* consists of the steps for developing strategy within an organization.

There are many different processes for developing strategy. We will highlight a strategic process for firms having little or no experience developing strategy—known as the *forced-choice model*. Afterwards, we present a comprehensive strategy development process that has been used in companies with mature quality processes.

Forced-Choice Model

The forced-choice model of strategic planning is one of several strategic planning models that could be adapted to demonstrate integrated quality planning. The forced-choice model is selected here because of its simplicity and its usefulness for firms that are beginning strategic planning. The forced-choice model is generic and is used simply for explanation purposes. It is simple when compared with other strategic planning models that are generally more complicated. Figure 4-2 provides an outline of the forced-choice strategic planning model.

FIGURE 4-2 Forced-Choice Model

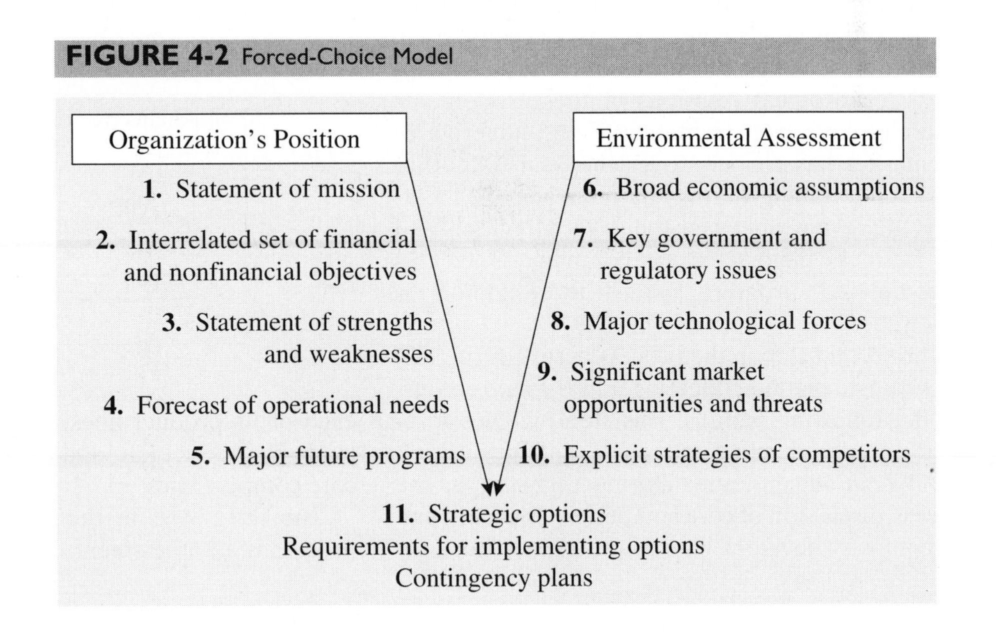

The forced-choice model is particularly useful for companies that are relatively inexperienced in strategic planning. The process begins by sequestering 6 to 12 members of upper management to a retreat. An organizational assessment is performed to identify strengths and weaknesses within the firm. Next, management performs an environmental assessment to evaluate the company's relative position in the marketplace. In the wrap-up session, alternative strategies are developed by executives.

A CLOSER LOOK AT QUALITY 4-2

A MATURE STRATEGIC PLANNING PROCESS

The strategic planning process contains quality concerns at every step of the process. XYZ Corporation is a nationally known organization that has been very successful in implementing quality strategy throughout the entire organization. (The company identity is disguised because of confidentiality reasons.) The company uses an annual strategy development cycle. However, part of the annual cycle involves "roll-ups" of five-year and ten-year planning projections. The strategic planning process for XYZ is coguided by the CFO and vice president of marketing, under direction of the corporate CEO.

As shown in Figure 4-3, the planning process begins with an annual management retreat in October to review the current year's mission statement, vision statement, values statement, and strategy. Using professional facilitators, the strategic management team reviews the current year's performance against projections and makes adjustments to the current year's plan. Once this is completed, the mission, vision, and values are reviewed and updated. Year-to-year revisions in these statements generally have been incremental and adaptive as the market has changed over the years. Another major undertaking during the management meeting is the review of key customer requirements and key business factors for the company. Once this is completed, progress against these factors is reviewed.

FIGURE 4-3 Mature Strategic Planning Process

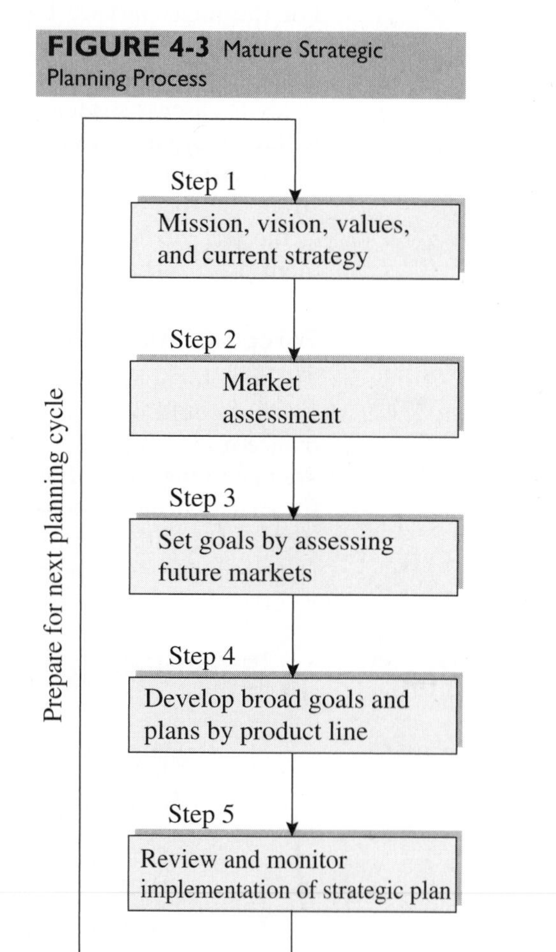

The second step in the process occurs during the fall strategic planning meeting and is finalized during the following months. This step involves forecasting market demand and reassessing the position of XYZ in the industry. During this step in the process, discussion of core competencies occurs, and the company develops visions and expectations for each of its product lines. Revenue projections are made by product line, and plans are made to extend core competencies.

The third step in the XYZ strategic planning process is an assessment of future markets in each of the product divisions. This occurs in January and involves an analysis of capacity needs for the coming year based on projections, changes in customer needs, technological needs assessment, profiling major competitors, assessment of the customers of major competitors, and the development of new marketing plans. Based on these marketing plans, information technology and operations decisions are made concerning capital investments for the coming year.

The fourth step involves the development of broad goals and plans by product line. These plans are used to develop separate missions, visions, values, and plans by product line on a very broad basis. Key business factors and key customer requirements are reviewed and updated at this point as management has evaluated formalized input from the customers. These plans are reviewed by upper management to ensure alignment with overall corporate goals. This step generally occurs during the first quarter of the year.

The fifth step is to develop the coming year's annual strategic plan. This plan includes market and product projections, plant and equipment projections, labor projections, projections of needed competencies and capabilities, as well as the development of Hoshin plans. Hoshin plans are deployment plans for the XYZ corporate objectives.

At this point we should mention that XYZ Corporation has many metrics, measures, and indicators in place to monitor performance as it relates to the strategic business plan. These measures and indicators, which are quality-related, financial, delivery-related, and so forth, are monitored by management on a daily basis using an enterprise resource planning (ERP) system. Indicators relating to key business factors are also monitored by the ERP. Feedback from the quarterly customer surveys is used to update key customer requirements.

Finally, the last stage of the annual strategic planning process is a review of the strategic planning process itself to see where potential improvement can be made. Once goals have been completed for the coming year, corporate, divisional, and departmental meetings are held to ensure that the corporate plan is distributed and discussed with employees at all levels.

Although this is a thumbnail sketch of a strategic planning process, the management of XYZ Corporation feels strongly that this process has set the stage for its global success. Financial results have been impressive for the company.

The formalized strategy is prepared subsequent to the retreat. Individual steps are listed in Table 4-6, emphasizing quality-related issues.

DEPLOYING QUALITY (HOSHIN KANRI)

Hoshin is Japanese for a *compass*, a *course*, a *policy*, or a *plan*. This is to indicate a vision or purpose to an existence. *Kanri* refers to *management control*. In English, this is generally referred to as *policy deployment*. Hoshin has been used in Japan since the 1960s as a means of implementing policy. Implicit in the Hoshin Kanri (or Hoshin for short) is the use of the basic seven tools of quality (these are discussed in Chapter 10), the new tools of quality, and quality function deployment.

Figure 4-4 gives an overview of the Hoshin process. The company develops a three- to five-year plan, and senior executives develop the current year's Hoshin objectives. Then the process of catchball occurs. **Catchball** is the term used to describe the interactive nature of the **Hoshin planning process**. Catchball involves reporting from teams and feedback from management. The development of the Hoshin plan results in the cascading of action plans that are designed to achieve corporate goals (see A Closer Look at Quality 4–2).

As shown in Figure 4-4, functional managers also should develop Hoshins in conjunction with upper management. This results in the development of specific plans for action and postimplementation review of Hoshins to evaluate the success of the process.

TABLE 4-6 Forced-Choice Steps

Step 1: Statement of Mission. The CEO presents the mission statement to the group of executives as a working document that is open to revision and discussion. The group discusses the company's mission with respect to quality and integrates quality-related language and objectives into the mission statement. Quality-related language might include an operational definition of quality.

Step 2: Interrelated Financial and Non-Financial Objectives. In addition to discussing and adopting financial goals, the group discusses quality financial issues, such as costs of quality, and documents obvious and not-obvious linkages between quality and financial results.

Step 3: Statement of Strengths and Weaknesses. Nominal group sessions for both strengths and weakness are performed. Over several hours, the group identifies significant operational strengths and weaknesses. In quality, major causes of quality problems, systemic failures, communications issues and other problems are brought to the fore and prioritized through nominal group voting techniques. The same process is utilized to identify quality strengths.

Step 4: Forecast of Operations. The financial forecast is used to gain an understanding of limitations of resources. Contribution of quality improvement to financial resources is evaluated and quantified.

Step 5: Major Future Programs. A review of current programs in place is undertaken. Ideas for improvement of current programs, identification of those that are not meeting expectation and potential future programs are evaluated. This limits available options. For example, long-term commitment to continuous improvement programs may limit choices concerning organizational redesign (such as reengineering).

Step 6: Broad Economic Assumptions. To initiate the environmental assessment, a marketing executive or economist presents the economic assumptions underlying the strategic plan. Among the economic factors discussed is the predicted importance of quality issues in winning orders in the marketplace.

Step 7: Key Government/Regulatory Issues. Emergent regulatory issues are presented by legal council. Quality related issues include environmental issues, product liability, safety, and employee relations.

Step 8: Major Technological Forces. Technologies that impact quality are discussed. Discussion includes in-house research and development, and externally emergent new production technologies. These technologies are discussed from a systemic perspective, recognizing interactions between multiple operating variables.

Step 9: Significant Marketing Opportunities/Threats. A nominal group session is performed asking questions such as: "What are the new opportunities for new or existing products?" "What role does quality play as an order-winning criterion?" and "What are likely changes to future quality related expectations?" This step can last up to one-half day of a retreat depending upon the number of questions posed.

Step 10: Explicit Competitor Strategies for Each Competitor. Each major competitor is identified by name and each competitive strategy assessed. The primary input for this activity is competitive intelligence gathered from benchmarking best-in-breed and world-class organizations. Approaches to quality management are identified and assessed for appropriateness, strengths, and weaknesses. This assessment provides a basis for identifying potential opportunities to obtain competitive advantage through improving quality processes.

Step 11: Generating Strategic Options. Along with developing important financial, marketing, and operational strategies, quality strategies are identified to address specific areas of weakness and to improve areas for strength. Gap analysis can be utilized at this point to identify the prioritization of quality emphases.

DOES QUALITY LEAD TO BETTER BUSINESS RESULTS?

One of the most common questions posed by executives seeking improvement is: "Will this quality effort pay off?" An answer to this question was provided by W. Edwards Deming with the Deming value chain, which we considered in Figure 1-8. Deming proposed a theoretical basis for concluding that quality will pay off.

The effects of quality on business results are mixed; some firms have been wildly successful with their quality efforts, and other companies have been unsuccessful in gaining bottom-line results. There are two primary reasons for this. *First, there are many variables that affect profitability besides quality*. You might produce the highest-quality, obsolete product in the world. If you produce a high-quality product or service that no one wants to buy, quality management systems likely will not save you. Macroeconomic factors, such as the recession-boom business cycle, affect profitability. American Motors products of the early 1980s were given awards for outstanding

FIGURE 4-4 Hoshin Planning

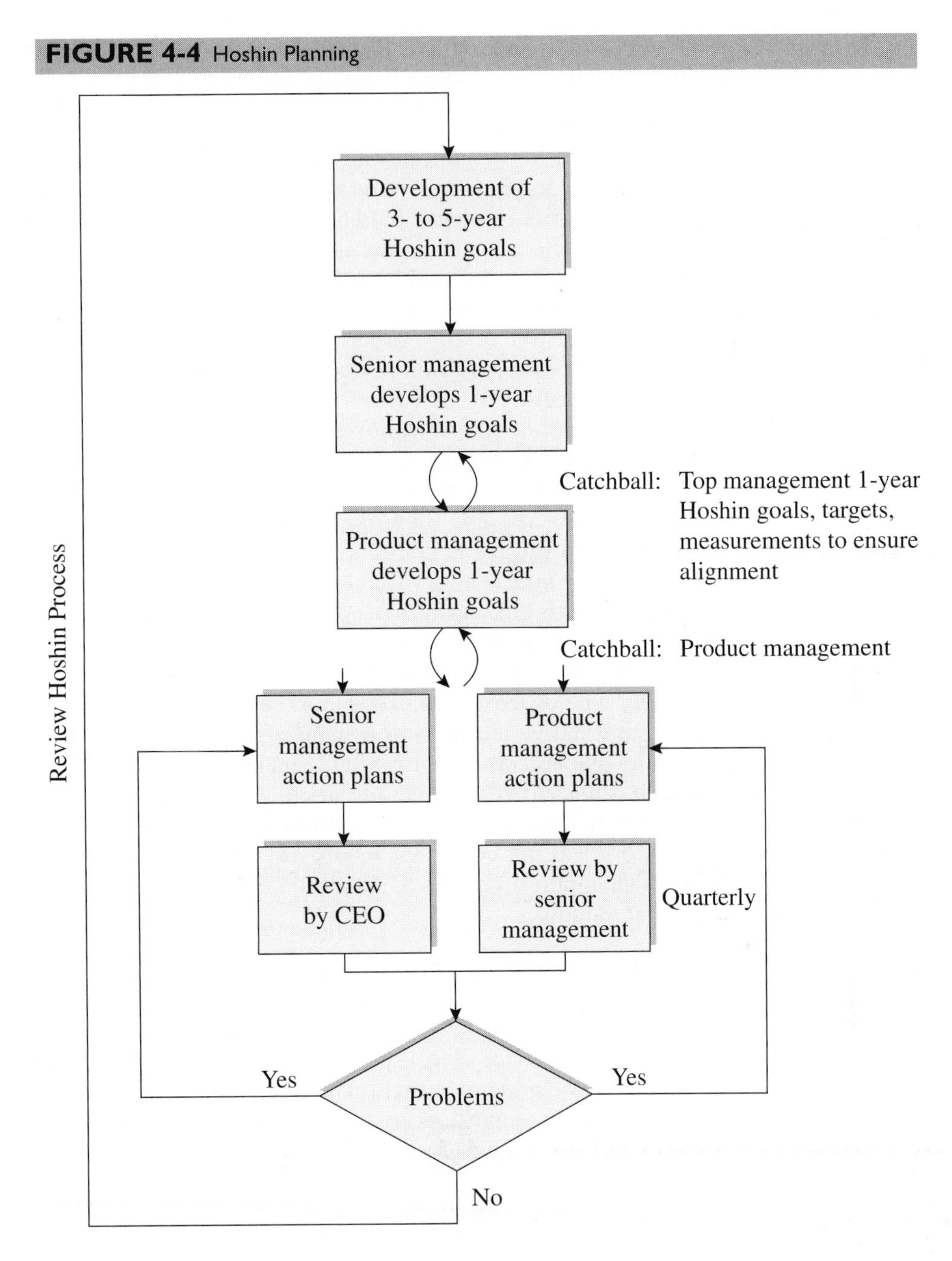

design (e.g., the Renault Alliance). At the same time, these products were not well accepted by their customers, and the company went out of business.

Second, many companies implement quality incorrectly. That you can claim you are implementing quality does not guarantee that you will be successful. Quality improvement takes a long time, and many firms desire quick returns on investment for quality training programs. When these returns are slow in coming, the companies give up in midstream and wonder why their quality efforts were ineffective. At the same time, quality programs have been shown to be effective in a variety of cultures and industries when implemented correctly. Hence the mixed results. As a result, we need to understand the relationships between quality and other variables.

Quality and Price

The relationship between price and quality has long been studied. Indeed, the laws of supply and demand lead to a natural ordering of competing products on a price scale such that a superior product would be the most highly priced. Another view is that high price is a psychological enabler allowing perceptions of high quality to result. However, the price-quality relationship becomes increasingly unclear when cultural differences in an international setting are considered. Kettinger et al.[12] found that in different cultures the relative ordering of customer priorities differed. The implication is that different cultures could perceive the price-quality relationship either positively or negatively, resulting in variations in financial performance. For example, the Mercedes is considered a workhorse car in much of Europe. In the United States, the Mercedes is perceived as a luxury vehicle.

Another reason that the price-quality relationship might be difficult to assess has to do with the increase in high-quality, low-priced goods over the past decades. It is expected that if this trend continues, the relationship between price and quality will decrease. Other intervening variables in the international markets are the prevalence of price supports for various products and "dumping" practices by international firms. Price supports often take the form of governmental subsidies that allow producers to sell products below cost internationally. Dumping occurs when producers sell products below cost to improve market share and to kill competition. An example of these alternative methods of pricing is apparent in the many Japanese consumer electronics that are less expensive in U.S. markets than in their native Japanese market. Research is unclear concerning the price-quality relationship. Although some higher-cost products clearly have better quality, such as a $60,000 BMW versus a $17,000 Ford Escort, it is not clear that a bottle of shampoo that costs $4.00 is better than a bottle of shampoo that costs $3.00 all the time. Therefore, at best, this relationship is somewhat tenuous.

Quality and Cost

A fundamental difference exists between a low-price strategy, which is based on competitive pricing, and a low-cost orientation that is based on continual learning and production competence. Because of the possible relationship between pricing and low-cost structure, we anticipate that quality will tend to provide a competitive advantage relative to other competitors by allowing firms with a high-quality strategy to incur lower costs they can pass along to the customers.

Quality and Productivity

The relationship between quality and productivity is clearer and has been demonstrated over time. The elimination of waste results in higher productivity. Simplification of processes also results in flows that are simpler and of higher productivity. Some care should be taken to recognize there are many different measures of productivity, such as labor productivity, productive use of energy, technological productivity, efficiency, multifactor productivity, and total-factor productivity. Total-factor productivity measures generally are considered the most robust means of measuring productivity. Several

---

[12]Kettinger, W., Lee, C., and Lee, S., "Global Measures of Information Service Quality: A Cross-National Study," *Decision Sciences* 26, 5 (1995):569–588.

measures have been developed that simultaneously monitor the relationship between quality and productivity.[13]

There is one caveat when evaluating the relationship between quality and productivity. It appears that changes to processes and procedures often will result in a temporary worsening of productivity. Hayes and Clark,[14] in a study of engineering change orders and productivity, found a temporary decrease in productivity as a result of too many simultaneous changes. Lillian Gilbreth[15] showed through a series of "tabletop experiments" that change results in a worsening of productivity that is usually not prolonged. Training is the variable that moderates the effects of too much change. If employees are properly trained in the new procedures, confusion caused by change can be reduced. At a minimum, changes resulting from process improvement must be managed and paced at the rate that the firm can absorb. The "rate of absorption" of change that different firms can handle is variable and can be managed by creating a learning organization.

Quality and Profitability

Many quality enthusiasts have declared that quality will always result in improved profitability. They will further state that those cases where quality does not lead to improved profits are the fault of the offending implementer. There are some facts to substantiate this claim. Every year the National Institute of Standards and Technology (NIST), the administrator of the Baldrige award, touts the outstanding stock performance of Baldrige winners. In addition, the U.S. General Accounting Office[16] performed a study of 30 Baldrige award winners and found the effects of quality improvement to include improved employee relations, lower cost, greater customer satisfaction, and improved market share. In a study of quality and profitability, Adam[17] found quality improvement in specific firms to be more long term rather than immediate. This makes some intuitive sense because quality improvement may not be recognized by the customers for some time. However, it should be stated that high quality is no guarantee of success. Firms must still successfully market, manage cash, and do the many other things that ensure profitability.

Quality and the Environment

Another strategic imperative receiving increasing attention is the effect of business on the environment. Because of regulatory pressures—both domestic and international—firms realize they must integrate environmental concerns into their strategic plans. When addressing this issue, Ed Woolard of DuPont Corporation stated the following:

> As we move closer to zero, the economic cost which society must ultimately bear may be very high (when considering reduction of pollution). Or the energy expenditure necessary to eliminate a given emission may have more

---

[13]See Adam, E., Hershaurer, J., and Ruch, W., *Productivity and Quality: Measurement as a Basis for Improvement* (Columbia, MO: Research Center, University of Missouri–Columbia, 1986).
[14]Hayes, R., and Clark, K., "Why Some Factories Are More Productive Than Others," *Interfaces* 64 (September–October 1986):66–73.
[15]Robinson, A., and Robinson, M., "The Tabletop Experiments," *Production and Operations Management* 3, 3 (1994):171–186.
[16]United States General Accounting Office, "Management Practices: U.S. Companies Improve Performance through Quality Efforts" (Washington, DC: GAO Report to the Honorable Don Ritter, House of Representatives, 1991).
[17]Adam, E., "Alternative Quality Improvement Practices and Organization Performance," *Journal of Operations Management* 12 (1994): 27–44.

> ecological impact than trade emissions themselves. Society will have to decide where the balance will be struck, and may conclude in some cases that zero emissions is neither in the environment's nor the public's best interest.[18]

It appears that the day when society will decide that the balance has been struck is still a long way off. As the United States becomes more urbanized, a greater percentage of the population worries that the entire world is becoming urbanized. This has led to an increase in the environmental movement.

Companies have to address many environmental issues. Besides the regulatory requirements, firms realize more and more that environmental friendliness is part of being a good corporate citizen. Therefore, tasks such as environmental protection, waste management, product integrity, worker health, government relations, and community relations compose the environmentally related strategic issues that must be addressed. As a result, firms are implementing quality-based environmental management systems, sometimes referred to as *total quality environmental management* (TQEM). These systems involve a holistic "systems" view of the processes causing environmental degradation. Measurements are implemented that identify indicators of environmental performance. This involves a focus on preventive rather than reactive cleanup.

Another technique that is used in TQEM is life-cycle costing. This uses value analysis to identify the total costs of products from a worldwide perspective. Figure 4-5 is an example of life-cycle costing for a typical cheeseburger. One might ask: "How much does a cheeseburger cost?" One answer is that a cheeseburger can be purchased at Burger King for less than a dollar. Indeed, this is the immediate price to the consumer. However, as proposed by some environmentalists, if one considers the environmental costs associated with deforestation, degradation to riparian habitats resulting from cattle grazing, and health problems because of the overconsumption of fatty foods, another cost emerges that is much higher than one dollar. Regardless of one's political stance concerning these issues, the debate will rage for many years, and regulation likely will not reduce over time. Therefore, it is in the best interest of management to address these issues in a proactive manner. Quality management philosophies and continuous improvement approaches can help in addressing these issues.

SUPPLY CHAIN STRATEGY

Strategic planning is important for effective supply chain management. Figure 4-6 shows various things to consider in supply chain strategy. First, you need to understand the different classes of sourced items. Next, you need to identify your supply chain's optimal performance levels. This means you should know your objectives and metrics for performance. Third, you should understand your current levels of performance and where you can improve. Fourth, organize team projects and plans for achieving desired supply chain objectives. Fifth, implement your improvement teams and plans. Finally, you monitor your results and take corrective action where needed. Notice that this process has its roots in Deming's plan-do-check-act cycle.

Among the questions to be answered in a supply chain strategy are[19]: How many plants are needed? Should we add new plants? Should we close plants? Should I stock inventory? For which products? In which locations? Should I outsource the task of

---

[18] *www.dupont.com/whatsnew/speeches/woolard/essu.html* (2006).

[19] Hicks, D., "The State of Supply Chain Strategy," *IIE Solutions*, 31, 8 (1999):24–30.

FIGURE 4-5 Life-Cycle Costs of a Hamburger

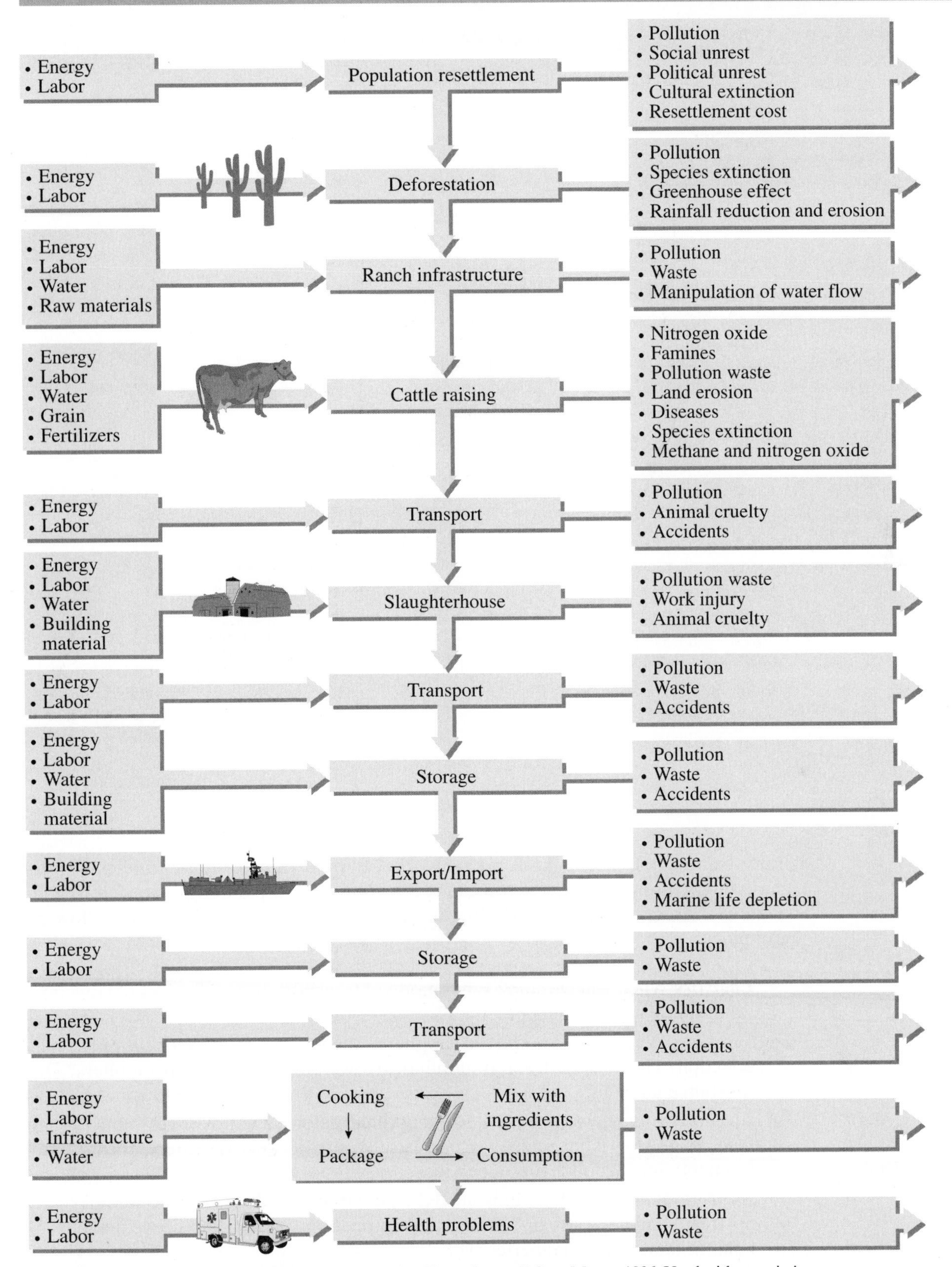

SOURCE: Thomas N. Gladwin, "Full Cost Pricing of a Cheeseburger" Case Memo, 1996. Used with permission.

FIGURE 4-6 Supply Chain Strategic Planning

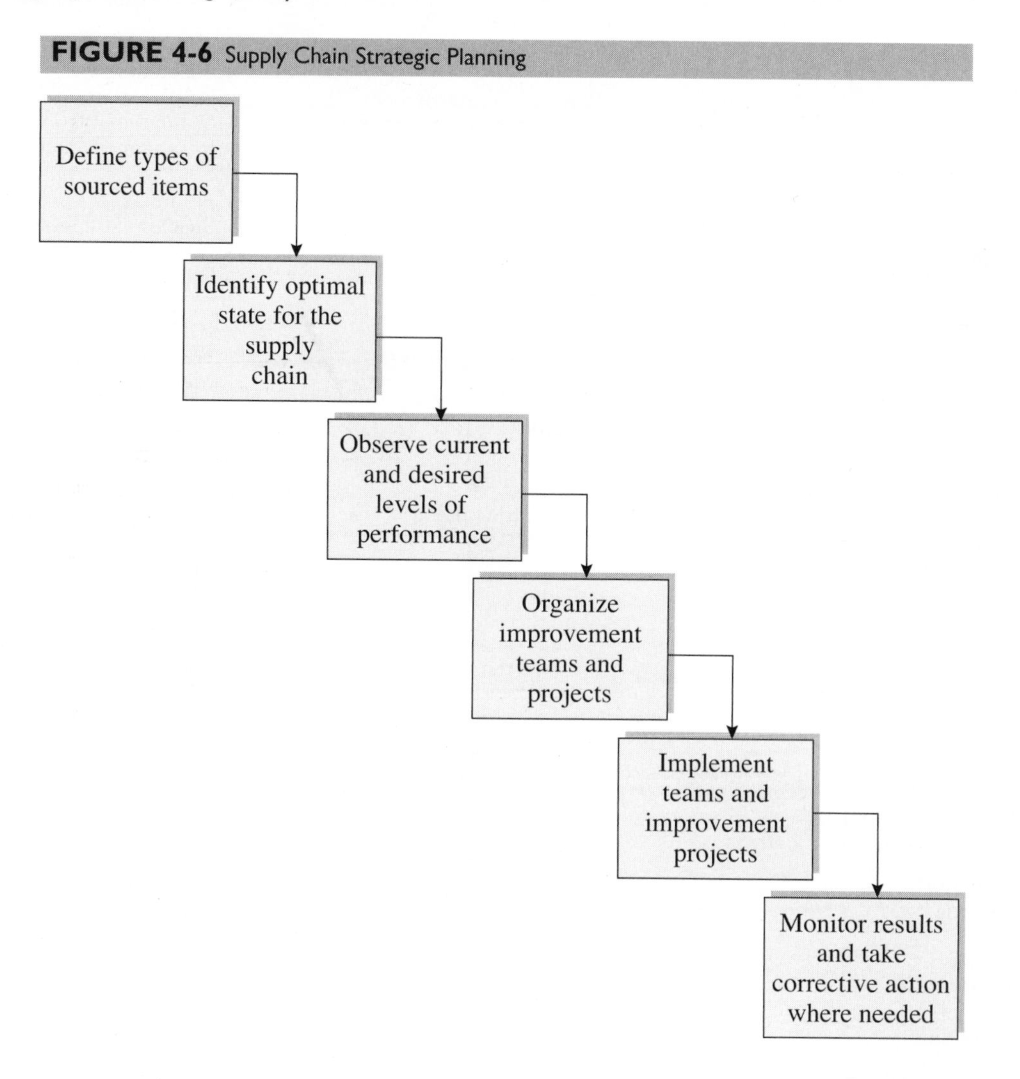

transporting goods throughout my networks? Should I make subassemblies or purchase them? From whom? As you can see, a good deal of planning is needed to adequately address these questions. In general, these questions fall into the following groupings:

- ***Logistics*** When will we ship? What mode of transportation will we use? How do we optimize our shipping practices?
- ***Suppliers*** Who are our preferred suppliers? What is our process for supplier selection? How do we develop suppliers? How do we link with our suppliers? Do we source globally?
- ***Inventory management*** Where do we optimally store inventory? How much? How long? Do we have perishable stocks? Where do we carry safety stocks? Are we maintaining good levels of services?
- ***Information flows*** What kinds of ERP systems are needed? How do we link upstream and downstream? What are data relations? What data do we need to manage our supply chain effectively?

- ***Products*** How many products do we stock? What product variety is necessary? How do we win orders? Where are our products in their life cycles?
- ***Service*** How do we define service along the supply chain? What do our customers want? Can we segment the supply chain? Who are our customers?

SUMMARY

This chapter discussed quality from the perspective of strategy. One of the fundamental themes of this text is that quality improvement is a managed process. Although this may seem obvious now, a framework is needed to help achieve desired results. This chapter provided a strategic planning framework for this to occur. Future chapters will focus on the implementation of these strategies.

We discussed various strategic issues relating to strategy content and process. Two processes were discussed: the forced-choice model and Hoshin plan. Finally, we discussed the results associated with quality. Results are of strategic importance as they help us to determine if we have accomplished what we wanted to.

KEY TERMS

- Appraisal costs
- Attrition
- Catchball
- External failure costs
- Failure costs
- Hoshin planning process
- Internal failure costs
- Leader behavior
- Leadership
- Leader skills
- Leading
- PAF paradigm
- Prevention costs
- Superordinate goals
- Trait dimension

DISCUSSION QUESTIONS

1. Recently, the Malcolm Baldrige National Quality Award changed the name of the "Strategic Quality Planning" category to simply "Strategic Planning." Why was this change made? Do you agree or disagree with the committee's rationale?
2. A study by the U.S. General Accounting Office reported that, on average, a firm takes 3.5 years to see significant results from a quality improvement program. Why do you think it takes so long to see significant results? Please make your answer as substantive as possible.
3. According to the quality literature, without top management leadership, quality improvement will not occur. Why do you believe this is the case?
4. Explain the hazards of the "management by dictate" model.
5. Trust has been identified as a very important attribute for leaders who are initiating quality improvement efforts. Why do you believe trust is such an important attribute?
6. Why is commitment an important variable in quality improvement initiatives?
7. Discuss some of the ways leaders resolve conflict in organizations. Which of these ways have you found to be most effective? Why?
8. Do you agree with the statement "quality is good ethics"? Please explain your answer.
9. Why do companies that focus on their customers often develop a set of ethics that includes valuing employees? Please make your answer as substantive as possible.
10. Discuss the concept of prevention cost. Why is prevention cost such a pervasive consideration in quality programs?
11. Describe the difference between external and internal failure costs. Is one cost more important than the other? Explain your answer.

12. What are the quality costs incurred when your hard disk crashes?
13. Think of a product you buy that is differentiated through quality. Do you believe the manufacturer's strategy to differentiate this product through quality is sustainable, or will the manufacturer eventually have to find other ways to attract you to the product? Explain your answer.
14. Describe the concept of core competency. Using the Prahalad and Hamel model, can quality be considered a core competency? Why or why not?
15. Describe the difference between "strategy content" and "strategy process." Describe examples of quality-related strategy content and strategy process issues.
16. Describe the benefits of strategic planning.
17. Why is the forced-choice model particularly useful for companies that are relatively inexperienced in strategic planning? Explain your answer.
18. Should the strategic planning process consider quality concerns at every step of the process? Why or why not?
19. Describe the concept of catchball.
20. Juran argues that both incremental (continuous) improvements and stepwise (breakthrough) improvements are needed in a strategic framework. Do you agree with Juran's assessment? Why or why not?

PROBLEMS

1. The Colorado Manufacturing Company of Boulder, Colorado, has gathered the following quality-related costs. You are hired as a consultant to evaluate these costs and to make recommendations to management.

Annual Quality Costs

Failure costs		
Defective products	$	4,234
Engineered scrap		21,265
Nonengineered scrap		224,123
Consumer adjustments		125,654
Downgrading products		2,125,328
Lost goodwill		Not evaluated
Customer policy changes		Not evaluated
TOTAL		
Appraisal costs		
Receiving inspection	$	24,138
Line 1 inspection		7,256
Line 2 inspection		8,543
Spot checking		2,766
TOTAL		
Prevention costs		
Quality training	$	25,500
Process engineering		
Corporate		132,678
Plant		44,124
Product redesign		10,422
TOTAL		

a. Compute the ratio of prevention and appraisal costs to failure costs.
b. Identify strategies for reducing failure costs.

2. The Aggie Remanufacturing Company of College Station, Texas, has gathered the following quality-related costs data:

Annual Quality Costs		
Failure costs		
Defective products	$	4,234
Engineered scrap		21,265
Nonengineered scrap		24,123
Consumer adjustments		25,654
Downgrading products		0
Lost goodwill		Not evaluated
Customer policy changes		Not evaluated
TOTAL		
Appraisal costs		
Receiving inspection	$	10,155
Line 1 inspection		9,225
Line 2 inspection		7,455
Spot checking		9,766
TOTAL		
Prevention costs		
Quality training	$	25,500
Process engineering		
Corporate		132,678
Plant		44,124
Product redesign		10,422
TOTAL		

a. Compute the ratio of prevention and appraisal costs to failure costs.
b. What would you recommend that this company do?

3. The Gorilla Manufacturing Company of Pittsburg, Kansas recently studied its expenditures and losses relative to quality for the month of October. They found that they had lost \$300,000 in scrap and rework. They had spent \$40,000 for inspection and \$25,000 in prevention-related activities. Evaluate their cost ratios and suggest whether or not the expenditures are warranted.
4. The Buffalo Machine Works of Boulder, Colorado was evaluating its cost structure relating to quality costs. In the prior year, the company had lost \$500,000 in warrantee and scrap. They didn't have good numbers for rework costs. In the prior year, they had spent \$100,000 training employees in quality tools and \$200,000 for quality assurance personnel. Inspection costs were 2% of sales of \$50 million. Evaluate these costs and recommend what actions should be taken to upper management. (Hint: The average company loses about 20% of sales due to poor quality.)

Cases

Case 4-1 Ames Rubber Corporation: Realizing Multiple Benefits through Improved Quality

Ames Rubber Homepage: ***www.amesrubber.com***

Ames Rubber, headquartered in Hamburg, New Jersey, produces rubber rollers for office machines and specialized parts for the assembly of front-wheel-drive vehicles. The company was founded in 1949, has annual sales of $45 million to $50 million, and employs 445 people at three New Jersey sites. At first glance, Ames Rubber appears to be a typical, small to medium-sized manufacturing firm. However, the company has some truly extraordinary aspects that make it deserving of a second look, particularly in the area of quality.

Unlike other companies that have used quality as part of a turnaround strategy, Ames Rubber has been successful throughout its corporate life. During its early history, the company flourished by providing component parts to the high-growth copier and printer industries. In the early 1980s, the global environment changed, and even though the company had achieved benchmark status in the copier industry, its customers were demanding products that met more stringent quality requirements at a lower cost. This challenge was exacerbated by the emergence of increased competition from foreign competitors. It was clear to the managers at Ames that "business as usual" was no longer sufficient to satisfy their customers' needs.

The company conducted an internal review of its operations. At this point, the company was unsure of how to improve its product quality in a cost effective manner. The review indicated that Ames was expending considerable effort to meet the quality, cost, and delivery requirements of its customers with no coherent quality plan in place to guide its efforts or anticipate customer requirements. As a result of this analysis, Ames decided to focus on the effectiveness of its manufacturing operations and quality efforts. Remembering that Xerox had launched a program called Leadership through Quality, Ames' CEO Joe Marvel put his entire management team through Xerox's supplier training program. Using Xerox as its benchmark, Ames announced that it was embarking on a new quality program entitled Excellence through Quality. The program was designed to achieve one common goal—the satisfaction of both internal and external customers.

Since it was introduced, the Excellence through Quality program has been effectively embraced by the entire Ames Rubber organization. The program involved the following key initiatives:

- **Involvement groups.** Everyone at Ames is a member of an involvement group, which meets a minimum of once a month. Involvement groups use team processes to improve product quality.
- **Cost of quality and reject tracking.** Reject tracking and cost of quality collection systems were established.
- **Yield improvement teams.** The company isolates the processes causing the greatest problems ("Pareto thinking").
- **Strategy review/operations review.** Strategy review and operations review meetings were established to review short- and long-term progress toward quality objectives.
- **Extensive training.** The company provides its employees extensive training in such areas as statistical quality control, communication skills, leadership skills, and quality management.

Cultural changes also were made to facilitate the Excellence through Quality program. All Ames' employees are called "teammates" to promote an egalitarian philosophy. Safety committees, made up of rank-and-file employees, were set up to ensure safety throughout the company's facilities. To demonstrate his personal commitment to satisfying internal and external customers, CEO Marvel redesigned the company's organization chart, with external customers on the top, the firm's employees in the middle, and himself at the bottom.

Ames Rubber has found that the benefits of its Excellence through Quality program have transcended its original objectives. At a minimum, the company has become more disciplined, knows its customers and employees better, and produces a better product than it has at any time in its history. In addition, according to the company's application summary for the Malcolm Baldrige National Quality Award, an emphasis on quality has produced the following benefits:

- Cultural change is happening, and it is being measured.
- Employee involvement groups are flourishing.
- Morale is at an all-time high because of involvement, improved communication, and improved recognition.
- A system for total quality and beyond is in place.
- The company has reorganized and decentralized.
- Prioritization of effort has become routine.
- Rejected products have been reduced by more than 50%.
- Financial results have improved.

The legacy of Ames Rubber Corporation and its experience with quality is twofold. First, regardless of how successful a company is, environmental change may result in pressures to improve quality. This includes a strong emphasis on employees as a conduit for change. As a result, maintaining a quality-minded culture is a prudent business practice. Second, there are multiple potential benefits to remaining attentive to quality issues that transcend improved operations management. ■

DISCUSSION QUESTIONS

1. In developing its Excellence through Quality program, Ames initially benchmarked against Xerox. Is benchmarking against another company's quality program a good idea? What are the potential hazards and benefits involved?
2. Discuss the manner in which Ames implemented its Excellence through Quality program. Did the company place its emphasis in the right areas? Please explain your answer.
3. Discuss the benefits that Ames Rubber achieved from its quality program. Are these benefits more encompassing than you would have expected? Why or why not?

Case 4-2 Make No Mistake—At Eastman Chemical, Quality is a Strategic Issue

Eastman Homepage: ***www.eastman.com***

Although some companies may struggle with the appropriate role of quality in their strategic thinking, there is no such ambiguity at Eastman Chemical. Eastman Chemical, a $5 billion, Ohio-based company, produces a wide range of products for the chemical, fiber, and plastics industries. The industry is very competitive, and the company's customers are becoming more and more demanding. As a result, although meeting customer needs has always been important to Eastman, the fervor attached to the issue has intensified over the past several years. Eastman knows that to achieve its vision of being the world's "Preferred Chemical Company," it must first meet its goal of producing high-quality products and delivering exemplary customer service.

Eastman's commitment to quality began in the early 1980s, after the company saw its market share decline in the 1970s. The quality issue was brought to the forefront when a major customer cited poor quality as the reason for finding a new supplier. At that point, Eastman knew it had to improve quality or continue to decline as a corporation. Early on, the company's efforts to integrate quality into its culture and operations were extensive, as illustrated by the following timeline:

1982 Initiated a customer emphasis program.

1983 Formally articulated a quality goal: "To be the Leader in Quality and Value."

1984 Invited and encouraged customers to visit Eastman plants and to provide input on process improvements.

1985 Reinforced the company's quality-minded culture by developing a set of principles referred to as the "Eastman Way."

1986 Developed a formal quality management process to improve the company's manufacturing and nonmanufacturing processes.

1987 Developed a formal customer satisfaction process.

1988 Implemented employee empowerment.

1989 Obtained ISO 9000 certification.

1990 Started using the Baldrige criteria to assess and guide future quality efforts.

1990 Initiated an employee development program.

1995 Eastman aggressively expands globally.

2000 Eastman expands e-commerce business.

2003 Realignment of organizational structure.

In the 1990s, quality was placed in a strategic context with the development of a values statement referred to as the company's "Strategic Intent." This statement, which is prominently displayed throughout the company's many sites, indicates the strategic intent of the Eastman Chemical Company is to be the "World's Preferred Chemical Company" through an emphasis on quality (both product and service), the Eastman Way (strong corporate culture), and responsible care (continuous improvement in health, safety, and environmental performance). The prominent role of quality in this statement elevates the role of quality in achieving the company's ultimate goals and objectives. As a result, quality has become a key strategic and performance variable at Eastman Chemical.

Eastman's decision to place quality in a strategic context has produced rich rewards. The company was the recipient of the Malcolm Baldrige National Quality Award in the manufacturing category. The company has won state quality awards in Tennessee, Texas, and Arkansas. *Industry Week* magazine named Eastman one of the world's 100 best-managed companies. More importantly, the company is working hard to achieve its goal of being the world's preferred chemical company. In customer surveys, more than 70% of Eastman's worldwide customers have rated the company their number one supplier. This level of success has opened new markets, and the company is expanding in the United States and abroad. ■

DISCUSSION QUESTIONS

1. Did Eastman make the right decision to define quality as a "strategic issue?" Defend your answer.
2. Eastman has made quality an important part of its overall mission or "Strategic Intent." As a way of drawing attention to the company's strategic intent, it is prominently displayed throughout the company's many sites. Why would Eastman display its strategic intent throughout the company?
3. Do you believe the majority of Eastman's employees know the role of quality in their firm? Why or why not?

CHAPTER 5

The Voice of the Customer

The customer is always right.
—MACY'S SLOGAN

We start with a list of our customers. "Tire-kicker. Mooch." The names come faster now, shouted out by the car dealership employees, "Dreamer. Stroker. Lookie-Lou." Then we ask, what do customers think of car salespeople? Silence. Then a few suggestions: "Snake oil salesman. Sleaze-bag. Crook."
—INFINITI BOOT CAMP FOR DEALERS

The customer is the enemy.
—REPORTEDLY A MOTTO AMONG CERTAIN MANAGERS AT ARCHER DANIELS MIDLAND CORPORATION

The quotes at the top of the chapter demonstrate that different employees have different views of their customers. We have all experienced instances of great or lousy customer service. Customer service is important because[1]

- Customers will tell twice as many people about bad experiences as good experiences.
- A dissatisfied customer will tell 8 to 10 people about the bad experience.
- Seventy percent of upset customers will remain your customer if you resolve the complaint satisfactorily.
- It's easier to get customers to repeat than it is to find new business.
- Service firms rely on repeat customers for 85% to 95% of their business.
- Eighty percent of new product and service ideas come from customer ideas.
- The cost of keeping an existing customer is one-sixth of the cost of attracting a new customer.

A **customer** is the receiver of goods or services. Typically, this involves an economic transaction in which something of value has changed hands.

Often customers are defined as internal or external customers. **Internal customers** are employees receiving goods or services from within the same firm. For example, management information systems (MIS) technicians and programmers view the users within their company as internal customers. Accounting departments and finance departments often have very little interaction with the bill-paying customer. However,

[1]Kabodian, A., "The Customer Is Always Right." Quotes from *http://customersatisfaction.com/book.html* (June 1997).

they have customers within the firm who use their services on a daily basis. In a sense, there is an economic transaction that takes place in internal services in that service providers are funded as a result of the service they provide to the organization as a whole. Some have used an abstraction of the term *internal customer* to include the person at the next step in the supply chain. Therefore, the person who works at workstation 3 can be considered the customer of the worker at workstation 2.

External customers are the bill-paying receivers of our work. The external customers are the ultimate people we are trying to satisfy with our work. If we have satisfied external customers, chances are we will continue to prosper, grow, and fulfill the objectives of the firm.

Another term that describes customers is **end user**. An end user is the final recipient of a product or service. The term is often used by software developers who program software solutions for customers. Service firms have many titles for customers. These titles include patient, registrant, stockholder, buyer, patron, and many others. As service providers and product producers, the customer is the focus of our activities.

CUSTOMER-DRIVEN QUALITY

Customer-driven quality represents a proactive approach to satisfying customer needs that is based on gathering data about our customers to learn their needs and preferences and then providing products and services that satisfy the customers. Customer-driven quality is one of the core values of the Malcolm Baldrige National Quality Award. A Closer Look at Quality 5-1 shows that companies have varying degrees of success in responding to customers.

Video Clip: Customer Satisfaction at Marriott

The Pitfalls of Reactive Customer-Driven Quality

Even though it is generally understood that listening to and understanding the customer is a good thing, there are some companies that implement customer feedback mechanisms incorrectly. As a result, these companies are placed in a reactive rather than a proactive mode with their customers.

One of the difficulties in satisfying customer requirements is that in a dynamic environment, customer needs are constantly changing. Consider the example of military suppliers. For many years, cost overruns and missed schedules were allowed by the military customers. When the purchasing standards were changed by the military, many suppliers such as McDonnell Douglas were incapable of adequately responding. The results have been layoffs, corporate restructuring, and mergers. Figure 5-1 shows a model of **reactive customer-driven quality (RCDQ)**. This model shows that a firm's quality performance is increasing while customers' expectations also are increasing. Problems occur when customer requirements increase at a faster rate than quality and service improvement. This places a firm in a reactive mode that may signal the need for major process and service redesign.

The RCDQ model demonstrates conceptually and graphically the primary pitfalls and dangers of RCDQ. In a sense, manufacturers and service organizations attempting to meet customer expectations are pursuing a moving target. As the supplier's competitors improve quality and competition increases, customers demand higher levels of quality and service. The difference between world-class and ordinary suppliers lies in whether suppliers stay ahead of the target or fall behind the target. Although a supplier to a customer might desire to provide high-quality service to the customer, the reactive posture engendered in the RCDQ approach will cause the supplier to fall farther and farther behind the moving target over time.

A CLOSER LOOK AT QUALITY 5-1

CUSTOMER SERVICE ON THE INTERNET[a]

Perhaps you have had questions that you would like to ask the producer of a product, such as whether Duracell batteries will last longer if stored in a refrigerator, how much corn is in a single Frito, or how McDonald's made round bacon for its egg McMuffins. To get answers to these questions, you can simply leave a message for these companies on the World Wide Web. However, you may not always get an answer in a timely fashion.

Many companies invite dialogue with their customers on Internet homepages, but they are often poorly prepared to answer queries. Tony Pittarese, a college professor from Florida, contacted Coca-Cola's Web site to help him plan a visit to the Summer Olympic Games in Atlanta, Georgia. The screen at Coca-Cola said "We're all ears." Sure, they were all ears, but no mouth. He never received a reply. He ended up dialing Coca-Cola's 1–800 number for the information. "If you're going to go on the Web and you can't do it right, then don't do it at all," says Pittarese. Coke apologetically stated that it had a large e-mail backlog at the time of the query.

Web sites offer a variety of approaches to e-mail. Some offer e-mail addresses and encourage inquiries. Others have a systems administrator Web address with no e-mail access.

In a research project, the *Wall Street Journal* sent e-mail inquiries to several major corporate Web sites with e-mail capabilities (see Table 5-1). Many never responded to the inquiries. Two took three weeks to reply. Others sent stock responses that failed to address the specifics of the query. Only three companies responded within a day with substantive responses.

When someone comes to your Web site, a customer is seeking you out," says Elizabeth Stites, marketing director of Matrixx Marketing, Inc. "If you're not talking back to them, you're crazy."

Why do companies have such a poor e-mail record? Many firms simply do not understand the maintenance hassle associated with placing a Web site on the Internet. Internet users often make inordinate demands and have high or unrealistic response expectations. That's why Saturn Corp. explicitly states on its Web site that it doesn't use e-mail. "You don't want to diminish the customer's passion by not responding when they take the time to contact you," according to Greg Martin, a Saturn spokesperson.

McDonald's Corporation has a typical Web site full of colorful animation, merchandise, and entertainment. In the McDonald's feedback section, users are questioned on their eating habits and asked how often they have eaten at McDonald's in the recent past. This is where the *Wall Street Journal* staff asked the question about the round bacon. No answer was ever received.

"I don't think we've taken our Web site to a high level of interactivity at this point," admitted a McDonald's spokesperson. An "appropriate response system does not exist."

[a]Weber, T., "Simplest E-Mail Queries Confound Companies," *Wall Street Journal* (October 21, 1996):B1.

WHAT IS THE VOICE OF THE CUSTOMER?

The **voice of the customer** represents the wants, opinions, perceptions, and desires of the customer. Firms perform a variety of activities to become familiar with the needs and wants of customers to better design products and services. Customers are also a source of knowledge concerning the performance of the production and service systems.

There is also a technical definition of the voice of the customer. It has to do with a standardized, disciplined, and cyclic approach to obtaining and prioritizing customer preferences for use in designing products and services. This definition of the voice of

TABLE 5-1 E-mail Responses

Site	*What We Asked*	*What Happened*	*Company Comment*
3M	Do Post-it notes get less sticky from just sitting around?	A few hours later: automated response promises an answer to follow. Twenty days later: another similar message. Two days after that: Post-it "adhesive does not become less sticky over time."	3M says it is working toward a 48-hour response time.
Budweiser	How did Bud come up with 110 days as the freshness standard for its new labeling?	No response.	Bud says it tries to answer e-mail in 72 hours or less. It believes our e-mail was either lost or was overlooked by staff.
Coca-Cola	How much caffeine is in Coke?	No response.	Coke says its records show that it wrote back the next day.
Compaq	Which PC is Compaq's lowest-priced multimedia machine with a phone-answering feature?	On the same day, Compaq sent a response asking that we visit their forum on America Online.	Compaq says it uses the forum to encourage customers to get in touch with local dealers.
Kellogg	Has the company discontinued its low-fat plain granola? Why do local stores only carry the raisin version?	Twenty days later: "Distribution may be limited in your area . . . may we suggest that you discuss this matter with your store manager."	Kellogg says a technical problem at its Internet service provider prevented it from receiving e-mail for more than a week; it took another week to catch up on messages.
Levi Strauss	Does Levi have a suggestion for softening new jeans?	One week later: "We do not recommend altering the finish of the fabric."	Levi says it tries to respond to all inquiries within two business days.
MCI	Why should someone switch to MCI from AT&T?	A few hours later: a form letter detailing benefits of MCI service, including information on a Mother's Day promotion three months after the holiday had passed. The company referred further requests to an 800 number.	MCI says it answers frequently asked questions with "consistent responses" but tailors answers when appropriate. It adds that it answers most inquiries within 24 hours.
Olean (olestra)	We didn't get a chance to ask anything.	We clicked on "Your Comments" but found only a multiple-choice quiz asking our sex, age, and modem speed. There was no space for comments and no e-mail address.	Procter & Gamble, which makes Olean, says it has a variety of sites for its brands but currently only fields e-mail inquiries at its Tide and Old Spice sites. P&G is working to better manage e-mail.
Reebok	Is it dangerous to wear running shoes to play basketball?	Four weeks later: "Sorry for the delay . . . running shoes are oftentimes lighter with not as much lateral support. I hope this helps!"	Reebok says it thoroughly vets all responses that concern the topic of injuries, resulting in a delay. In general, Reebok's goal is 24-hour turnaround.
Sony	For a malfunctioning television set, is there anything Sony recommends trying before calling a repair technician?	Six days later: "Try leaving the TV unplugged for a while." Sony also offered to track down the nearest service center.	Sony says some questions must be referred to an engineer, resulting in delays. It says it answers 80% of its e-mail within 12 business hours.

SOURCE: T. Weber, "Simplest E-mail Queries Confound Companies," *Wall Street Journal* (October 21, 1996):B1.

FIGURE 5-1 Reactive Customer-Driven Quality Model

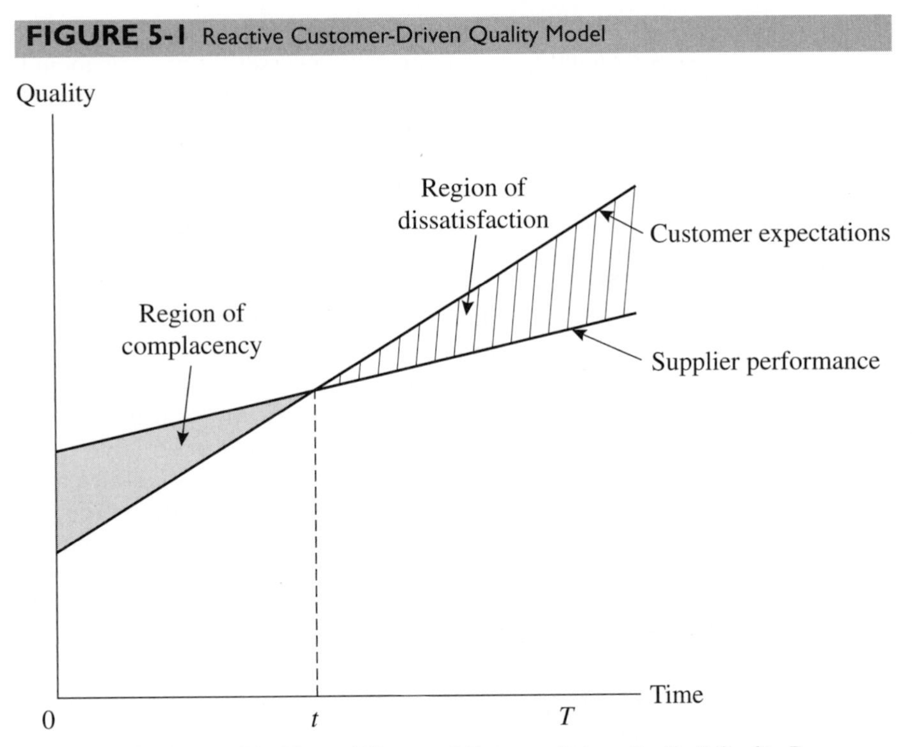

SOURCE: S. T. Foster, "The Ups and Downs of Customer-Driven Quality," *Quality Progress* (October 1998):70. © 1998 American Society for Quality. Reprinted with permission.

the customer is sometimes associated with **quality function deployment (QFD)** or the **house of quality**.[2] QFD translates customer wants into a finished product design. The Japanese developed this approach in the 1960s,[3] and it has been used in the United States since the 1980s. The QFD approach is discussed in Chapter 7.

As we stated in Chapter 1, *quality is as the customer sees it*. In spite of all of our efforts and work, if we do not adequately please the customer, we will cease to be economically viable. Therefore, companies spend a great deal of resources attempting to understand the customer, which is the focus of this chapter: How do we get to know our customers, and how do we develop systems from this information to constantly and forever improve our systems?

CUSTOMER-RELATIONSHIP MANAGEMENT

Much of the focus in marketing today is on maintaining the existing customer base that a firm has established. If it is true that 90% of the business for many service firms is in the form of repeat business, the focus of process and system design must be on developing relationships with customers rather than simply providing clean transactions at each stage of the process.

[2]Hauser, J. R., and Clausing, D., "The House of Quality," *Harvard Business Review* 66, 3 (1988):63–73.
[3]Akao, Y., *Quality Function Deployment* (Cambridge, MA: Productivity Press, 1990).

Process design in services often has focused on the transaction. For example, a university might focus on discrete processes for improvement in areas such as registration, financial aid, test taking, and so forth. However, some universities have learned that focusing on these internal processes does not help in the retention of students. Therefore, new programs for student retention focus on familiarizing the student with the university and developing the skills students need to be successful in a college setting. Some of these skills might involve study skills, social skills, or managing on a limited budget. In this way, the university begins to look at the whole system relating to the student and not just internal university processes.

Many times when you purchase products you are asked for your name and other personal information. Sometimes providing this information seems an intrusion on your privacy. Indeed, sometimes it is, because firms sell their mailing lists to other entities. For example, did you know that many states in the United States sell driving records to commercial firms to earn money? This is often so that other firms can direct market materials to you based on the kind of automobile you drive. These same firms may use the information they gather from you to develop databases to better serve you as a customer. This knowledge about customers is a very powerful marketing tool.

For many firms, the focus on process design includes the aspect of **customer-relationship management**. This view of the customer asserts that he or she is a valued asset to be managed. According to the introduction to the show "Cheers": "Sometimes you want to go where everybody knows your name." This is relationship management. The tangibles (such as facilities and machinery) meet the intangibles (such as professionalism and empathy) to provide a satisfying experience for the customer. There are four important design aspects (see Figure 5-2) to customer-relationship management that will be addressed here: complaint resolution, feedback, guarantees, and corrective action or recovery. For practical reasons, I will distinguish between CRM and Customer Relationship Management Systems (CRMS). CRMS are systems used for capturing customer-related data. CRMS are discussed later in the chapter.

FIGURE 5-2 Four Components of a Customer-Relationship Management Process

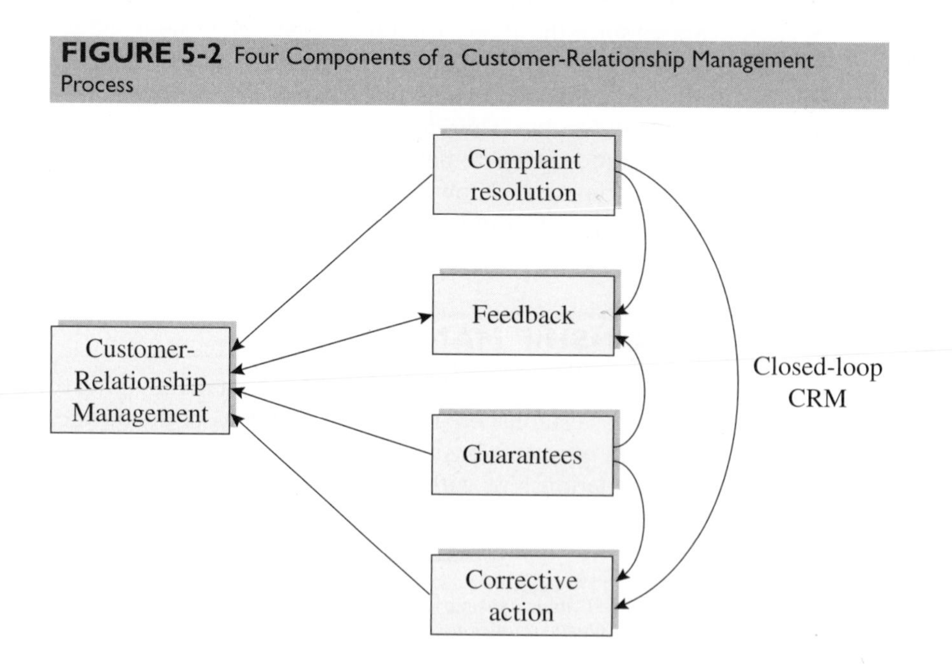

Complaint Resolution

As the famous saying goes, "you can please some of the people some of the time, but you can't please all of the people all of the time." As a result, complaint resolution is an important component of a quality management system. Complaints come in many forms. For our discussion, we will focus on three types of complaints that need to be resolved: regulatory complaints, employee complaints, and customer complaints. Although the focus of this chapter is on the customer, it is important to recognize all three types of complaints as potential sources of information for improvement. Donald Beaver, the owner of New Pig Corporation of Tipton, Pennsylvania, has the right attitude about complaints.[4] He states, "You should love complaints more than compliments. A complaint is someone letting you know that you haven't satisfied them yet. They have gold written all over them." Complaints should be viewed as opportunities to improve. Because only a small percentage of customers ultimately will complain, they should be taken very seriously. This small percentage of customers may represent a much larger population of dissatisfied customers.

The complaint-resolution process involves the transformation of a negative situation into one in which the complainant is restored to the state existing prior to the occurrence of a problem. In extreme cases, the complainant has incurred a loss, as in the case of a malfunctioning product leading to injury and liability. In the case of personal injury, if the complainant was injured as a result of the malfunction of a product, it is ethical that the company should restore the person by reimbursing him or her for the product, any medical expenses, and other costs such as lost time from work.

Typically, losses incurred by customers are not quite so dramatic. The losses are smaller, such as lost time, lost money, or lost patience. The first component of a complaint-resolution process is to **compensate** people for losses. This may be as small as an easy return policy with no questions asked. The second component to complaint resolution is **contrition**. The firm should apologize to the customer for the mistakes made and invoke the Macy's mantra, "The customer is always right." (See A Closer Look at Quality 5-2.) Third, the complaint-resolution process must be designed to make it easy for complainants to reach resolution to simple complaints.

The process associated with resolving complaints is called the **complaint-recovery process**. Recovery design is an important activity for many firms. Complaints can come from a variety of sources, such as questionnaires (low scores on key quality indicators can be considered complaints), formal direct inquiries, or informal channels. The recovery process must be developed for documenting complaints, resolving the complaint, documenting recovery, and feedback for system improvement.

Feedback

To understand customer behavior, wants, and needs, data about the customer are necessary. Some of these data come directly from the customer. Some customer data are solicited, and other data are provided without solicitation. The following pages discuss different approaches to collecting and analyzing customer data. One way to gather data is to receive customer feedback. There are two main types of feedback—feedback to the customer and feedback to the firm as a basis for process improvements. The customer-feedback loop includes reporting the resolution of the complaint to the customer. Many times this requires a data-gathering mechanism, such as a computerized information system, to ensure that the customer complaint has been resolved adequately. Feedback

---

[4]Whitely, R., *The Customer-Driven Company: Moving from Talk to Action* (Boston: Addison Wesley, 1991).

A CLOSER LOOK AT QUALITY 5-2

THE CUSTOMER IS ALWAYS . . . THWARTED?[b]

It all started when Dale Feinstein returned a cutlery set that his brother had purchased to the Macy's store in the Ocean County Mall in Toms River, New Jersey. The cashier in the housewares department accepted the unopened cutlery set and the total refund came to $286.19. The cashier stated that it would take about two weeks for the refund check to arrive in the mail. After waiting three weeks for the check to arrive, Feinstein telephoned Macy's to inquire about the status of his refund. The operator stated that the store could not assist him and advised that he call Macy's toll-free number. He promptly contacted their automated system and was connected to the department that deals with check returns. After a short wait, he spoke to a customer service representative who retrieved information about his refund from the computer. She told him that he would have to contact another department and connected him to an answering machine. Feinstein left a message with the relevant information.

A few days later, someone from the department called and stated that in order to mail the check, they would have to "verify the purchase." They asked him where he bought it, and Feinstein explained that his brother had purchased it from a Macy's store, but neither he nor his brother could recall which location. He asked them to check the two closest stores, and the representative stated that they would telephone him in a few days with the results of their investigation. About two and a half weeks passed, and Feinstein still had not received a call from Macy's. He telephoned the toll-free number again and explained that he was never contacted. They stated that there was nothing they could do, and he would have to leave a message on the answering machine again. Feinstein told the customer service representative that he wished to speak with a human, not a machine. Feinstein explained that because they did not return his call as they promised, he could not rely on them to respond to another message. The representative again stated that there was nothing he could do and said that Feinstein was lucky that they called once. The customer service rep said that he had heard from many other customers that they had never been called back once. Begrudgingly, Feinstein asked him to connect him to the answering machine. The employee told Feinstein that he normally did not work in this department and was unaware of how to connect Feinstein. Feinstein asked to speak to the manager and was put on hold before being told that there was no manager present at the time.

A few weeks later Feinstein called the toll-free number again and explained the problem. The customer service representative connected him to the machine, and he left a message. The next day, Macy's called late in the afternoon, but unfortunately, Feinstein was not home.

Feinstein called the toll-free number, but before he was connected to the answering machine, he told the operator how frustrating this experience had been and that he should be able to speak to a person during normal business hours. She reiterated how their department could not help him and that he would have to leave a message on the answering machine. She also offered to send the department an e-mail message indicating that if a check was not received soon, Feinstein would begin to take legal action against Macy's.

After some additional time had passed without a check arriving or a return telephone call, Feinstein called the toll-free number again. He spoke to a representative who stated that the "check had been released."

Two weeks later the check still had not arrived. Feinstein called and was put on hold before being told that the operator from a couple weeks prior denied telling him that the check had been released. Feinstein would have to speak with the machine again. Feinstein asked to speak with the manager and was placed on hold before being told that the

[b]Used by permission of Mr. Feinstein. This was originally published on the Web site *http://www.cybercomm.net/~dale/macys* (2002).

manager was unavailable and would just send any information to the answering machine people. Feinstein told the operator that he would take measures to have his money returned and ended the call.

A couple of weeks later Feinstein filed a small claims suit against the Toms River Macy's; the filing fee was $14.00. The next day the refund check arrived. Feinstein once again contacted Macy's, this time to ask whether they would compensate him for the filing fee. Macy's legal department refused, so he proceeded with the legal action in hopes of recovering court costs, interest, and punitive damages.

Approximately four months later the case was mediated. At first, the individual representing Macy's, an assistant manager of security, did not want to reimburse the $14.00 court filing fee. Feinstein threatened to have the case tried before Macy's eventually authorized the settlement.

to the firm should occur on a consistent basis with a process to monitor changes resulting from the process improvement.

Guarantees

Another important aspect of customer service is the guarantee. Many firms offer service guarantees. A guarantee outlines the customer's rights. Even with high-quality companies, such as Motorola, there are product and service failures. To design a process that ignores this fact is a form of denial. The guarantee is both a design and an economic issue that must be addressed by all companies before the first sale occurs. Guarantees should be designed prior to beginning business so that employees can be trained to implement the guarantee and marketing can advertise the guarantee properly. This is an important economic issue because of the sales potential that is created by a guarantee and the costs associated with fulfillment of the guarantee.

High Street Emporium sells products on United Airlines flights via catalog and extends a simple guarantee: "The Best Products from the Best Catalogs at the Best Prices—Guaranteed." This is a nice example of a simple, understandable guarantee. To be effective, a guarantee should be[5]

- *Unconditional.* No "small print." Inconsistent application can make the guarantee less compelling.
- *Meaningful.* To be meaningful, customer grievances must be fully addressed. For example, any financial loss must be fully recovered by the complainant.
- *Understandable.* The customer must be able to understand easily the parameters of the guarantee and how to achieve resolution quickly.
- *Communicable.* The phrasing of a guarantee should resonate with the customer. "The best quality at the lowest prices—I guarantee it," is the famous motto at the Men's Wearhouse. Not only is this a great guarantee, but it also is a great marketing line.
- *Painless to invoke.* The customer must not be inconvenienced too much. Wal-Mart has a "no questions asked" return policy that allows customers to return or exchange merchandise with a minimum of hassle. This return policy was an early service linchpin that differentiated its service from other retailers.

[5]Chase, R., *Operations Management for Competitive Advantage* (Homewood, IL: Irwin, 2005).

Corrective Action

An important aspect of customer-relationship management is corrective action. This means that when a service or product failure occurs, the failure is documented and the problem is resolved in a way that it never happens again. Corporate teams or committees should be in place to regularly review complaints and to improve processes so that the problems don't recur.

Referring to Figure 5-2, the corrective action results in gathering complaint and warranty data and determining the causes of these problems. Often, teams of employees and managers study complaints, do Pareto cost analysis of the complaints, and recommend improvements to the customer service delivery system. Such systems are often referred to as **closed-loop corrective action**. The loop is in effect *closed* because a process is in place that ensures that this information is used for improvement. This is why complaint and field repair data are so golden.

THE "GAPS" APPROACH TO SERVICE DESIGN

The **gap** has been recognized in the quality literature for some time. Typically, the gap refers to the differences between desired levels of performance and actual levels of performance. This could be something like the difference between the desired conformance level versus the existing conformance level in a manufacturing environment. In services, this is the difference between the expected and the actual level of service provided. Gaps are important in that once a gap is identified, it is a candidate for corrective action and process improvement. The formal means for identifying and correcting these gaps is called **gap analysis**.

One of the differences studied by gap analysis identifies the difference between managerial and customer perceptions of what the customer wants (gap 1). When this gap is large, service providers are likely producing excellent services that no one wants. The other gaps and gap analysis are discussed in depth in Chapter 8.

In addition to the gaps model, Parasuraman, Zeithamel, and Berry contributed a number of important concepts to managing service quality. These include 10 determinants of service quality (Table 5-2), service quality dimensions discussed in Chapter 1, and a research instrument (questionnaire) to test the hypothesized relationships in Figure 5-3. The survey instrument is called SERVQUAL. It is available for quality managers to use in assessing quality of service (see Chapter 8).

Another approach to identifying and measuring gaps in quality service is found in the two-dimensional gaps model shown in Figure 5-4. Surveys using this approach measure customer satisfaction perceptions on the *y* axis and ratings of relative importance on the *x* axis. This approach is presented in depth in Chapter 8, but it is presented here in the context of gathering customer data to better understand the customer.

SEGMENTING CUSTOMERS AND MARKETS

One of the preliminary steps that many analysts overlook is segmenting data. Data about customers can be gathered from a number of sources, such as industry groups, external sources, the Internet, commercial CD-ROM databases, or questionnaires. The Baldrige criteria emphasize data segmentation of customers and customer markets. The segmentation is simple (e.g., consumer, commercial, or wholesalers).

To segment markets means to distinguish customers or markets according to common characteristics. Personal computer markets are segmented into home and

TABLE 5-2 Determinants of Service Quality

Reliability involves consistency of performance and dependability. It means that the firm performs the service right the first time. It also means that the firm honors its promises. Specifically, it involves
- —Accuracy in billing
- —Keeping records correctly
- —Performing the service at the designated time

Responsiveness concerns the willingness or readiness of employees to provide service. It involves timeliness of service
- —Mailing a transaction slip immediately
- —Calling the customer back quickly
- —Giving prompt service (e.g., setting up appointments quickly)

Competence means possession of the required skills and knowledge to perform the service. It involves
- —Knowledge and skill of the contact personnel
- —Knowledge and skill of the operational support personnel
- —Research capability of the organization (e.g., securities brokerage firm)

Access involves approachability and ease of contact. It means
- —The service is easily accessible by telephone (lines are not busy, and customers are not put on hold)
- —Waiting time to receive service (e.g., at a bank) is not extensive
- —Convenient hours of operation
- —Convenient location of service facility

Courtesy involves politeness, respect, consideration, and friendliness of contact personnel (including receptionists, telephone operators, and so on). It includes
- —Consideration for the consumer's property (e.g., no muddy shoes on the carpet)
- —Clean and neat appearance of public contact personnel

Communication means keeping customers informed in language they can understand and listening to them. It may mean that the company has to adjust its language for different consumers—increasing the level of sophistication with a well-educated customer and speaking simply and plainly with a novice. It involves
- —Explaining the service itself
- —Explaining how much the service will cost
- —Explaining the tradeoffs between service and cost
- —Assuring the consumer that a problem will be handled

Credibility involves trustworthiness, believability, honesty. It involves having the customer's best interests at heart. Contributing to credibility are
- —Company name
- —Company reputation
- —Personal characteristics of the contact personnel
- —The degree of hard sell involved in interactions with the customer

Security is the freedom from danger, risk, or doubt. It involves
- —Physical safety (Will I get mugged at the automatic teller machine?)
- —Financial security (Does the company know where my stock certificate is?)
- —Confidentiality (Are my dealings with the company private?)

Understanding/knowing the customer involves making the effort to understand the customer's needs. It involves
- —Learning the customer's specific requirements
- —Providing individualized attention
- —Recognizing the regular customer

Tangibles include the physical evidence of the service
- —Physical facilities
- —Appearance of personnel
- —Tools or equipment used to provide the service
- —Physical representations of the service, such as a plastic credit card or a bank statement
- —Other customers in the service facility

SOURCE: A. Parasuraman, V. Zeithamel, and L. Berry, "SERVQUAL: A Multiple-Item Scale for Measuring Customer Perceptions of Service Quality," *Journal of Retailing*, Spring 1988, pp. 12–40.

FIGURE 5-3 Gaps and the Service Quality Model

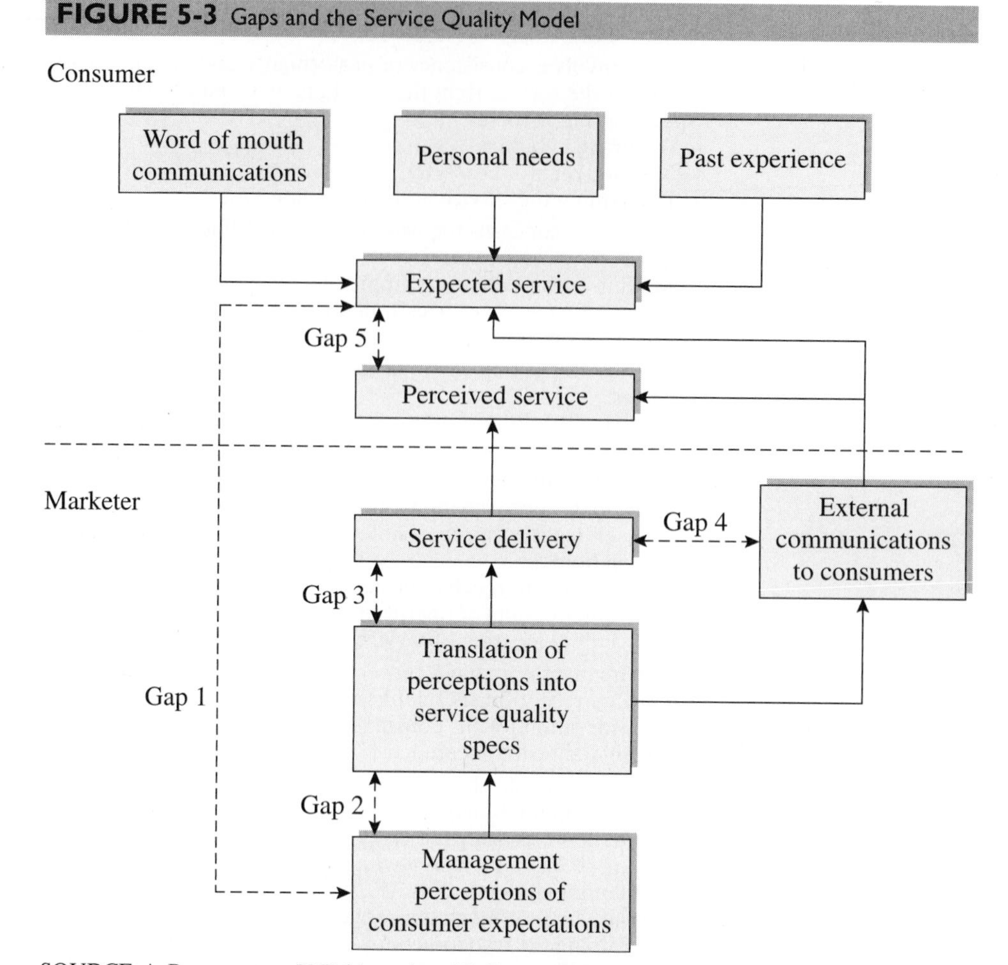

SOURCE: A. Parasuraman, V. Zeithamel, and L. Berry, "SERVQUAL: A Multiple-Item Scale for Measuring Customer Perceptions of Service Quality," *Journal of Retailing*, Spring 1988, pp. 12–40.

business markets. Sometimes segmentation is more complex, involving customer characteristics and demographics, psychographics, or benefits desired. Table 5-3 shows examples of segments for consumer markets. Notice how consumer markets can be segmented. This segmentation implies that data are gathered separately for each of these segments and analyzed separately. Table 5-4 shows examples of segments for commercial markets. Therefore, you may segment your markets according to consumer markets or commercial markets. You also may segment further within these markets.

STRATEGIC SUPPLY CHAIN ALLIANCES BETWEEN CUSTOMERS AND SUPPLIERS

Traditionally, the customer–supplier relationship has been viewed as a relationship in which one party attempts to gain advantage over the other. The customer obtains advantage over the supplier by exercising the ability to change suppliers. At the same time, the supplier attempts to gain power over the customer by increasing switching costs, thereby making it difficult for customers to switch to another supplier.

FIGURE 5-4 Gaps Model

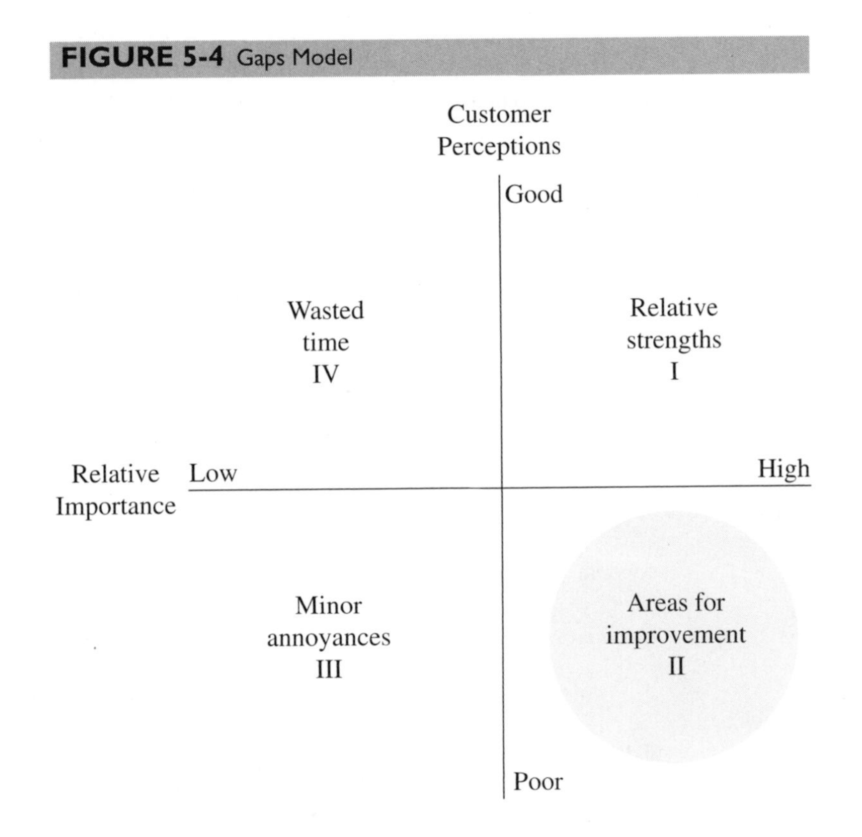

The theory behind this arrangement is essentially a competitive model because competition among the suppliers drives costs lower and quality higher. However, this theory ignores the costs associated with variability created by using multiple suppliers. This variability can be seen in process industries such as metallurgy, where customers use sheet steel raw materials from suppliers. Even though the dimensions and specifications for sheet metal are the same, if different suppliers are used, there will be increased variability in the physical properties of the materials, such as tensile strength. For example, one manufacturer found that by limiting the number of suppliers of sheet steel, it reduced its defects by 40%. This demonstrated that variability in sourced materials was a major cause of defects.

Today many Japanese and American companies use **single sourcing** as a way to reduce the number of suppliers. Single sourcing is a process for developing relationships with a few suppliers for long contract terms. In the late 1970s, this was the Japanese method of purchasing. Traditional American and Japanese approaches to purchasing were quite different. Japanese just-in-time (JIT) purchasing methods are now synonymous with world-class purchasing practice.

Increasingly, single-sourcing arrangements are developing into **strategic partnerships** where the suppliers become de facto subsidiaries to their major customers. In these arrangements, not only are suppliers single-source providers, but they also integrate information systems and quality systems that allow close interaction at all levels. Single-source suppliers to Ford are increasingly trained by their customers concerning the preferred and required organizations, processes, and delivery systems. Suppliers also enter into agreements to reduce costs and improve productivity and are graded on an annual basis concerning the attainment of these targets.

TABLE 5-3 Sample Consumer Market Segments

Respondent-related	
Geographic region	Pacific, Mountain, West North Central, West South Central, East North Central, East South Central, South Atlantic, Middle Atlantic, New England, Midwest
City, county, area	Under 5,000; 5,000–19,999; 20,000–49,999; 50,000–99,999; 100,000–249,999; 250,000–499,999; 500,000–999,999; 1,000,000–3,999,999; 4,000,000 or more
Demographic age	Infant, under 6; 6–11; 12–17; 18–24; 25–34; 35–49; 50–64; 65 and over
Sex	Male; female
Family size	1–2; 3–4; 5+
Family life cycle	Young, single; young, married, no children; young, married, youngest child under 6; young, married, youngest child 6 or over; older, married, with children; older married, no children under 18; older, single; other.
Income	Under $5,000; $5,000–$7,999; $8,000–$9,999; $10,000–$14,999; $15,000–$24,999; $25,000–$39,999; $40,000 or over
Occupation	Professional and technical; managers, officials, and proprietors; clerical, sales; craftspeople, supervisors; operatives; farmers; retired; students; homemakers; unemployed
Education	Grade school or less; some high school; high school graduate; some college; college graduate; postgraduate
Religion	Catholic; Protestant; Jewish; Latter-Day Saint; other
Race	White; African American; Asian American; other
Nationality	American; British; French; German; other
Social class	Lower-lower; upper-lower; lower-middle; upper-middle; lower-upper; upper-upper
Product-related	
Benefits offered	
Need satisfiers	Motives: economic and more detailed needs
Product features	Situation specific, but to satisfy specific or general needs
Consumption/use patterns	
Rate of use	Heavy; medium; light; nonusers
Use with other products	Situation specific (e.g., gas with a traveling vacation)
Brand familiarity	Insistence; preference; recognition; nonrecognition; rejection
Buying situation	
Kind of store	Convenience; shopping; specialty
Kind of shopping	Serious versus browsing; rushed versus leisurely
Depth of assortment	Out of stock; shallow; deep
Type of product	Convenience; shopping; specialty; unsought

Toyota employs an extensive supplier development program for each of its suppliers. As a result, variability to Toyota is reduced as the relationship between customer and supplier is enhanced over time. It is to the benefit of both parties to continue this relationship over a long period of time. In addition, some Japanese companies actually include their suppliers on their organization charts. Therefore, the task of managing suppliers is simplified over time. Variability and complexity are reduced.

TABLE 5-4 Sample Commercial Market Segments

Type of organization	Manufacturing, institutional, government, public utility, military, farm, other
Demographics	Size Employees Sales volume SIC code Number of plants Geographic location East, Southeast, South, Midwest, Mountains, Southwest, West Large city or rural area
Type of product	Installations, accessories, components, raw materials, supplies, services
Buying situation	Decentralized versus centralized Multiple buying influence Straight rebuy, modified rebuy, new buy
Source loyalty	Weak versus strong loyalty Last resort, second source, first source
Kind of commitments	Contracts, agreements, financial aids
Reciprocity	None versus complete

Bose Company, a speaker and sound systems manufacturer, broke new ground in dealing with its suppliers. Bose actually has delegated many purchasing responsibilities to its suppliers. Using this approach, major suppliers are provided office space and are authorized to make purchases for their own accounts. Using the JIT II approach, Bose suppliers provide only what is needed, when it is needed. Single sourcing is discussed further in Chapter 7.

The Role of the Customer in the Supply Chain

The goal of supply chain management is customer satisfaction. But who is the customer? Figure 5-5 shows a picture of a supply chain for yogurt. Notice that the supply

FIGURE 5-5 Segmenting the Supply Chain to Define Customers

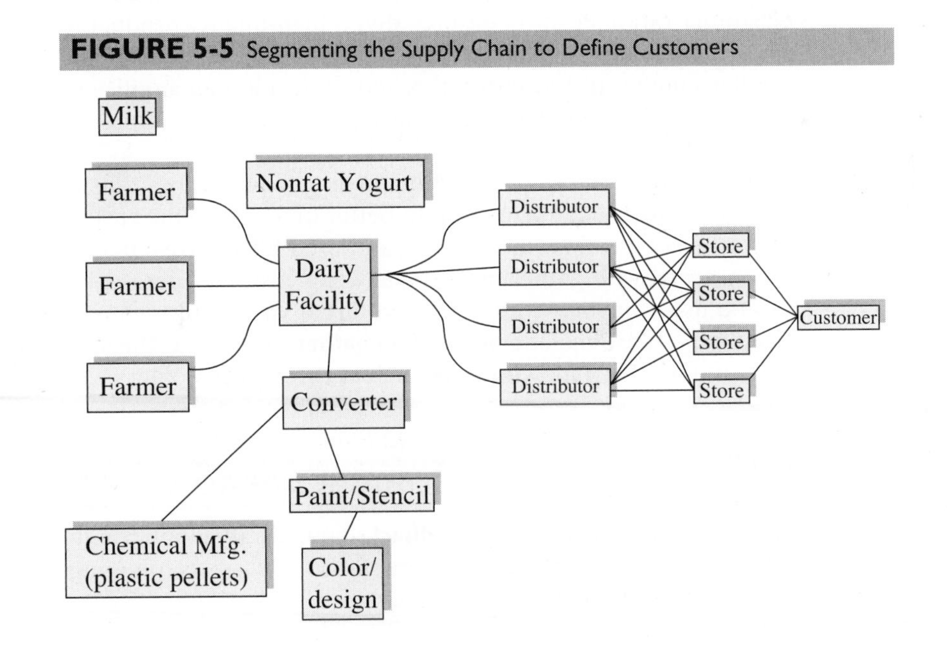

chain consists of several parts. In one, the farmers are providing product to a dairy. On another part of the supply chain, a chemical company provides plastic pellets to a converter who molds the cups. The label maker also provides labels to the converter. Do you suppose the converter is primarily concerned with the final consumer who buys the yogurt? Likely not. The supply chain can be segmented. The chemical company is primarily focused on the converter. The dairy company schedules shipments to the distributors and the distributors handle shipments to the store. The dairy works with the distributors and retailers to understand consumer preferences and to provide the flavors desired by the customers.

Customer satisfaction from a supply chain perspective is made more difficult because of the realities of the market. There are over 200 flavors of yogurt today. Thirty years ago, there was only one flavor. In many industries there are now many more high-priced premium items that must be considered. Shipping costs are also higher now. In addition, as we have discussed previously, customers are now more demanding than ever. It is clear that while much of the focus in supply chain management has been on cost reduction, supply chain managers must now focus more on service and product quality.

COMMUNICATING DOWNSTREAM

Marketing views every dollar of income equally, but operations does not. Operations views the costs and confusion associated with trying to satisfy diverse customer groups. It is less expensive to produce one item that satisfies a larger segment of the market than it is to produce several products that please several niche markets. Therefore, the operations professional does not view a dollar of income from diverse customers equally. **Customer rationalization** results from agreement between marketing and operations as to which customers add the greatest advantage and profits over time. This does not necessarily mean pursuing customers who are currently the most profitable. It could mean pursuing customers that cause the company to improve in ways necessary for continued survival. For example, consulting firms often have to turn away customers desiring services in areas outside the expertise of the company. Customer rationalization ensures that a high-quality product is provided and that the service provider stays within its field of expertise. Also, this allows firms to focus on a smaller number of key customers and to develop an **annuity relationship**. An annuity relationship is one in which the customer provides a long-term, steady income stream to the provider.

As suppliers focus on satisfying their customers, these customers are recognized as primary sources of information. To better understand the customer, data about the customer must be gathered, analyzed, and used for improvement. The rest of this chapter is concerned with gathering and analyzing customer data in such a way that the data can be used for improvement. There are a variety of means for gathering data from customers. These include **active data gathering**, such as through focus groups and surveys, and **passive data gathering**, such as through customer comment cards.

ACTIVELY SOLICITED CUSTOMER-FEEDBACK APPROACHES

Actively solicited customer feedback includes all supplier-initiated contact with customers. The three most common arenas for this are telephoning customers, conducting focus groups, and sending out surveys.

Phone contacts, focus groups, and survey results are referred to as **soft data**. As opposed to soft data, **hard data** are measurement data such as height, weight, volume, or speed that can be measured on a continuous scale. Soft data are not continuous and are, at best, ordinal. **Ordinal data** are ranked so that one measure is higher than the next. For example the continuum of "strongly disagree–disagree–neutral–agree–strongly agree" is a five-point scale that is ordinal. However, the interval between disagree and neutral is not equal to the interval between agree and strongly agree in the same way that the interval between 50 and 60 pounds is equivalent to the interval between 667 and 677 pounds. The weight measure of pounds is a hard measure. Because the ordinal scale is based on perception, measurements using ordinal data are subject to greater error than hard measurement data.

In spite of the error associated with soft data, soft data are useful in measuring the perceptions of customers. One use of this soft data is to compare employee and customer perceptions of quality. In a recent survey of hospital patient and employee perceptions of quality, it was found that patients consistently rated service quality higher than did the nurses and other employees in the survey. This can be explained in a services environment by the fact that employees have a view of the whole system, including customer contact activities and background activities that the patients do not see. Also, extremely dissatisfied customers of hospitals may not be able physically to respond to the survey.

Another important use of soft data is to provide an *external* source of data. It is difficult to gauge customer satisfaction. Hard external measures include returns, refunds, and warranty work. Sales data can be considered an external indicator of customer satisfaction. However, as you recall, when measuring quality, we are interested in understanding the many dimensions of quality that are important to our customers. Once an instrument is developed with valid measures of different dimensions, such as empathy or professionalism, then results can be baselined over time to determine whether things are improving. Once this baselining system is in place, it is possible to determine if improvements in the systems of service and production are improving customer satisfaction. If no improvements in customer satisfaction occur, it could be that the processes improved under the guise of quality improvement are really internally oriented and not customer oriented. In this way, survey instruments are useful tools for developing more of an external orientation in setting priorities for process and service improvement.

Telephone Contact

Telephone surveys are often used to gather information related to customers. Most often these are used for surveys or structured interviews. This is a type of convenience survey method. The major problem with telephone surveys is bias. Often major segments of the population of interest are not available via telephone at certain times. This makes random sampling difficult. Also, customers resent being called at inconvenient times. With recent changes in laws, service providers can be asked to remove customers from contact lists if this method is used. In addition, phone surveyors must respect state and federal do-not-call lists. However, it is often helpful to call respondents prior to sending a survey to see if they are willing to respond to a survey.

Focus Groups

A **focus group** allows a supplier to gather feedback from a group of consumers at one time. The groups are focused in two ways: First, focus groups narrowly address a single topic or group of topics. Second, focus groups draw individuals with similar characteristics

FIGURE 5-6 Focus Group Steps

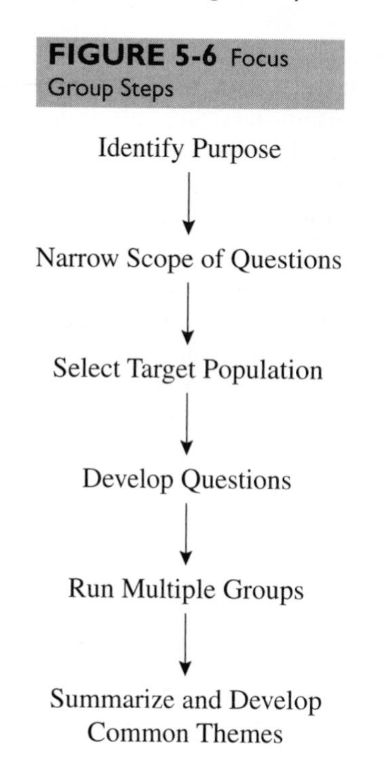

or demographics. This limits the discussion to subjects and market segments that are of particular interest to the firm. Figure 5-6 shows the steps included in performing a focus group session. After identifying the purpose of the focus group, a set of questions is developed that sequences from broad to specific. After the focus group sessions are completed with multiple groups, the notes from the sessions are reviewed to find common themes. These common themes become the basis for planning and improvement.

Focus groups are often used by business and government agencies to gauge topics or issues that elicit the strongest emotional responses from the subject. Focus groups need to be carefully facilitated so that the objectives of the research are met.

Customer Service Surveys

In Chapter 1 we said that quality ultimately is as the customer perceives it. The customer service survey is an important tool for determining customer perceptions of customer service and quality and is used by marketers and quality professionals in defining areas of strength and areas for improvement in quality systems. It is disappointing that surveys are sometimes misused as is shown in A Closer Look at Quality 5-3. A survey (or instrument) consists of a series of items (or questions) designed to capture perceptions. The number of items is determined by the purpose of the instrument and the willingness of respondents to spend time filling out the survey.

There are four steps to developing a useful survey: identifying customer requirements, developing and validating the instrument, implementing the instrument, and analyzing the results.

1. ***Identifying customer requirements.*** Customer requirements include the dimensions of quality, service, and performance that are necessary to satisfy the customer. Identifying customer requirements initially involves reviewing the purchase orders and contracts established when the relationship with the customer

A CLOSER LOOK AT QUALITY 5-3

MISUSING SURVEYS

Figure 5-7 shows a letter recently received by this author from a Chrysler dealer. The representative from the Chrysler dealer understands that surveys are used to evaluate dealer performance. It is unclear if financial incentives are linked to the surveys. However, it is clear that there must be a "hammer" that comes down on the dealer when less-than-perfect responses are received by Chrysler. Therefore, this represents both a misuse of the survey data by Chrysler and a feeble attempt by the dealer to distort the process for collecting data in a way that is unhelpful to them. Remember that survey information is only useful if the data is unbiased. This is an attempt by the dealer to bias their results. This is not unusual. Hilton Corporation does the same thing with signs in the elevators encouraging guests to "check excellent" on questions on their surveys. Both Chrysler and Hilton need to reevaluate how they use survey data to improve service performance.

FIGURE 5-7 Misusing Surveys

July 2, 2004

Tom Foster

Dear Tom,
Thank you for your recent visit to our service department. I am following-up to ensure that you were completely satisfied with the service.

Chrysler may contact you requesting your participation in a survey. Thank you in advance for taking your time to complete their report card on Jim's performance during your last visit. Hopefully you can give him an excellent as anything less means he failed in the eyes of the manufacturer.

In our continuing effort to find new ways to improve, we value your opinion or suggestions that would allow us to better serve our customers. May we ask you, what was the one thing we could have done better on your last service visit?

If for any reason you were not completely satisfied, please feel free to rectify any concerns so you would feel comfortable giving us a perfect score. If I am not available, please leave me a message and I will be glad to get back to you. Or, you can email me.

Sincerely,

Customer Care Manager

FIGURE 5-8 Timeliness at Henry's Fast Food—Specific-example Approach

Timeliness Dimension

_____ I received my food quickly.
_____ The server responded in a timely manner.
_____ The line moved quickly.
_____ I was served rapidly.
_____ The food service at Henry's is quick.

begins. Second, customer needs are reviewed with marketing and production. Third, interviews are conducted with a sampling of customers to determine what to add to the list of customer requirements.

2. ***Developing and validating the instrument.*** Once dimensions of customer requirements are identified, specific examples are developed to measure the particular dimensions. As shown in Figure 5-8, the dimension of timeliness is important for a fast-food restaurant. In a survey of their patrons, they had several items relating to the timeliness dimension (among others). Notice that the survey items are not questions. They are simple declarative sentences that use action verbs. Declarative statements are easy to understand and fit well with five- or seven-point scales. Each statement is a specific example of the quality dimension being measured.

An alternative means of developing survey items is the *critical-incident approach*. The critical-incident approach involves obtaining information from customers about the process they use to receive goods and services. The critical incidents are aspects of organizational performance with which the customers come in direct contact. These are important for monitoring and measuring process performance as it relates to customer service. In your baselining system, this approach is important in determining whether your process performance is improving. Figure 5-9 shows the same dimension of timeliness for the fast-food restaurant. Notice that the items relate to specific steps in the process of purchasing product in a fast-food restaurant. As with the specific-example approach, the items relating to timeliness can be averaged to determine if the process is performing well on some dimension.

3. ***Implementing the survey.*** Reliability and validity are two different but interrelated issues of survey development. We use the traditional target approach to explain this relationship. If the target in Figure 5-10 is the dimension of customer

FIGURE 5-9 Timeliness at Henry's Fast Food—Critical-incident Approach

_____ I was greeted on entering Henry's.
_____ There was a server available when I approached the service counter.
_____ My line had less than three people when I arrived.
_____ As soon as I reached the counter, the server requested what I wanted immediately.
_____ My food was delivered within 60 seconds of entering Henry's.

FIGURE 5-10 Reliability and Validity

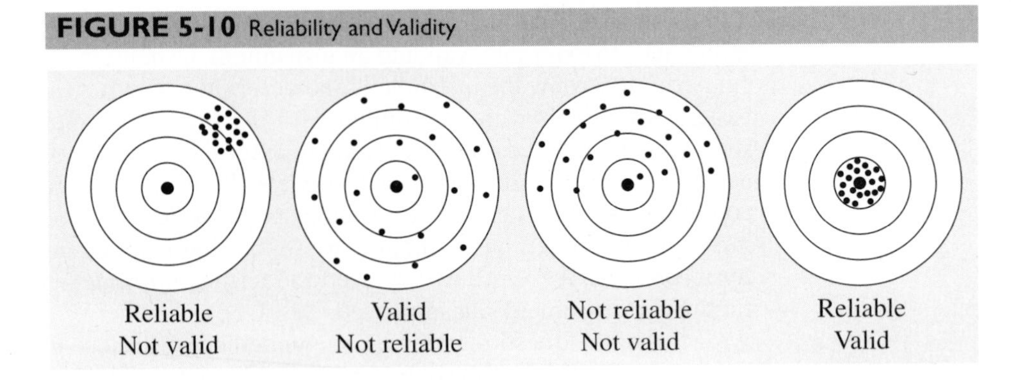

service that you are trying to measure, each arrow mark represents a single response using the survey instrument. If the measure of the dimension ascertains the dimension perfectly, the shot will be right in the center of the target. If you don't hit the center, you are not perfectly measuring the quality dimension. The more imperfect the measurement, the farther the shot will be from the center of the target.

Figure 5-10 shows four different situations. In target one, the arrows consistently hit the target in the same area but were off center. The instrument developed was reliable but not valid. In other words, the responses were consistent, but they were not measuring the right thing. In target two, the responses were all over the target. The average responses will hit the exact center of the target; however, there is a great deal of variability in responses. In this case, the group estimate is valid, but the measurement is not reliable. This shows that reliability is directly related to variability in responses from the respondents. Target three is neither reliable nor valid because the group averages will be off center and the variability is high. Target four is both reliable and valid because the survey is both centered and there is little variability in the responses. We briefly address these two issues separately.

An instrument with low reliability is a problem; the great deal of unwanted noise or variability in the responses hides the data's message. Two approaches to testing for reliability are test/retest reliability and interjudge assessment. With test/retest, the instrument is administered to a group of respondents randomly selected from the population of interest in a pretest. The same instrument is administered to the same group of individuals at a later point after some time has passed (say two weeks). The responses are then analyzed statistically.

Sloppy or unreliable terms, such as *quality, outstanding, acceptable*, and other adjectives, should be avoided as much as possible in designing surveys. With interjudge assessment, the survey is administered to multiple respondents and analyzed to gauge consistency of response among the respondents.

Validity is related to reliability, but a reliable instrument is not guaranteed to be valid. There are different types of validity; for example, *construct validity* refers to the use of certain terms and whether terms really measure what it is we want to measure. For example, self-reported measures of percentage growth in sales may not be a valid measure of success in customer satisfaction. Sales increases may instead reflect favorable market conditions. *Criterion validity* indicates that your measuring instrument has the ability to predict or agree with constructs external to that which you are measuring. *Content validity* refers to whether the item really measures what we want to measure.

To help ensure (but not guarantee) content validity, it is helpful to ask some outside individuals to externally validate an instrument. Usually, this includes asking five or six "experts" to review the instrument and determine whether the instrument is valid. The experts are people who are familiar with the firm's customers or have done previous surveying in a related area. Pretesting to externally validate the instrument can include asking managers and customers to review the instrument for understandability and completeness. Ask them what questions they would change, delete, or add. In validating a quality-related questionnaire, we found that respondents did not understand the term *conformance*. Once we replaced the term *conformance* with *meets specifications*, the survey instrument was improved.

Most of the questions will be close-ended because close-ended questions provide a better basis for data analysis. It is preferable to have at least one open-ended question in the customer service survey to allow customers to vent frustrations or make suggestions or other comments. Open-ended questions allow respondents to offer extemporaneous responses and comments.

4. ***Analyzing the results.*** For business purposes, data analysis generally should be kept simple. Means, histograms or numerical responses, and simple correlations are best for analyzing survey responses. Open-ended questions are analyzed with Pareto analysis using bar charts of the various categories of responses. More extensive data analysis using advanced statistical techniques, such as multiple regression, analysis of variance, or other procedures, should be performed if necessary. Explanations of these statistics can be found in introductory statistics textbooks. However, business experience has shown that simpler analysis is better because simple statistical results are easy to communicate to managers and coworkers. Quality Highlight 5-1 shows how ADAC Industries serves its customers.

QUALITY HIGHLIGHT 5-1

ADAC INDUSTRIES[c]

www.iotech.com

Focusing on the customer to improve quality has paid handsomely for ADAC Industries (recently acquired by IOtech), a Silicon Valley–based maker of high-technology health care products. During the past decade, revenue per employee increased from $200,000 to $330,000. This level of performance was 65% better than that of the competition. At the same time, ADAC increased market share in its core business, nuclear medicine, to more than four times that of its nearest competitor and became the market leader in Europe, Asia, and Latin America.

ADAC's whole-organization approach to increasing customer satisfaction and improving quality is illustrated by a decision to eliminate the quality council, a body composed of executives and managers and charged with overseeing the company's quality management process. As a result of benchmarking a Baldrige-winning company, ADAC replaced the council with two weekly meetings that were open to all employees as well as customers and suppliers. During these meetings, numerous employees presented data on key measures of customer satisfaction, quality, productivity, and operational and financial performance.

The company's corporate planning process, known as DASH, resulted in a strategic plan for three to five years and an annual business plan. Consistent with ADAC's primary core value, "customers come first," the DASH process begins with a comprehensive, fact-based analysis of customer

[c]Adapted from the MBNQA Profiles of Winners, 2003.

requirements for both the near and long terms. This analysis includes data from many sources, such as surveys, lost-order information, interviews conducted by customer-contact employees, logs of service calls, and focus groups. Results are integrated with analyses of competitive forces, risks, company capabilities, and supplier capabilities.

Most ADAC employees are part of empowered teams, and all manufacturing employees are members of self-directed work teams. All employees are trained in customer and supplier models, problem solving, and basic statistics.

As a result of improvements, the volume of service calls during the first 30 days after installation—an especially critical time when customers are forming their perceptions of quality—has been cut in half. An independent survey rated the first-month reliability of ADAC's cameras as the best in the industry. If customers do encounter serious problems, they can expect a quick and effective response. For example, if a system breaks down, ADAC technicians will have it back in operation within an average of 17 hours after receiving the customer's call. This is a third of the time that it took five years earlier.

ADAC's business system is delivering increases in customer satisfaction as ascertained through surveys. Customer retention rates have increased from 70% to 83% in five years, and service contract renewals have risen from 85% to 95% during the same period. In independent surveys of nearly 2,000 clinics and hospitals, nuclear medicine customers consistently have rated ADAC best at meeting customer needs.

PASSIVELY SOLICITED CUSTOMER-FEEDBACK APPROACHES

Customer-initiated contact, such as filling out a restaurant complaint card, calling a toll-free complaint line, or submitting an inquiry via a company's Web site, is considered **passively solicited customer feedback**.[6] Table 5-5 outlines several of the differences between actively and passively collected data. According to recent research in the area of passive data collection, it was found that passive collections resulted in lower ratings of quality than active collections. It is not clear which approach is more

TABLE 5-5 Differences between Actively and Passively Collected Data

Issue	*Active Data Collection*	*Passive Data Collection*
Sampling	Controlled by researcher	Controlled by respondent
Response rates	High rates (>50%) generally expected	Low rates (<20%) generally expected
Response bias	Controlled by researcher	Extreme response bias expected
Extent of questioning	Multipage questionnaires acceptable	Each question decreases response rate, increases nonresponse bias
Time frame	Discrete studies	Continuous data collection
Data use	Data useful as representation of target population	Data useful to track quality on continuous basis
Cost	High to medium	Low

SOURCE: S. Sampson, "Ramifications of Monitoring Service Quality through Passively Solicited Customer Feedback," *Decision Sciences* 27, 4 (1996):601–622.

[6]Sampson, S., "Ramifications of Monitoring Service Quality through Passively Solicited Customer Feedback," *Decision Sciences* 27, 4 (1996):601–622.

biased. However, it is expected that people who fill out customer response cards based on their own initiative probably have issues they would like resolved.

Customer Research Cards

Figure 5-11 shows an example of a customer research card. As you might guess, the card is from a local pizzeria catering to a college student population. We are all familiar with customer cards. We see them at many services companies and receive them with products. They are often a cheap way to involve the customer in the process. Research shows that respondents to these cards tend to be expressing extreme responses—either very highly pleased or extremely displeased. Response cards provide an opportunity for the service provider to develop a relationship with a customer through properly recovering from an extremely poor service encounter. Companies should have a method for logging, resolving, and tracking complaints. Resolved complaints also should be used for future feedback for systems improvement.

Customer Response Lines

Many companies provide toll-free phone lines for customer complaints, questions, and inquiries. These services are offered by many third parties or can be offered in-house. For example, Heinz Frozen Food Company uses a third party to handle and resolve

FIGURE 5-11 Pizzeria Complaint Card

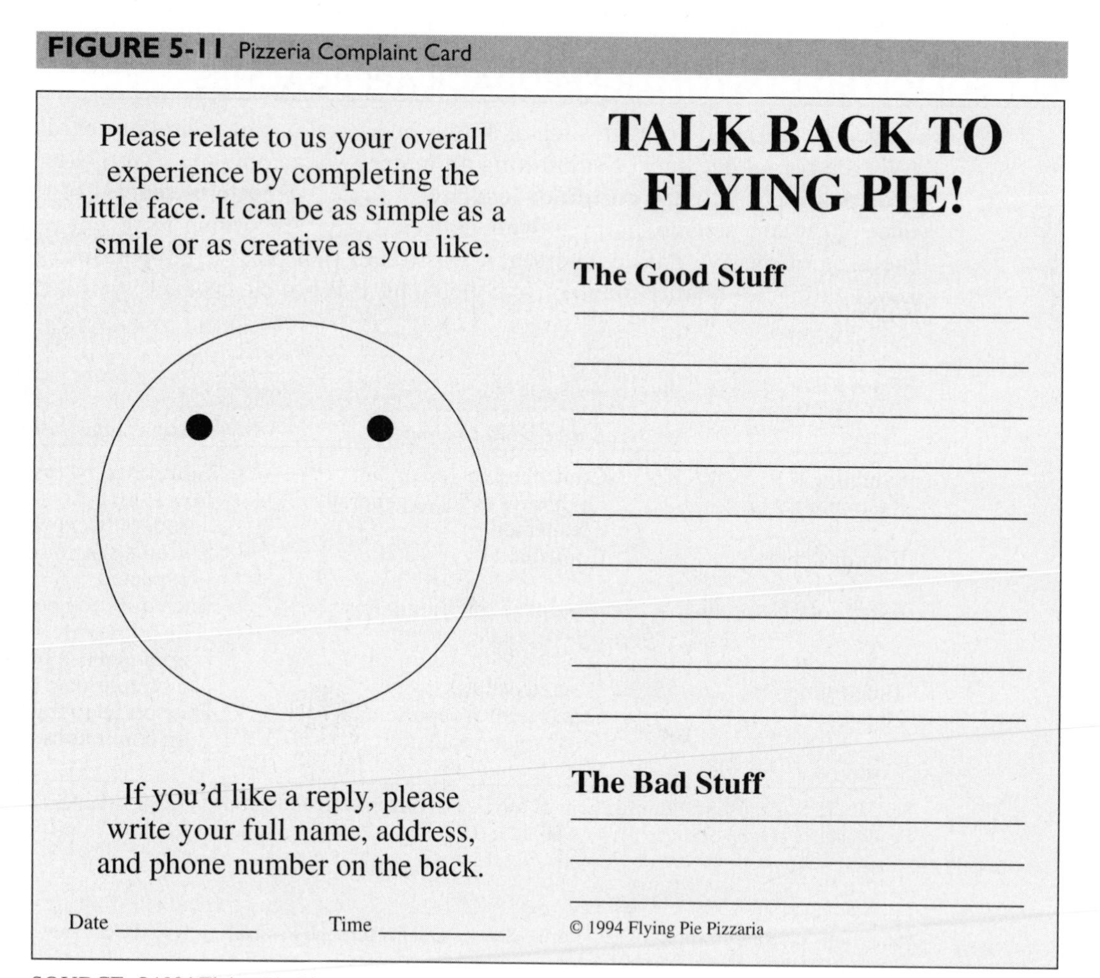

Please relate to us your overall experience by completing the little face. It can be as simple as a smile or as creative as you like.

TALK BACK TO FLYING PIE!

The Good Stuff

The Bad Stuff

If you'd like a reply, please write your full name, address, and phone number on the back.

Date ____________ Time ____________

© 1994 Flying Pie Pizzaria

SOURCE: ©1994 Flying Pie Pizzeria. Used by permission of Howard Olivier, owner (*www.flyingpie.com*).

complaints. If complaints are not resolved to the customer's satisfaction by the third party, the quality manager resolves complaints personally. Common problems with complaint lines occur when there are insufficient phone lines, long waits, poorly trained personnel, or unresponsive personnel.

Web Site Inquiries

Figure 5-12 (in two parts) shows Web pages that gather passively solicited customer feedback. The first is for the film *Bela Fleck* and the second is for Lands' End, a clothing retailer. Here the design is appropriate for the market segment. Both sites offer information and feedback mechanisms. However, the designs of the pages are very different, serving different market segments. The Lands' End feedback is handled through the "Customer Service" hot link on their homepage. As was discussed previously, many companies have underestimated the cost and resources needed to manage Internet inquiries.

MANAGING CUSTOMER RETENTION AND LOYALTY

An important indicator of customer satisfaction is **customer retention**. Customer retention is measured as the percentage of customers who return for more service. Customer retention will increase by application of the service tools and concepts contained in this chapter, such as tools for data gathering and analysis. This is an important indicator that every company should track. In services, where the customers are an input to the process, variability can be reduced by maintaining a stable pool of customers who are familiar with the transaction processes.

FIGURE 5-12 Customer Feedback: Bela Fleck and Lands' End Web Sites

SOURCE: Flecktones Web site, *www.flecktones.com/frameset.htm*. Used with the permission of Bela Fleck & The Flecktones and Richard Battaglia. Lands' End Web site, *www.landsend.com*. © 2002 Lands' End, Inc. Used with permission.

FIGURE 5-12 Continued

Customer loyalty can be instilled by offering specialized service not available from competitors. This can take many forms, including high customer contact or technology advancements. If a customer and supplier are linked through electronic data interchange (EDI), it is more difficult to switch to an alternative supplier because information systems have to be upgraded in order to work with new suppliers. This also speeds up data transmission between supplier and customer, reducing cycle times, and lead times for delivery of products and services.

There is an intangible aspect to customer loyalty. Harley-Davidson is probably the best example of brand loyalty. After all, how many products induce the kind of loyalty that causes people to tattoo the company logo on their bodies or buy garish clothing reflecting their love of the product. It is difficult to isolate the ethos that results in this type of customer loyalty. Honda, Yamaha, and Kawasaki don't elicit the same passion. Some automobile brands such as Volkswagen Beetle, Ford Mustang, Chevrolet Corvette, and more recently, Saturn, create such customer loyalty that people travel great distances to go to national expositions and jamborees that center on these products. There is a certain intrigue about these products that results in this level of excitement and loyalty.

CUSTOMER-RELATIONSHIP MANAGEMENT (CRM) SYSTEMS

With business information systems—especially over the Internet—companies are receiving volumes of customer-related data. These data include personal, Internet, process, and customer-preference information. As a result, systems have been created to mine these data to improve customer service and retention. These systems are called **customer-relationship management systems** (CRMS), CRMS systems use data to manage the three phases of customer-relationship management. These three phases are *acquisition, retention*, and *enhancement*.

CRMS technologies are used in customer data-acquisition and data mining efforts. All firms desire profitable customers. Customer self-service and product customization are ways to acquire new customer data. Another means is to provide customer access to information technology (IT) systems for configuring orders or researching information online. Customer retention is enhanced through a variety of activities. Since it costs 6 times as much to acquire customers, it is cost-effective to retain customers. Frequent-flier programs and grocery discount cards are examples of information-based methods for retaining customers. Enhancement involves improving service to the customer through the use of information systems. On Amazon.com, users can customize their desktops. Also, advertisements for new books in the area of interest for the customer are created based on the customer's historical purchase patterns. This is accomplished through CRMS.

In terms of functionality, CRMS allow for providing customer contact, product configuration, campaign management, dealer/distribution management, pipeline management, telemarketing, customer interaction centers, customer analysis, field service management, self-service, personalization, and supply-chain management. Table 5-6 provides a more detailed listing of CRMS functions. These are listed by category.

CRMS is utilized to monitor customer interactions, preferences, and relationships through media such as customer transaction records, call center logs, searches, and Web site clicks. Among the activities monitored include **customer defections**. These are customers who do not repeat business in some fixed period of time. This involves determining who are active versus inactive customers. Personalized service to those who are in danger of becoming inactive is used to achieve **churn reduction**—that is, reduction of the loss of customers.

CRMS are used to determine which customers are profitable and those who are unprofitable. Personalized communication can take place with customers, which serves them better and helps to maintain them as customers. For companies such as Amazon.com, **clickstream** information is kept that demonstrates how a customer navigates the Web site. This allows tailoring of the Web site to the preferences of the customer.

TABLE 5-6 CRMS Functions by Category

Customer-centric activities
Consolidate customer information from multiple channels—including e-mail, call centers, mobile devices, the Internet, and in-person encounters.
Give all departments a composite image of the customer's purchasing and service history, as well as the customer's buying, delivery, and contact preferences.
Support coordinated interactions throughout all customer touch points, including field sales, telesales, customer service, billing, and order fulfillment.
Analyze data to determine your most valuable customers, target services to them, and use their behavior to predict new products that will succeed in their marketplace.
Enterprise capabilities
Marketing automation to let marketers analyze customer purchasing trends, design targeted sales and marketing campaigns, and then measure results.
Customer service software to create customer profiles and to provide scripts to help representatives solve customer's problems and cross-sell to promote new purchases.
Sales automation tools to help the team manage accounts and prospects, as well as check metrics and inventories.
Partner relationship management solutions that link together vendors and other partners with your systems.
Customer acquisition
Marketing automation
Campaign management
Customer analysis
Web measurement tools
Advertising management
Sales management
Sales process automation
Configuration tools
Order management
Sales compensation
Channel management
Sales planning and analysis
Wireless device support
Customer retention and enhancement
Customer contact center
Web/telephone self-service
E-mail management
Web/interactive chat
Workflow analysis
Field service

SOURCE: Microsoft Corporation, 2002.

Maintaining these types of information is referred to as **knowledge management**. This involves managing the mountain of information generated by Web site usage in a way to improve marketing to key customers. The sales activities are managed specifically to each customer's preferences. This includes **transactional analysis**, which consists of customer service policies, sales processes, service process design, and after-sales service.

Like many newer technologies such as enterprise resource planning systems, CRMS implementations often result in failure. To properly implement CRMS requires an understanding of your business, your customers, competitors, culture, and processes. Many of the project management, team management, and change management techniques discussed in this book can be extremely helpful for CRMS implementation. This requires strong planning and project management skills. There are many vendors of CRMS, such as Oracle, IBM, Siebel, and others.

A WORD ON EXCELLENT DESIGN

It should be noted that not all good ideas come from customers. Some excellent products arise from advances in technology resulting from good engineering. IBM, in the early days of the computer, would not have pursued the computer as a viable product if it had relied on customers to tell the company their needs. For this purpose, the **ready–fire–aim** approach has been adopted by many high-technology companies as the best way to market their goods. An example of this was the Sony Walkman. Prior to the introduction of the Walkman radio, there probably was not much need for a portable radio with headphones and no speakers. In fact, when the Walkman first entered the market, many people commented that these would ruin the hearing of young people and cause alienated youth to further retreat from society. However, others loved the new product, and Sony was successful in marketing the Walkman to sports enthusiasts and young people. More recently, the MP3 technology for downloading music was available for a long time before it entered the mainstream for consumer use. (see Figure 5-13)

FIGURE 5-13 Apple iPod

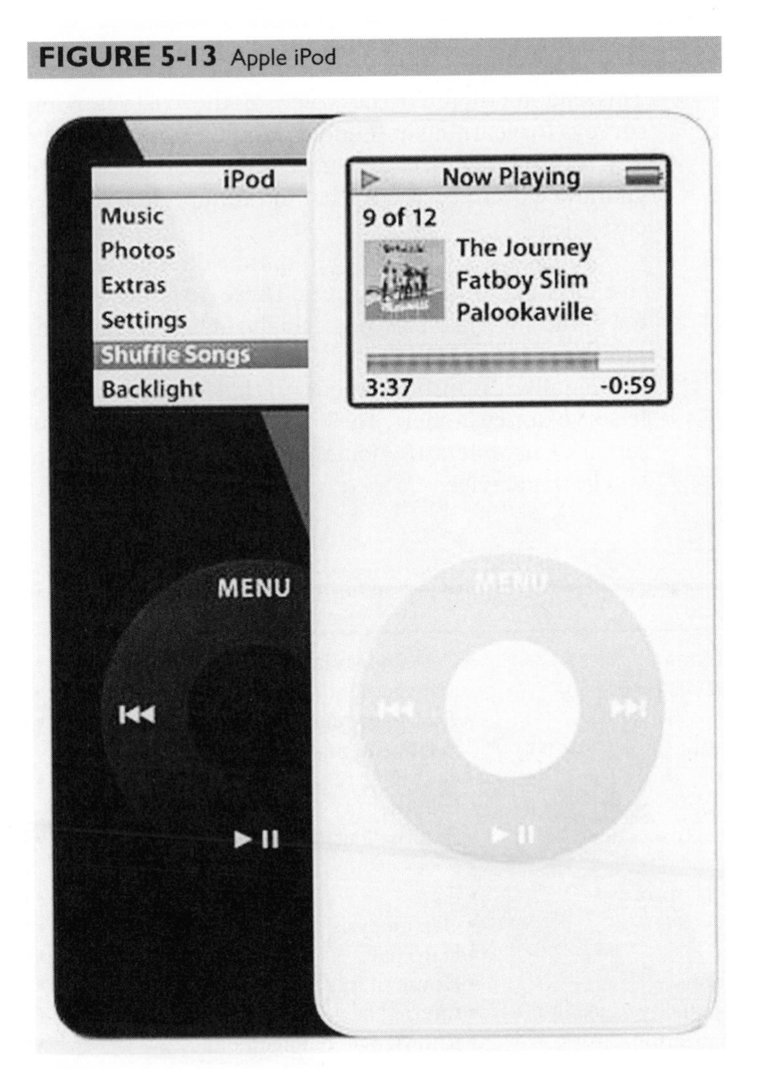

Used with permission.

Other new products are developed by identifying a need that customers do not necessarily recognize. There is much work being performed in alternative technologies to be used in third world countries that do not have the infrastructure of developed countries. In these countries, short-wave radios or personal computers that create their own electricity using a hand crank are being developed to help educate people about sanitation and disease prevention. BayGen has developed a radio that, after 20 seconds of winding, will play for one-half hour. The inventor, Trevor Bayless, invented the radio as he pondered the need for a technology to help stem the growth of AIDS on the African continent. The company now produces 1,000 radios a day and donates 10% of the radios to charity. This alternative technology was developed by an engineer. Others are discovering that this type of technology is friendly to the environment and demand is increasing. The BayGen radio has been called the environmental radio for the coming decade. Good customer intelligence coupled with innovative research and development programs appears to be the best marriage of resources.

SUMMARY

This chapter defined the voice of the customer and presented techniques such as surveys for learning and understanding this voice. Customer focus and satisfaction are key for companies to be successful. Tools such as complaint mechanisms, feedback, guarantees, and corrective action are necessary to develop annuity relationships with customers.

There are several ways to gather data about customers, including active and passive data-gathering techniques. These activities should be a part of an ongoing process for gathering customer data, analyzing the data, and implementing improvements to processes and design.

Finally, it is through focus on the customer and ingenuity that we find better ways to serve our customers. The corporate battleground in the new century is in the area of service. Customers are demanding more, and suppliers are responding by giving better and better service.

KEY TERMS

- Active data gathering
- Actively solicited customer feedback
- Annuity relationship
- Churn reduction
- Clickstream
- Closed-loop corrective action
- Compensate
- Complaint-recovery process
- Contrition
- Customer
- Customer defections
- Customer-driven quality
- Customer rationalization
- Customer-relationship management
- Customer-relationship management systems (CRMS)
- Customer retention
- End user
- External customers
- Focus group
- Gap
- Gap analysis
- Hard data
- House of quality
- Internal customers
- Knowledge management
- Ordinal data
- Passive data gathering
- Passively solicited customer feedback
- Quality function deployment (QFD)
- Reactive customer-driven quality (RCDQ)
- Ready–fire–aim
- Soft data
- Sole sourcing
- Strategic partnerships
- Transactional analysis
- Voice of the customer

Discussion Questions

1. Describe the difference between the internal and the external customers of a business organization. Why is it important to distinguish between internal and external customers?
2. Describe some of the potential pitfalls of customer-driven quality. Can you think of any ways to avoid or lessen the impact of these potential pitfalls?
3. How can a supplier avoid settling into a reactive customer-driven quality (RCDQ) mode?
4. What industries pose the greatest challenge for suppliers in terms of anticipating customer needs and requirements? What are the distinctive characteristics of these industries?
5. When was the last time you purchased a product and were asked to provide the seller or manufacturer of the merchandise information about yourself, such as your name, address, and telephone number. Did the request for information seem intrusive to you? Did you have any idea why you were being asked for the additional information?
6. Reflect on the last time that you complained to the manager of a store, restaurant, or other business about something that dissatisfied you. Was the complaint resolved to your satisfaction? Did the complaint-resolution process tell you something about the quality of the organization that you were dealing with?
7. Can you think of an example of an experience that you have had with a firm in which the difference between the espoused and the actual level of service provided was great in either a positive or negative way? If so, did this experience influence your perception of the business? Has this experience affected your willingness to do business with this company again?
8. Describe the basic concept behind strategic partnerships. In what ways can strategic partnerships facilitate a firm's quest for quality?
9. Describe some great service experiences and some horror stories that you have experienced. What were the variables that in your mind differentiated between the great and the horrible companies?
10. Suppose that the marketing department of a large manufacturing firm decided to adopt the motto, "We will build a product to suit any buyer's needs." What type of difficulties could this philosophy impose on the operations department? Through what process could the marketing department and the operations department determine which customers add the greatest advantage and profits over time?
11. Describe the basic idea behind a focus group. Are focus groups an effective way of gathering data about customer preferences and tastes?
12. Should focus group settings be formal and highly structured? Explain the rationale for your answer.
13. How can firms gain an overall understanding of the market segments they serve? Please make your answer as substantive as possible.
14. Why is it important that the facilitator of a focus group not bias the discussion in any manner? How could the results of a focus group analysis be tainted if the facilitator biased the discussion?
15. Describe a situation in which the use of an Internet customer response might be appropriate.
16. Describe the difference between actively solicited customer feedback and passively solicited customer feedback. Which type of feedback results in a lower rating of quality? Explain why.
17. Describe the difference between hard data and soft data. What are the unique advantages of each type of data in terms of obtaining information about customer perceptions?
18. Are customer surveys better suited for accessing customer perceptions in services or manufacturing? Explain your answer.
19. Explain the concepts of reliability and validity. Why is it important that survey instruments be both reliable and valid?
20. Why is it important to have open-ended questions in survey instruments?

PROBLEMS

1. One of the problems encountered by universities is developing reliable and valid course evaluation survey instruments. Choose a class you took last semester. For that class, identify two dimensions relative to course delivery. Now, develop five valid survey items for each of your two dimensions. Defend why you think these items are valid.
2. For the above class (Problem 1), develop five valid critical-incident survey items for each dimension. Defend why you believe your items are valid.
3. A manager for the Golden Bear publishing company found out that you are taking quality management. The manager desires to improve her supply chain performance by employing you as a consultant to provide quality improvement training to the supply chain employees. As a result, you now have to create a survey instrument to gauge the effectiveness of your training sessions. Choose four dimensions relative to your quality management training and develop five valid survey items for each dimension. Defend why you feel your survey items are valid.
4. For your above performance dimensions (Question 3), develop five critical-incident survey items for each dimension. Explain how you would demonstrate that these items are reliable.

Cases

Case 5-1 Customer Quality Feedback at Apple Computer

Apple Computer Homepage: *www.apple.com*

In the fast-paced personal computer industry, it would be very tempting for a computer company to rush a new product to market without taking the time to solicit customer input and feedback during the product development cycle. To avoid this temptation and to highlight its commitment to customer satisfaction, Apple Computer has developed a program called Customer Quality Feedback (CQF). CQF is a "hands-on" program providing Apple engineers with the ability to communicate with potential end users during the entire development cycle of an Apple product. The program integrates many of the features of a focus group but is sustained on an ongoing basis. It is also a very substantive and useful tool for Apple because it keeps the company attuned to the needs, preferences, and desires of its end users.

For people interested in participating in the program, Apple has posted an application form on its Web site. The application form is fairly comprehensive and outlines the terms and conditions of participation. Although the program is open to anyone, it is clear that Apple wants well-informed participants who will stick with the program. Participants are selected based on their interest, ability to provide timely information, commitment to working with Apple personnel, and the suitability of their computing environment as it relates to Apple's current needs. Once selected, the participants become an integral part of the development process for the products they are evaluating. They are provided early prototypes of Apple products and are asked to provide feedback pertaining to the product's features, interaction with employees, ease of use, performance, compatibility with third-party software, and other topics. The participants are also asked to provide suggestions as the product development cycle matures. The information provided by the participants is fed directly to the Apple engineers who are developing and testing the products. The overriding objective of the program is to incorporate customer input into the development of Apple products before they are shipped, rather than waiting for customers to react to the company's products after they are made available for sale. Prior to a product launch, the CQF participants involved with the product are asked to write testimonials about their input into the product's final design. These testimonials are used by Apple to demonstrate to other potential end users how Apple incorporates user feedback into the design and development of its products.

Apple's CQF program is a good example of a proactive approach to satisfying customer needs. It is also evidence of the company's willingness to "listen to the voice of the customer" in its product development and design. These are important steps in the development of a customer-driven approach to quality. ■

DISCUSSION QUESTIONS

1. Explain how Apple's Customer Quality Feedback program helps the firm hear the voice of the customer.
2. In your opinion, what are the most important aspects of Apple's program? Would you make any changes or modifications?
3. If you were an Apple user, would you enjoy participating in the Customer Quality Feedback program? Why or why not?

Case 5-2 Chaparral Steel: Achieving High Quality through a Commitment to Both External and Internal Customers

Chaparral Steel Homepage: ***www.chaparralsteel.com***

Chaparral Steel Company, located near Dallas, Texas, is a steel minimill producing a variety of steel products by recycling scrap steel. The company, founded in 1973, sells to a diverse group of industries, including construction, railroad, defense, mobile homes, and appliances. Chaparral has received a great deal of attention in the media and among business leaders because it has been relatively successful in an industry beset by a multitude of problems. Much of the company's success can be attributed to a focus on customer service and product quality through a commitment to both its external customers and internal customers.

To ensure that external customers are satisfied, Chaparral routinely conducts customer surveys, sends employees on site visits, and listens carefully to customer comments and suggestions. In addition, the company practices a number of quality-minded manufacturing techniques to reduce defect rates and prevent problems from occurring. Chaparral is very efficient, and new ideas and manufacturing techniques are transferred very quickly to the factory floor. As a result, the company remains on the leading edge of steel manufacturing technology and can adapt as customer requirements change. The company continues to improve its products and operations by benchmarking against world-class producers and giving employees paid sabbaticals where they learn about new work practices and technologies from academic institutions and industry leaders.

Although these efforts are commendable, the company's commitment to its internal customers is equally important. At any one time, approximately 85% of Chaparral's employees are enrolled in some type of class ranging from electronics to Spanish. If the training is off-site, the company reimburses employees for their tuition costs. In the plant, the majority of the employees are cross-trained, which enhances their individual job skills. The company benefits through consistency in operations because one employee can step in and perform the job of another employee if the need arises.

In many instances, Chaparral's commitment to its employees directly contributes to its employees' ability to contribute to company objectives of customer service and product quality. All of Chaparral's employees, from the CEO to the maintenance crew, have business cards they can give to customers to promote interaction. Frontline employees are periodically sent on customer site visits to answer questions, observe the customer's manufacturing process, or simply to see how Chaparral products perform. Every employee is salaried, and merit increases are tied to a variety of factors, including individual performance, versatility of skills, and training credits earned through the company's continuing education program. All these practices serve the dual purpose of increasing quality while at the same time enriching the jobs of the employees. Through its commitment to its internal customers, Chaparral has obtained some remarkable results. For example, a group of Chaparral employees recently developed a proprietary system for manufacturing wide-flange steel beams resulting in a substantial cost savings for the company. Chaparral credits its training program, along with individual effort on the part of its employees, for this accomplishment.

Overall, Chaparral's commitment to both its internal and external customers has produced impressive results, particularly as it relates to firm productivity and product quality. The company's daily absentee rate is less than 1%, and the average yearly turnover rate is only 5%. These rates are far lower than industry averages. These low rates provide for continuity in operations. Partially as a result of this, Chaparral produces steel at a lower cost per ton than its rivals. At Chaparral, it takes 1.4 hours to produce a ton of steel, as compared with an average of 2.4 hours per ton at other steel minimills. Chaparral was awarded the Japanese Industrial Standard Certification on its general structural steel products, becoming the only steel company outside Japan given that recognition. Similarly, Chaparral is the only steel company of its type to be certified by the American Institute of Mining, Metallurgical, and Petroleum Engineers to manufacture its products for nuclear applications. ■

DISCUSSION QUESTIONS

1. For a company like Chaparral Steel, why is a commitment to both its internal and external customers necessary?
2. As mentioned in the case, Chaparral periodically sends frontline employees on trips to visit the manufacturing sites of the company's customers. In your opinion, is this a justifiable expense? Why or why not?
3. Compare Chaparral's employee commitment to your current employer or a recent employer.

CHAPTER 6

The Voice of the Market

And he said unto them, What have I done now
in comparison of you? [Is] not the
gleaning of the grapes of Ephraim better
than the vintage of Abiezer?
—JUDGES 8:2

In the movie *The Paper Chase* a group of law students forms a study group. One student chooses real estate law because that is the specialty of his family practice. Another student chooses contracts because he enjoys the topic and is intrigued by the professor who teaches the course. In the end, the group benefits as each individual focuses on his expertise and coaches the others in that area.

Different people have different ideas about how to perform similar work. As a result, we can benefit by exploring different perspectives in designing products and processes. In the same sense, different organizations solve problems differently and take different approaches toward their work. For this reason, it can be helpful to observe how different firms perform tasks.

For example, a small computer software firm must consider many things when establishing a customer service unit. Some questions they must ask themselves include: What type of equipment would be needed? How should customer service specialists be trained? How would customer complaints be settled? How should refunds be handled? What procedures should be established to resolve customer issues? And what is an acceptable response time for returning customer calls? One method could be for the firm to blaze forth with the new customer service department and make mistakes as it goes along. A wiser approach would be to benchmark external customer service units in other companies first.

WHAT DO WE MEAN BY THE VOICE OF THE MARKET?

Strategy formation results from understanding the customers and the marketplace (see Figure 6-1). The marketplace includes immediate customers as well as competitors, the customers of competitors, potential competitors, and potential customers. In every market, advances shape and reshape the markets. Each firm strives to introduce new products, develop innovative processes, and find better ways to satisfy the customer. Customers can be good sources of information about competitors. This type of data gathering should include both strengths and weaknesses of your competitors. Some customers might be reticent to share this information. However, most customers realize that it is in their best interest to improve the performance of suppliers and will eagerly help. Information from lost customers also can be extremely useful for targeting weaknesses and improving products and services.

FIGURE 6-1 Strategic Quality Planning Model

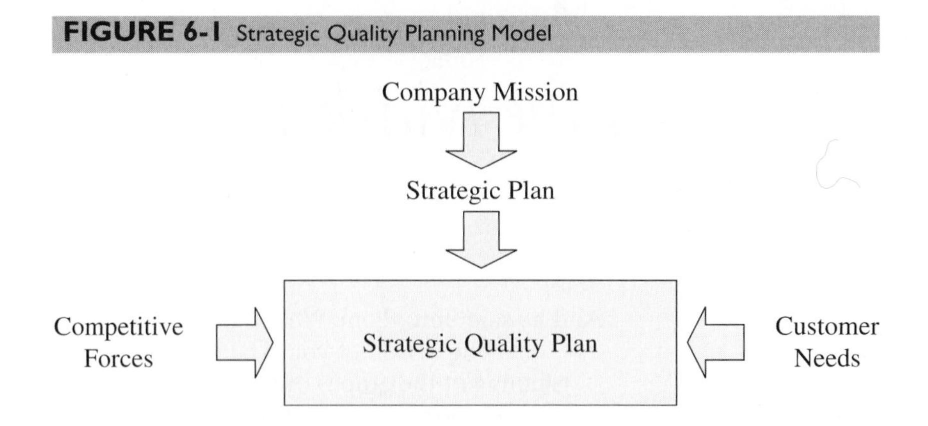

In Chapter 5 we considered how customers shape markets and how information about customers is obtained. However, customers are not the only source of information about the market. One of the best sources of information can be other companies. By understanding our competitors, we begin to understand the marketplace better and what it takes to compete successfully in the marketplace.

GAINING INSIGHT THROUGH BENCHMARKING

Suppose that you wanted to learn how to snow ski during your winter break. There are a few options available to you: You could teach yourself how to ski, you could enlist the help of your friend who learned how to ski last week, or you could take ski lessons with a certified instructor. Perhaps you could become a decent skier using any of these methods; however, your odds of success in learning quickly, avoiding injury, and gaining an appropriate respect for the sport would be greatest if you opt for the lessons.

The same is true in business. For a start-up firm, a rapidly growing company, or an organization in need of some improvements, the opportunity to observe and learn from a master could be invaluable. A **benchmark** is an organization recognized for its exemplary operational performance. There are many benchmarks in the world, including Toyota for processes, Intel for design, Motorola for training, Scandinavian Airlines for service, and Honda for rapid product development.

To be a benchmark, a company must be willing to open its doors and allow others to view its operations and tour its facilities. Thus a distinguishing feature of benchmarks is their amazing openness to other firms. Consider Toyota Motor Company. From the 1960s to the 1980s, Toyota developed the world-class production system known as just-in-time (JIT). This production system resulted in previously unseen levels of productivity, minimal cost, and a source of competitive advantage. Some companies might have put barbed wire and guard dogs around their JIT facilities to keep this technology to themselves and maintain their competitive advantage. Instead, Toyota allowed employees such as Taiichi Ohno and Shigeo Shingo to write books explaining the JIT concept to the rest of the world. Toyota also allowed thousands of visitors to tour its facilities and learn about the JIT system. The company even entered into strategic alliances with chief competitors such as General Motors so that GM could learn about JIT production.

Two rationales explain why benchmarking is good business. The first originates from Deming's thought that "the worst thing for a business is a weak competitor." Therefore, strengthening the competition forces everyone to improve in order to maintain a competitive edge. As such, *openness provides an impetus to continual improvement.* An opposite view is that openness can create a competitive advantage through creating *psychological barriers to competition*. A manager of a high-technology firm marketing a new product once said, "We do not mind if the others come to see how we produce our product. Once they see what we can do, they will not want to compete against us." In other words, the large amount of work it would take to establish and develop the systems that a truly outstanding competitor already has in place can be daunting and discouraging to its competitors.

Benchmarking is the sharing of information between companies so that both can improve. The first step a benchmarking firm must take is to document current performance. This activity will allow the company to pinpoint its goals and find a company (inside or outside the industry) that already excels at what it is trying to accomplish, study what it does, and gather ideas for improvement. Benchmarking is useful for externally validating an organization's approach to its business. If the managers in a firm are unsure that they are pursuing a useful plan of action, benchmarking can help them understand how what they are doing stacks up against the masters.

There are two parties to each benchmarking relationship: an **initiator firm** and a **target firm**. The initiator firm is the firm that initiates contact and studies another firm. The target firm is the firm that is being studied (also called a *benchmarking partner*). These are not static roles. Often the target firm enters into a reciprocal agreement to observe the initiator firm. As is shown in Table 6-1, there are several types of benchmarking. Note that they are not all mutually exclusive.

Process Benchmarking

In **process benchmarking**, the initiator firm focuses its observation and investigation on business processes. This can involve studying process flows, operating systems, process technologies, and the operations of target firms or departments. The goal is to identify and to observe the best practices from one or more benchmark firms. By improving core processes, overall business performance is enhanced.

Financial Benchmarking

The goal of **financial benchmarking** is to perform financial analysis and compare the results in an effort to assess your overall competitiveness. This type of benchmarking need not involve direct interaction between the initiator firm and the target firms. There is, however, interaction between the initiator and a third party that gathers this information. Usually the information can be gathered using CD-ROM databases such as Lexis/Nexis or Compact Disclosure. As more companies place annual reports on-line, the Internet has become an important tool for benchmarking financial performance.

Performance Benchmarking

Performance benchmarking allows initiator firms to assess their competitive position by comparing products and services with those of target firms. Performance issues may include such things as cost structures, various types of productivity performance, speed of concept to market, quality measures, and other performance evaluations. For example,

TABLE 6-1 Benchmarking Types

Process benchmarking—comparing processes
Financial benchmarking—comparing business results
Performance benchmarking—comparing cost structures, speed, quality levels, etc.
Product benchmarking—comparing product attribute and functionality
Strategic benchmarking—comparing firm competitiveness along several dimensions
Functional benchmarking—comparing or learning how another firm performs a particular function

an initiator firm may be interested in identifying other firms that have implemented effective cost accounting practices such as activity-based costing systems to observe and compare the performance of various cost drivers.

Product Benchmarking

Many firms perform **product benchmarking** when designing new products or upgrades to current products. Product benchmarking often includes **reverse engineering**, or dismantling competitors' products to understand the strengths and weaknesses of their designs. By observing the designs of others, the initiator firm can develop new ideas for product and service design. Micron Technologies maintained a cost advantage in producing DRAM computer chips because of its ability to produce chips with fewer mask levels than its competitors. To compete with Micron, competitors would have to analyze their chips and try to understand their chip-making processes.

Strategic Benchmarking

Strategic benchmarking involves observing how others compete. This often is not industry-specific because firms go outside their own industries to learn lessons from companies and organizations in different industries. This typically involves target firms that have won prestigious honors such as the Malcolm Baldrige National Quality Award, the Shingo Prize, or the Deming Prize. The focus of this type of benchmarking is to identify the mix of strategies that makes these firms successful competitors. Such benchmarking can be very time-consuming and costly. At Boise Cascade Company, when establishing a quality management process, the firm took a team of executives to Japan and visited several firms to get a sense of what a high-quality firm looked like. Though firms were from other industries, the executives returned with a very realistic idea of the task they faced in establishing quality processes. Pal's Sudden Service (Quality Highlight 6-1) uses strategic benchmarking.

QUALITY HIGHLIGHT 6-1

PAL'S SUDDEN SERVICE

www.palsweb.com

A privately owned, quick-service restaurant chain, Pal's Sudden Service, serves primarily drive-through customers at 17 locations, all within 60 miles of Kingsport, Tennessee, where its first restaurant opened in 1956. Carefully following its formula for standardizing high levels of product and service quality, Pal's has since grown to become a major regional competitor.

Today, Pal's employs about 465 people, 95% of whom are in direct production and service roles. The company competes directly with national fast-food chains, earning a steadily increasing—and, now, second best in its region—market share of almost 19%, doubling since 1994. In 2004, sales totaled about $32 million.

Featuring hard-to-miss exteriors festooned with larger-than-life menu items, Pal's restaurants sell hamburgers, hot dogs, chipped ham, chicken, French fries, and beverages, as well as breakfast biscuits with country ham, sausage, and gravy. The company aims to distinguish itself from fast-food competitors by offering competitively priced food of consistently high quality, delivered rapidly, cheerfully, and without error. The majority of customers live or work within 3 miles of Pal's locations, and nearly two-thirds are women.

Pal's is the first business in the restaurant industry to receive a Malcolm Baldrige National Quality Award. For everything organizational and operational, Pal's has a process. Almost nothing—

from new product introductions to hiring decisions to the design of support processes and work systems—is done without thorough understanding of likely impacts on customer satisfaction.

The company's Business Excellence Process is the key integrating element used in every transaction. Carried out under the leadership of Pal's two top executives and its 17 store owner/operators, the Business Excellence Process spans all facets of the operation—from strategic planning (done annually with two-year horizons) to on-line quality control. Every component process, including those for continual improvement and product introduction, is interactively linked, producing data that directly or indirectly inform the others.

Benchmarking underpins the entire Business Excellence Process. Managers are continually on the lookout for benchmarking candidates, and each one compiles a running list of potential subjects. For the leadership team, benchmarking yields meaningful competitive comparisons, new best practices for achieving higher performance goals, or new organizational directions. For the entire organization, benchmarking results are a constant reminder that performance always can be improved.

Pal's is exhaustive in its pursuit of useful data, the basis for sound planning and decision making. In particular, customer, employee, and supplier feedback is central to all processes, and it is gathered in numerous formal and informal ways. For example, Pal's owner/operators must devote part of every work day to "marketing by wandering around." A portion of this period is spent engaging employees and customers to hear their views on how a location is performing and to solicit ideas for improvement.

Owner/operators also go door-to-door within a 3-mile radius of their restaurants, seeking direct input on customer requirements and satisfaction levels. Answers to predesigned questions are recorded, compiled, and later analyzed at the store and corporate levels.

Owner/operators also maintain a communications log. They record what they have learned about sales, expenses, customers, staff, products, services, equipment, and suppliers, and they list ideas for improvement. Weekly logs are sent to senior Pal's executives, who comb the entries for issues and opportunities to be addressed at formal monthly management reviews of organizational and business results.

Data are gathered systematically at all levels—process, shift, individual store, and entire business. The company's enterprise resource planning system, SysDine, is a key tool, generating store-level and company-wide data on sales, customer count, product mix, ideal food and material cost, and turnover rates. This information supports daily operational decisions. It also is used to update Pal's Balanced Scorecard of Core Performance Measures, which links directly to its key business drivers: quality, service, cleanliness, value, people, and speed.

Managers regularly review the value of the data collected, and the company employs an outside statistician to evaluate the type of information tracked, how it is used, and how it is collected.

Developed with the aid of benchmarking studies, the company's training processes support improvement in operational and business performance. Owner/operators and assistant managers have primary responsibility for staff training. They use a four-step model: show, do it, evaluate, and perform again. Employees must demonstrate 100% competence before they are certified to work at a specific work station.

In customer satisfaction, including food quality, service, and order accuracy, Pal's is outperforming its primary competitor. For example, customer scores for quality averaged 95.8%, as compared with 84.1% for its best competitor.

Pal's order handout speed has recently improved more than 30%, decreasing from 31 to 20 seconds, almost four times faster than its top competitor. Errors in orders are rare, averaging less than 1 for every 2,000 transactions. The company aims to reduce its error rate to 1 in every 5,000 transactions. In addition, Pal's has consistently received the highest health inspection scores in its market and in the entire state of Tennessee.

SOURCE: NIST, Profiles of Baldrige Winners, 2006.

Functional Benchmarking

Another type of benchmarking is referred to as **functional benchmarking**, where a company will focus its benchmarking efforts on a single function to improve the operation of that function. An example of functional benchmarking occurs in purchasing. Often purchasing managers use their networks to share information about the purchasing function in many different organizations. The Institute for Supply Management (ISM) provides a framework for networking purchasing professionals that facilitates the sharing of information about the purchasing function. In addition, the ISM gathers information about purchasing departments and makes this information available to members of the organization.

PURPOSES OF BENCHMARKING

There are seven primary purposes for benchmarking (see Figure 6-2). These purposes require different levels of involvement in the benchmarking activities. Time consumption and costs may vary according to the purpose of benchmarking.

The purposes of benchmarking range from basic learning to achieving world-class leadership. The life cycle in Figure 6-2 shows that benchmarking purposes evolve as the firm becomes more mature in its quality journey.

FIGURE 6-2 Benchmarking Purpose and Quality Maturity

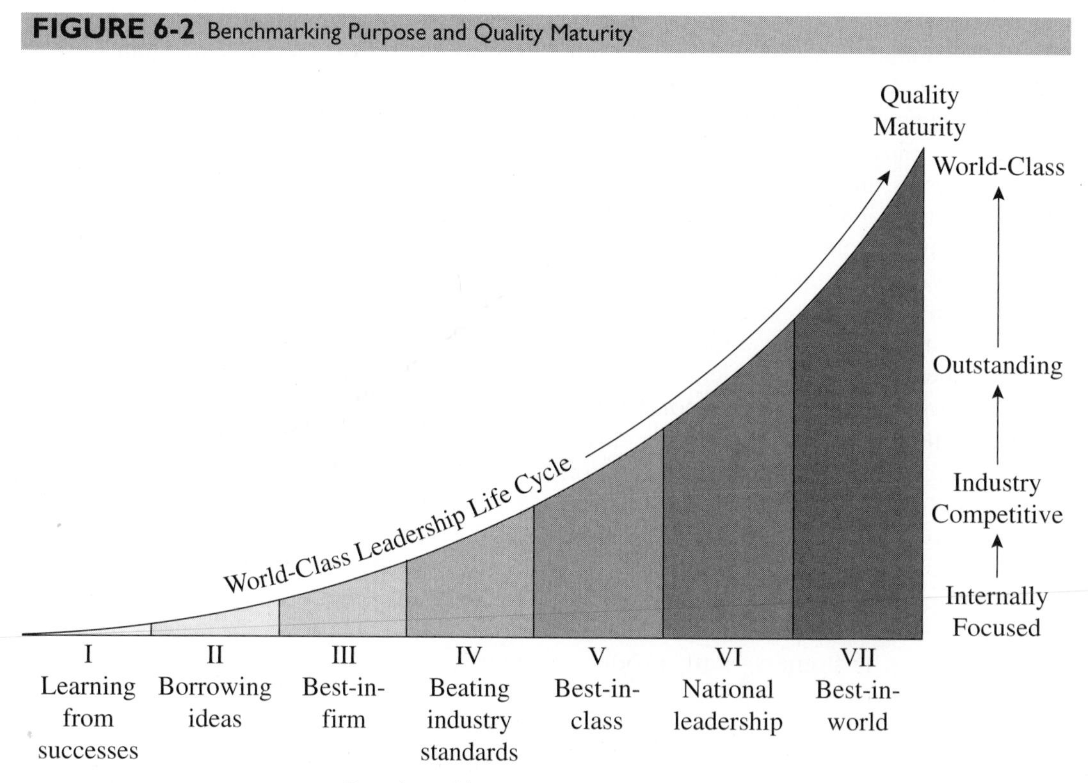

DIFFICULTIES IN MONITORING AND MEASURING PERFORMANCE

Many times firms desire to compare financial measures between companies when benchmarking. This can be a useful activity; however, there can be problems. One of the most significant problems stems from limitations of accounting systems. Often companies have variations in the way they compute their measures that affect the results.

Consider the computation of scrap in which a company computes the ratio of cost of goods sold to scrap. The formula for this computation is

$$\text{Scrap efficiency} = \text{cost of goods sold/scrap} \tag{6.1}$$

This formula normalizes the cost of scrap based on the volume of business that a firm does. The resulting ratio is the proportion of the material inputs to production that is wasted as scrap. The higher the ratio, the more efficient is the use of these materials.

Now consider two companies, A and B. Company A uses Equation 6.1 and computes scrap by weighing discarded materials at a standard cost of \$.15 per pound of scrap. However, Company B has a variation in its measures. Company B computes scrap at the industry standard of \$.15 per pound, less \$.03 per pound (the amount it receives from a recycling company that purchases its scrap). Therefore, Company B's account equation would be

$$\text{Computed ratio for company B} = \text{cost of goods sold/(scrap} - \text{recovery)} \tag{6.2}$$

The resulting equations are

$$\text{Company A ratio} = \text{cost of goods sold/.15}$$

$$\text{Company B ratio} = \text{cost of goods sold/.12}$$

If the cost of goods sold is the same for both companies, the ratio will be higher for Company B. In this case, Company A is at a disadvantage because of the differences in the ways that scrap is costed. These differences might be more apparent if a careful benchmarking study is performed in which the participants know exactly what the numbers mean and what the differences in accounting systems are. However, if a great number of data, ratios, measures, and numerical statistics are shared between the companies, the differences in accounting methods might not be as obvious.

The ratio we have focused on is a type of productivity ratio. The cost of goods sold to scrap ratio measures the efficient use of materials. One remedy for the effects of accounting differences on productivity ratios is to compute total-factor productivity measures. To understand total-factor productivity measures, we first must understand single-factor productivity measures. Single-factor productivity measures are computed as

$$\text{Single-factor productivity} = \text{output/(a single input)} \tag{6.3}$$

The cost of goods sold to scrap ratio is a single-factor ratio in that it focuses on scrap material alone. Multiple-factor productivity measures use multiple inputs in their computation. For example

$$\text{Multiple-factor productivity} = \text{output/(the sum of multiple inputs)} \tag{6.4}$$

Multiple inputs might include scrap, labor, energy, materials, equipment, and other measures of inputs. Hayes and Wheelwright[1] suggest that it is better for firms to use

[1]Hayes, R., and Wheelwright, S., *Dynamic Manufacturing* (Boston: Free Press, 1988).

multiple-factor productivity measures in making comparisons because these measures are more robust. The total-factor productivity measure is similarly computed as

$$\text{Total-factor productivity} = \text{output}/(\text{sum of all inputs to production}) \quad \textbf{(6.5)}$$

Total-factor productivity, although still subject to the "apples and oranges" problems of comparisons, is the least sensitive to differences in costing conventions and accounting practices.

Notice that if you focused on plant and equipment productivity, the conclusion would be that the competitor is lagging in productivity. Also notice that labor productivity may be a small part of total productivity.[2]

Example 6-1: Computing Productivity

A company has gathered the following financial information for itself and a competing firm. The company wishes to compare productivity for the two firms (all the following numbers are in 000s).

	Firm A	Competition
Labor	$ 20,000	$ 15,000
Plant and equipment (P & E)	100,000	145,000
Energy	5,500	10,500
Materials	220,000	190,000
Sales	425,000	500,000

Labor and other productivity ratios for the two firms are computed as

Firm A labor productivity = (sales/labor) = 425,000/20,000=21.25
Competitor labor productivity = (sales/labor)500,000/15,000 = 33.33
Firm A P&E productivity = (sales/P&E) = 425,000/100,000 = 4.25
Competitor P&E productivity = (sales/P&E) = 500,000/145,000=3.45
Firm A energy productivity = (sales/energy) = 425,000/5,500=77.27
Competitor energy productivity = (sales/energy) = 500,000/10,500=47.62
Firm A materials productivity = (sales/materials) = 425,000/220,000=1.93
Competitor materials productivity = (sales/materials) = 500,000/190,000 = 2.63
Total-factor productivity for Firm A = (sales/total of inputs) = 425,000/345,500 = 1.23
Total-factor productivity for the competitor = (sales/total of inputs) = 500,000/360,500 = 1.39

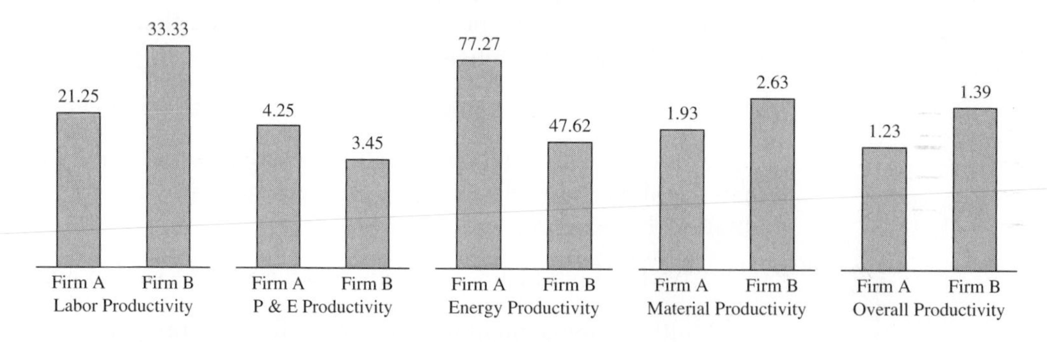

[2]Some firms are using a method called *data envelopment analysis* (DEA) to improve comparisons between firms. This is a linear programming approach that can be used to eliminate the "apples and oranges" problem.

As can be seen, the total-factor productivity measure provides a more complete picture of firm productivity.

Another caveat for comparing measures between companies is based on the Deming arguments against such things as work measurement and management by objective. To concentrate too much on comparative measures might focus managers on results and not causes. This could result in the unfortunate development of numerical goals that ignore the necessity of improving the system of production.

Because of the possibility of serial correlation of data, even greater care should be taken when analyzing longitudinal (time-series) data. For example, company researchers might want to compare *rates of growth* for two companies in assets, sales, and other measures. If the data have serial correlation, they might reach the wrong conclusions.

Problems in comparisons are even more pronounced when comparing U.S. firms with foreign companies. The cost accounting conventions and accepted accounting principles can vary greatly between countries. For this reason, it is best not to compare accounting figures with foreign companies. However, if you must, take great care as to how the results of the comparisons are used in formulating policy.

COMMONLY BENCHMARKED PERFORMANCE MEASURES

Different firms in different industries use thousands of different benchmarking measures. The measures a firm chooses depends on the **key business factors (KBFs)**, or critical success factors, of each particular firm. The key business factors are factors that are significantly related to the business success of the firm. For example, in a given firm, customer satisfaction might be significantly related to market share. If this is the case, then a company would be very interested in keeping score of its customer satisfaction.

Table 6-2 shows the categories of measures that are often gathered in benchmarking studies. **Financial ratios** such as return on assets (ROA) or return on investments (ROI) are probably the easiest to obtain and compare. For many financial ratios, all that is needed is an income statement of a firm and a balance sheet. Many of these statistics are available in annual reports and on the Internet. Generally, senior management is keenly interested in these measures to guide their decision making.

Productivity ratios are useful in measuring the extent to which a firm effectively uses the scarce resources that are available to the firm. As we have previously discussed, these include single-factor, multifactor, and total-factor productivity measures.

TABLE 6-2 Benchmarking Data
Financial ratios
Productivity ratios
Customer-related results
Operating results
Human resource measures
Quality measures
Market share data
Structural measures

Customer-related results include customer satisfaction, customer dissatisfaction, and comparisons of customer satisfaction relative to competitors. These measures may be in the form of retention, gains, losses, customer perceived value, competitive awards, competitive customer ratings, and independent organization evaluations. Customer satisfaction measures are important for gauging the effectiveness of quality improvement because they are good indicators of financial performance.

Operating results are important for monitoring and tracking the effectiveness of company operations. These might relate to cycle times, waste-reduction measures, value-added measures, lead times, time from concept to market, setup times, percent reduction in setup times, and myriad other operating results.

Often, one of the key aspects of running a business involves the people employed by the organization. Therefore, **human resources measures** provide important insights into how effectively the business is being run. Employee satisfaction is significantly related to business performance in many firms. Therefore, employee satisfaction measures, training expenditures, training hours per year, work system performance, employee effectiveness measures, turnover, safety statistics, absenteeism, and many other data might be important measures for benchmarking.

Of course, **quality measures** are often compared between firms. These can include conformance-based quality information such as reject rates, capability information, performance information, or other measures. These quality measures also can include scrap and rework measures, percent defectives, field repairs, costs of quality, and many other metrics. The quality measures also may include data concerning the performance of processes and time-related statistics. Where customer service quality is important, the types of data that are captured often include such things as percentage of customers whose phone calls are answered within seven seconds, average response time for phone inquiries, number of people a caller must contact to get a problem resolved, and many other metrics (or measures).

Market-share data are an essential indicator of business success. This has resulted from the change in competition because of increased global competition. However, market share does not encompass only a single measure. Since markets are segmented, market share includes shares in the different markets served by the firm. Market-share comparisons also are made to determine where the initiator firm ranks in the market.

Finally, with the advent of ISO standards and a move to formalized production systems, **structural measures** often are benchmarked. Structural measures include objectives, policies, and procedures followed by a firm. They may include safety, production, accounting, financial, engineering, and other types of structural measures that are used in determining competitiveness.

Why Collect All These Measures?

Management by fact dictates that decisions are made based on the sound collection and analysis of data. The Malcolm Baldrige criteria are clear about this:

> Data and analysis support a variety of company purposes, such as planning, reviewing company performance, improving operations, and comparing company performance with competitors or with "best practices" benchmarks.[3]

[3]*2006 Criteria for Performance Excellence* (Gaithersburg, MD: National Institute of Standards and Technology, 2006), p. 4.

A major consideration in performance improvement involves the creation and use of performance measures or indicators. Performance measures or indicators are measurable characteristics of products, services, processes, and operations the company uses to track and improve performance. The measures or indicators should be selected to best represent the factors that lead to improved customer, operational, and financial performance. A comprehensive set of measures or indicators tied to customer and/or company performance requirements represents a clear basis for aligning all activities with the company's goals. Through the analysis of data from the tracking processes, the measures or indicators themselves may be evaluated and changed to better support such goals.

Key Business Factors

When benchmarking, it is important to understand your target firm's key business factors (KBFs) as well as your own. Key business factors are important attributes of a business that influence its operations and decision making. Examples of KBFs include mission, vision, values, key customer segments, core capabilities, culture, governance, and other facets of a business. KBFs are not to be confused with **key measures** (the metrics that need to be monitored to gauge the health of the business) and **critical success factors** (factors that help to determine the success of the firm).

BEST-IN-CLASS BENCHMARKING

When beginning the benchmarking process, the question is often asked, "Whom do we benchmark?" This is not an easy question to answer. It is somewhat daunting to consider benchmarking against direct competitors because of the dog-eat-dog nature of competition. However, a surprising number of firms are participating in benchmarking studies with competitors.

Generally, initiator firms will choose to benchmark the **best in class** (or best in breed). Best in class refers to firms or organizations that have been recognized as the best in an industry based on some criterion. There might be firms that have excellently engineered product designs or excellent customer service. Often industry-related professional societies publish data concerning the major players in an industry. These societies can be excellent sources for obtaining names of firms to be benchmarked.

The objective of best in class is to provide a basis for continual improvement. By observing the processes of a firm, another firm generates ideas for improvement. Typically, these are reciprocal agreements that are well defined where the target firm also visits the initiator firm. Although it may seem that an established best-in-class firm would have little to gain from a benchmarking agreement, in many cases it is up to the initiator firm to demonstrate how the target firm also could benefit from benchmarking. By continually improving, initiator firms can become best in class too.

BEST-OF-THE-BEST BENCHMARKING

After becoming a best-in-class firm, it may be difficult to gain new insight and information from direct competitors. It is usually most helpful to begin by performing best-in-class benchmarking prior to doing this out-of-industry benchmarking. Therefore, the next level of improvement is called **best of the best** or best in the world. These are the outstanding global benchmark firms—some of which we have already mentioned in this book such as Baldrige winners and other world-class firms. It has been suggested that best-of-the-best benchmarking can lead to breakthrough improvement by causing individuals within firms to look at other industries and to visualize the business in new and different ways.

An example of best-of-the-best benchmarking involved a visit by upper management of a large hotel chain to Scandinavian Airlines (SAS) to observe how SAS served customers. By doing this, they also were able to generate ideas for efficient use of space in hotel rooms (e.g., design of kitchenettes and rest rooms). Such synergy may lead to radical improvement and change.

One of the greatest impediments to seeking a benchmarking relationship is the belief that an initiating company or firm is so unique that it cannot learn from other firms because they are different. However, employees entrenched in a process may have trouble envisioning another way to perform work. Therefore, process improvement may be based on the same old ingrained philosophy of production for the particular firm. To break this mold, it is useful to see other philosophies and ideas. This is not unlike going to a university, where students are faced with new ideas and have to process various viewpoints to mature. Just as the student benefits from exposure to a variety of disciplines, such as psychology, philosophy, engineering, mathematics, sociology, anthropology, and the humanities, a business should benefit by exposing its employees to other business entities.

BUSINESS PROCESS BENCHMARKING

Many benchmarking professionals believe that the most important type of benchmarking is business process benchmarking. There is good reason for this. Suppose that you compare your customer service quality with a competitor's. You find out that your competitor is using a standard survey instrument to gauge customer satisfaction. You then administer the survey to your customers and find out that on a five-point scale your company rates a 4.3 and your chief competitor rates a 4.7 (with equal dispersion). Now what do you do with this information? The fact that your score differs from your competitor's tells you nothing about how the competitor is achieving these higher scores. To understand how your competitor has achieved these scores, business process benchmarking is necessary.

Business process benchmarking is based on the concept of **5w2h** developed by Alan Robinson.[4] The 5w2h concept is labeled as such because a business process benchmarking project should result in the answers to seven questions (see Table 6-3). Five of these questions begin with the letter *w* (*w*ho, *w*hat, *w*hen, *w*here, *w*hy), and the remaining two questions begin with the letter *h* (*h*ow and *h*ow much).

The 5w2h concept is a good starting point because it focuses the participants in the benchmarking process on the nuts and bolts of what is being done. If the initiator organization can answer the 5w2h questions at the end of a benchmarking process, then information will be in place that could, for instance, help a company improve its customer satisfaction from 4.3 to 4.7 or beyond.

The 5w2h questions should be viewed in the context of a process. Figure 6-3 contains a diagram of a generic process. In a broad sense, inputs include the equipment, people, machines, materials, and design that combine to form a product or service. The inputs are combined in what is known as the **conversion process**. In the conversion process, we align the inputs together to form the product or service. Conversion

[4]Robinson, A., *Continuous Improvement in Operations: A Systematic Approach to Waste Reduction* (Cambridge, MA: Productivity Press, 1991), p. 245.

TABLE 6-3 Business Process Benchmarking Questions

Type	*5w2h*	*Description*	*Countermeasure*
Subject matter	What?	What is being done? Can this task be eliminated?	Eliminate unnecessary tasks.
Purpose	Why?	Why is this task necessary? Clarify the purpose.	
Location	Where?	Where is it being done? Does it have to be done there?	Change the sequence or combination.
Sequence	When?	When is the best time to do it? Does it have to be done then?	
People	Who?	Who is doing it? Should someone else do it? Why am I doing it?	
Method	How?	How is it being done? Is this the best method? Is there some other way?	Simplify the task.
Cost	How much?	How much does it cost now? What will the cost be after improvement?	Select an improvement method.

A number of simple guidelines have been developed to help people or groups generate new ideas. In general, these guidelines urge you to question everything, from every conceivable angle.
SOURCE: Reprinted from *Continuous Improvement in Operations: A Systematic Approach to Waste Reduction*, edited by Alan Robinson. Copyright © 1991 by Productivity Press, 800-394-6868, *www.productivityinc.com*.

processes might include turning sheet steel into chimney pipes or turning wood pulp into paper or, for a shipping service, moving materials from point A to point B.

The conversion process results in outputs that are eventually sold to customers. Notice in Figure 6-3 that there are two feedback loops in the process. The first feedback loop results from gathering data from the process. This is known as the **control process**. As was explained in Chapter 1, the control process involves gathering, analyzing, and using the data to adjust the process. This is often the result of using process control charts.

The second feedback loop in the process is the customer feedback loop and results from gathering data from the customers. Using these data, the control process is improved to give the customer greater satisfaction. A third feedback process can be added by gathering information from competitors through benchmarking.

Figure 6-4 shows what to benchmark at each stage. Again, the questions asked are similar to the 5w2h questions posed earlier.

FIGURE 6-3 Process Model

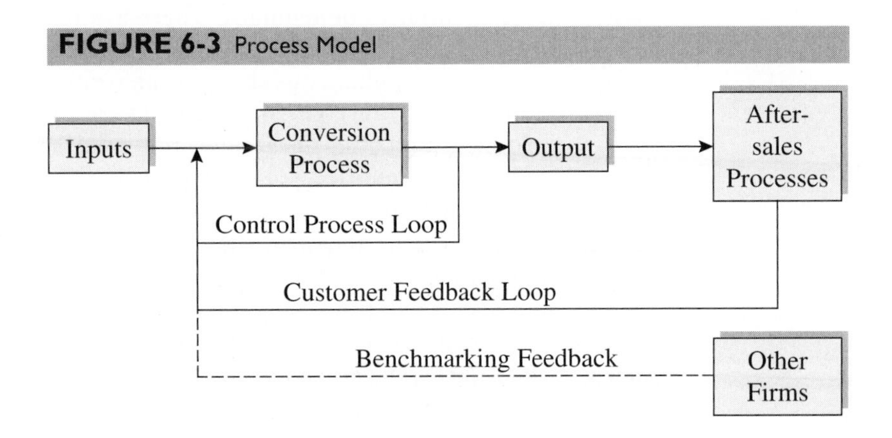

FIGURE 6-4 Process Information

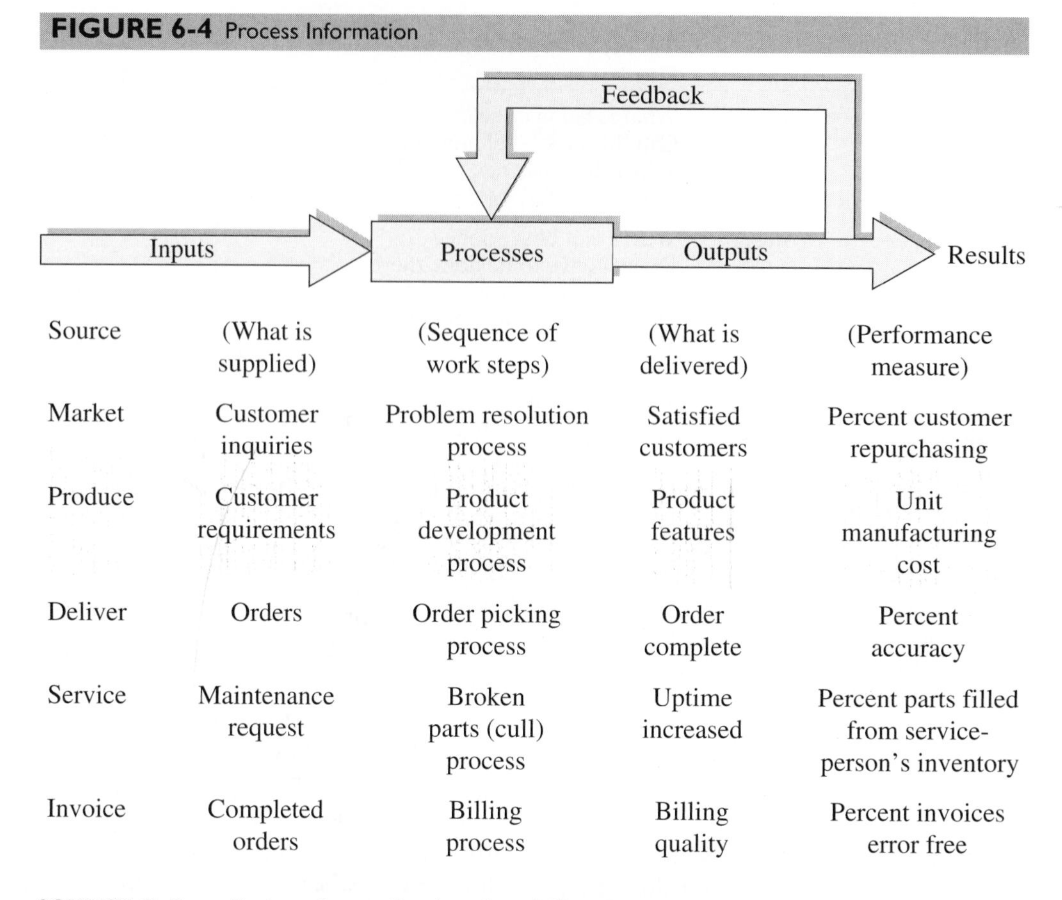

Source	(What is supplied)	(Sequence of work steps)	(What is delivered)	(Performance measure)
Market	Customer inquiries	Problem resolution process	Satisfied customers	Percent customer repurchasing
Produce	Customer requirements	Product development process	Product features	Unit manufacturing cost
Deliver	Orders	Order picking process	Order complete	Percent accuracy
Service	Maintenance request	Broken parts (cull) process	Uptime increased	Percent parts filled from service-person's inventory
Invoice	Completed orders	Billing process	Billing quality	Percent invoices error free

SOURCE: R. Camp, *Business Process Benchmarking* (Milwaukee, WI: ASQ Quality Press, 1995), p. 26. Reprinted with permission from the Quality Press. © 1995 American Society for Quality.

Video Clip: Benchmarking at Xerox

Robert Camp's Business Process Benchmarking Process

We have already defined process benchmarking. We now discuss the benchmarking process developed by Robert Camp. Xerox was an early adopter of benchmarking and has used benchmarking effectively to improve processes. This approach includes a formal 10-step process to benchmarking, as shown in Table 6-4

Step 1: **Decide what to benchmark.** There are innumerable areas for improvement in every company. Obviously, not all of these can be tackled at once. Too many simultaneous changes can result in a confused organization and actually can hurt performance. Therefore, one must prioritize those processes that offer the greatest potential for improvement. To do this, you must have identified your key processes and charted those processes for future analysis.

Step 2: **Identify whom to benchmark.** This involves identifying those competitors in your industry and those firms outside your industry that have outstanding results and processes for study. Those companies that have developed best practices are prime candidates for benchmarking.

Step 3: **Plan and conduct the investigation.** To complete this step, identify data to be collected. Next, a method is developed with the target firms to determine

TABLE 6-4 Benchmarking Steps at Xerox

1. Decide what to benchmark	7. Revise performance goals
2. Identify whom to benchmark	8. Develop action plans
3. Plan and conduct the investigation	9. Implement specific actions and monitor progress
4. Determine the current performance gap	10. Recalibrate the benchmarks
5. Project future performance levels	
6. Communicate benchmarking findings and gain acceptance	

how the data are to be collected. Once the data are selected, observe the practices of the target firm and document the best practices that are observed.

Step 4: **Determine the current performance gap.** Once you have collected the data about the processes, a determination is made as to which processes in the initiator company have the greatest performance gap with the target company. Brainstorming can help make this determination. This step helps to prioritize which areas are the first candidates for change and improvement.

Step 5: **Project future performance levels.** Predict whether the performance gap for the benchmarked processes will narrow or widen in the coming years. If the performance gaps are likely to widen in the future, project how this will affect the company.

Step 6: **Communicate benchmarking findings and gain acceptance.** This step begins the communication of the benchmarking findings to those who will be affected by the results. This can be done through meetings and written media. Communication keeps the process as open as possible, which minimizes uncertainty and fear.

Step 7: **Revise performance goals.** Once best practices are identified, operational goals are revised. This establishes measures for evaluating the results of improvements based on benchmarking.

Step 8: **Develop action plans.** Action plans are the specific steps and objectives for implementation. Assignments to personnel and timetables for implementation are used with a project management approach.

Step 9: **Implement specific actions and monitor progress.** As implementation proceeds, progress is reported to management and stakeholders.

Step 10: **Recalibrate the benchmarks.** Continued benchmarking with the best firms helps to identify new best practices. This "raises the bar" for higher levels of future performance.

Figure 6-5 combines these ten benchmarking steps into five phases of development: *planning, analysis, integration, action*, and *maturity*. The steps they contain define all these phases. The maturity phase is characterized by achieving market and world leadership in some aspect of your processes. At this stage, benchmarking is an ongoing process that supports continual improvement efforts.

LEADING AND MANAGING THE BENCHMARKING EFFORT

Like other quality management efforts, benchmarking is a managed process. Therefore, management must have an understanding of the benchmarking process, the participants involved, and the objectives of the exercise. Robert Camp specifies a management

FIGURE 6-5 Camp's Five Phases

Phase 1: Planning—Identify what to benchmark; identify whom to benchmark; and gather data.
A plan for benchmarking is prepared.
- Decide what to benchmark
- Identify whom to benchmark
- Plan the investigation and conduct it
  - —Gather necessary information and data
  - —Observe the best practices

Phase 2: Analysis—Examine the performance gap and project future performance.
The gap is examined and the performance is assessed against best practices.
- Determine the current performance gap
- Project future performance levels

Phase 3: Integration—Communicate the findings and develop new goals.
The goals are redefined and incorporated into the planning process.
- Communicate benchmarking findings and gain acceptance
- Revise performance goals

Phase 4: Action—Take actions, monitor progress, and recalibrate measures as needed.
Best practices are implemented and periodically recalibrated as needed.
- Develop action plans
- Implement actions and monitor progress
- Recalibrate the benchmarks

Phase 5: Maturity—Achieve the desired state.
Leadership may be achieved.
- Determine when leadership position is attained
- Assess benchmarking as an ongoing process

SOURCE: R. Camp, *Business Process Benchmarking* (Milwaukee, WI: ASQ Quality Press, 1995), p. 26.

process for business process benchmarking that can be generalized to the different approaches to benchmarking specified in this chapter (see Chapter 1).

Managing the benchmarking process involves establishing, supporting, and sustaining the benchmarking program. To begin the management process, a strategy statement outlining the goals and strategies to be used is developed (see A Closer Look at Quality 6–1).

With the strategy statement in place, management sets expectations for performance relating to the benchmarking project. At a minimum, the expectations for benchmarking are that this is an ongoing process that serves as a basis for improvement (not a one-time event) and that there are specific deliverables that are identified by management that must be fulfilled.

Other activities for management include providing management awareness training, establishing a benchmarking competency center, developing guidelines for information

A CLOSER LOOK AT QUALITY 6-1

TOYOTA: BENCHMARKING AMERICAN AUTO COMPANIES (BEFORE IT WAS CALLED BENCHMARKING)[a]

We talk a lot about how we have learned much from Toyota. It is interesting to look to the past and find that Toyota used benchmarking to get started. This included reverse engineering of American automobiles.

One was the zealous mentor, the scion of a prominent clan with a vision that stretched far beyond the family business. The other was the younger cousin, a protégé who knew how to make that frustrated vision come alive. Both thought they had the makings of a great company. But on that day in 1938 when Kiichiro Toyoda, the founder of Toyota Motor Corporation, instructed his understudy, Eiji, to build a factory on land cleared from a red pine forest in central Japan, neither realized they were about to make history. That plant, located in what is now called Toyota City, pioneered concepts such as lean manufacturing, *Kaizen*, and *kanban*. All of these approaches are now used from Detroit to Sweden to Korea.

Early on, Kiichiro Toyoda spent his days reverse engineering Chevrolet engines. By the late 1930s he had launched his own company. By 1935 the A1 was launched (a copy of the Chrysler DeSoto Airflow). Kiichiro shortened his supply chain so parts were delivered just in time for assembly.

In 1947, Eiji became managing director and was sent to the United States to study Ford at the Rouge River plant. He came back impressed by Ford's scale but scornful of their inefficiencies. So he and veteran machinist Taiichi Ohno came up with the *kanban* system of labeling, an early precursor to bar codes, to keep the flow of parts smooth. They also perfected the art of *kaizen*. Coming full circle, it is fair to say that every American automotive executive has made a pilgrimage to Toyota City to see how they do it.

[a]Adapted from C. Dawson, "Blazing the Toyota Way," *Business Week*, May 24, 2004, p. 22.

sharing, and overseeing the development of a visit protocol. Once this benchmarking infrastructure is in place, benchmarking teams are commissioned and trained to perform the benchmarking project. Table 6-5 outlines additional management process steps that are included in the Camp benchmarking model. Some or many of these steps may be collapsed or simplified for smaller firms involved in benchmarking processes. Admittedly, this process is a very mature and successful benchmarking process.

Training

Training is the key to success in all quality management approaches. This is especially true for benchmarking. Participants must have project management skills and be familiar with benchmarking approaches and protocols. Benchmarking carries with it legal liabilities that should be addressed during the training (see A Closer Look at Quality 6-2). Training should include managerial training, cross-functional benchmarking skills training, team training, and documentation training (flowcharting). Many of these training courses are available from many different consulting organizations. It is best to obtain training from organizations that are experienced with benchmarking. Many companies have established external training arms that can be hired by competing firms.

Outside consultants can be useful in providing training and coaching on benchmarking and other quality-related efforts. However, it is best for the firm that is interested in benchmarking to perform the benchmarking with its own employees rather than through consultants because a self-perpetuated approach creates the best plat-

TABLE 6-5 Process Benchmarking Management Process

1. Strategy statement	10. Document process
2. Set expectations	11. Qualify partners
3. Management and training	12. Follow 10-step process
4. Benchmarking competency center	13. External assistance
5. Guidelines and protocol	14. Benchmarking handbook
6. Network	15. Share successes
7. Champions identification	16. Inspect for benchmarking use
8. Establish benchmarking teams	17. Rewards
9. Team skills training	

form for organizational learning. It may take longer; however, it probably will be more useful to do it yourself.

BASELINING AND REENGINEERING

Many firms are facing the important decision to reengineer business processes to enhance productivity, to reduce cost, to improve quality, and to achieve better customer service. *Reengineering* (see Chapter 2) is defined as a fundamental rethinking and redesign of business processes. Such change is often accompanied by the automation of business processes. Two factors are critical to achieving success through reengineering: breadth and depth. *Breadth* refers to the impact of the reengineering process to the entire organization. *Depth* refers to organizational elements such as responsibilities, measurements, information technology, and skills. These dichotomous factors imply that business process reengineering will result in both localized (such as the affected work areas) and generalized (or organization-wide) performance impacts. However, these impacts are difficult to assess. Therefore, appropriate, effective methodologies are needed to assess the impacts of business process reengineering.

A methodology that can be applied in assessing business process reengineering impacts is baselining. **Baselining**, which has been discussed in other chapters, requires the monitoring of key internal firm performance measures over time to identify trends such as improvement (or decline) to inform managerial decision making. The baselining process involves identifying measures, establishing time frames for future data collection, gathering data, and analyzing data on an ongoing basis to identify performance trends and changes (see Figure 6-6).

FIGURE 6-6 Baselining Process

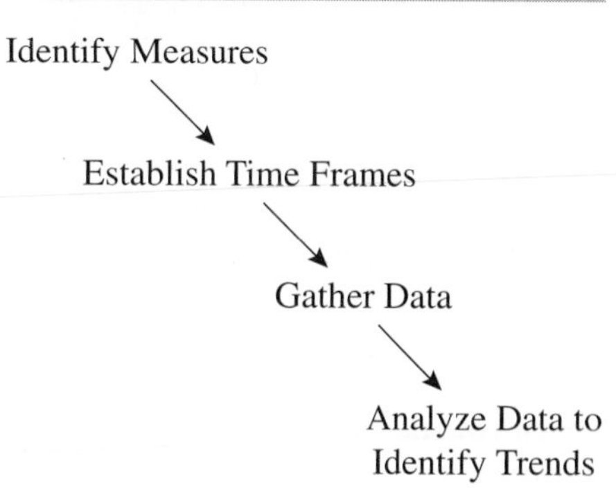

A CLOSER LOOK AT QUALITY 6-2

THE LEGAL ENVIRONMENT OF BENCHMARKING

Legal issues such as antitrust, intellectual property, and trademark issues are central concerns for firms entering into benchmarking agreements. To overcome these issues, benchmark practitioners should follow a code of conduct. One such code of conduct for benchmarking has been proposed by the Strategic Planning Institute Council on Benchmarking.[b]

THE BENCHMARKING CODE OF CONDUCT

Principles

To contribute to efficient, effective, and ethical benchmarking, individuals agree for themselves and their organization to abide by the following principles for benchmarking with other organizations:

- **Principle of legality.** Avoid discussions or actions that might lead to or imply an interest in restraint of trade: market or customer allocation schemes, price fixing, dealing arrangements, bid rigging, bribery, or misappropriation. Do not discuss costs with competitors if costs are an element of pricing.
- **Principle of exchange.** Be willing to provide the same level of information that you request in any benchmarking exchange.
- **Principle of confidentiality.** Treat benchmarking interchange as something confidential to the individual and organizations involved. Information obtained must not be communicated outside the partnering organizations without prior consent of participating benchmarking partners. An organization's participation in a study should not be communicated externally without their permission.
- **Principle of use.** Information obtained through benchmarking partnering should only be used for the purpose of improving operations within the partnering companies themselves. External use or communication of a benchmarking partner's name with their data or observed practices requires permission of that partner. Do not, as a consultant or client, extend one company's benchmarking study findings to another without the first company's permission.
- **Principle of first-party contact.** Initiate contacts, whenever possible, through a benchmarking contact designated by the partner company. Obtain the permission of the contact before relaying any information or delegating any responsibility to other parties.
- **Principle of third-party contact.** Obtain an individual's permission before providing their name in response to a contact request.
- **Principle of preparation.** Demonstrate commitment to the efficiency and effectiveness of the benchmarking process with adequate preparation at each process step; particularly, at initial partnering contact.

Remember:

- Keep it legal.
- Be willing to give what you get.
- Respect confidentiality.
- Keep information internal.
- Use benchmarking contacts.
- Don't refer without permission.
- Be prepared at the initial contact.

Etiquette and Ethics

Benchmarking partners must make themselves vulnerable to one another in order to gather useful information. Therefore, a level of trust must be established. The following general guidelines should be followed by both partners in a benchmarking encounter:

- Establish specific ground rules up front (e.g., "We don't want to talk about those things that will give either of us a competitive advantage, rather, we want to see where we both can mutually improve or gain benefit").
- Do not ask competitors for sensitive data or cause the benchmarking partner to feel that sensitive data must be provided to keep the process going.
- Use an ethical third party to assemble and blind review competitive data, with inputs from legal counsel, for direct competitor comparisons.

[b] *www.orau.gov/pbm/pbmhandbook/apqc.pdf*, 2005.

Continued

- Consult with legal counsel if any information gathering procedure is in doubt before contacting a direct competitor.
- Any information obtained from a benchmarking partner should be treated as internal, privileged information.

There are also some general rules of etiquette to be followed. On the one hand, you should never disparage a competitor's business or operations to a third party, nor should you attempt to limit competition or gain business through the benchmarking relationship, nor misrepresent yourself as working for another employer.

On the other hand, you should always know and abide by the benchmarking code of conduct; have a basic knowledge of benchmarking and follow a benchmarking process; have determined what to benchmark, identified key performance variables, recognized superior performing companies, and completed a rigorous self-assessment; have developed a questionnaire and interview guide and shared these in advance if requested; have authority to share information; work through a specified host; and have mutually agreed on scheduling and meeting arrangements. For face-to-face visits, the following guidelines are essential: Provide a meeting agenda at least a week in advance; be professional, honest, courteous, and prompt; introduce all attendees and explain why they are present; adhere to the agenda and maintain the focus on benchmarking issues; use language that is universal, not a company's specialized jargon; do not share proprietary information without prior approval from the proper authority of both parties; and share information about your benchmarked processes.

Because business process reengineering affects multiple levels in a firm, abrupt organizational changes are reflected in baselining results. If changes are effective in improving the organization, the data collected will reflect this improvement.

PROBLEMS WITH BENCHMARKING

Benchmarking is not a simple activity, and it can be difficult to implement any of these tools and concepts. There are four key problems with benchmarking:

1. There may be substantial difficulty obtaining cooperation from other firms in your own industry—unless you happen to be a Fortune 100 firm. Most organizations have much less clout. The thing to remember is *reciprocity*. To be effective, you must have something to offer the target firm in return for sharing information. For example, a small firm may feel that it has little to offer a larger firm. If the small firm is flexible, the Goliath firm may want to learn how to achieve that flexibility.
2. The predominance of functional benchmarking with firms in noncompeting industries makes it difficult to benchmark with these firms. It takes much ingenuity to properly identify benchmarks from noncompeting firms. This is where the business and industry literature is very helpful in trying to identify benchmarking firms in noncompetitive industries.
3. Your efforts will be wasted unless you fully understand your own processes before you benchmark someone else. Using tools such as business process maps, it is possible to identify the exact performance measures and metrics needed from the target firm.
4. Benchmarking is time-consuming and costly. A recent benchmarking project between two firms took nearly two years to complete and cost each of the firms more than $250,000. Of course, this was a major benchmarking project. Other projects have been completed at lower costs. However, some have been much

more expensive. Costs of benchmarking include things like time for planning, travel, documentation, and implementation. By far, the largest costs are associated with implementation. The investment is lost if benchmarking data are not used to drive improvement.

SUMMARY

In this chapter we discussed an important method for listening to the voice of the market. We discussed the purposes of benchmarking. Most often companies benchmark processes. They also benchmark many performance measures such as productivity.

The goal of benchmarking is to become best in class and then best of the best. Benchmarking is more effective for firms that have been pursuing quality and process improvement over time. This is certainly not a starting point for quality improvement efforts.

Remember that the use of data and measures can result in undesired outcomes as individuals attempt to exploit the measurement system to reflect well on them. Baselines and other measures should be implemented carefully with attention to the possibility of unintended outcomes.

KEY TERMS

- Baselining
- Benchmark
- Best in class
- Best of the best
- Control process
- Conversion process
- Critical success factors
- Customer-related results
- Financial benchmarking
- Financial ratios
- 5w2h
- Functional benchmarking
- Human resources measures
- Initiator firm
- Key business factors (KBFs)
- Key measures
- Management by fact
- Market-share data
- Operating results
- Performance benchmarking
- Process benchmarking
- Product benchmarking
- Productivity ratios
- Quality measures
- Reverse engineering
- Strategic benchmarking
- Structural measures
- Target firm

DISCUSSION QUESTIONS

1. Describe the concept of benchmarking. Provide an example of how a restaurant that you are familiar with could use benchmarking to improve its performance.
2. What is a benchmark firm? Why is it good practice for a benchmark firm to open its doors and allow others to view its operations and tour its facilities?
3. What are the pros and cons of becoming a benchmark firm? If you were the manager of a highly successful company, would you want other companies benchmarking against your firm? Why or why not?
4. In the context of benchmarking, describe the distinction between an initiator firm and a target firm.
5. Describe how benchmarking can be used by a firm to externally validate the value of its present business practices.
6. Describe the concept of process benchmarking. How does process benchmarking improve a company's overall business performance?
7. Is the growing popularity of the Internet a positive development or a negative development for the future of benchmarking? Please explain your answer.

8. Compare and contrast process benchmarking, product benchmarking, and strategic benchmarking.
9. When benchmarking, what is the primary hazard in comparing measures across companies to gauge performance differences?
10. How does a firm's key business factors help direct its benchmarking program?
11. Provide several examples of the types of measures that are often gathered in benchmarking studies. How does a firm determine which measures should be included in its benchmarking program?
12. Imagine yourself in the role of the CEO of a medium-sized auto parts company. Briefly describe how you would set up a benchmarking program. Include in your description an analysis of how you would determine "what" to benchmark and how you would determine "whom" to benchmark against.
13. What does the term *best in class* mean? Describe a firm that you consider best in class. Would other firms benefit from benchmarking against the firm that you selected in this particular area? Please explain your answer.
14. What does the term *best of the best* mean? Why can best-of-the-best benchmarking lead to breakthrough improvements?
15. Consider the following: "Despite its many advantages, benchmarking is not an appropriate technique for all firms. Some firms are so unique that it is impossible for them to find benchmark firms that experience the same challenges they do. As a result, it is impractical for these firms to participate in benchmarking." Do you agree or disagree? Please explain your answer.
16. Many scholars believe that business process benchmarking is the most important type of benchmarking. Do you agree with this particular belief? Why or why not?
17. Describe the concept of 5w2h.
18. Briefly describe Camp's 10-step process to benchmarking.
19. What management activities can be undertaken to support a firm's benchmarking activities? Please make your answer as substantive as possible.
20. What are some of the pluses and minuses of the benchmarking code of conduct? Would you add any other points to the code? If so, what?
21. Although benchmarking is a popular management technique, many firms are not engaged in the benchmarking process. Why do you think that some firms avoid benchmarking? Are any of the reasons valid? Why or why not?

PROBLEMS

1. A company has gathered the following financial information for itself and a competing firm. They wish to compare productivity for the two firms (all numbers in 000s).

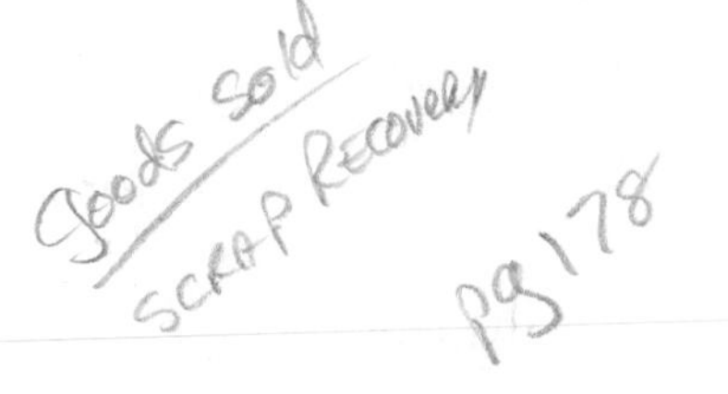

	Firm A	Firm B
Labor	$30,000	$16,000
Plant and equipment	200,000	150,000
Energy	17,500	20,500
Materials	200,000	180,000
Sales	1,200,000	900,000

a. Compute partial and total factor productivity measures for Firms A and B.
b. What is the picture you get of the two firms?
c. What would you suggest to the management of Firm B?

2. For the firms in Problem 1, you have been asked to make a report to the management of Firm A. What are some of the caveats relating to accounting practices that you would include in the report? Which numbers should be interpreted cautiously?

3. For the data in Problem 1, suppose that Firm B is a foreign firm. What additional caveats would you place on interpretation of the data?
4. A domestic company operating a subsidiary in an LDC (less-developed country) has shown the following financial results:

	Parent (Domestic)	Subsidiary (LDC)
Sales (units)	200,000	80,000
Labor (hours)	40,000	60,000
Materials (currency)	$40,000	FC40,000
Equipment (hours)	120,000	10,000

   a. Calculate partial labor and capital productivity numbers for the parent and subsidiary. Interpret the results.
   b. Compute total labor productivity figures. Does your interpretation change?
   c. If $1 = 10 FC units, calculate materials productivity figures. Explain your finding

5.

	Marketing	Manufacturing Engineering	Supply Chain
Aerospace	4.8	12.5	10.0
Automotive	8.5	5.0	6.0
Communications	11.6	11.8	20.0
Components	3.0	2.4	5.5
Computer	5.0	8.0	12.5
Electromechanical	6.3	6.0	10.0
Electronic Subsystems	4.2	4.5	8.0
Heavy Mechanical	6.0	9.0	9.0
Instrumentation	7.5	4.8	6.1
Light Assembly	2.8	2.5	3.3
Medical	3.8	4.0	13.4
Cross-Industry	4.9	6.0	8.2

   Above are industry comparisons of the percent of sales spent on salaries for marketing employees, manufacturing engineering employees, and supply chain employees for several different industries. Rank these industries by the amount spent in each of the areas and report your findings to management. What can you infer from your findings about each of the industries?
6. Using the data from Problem 5, suppose you are a computer company who spends 9% on marketing staff, 6% on manufacturing engineering, and 8% on supply chain staff. Comparing your expenditures with the above averages, what would you recommend to management? What more information would you want?
7. Following is an announcement from a study from The Benchmarking Network, Inc.

5th Annual Shared Services Measures

The Benchmarking Network and the Shared Services Benchmarking Association announced they will be kicking off **the fifth Annual Shared Services Measures** benchmarking study. Now is the time to join and become involved in setting the focus and direction of the study by attending the kickoff meeting. The study will review **Shared Services Measures** including research into:

- Accounting;
- Finance and Treasury;

- Human Resources;
- Information Technology;
- Procurement and Supply Chain;
- Legal;
- Regulatory;
- Auditing;
- Corporate Communications/External Affairs;
- Facilities, Real Estate, Security;
- Environmental;
- Fleet; and
- Other Shared Services measures.

Put yourself in the place of the Benchmarking Network researchers. How would you design this study? What would you have to consider in undertaking such a study?

8. Below are enrollment data from several British universities.

	Change in student numbers					*Supporting data 2000/01 Student numbers*	
	99/2000 to 2000/01		*98/99 to 99/2000*				
	FT	*PT*	*FT*	*PT*	*ALF*		
Institution	%	%	%	%	£	FT	PT
Abingdon and Witney College	b	b	0	−7	b	b	b
Accrington and Rossendale College	−16	−1	7	−1	17.22	1246	8896
Alton College	2	34	11	10	17	1478	1522
Amersham and Wycombe College	−4	−6	5	21	17.07	1762	5064
Aylesbury College	−14	11	−5	−2	22.63	800	3117
Barking College	−6	6	2	−68	17.05	2193	8638
Barnet College	−7	−2	5	−22	17	4383	13189
Barnfield College	3	7	7	24	17	2601	15480
Barnsley College	−3	−85	−9	−24	17	3084	6029
Basildon College	−8	−43	−12	41	17	583	5695
Basingstoke College of Technology	−4	2	−6	8	17	1463	7670
Bedford College	−1	19	−3	−7	17.32	1855	8524
Beverley College of Further Education	17	6	31	4	17	851	3972

ALF = Average Level of Funding
FT = Full Time
PT = Part Time
b = not available

Compare the data from these schools and present them in a way that is useful for the administrators in these schools.

9. Suppose the chancellor of Bedford College asks you to infer meaning from the data in Problem 8. Put together a report for the chancellor explaining how Bedford is performing relative to the other schools.
10. Choose a company in your local area. Develop a list of five companies among the best in class in their industry and the best of the best. Explain how you chose the benchmarking

targets you chose. Identify a list of 20 benchmarking questions. What kind of data would you select? How would you contact the target firms?

11. Following are financial statements from American Ecology. Studying these figures, what are some possible financial benchmarks this firm might want to develop?

American Ecology Corporation Consolidated Statements of Operations: Proforma and Unaudited ($ in 000s except per share amounts)

	2003	2004	2005
Revenues	38,960	41,522	49,972
Operating costs	23,545	23,219	33,571
Gross profit	15,415	18,303	16,401
Selling, general, and administrative expenses	15,702	21,909	24,187
Impairment loss on long-lived assets	—	—	7,451
Loss from operations	(287)	(3,606)	(15,237)
Investment income	(618)	(1,203)	(932)
Gain (or loss) on sale of assets	(72)	(136)	(55)
Other expense	(414)	(1,723)	(1,326)
Gain (or loss) before income taxes	817	(544)	(12,924)
Income tax expense (benefit)	55	132	(1,517)
Net income (or loss)	762	(676)	(11,407)
Preferred stock dividends	417	760	465
Net income (or loss) available to common shareholders	345	(1,436)	(11,872)
Basic earnings per share	0.03	(0.17)	(1.47) actual dollars
Diluted earnings per share	0.03	(0.17)	(1.47) actual dollars

12. Baseline the data from Problem 11. Is the company's performance improving or worsening?

Cases

Case 6-1 Amgen Corporation: Using Benchmarking as a Means of Coping with Rapid Growth

Amgen: ***www.amgen.com***

American Productivity and Quality Center: ***www.apqc.org***

Amgen Corporation is the largest biotechnology company in the world. Founded in Thousand Oaks, California, in 1980, the company produces life-saving pharmaceutical products based on advances in cellular and molecular biology. One of the company's products, EPOGEN, is a product that is used for the treatment of anemia associated with chronic renal failure and is one of the top-selling pharmaceutical products in the world.

Although Amgen is very good at developing cutting-edge pharmaceutical products, the company has struggled at times with the demands of managing a rapidly growing workforce. In the past 10 years, the company has grown from a workforce of 400 employees to 4,600. In particular, Amgen has faced challenges in terms of making the transition from a single-product company to a multiproduct company and in staffing and training its growing divisions. To deal with these challenges, Amgen decided to turn to benchmarking. The first department involved with the benchmarking initiative was sales training and development because the company's employees believed that improvement was needed in that area. There also was recognition within Amgen that training and development needed to be strengthened. Commenting on the merits of the first benchmarking study, Ellen Nichols, the director of the sales training and development department, said, "We'd been clearly focused on our products and customers, but perhaps we haven't had as focused an effort on developing our people."[5]

To conduct its first benchmarking study, Amgen solicited help from the American Productivity and Quality Center (APQC) in Houston, Texas. APQC is a nonprofit organization that provides benchmarking training, maintains a best-practice database, and helps firms locate businesses to benchmark against. APQC helped Amgen develop a list of 40 potential benchmark targets, based on Amgen's criteria for selection. Amgen wanted to benchmark against companies that were high-growth, high-tech, successful, and had a geographically dispersed sales force. Eventually, Amgen narrowed the list to seven companies and benchmarked its sales training and development department against similar departments at Dow Chemical, Lexus, Lucent Technologies, Motorola, Anheuser-Busch, Eastman Chemical, and IBM. Four of the companies invited Amgen personnel to visit their facilities, and the other three participants were interviewed over the phone. The study was successful, and Amgen redesigned its sales training and development department as a result of the benchmarking initiative.

Amgen has found other opportunities to use benchmarking to help cope with the challenges of rapid growth in a high-technology industry. For example, the company used benchmarking to study the way it moves its products from production to the end user, and it conducted a benchmarking study that examined its marketing practices. Through its experiences, Amgen has learned some of the barriers to effective benchmarking and some of the global benefits. In terms of barriers, Amgen learned that a benchmarking study is only as good as its implementation. If the implementation stage is lengthy or is not taken seriously by the people involved, the study will not be useful. Another potential barrier is finding companies that are similar enough to your company to benchmark against and that are willing to participate. Having a partner like the American Productivity and Quality Center that maintains a best-practice database to draw from helps address this concern. Amgen has learned there are also global benefits to benchmarking, beyond the specific ideas and techniques that are gleaned from a particular benchmarking study. Benchmarking helps a firm gain visibility and sends a clear message to its stakeholders that the firm is interested in continuous improvement. In addition, benchmarking helps a firm determine whether its best-in-class activities are truly best in class. Often, it is difficult for a firm to know just how good (or poor) it is at some activity until it measures its activity against a similar firm.

Amgen has been successful in its early experiences with benchmarking and serves as a model for how to initiate a benchmarking program. Although Amgen excels in the development of high-potential pharmaceutical products, it recognizes that it has a lot to learn about other management issues. Benchmarking has been a useful tool for the company in its learning efforts. ■

[5]Powers, V. J. (1997). "Amgen Succeeds with Benchmarking through Outside Facilitation, Hard-Working Teams," *American Productivity and Quality Center* 8 (October 1997): 1.

DISCUSSION QUESTIONS

1. Why was benchmarking so important for Amgen at the point in the company's history when benchmarking was initiated? Do you believe that benchmarking will contribute to Amgen's long-term success?
2. Was it a good idea for Amgen to solicit the help of the American Productivity and Quality Center? Do you think that Amgen would have been successful without the APQC's involvement?

Case 6-2 Ameritech: Making Benchmarking a Part of the Process Improvement Toolkit

Ameritech: *www.Ameritech.net*

Ameritech, now merged with SBC, is a regional phone operating company that offers local telecommunications services to people in Illinois, Indiana, Ohio, Michigan, and Wisconsin. The company has a rich history of innovation and customer service. For example, Ameritech was the first U.S. company to offer commercial cellular service. Today, the company offers a broad array of services to its customers and is an industry leader in productivity and financial performance.

Ameritech is firmly committed to benchmarking. As evidence of this, the company is a founder and active participant in the Telecommunications Industry Benchmarking Consortium, a group of 18 telecommunications companies that uses benchmarking to improve quality. To maximize the potential of its benchmarking efforts, Ameritech employs internal benchmarking experts and conducts benchmarking forums for its employees. Stories about successful benchmarking efforts in the company are always included in the forums. According to Orval Brown, the company's manager of business process architecture and benchmarking, "Other people's success stories (about benchmarking) get people interested and excited about the possibility of improvement."

There are three types of benchmarking that take place at Ameritech. The company has found each of these approaches helpful in the attainment of quality improvements:

Internal (best-of-breed) benchmarking involves a comparison of processes between different business units within Ameritech.

External (best-in-class) benchmarking involves finding companies to benchmark against, even if the company is in an unrelated industry.

Competitive (industry best) benchmarking involves benchmarking against a leader in Ameritech's industry.

Each of these types of benchmarking has unique challenges and rewards. Internal benchmarking is the simplest because there is typically no problem getting access to information. External benchmarking is more challenging because of access and confidentiality issues, but the rewards can be quite good. Competitive benchmarking is the toughest because a direct competitor typically will go only so far in terms of sharing information. The rewards, however, can be substantial.

Because Ameritech is multifaceted in its benchmarking efforts and because a lot of employees are involved, the firm uses a benchmarking code of conduct to maintain strict control of its benchmarking efforts. The benchmarking code of conduct (reflecting the guidelines from The Strategic Planning Institute Council of Benchmarking) is as follows:

- Keep it legal.
- Be willing to provide the same information you request.
- Respect confidentiality.
- Keep information internal for your use only.
- Initiate contact through benchmarking contacts.
- Don't refer possible benchmarking candidates without their permission.
- Be prepared at initial contact.
- Have a basic knowledge of benchmarking and follow the process.

This code of conduct provides a measure of continuity across the company's benchmarking efforts that is very helpful in developing a consistent set of benchmarking behaviors. It also helps ensure that benchmarking will in no way impinge on the ethical standards of the firm.

One component of the benchmarking process that Ameritech has become very good at over the years is finding suitable firms to benchmark against. Ameritech typically looks for the following types of companies to include in its benchmarking studies: companies that have received quality or business awards, companies with excellent financial results, companies with success stories published in major periodicals, and companies that are top-rated in their industries. These criteria help Ameritech select firms that have the highest potential to provide successful benchmarking results.

Ameritech is also keenly aware of the fact that the litmus test of any management initiative is effective implementation. The company typically takes the following steps in implementing what it learns from a benchmarking initiative:

Select implementation alternatives.
Assign resources and create a schedule.
Establish goals.
Develop a monitoring plan.
Gain appropriate approval to alter current practices.
Implement the plan.
Communicate the benchmarking findings.

This formalized process of implementation helps ensure that a benchmarking effort translates into actual change. There is nothing more discouraging for a progressive firm such as Ameritech than to complete a lengthy study only to have the results sit in a three-ring binder and never be used.

As the telecommunications industry continues to become more competitive, the company's reliance on benchmarking as part of its process improvement tool kit will probably become even more pronounced. ■

DISCUSSION QUESTIONS

1. Is Ameritech's benchmarking code of conduct a good idea? What types of ethical abuses potentially could occur in benchmarking if a firm did not follow a strict code of conduct?
2. This case describes benchmarking as an important part of Ameritech's "tools" of quality. What other tools of quality does benchmarking complement?
3. Ameritech is in a fast-paced, rapidly changing industry (i.e., telecommunications). Do you believe that benchmarking is particularly important in a fast-paced industry? Why or why not?

CHAPTER 7

Quality and Innovation in Product and Process Design

We all prefer copiers whose copies are clear under low power;
we all prefer cars designed to steer safely and predictably, even on
roads that are wet or bumpy. . . . We say the products are robust.
They gain steadfast customer loyalty.
— GENICHI TAGUCHI AND DON CLAUSING[1]

Have you ever needed a copy quickly but the copy machine was jammed? Have you ever worked against an impending deadline only to have a computer or a printer go haywire? Have you ever gotten into your car on a cold day to find it would not start? These annoyances are relatively minor. However, other examples of product failures can be catastrophic. If a lumberjack uses a defective chainsaw, he or she might lose an arm. If a heart monitor malfunctions, the results might be fatal to a patient. A race car driver's tires blow at 210 miles per hour, and a spinout results. A faulty fire alarm fails to alert the home's occupants until it is too late. These are major product failures that can result in severe injury or death.

In this chapter we focus on quality assurance. We have already learned that we cannot ensure quality at the final stages of inspection. After all, assurance is best achieved at the design stage.

DESIGNING PRODUCTS FOR QUALITY

In designing products, we must first answer many questions. For example, what are the functions the customer wants? What are the capabilities of current products? What are the limitations of the materials we have selected for the product? Are there better materials available? How much will the product cost to make? How much must the product cost to make it successful in the marketplace?

What does it mean to design products for quality? As we have already said, quality has many different dimensions. Reflecting on David Garvin's dimensions of quality from Chapter 1, it is clear that each dimension poses different design problems. Take the first dimension of performance. What are the critical attributes of performance? How much performance does the customer want? How much performance is overkill? How do we balance competing dimensions of performance (e.g., audio output versus distortion)? Similar questions could be constructed concerning any of Garvin's product quality dimensions. All these questions must be answered early in the design

[1]Taguchi, G., and Clausing, D., "Robust Quality," *Harvard Business Review* 68 (1990):65–75.

process. For example, if durability is an important quality dimension for the producer of an electronic calculator, then the design team might actively investigate new polymers to find a durable housing for the internal electronics. Computer chips are made of silicon because silicon is cheap and reliable. More recently, researchers have discovered an impressive array of materials that support transistor functions, including conducting polymers, organic crystal structures, compound inorganic systems such as gallium arsenide, and complex rare-earth ceramics that are high-temperature superconductors. All offer some unique advantage over silicon—either high-speed, low-power operation or ease of handling in manufacture. Time has shown the choice of silicon to be a good selection because of its low cost and wide availability. Someday computer memory might be held in electromagnetic fields known as "bubbles" or in some other state.

It might seem that materials choices are technical and should be made by engineers. However, engineers need input from marketing and operations to understand customer needs, marketing requirements, and the realities of production. Given free reign, engineers would design many products to the *n*th degree. If you don't believe this, look at a remote control for a VCR or a television set. You will never use many of the buttons on the remote. A study was performed by Sony Corporation to determine which buttons were actually used by the customer. Based on the results of the study, the remotes for Sony televisions were simplified.

THE DESIGN PROCESS

There are many different approaches to designing products. Even within the same industries, the approaches will vary in some important ways. Yet there are some similarities across the board. For example, design projects are likely to involve a project team rather than a single designer working independently. Preferably, these teams will work closely with customers to ensure that customer needs are met.

Figure 7-1 shows a generic approach to designing products. The design process includes nine phases that are interrelated. These stages begin with product idea generation and end with manufacture, delivery, and use. Project managers monitor design projects at each stage for cost and adherence to schedules.

Product idea generation is the first step. During this stage, external and internal sources brainstorm new concepts. Internal sources include marketing, management, research and development (R&D), and employee suggestions. The primary source for external product ideas is the customer. Original equipment manufacturers (OEMs) and contract manufacturers work closely with customers to develop new products. In other circumstances, customer needs are identified to generate product ideas. Other external sources for product ideas can be market-related sources such as industry experts, consultants, competitors, suppliers, and inventors. There are fundamental differences between R&D-generated ideas (known as *R&D push*) and marketing-generated ideas (known as *marketing pull*). R&D-generated ideas tend to be groundbreaking, risky, and technologically innovative. An example of R&D-based development was the Altair microcomputer. In the mid-1970s, the MITS (Micro Instrumentation and Telemetry Systems) Altair 8800 appeared on the cover of *Popular Electronics*. At the time, there was a very small market for this product. However, the article inspired two computer whizzes named Paul Allen and Bill Gates to develop a BASIC Interpreter for the Altair. The rest is history. Although there was not a large established market for personal computers, they have radically affected business and home life since their introduction.

FIGURE 7-1 Product Development Process

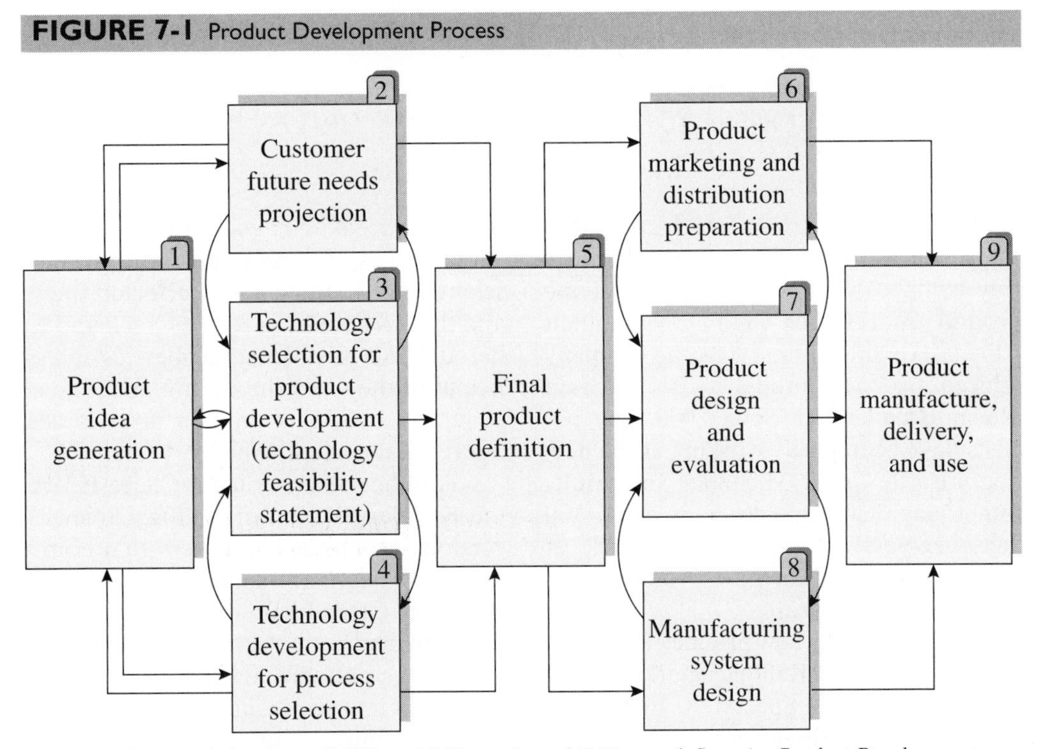

Adapted with permission from: C. Wilson, M. Kennedy, and C. Trammel, *Superior Product Development* (Blackwell Publishing, 1995).

Marketing-generated ideas tend to be more incremental—that is, they build on existing designs—and are better aligned with customer needs. For example, at the product idea-generation stage, a gap in the market or a customer need should be identified. Preliminary assessment of the marketability of the product is performed and funding provided for beginning development of a prototype of the product. Recent developments in computers have included technological developments such as improved multimedia capabilities and faster speeds as well as cosmetic changes in casings such as tablet designs and the use of clear plastics. These are marketing-oriented changes. As is shown in Quality Highlight 7-1, marketing-driven ideas have enhanced the bottom line.

Stage 2 is **customer future needs projection**. This uses data to predict future customer needs. Designers for Intel, the maker of the microprocessors for personal computers, have been masters at this. They have been able to project and introduce new products that are well timed to fit with changes in the technology requiring them. With the explosion of graphics in programs and on the Internet, Intel developed new chips to fit these needs. The company also has introduced these microprocessors at a rate that has not outstripped the ability of the market to absorb the new technology. At the same time, the company has been able to outpace competing microprocessor developers by staying slightly ahead of the technological curve.

The task of the product designer is to offer products with value that exceed customer needs at any point in time by careful planning and thought as to what future customer needs will be. There is no single approach to gathering information about future customer needs. Surveys might give insights, but they are usually insufficient to uncover emerging customer needs.

QUALITY HIGHLIGHT 7-1

A TURNAROUND AT KELLOGG'S CEREALS: DRIVEN BY DESIGN

www.KelloggCompany.com

Due to the leadership of Carlos Gutierrez, former CEO, Kellogg had regained the top spot in sales in the highly competitive cereal industry. Driving the turnaround is research and development (R&D). In the last decade of the 1900s, the company only introduced two new products—Nutri-Grain Bars and Raisin Bran Crunch. Since 2000, over 100 food scientists have been busily working at the new R&D center in Battle Creek, Michigan. The results are evident—today Kellogg generates over 100 new products per year! In addition, over 20% of Kellogg's sales are now from new products—up from less than 5% in prior years.

Here is one example of a new product: Special K with Red Berries. When Kellogg marketers in France decided to jazz up Special K by adding freeze-dried berries, headquarters took notice. Cereal makers had tried using freeze drying in the 1970s, but freeze drying was more primitive back then and wreaked havoc with moisture levels in the cereal boxes, turning flakes into mush. Now, food scientists at Kellogg have perfected the technology and the R&D folks in Battle Creek and Europe tinkered with a mix of tart raspberries and sweet cherries until they had the right blend. The cereal now generates over $100 million in sales according to figures that exclude Wal-Mart sales.

Another new product is Cheez-It Twisterz from the Keebler division of Kellogg. Spicier, crunchier versions of Cheez-It crackers that come in twisty shapes, Twisterz are packaged in new cube-shaped boxes that stand out on crowded snack shelves. They are made out of a type of extrusion technology, which uses pressure and heat to turn water inside the dough into steam, making it expand and turning the dough into a light and airy snack. Interestingly, this is the same process historically used to make Froot Loops cereal.[a]

[a]Adapted from "The Man Who Fixed Kellogg," *Fortune*, September 6, 2004, pp. 218–226.

During **technology selection for product development**, designers choose the materials and technologies that will provide the best performance for the customer at an acceptable cost. A **technology feasibility statement** is used in the design process to assess a variety of issues such as necessary parameters for performance, manufacturing imperatives, limitations in the physics of materials, special considerations, changes in manufacturing technologies, and conditions for quality testing the product. At this stage, preliminary work can be performed to identify key quality characteristics and potential for variability with each of the different materials.

Technology development for process selection means choosing those processes used to transform the materials picked in the prior step into final products. Careful technology selection of both automated and manual processes is key from a quality perspective because machinery, processes, and flows need to be developed that will result in a process insensitive to variations in ambient and material-related conditions.

Final product definition results in final drawings and specifications for the product with product families by identifying base product and derivative products.

Product marketing and distribution preparation are marketing-related activities such as developing a marketing plan. The marketing plan should define customers and distribution streams. The production-related activities are identifying supply-chain activities and defining distribution networks. Nowadays, this step often requires the design of after-sales processes such as maintenance, warrantees, and repair processes that occur after the customer owns the product.

Product design and evaluation requires definition of the product architecture, the design, production, testing of subassemblies, and testing of the system for production. A product design specification (PDS) demonstrates the design to be implemented with its major features, uses, and conditions for use of the product. The PDS contains product characteristics, the expected life of the product, intended customer use, product development special needs, production infrastructure, packaging, and marketing plans.

Manufacturing system design is the selection of the process technologies that will result in a low-cost, high-quality product. The selection of process technology is a result of projected demand and the finances of the firm. Processes must be stable and capable of producing products that meet specification. One of the major developments in this area is that firms now desire the ability to change over to new products with a minimum of cost associated with defects. In the past, it was considered standard operating procedure to produce a certain amount of bad product to prove that the system works. For example, a producer of stove pipe would process a small batch of pipe, inspect the pipe and then adjust the line, produce another small batch and reinspect, and so forth until they *proved* the process. This is no longer considered a cost-effective means of introducing new products.

Finally, **product manufacture, delivery, and use** finish this process. The consumer then enjoys the result of the design process.

QUALITY FUNCTION DEPLOYMENT (QFD)

When you have determined customer needs, those needs must be translated into functional product design. Quality function deployment (QFD) describes a method for translating customer requirements into functional design. Sometimes this process of translation is referred to as the *voice of the customer*. The quality function deployment approach was developed by Dr. S. Mizuno,[2] a former professor of the Tokyo Institute of Technology. Since then, this approach has been used extensively throughout the world. In the United States, Hauser and Clausing,[3] two MIT professors, were central in researching and publishing articles describing this approach.[4]

Designers need a means for implementing customer requirements into designs. The house of quality illustrated in Figure 7-2 shows how QFD is used to accomplish this. The left wall on the house of quality contains a listing of customer requirements. The roof on the house of quality lists technical requirements. We introduce QFD step by step so that you can see how a house of quality is developed and analyzed. Following are steps in performing a QFD:

1. ***Develop a list of customer requirements.*** The list of customer requirements includes the major customer needs as they relate to a particular aspect of a process. In Figure 7-3, a part of a QFD house of quality is shown with customer requirements for a restaurant. Customers want to have a clean restaurant, a comfortable seating arrangement, delicious food, and responsive servers.

---

[2]Mizuno, S., and Akao, Y., "QFD: The Customer-Driven Approach to Quality Planning and Development," Asian Productivity Organization, Tokyo, Japan, 1994.
[3]Hauser, J., and Clausing, D., "The House of Quality," *Harvard Business Review* (May–June, 1988):63–73.
[4]Ibid.

FIGURE 7-2 QFD Layout: The House of Quality

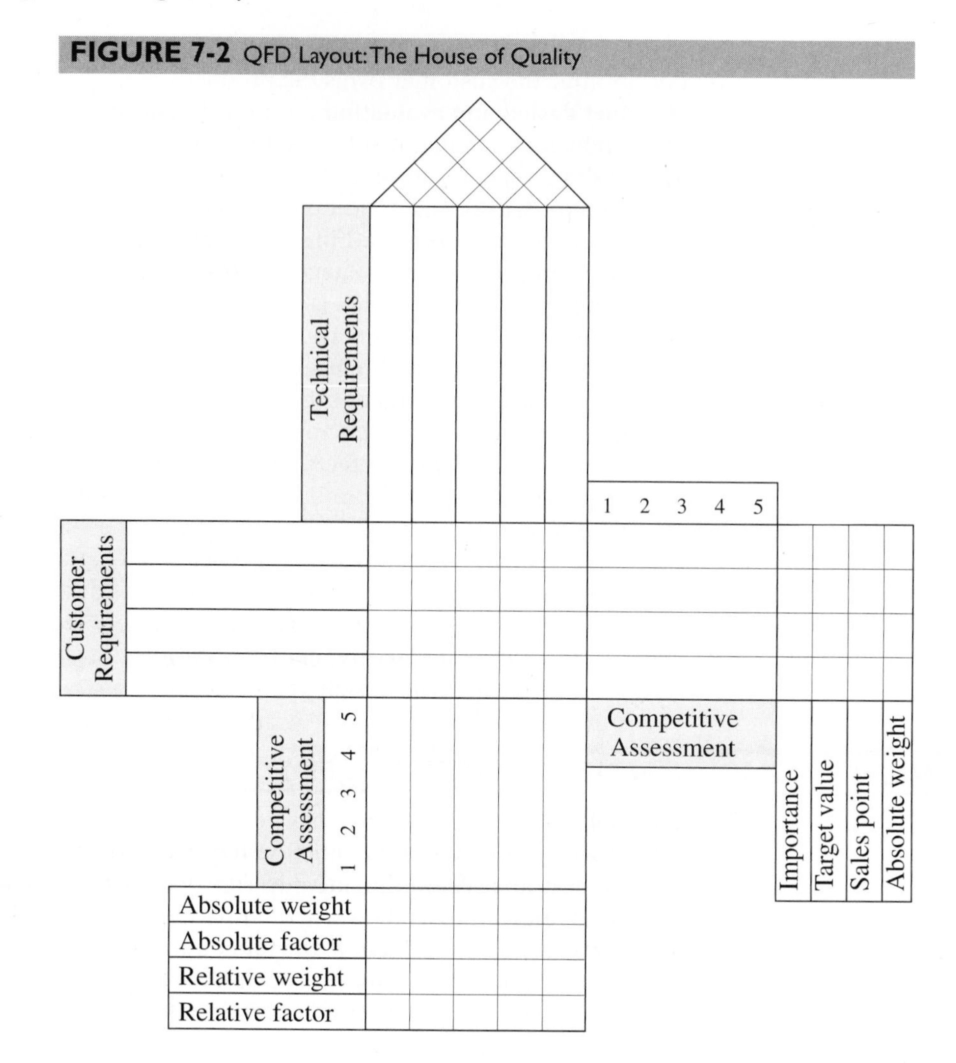

2. ***Develop a listing of technical design elements along the roof of the house.*** These are the design elements that relate to customer needs. Figure 7-4 shows the design elements for the restaurant that may affect the customers' requirements. These design elements are building materials such as type of tile, dirt resistance of floor tiles, material used in making seats, training for servers, and standardization of menu.

FIGURE 7-3 QFD Customer Requirements

Customer Requirements	
	Clean facilities
	Comfortable seating
	Delicious food
	Responsive servers

FIGURE 7-4 QFD Technical Requirements

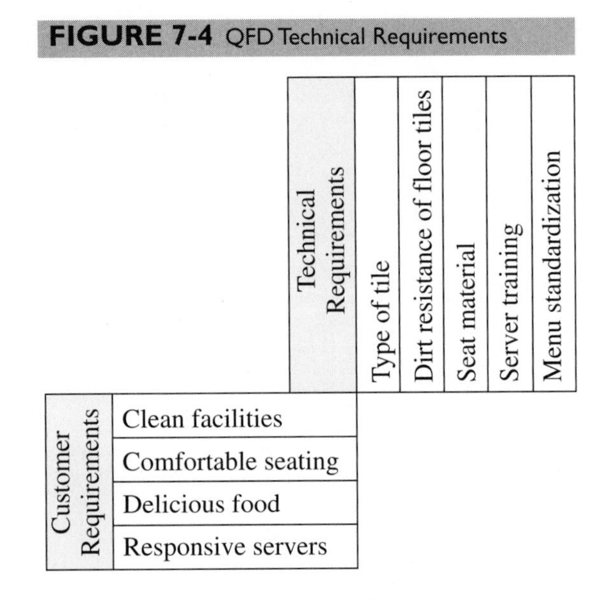

3. ***Demonstrate the relationships between the customer requirements and technical design elements.*** A diagram can be used to demonstrate these relationships. The symbols shown in Figure 7-5 are used, and scores are assigned relating to these symbols (i.e., 1, 3, and 9). Where 9 means strongly associated, 3 is somewhat associated, and 1 is weakly associated. Notice that tile and dirt resistance are strongly associated to clean facilities.
4. ***Identify the correlations between design elements in the roof of the house.*** Using the symbols identified in Figure 7-6, show whether different design elements are positively or negatively correlated. Positive and negative scores are assigned to each symbol as shown. Notice that seat material and type of tile are negatively related, whereas, type of tile is strongly positively related to dirt resistance. Server training and menu standardization are also strongly positively related.

FIGURE 7-5 QFD Technical Requirements and Customer Requirements Relationships

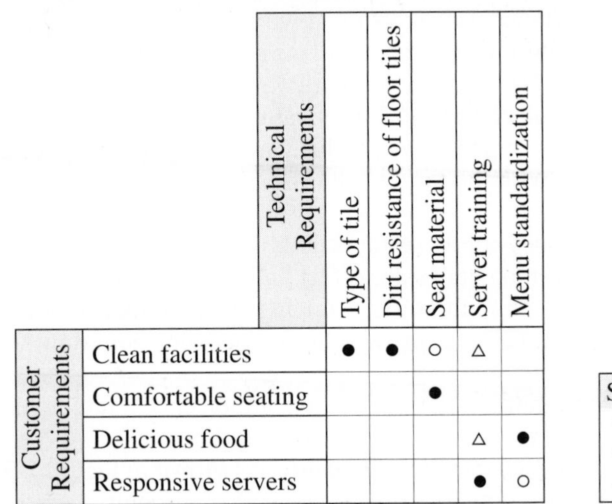

FIGURE 7-6 QFD Technical Requirements Interrelationships

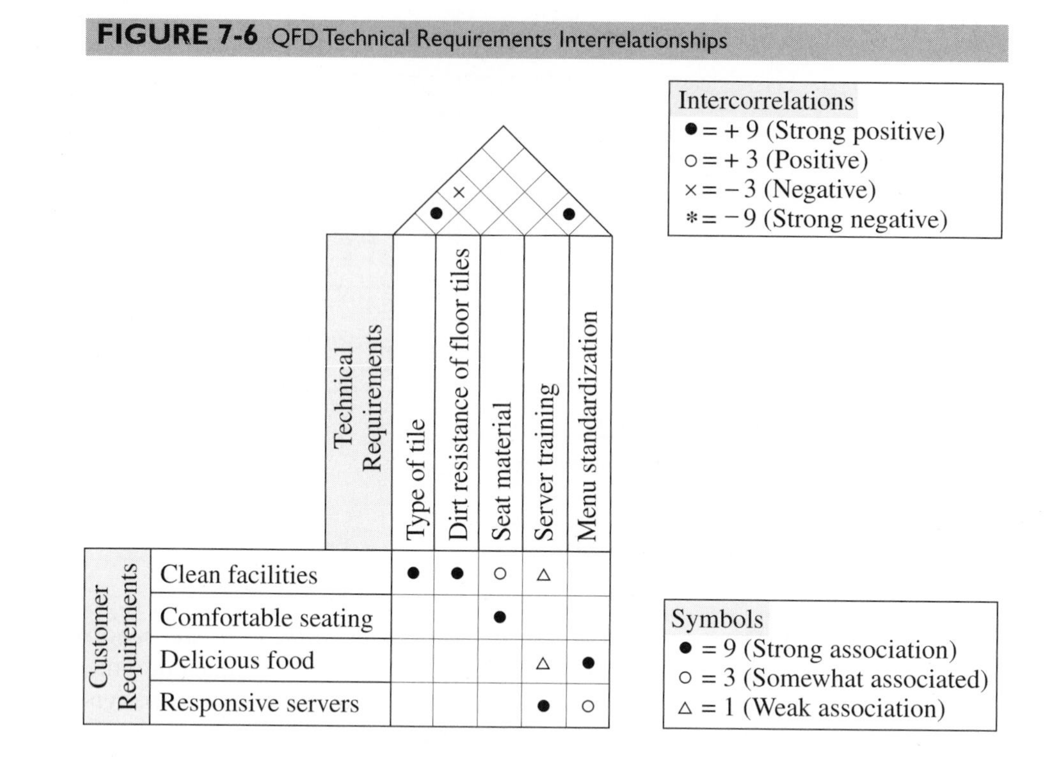

5. ***Perform a competitive assessment of the customer requirements.*** On both the right side and in the lower middle portion, of Figure 7-7, there is an assessment of how your product compares with those of your key competitors. These comparisons are on a five-point scale with five being high. A stands for competitor A, B means competitor B, and Us stands for the company in question. Note that there are two assessments, one for customer requirements and another for technical requirements.
6. ***Prioritize customer requirements.*** On the far right side of Figure 7-8 are customer requirements priorities. These priorities include importance to customer, target value, sales point, and absolute weight. A focus group of customers assigns ratings for importance. This is a subjective assessment of how critical a particular customer requirement is on a 10-point scale, with 10 being most important. Customer requirements with low competitive assessments and high importance are candidates for improvement. Target values are set on a 5-point scale (where 1 is no change, 3 is improve the product, and 5 is make the product better than the competition). With the target value, the design team decides whether to change the product.

   The sales point is established by the QFD team members on a scale of 1 or 2, with 2 meaning high sales effect and 1 being low effect on sales. The absolute weight is then found by multiplying the customer importance, target factor, and sales point. This is expressed in the following equation:

$$\text{Absolute weight} = \text{customer importance} \times \text{target value} \times \text{sales point} \qquad \textbf{(7.1)}$$

7. ***Prioritize technical requirements.*** As shown in Figure 7-9, technical requirements are prioritized by determining degree of difficulty, target value, absolute weight,

FIGURE 7-7 Competitive Assessment

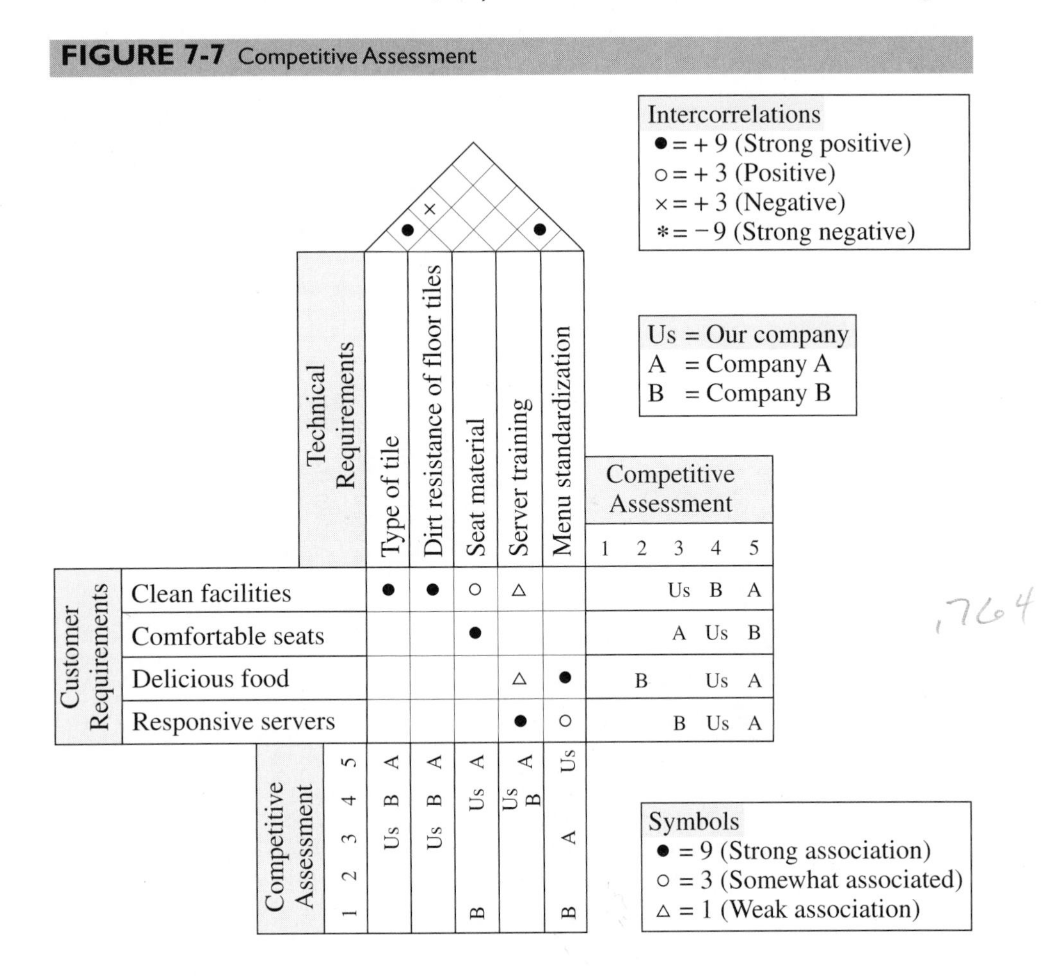

and relative weight. The degree of difficulty is assigned by design engineers on a scale of 1 to 10, with 1 being least difficult and 10 being most difficult. The target value for the technical requirements is defined the same way the target values for the customer requirements were assigned.

The values for absolute and relative weights are now established. The value for the absolute weight is the sum of the products of the relationship between customer and technical requirements and the importance to the customer columns (fourth column from the right). The value for relative weight is the sum of the products of the relationship between customer requirements and technical requirements and the customer requirements absolute weights (the farthest right column).

8. ***Final evaluation.*** The relative and absolute weights for technical requirements are evaluated to determine what engineering decisions need to be made to improve the design based on customer input. This is performed by computing a percentage weight factor for each of the absolute weight and relative weight numbers (see Figure 7-10).

As can be seen in this example, the standardized menu has a very high relative importance. This gives the restaurant a focus for the coming period.

FIGURE 7-8 Competitive Assessment Requirement Priorities

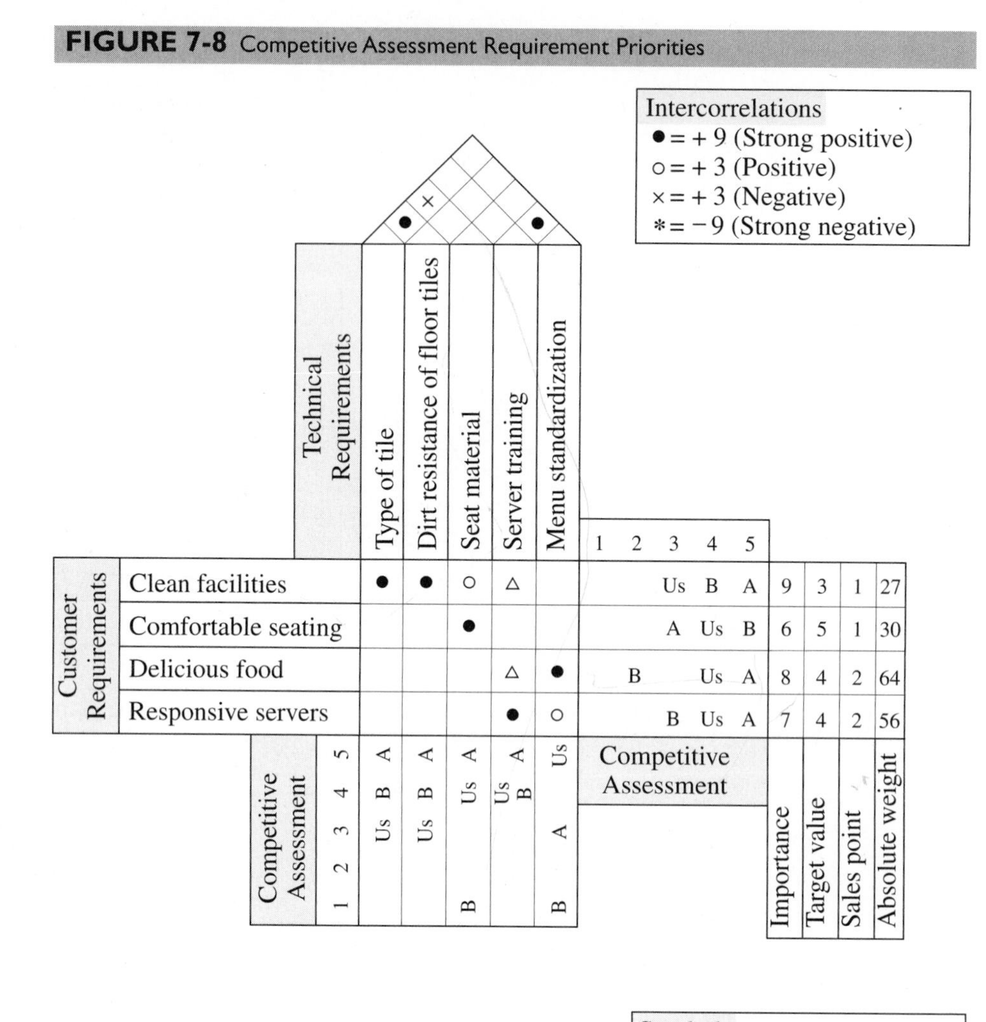

TECHNOLOGY IN DESIGN

No longer are the tools of the designer a square, a pencil, and a drafting table. Today, a designer is much more likely to use a **computer-aided design (CAD) system**. These systems are used in designing anything from an ultralight airplane, to a hamburger, to a home, or to a new intersection that can handle higher volumes of traffic. Computer-aided tools greatly improve the ability of designers to generate new and varied designs. In addition, they simplify the design process. For example, auto designers once had to place mock-ups of automobiles into wind tunnels to test the aerodynamics of a design. However, now the wind resistance coefficients for automobiles can be simulated on computers, cutting costs and design times and allowing for quick adjustments to the design. CAD systems help to develop more reliable and robust designs.

FIGURE 7-9 QFD Example

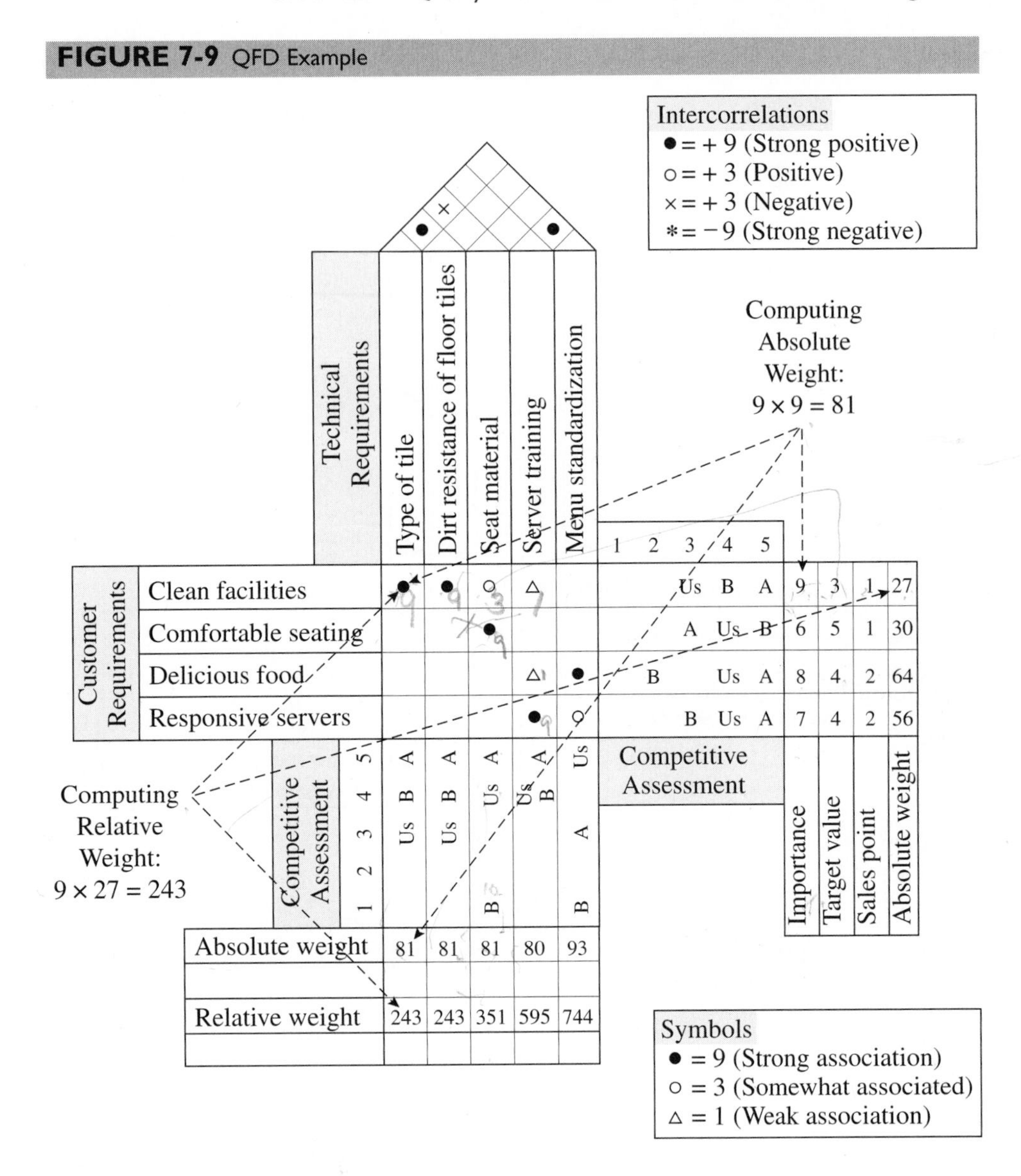

An important advance in CAD systems has been the advent of **multiuser CAD systems**. Using a common database in a network, multiple designers in locations worldwide can work on a design simultaneously around the clock. Consider a multinational corporation developing a new product. When the U.S. designers sleep, Asian and European designers work. When the U.S. designers return to work, they can see the progress that has been made overnight. When developing the Boeing 777, Boeing used hundreds of designers on the project simultaneously. These designers used their CAD systems to ensure there were no inconsistencies in design that would render the airplane unusable.

CAD systems are used in geometric design, engineering analysis, design review and automation, and automated drafting. **Geometric modeling** is used to develop a

FIGURE 7-10 QFD Example

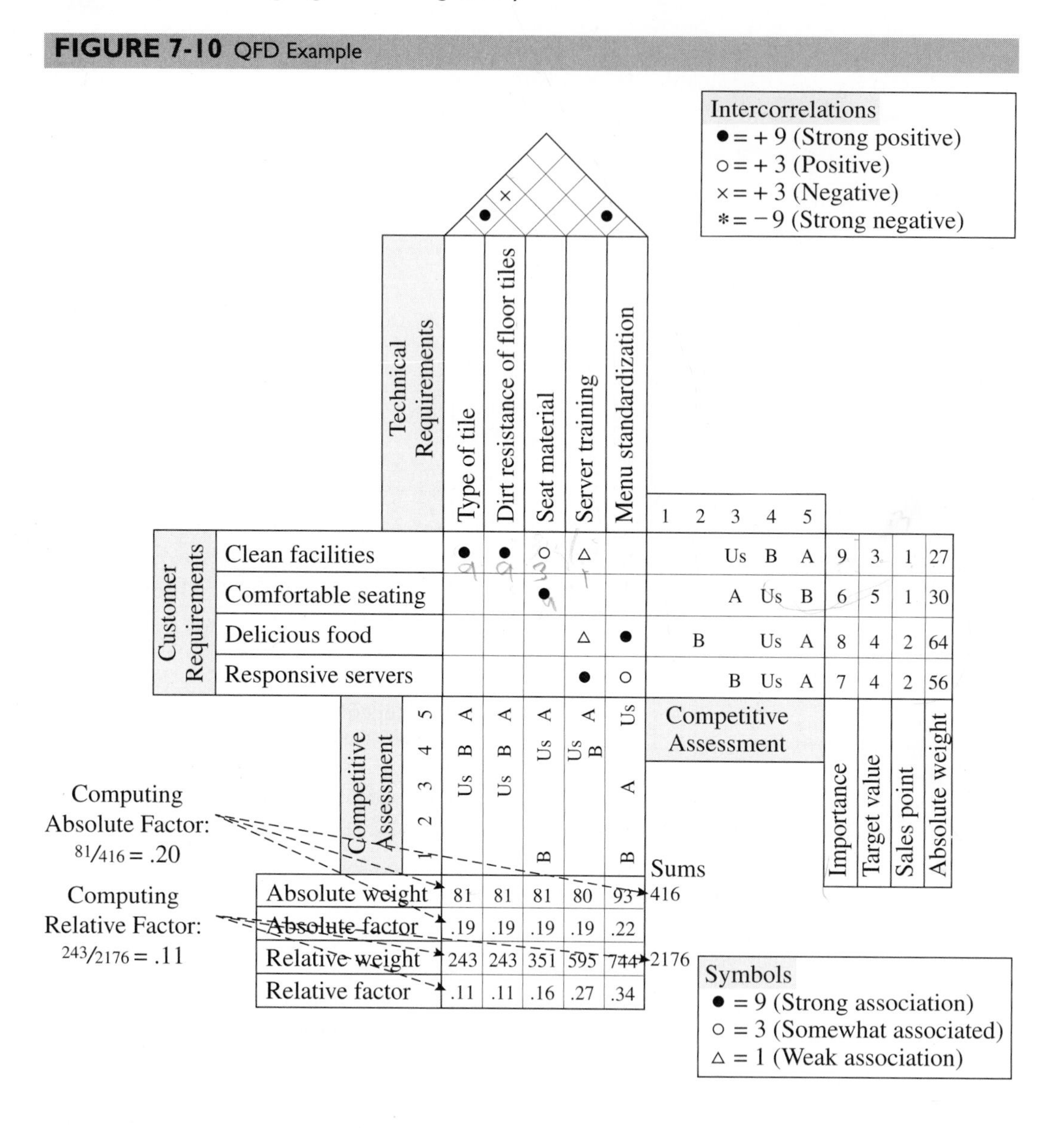

computer-compatible mathematical description of a part.[5] The image developed is typically a wire-frame drawing of a component. This part may appear in two dimensions, as a two-dimensional drawing of a three-dimensional object, or in full three-dimensional view with complex geometry.

Engineering analysis may involve many different engineering tests such as heat-transfer calculations, stress calculations, or differential equations to determine the dynamic behavior of the system being designed. Analysis-of-mass-properties features in CAD systems automatically identify properties of a designed object such as weight, area, volume, center of gravity, and moment of inertia. CAD systems allow for the automatic calculation of these properties.

---

[5]Groover, M., and Zimmers, E., *CAD/CAM: Computer Aided Design and Manufacturing* (Englewood Cliffs, NJ: Prentice Hall, 1987).

Designs are checked for accuracy during **design review**. Using CAD, the designer can zoom in on any part of design detail for close inspection of a part. Layering also is performed during design review by overlaying the geometric images of the final shape of a part over an image of a rough casting. This validates the design by ensuring that enough material is available on the casting to accomplish the final machined dimensions of the part.

Examining a design to see if different components in a product occupy the same space is called **interference checking**. Interference checking was of major importance in the design of the Boeing 777. Hundreds of pipes and thousands of wires occupy the walls of the aircraft. Interference checking in design review ensured that designs were feasible. This was especially important for Boeing because so many engineers were participating in the design. Automated drafting results in the creation of a final drawing of the designed product and its components. Some of the features of an engineered drawing include automated dimensioning, generation of cross-hatched areas, scaling of the drawing, development of sectional views, and enlarged views of particular part areas.

CAD systems can be stand-alone or tied into computer-assisted manufacturing (CAM) systems that are used in automated production systems. Another important component of a CAD system is the **group technology** component that allows for the cataloging and standardization of parts and components for complex products. Standard parts can result in fewer suppliers, simpler inventory, and less variability in processes.

CAD/CAM systems are often tied together in a closed-loop system with **computer-aided inspection (CAI)** and **computer-aided testing (CAT)** quality control systems. CAI and CAT allow for 100% inspection of products at a relatively low cost. Inspection is performed by infrared and noncontact sensors that allow for parts to be inspected without handling, thereby reducing the chance of damage to products.

PROTOTYPING METHODOLOGIES

With the increase of CAD systems, the approaches to prototyping products have expanded. **Prototyping** is an iterative approach to design in which a series of product mock-ups is developed until the customer and the designer agree on the final design. In some cases, the customer might not be an external user but upper management that approves the final designs of products.

To get a better idea of how prototypes can be used, we will define the types of prototypes. The first is the **basic prototype.** The basic prototype is a nonworking mock-up of the product that can be reviewed by customers prior to acceptance. Sometimes simple prototypes are developed prior to trade shows. For the auto companies, these are called *concept cars* that might not hit the market for several years, if ever.

Paper prototypes consist of a series of drawings developed by the designer on CAD systems and reviewed by decision makers prior to acceptance. Again, this can be an iterative process. In Windows- and Apple-based computer applications, graphical-user interface (GUI) prototypes are developed using sticky note pads and flip-chart paper to allow the user to view mock-ups of the program's proposed computer screens.

At times, companies build **working prototypes.** These are fully working models of the final product. However, depending on the product, working prototypes can be cost prohibitive because the complete design cycle must be completed to create them.

Organizing the Design Team

If the design process steps discussed previously are performed sequentially, the design process will be very time-consuming. Therefore, the steps are performed simultaneously as often as possible. This approach is called **concurrent engineering** and has been very helpful in speeding up the design life cycle. Products such as John Deere tractors and all-new automobiles have been designed using this strategy. Teams are a primary component of concurrent engineering and include program management teams, technical teams, and design-build teams.

The benefits of concurrent engineering primarily include communication among group members and speed. By working on products and processes simultaneously, the group makes fewer mistakes, and the time to get the concept to market is reduced drastically. The team concept joins people from various disciplines, which enhances communication and the cross-fertilization of ideas.

Another benefit of concurrent engineering is increased interaction with the customer. Often customers are included in concurrent engineering teams to give immediate feedback on product designs. This requires contractual agreements between suppliers and customers because the customer representatives may work for the design team on a contract basis. However, the immediate feedback is very helpful.

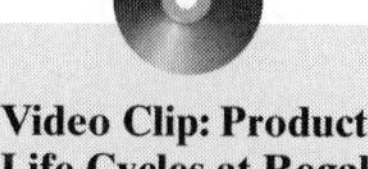

Video Clip: Product Life Cycles at Regal Marine

The Product Life Cycle

As is shown by the ski industry (see A Closer Look at Quality 7-1), product development is not a static process. Once new products are developed, work may already be underway to introduce the next generation of products. The product life-cycle concept demonstrates the need for developing new products by showing product design, redesign, and complementary product development on a continuum. Figure 7-11 shows a product life cycle for a typical product. As soon as a product is developed, it is on its way to decline.

Product Families and the Product Life Cycle

Two imperatives have come to the forefront in the study of product life cycles. The first is that product life cycles are becoming shorter. This means that obsolescence is a greater problem for designers and that the speed at which new product concepts are delivered to market is becoming much more important for companies around the world. The second imperative is that as product life cycles shorten, product variety and change become much more important to the successful competitor because complementary products are needed to consume productive capacity. Complementary products are needed for two reasons. First, as discussed, product obsolescence requires that products be updated. Second, some products have seasonal demand necessitating counterseasonal products. **Variety** refers to the differences in products that are produced and marketed by a single firm at any given time. **Change** is the magnitude of the differences in a product when measured at two different times.[6] Using the framework for variety and change developed by Sanderson and Uzumeri,[7] variety is the range of different items produced by a firm. Variety is related to a specific family of products. Change can occur as a result of evolutionary small changes to a product or drastic big changes to a product.

---

[6]Ashby, W. R., *An Introduction to Cybernetics* (New York: Metheun, 1956).
[7]Sanderson, S. W., and Uzumeri, M., *Managing Product Families* (Chicago: Irwin, 1997).

A CLOSER LOOK AT QUALITY 7-1

SKI DESIGN

www.rossignol.com

During the 1930s and 1940s, skiing in the United States was a sport for the very wealthy who could afford to take a train to Sun Valley, Idaho, or a resort in Colorado. By the 1950s and 1960s, middle-class people had taken up skiing. By the 1970s, skiing was at the peak of its popularity. People in the 1960s and 1970s began spending more time outside and were living a less sedentary lifestyle. However, by the 1980s, skiing had begun to decline. Other winter sports were becoming popular. The baby boomers were aging, and many downhill skiers were turning to cross-country skiing. This resulted in a decline of sales for ski makers such as K2 and Rossignol. These companies responded by pouring more money into research and development. One development was the introduction of snowboards. Snowboards appealed to skate boarders and young people. This market grew steadily, and now almost all ski areas in the United States have snowboard parks with half-pipes (a snowboarding area) and other amenities for snowboarders. In the early 1990s, ski sales were still flat, and "capped" skis were introduced. The theory behind capping of skis was that shock was more evenly dispersed throughout the skis. However, it was suspected that the real appeal of the capped ski was that it could hold 30% more graphics than the traditional models. In the mid-1990s, "fat boy" skis were introduced. These were skis that were much shorter and wider than traditional skis, and they made it much easier to ski in deep-powder snow and frozen, chunky, icy snow.

In the late 1990s, the industry introduced a radical change in ski design known as *parabolic* or shaped skis. The shaped skis have wider tips and tails that provide ease in turning and greater stability in the snow. One interesting variation of this growth in ski designs is the "parabolic fat boy" manufactured by Rossignol (see Figure 7.11). The parabolic skis have adapted snowboard technology into skis and make skiing much easier. Although this ski is recommended for expert skiers, it is great for intermediate skiers who want to improve. This is a pleasure to aging baby boomers whose legs enjoy the relief provided by these technological design changes. These design changes have reinvigorated the ski industry.

FIGURE 7-11 Rossignol B3-Ski

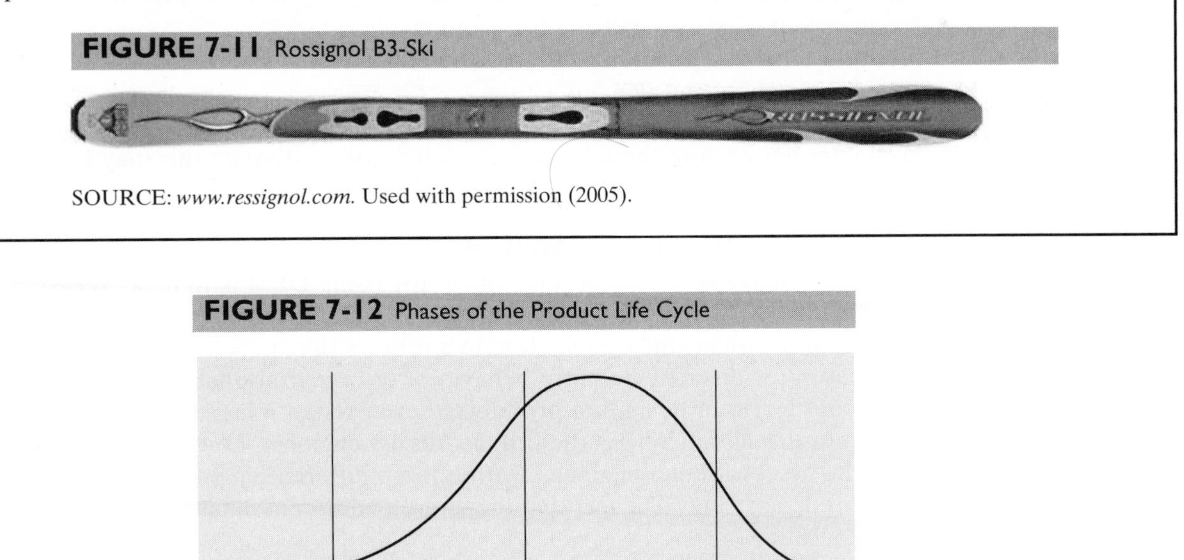

SOURCE: *www.ressignol.com.* Used with permission (2005).

FIGURE 7-12 Phases of the Product Life Cycle

Start-up Growth Maturity Decline

Complementary Products

What do we mean by managing product life cycles? Don't they just occur naturally? Well, yes, they do occur naturally. There probably isn't a lot you can do to control the rate at which the life cycle occurs. However, we can plan to introduce new and **complementary products.** Complementary products are new products using similar technologies that can coexist in a family of products. These products extend the life of the product line by offering new features or improvements to prior versions of the product. At times, these improvements are cosmetic, and at times, they are substantive. One example of a complementary product is a product that has a counterseasonal demand when compared with a base product such as motorcycles and snowmobiles. Arctic Cat produces ATVs for summer use and snowmobiles for winter use. This allows for level production rates throughout the year.

When we study issues such as variety and change, it becomes clear that at a strategic level, the problem of the product life cycle is not that a single product is becoming obsolete. Rather, if a variety of products are produced by a given company, management must be aware that several product life cycles must be managed simultaneously.

Designing Products That Work

As A Closer Look at Quality 7-2 shows, there are many things to consider when designing products. One of the biggest considerations is **design for manufacture (DFM)**—"Now that we have designed it, can we make it?" Loosely speaking, design for manufacture means to design products so that they are cost-effective and simple to build. However, there are many other considerations in a design. One consideration is how we design the product so that it is easy to maintain. After all, maintenance, if required, can be very expensive. Another aspect is designing for reliability. It makes little sense to design a product that is capable and stable but not reliable. Another issue relating to design is speed. *Speed* refers to the time it takes for a concept to reach the market. If it takes a long time for products to reach the market, competitiveness may be hampered. Product designs must be simple. Designing for simplicity means standardizing parts, modularizing, and using as few parts as possible in a design. Environmental issues also have become key considerations for companies designing products. With changes in regulations around the world, products must be designed for reuse, disassembly, and remanufacture.

Design engineers, operations managers, and others involved in the design process must consider each of these topics simultaneously. Although this may seem complicated, in fact, design's cycle times have improved for many companies.

Design for Manufacture Method

The overriding concept to consider when discussing DFM is to *make it easy to build* (see Quality Highlight 7-2). This may seem intuitive; however, it is sometimes difficult to be intuitive when you are too close to a process (like people who design products). The reason for this may be more behavioral or organizational rather than technical. In the old world of designing products, there existed a hierarchy of engineers. At the top of this hierarchy was the product design engineer. Lower down the hierarchy were the process design engineers. Often these different engineers worked in totally different departments. The fact that they were in different departments often impeded communication.

This organizational problem has been referred to as the **over-the-wall syndrome**. The over-the-wall syndrome is demonstrated by looking at the design process sequentially. First, the product design engineers developed a design. This product design would

A CLOSER LOOK AT QUALITY 7-2

WHY IT TAKES A ROCKET SCIENTIST TO DESIGN A GOLF BALL[b]

www.wilsonsports.com/golf

If you think that golf is unrelated to rocket science, you should meet Robert Thurman, a rocket scientist who designs golf ball dimples for a living (see Figure 7-13).

Thurman, a 35-year-old aeronautical engineer, once helped build fuel tanks for the space shuttle. Now he works for Wilson Sporting Goods Co., a unit of the Finnish company Amer Group Ltd. He is one of a growing cadre of specialists designing dimples that can make golf balls travel faster or slower, higher or lower, longer or shorter.

It is no fly-by-night avocation. With the global golf boom, golfers now spend some $800 million per year on balls and tend to flock to the latest wrinkle—or dimple—in ball design. Wilson has spent nearly $1 million on Thurman's dimple lab in Humboldt, Tennessee, including installing a 340-yard driving range complete with sand traps and a water hazard.

Steven Aoyama is an aeronautical engineer who designs dimples for Fortune Brands, Inc.'s Titleist & Foot-Joy Worldwide unit in Fairhaven, Massachusetts. He has his own computerized wind tunnel and a three-hole course with a full-time superintendent. Spalding Sports Worldwide, the world's biggest golf ball maker, estimates that it spends about $2 million a year on two dimple-designing facilities in Massachusetts and Florida. Spalding employs an aeronautical engineer and a physicist to help with the designs.

"Dimples," says Frank Thomas, technical director of the U.S. Golf Association in Far Hills, New

FIGURE 7-13 Anatomy of a Dimple

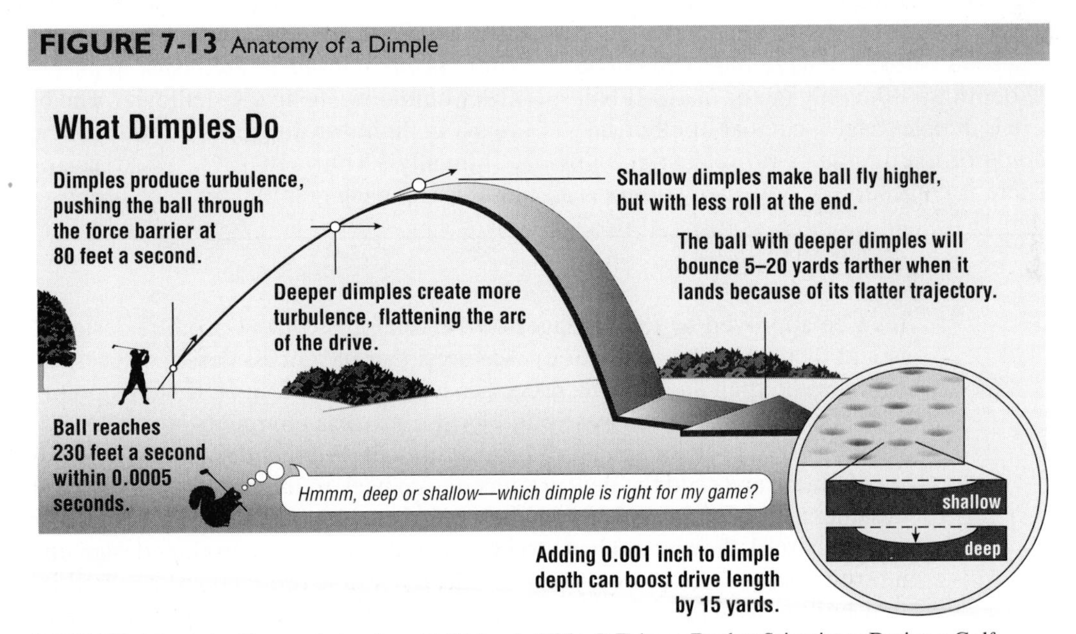

SOURCE: Adapted with permission from B. Richards, "Why It Takes a Rocket Scientist to Design a Golf Ball," *Wall Street Journal* (November 7, 1997): B1.

[b]Richards, B., "Why It Takes a Rocket Scientist to Design a Golf Ball," *Wall Street Journal* (November 7, 1997):B1. Copyright 1997 by Dow Jones & Co. Inc. Reproduced with permission of Dow Jones & Co. Inc. in the format Textbook via Copyright Clearance Center.

Continued

Jersey, "are very important, and very, very complex." A dimpleless golf ball in the hands of even a good golfer would travel an anemic 130 yards or so off the tee. The same ball, properly dimpled, could fly twice that far.

But dimpled how?

"Ah," says Thurman, as he tees up one of his dimpled creations on an Iron Byron. It's a contraption designed to mimic the swing of Byron Nelson, the legendary golf pro whose swing in the 1930s and 1940s was reputed to be close to perfect. The device winds up and whacks Thurman's ball about 270 yards, straight down the fairway toward the pin.

The key, Thurman explains, is to create a dimple with enough depth to produce air turbulence around the ball as it rotates in flight. That helps the ball penetrate the force barrier—the resistance that builds up as the ball moves through the air. But too much turbulence increases the drag on the ball, causing it to drop too fast.

Most dimpled balls crack the force barrier when they reach a speed of about 80 feet per second. Once they get through, air resistance drops by half, and the ball goes farther. Thurman estimates that the same ball, undimpled, would have to travel more than 200 feet a second before breaking the force barrier.

Dimple depth isn't the only factor in a golf ball's flight. There is dimple shape—circular or elliptical—and the angle of the dimple as it breaks the ball's surface. There is a question of how many dimples a ball should have—the consensus seems to run somewhere between 300 and 500.

Perhaps the most daunting factor is how to align those dimples. Thurman says he'll never forget his wedding anniversary, October 8, because it was on that date that he received his first U.S. patent for a dimple pattern.

Aoyama holds about a half-dozen patents and estimates that he has designed hundreds of dimple shapes over the past 20 years. "I'm always thinking about dimples when I play," he says.

It's all a long way from early golf balls, which were leather bags stuffed with feathers. Thurman and Aoyama feel that with the help of computers and gadgets like Iron Byron, they have pretty much figured out the geometry of dimples. So is the perfect golf ball on the way?

Not by a long shot.

The perfect ball, Aoyama says, would have to morph through a series of dimple changes during its flight. "You'd start with dimples as they are now, then they'd get a little shallower for more lift toward the middle of the flight." At the peak of the ball's trajectory, when it was slowing down, the dimples would suddenly get deeper to reduce drag. Then, as the ball accelerated back to earth, the dimples would once again get shallower. The ball would adjust in flight for the quality of the golfer, too.

"Ideally," Thurman says, "we'd customize the dimples on each golf ball for each user."

then be approved by the manager or the vice president of product design. The design would then go to the manager or vice president of process design. Process design engineers would then develop the processes to make the product. If at any point a problem with the product design was found by the process engineers, a request for redesign would be sent to the manager of product design. Product design engineers would solve the problem and send the new product design back to the process designers. The process designers would then continue their work of developing the process. When other problems occurred, they would have to be referred back to the product design engineers. As a result, many times it took years to develop a new product and resulted in processes that built poorly functioning products. DFM methods are designed to eliminate the over-the-wall syndrome and radically reduce design cycle times.

Many firms use **enterprise resource planning (ERP) systems** to integrate financial, planning, and control systems into a single architecture. As a result, there is an effort in the business world to include computerized design systems in these ERPs. An important component of such design software is the **product data management (PDM)** tool. PDM is a general extension of techniques commonly known as *engineering data*

QUALITY HIGHLIGHT 7-2

FOCUS ON DFM AT GENERAL MOTORS

www.gm.com

Design for manufacture (DFM) is not a new concept. Applications of DFM in the United States occurred as early as the late 1970s. Today, all auto companies use DFM methodologies. At General Motors, the DFM process includes team activities at every level and from every discipline. These multidisciplinary teams are called *product development teams*. In the early days at General Motors, DFM focus was on simpler designs with fewer parts and did not demonstrate an overall concern for the complete design. As time passed, DFM focused more on the product as a whole and the ease of making the product. General Motors used a workshop approach to begin DFM initiatives. In these workshops, design engineers would work together to begin the development of designs for new products. As time passed, these workshops became the modus operandi for the design people at General Motors. Concurrent engineering concepts were integrated into the product development teams and became a formal part of the product development process.

The DFM process at General Motors began with a detailed plan (known as the *total-vehicle DFM plan*) listing all the requirements in a plan that included all the DFM tools that would be used and timing relative to the product development life cycle. Typically, the plan began with quality function deployment (QFD), the process that captured the voice of the customer and helped to develop a plan to verify that customer needs were incorporated into final product designs. The total-vehicle DFM plan included all new product content and selected carryover content from previous vehicle designs. Potential DFM opportunities were identified and prioritized. Design priorities were established and were translated into account program imperatives, customer needs, and development schedules. DFM projects were then constantly reviewed and approved by the DFM team, functional area managers, and engineering directors.

The major component of DFM at General Motors is a requirement to assemble parts easily and without defects. To accomplish this, process variation must be held to a minimum. Dimensional variation management requires a structured process of analysis, variation management in product design, and manufacturing processes to optimize the vehicle quality.

DFM teams often encounter conflicting requirements. General Motors uses tools such as the **Pugh matrix**, a method of concepts selection, to identify conflicting requirements and prioritize design tradeoffs. The sooner design conflicts are identified, the better so that they can be resolved quickly by the team and not delay design schedule.

The final step in DFM is to verify manufacturability. The design team develops a working prototype to verify that the design has met all predetermined requirements. As a result, several lessons learned from this entire process are fed to the next vehicle development program to ensure continuous product/process improvements.

management, document management, and other similar names. PDM helps manage both product data and the product development process by tracking the masses of data needed to design, manufacture, support, and maintain products. It does this in part using a bill of materials that is later transferred to manufacturing, planning, and control systems after the design phase of the product life cycle.[8]

---

[8]Bourke, R., "PDM and ERP Continuing to Converge," APICS: The Performance Advantage (August 1997):66–67.

Design for Maintainability

One of the major concerns with new products is ease of maintainability. It often seems cheaper to replace a product than to repair it. This certainly is true for inexpensive products such as electric can openers, transistor radios, and other small appliances. However, the cost of repairing relatively expensive products such as personal computers, automobiles, and large appliances is becoming prohibitive. Consider a broken video camcorder. A new, relatively inexpensive camcorder may cost $250. If the camcorder breaks, a repair may cost half to two-thirds that. The decision to repair is essentially an economic decision involving costs, benefits, and tradeoffs. This decision becomes particularly difficult when the product life cycle is short. Suppose you owned a personal computer for the past year. You purchased the computer for $1,000, and it satisfied your needs at the time. Now you realize that you need more memory, your sound card is inadequate, and you need a larger hard drive. If you go to a discount store, you can purchase the upgrades you need for about $500. At the same time, for a little more money you can purchase a new computer that has all the desired features as well as many others. For example, the new computer will be faster than your old computer. Should you spend $1,000 on a new computer or $500 to upgrade your old computer? One solution to this problem is design for maintainability. Design for maintainability concepts include

Components that are easily replaced

Components that are easily removed with standard tools

Adequate space to perform the maintenance function

Nondestructive disassembly

Safe maintenance

Available adequate owners manuals and documentation (e.g., wiring diagrams, help facilities, or videos showing how to perform minor repairs)

Many personal computer manufacturers include how-to videos in their memory that demonstrate maintenance functions such as adding memory, connecting interfaces, and other simple maintenance functions. Craftsman tractors sold by Sears include videos that demonstrate how to change oil, how to operate the tractor, and how to perform other service functions. The bottom line is that customers should be provided with the necessary information and ease of access to the product that allow for simple or preventive maintenance.

An important issue is ease of delivery of more serious maintenance. There are many repairs that can be performed only by trained professionals. Many personal computer companies offer at-home maintenance for their personal computers. It is common for auto repair facilities to offer rides and loaner cars for customer use. Car rental companies typically offer on-the-road repair and towing service when their cars break down. It is important to recognize that service is also a design issue. At the design phase, after-sale processes must be designed such that maintenance is received simply, rapidly, and cost-effectively. Experience has shown that consumers are willing to pay more for products that are supported by outstanding service.

DESIGNING FOR RELIABILITY

Reliable products are always available when you need them, and you can depend on them to work properly. Reliability, as it relates to products, results from the interaction of multiple components in a system. Quality Highlight 7-3 shows how a luxury product

QUALITY HIGHLIGHT 7-3

DESIGNING RELIABLE LUXURY AT VUITTON[c]

www.Vuitton.com

Behind a locked door in the basement of Louis Vuitton's elegant Paris headquarters, a mechanical arm hoists a brown and tan handbag a half meter off the floor and drops it. The bag, loaded with an eight-pound weight, will be lifted and dropped, over and over again, for four days. This is a type of rapid-life-testing usually reserved for designing reliability into machinery and vehicles. Here it is being used on elegant handbags.

Vuitton has designed a high-tech torture chamber for testing its products. A piece of lab equipment bombards handbags with ultraviolet rays to test resistance to fading. Still another machine tests zippers by tugging them open and shutting them 5,000 times. There's even a mechanized mannequin hand, with a Vuitton charm bracelet on its wrist, being shaken vigorously to make sure none of the charms falls off.

If you think about Louis Vuitton, you likely don't think about robots beating up on bags. Likely, you think of waiflike supermodels or lithe Hollywood celebs toting Vuitton luggage to Palm Springs. However, to understand what makes Vuitton tick, you have to look behind the glittery façade and look closely at the world's most profitable luxury moniker.

Vuitton focuses relentlessly on quality in design and performance. Remember that the robot makes sure that Vuitton rarely has to make good on its lifetime repair guarantee. The supply chain is rigorously controlled and no bag is ever discounted. Above all, there's the efficiency of a finely tuned machine, fueled by ever-increasing productivity in design and manufacturing—and, as Vuitton grows bigger, the ability to step up advertising and global expansion without denting the bottom line.

Following are some examples of how attention to design detail sets the Vuitton bag apart:

- Zipper: Laboratory equipment randomly tests zippers by opening and closing them 5,000 times.
- Production: Manufacturing methods adopted from automakers and other industries are boosting productivity by 5% per year.
- Handle Metal Ring: To cut costs, Vuitton pressured supplier of metal rings to improve production efficiency.
- Leather Trim: Vuitton uses hides from northern European cattle, which have fewer blemishes from insect bites.
- Price Tag: Forget bargains: Vuitton never holds sales and price increases are common.

[c]Adapted from C. Matlack, "The Vuitton Machine", *Business Week*, March 22, 2004, pp. 98–102. Used with permission.

is designed to be reliable at Vuitton. Reliability has two dimensions, *failure rate* and *time*, both of which can be applied to components and to systems. **Component reliability** is defined as the propensity for a part to fail over a given time. **System reliability** refers to the probability that a system of components will perform the intended function over a specified period of time. It is important to recognize the difference between component reliability and system reliability. The levels of measurement are different for system and component reliability. When we talk of component reliability, we refer to a finite aspect of the overall product. System reliability is computed from the aggregation of multiple components. Reliability models are discussed in Chapter 13.

Reliability Analysis Tools

There are many ways to make designs more reliable. These methodologies include failure modes and effects analysis; fault-tree analysis; and failure modes, effects, and criticality analysis.

Failure Modes and Effects Analysis (FMEA)

Failure modes and effects analysis (FMEA) systematically considers each component of a system, identifying, analyzing, and documenting the possible failure modes within a system and the effects of each failure on the system. It is a bottom-up analysis beginning at the lowest level of detail to which the system is designed and works upward. The FMEA process results in a detailed description of how failures influence system performance and personnel safety. FMEA answers the question, "How do the systems or components fail?"

Failure modes and effects analysis was created by the aerospace industry in the 1960s and is used extensively in Six Sigma (see Chapter 14). An early application of FMEA occurred in 1972 when Ford Motor Co. used it to analyze engineering design. Ever since, Ford has tried to refine FMEA through continued use in its operations as many software applications have evolved recently that aid in the use of FMEA. Some benefits that can be derived through the use of FMEA include

1. Improvement of the safety, quality, and the reliability of product
2. Improvement of a company's image and its competitiveness
3. Increased satisfaction from a user standpoint
4. Reduction in product development cost
5. Record of actions taken to reduce a product risk

There are five basic areas where FMEA can be applied. These are concept, process, design, service, and equipment.

Concept. FMEA is used to analyze a system or its subsystems in the conception of the design.

Process. FMEA is applied to analyze the assembly and manufacturing processes.

Design. FMEA is used for analysis of products before mass production of the product starts.

Service. With respect to services, FMEA is used to test industry processes for failure prior to their release to customers.

Equipment. A company also can use FMEA to analyze equipment before the final purchase.

How FMEA Works

As is shown in Figure 7-14, failure modes and effects analysis uses a nine-step process.

1. The first FMEA step is to give each component in the system a unique identifier; this is so that none of the parts will be overlooked in the analysis.
2. In the second step, list all the functions each part of the system performs. This step requires you to develop a block diagram for the description of your design.
3. List the one or two failure modes for each function from the second step. The best description of a failure mode is a short statement of how a function may fail to be performed. What a product does or does not do when it fails describes the failure mode.
4. The fourth step describes what effects each failure mode of the component will have, especially the effects perceived by the user or operator. Analysis of the effects should follow a hierarchical order because any effect should be fairly detailed so that the severity of each effect can be judged. Some of these effects measure the consequence of failures on a component or part of a device, the whole system, the user, and/or the public.

FIGURE 7-14 FMEA Steps

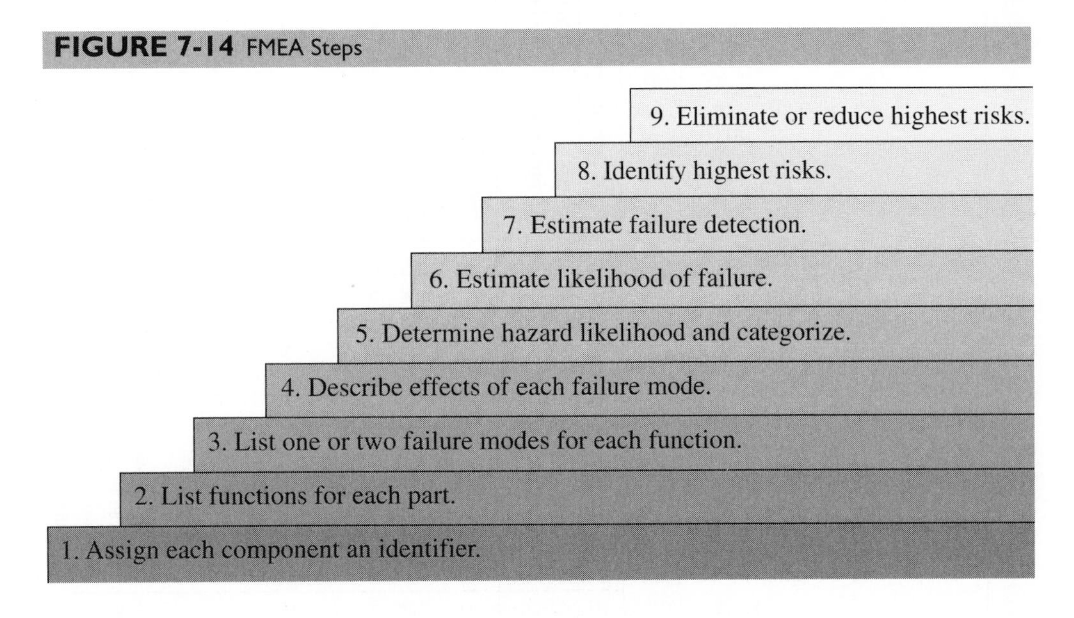

5. Determine whether the failure will result in a potential hazard to personnel or the system. Then categorize how severe each hazard will be. There are four basic categories that hazards fall into: catastrophic, critical, marginal, and negligible.
6. Estimate the relative likelihood of occurrence for each failure. The likelihood of occurrences is estimated using a 10-point scale and described in steps 4 and 5, ranging from highly unlikely (1) to very likely (10).
7. Estimate the ease with which the failure may be detected. If the failure takes so long to be detected that it becomes too late to replace or repair, the magnitude of the problem is likely to be much greater than if the failure can be easily detected.
8. Use the estimates from steps 5, 6, and 7 to identify the highest risks related to the system.
9. Decide what action will be taken to eliminate or reduce the highest risks in the system. The most common decision made is to alter the design to reduce the likelihood of occurrence and failure severity or simply to bring about easy failure detection. Figure 7-15 shows an example of an FMEA form.

Fault-Tree Analysis

Fault-tree analysis (FTA) is an analytical tool that graphically renders the combinations of faults that lead to failure of a system. This technique is useful for describing and assessing the events within a system. Such events can be either normal or abnormal, but it is their sequence and combination that are important. FTA shows the probabilities of systems failure caused by any event and is widely used in the aerospace, electronics, and nuclear industries. A fault tree is a qualitative model that also can be evaluated quantitatively. FTA is used for reliability, maintainability, and safety analysis and was used originally in 1961 at Bell labs to evaluate Minuteman Launch Control Systems to avoid inadvertent missile launches.

Failure Modes, Effects, and Criticality Analysis (FMECA)

FMECA is an extensive but simple method for identifying ways in which an engineered system could fail. As in FMEA, failures, effects, and causes are identified. FMECA rates failure modes by ranking each possible mode according to both the

FIGURE 7-15 FMEA Form

Potential
Failure Modes and Effects Analysis
(Concept FMEA)

_X_System
___Subsystem
___Component: 000000/COMPLETE VEHICLE SYSTEM.
Model Year/Vehicle(s): / (3)
Core Team: (0)

Design Responsibility: (3)
Key Date: (6)

FMEA Number: (1)
Page 1 of 1
Prepared by: (4)
FMEA Date (Orig.): 94.06.06 (Rev.): 94.06.06

Item / *Function*	*Potential Failure Mode*	*Potential Effect(s) of Failure*	*Sev*	*Class*	*Potential Cause(s)/Mechanism(s) Failure*	*Occur*	*Current Design Controls*	*Detec*	*R.P.N.*	*Recom-mended Action(s)*	*Responsibility & Target Completion Date*	*Action Results*				
												Actions Taken	*Sev*	*Occ*	*Det*	*R.P.N.*
(9) Enter a system function. Use the verb-noun format. If known, enter the Engineering require-ments and constraints associated with each function.	(10) Enter the potential failure mode(s) for the system function. Describe the failure mode in terms of "loss of function," or as the negative of the function.	(11) For each failure mode, list its con-sequences on: —Other Systems —The Vehicle —The Customer —Government Regulations			(14) From the block diagram, determine if/how each element can cause the System failure mode. (The cause will be a failure mode of the element.) Typical causes (element failure modes) will be: —fails to operate —operates prematurely —operates intermittently —fails to stop operating —loss of signal to next element		(16) Enter the analytical method, test, or technique used to detect the cause of the System failure mode. If no detection methods are known, enter "None identified at this time"		0	(19) Enter the recommended design actions intended to reduce one or more of the Severity, Occurrence, and/or Detection ratings. If no actions can be listed, enter "None at this time."	(20) Enter —System design Dept —System design Eng —Target Completion date	(21) Enter a brief description of the action taken and its completion date.				
		Severity (12) —> For each failure mode, rate the most serious effect. Enter rat-ing in column 12. Use Severity Rating Table for System FMEA.			Occurrence (15) —> Estimate the rate at which a cause is expected to occur over the design life of the element. Use Occurrence Rating Table for System FMEA. If no information is available to estimate the Occurrence rating, enter a rating of 10. <— (13) Reserved for future use		Detection (17) —> Estimate the likelihood the Detection method(s) will detect the cause of the System failure mode. If several methods are listed, enter the lowest (best) rating. Use the Detection Rating Table for System FMEA. If no methods can be listed, enter a rating of 10.		0	<—(10) RPN Risk Priority Number		Revised (22)—> RPN After actions have been taken, re-estimate the rat-ings for Severity, Occurrence, and Detection. Enter the revised ratings in the columns to the right. If no actions are listed, leave these columns blank.				

SOURCE: *FMEA Handbook, Environmental and Safety Engineering* (Dearborn, MI: Ford Automotive Operations, 1995).

probabilities of its occurrence and the severity of its effects. The primary goal of FMECA is to develop priorities for corrective action based on estimated risk. FMECA is used to analyze a probable cause of a product failure, to determine how the problem affects a customer, to identify the probable manufacturing or assembly processes responsible, to identify which process control variable to focus on for prevention detection, and to quantify the effects on the customer.

Criticality in FMECA is important because it prioritizes how the design team should be spending its resources. In general, criticality refers to how often a failure will occur, how easy it is to diagnose, and whether it can be fixed. Criticality assessment is somewhat subjective because it depends on the viewpoint of a service or field engineer. This view is markedly different from the designer or marketing manager. All members of an interdisciplinary team should participate in ranking criticality so that their concerns are factored into rankings. As a design team considers the various failure modes in the ramifications, one or more of the team members fills out a structured FMECA form that summarizes all that is involved in what can go wrong. In general, a design FMECA includes

1. A description of the product's function
2. Listings of the potential failure modes
3. Potential effects each failure mode could have on the end user
4. Potential causes of each failure mode with the likelihood ranking for each
5. Preventive measures in place for firmly scheduling by the time production starts
6. Ranking of the effectiveness of each preventive measure
7. A ranking of the difficulty of detection
8. An estimate of the probability that the cause of a potential failure will be detected and corrected before the product reaches the end user

Product Traceability and Recall Procedures

Although FMEA and FTA help predict where defects will occur and what their effects will be, from time to time unforeseen defects will occur that can result in dangerous and costly results that can subject the firm to liability.

For example, one cool summer morning a young pilot was asked to fly some hunters over some extremely remote wilderness territory. The pilot was rather inexperienced at flying in these types of conditions; after all, backcountry flying is dangerous, and experience is hard to obtain. However, the temperature and weather conditions were perfect. The adventure proceeded without incident until the final leg of the trip over some of the most extreme country. The wind began to pick up and gusted to more than 100 miles per hour. The light aircraft had never experienced these types of winds before. The young pilot pulled out of the mountainous canyon to get well above the mountaintops. This caused the aircraft to pitch and roll, and the wind shear coming over the mountaintops stressed the plane. The plane's vertical stabilizer was rendered useless by the powerful winds, but the team was able to return to the airport safely.

Subsequent investigation showed that an alloy structural support had been stressed to the breaking point in the plane's vertical stabilizer. The plane's manufacturer issued a recall of 1,200 planes fitted with the same special alloy in the vertical stabilizer. The problem occurred in one piece used in the manufacturing process. This piece had been produced by another vendor. An identification number allowed the plane manufacturer to track the purchase of this part back to its supplier. All aircraft built with the weak structural piece were recalled to replace the defective vertical stabilizer. Without proper identification techniques and sufficient tracking systems, potentially hazardous products could remain in use without any way of recalling or repairing them.

This characteristic is called **product traceability.** Product traceability and **recall procedures** are important aspects of product design. Because companies are liable for the products they create, it is important to be able to identify the origins of defective products or components through product traceability procedures.

In 1972, Congress created the **Consumer Product Safety Commission (CPSC)** to protect citizens from unreasonable risks of injury and death. To avoid being listed on the CPSC list of hazardous products, a company must have a system in place for tracing components. When a recall is demanded by the CPSC, a company needs to narrow its recall to a particular identification (ID) number or product line. Therefore, a good ID system can help isolate where the breakdown in the product occurred.

A major goal of product traceability and recall procedures is to be able to trace products with a minimum of cost. Product traceability also helps limit product liability relating to safety hazards.

ENVIRONMENTAL CONSIDERATIONS IN DESIGN

Currently, society demands much more from product designers than just high-quality products. Many manufacturers have turned to a more environmental form of manufacturing that offers positive returns on investment. Many companies, such as Siemens, Caterpillar, Xerox, Eastman Kodak, Hewlett Packard, and others, are using environmentally friendly forms of manufacturing.

The move to **green manufacturing** began in Germany with requirements for importers to remove packaging materials. Using a life-cycle approach to product design causes designers to focus not only on incoming materials, manufacturing processes, and customer use but also on the eventual disposal of the product. This life-cycle approach has led to practices known as **design for reuse, design for disassembly,** and **design for remanufacture.**

Design for reuse refers to designing products so that they can be used in later generations of products. One example is the Kodak FunSaver camera.[9] Initially, the camera was made so that it could be disposed of after use. Although Kodak had experimented with recycling the cameras, the cameras really ended up in landfills. Kodak received a wake-up call when it received the dreaded "wastemaker of the year" award for the disposable cameras, and it responded by converting the design from disposable cameras to recyclable cameras. Initially, the camera had been ultrasonically welded. Through design-for-disassembly processes, the camera case is now made so that it snaps apart easily. The customer would deliver the camera to the photofinisher, who would return the camera to Kodak. Kodak subcontracted with a company named OutSource, a New York state sponsoring organization that employs handicapped people, to take the cameras apart. Camera covers, lenses, and other parts were removed and reused. Miscellaneous plastic parts were ground into pellets that are used in molding new camera parts. Today, 87% of the FunSaver camera is either reused or recycled. There is great potential for reuse of products. Consider that currently two computers are discarded for every three computers purchased. The method for designing for reuse involves analyzing existing products for materials, identifying other uses for these materials, and developing a disassembly process to sort out these materials. This is good business for the producers of personal computers because if chemicals used in

[9]Bylinsky, G., "Manufacture for Reuse," *Fortune* (February 6, 1995):102–112.

making PCs were to find their way into groundwater, the manufacturers could be held liable. The resulting costs could be in the billions of dollars.

The principles for design for disassembly include using fewer parts and fewer materials, using snap-fits instead of screws, making assembly efficient and improving disposal, using design for disassembly experts in concurrent design teams, and eliminating waste through better design.

SUMMARY

As life cycles for products become shorter, a focus on quality in the product design process is necessary to remain competitive. Many of the dimensions of quality that we discussed in Chapters 1 and 2 are addressed in the design phase of the product life. By focusing on issues such as maintainability, assembly, reliability, and product traceability, we are able to continually improve our ability to make things.

We have said that you should design products so that they are easy to build. By simplifying design processes (through concurrent design teams, by standardizing, and through the use of modular designs), we make products that are more reliable.

Companies have implemented these processes with great results. These results have facilitated huge increases in production capacity, coupled with a reduction in cost. These cost savings do not always result in higher profits. As we have seen, the costs of computer chips dropped consistently. However, a company that does not become better at design will simply not be competitive in the future.

KEY TERMS

- Basic prototype
- Change
- Complementary products
- Component reliability
- Computer-aided design (CAD) system
- Computer-aided inspection (CAI)
- Computer-aided testing (CAT)
- Concurrent engineering
- Consumer Product Safety Commission (CPSC)
- Criticality
- Customer future needs projection
- Design for disassembly
- Design for manufacture (DFM)
- Design for remanufacture
- Design for reuse
- Design review
- Engineering analysis
- Enterprise resource planning (ERP) systems
- Failure modes, effects, and criticality analysis (FMECA)
- Failure modes and effects analysis (FMEA)
- Fault-tree analysis (FTA)
- Final product definition
- Geometric modeling
- Green manufacturing
- Group technology
- Interference checking
- Manufacturing system design
- Multiuser CAD systems
- Over-the-wall syndrome
- Paper prototypes
- Product data management (PDM)
- Product design and evaluation
- Product idea generation
- Product manufacture, delivery, and use
- Product marketing and distribution preparation
- Product traceability
- Prototyping
- Pugh matrix
- Recall procedures
- System reliability
- Technology development for process selection
- Technology feasibility statement
- Technology selection for product development
- Variety
- Working prototypes

DISCUSSION QUESTIONS

1. Product idea generation initiates the process of designing a product by generating ideas from external and internal sources. What are some examples of external and internal sources that are used in this process?

2. Discuss the concept of consumer future needs projection. Does a firm that excels in this area have a competitive advantage? Please explain your answer.
3. What is a technology feasibility statement? Why is it important?
4. Briefly describe the role of computer-aided design (CAD) in the product design process. How has CAD changed the way that product designers go about their jobs?
5. What role does prototyping play in the product design process? What is the difference between a basic prototype, a paper prototype, and a working prototype?
6. Describe the concept of concurrent engineering. How does concurrent engineering improve the product design process?
7. The product life cycle for many products is getting shorter. In what ways does this trend complicate the product design process? Can you think of any advantages to shorter product life cycles for firms that have exemplary product design processes?
8. What is the role of complementary products in managing the product life cycle?
9. What is meant by design for manufacture?
10. The design for maintainability concept states that a product should be designed in a way that makes it easy for a consumer to maintain it. What product attributes make it easy for a product to be serviced or maintained?
11. What is the over-the-wall syndrome? How can the over-the-wall syndrome be avoided?
12. Define component reliability and system reliability. What is the major difference between these two concepts?
13. Describe the concept of failure modes and effects analysis (FMEA). What is the end result of an FMEA analysis? What are some of the ancillary benefits that can be derived through engaging in FMEA?
14. What is the primary purpose of conducting a fault-tree analysis?
15. Describe a method for identifying ways in which an engineered system could fail. What is the primary goal of this method of analysis?
16. Discuss the importance of product traceability and recall procedures. Why is product traceability considered an important consumer safety issue?
17. What environmental considerations are important for product designers? Do you believe that environmental considerations will become more important or less important in the future? Explain your answer.
18. Compare the job of a product designer 20 years ago to the job of a product designer today. In your opinion, what has been the single most significant technological advancement that has changed the job of a product designer?

PROBLEMS

1. Flowchart the design and production processes for writing a book such as *Managing Quality: Integrating the Supply Chain.* Use the standard process for designing products in the chapter.
2. Define key customer requirements for a pen. Next, define key technical requirements for the pen. Create a matrix showing the relationships between technical and customer requirements using the QFD format.
3. Define key customer requirements for an automobile windshield. Next, define key technical requirements. Create a matrix showing the relationships between technical and customer requirements using the QFD format.
4. For the QFD Problem 4 Matrix, compute the
   a. Customer requirements absolute weight.
   b. Technical requirements absolute weight and factor.
   c. Technical requirements relative weight and factor.
   d. Which design and technical factors should be emphasized? Why?

Problem 4 Matrix

Customer Requirements	Technical Requirements 1	2	3	4	5	Competitive Assessment 1 2 3 4 5	Importance	Target	Sales point	Absolute weight
A	○		●				3	5	2	
B				○			4	4	2	
C		△					7	3	2	
D			△		△		2	2	1	
E				○	●		1	1	1	
Difficulty										
Target										
Absolute weight										
Absolute factor										
Relative weight										
Relative factor										

5. For the QFD Problem 5 Matrix, compute the
   a. Customer requirements absolute weight.
   b. Technical requirements absolute weight and factor.
   c. Technical requirements relative weight and factor.
   d. Which design and technical factors should be emphasized? Why?

Problem 5 Matrix

Customer Requirements	Technical Requirements 1	2	3	4	5	Competitive Assessment 1 2 3 4 5	Importance	Target	Sales point	Absolute weight
A	△	●	△		●		2	1	2	
B	△	○		○			8	1	1	
C	●	△	●		●		5	5	2	
D			●	○	△		3	4	2	
E	○	△		△	△		2	3	1	
Difficulty										
Target										
Absolute weight										
Absolute factor										
Relative weight										
Relative factor										

6. For the QFD Problem 6 Matrix, compute the
   a. Customer requirements absolute weight.
   b. Technical requirements absolute weight and factor.
   c. Technical requirements relative weight and factor.
   d. Which design and technical factors should be emphasized? Why?

Problem 6 Matrix

Customer Requirements	Technical Requirements 1	2	3	4	5	Competitive Assessment 1 2 3 4 5	Importance	Target	Sales point	Absolute weight
A	●	△	●		△		6	4	1	
B		○	●	○			9	5	2	
C	△	●		○	●		9	3	2	
D	○	△	△	○			4	4	1	
E	●				●		2	2	2	
Difficulty										
Target										
Absolute weight										
Absolute factor										
Relative weight										
Relative factor										

7. What are important design elements for a pair of pants?
   a. Define the customer requirements.
   b. Define technical requirements.
   c. Using the QFD format, show the relationships (with strengths, i.e., 1, 3, or 9) between a and b above.
8. Using the format in Figure 7-15, develop an FMEA for a pair of women's panty hose.

Cases

Case 7-1 Designing the Ford Taurus: Why Quality Had to Be Job 1

Ford: ***www.ford.com***

Ford Motor Company was started in a converted wagon shop by Henry Ford in 1902. In 1908, the company produced the Model T, the first car designed to be produced in great volume at a low cost. Because of its affordability, the Model T was called the car that "put the world on wheels." Since the introduction of that grand old car, the Ford Motor Company has treated the American public to many popular cars, including the Mercury, the Thunderbird, the Mustang, and the Taurus. Although all these cars have been important to the Ford Motor Company, perhaps the most important car has been the Taurus. The story of the Taurus vividly illustrates the importance of quality in all aspects of product design and development.

Despite Ford's rich history, the company was suffering badly in the 1970s. The American auto market was flooded with good-quality cars from Japan, and Ford's competitive position was in decline. By the end of the 1970s, Ford faced declining sales, increasing global competition, major layoffs, and the real possibility of permanent financial damage. Management knew that business as usual with minor modifications would lead the company to bankruptcy. The only hope was offering the American public a new kind of car, a car designed with quality in mind. The car was the Ford Taurus.

The Taurus wasn't just a new car for Ford; it represented an opportunity for the company to design a new way to build cars. Instead of organizing the development of the Taurus along the lines of functional expertise, Ford used integrated teams of people from various functions. Ford also knew that the Taurus would not appeal to all customer groups, so the Taurus was designed to be part of a broader product line, rather than the single car to save the company. Ford carefully documented what it did in the Taurus project and learned from its successes and failures. As the Taurus was coming to market, Ford announced a new set of product development principles it called the "Concept-to-Customer" design paradigm. Some of the basic principles of the new approach to product design, which were pioneered during the Taurus project, were as follows:[10]

Senior management review of a development program should be driven by substantive milestones rather than by calendar schedules.

Suppliers of production parts should be responsible for the prototypes of those parts; they may subcontract production of prototypes, but the production suppliers remain responsible for quality, performance, cost, and delivery.

Parts used to manufacture pilot vehicles should be made with production tools.

The development of these principles, along with the other principles contained in Ford's "Concept-to-Customer" paradigm, represented a new concept for product development at Ford. Fortunately for Ford, the Taurus sold well and established a new standard of excellence for American cars. The improved engineering of the Taurus and its design features were enthusiastically accepted by the American public. The car also set a new standard for the way Ford builds cars. Subsequent to the design of the Taurus, Ford experimented with a number of additional innovative management techniques, including concurrent engineering, early manufacturing input, improved prototyping, and benchmarking. The company's new attitude toward product design, obtained during the development of the Taurus, was instrumental in helping the company restore its leadership position in the global automobile industry. ■

[10]Bowen, H. K., Clark, K. B., Holloway, C. A., and Wheelwright, S. C., *The Perpetual Enterprise Machine* (New York: Oxford University Press, 1994).

DISCUSSION QUESTIONS

1. What would probably have happened to Ford if the company had not developed a new concept of product design? Do you believe that Ford would have survived? Why or why not?
2. What can other companies learn from Ford's experiences with product development and design? Make your answer as substantive as possible.
3. Can you think of any companies in other industries that have significantly improved the way they design their products in the past 10 to 20 years? If so, please describe how the company changed its product design process.

Case 7-2 Nucor Corporation: Producing Quality Steel by Stressing Sound Management Practices

Nucor Corporation: ***www.nucor.com***

Nucor Corporation, headquartered in Charlotte, North Carolina, is the largest manufacturer of steel and steel products in the United States. The company received a great deal of attention because of its impressive performance in an industry plagued by a multitude of problems, especially in recent years. Since the 1970s, Nucor pioneered the minimill concept, which is a method of making steel by melting scrap metal in electric arc furnaces at a fraction of the cost of conventional steelmaking. Nucor is admired for its quality products, its state-of-the-art manufacturing processes, and its industry-leading productivity ratios.

It is difficult to find a single reason that explains Nucor's success. Although the company has recently made key acquisitions and has modern facilities and equipment, competitors that have the same level of technology do not fare as well. What Nucor does have that is unique is a set of sound management principles and a somewhat novel approach to employee relations. Although Nucor is a $4.8-billion-per-year company, there are only four management layers between the CEO and frontline employees, and the general managers on the plant floor make the day-to-day decisions. Rank-and-file employees are involved in devising methods to improve operations. The company has a very egalitarian culture. There are no company cars, company planes, assigned parking spaces, hunting lodges, or other indications of status. All the employees wear the same color hard hat (with the exception of maintenance workers and visitors, who must be easily recognizable in case of an emergency), have the same group insurance program, have the same holidays, and have the same vacation plan.

There are other areas in which Nucor is distinct. The company has a well-developed employee incentive plan that aligns the interests of the employees with the interests of the firm. The typical millworker at Nucor receives a base pay that is slightly below the industry standard, but the firm's bonus plan is very generous when the company is doing well. Two distinctive features of Nucor's bonus system are that it is all written down and is totally objective, based on firm performance criteria. There is no subjectivity involved. If the firm reaches certain performance levels, a bonus will be paid, period. With bonuses figured in, Nucor employees typically lead the steel industry in terms of average pay. Yet the company's total cost per ton of steel produced is lower than that of other integrated producers.

In return for the generous compensation package, Nucor holds its employees to a high standard. Decision making is pushed down to the factory floor in many instances, requiring mental toughness and continuous education on the part of the company's employees. The company also asks its employees to be prompt and fully engaged in their jobs. For example, if an employee is late to work, he or she loses his or her bonus for the day. If the employee is more than 30 minutes late, the bonus is lost for a week. In return for this level of employee commitment, Nucor has not laid off a single employee for lack of work in 20 years. A very unusual indication of what Nucor thinks of its employees is evidenced in the company's Annual Report for 2006[11] (and in many previous years). The name of each of the company's 10,600 employees is written on the front and back cover of the Annual Report. Nucor produces high-quality products by stressing sound management techniques. Commenting on this issue in a book about Nucor, Jeffery L. Roengen wrote, "The amazing thing about Nucor's success is that it is so simple: Give employees a stake in the company's growth; focus on the business at hand; keep red tape and bureaucracy to a minimum." Apparently, this formula has continued to work for Nucor. ■

[11]Nucor Annual Report, 2006.

DISCUSSION QUESTIONS

1. This chapter has emphasized process design. At Nucor, do human resources processes affect product quality?
2. How do Nucor's management practices affect its ability to produce high-quality products? Make your answer as substantive as possible.
3. Would you enjoy working at Nucor? Why or why not?

CHAPTER 8

Designing Quality Services

Encouraging employees to solve customer problems and eliminate the source of complaints allows them to be "nice," and customers treat them better in return. Not just customers but also employees want to continue their relationship with the business.

—Frederick Reichheld and W. Earl Sasser[1]

High-quality service is essential for competitiveness and can even improve employee satisfaction. However, service, like quality, is a multidimensional term. To provide high-quality service, we need a profound understanding of the needs, wants, and desires of the customer and an understanding of who the customer is.

Quality service is not only an imperative for competitiveness but also a sign of quality maturity. As we have discussed previously, even manufacturing firms—after reliability, conformance, design, and other requirements have been met—eventually focus on service throughout the supply chain. In today's economy, service still is a major differentiator that allows firms to beat competitors in the marketplace.[2]

Figure 8-1 shows the power of satisfied customers. If customers are satisfied, they will be loyal. Revenue streams will increase—as will profits. If a credit card customer leaves after one year, the credit card firm will lose $51 on average. Notice that if the same customer stays, the revenues increase as each year passes. Profit per customer for the laundry industry increases steadily year after year as well. This principle is the same for other service industries.

In this chapter we discuss services in general first and then services from a quality perspective. Tools such as the SERVQUAL, gap analysis, and services blue printing will be developed. The central theme of this chapter is to understand customers' needs and to use that understanding to design services that will satisfy customers.

DIFFERENCES BETWEEN SERVICES AND MANUFACTURING

In Chapter 1 we talked about the multidimensional nature of quality. If quality is multidimensional for manufactured products, it will be more so for services. Understanding some of the differences between manufacturing and services helps to design useful approaches to quality improvement in services. Using a contingency perspective, we understand that the nature of services causes us to approach service quality improvement from a different direction from manufacturing.

[1]Reichheld, F., and Sasser, E., "Zero Defections: Quality Comes to Services," *Harvard Business Review* (September–October 1990):105–113.

[2]Heskett, J. L., et al., "Putting the Service-Profit Chain to Work," *Harvard Business Review* (March–April 1989):164–174.

FIGURE 8-1 How Much Profit a Customer Generates over Time

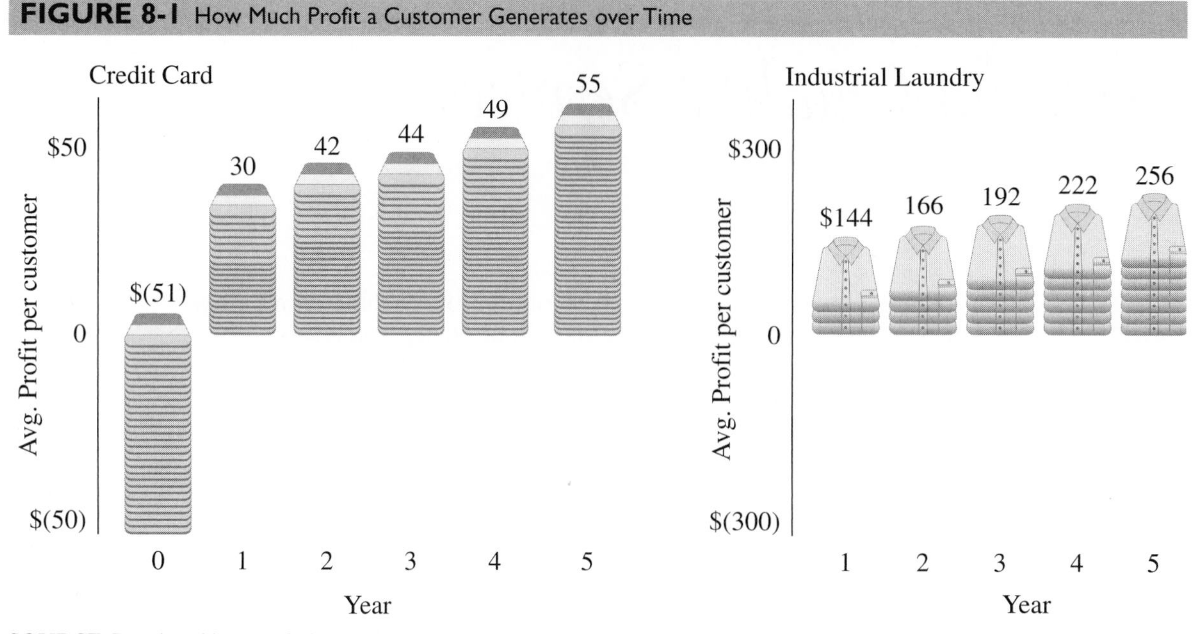

SOURCE: Reprinted by permission of Harvard Business Review. Figure from F. Reichheld and E. Sasser. "Zero Defections: Quality Comes to Services." Harvard Business Review (September–October 1990): 106.

Services are distinguished from manufacturing on several dimensions. First, many service attributes are **intangible**. This means that they cannot be inventoried or carried in stock over long periods of time. However, all services have some tangible aspects as well. The outputs of services are also **heterogeneous**. This means that for many companies, no two services are exactly the same. Consider, for instance, an advertising firm. No two advertising campaigns are alike. Customer requirements are different, and different campaigns are launched for different customers according to their needs. A third factor is that production and consumption of services often *occur simultaneously*. If you hire someone to mow your grass, you'll receive the service exactly at the same time it is produced.

The term *service* is very broad and covers many diverse industries, such as hotels, hospitals, financial services, and even prisons. Are financial services distinctly different from transportation, health care, or law firms? Certainly, there are similarities among the problems faced by these different categories of firms. However, there are also many dissimilarities.

One useful distinction between services and manufacturing centers on the aspect of **customer contact**. Customers tend to be more involved in the production of services than they are in production of goods. For instance, you probably have never seen anything you own during its manufacturing stage. In fact, many of the products you own were manufactured overseas. However, you probably are actively involved in the production of services you receive. If you work for a firm that hires a consultant, you will work closely with the consultant as he or she provides the services you purchased. When you receive a haircut, you are actively involved in the service by providing information and sitting still. In many restaurants it is not uncommon for the customers to fill their own drinks. This is called **customer coproduction**.

Because customers are actively involved in producing the services they consume, they create problems for service providers. For example, the time required to serve different customers can vary widely, making it difficult to plan capacity. The varying demands of customers also contribute to process variability that makes quality production of services difficult. Therefore, even though customers are the reason for the existence of services firms, they also make providing good service difficult.

The good news for customers is that by being actively involved in the production of the service, they can exert great control over the service provider and achieve great customization. This control can be manifested in a variety of different ways. For instance, if you have never visited the producer of a food product you purchase, you may not be aware of many issues concerning the products, such as sanitation or environmental pollution. However, you are not likely to remain a customer of a restaurant that is unclean or creating environmental problems. As a result of this greater customer control, service facilities, processes, and interactions must be designed in a way that promotes a positive encounter with the customer.

Internal versus External Services

An aspect of services that affects the definition of quality is whether a service is internal or external. **External services** are those whose customers pay the bills. **Internal services** are in-house services such as data processing, printing, and mail. Typically, these services are separate from the external customer. However, customer service to internal customers is very important to internal service because their services often can be outsourced. There is a trend in companies to outsource internal services. In a sense, this presents a competitive pressure on internal services. Although internal and external services may be very different, they do have many similarities. In both cases, competitive pressures can result in the possible loss of customers.

Voluntary versus Involuntary Services

Another way services differ is by being voluntary or involuntary. **Voluntary services** are services that we actively seek out and employ of our own accord. Generally, we research a voluntary service, such as a gas station, a restaurant, or a hotel, and have certain expectations when we engage its services. Even doctors often provide many voluntary services because patients can chose among different doctors to some extent.

The quintessential example of an **involuntary service** is a prison. Other involuntary services include hospitals, the IRS, the police department, the fire department, and other services that you do not choose. If you have the chance to engage this type of service at some point, you likely will have vague expectations about the experience. It is generally more difficult to achieve high levels of customer satisfaction in involuntary services. For example, does anyone really enjoy visiting a hospital or a dental clinic? A customer service survey would be laughable for a county jail. Yet employees of these organizations often desire to provide better service to the patrons. For example, many IRS agents are involved in quality improvement activities. Certainly, our perceptions and expectations of service quality can be affected by whether the service is voluntary or involuntary.

How Are Service Quality Issues Different from Those of Manufacturing?

We've identified three major realities in services that affect the approaches to quality adopted by service providers; these are intangibility, simultaneous production and consumption, and customer contact. Not surprisingly, they lead us to the major differences between services and manufacturing when it comes to quality.

Because services attributes can be intangible, it is sometimes difficult to obtain hard data relating to services. In manufacturing, dimensions such as height, weight, and width are available for measurement. Conformance to these measurements implies a certain dimension of quality. However, in services, such measurable dimensions are often unavailable. For this reason, many services organizations that use quality control charts encounter difficulty in using them, or they use them incorrectly. This is not to say that control charts cannot be used in services. However, compared with manufacturing, their use in services is quite low. Generally speaking, time (such as cycle time or response time) is a primary measurement available in service environments.

Simultaneous production and consumption of services means that you have to do it right the first time. You can't easily inspect and rework defects in a haircut the way you can in manufacturing.

Customer contact leads to an increase in variability in the process. This leads to a high degree of customization in services as well as great variability in the time required to perform services. In manufacturing, repetitive tasks are easily measured, and cycle times are generally consistent. When customers are intimately involved in processes, there is much more customization and much more variability than in manufacturing.

Services design, as is discussed in this chapter, is also very different from design in manufacturing. Because services are intangible, warranty or repair processes are not as important as recovery or reimbursement processes (see A Closer Look at Quality 8-1). Also, the design of the services must take into account such variables as customer moods and feelings because these affect customer perceptions of service quality.

Product liability issues in services are very different from manufacturing. Whereas in manufacturing liability issues center around safety concerns, in services liability issues often relate to **malpractice**, which refers to the professionalism of the service provider and whether reasonable measures were taken to ensure the customer's well-being. However, services also may have liability issues. In the Rocky Mountains, ski areas are sued regularly by customers who are injured. In many states, laws protect ski areas from such lawsuits by limiting liability for injuries.

Services do not have as long a history of quality practice as does manufacturing. Although many quality techniques such as control charts have been adopted by services companies, this trend is still new. Certainly, as time passes, more quality techniques are being developed specifically for services.

How Are Service Quality Issues Similar to Manufacturing?

For both manufacturing and service firms, the customer is the core of the business, and customer needs provide *the* major input to design. By focusing on the customer, many manufacturers and services firms have come to view themselves as service providers. Companies from Harley-Davidson to Hewlett Packard have spent extraordinary amounts of time designing services for their customers.

WHAT DO SERVICES CUSTOMERS WANT?

In Chapter 1 we considered the different quality dimensions relating to services. These were

Tangibles
Reliability
Responsiveness
Assurance
Empathy

Zeithamel, Parasuraman, and Berry[3] provided this list of dimensions of service quality after extensive research in a number of services sectors, and they have become widely

[3]Zeithamel, V., Parasuraman, A., and Berry, L., *Delivering Quality Service* (New York: Free Press, 1990).

A CLOSER LOOK AT QUALITY 8-1

SERVICE WARRANTIES: PROFITABLE OR A RIP-OFF—YOU DECIDE

Here's a secret two of the nation's largest consumer electronics retailers don't want you to know: Many times, they make more money off of service contracts than they do selling products. Best Buy Co. and Circuit City, Inc. aren't banking on sales of TVs, computers, and DVD players to make profits. They are counting on the service contracts to make them profitable.

Just look at the numbers. At Best Buy, service contracts are 4% of sales but provide 45% of profits. At Circuit City, the numbers are 3.3% and 100%! The profit margins on these contracts are between 50% and 60%. If you spend $400 on a service contract, Best Buy keeps $240 itself and gives $160 to the insurer. As profitable as these are, it is interesting that Wal-Mart has been slow to jump onboard. However, unlike Best Buy, they don't have as many salespeople to pitch these contracts.

As service contracts have become more profitable, Best Buy has cut back on disclosure. The company no longer reports its warranty profits separately. Best Buy's spokesperson says the products and contracts should be seen as inseparable. A Circuit City spokesperson adds: "We feel we give an appropriate amount of information."

So, as a consumer, when do service contracts make sense? Most often, the correct answer is, "Never!" Typically, only 20% goes to repair or replacement of products. That's why consumer organizations generally counsel against service contracts. According to one consumer advocate, "The worst rip-off is on appliances because they have gotten so reliable."

Consumer Reports cites four products for which contracts *might* make sense—these are treadmills, elliptical trainers, plasma TVs, and laptop computers. Remember, most products come with manufacturer warranties. Many times, extended warranties can be purchased more cheaply directly from the manufactuer.[a]

[a]Adapted from: W. Symonds, "The Warranty Windfall," *Business Week*, December 20, 2004, pp. 84–89.

accepted. However, this does not mean that your particular services industry does not include other dimensions. Therefore, the adoption of these dimensions in your service should include a careful consideration of the applicability of these and other dimensions.

As in any industry, the concept of leadership is one that Zeithamel, Parasuraman, and Berry believe is the key to success. However, they define this leadership role in a way that is quite interesting. The key aspects of a leader in services are given in Table 8-1.

First, a leader has a *service vision*. Such leaders really view service quality as the force underlying profitability and business success. When selecting strategies for improvement, leaders see quality as the winning strategy. Such a vision can be translated into action and excitement for others in the company. To win in services, a firm must develop a passion for service quality within the entire workforce. When there is intense interaction between customers and service providers, the attitude of employees is the key element in achieving service success. Active and involved leadership is very important to attaining this important organizational attribute.

TABLE 8-1 Attributes of Service Leaders

Attributes of Service Leaders
Service vision
High standards
In-the-field leadership style

Services leaders have *high standards*. In services, you will notice that some firms are better equipped and maintained than others. Sometimes this is evident in the small details. Some doctors' offices have a better selection of magazines than others; some restaurants are more comfortable than competitors; some grocers have a better selection of products; the list goes on and on. Those things don't happen by magic. They are the result of a leader with high standards and a focus on details. Have you noticed that some professors come to class impeccably prepared and others appear somewhat disorganized? This is so because the student/customer-oriented professor has higher standards for preparation and presentation than other professors. Think about yourself in a work situation. Do you provide a high level of service that reflects a high standard?

Outstanding services leaders have an *in-the-field style of leadership*. Because there is so much contact with the customer in a service system, the field is where the action is. Sam Skaggs, the founder of American Stores Corporation in Salt Lake City, Utah, was famous for stopping by his stores to make sure that things were in order. He viewed this as an important way of keeping the management on its toes. If Skaggs showed up at a single store in Kansas City, Missouri, for example, the manager immediately contacted all the other managers in town to give them a "heads up" that Skaggs was in town. Too often owners can become isolated from their businesses. By being in the field, they gain a better understanding of the business and how to serve the customer. Quality Highlight 8-1 shows how Ritz-Carlton uses a gold standard to keep its managers in touch with its customers.

SERVQUAL

An important tool developed by Parasuraman, Zeithamel, and Berry for assessing services quality is **SERVQUAL**. These researchers developed this survey instrument for assessing quality along the five service quality dimensions discussed in detail in Chapter 1. The SERVQUAL survey has been used by many firms and is an off-the-shelf approach that can be used in many services situations. The SERVQUAL instrument, a survey, has many advantages. Among these are

- It is accepted as a *standard* for assessing different dimensions of services quality.
- It has been shown to be *valid* for a number of service situations.
- It has been demonstrated to be *reliable*, meaning that different readers interpret the questions similarly.
- The instrument is *parsimonious* in that it has only 22 items. This means that it can be filled out quickly by customers and employees.
- Finally, it has a standardized analysis procedure to aid interpretation and results.

One of the benefits of statistical quality control (SQC) is that it is an accepted procedure for assessing process variability. One of the comforts of implementing SQC is knowing that many other firms have used this approach and benefited from it. Although the SERVQUAL survey is not as widely used as SQC, it is a standardized approach to gathering information about customer perceptions of service quality. As such, it provides a base, or a means, to get started in assessing customer perceptions of quality.

Expectations

The SERVQUAL survey has two parts—**customer expectations** and **customer perceptions**. Before discussing SERVQUAL expectations, we should first discuss the reasons for assessing both expectations and perceptions.

QUALITY HIGHLIGHT 8-1

RITZ-CARLTON HOTELS[b]

www.ritzcarlton.com

The Ritz-Carlton hotel company is a success in one of the economy's most logistically complex service businesses. Targeting primarily industry executives, meeting and corporate travel planners, and affluent business travelers, the Atlanta-based company manages more than 60 luxurious hotels that pursue the goal of being the very best in each market. Ritz-Carlton does so on the strength of a comprehensive service quality program that is integrated into marketing and business objectives.

Hallmarks of the program include participatory executive leadership, thorough information gathering, coordinated planning execution, and a trained workforce that is empowered to satisfy customers. Quality planning begins with the president, the CEO, and the 13 senior executives who make up the corporate steering committee. This group, which doubles as the senior quality management team, meets weekly to review the quality of products and services, satisfaction, market growth and development, organizational indicators, profits, and competitive status. Each year executives devote about one-quarter of their time to quality-related matters.

The company's business plan demonstrates the value placed on goals for quality products and services. Quality goals draw heavily on consumer requirements derived from extensive research by the travel industry and the company's customer reaction data, focus groups, and surveys. The plan relies on a management system designed to avoid the variability of service delivery traditionally associated with hotels. Processes are well defined and documented at all levels of the company.

Key products and service requirements of the travel consumer have been translated into Ritz-Carlton *gold standards*, which include a credo, three steps of service, and 20 "Ritz-Carlton basics." Each employee is expected to understand and adhere to the standards with defined processes for solving guests' problems as well as detailed grooming, housekeeping, safety, and efficiency standards. Company studies prove that this emphasis is on the mark, paying dividends to customers and, ultimately, to Ritz-Carlton.

The corporate motto is "ladies and gentlemen serving ladies and gentlemen." To provide superior service, Ritz-Carlton trains employees with a thorough orientation, followed by on-the-job training, and then job certification. Ritz-Carlton values are reinforced continuously by daily "line-ups," frequent recognition for extraordinary achievement, and a performance appraisal based on expectations explained during the orientation, training, and certification processes.

To ensure that problems are resolved quickly, workers are required to act at first notice regardless of the type of problem or customer complaint. All employees are empowered to do whatever it takes to provide "instant pacification." No matter what their normal duties are, other employees must assist if aid is requested by a coworker who is responding to a guest's complaint or wish.

Much of the responsibility for ensuring high-quality guest services and accommodations rests with the employees. Surveyed annually to ascertain their levels of satisfaction and understanding of quality standards, workers are keenly aware that excellence in guest services is a top hotel and personal priority. At each level in the company—from corporate leaders to managers and employees in individual work areas—teams are charged with setting objectives and devising action plans, which are reviewed by the corporate steering committee. In addition, each hotel has a "quality leader" who serves as a resource and advocate as teams and workers develop and implement their quality plans.

Teams and other mechanisms cultivate employee commitment. For example, each work area is covered by three teams responsible for setting quality-certification standards for each position, solving problems, and planning strategy.

[b]NIST, Profiles of Baldrige Winners, 2006.

Let's say that you desire to improve service quality along some dimension—either tangibles or reliability. The natural question is, "Which will create the *greater* improvement to the system for service?" If we understand both customer expectations and perceptions, we can assess the **gap** in these areas. For example, if customers have higher expectations for tangibles than for reliability, and customers perceive tangibles as poor, then a large gap or disconnect exists between the expected and delivered performance on tangibility. Given that this gap is larger, the greater potential for increasing customer satisfaction lies in addressing tangibles first. This type of analysis provides a good way to understand how best to improve customer satisfaction. Figure 8-2 shows the 22 survey items for expectations. The wording of the statements in the expectations survey relates to a generic firm in an industry that interests you. For example, if you were assessing customer service for a given grocery store, you might first administer the expectations survey to customers concerning a grocery store in general. Later, the perceptions survey might be administered to the customers of the particular grocery store. Table 8-2 shows the items that address specific service quality dimensions. The averaged scores for these dimensions provide SERVQUAL difference scores (to be demonstrated later).

Perceptions

The SERVQUAL perceptions survey shown in Figure 8-3 is administered to customers in the same way that the expectations survey was administered. Notice that the perceptions survey also contains 22 items that are matched with the same five service quality dimensions as the expectations survey (the dimensions are listed in Table 8-2).

Gap Analysis

The SERVQUAL instrument is useful for performing what is called **gap analysis**. Because services are often intangible, gaps in communication and understanding between employees and customers have a serious negative affect on the perceptions of services quality. The model in Figures 8-4 through 8-8 shows the gaps that commonly occur and can affect the perceptions of services quality.

Each of the gaps in the model demonstrates differences in perceptions that can have a detrimental effect on quality perceptions in services. The SERVQUAL survey instrument can be administered in a variety of ways that examine each of these gaps. For example, SERVQUAL can be used to explore differences in perceptions between customers, between managers, between managers and customers, and between employees. We will briefly examine each of the different gaps in the next paragraphs.

Gap 1 shows that there can be a difference between *actual* customer expectations and *management's idea* or perception of customer expectations. As a customer, have you ever wanted to tell a service provider, "I don't want you to do that; I want you to do something else!" It is very difficult for managers or employees to break out of the internal, process-oriented view of the business. Many times, improving processes does not equal improving customer service. To truly improve customer service, we must understand what the customer wants. The SERVQUAL instrument can be used to help in this understanding (see Figure 8-4).

Managers' expectations of service quality may not match service quality specifications. This mismatch is demonstrated in *gap 2* (see Figure 8-5). Once managers truly understand what the customer wants, then a system can be developed to help provide exactly what the customer wants. Often, because firms do not specify customer requirements according to a well-defined process, there is no way to know whether customer specifications and management expectations are aligned.

FIGURE 8-2 SERVQUAL Expectations Survey

	Strongly Disagree						*Strongly Agree*
1. Excellent ______ companies will have modern-looking equipment.	1	2	3	4	5	6	7
2. The physical facilities at excellent ______ companies will be visually appealing.	1	2	3	4	5	6	7
3. Employees at excellent ______ companies will be neat-appearing.	1	2	3	4	5	6	7
4. Materials associated with the service (such as pamphlets or statements) will be visually appealing in an excellent ______ company.	1	2	3	4	5	6	7
5. When excellent ______ companies promise to do something by a certain time, they will do so.	1	2	3	4	5	6	7
6. When a customer has a problem, excellent ______ companies will show a sincere interest in solving it.	1	2	3	4	5	6	7
7. Excellent ______ companies will perform the service right the first time.	1	2	3	4	5	6	7
8. Excellent ______ companies will provide their services at the time they promise to do so.	1	2	3	4	5	6	7
9. Excellent ______ companies will insist on error-free records.	1	2	3	4	5	6	7
10. Employees in excellent ______ companies will tell customers exactly when services will be performed.	1	2	3	4	5	6	7
11. Employees in excellent ______ companies will give prompt service to customers.	1	2	3	4	5	6	7
12. Employees in excellent ______ companies will always be willing to help customers.	1	2	3	4	5	6	7
13. Employees in excellent ______ companies will never be too busy to respond to customers' requests.	1	2	3	4	5	6	7
14. The behavior of employees in excellent ______ companies will instill confidence in customers.	1	2	3	4	5	6	7
15. Customers of excellent ______ companies will feel safe in their transactions.	1	2	3	4	5	6	7
16. Employees in excellent ______ companies will be consistently courteous with customers.	1	2	3	4	5	6	7
17. Employees in excellent ______ companies will have the knowledge to answer customers' questions.	1	2	3	4	5	6	7
18. Excellent ______ companies will give customers individual attention.	1	2	3	4	5	6	7
19. Excellent ______ companies will have operating hours convenient to all their customers.	1	2	3	4	5	6	7
20. Excellent ______ companies will have employees who give customers personal attention.	1	2	3	4	5	6	7
21. Excellent ______ companies will have the customer's best interests at heart.	1	2	3	4	5	6	7
22. The employees of excellent ______ companies will understand the specific needs of their customers.	1	2	3	4	5	6	7

Adapted from: V. Zeithamel, A. Parasuraman, and L. Berry, "SERVQUAL: A Multiple-Item Scale for Measuring Customer Perceptions of Service Quality," *Journal of Retailing*, Spring 1988, pp. 12–40.

TABLE 8-2 SERVQUAL Items and Dimensions

Dimension	*Items*
Tangibles	1–4
Reliability	5–9
Responsiveness	10–13
Assurance	14–17
Empathy	18–22

FIGURE 8-3 SERVQUAL Perceptions Survey

	Strongly Disagree						*Strongly Agree*
1. XYZ Co. has modern-looking equipment.	1	2	3	4	5	6	7
2. XYZ Co.'s physical facilities are visually appealing.	1	2	3	4	5	6	7
3. XYZ Co.'s employees are neat-appearing.	1	2	3	4	5	6	7
4. Materials associated with the service (such as pamphlets or statements) are visually appealing at XYZ Co.	1	2	3	4	5	6	7
5. When XYZ Co. promises to do something by a certain time, it does so.	1	2	3	4	5	6	7
6. When you have a problem, XYZ Co. shows a sincere interest in solving it.	1	2	3	4	5	6	7
7. XYZ Co. performs the service right the first time.	1	2	3	4	5	6	7
8. XYZ Co. provides its services at the time it promises to do so.	1	2	3	4	5	6	7
9. XYZ Co. insists on error-free records.	1	2	3	4	5	6	7
10. Employees in XYZ Co. tell you exactly when services will be performed.	1	2	3	4	5	6	7
11. Employees in XYZ Co. give you prompt service.	1	2	3	4	5	6	7
12. Employees in XYZ Co. are always willing to help you.	1	2	3	4	5	6	7
13. Employees in XYZ Co. are never too busy to respond to your requests.	1	2	3	4	5	6	7
14. The behavior of employees in XYZ Co. instills confidence in you.	1	2	3	4	5	6	7
15. You feel safe in your transactions with XYZ Co.	1	2	3	4	5	6	7
16. Employees in XYZ Co. are consistently courteous with you.	1	2	3	4	5	6	7
17. Employees in XYZ Co. have the knowledge to answer your questions.	1	2	3	4	5	6	7
18. XYZ Co. gives you individual attention.	1	2	3	4	5	6	7
19. XYZ Co. has operating hours convenient to all its customers.	1	2	3	4	5	6	7
20. XYZ Co. has employees who give you personal attention.	1	2	3	4	5	6	7
21. XYZ Co. has your best interests at heart.	1	2	3	4	5	6	7
22. Employees of XYZ Co. understand your specific needs.	1	2	3	4	5	6	7

Adapted from: V. Zeithamel, A. Parasuraman, and L. Berry, "SERVQUAL: A Multiple-Item Scale for Measuring Customer Perceptions of Service Quality," *Journal of Retailing*, Spring 1988, pp. 12–40.

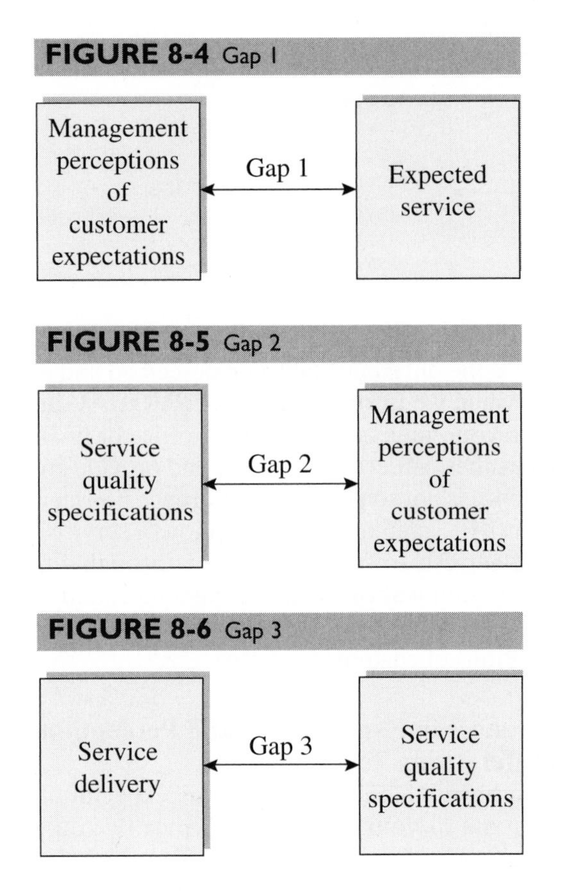

FIGURE 8-4 Gap 1

FIGURE 8-5 Gap 2

FIGURE 8-6 Gap 3

Once services specifications have been established, the delivery of perfect services quality is still not guaranteed. Inadequate training, communication, and preparation of employees who interact with the customer, referred to as **contact personnel**, can lower the quality of service delivered. This mismatch is shown in Figure 8-6 as *gap 3*.

Gap 4 (see Figure 8-7) shows the differences between services delivery and external communications with the customer. Companies influence customer expectations of services through word of mouth and through other media such as advertising. As a result, there could be a difference between what customers hear you say you are going to deliver as a service provider and what you actually deliver. Have you ever heard someone say, "They promised me one thing and gave me another." This gap can lead to seriously negative customer perceptions of service quality.

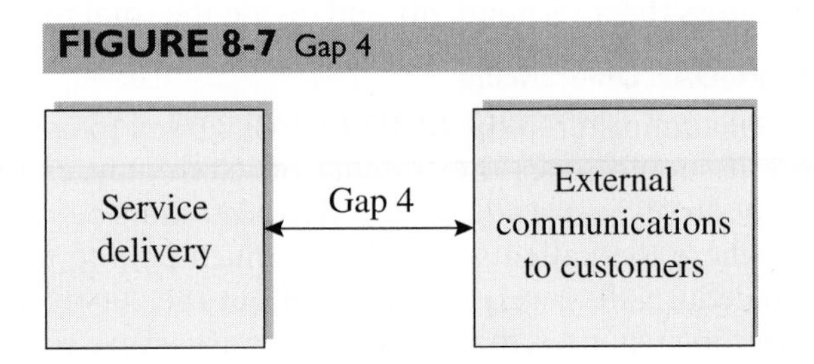

FIGURE 8-7 Gap 4

FIGURE 8-8 Gap 5

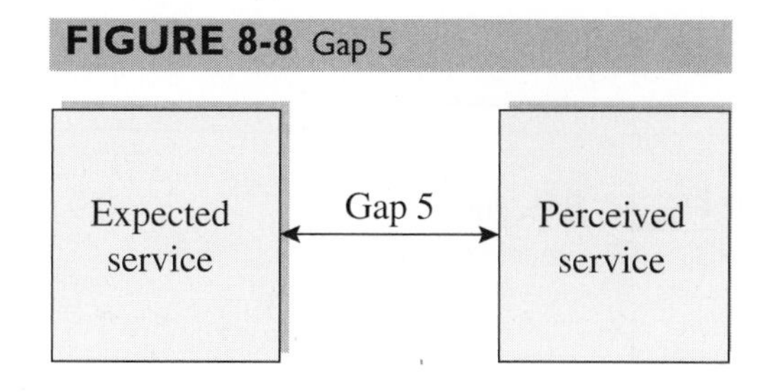

Gap 5 (Figure 8-8) is the difference between perceived and expected services, which we considered briefly when we introduced the SERVQUAL instrument. Think of the first time you dealt with your university admissions office or financial aid office. In many universities and colleges, these offices are well run and provide great service. However, in other colleges, their service is not so good. The difference between your expectations and your perceptions is directly related to your perception of service quality.

The key to closing gap 5 is to first close gaps 1 through 4 through thoughtful systems design, careful communication with the customer, and a workforce trained to provide consistently outstanding customer service. As long as these gaps exist, there will be lowered perceptions of customer service.

Assessing Differences in Expectations and Perceptions by Using the Differencing Technique

Let's suppose that you have administered both the expectations and the perceptions SERVQUAL instruments to your customers. Typically, you need a sample size of between 50 and 100 for each of the surveys (i.e., $50 < n < 100$, where n is the sample size). The difference score for SERVQUAL is computed using the following steps. Separate the SERVQUAL dimensions as follows:

Dimension	Items
Tangibles	1–4
Reliability	5–9
Responsiveness	10–13
Assurance	14–17
Empathy	18–22

For each respondent, sum your SERVQUAL scores for each item relating to a given dimension. Sum across the n respondents and divide the total by n.

Example 8.1: SERVQUAL Differencing

Recently, a hospital administered the SERVQUAL survey to its customers as a way to determine where it should focus the training of its employees to best improve customer service. Fifty surveys were administered to customers before and after they were treated. In cases where the patients were in too much pain to fill out the perceptions survey after a procedure, they were asked to fill out the survey at the follow-up visit. On the basis of the 50 responses, the following averages were computed for each item:

Item Number	Average Perception	Average Expectation
1	6.5	6.3
2	6.4	6.4
3	6.9	6.2
4	6.8	6.8
5	3.2	5.2
6	3.4	6.1
7	3.3	6.3
8	3.5	5.9
9	3.6	6.6
10	5.2	2.4
11	5.5	2.2
12	5.6	2.4
13	5.8	2.6
14	4.1	3.2
15	5.5	3.3
16	4.3	3.4
17	4.1	3.2
18	4.2	3.5
19	2.6	6.5
20	2.8	6.6
21	2.5	6.4
22	2.4	6.3

On the basis of these means, the following overall averages were computed for the different dimensions:

Perception Averages		Average Perception	Average Expectation	Expectation Averages	
Tangibles:	6.650	6.5	6.3	Tangibles:	6.425
(Avg. items 1–4)		6.4	6.4	(Avg. items 1–4)	
		6.9	6.2		
		6.8	6.8		
Reliability:	3.400	3.2	5.2	Reliability:	6.020
(Avg. items 5–9)		3.4	6.1	(Avg. items 5–9)	
		3.3	6.3		
		3.5	5.9		
		3.6	6.6		
Responsiveness:	5.525	5.2	2.4	Responsiveness:	2.400
(Avg. items 10–13)		5.5	2.2	(Avg. items 10–13)	
		5.6	2.4		
		5.8	2.6		
Assurance:	4.500	4.1	3.2	Assurance:	3.275
(Avg. items 14–17)		5.5	3.3	(Avg. items 14–17)	
		4.3	3.4		
		4.1	3.2		
Empathy:	2.900	4.2	3.5	Empathy:	5.860
(Avg. items 18–22)		2.6	6.5	(Avg. items 18–22)	
		2.8	6.6		
		2.5	6.4		
		2.4	6.3		

The averages for each of the dimensions of service quality were computed by averaging the items pertaining to each dimension. Finally, differences for the dimensions were computed as follows:

	Perception		Expected		
Tangible difference =	6.65	less	6.425	equals	0.225
Reliability difference =	3.4	less	6.02	equals	−2.62
Responsiveness difference =	5.525	less	2.4	equals	3.125
Assurance difference =	4.5	less	3.275	equals	1.225
Empathy difference =	2.9	less	5.86	equals	−2.96

The differences show that the greatest negative mismatch exists in the dimension of empathy, with reliability as a close second. Therefore, the training program should focus on teaching employees to be empathetic. Also, the process improvement efforts should focus on improving reliability. These changes will lead to the greatest improvements in customer service.

Example 8.2: SERVQUAL Two-Dimensional Differencing

If there is enough variation in the responses given to different dimensions, the two-dimensional differencing technique is very useful for evaluating SERVQUAL responses. Note that this technique is also used for specific questionnaires relating to specific services offered by companies. For example, St. John's Hospital administers surveys to patients asking about several specific services such as food, laundry, nursing, and many other services. The two-dimensional differencing technique allows the hospital to determine which services it should emphasize to improve customer perceptions and those that make little difference.

Using the information from Example 8.1, it is fairly simple to develop a two-dimensional services plane. The vertical axis reflects the expectations score and the horizontal axis relates to the perceptions score (Figure 8-9) using 4 (the neutral response) as the origin. The hospital analyst learns that emphasis is needed in the areas of reliability and empathy, as these are areas where expectations are high and perceptions are relatively low.

DESIGNING AND IMPROVING THE SERVICES TRANSACTION

So far in this chapter we have discussed customer perceptions of quality. One of the ways to improve customers' perceptions of quality is to improve the process of delivery of the service. Just as teams can succeed in manufacturing, teams in services can develop ways to improve processes and customer satisfaction.

Other concepts and tools include services blueprinting, moments-of-truth concept, and the Japanese method known as poka-yoke. Each of these is discussed in the following paragraphs.

Services Blueprinting

Lynn Shostack, CEO of Joyce International, Inc., is known for the statement, "The process is the service." Shostack also developed the process known as **services blueprinting.**[4] A services blueprint is a flowchart that isolates potential fail points in a

[4]Shostack, G. L., "Designing Services That Deliver," *Harvard Business Review* 62, 1 (1984):135.

FIGURE 8-9 Two-Dimensional Differencing Plane

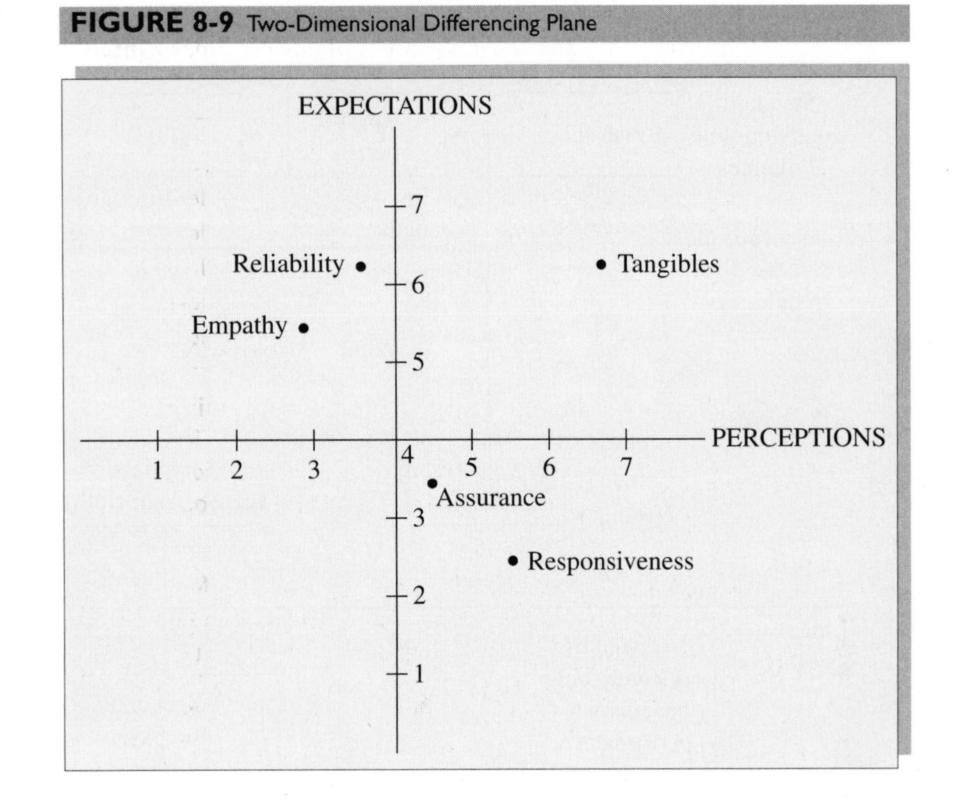

process. She recommends that blueprints be kept on every process in a service and that a "keeper of the blueprint" make the blueprints available for others in the firm. If possible, the blueprint also should be available on a computer network for all to view. There are four steps to developing a services blueprint.

1. ***Identify processes.*** In this step, processes are flowcharted so that the bounds of the process are identified. Figure 8-10 shows a simple process used by Shostack to demonstrate services blueprinting, in this case for a shoe-shine process. The steps include brushing shoes, applying polish, buffing the shoes, and collecting the payment.
2. ***Isolate fail points.*** Notice in Figure 8-10 that the applying polish stage is a possible fail point. What can happen here? The wrong color of polish could be applied, and the shoes will be ruined. This would be a very expensive mistake.
3. ***Establish a time frame.*** In a shoe-shining operation, time is a major determinant of profitability. As a result, those steps that waste time result in lost income. The analyst observing this process should establish a standard time for each step in the process.
4. ***Analyze profits.*** The customer spends about three minutes in the process. As errors occur in the process, the shoe shiner becomes liable for replacing ruined shoes, and other business is lost. Because delays and errors affect profitability, the figure shows that after five minutes customers begin to be lost and the business person loses money.

FIGURE 8-10 Services Blueprinting Example

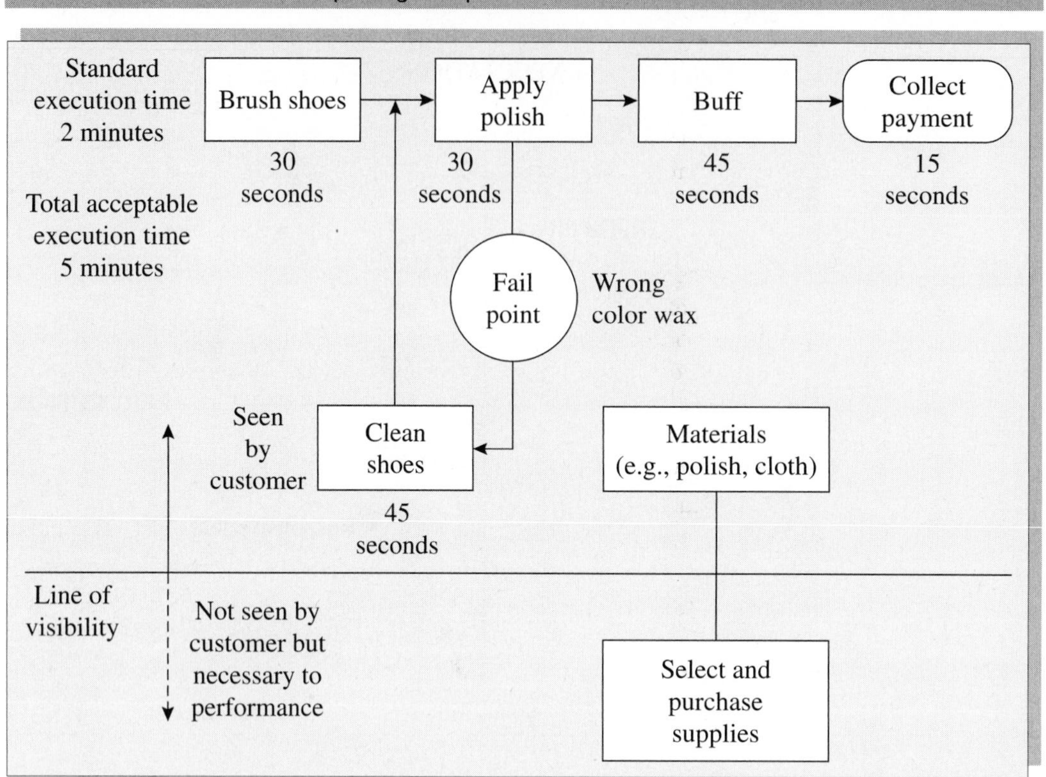

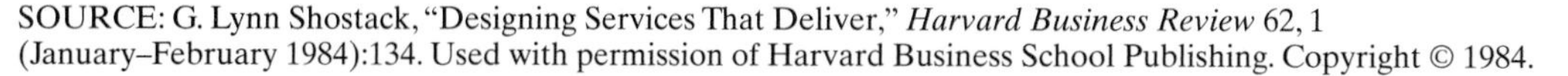

SOURCE: G. Lynn Shostack, "Designing Services That Deliver," *Harvard Business Review* 62, 1 (January–February 1984):134. Used with permission of Harvard Business School Publishing. Copyright © 1984.

Notice also that Figure 8-10 includes a line of visibility. The activities below the line of visibility are not seen by the customer, but they influence performance. This is true in many organizations. Often, the area above the line of visibility is referred to as the *front office*, and the area below the line of visibility is referred to as the *back office*. Many times process improvements focus on back-office activities, whereas front-office activities that involve high customer interaction are ignored. Services process blueprinting places the focus on front-office activities.

To understand how you could apply services blueprinting, think about a restaurant. Typically, when you first enter a restaurant, you expect to be greeted at the door. Can you remember a time when you weren't? This has happened to all of us at some point. A restaurant can install sensors or provide backups for the greeter so that this breakdown never occurs. Services blueprinting is a tool to help with brainstorming activities that lead to customer service improvement.

Moments of Truth

The fail points in the services blueprints are often referred to as **moments of truth**. These are the times at which the customer expects something to happen. Remember the SERVQUAL items? Expectations are a major determinant of customer perceptions of service quality. Therefore, when the customer expects something to happen, it has to happen. It is that simple! Some companies list these moments of truth and define fail-safes and procedures to see that they result in satisfied customers.

Customers' contact with the business can occur in many different ways—face to face, over the Internet, by phone, through a machine such as an ATM, or through the mail. All these moments of truth result in either happy customers or lost customers. Moments of truth also can happen at various stages of the product life cycle, such as when the product is being used, when customer service queries arise, when the product needs repair, and when it is eventually disposed of. A Closer Look at Quality 8-2 considers one firm that used the moment-of-truth concept to improve service.

Poka-yoke

Dr. Richard Chase and Dr. John Grout have been influential in promoting the use of **poka-yokes** (fail-safes) in services.[5] The idea behind fail-safing is to ensure that certain errors will never occur. Just as many processes seem to be designed to fail, they also can be designed *not* to fail. In services, Chase defines different classifications for fail-safe devices. These are

Warning methods
Physical contact methods
Visual contact methods

Fail-safe methods can also be defined using the "**three Ts**" (see Figure 8-11):

Task to be performed
Treatment provided to the customer
Tangibles provided to the customer

A CLOSER LOOK AT QUALITY 8-2

MOMENTS OF TRUTH IN ACTION

The Kroger Company: www.kroger.com

To achieve high levels of customer satisfaction, Brian Caldwell, a Kroger manager in Seymour, Tennessee, bases training programs for his store on the moments-of-truth philosophy. This philosophy is reinforced by a series of tools to stimulate and encourage an "atmosphere of employee and customer ownership" throughout all store operations. The program involves building a team ethic, eliciting customer feedback, and understanding employees' self-perceptions. The notion at Kroger is that satisfied employees are better able to satisfy customers. "This empowered, positive outlook creates employee excitement, especially when they see their idea go full circle from beginning to end in a supportive atmosphere."

Mr. Caldwell has seen significant improvement in the stores where this has been implemented. His personal mission statement is instructive: "Investing in people who in turn will invest in me." This means that if he helps employees to grow as people, they will, in turn, help him to grow as a person and a manager.

[5]Chase, R., and Stewart, D., "Make Your Service Fail-Safe," *Sloan Management Review* 35, 3 (1994):35–44.

FIGURE 8-11 The Three Ts

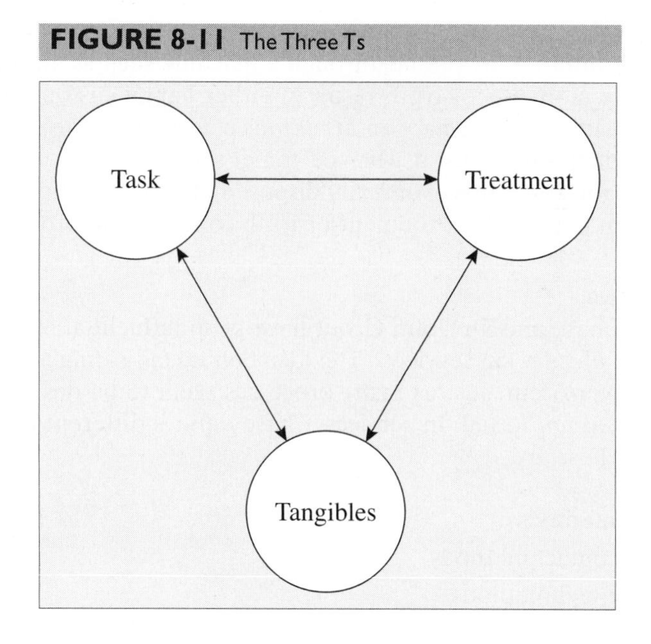

These poka-yoke classifications and Ts occur in many different forms. Some examples include[6] beepers in ATM machines that warn you to remove your card, toilets and sinks that automatically flush and shut off, the mechanism that stops you from inserting a disk upside-down in a computer, surgical trays that have indentations for different instruments, needle removers that prevent accidental needle pokes, requirements that bank tellers enter a customer's eye color before beginning a transaction so that identity is confirmed, or a file cabinet that locks the other drawers when any one drawer is opened so that the cabinet doesn't fall over.

Poka-yokes such as these represent a good amount of creativity and are very often used by Japanese and American companies to help ensure quality service. In a nutshell, you should isolate fail points in a process and then fail-safe the process to make sure that errors don't occur. Thinking back to Lynn Shostack's shoe-shine process, how would you fail-safe the process so that the wrong color polish would never be applied?

THE CUSTOMER BENEFITS PACKAGE

Just as many organizations have employee benefits packages, services firms can develop **customer benefits packages (CBPs)**. A customer benefits package consists of both **tangibles** that define the service and intangibles that make up the service. The tangibles are known as *goods-content*. Intangibles are referred to as *service-content*. The only difference between an employee benefits package and a services benefits package is the ultimate recipient of the benefits package.

CBPs are important not only in that they help define what it is that your service firm *will* provide to the customer but also in helping to define what *will not* be provided

[6]For an extensive literature review and interesting information about poka-yoke, you should check out Dr. John Grout's homepage on the Internet (*www.campbell.berry.edu/faculty/jgrout/pokayoke.shtml*). He has devoted a great deal of time to developing this resource, and it is the most concise resource on poka-yoke available. Grout is currently a professor of operations management at Berry College in Mount Berry, Georgia.

FIGURE 8-12 CBP Design Process

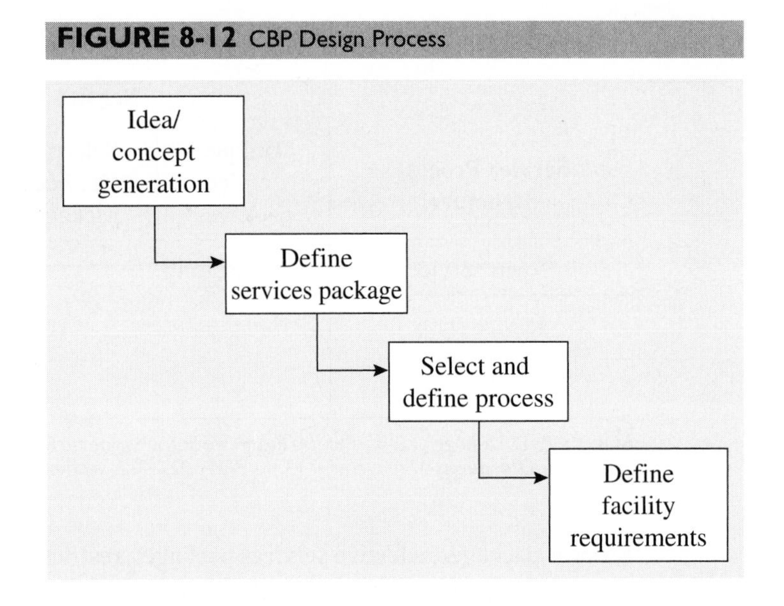

to the customer. More and more, firms are focusing on better defining the niches that they serve. As a result, the question of what they *will not* provide the customer is often as important as what they *will* provide the customer. By helping to answer this question, CBPs provide a foundation for developing a service strategy. The four stages of the service benefit package design process are as follows (see Figure 8-12):

1. Idea/concept generation
2. The definition of a services package
3. Process definition and selection
4. Facilities requirement definition

As defined by David Collier,[7] a professor at Ohio State University, the objectives of customer benefits package design are to

- Make sure the final CBP attributes you are using are the correct ones.
- Evaluate the relative importance of each attribute in the customer's mind.
- Evaluate each attribute in terms of process and service encounter capability.
- Figure out how best to segment the market and position CBPs in each market.
- Avoid CBP duplication and proliferation.
- Bring each CBP, and associated process and service encounters, to market as quickly as possible. Use the CBP framework and final attributes to design facilities, processes, equipment, jobs, and service encounters.
- Maximize customer satisfaction and profits.

The CBP is defined largely by the degree of freedom allowed by the firm in the customization of the services package. Deborah Kellogg and Winter Nie[8] provided a services process/services package matrix. As shown in Figure 8-13, firms will offer **unique**

[7]Collier, D., *The Service/Quality Solution* (Milwaukee, WI: Irwin/ASQC, 1994).
[8]Kellogg, D., and Nie, W., "A Framework for Strategic Service Management," *Journal of Operations Management* 13, 9 (1995):323–337.

FIGURE 8-13 Service Process/Service Benefits Package Matrix

	Service Package Structure			
Service Process Structure	Unique service packages	Selective service packages	Restricted service packages	Generic service packages
Expert service	Consulting			
Service shop		Higher education		
Service factory				Package delivery

SOURCE: D. Kellogg and W. Nie, "A Framework for Strategic Service Management," Reprinted from *Journal of Operations Management* 13, 4 (1995): 323–327, with permission from Elsevier Science.

services packages, selective services packages, restricted services packages, or **generic services packages**. Generic services packages are of the one-size-fits-all variety. Unique services packages are especially tailored for each customer. Your firm's ability to custom tailor a benefits package depends on the amount of flexibility you have as a service provider. Strategic issues such as organizational flexibility, top-management skill, employee motivation, training, hiring practices, culture of the service customer, nature of the service, and technological choice affect your ability to provide unique services packages.

Table 8-3 shows a customer benefits package from the Slide-Master firm. Notice that Slide-Master has taken great care in defining tangibles and intangibles for its CBP. Slide-Master evaluates its CBP performance using monthly surveys of employees and customers.

SERVICE TRANSACTION ANALYSIS

Since we have now discussed design of services processes and improving services processes, we can now present **service transaction analysis (STA)**.[9] This is a service improvement technique that allows managers to analyze their service processes at a very detailed level. As we stated, Crosby views service encounters as a series of transactions (or moments of truth). STA is a method for identifying these transactions and evaluating them from the customer's perspective to determine if there is a gap between service design and what the customer perceives as the service.

Figure 8-14 shows a service transaction analysis sheet. This sheet is a tool in STA. Once you have specified the service process to be studied, "mystery shoppers," or independent consultant-customers, walk through the entire process. After receiving the service, they then rate each transaction in the process with either a "+" (delighting), "0" (satisfactory), or "−" (unsatisfactory). The rationales for these scores are entered into the right side of the sheet, and an overall evaluation is provided in the bottom of the sheet. Using these sheets, service designers, managers, and staff can attempt to understand why the customer did not like certain aspects of the service and use this as an input into improving the process.

[9]Johnston, R., "Towards a Better Understanding of Service Excellence," *Managing Service Quality* 14, 2/3 (2004): 129–134.

TABLE 8-3 Final CBP Attributes for Slide-Master

Tangible (Goods-Content)	***Intangible (Service-Content)***
Slides	*Convenience and timeliness*
Quality, durable, clean, new film, and cardboard	Facility close to downtown
High-resolution pictures and slides	Pickup and delivery service
Multiple color combinations (6 million)	Telephone/fax order capability
Slide properly centered and focused	Standard 3-day service
	Rush 1-day service
Equipment/technology	Really try to handle last-minute customer changes
Latest computer hardware	Accurate, itemized billing by fifth working day of each month
State-of-the-art software	
High-quality camera	
Superior maintenance of all equipment	*Professionalism*
Well-maintained delivery vehicles with ads on the sides	Absolutely confidential services
Flexible for custom designs	Emphasis on telephone courtesy
	Refer to client by name
People	Client order documentation correctly filled out and processed
Clean and very well groomed	Flexible to customer needs
Attractive uniforms	
Packaging	*Consulting services*
Heavy, clear, high-quality sleeves	Artistic expertise
Sequenced and numbered properly	History of previous jobs
Loaded in attractive boxes or slide carousels	Technical knowledge
	Nonbusiness hours on-call professional
Facilities	
Ample parking spaces	*Service attitude*
Secure parking lot	User-friendliness
Attractive signs	Polite, responsive attitude
Clean, attractive building—outside	Do exactly what we promise
Up-scale indoor wall pictures and decor	Script dialogues for order taking and postpurchase callbacks
Reception area and service counter—clean and professional appearance	Confident but relaxing behavior
Soft, relaxing background music	
Plush furniture	
Bright lighting	
Complimentary coffee and soft drinks	
Restrooms, soap, etc.—clean	

SOURCE: Adapted from David Collier, *The Service/Quality Solution* (Milwaukee, WI: Irwin/ASQC, 1994).

Example 8.3: Service Transaction Analysis in Action

In the transaction analysis sheet in Figure 8-14, the large oak door, while professional, looks foreboding. When going to this lawyer's office, the client is probably nervous about approaching the attorney and can find the décor daunting and scary. The message on the second door made it appear that the attorneys may not provide impartial advice due to their relationships with a lending company. The not-so-subtle message sent by the rude receptionist demonstrated a mismatch between the service design and execution . . . and so forth. The application of this technique alerted the partners to what was really going on in their own office. Changes were made to address these problems.

FIGURE 8-14 STA Sheet

SERVICE TRANSACTION ANALYSIS SHEET

Organization:	lawyer			Service concept: general legal services for personal customers in a user-friendly, sympathetic and non-intimidating way
Process:	reception			
Customer type:	personal client			
	Score :			
Transaction	+	0	–	Message
Imposing oak entrance door, firmly shut				"trustworthy, professional but a little formidable"
Second door with advertisement for a lending company				"they like constructing barriers" "they may not be impartial"
Carpeted corridor but no sign of receptionist				"homely but is this the right place?" "unhelpful"
Receptionist behind desk ignores customer and continues typing				"they don't seem to care about me" "they don't think I am important"
She says "Yes?"				"not very welcoming" "I feel like I am intruding on her work"
Phone rings which receptionist answers				"I am not important" "other people have prority on her time"

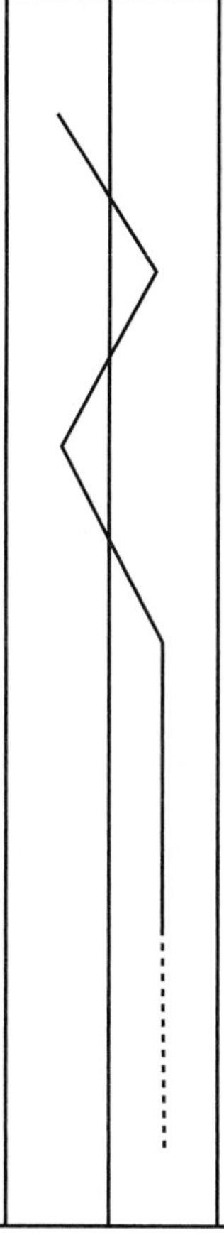

Overall evaluation
Poor service design. Little thought or concern for clients. Unfriendly and intimidating service.

THE GLOBALIZATION OF SERVICES

Just as CBPs may vary according to the culture of the area where the service is delivered, the trend toward globalization in services will alter the way we manage service quality. Economies such as those in eastern European countries and eastern Asian countries are following the lead of the United States by transferring labor and GDP

TABLE 8-4 Japan's Economy in the Year 2000

	Composition of Gross Domestic Product (%)				*Structure of Employment (in thousands of persons)*				*Annual Change (%)*
	1980		*2000*		*1980*		*2000*		
Primary sector	3.7		4.2		5,770		3,080		–3.1
Secondary sector	38.2		31.5		19,250		21,110		0.5
Manufacturing		29.3		21.6		13,770		14,200	0.2
Chemicals		5.4		1.5		1,750		1,450	–0.9
Primary metals		3.6		0.8		670		540	–1.1
Machinery		11.9		15.7		5,380		8,930	2.6
Other		8.4		3.6		5,970		3,280	–3.0
Construction		8.9		10.0		5,480		6,900	1.2
Tertiary sector	58.1		64.2		30,190		39,120		1.3
Utilities		3.0		1.5		300		330	0.4
Finance, insurance, real estate		15.5		8.5		1,910		2,410	1.2
Transport and communications		6.6		5.6		3,500		3,550	0.1
Service, and so on		33.0		48.6		24,480		32,830	1.5
Total	100.0		100.0		55,360		63,290		0.7

SOURCE: Adapted from Keizai Kikaku Hen (Economic Planning Agency), *2000 Nen No Nihon* [*Japan in the Year 2000*], (Tokyo: Okurasho, 1982), pp. 64–65, 72.

into the services sector. Table 8-4 shows the transition to the services economy by the Japanese between 1980 and 2000. The implication is that service competition will increase on a global scale, as has been the case in manufacturing for the past 40 years.

IMPROVING CUSTOMER SERVICE IN GOVERNMENT

If customer service is the battlefield for business in the twenty-first century, then government is probably the last frontier. There are some evidences of improvement in several aspects of government. The National Productivity Review[10] reports that some federal government agencies have adopted quality management (see A Closer Look at Quality 8-3). Quality professionals know that the military has long been an early adopter of statistical quality techniques. Many standards have been established. As of 1998, the government had developed a searchable list of 4,000 customer service standards for 570 federal departments and agencies. This effort started with a 1993 executive order for a "customer-driven government that matches or exceeds the best service available in the private sector." In the first year, only three agencies responded with agreements to commit to service standards. These early adopters were the Social Security Administration, the IRS, and the Postal Service.

States are also jumping on the bandwagon. By 2005, 41 states had established quality award programs. Many state agencies around the country have adopted quality techniques. However, many of these implementations appear to be in the early stages. Overall, government is lagging behind the private sector in quality adoption. Although

[10]Milakovich, M., "The State of Results-Driven Customer Service Quality in Government," *National Productivity Review* 17, 2 (1998):47–54.

A CLOSER LOOK AT QUALITY 8-3

GOVERNMENT SERVICE QUALITY: A STOP-AND-GO PROCESS[c]

Quality improvement in the U.S. government has been a hit-or-miss proposition since the 1980s. There have been serious and significant efforts that have been implemented since the Reagan administration. While the Reagan and Bush (GHW) administrations promoted the Baldrige and Deming approaches to improvement, Clinton emphasized improvement via Executive Order. The GW Bush administration has done little to improve government service quality.

During the Clinton administration, the major improvement effort was called "Reinventing Government." Among the triumphs of **reinventing government** were the IRS TeleFile program and the Social Security Administration telephone answer lines. However, the reinventors also avoided contact with traditional good-government groups, such as the National Academy of Public Administration and the Council for Excellence in Government, and wanted little to do with the Office of Management and Budget, the Office of Personnel Management, the federal inspectors general, and the General Accounting Office, all of which they viewed more as part of the problem than of the solution. Created as a quasi-independent unit appended to the vice president's office, the campaign to reinvent the government operated with enormous autonomy. Its leaders could not, and would not, testify before Congress, and its directives were kept in virtual space at a Web site whose days were numbered when the Bush administration arrived.

Convinced that statutory reform was either impossible or unnecessary, the reinventors were left with a number of weak devices for spurring change. Reinventing was poorly linked, if at all, to the employee performance appraisal process and carried few, if any, budgetary or personnel consequences. It simply did not matter to the things that matter to agencies—money and headcounts. Federal employees could earn Hammer awards for reinventing government but not bigger budgets or a more forgiving congressional committee.

Customer satisfaction surveys might be the one piece of reinvention to survive in the Bush administration. Few would expect the administration to interview the same customers, however. The Environmental Protection Agency surveyed reference librarians under Clinton, for example, but a survey of regulated parties was more plausible under Bush.

According to Paul C. Light, vice president and director of governmental studies at the Brookings Institution, the experience from reinventing government suggests at least three ingredients needed for more vibrant efforts in future administrations:

- ***Statutes.*** Although congressional action is always difficult, it outlives the ends of administrations in a way that executive directives cannot.
- ***Structure.*** The federal hierarchy is just as thick today as it was prior to reinventing government. Indeed, considerable evidence suggests that the most senior levels of the hierarchy are even thicker. This is the cost of letting a thousand flowers bloom without doing any weeding.
- ***Bipartisanship.*** Reinventing government would be alive today if it had been rooted in a bipartisan agreement on the need for big-government reforms. It is a lesson well worth remembering for future efforts.

[c]Adapted from "*Requiem for Reinvention, Government Executive*," *www.govexec.com/dailyfed/0201/021201ff.htm*, February 12, 2001.

the results are mixed, it is clear that private-sector quality management practices are being adopted in government. Why is this progress occurring? Many used to question whether the government had the inclination to adopt quality techniques, given the lack of profit motive. However, several factors seem to be driving this change:

- People want and desire to do good quality work.
- Because quality management is associated with improved employee satisfaction, there is a major impetus to improve.

- Government leaders are mandating standards, strategic plans, and new levels of performance at all levels of government. These standards are being adopted in government agencies because of the mandates.
- Demand for government services is growing at a faster rate than funding for them. The natural reaction is to simplify processes that have become bloated.
- Finally, the threat of privatization in government has led to an improvement in service in many areas.

QUALITY IN HEALTH CARE

Another area of services that is receiving much attention is health care. Several factors have contributed to this phenomenon:

- Health care is facing the same "cost squeeze" that government is facing.
- A move toward health maintenance organizations (HMOs) is causing hospitals to streamline operations.
- There is increasing diversity in health care.
- Calls for a nationalized health care system threaten the status quo and provide the competitive pressures that spur the impetus to improve.

In some cases, insurers such as Blue Cross are encouraging the use of quality management approaches. Health care workers are becoming increasingly knowledgeable about quality management practices and concepts. In fact, the very nature of health care requires careful and well-planned procedures.

Many health care customers, however, are uncomfortable with these changes to government and health care. If quality approaches are applied, it is probably best that efforts not focus entirely on efficiency. Reliability and empathy are dimensions that can only be good for health care.

SUPPLY CHAIN QUALITY IN SERVICES

As we have talked about before, one of the major differences between services and manufacturing is the involvement of the customer in the process. As is shown in Figure 8-15, service supply chains are bidirectional. This means that service customers actually provide inputs to the supply chain. Many times, these inputs can be information or, in some cases, labor—as in the case of a self-service gas station. Therefore, effective communication is necessary between customers and suppliers to prevent inadequate

FIGURE 8-15 Bidirectional Services Supply Chain

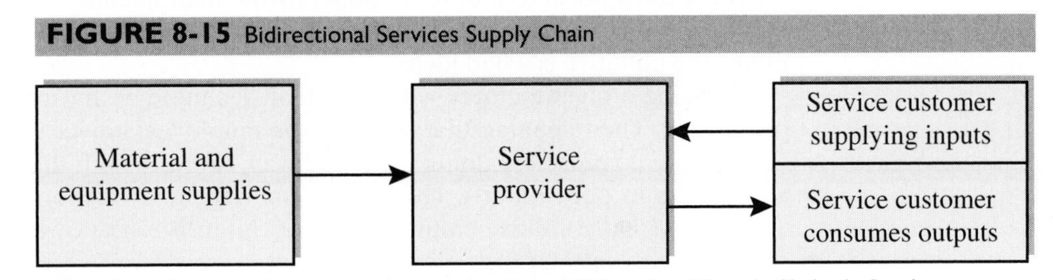

SOURCE: S. Sampson, "Customer-Supplier Duality and Bidirectional Supply Chains in Service Organizations," *International Journal of Service Industry Management* 11, 4 (2000), 348–364.

fulfillment of customer expectations. The concept of garbage in, garbage out implies that the quality of a service supply chain will be limited by the quality of the supplied inputs. Even if the customer provides poor inputs (inaccurate information), they may still expect accurate outputs. This has implications for service design. In an integrated supply chain, the service provider still has greater responsibility for verifying customer inputs to ensure they are accurate. In some cases, the service provider may be able to initiate communication of customer-input delivery expectations, such as through reservation systems. This may be complicated by the fact that many times, service delivery systems are inherently just-in-time. Of course, in services, there are other suppliers besides the customers. Often, these are managed in a more traditional supply-chain manner. However, the customer-supplier linkage makes services unique.

A THEORY FOR SERVICE QUALITY MANAGEMENT

As we studied in Chapter 2, theory development in the area of quality is an important work that continues. Dr. Scott Sampson of Brigham Young University is a researcher who is developing theory in services management. His **unified theory for services management** provides interesting insights for quality management.[11] This theory consists of several propositions. These propositions are based on the definitions of services that were introduced early in this chapter. Some of the propositions are as follows:

Proposition 1: The Unified Services Theory. "With services, the customer provides significant inputs into the production process. With manufacturing, groups of customers may contribute ideas to the design of the product; however, individual customers' only part in the actual process is to select and consume the output. Nearly all other managerial themes unique to services are founded in this distinction."

Proposition 2: The Unreliable Supplier Dilemma. "With services, the customer-suppliers often provide unreliable inputs."

"The Unreliable Supplier Dilemma" occurs because service customers provide themselves, their belongings, and/or their information as process inputs (by the Unified Services Theory). This simultaneous relationship as supplier and customer makes it difficult for the service provider to control the supplied inputs.

Proposition 3: Capricious Labor. "With services, customer-labor may ignore, avoid, or reject technologies or process improvements which are intended to increase quality and productivity. As a result, customer buy-in to process changes must be carefully addressed."

Capricious labor occurs because many services customers provide themselves as labor inputs into the production process. In manufacturing organizations, labor is expected to conform to corporate policy. If the manufacturer mandates that a quality initiative be implemented, labor is generally expected to conform, even when labor thinks the initiative is a bad idea.

Imagine a manufacturer who invests in technology that will improve quality of production. Then imagine that some of the employees reject the technology stating, "I am more comfortable doing it the old way," even though the old way is inefficient and results in poor quality. How would management respond to those self-willed employees? Unless those employees were children of the owner, one might suspect

[11]Sampson, S., "The Unified Theory Approach to Service Quality," DSI 1998 Proceedings, Las Vegas.

their jobs would be in jeopardy. In services, much labor is performed by customers who coproduce. Therefore, they can adopt or reject what they don't like. For example, customers don't have to stand in line if they don't wish.

Proposition 4: Everyone Presumes to be an Expert. "With services, the customer often provides product specifications (what to make) and process design (how to make it), often without the invitation of the service provider."

"Everyone thinks they're an expert" occurs because the necessity for customer inputs in service processes means that most customers have extensive experience with the service process. This experience breeds process knowledge and ideas for improvement. The words of Richard Chase capture this idea well:[12] "Everyone is an expert on services. We all think we know what we want from a service organization and, by the very process of living, we have a good deal of experience with the service creation process."

SUMMARY

In this chapter we have studied quality in services. Because services involve intangibles, they are different from manufacturing. Because of the lack of hard measures, statistical quality control techniques are not always as successful in services as they are in manufacturing. This doesn't mean that statistical thinking is not extremely useful in services.

Services definitions and classifications were presented that help us better understand services. We have provided tools for services such as SERVQUAL, services blueprinting, moments of truth, poka-yokes, and customer benefits packages.

The bottom line is a satisfied customer. Customers pay the bills. They are the object of our efforts. At times, all the customer wants is a caring ear to bend. In our race for profits, efficiencies, and better processes, let's not forget the human touch.

KEY TERMS

- Contact personnel
- Customer benefits packages (CBPs)
- Customer contact
- Customer coproduction
- Customer expectations
- Customer perceptions
- External services
- Gap
- Gap analysis
- Generic services packages
- Heterogeneous
- Intangible
- Internal services
- Involuntary service
- Malpractice
- Moments of truth
- Poka yoke
- Product liability
- Reinventing government
- Restricted services packages
- Selective services packages
- Service transaction analysis (STA)
- Services blueprinting
- SERVQUAL
- Tangibles
- Three Ts
- Unified theory for services management
- Unique services packages
- Voluntary services

DISCUSSION QUESTIONS

1. Discuss the ways in which services are unique in comparison with manufactured goods. How do these differences affect the management of service quality?
2. Provide an example of customer coproduction other than the example provided in the text. What are the advantages and disadvantages of customer coproduction for service providers?

---

[12]Chase, R., Jacobs, H., and Aquilano, N., *Operations Management for Competitive Advantage* (Homewood, IL: Irwin, 2003), p. 142.

3. Are quality techniques in the service industry well developed or still fairly immature? If you believe that they are immature, why do you think that this is the case?
4. Why do you believe that quality techniques in the service industry are less mature than quality techniques for manufactured products? What can be done to bring quality techniques for the services industry up to a higher level?
5. Discuss the distinction between voluntary services and involuntary services. Why is this distinction important in our understanding of service quality?
6. Are the Baldrige criteria applicable to service situations? If so, how?
7. List Zeithamel, Parasuraman, and Berry's five dimensions of service quality. Is the list identical for every service provider, or does it vary from company to company? Explain your answer.
8. Discuss some of the qualities of an effective leader in a service context.
9. What is SERVQUAL? How does SERVQUAL help a firm assess its service quality?
10. What are the advantages of the SERVQUAL instrument?
11. Discuss the concept of gap analysis in the context of a SERVQUAL assessment.
12. What is a service blueprint? How is a service blueprint developed?
13. Describe the concept of moment of truth in a service context. Describe a moment of truth that you recently experienced as a consumer. Was your service experience satisfactory or unsatisfactory? Explain your answer.
14. How can the moment-of-truth concept be used as a training tool in a service setting?
15. What is a customer benefits package? What is the purpose of developing customer benefits packages in a service context?
16. In what ways will the globalization of services alter the way that businesses manage their service quality?
17. Discuss the initiatives that the U.S. government has taken to increase its emphasis on service quality. Are they effective?
18. Discuss several of the factors that have contributed to an increase in attention directed toward quality in health care.
19. Discuss the unified theory for services management developed by Dr. Scott Sampson. Do you agree or disagree with the principles underlying Sampson's theory? Explain your answer.
20. In your judgment, will the management of "service quality" ever progress as far as the management of "manufactured goods quality"? Please make your answer as substantive as possible.

PROBLEMS

1. A national electronics retail chain charges $350 for a service contract. Of this, the company sends $120 to an insurer. Calculate the profit margin for the service contract.
2. Using the example of the service transaction analysis (STA) worksheet in Figure 8-14, chart and evaluate the transactions for your university or college advising office. Report your findings and overall evaluation.
3. Develop a consumer's benefits package for a service business in your community. Be exhaustive and explicit in your package identification.
4. Develop a services blueprint for a local hair salon or barber. Identify possible fail points, back-office processes, and fail-safes.
5. Recently, a medical office administered the SERVQUAL survey to its customers as a way to determine where it should focus the process improvement. Forty surveys were administered to customers before and after they were treated. On the basis of the 40 responses, averages were computed for each item. Using the averages in the table that follows, compute dimension averages. Based on your findings, which dimensions should be emphasized?

Item	Perception Average	Expectation Average	Differences
1	5.5	2.3	3.2
2	5.4	2.4	3.0
3	5.9	2.2	3.7
4	5.8	2.8	3.0
5	3.2	3.2	0
6	4.4	4.1	0.3
7	4.3	5.3	−1.0
8	2.5	4.9	−2.4
9	4.6	5.6	−1.0
10	6.2	3.4	2.8
11	6.5	3.2	3.3
12	6.6	3.4	3.2
13	6.8	3.6	3.2
14	3.1	3.2	−0.1
15	4.5	3.3	1.2
16	3.3	3.4	−0.1
17	3.1	3.2	−0.1
18	3.2	3.5	−0.3
19	1.6	6.5	−4.9
20	1.8	6.6	−4.8
21	1.5	6.4	−4.9
22	1.4	6.3	−4.9

6. For the data in Problem 5, perform a two-dimensional differencing analysis. Do your results differ from your answer in Problem 5?
7. The averages for different dimensions of service quality were computed by averaging the items pertaining to the dimension. Use the following data to determine which dimensions to emphasize.

	Perceptions	Expectations
Tangibles =	5.40	1.42
Reliability =	3.20	6.40
Responsiveness =	2.45	2.30
Assurance =	5.60	3.30
Empathy =	1.90	6.40

   a. Using simple differencing, determine which dimensions should be emphasized.
   b. Use two-dimensional differencing to determine which dimensions should be emphasized.
   c. Based on your findings, choose the most important dimension and describe how you would develop a process improvement program to address the dimension that needs to be improved.

8. A state university wants to perform a gap analysis to determine what student traits corporate CEOs find most important. The exercise is to be administered to CEOs and involves two surveys—an expectations survey and a perceptions survey. All questions are answered on a 10-point scale. The attributes the CEOs are asked to rate are propensity for lifelong learning, ability to work in teams, innate ability, and cognitive ability. The results of nine surveys are synopsized in the following table:

	Lifelong Learning		Teamwork		Innate Abilities		Cognitive Abilities	
Respondent	Expectations	Perceptions	Expectations	Perceptions	Expectations	Perceptions	Expectations	Perceptions
1	8	0	8	10	2	5	3	9
2	7	4	9	9	6	5	4	8
3	9	2	9	10	4	5	2	7
4	10	5	10	9	7	4	3	8
5	5	1	8	7	5	6	4	9
6	9	1	7	9	2	7	2	10
7	9	3	9	10	2	5	3	10
8	8	2	10	10	3	6	2	10
9	7	0	9	10	2	5	4	9
Average	8.00	2	8.78	9.33	3.67	5.33	3.00	8.89

Perform a gap analysis by developing a two-dimensional differencing plane and evaluate the results.

Cases

Case 8-1 Yahoo! Designs Quality Services with Customers in Mind

Yahoo!: ***www.yahoo.com***

Millions of people log onto the Internet every day and use the services of Yahoo!, the World Wide Web's most popular portal. Although Yahoo! is a young company, it has evolved quickly into a firm that provides comprehensive Internet services to a global clientele. The story of how Yahoo! started and how it has attracted such a loyal clientele in an extremely competitive industry is quite amazing.

Yahoo! started in 1994 as a hobby of its co-founders, Jerry Yang and David Filo. Both individuals were doctoral students in electrical engineering at Stanford University who took time off from writing their dissertations to surf the Web, classify the content, and create categories. As the two students started classifying more and more Web sites, the product that they were developing started to attract the attention of other people. This provided the two individuals the motivation to continue to expand their efforts, and Yahoo! as a company was born.

Yahoo! is free to its users. The company generates revenue by selling advertising space on the Yahoo! search engine. What is particularly remarkable about Yahoo! is the customer base that the company established in only a few short years. Yahoo! serves more than 237 million unique users and has become the most recognized and valuable Internet brand globally.

How has Yahoo! established such a large customer base? Largely by trying very hard to determine what its customers want and then designing quality services to meet its customer's needs. For example, Yahoo! follows the traffic patterns of its search engine very carefully in an attempt to determine the types of information its users are seeking. Early on, the company noticed that many of its users were searching for stock quotes by typing in either a company's name or its ticker symbol. Yahoo! created a financial site on its search engine and partnered with the major stock exchanges to get direct feeds of stock quotes. The result—Yahoo! now gets more than one million queries per day just for stock quotes. Yahoo! offers similar levels of service for news, weather, and sports.

Particularly striking is the way that Yahoo! has customized its search engine to appeal to different demographic groups. Yahoo! now has search engines specifically tailored to its customers in 25 foreign countries and 13 languages. Also, Yahoo! is tailored to specific cities. If you click on Yahoo! Seattle, for example, you instantly have at your fingertips a vast amount of information specifically about the Seattle area. Yahoo! also has segmented its market by age and area of interest. Yahooligans is a search engine designed specifically for children, with kid-safe content presented in a manner that they can use. Similarly, Seniors' Guide is a directory designed with information of interest to older users.

A challenge for Yahoo! is staying current with its customers' preferences and demands. To accomplish this, the company encourages input from its users and gets thousands of e-mail messages per day. The users simply tell Yahoo! what they like and what they don't like. Another thing that Yahoo! does is move very quickly to get a product to market. The corporate culture does not demand that a product be perfect before it is placed on the search engine. The company is willing to take chances and will simply pull a product from its Web site if its users don't like it.

As a result of its success, Yahoo! has attracted many competitors, such as Excite, Lycos, Infoseek, and WebCrawler. Several of these companies have now partnered with large firms (e.g., Infoseek with Disney and AOL with Time Warner), so the heat will remain on Yahoo! to continue to design high-quality Internet services and products. What other companies can learn from Yahoo! is that a thorough understanding of customer needs is the first step toward designing high-quality service products. Also, a strong follow-through and a willingness to listen to customer suggestions and complaints are key attributes to a service company's success. ■

DISCUSSION QUESTIONS

1. Think about the Internet search engine that you use the most often. If it is Yahoo!, what is it about Yahoo! that attracts you as a user? If it is not Yahoo!, what could Yahoo! learn from the search engine that you use that could help make it better?
2. What parallels do you see between developing a high-quality service product and a high-quality manufactured product? Make your answer as substantive as possible.
3. Is Yahoo! a company that was simply at the right place at the right time, or are many of its service innovations truly unique? Explain your answer.

Case 8-2 UPS: Delivering the Total Package in Customer Service

UPS: ***www.ups.com***

In 1907 there was a great need in the United States for private messenger and delivery services. The U.S. Postal Service was not yet offering parcel delivery, and few offices and private homes had telephones, so messages had to be delivered by hand and packages by courier. To help meet customers' communication needs, an energetic 19-year-old, James (Jim) E. Casey, started the American Messenger Company in Seattle. Although the company began with a small staff and faced stiff competition, it did fairly well, primarily because of Casey's strict policies. He built his business on four principles: customer courtesy, reliability, around-the-clock service, and low rates.

Casey's company eventually became United Parcel Service, or UPS. The name United Parcel Service was chosen to draw attention to the words *United*, to emphasize the fact that shipments were consolidated to increase efficiency, and *Service*, because the company recognized that service was all that it had to sell. UPS grew quickly through the years and became well known for its chocolate-colored delivery vans and courteous drivers. The public also liked UPS's business concept. It was convenient to send packages by

UPS, and people trusted UPS to deliver packages safely to their destinations. All kinds of people and businesses used UPS's services, from pharmaceutical companies that shipped life-saving drugs across country to grandparents who sent their grandchildren birthday presents and boxes of candy at Christmas.

Although UPS has always been a friendly company, until the mid-1980s it relied primarily on technology to maintain efficiency, keep prices low, and provide new services. A major internal change took place at UPS in the mid-1980s when the company decided to shift its emphasis from technology to satisfying customer needs. This shift represented a recognition that UPS customers were becoming more sophisticated and had a variety of needs the company was uniquely equipped to satisfy. Paramount among these were an increased need for information, a desire to move packages even more quickly and efficiently, tremendous competitive pressure from Federal Express, and a demand for customized prices and services.

UPS moved quickly to satisfy its customers' needs by developing new service products. For example, TotalTrack, which is available at UPS's Web site, can instantly provide customers with tracking information on all bar-coded UPS packages. This service helps vendors know when their buyers have received their shipments. Inventory Express is a contract logistics management service in which UPS stores a customer's merchandise and then ships it when it is needed, often on a just-in-time basis.

UPS also has improved its basic package pickup and delivery services. Customers with urgent shipments can telephone UPS to take advantage of On-Call Air Pick Up, which provides fast pick up at the customer's home and overnight delivery of packages. To accommodate customers who ship to sparsely populated areas in the United States and abroad, UPS has improved its geographic reach to every address in the United States and locations in more than 185 countries and territories. ■

DISCUSSION QUESTIONS

1. Based on the description of UPS, what do you believe are UPS's strengths and weaknesses?
2. How has UPS used technology in its design of quality services? Make your answer as substantive as possible.
3. Describe a positive or negative experience that you have had with UPS (or one of its competitors such as FedEx or the U.S. Postal Service). If the experience was positive, reflect on whether the experience is consistent with UPS's new emphasis on customer needs. If the experience was negative, what could UPS have done to better satisfy your needs?

CHAPTER 9

Managing Supplier Quality in the Supply Chain

We define a supply chain as a network of facilities that procures raw materials, transforms them into intermediate subassemblies and final products and then delivers the products to customers through a distribution system.
—COREY BILLINGTON OF HEWLETT-PACKARD[1]

As you can tell from the introductory quote, a great deal of thought has gone into managing the supply function at Hewlett Packard Corporation (HP). HP uses analytical, accounting, and managerial tools to improve its performance. HP is known for its commitment to its customers and for understanding that quality performance is closely related to supply-chain activities. HP, therefore, focuses a lot of attention on those supply-chain activities.

In this chapter we discuss the roles of purchasing, supplier development, logistics, and other supply-chain functions. We also introduce some statistical quality control tools that are used to evaluate the inputs provided from suppliers. Remember that the performance of your suppliers directly affects your reliability and your ability to satisfy customers. As a result, the supplier is key. This connective relationship between suppliers, producers, and consumers is both important and timely as firms are attempting to improve quality along with on-time performance. This is a major theme of this chapter.

THE VALUE CHAIN

To understand the **supply chain**, we first discuss the economic concept of the value chain. Michael Porter,[2] the noted economist and author, identified a systematic means for examining all the activities a firm performs and how those activities interact. The **value chain** is a tool that disaggregates a firm into its core activities to help reduce costs and identify sources of competitiveness. It is part of the **value system** that consists of a network of value chains. The value chain and core activities that are performed by any company include inbound logistics, operations, outbound logistics, marketing and sales, and service.

Figure 9-1 shows Porter's value chain. Notice that this is a chain for a single firm; however, the firm's suppliers have value chains also. The core activities shown in the figure are termed **value-chain activities** because they are the tasks that add value for

[1]Billington, C., "Strategic Supply Chain Management," *OR/MS Today* (April 1994):20–26.
[2]Porter, M., *Competitive Advantage* (New York: Free Press, 1995).

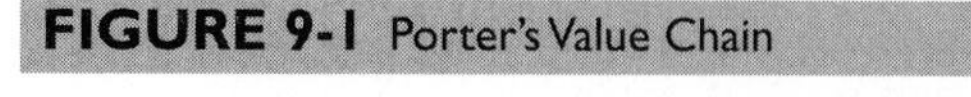

FIGURE 9-1 Porter's Value Chain

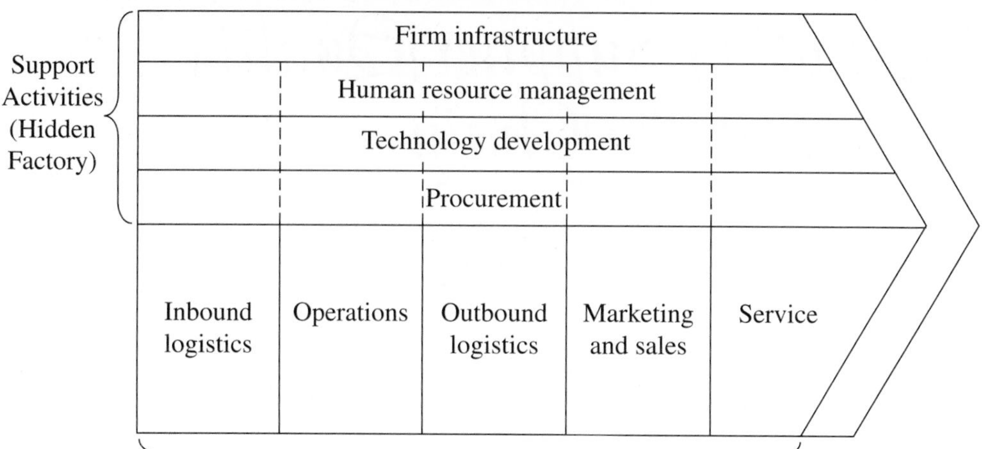

SOURCE: M. Porter, *Competitive Advantage* (New York: Free Press, 1995). Reprinted with the permission of The Free Press, a Division of Simon & Schuster Adult Publishing Group. All rights reserved.

the customer. If a firm performs these core functions well, the result is high customer satisfaction. Non-value-chain activities typically have costs but no effect on the customer and are referred to as the **hidden factory.** The hidden factory contains all the bureaucratic processes that are not part of the core activities in Figure 9-1.

The Chain of Customers

From a quality perspective, an interesting variation of the value chain is the concept of the **chain of customers.**[3] Looking at the activities along the value chain sequentially, we see that the links in the value chain are really people performing different functions. The chain of customers is revealed when you view the step in the chain after you as your own customer. This means that if you work at workstation 4 in a process at the core of the value chain, you will make sure that the work you do is absolutely impeccable before you release it to your "customer" in workstation 5. This chain extends from raw materials through supplier firms to the producing firm, with the final link in the chain being the ultimate consumer of the product. The notion is that if each of us along a chain works to satisfy our own customer, the final customer will be very satisfied, and our products and services will be free of defects and mistakes.

Managing the Supply Chain

The concept of supply-chain management extends the economic concept of the value chain. Figure 9-2 shows a rendering of a supply chain. Notice that it includes several suppliers, plants, distribution centers, and customer groups. This is a useful extension of the value chain because it provides a more realistic picture of the value chain. Notice that the value chain focuses on activities such as inbound and outbound logistics. These are supply-chain activities. One of the most significant aspects of the value chain is the *linkage* between a series of suppliers and consumers. This linkage is especially tenuous because it involves the complex interaction of logistics, systems, and human behavior. These linkages and relationships between suppliers and customers have undergone radical changes in the past decade. Much of this chapter focuses on this linkage.

---

[3]Schonberger, R., *Building a Chain of Customers* (New York: Free Press, 1990).

FIGURE 9-2 A Supply Chain

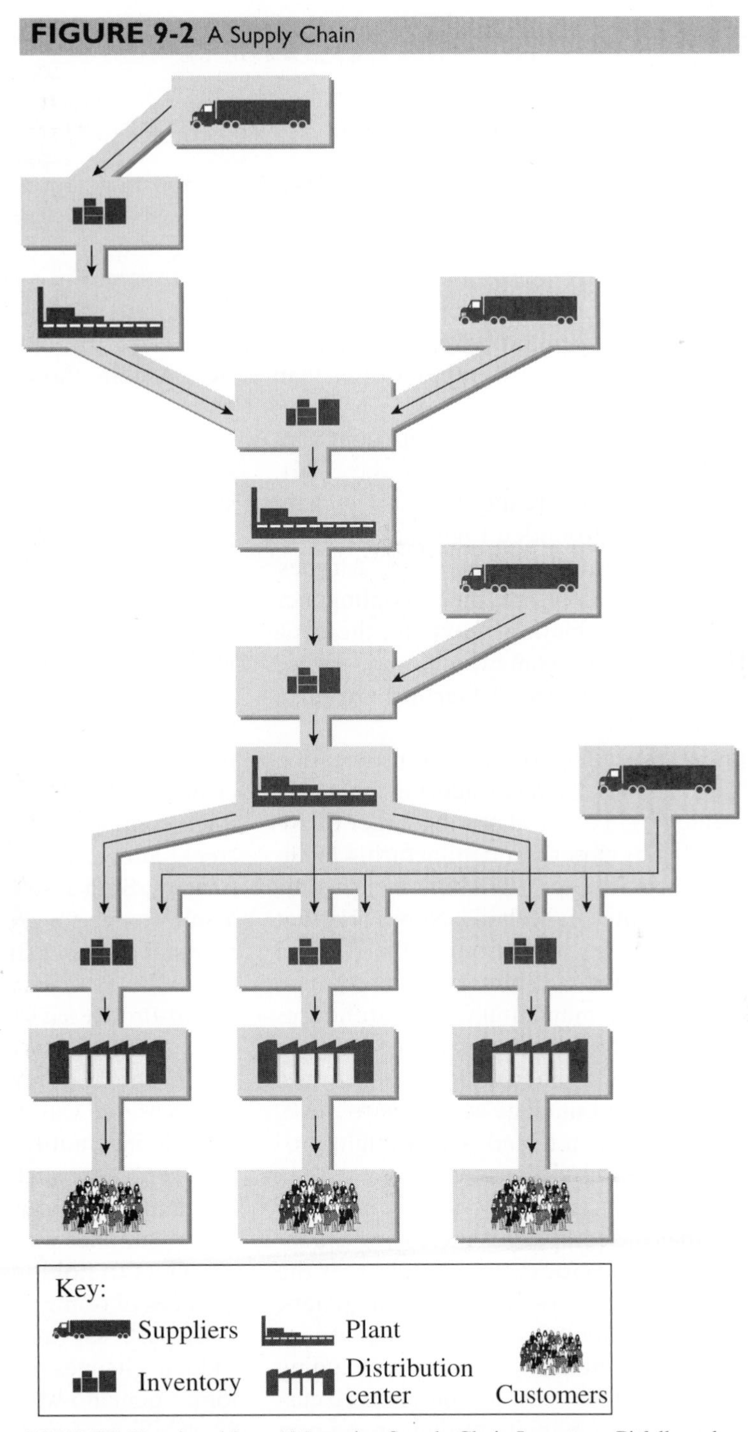

SOURCE: Reprinted from "Managing Supply Chain Inventory: Pitfalls and Opportunities," by H. Lee and C. Billington, *MIT Sloan Management Review* (Spring 1992): 66 by permisison of publisher.

A CLOSER LOOK AT QUALITY 9-1

SUPPLY CHAINS AND TERRORISM[a]

Since the terrorist attacks on New York and Washington, many companies are shifting their supply-chain priorities from squeezing costs through inventory reduction to limiting the consequences of transportation disruptions on production. The concept of lean-inventory management systems feeding directly into production lines—in other words, lean manufacturing—has been the goal of companies looking to save money through more efficient supply chains. However, the September 11, 2001, tragedy introduced a new reality for supply-chain managers faced with grounded flights, closed borders, and lengthy inspections of truck cargoes. "People are coming to see they need contingency plans so that in case something happens, there's a stockpile of critical parts somewhere," says Bruce Bond, a supply-chain analyst at Gartner Group, a tech-consulting leader.

To minimize the effect of such transportation disruptions, companies are reconsidering their lean inventory strategies in favor of storing larger quantities of critical parts closer to manufacturing facilities. Arrow Electronics, Inc., a $13 billion electronic components distributor in Melville, New York, that has 13 distribution centers around the United States, is working with customers to determine whether it can help them maintain miniwarehouses near production lines through the use of Arrow's proprietary inventory management software. Arrow might open and manage such a warehouse, or a customer might open it, and Arrow might provide the management services.

Bill Forster, Arrow's vice president of worldwide logistics, says that the company is increasing the inventory of parts it holds for customers and sees the same trend among some of its largest customers. For Arrow, decisions about whether to increase inventory depend on a number of factors, including the importance of a given part to one or more customers and whether an inability to deliver the part could halt production.

Boosting stores of easily accessible inventory may become necessary as tighter security results in more frequent—and longer—cargo inspections by U.S. Customs at airports, on highways, and at shipping ports. "Every day we hear different requirements for the inspection of materials," Forster says. Among those requirements are hand searches and x-ray examinations of cargo, as well as additional testing, such as passing shipments through compression chambers to make sure that no device being shipped is set to explode at reduced atmospheric pressure. One airline that before September 11 required no waiting time for cargo inspection now requires at least 24 hours, Forster says.

Ford Motor Company is one company that's adjusting its lean-inventory model. The automaker has stockpiled engines from Canada and other critical parts manufactured outside the United States. Ford is also reevaluating its global-sourcing strategy for critical parts. Rival General Motors Corp. is developing a contingency plan that likely will include some stockpiling and a reevaluation of its global-sourcing strategy. Both automakers are reconsidering whether a sole source for a part should be a supplier whose manufacturing facility is in another country.

AMR Research automotive industry analyst Kevin Prouty says that DaimlerChrysler, Ford, and General Motors are all asking suppliers to increase the inventory of non-U.S.-made parts in the United States as one way to mitigate risk in the face of less reliable international shipping and potential recurring problems such as border slowdowns or closings and air shipment delays or stoppages. He notes that a supplier-managed inventory strategy forces suppliers to hold inventory and to bear the consequences of doing so.

Experts say that the ability to adapt quickly to sudden changes in supply-chain activity and customer demand will be imperative. Hon Industries, a $2.04 billion office furniture manufacturer in Muscatine, Iowa, is an example of this adaptability. The day

[a]Adapted from Gonsalves, A., and Konicki, S., "In Search of the Big Picture: Supply Chains," *Information Week* (October 8, 2001): 34–40.

of the attacks, the company received a 20-truckload rush order for office furniture from a customer in the Northeast. The customer that placed the order was setting up offices for companies affected by the World Trade Center attack, and it needed the furniture within five days. Jim McKeone, Hon's investor relations manager, says filling the order was a daunting task, considering that the company is a lean manufacturer, generally requiring two weeks to build orders and not holding inventory of its products.

Many CFOs, says Gartner's Bond, are starting to think that it is more economical to minimize supply-chain disruptions by building up inventory selectively, deploying supply-chain software that allows flexibility in production scheduling, and working closely with suppliers to meet unexpected needs rather than to be forced to halt production at a factory with a lean manufacturing model. Agility once meant the ability to respond to whatever a customer wanted. Now, Bond says, people are saying it "has to include the idea of planning for disruptions in the supply chain that could keep you from serving your customer at all."

SUPPLIER PARTNERING

Managing inbound logistics in the supply chain involves working with suppliers who provide parts, raw materials, components, and services. As we have already discussed, there has been a trend toward developing closer working relationships with fewer suppliers. Given this new approach to suppliers, a big part of quality improvement requires developing and assisting suppliers so that they can provide needed products with low levels of defects, in a reliable manner, while conforming to requirements. Several approaches to improving suppliers result in what is called **supplier partnering**. Inspired by lean purchasing approaches learned from Japanese industry, supplier–partner relationships have emerged that treat suppliers as de facto subsidiaries of the customer organization. We say de facto subsidiaries because as information is shared and communications are improved, the relationship begins to resemble a parent/subsidiary instead of separate firms.

As is shown in Table 9-1, a number of systems are used to help develop suppliers. **Single sourcing** refers to narrowing down the list of approved suppliers for a single component to just one supplier. Companies that are uncomfortable with using a single supplier may use **dual sourcing**, where the number of approved suppliers is reduced to just a few. Dual sourcing reduces the exposures of having a single supplier.

Supplier evaluation is a tool used by many firms to differentiate and discriminate between suppliers. Supplier evaluations are often recorded on *report cards* in which potential suppliers are rated based on criteria such as quality, technical capability, or ability to meet schedule demands.

TABLE 9-1 Supplier Development Approaches

Single sourcing
Dual sourcing
Supplier evaluation
Sourcing filters
ISO 9000:2000
MBNQA
Supplier certification or qualification programs
Supplier development programs
Supplier audits
Partnering

Sole-source filters that are used in many companies rely on external validation of quality programs. The external validation comes from outside examiners and registrars that are used in these processes. This gives customers the comfort that outside authorities have given your company a sort of seal of approval. Two of the most commonly used filters are the Baldrige criteria and ISO 9000:2000 (Chapter 3). In these cases, companies must show either that they are using the Baldrige criteria to improve or that they have become ISO 9000:2000 registered. The ISO 9000:2000 filter is used commonly in the international community.

Video Clip: Supplier Development at Nylamode

Many companies perform lengthy inspections of their suppliers that involve long-term visits and evaluations. These programs are often called **supplier certification of qualification programs** if the focus is entirely on evaluation. If the focus is on helping the supplier to improve by training the supplier over long periods of time, they are termed **supplier development programs**. Supplier development programs are sort of like mowing Mrs. Gunther's yard. The first time I mowed Mrs. Gunther's yard as a youth, she was not especially satisfied. After I had finished, she brought me a glass of lemonade and explained to me gently how she really wanted her yard to look. A week later, I returned and did a better job. Over lemonade, Mrs. Gunther explained where I had done better and where I needed to improve further. With each successive week, I did a better job. As my work improved, she paid me more. By the end of the summer, I realized how much better I had become at caring for her yard. She patiently developed me as a service provider. Over time she became more satisfied, as did I.

Another tool that is used often is the **supplier audit**. This is similar to supplier certification except that a team of auditors visits the supplier and then provides results of the audit to the customer. The audits are performed to ensure that product quality and procedural objectives are being met. Supplier audits tend not to have the developmental component that is found in supplier development programs.

We should mention that there are drawbacks to single sourcing. When there are few suppliers, there is more exposure to interruption of supply. For example, General Motors experienced a major shutdown as a result of single sourcing from a single supplier named Delphi. If labor relations are not solid, single sourcing can have the effect of shifting negotiating power to unions in supplier plants. Other problems include possible interruptions because of transportation problems, quality problems, disagreements concerning pricing, or global security problems (see A Closer Look at Quality 9-1).

Single-Sourcing Examples

In the 1980s, a defective rate of 5% for a supplier was acceptable. In this new century, parts-per-million levels of quality are expected from suppliers. In addition, many companies such as Mercedes-Benz are moving to single-source suppliers (see Quality Highlight 9-1). Other changes have occurred as well. Purchasing groups were viewed in the past as in-house experts who expedited orders and solved materials supply problems. The dollars spent on supply were not critical as long as parts were available for manufacture. Strategies are different now. The way to develop a supplier is to have adequate communications, linear production schedules, and time to make necessary changes. Supplier contacts are one way to ensure adequate communication. Assigning one person or a team to each supplier can reduce the potential for miscommunication. Another way to communicate is through supplier programs where the product or service producer ensures supplier access to information. This provides open communication on mutually critical issues between the customer and the supplier. Another issue of communication between suppliers and customers is that production schedules must match. Suppliers constantly must be updated as to when the customers need products with lead times becoming shorter.

QUALITY HIGHLIGHT 9-1

SUPPLIER PARTNERSHIPS AT MERCEDES-BENZ

Mercedes-Benz was recently able to manufacture and sell a sports utility vehicle (SUV) in a price range only slightly higher than a Grand Cherokee or Ford Explorer. The M-Class, classified by Mercedes-Benz as an "all-activity vehicle" rather than an SUV, was manufactured using the basics of JIT production. When a vehicle is put through the paint shop, a computer system sends an order to a supplier, such as Johnson Controls, Inc., in Milwaukee, which delivers a dashboard to the Vance, Alabama, assembly line in a few hours. Therefore, Mercedes-Benz does not have to store that part before it is needed. As the vehicle nears readiness for dashboard installation, that part arrives. The key to success here is the communication link between the two firms. If the electronic data interchange is interrupted for some reason, a supplier will not be aware of an order, and the manufacturer's effectiveness is hampered. Even with the chance of a flaw, the payoffs are significant. Robert Sigler, an automotive analyst, says, "With 70% of its components being developed by other suppliers, Mercedes-Benz could slash millions of dollars in inventory management costs."[b]

One of the key elements in improving quality is the supplier. As illustrated by Mercedes-Benz, if the electronic data interchange is not functioning, its delivery system is not operating on schedule, or the production of the M-Class components is behind, the overall reputation of the company will be affected. As a result, supplier selection is a key issue for many companies. Is it best to take the time to inspect incoming components before putting them into the final product? To develop a solid partnership between customers' suppliers, a mutually beneficial program must be developed where both sides contribute to the stability and capability of the production process.

[b]Hoffman, T., "Mercedes-Benz Introduces M-Class All-Activity," *Computerworld* (November 17, 1997):42–53.

Electronic data interchange (EDI) is a system that aids customer and supplier communication by linking together supplier and customer information systems. Customers now are helping suppliers to isolate bottlenecks in the operation, balance production systems, and reduce setup times in an effort to reduce lead times. For example, suppliers of seats to Chrysler Motor Corporation must be able to meet the schedule changes within 36 hours. Schedules are communicated through an electronic data interchange link on a real-time basis.

Single sourcing has changed the landscape of purchasing. Prior to single sourcing, Xerox used 5,000 suppliers. Since implementing single sourcing, Xerox uses only 300 suppliers. Suppliers were chosen for their current quality practices and their willingness to work with Xerox to implement a quality improvement program. In Britain, dual or multisourcing has been chosen by many major customers predominantly to avoid unfair pricing and the possibility of the customer's production being disrupted by suppliers' labor disputes. Having a limited set of suppliers also reduces shopping around for the best price.

The Dell Computer Company's goal was to work with suppliers to figure out how to minimize the supply chain and hold the least amount of inventory. Personal computer production is characterized by a tremendous inventory control problem. Suppliers who win in this industry are suppliers who are able to help the personal computer producers to overcome this inventory control problem. Dell is responding by bringing in suppliers who understand the personal computer production business. If suppliers don't understand your business, you end up creating buffers that translate into inventory.

Lockheed-Martin was a defense contractor that worked closely with its suppliers. At its Aeronautics Materials Management Center, a special program for suppliers was developed. This program was called the Star Supplier Program. Lockheed-Martin developed criteria that each supplier had to meet. Quality was a top requirement. Each Star Supplier was required to use statistical process control, have a 0% rejection rate at the point of inspection for six months, and achieve zero nonconformance—documented at the Aeronautics Materials Management Center. The next requirement was to meet scheduled delivery dates. A supplier had to maintain a 98% concurrency to contract delivery schedules. Finally, the cost criteria were used, which showed favorably improving price trends and favorable purchase order administration.

Perhaps the most extreme example of supplier partnering comes from the Bose Corporation. Bose has implemented what it terms JIT II. In this effort, Bose eliminated a large part of its purchasing department and empowered suppliers to write their own purchase orders. This in effect made the suppliers responsible for managing inventories and keeping inventory costs low. It will be interesting to see if JIT II catches on. This is now called **vendor-managed inventory (VMI)**.

SUPPLIER DEVELOPMENT

Supplier Development has to do with the activities a buyer undertakes to improve the performance of its suppliers. Some of these activities may include supplier evaluation, supplier training, consultation, sharing data, and sharing processes. Companies such as Toyota and Honeywell have become very good at developing suppliers. However, recent data suggests that over 50% of U.S. companies do not have adequate supplier development programs. There is much work left to be done in this area.

There are seven steps for supplier development. First, you *identify critical products and services*. This involves identifying strategic products and components (those that are difficult to obtain, high costs, or high volume). Second, *identify critical suppliers*. These may be suppliers who provide strategic components but do not meet quality or reliability objectives or suppliers who do not meet schedules. Third, *form cross-functional teams*. The buyer forms a cross-functional team to work with the supplier. Fourth, *meet with supplier top management*. This is to discuss details of strategic alignment, performance expectations and measurement, and processes for improving. Fifth, *identify key projects*. These occur when there is agreement about how the supplier needs to improve and where. Projects are selected in the same way six-sigma projects are selected, utilizing criteria such as impact, ROI, feasibility, and required investments. Sixth, *define details of agreement*. This involves cost (and benefit sharing), commitments of resources, metrics for improvement, project charters, accountability, and deliverables. Finally, *monitor status and modify strategies*. To ensure success, management must actively monitor progress and revise strategies as needed.

It should be noted that many companies confuse supplier evaluation with supplier development. These are not synonyms. Implicit to supplier development is the expenditure of resources designed to improve the performance of the supplier. This may occur over a long period of time—sometimes months or years. Many companies couple this with expectations for shared cost reductions. For example, Toyota sets goals for cost reductions with its suppliers. If the target is 10%, Toyota may ask for a cost reduction of 5% and provide the other 5% benefit to the supplier. Suppliers who successfully complete development activities are often designated as preferred suppliers due to their alignment with customer needs.

Supplier Awards

Many times, companies will provide awards to outstanding suppliers. This provides an opportunity to celebrate supplier performance that is best of the best. Some of these awards are based on the Baldrige criteria or are decided by a committee within the buyer's company. An example of a supplier award is the Ford World Excellence Award. In 2004, Ford provided this designation to over 50 companies around the world. Their program includes Gold, Silver, and Recognition level awards.

Supplier Relationship Management Systems (SRMS)

Elsewhere in this text, we discuss customer relationship management systems (CRMS). For upstream activities, there are similar systems called SRMS. These systems include spend analytics, sourcing execution, procurement execution, payment, supplier scorecarding, and performance monitoring. In SAS ERP systems, the SRMS have the following capabilities:

- Create complete spend transparency.
- Develop a comprehensive, accurate profile of the supplier base.
- Identify opportunities for optimal sourcing of materials, equipment, and services.
- Consolidate and prioritize suppliers based on quality, performance, and on-time delivery.
- Ensure contract compliance and reduce maverick spending.
- Ensure the quality of purchased items.
- Ensure appropriate levels of supply.

APPLYING THE CONTINGENCY PERSPECTIVE TO SUPPLIER PARTNERING

Different firms take different approaches to supplier development. Remembering the contingency perspective discussed in Chapter 1, you should not be surprised. Apparently, one variable that affects what customers want from their suppliers is the customer's position in the supply chain. A study of the auto industry supply chain found that the auto manufacturers, direct suppliers, and indirect suppliers had somewhat different expectations from their suppliers. Table 9-2 shows these priorities and

TABLE 9-2 Supply Chain Priorities

	Rankings		
Factors	*Auto Assemblers' Importance*	*Direct Suppliers' Importance*	*Indirect Suppliers' Importance*
Consistency	1	1	1
Reliability	2	3	3
Relationship	3	2	2
Technological capability	4	4	4
Flexibility	5	5	4
Price	6	6	6
Service	7	7	7
Finances	8	8	8

SOURCE: T.Y. Choi and J.L. Hartley, *Journal of Operations Management* 14 (1996):340.

their rankings. This table shows relative priorities between auto assemblers, their direct suppliers, and their indirect suppliers. Notice that suppliers are marginally more interested in relationships than customers. This makes sense because it is to their advantage to develop these relationships to gain sales. Perhaps customers should pay more attention to relationship building.

A SUPPLIER DEVELOPMENT PROGRAM: ISO/TS 16949

Now that we have discussed supplier development conceptually, let's look at a specific example of a supplier development program—ISO/TS 16949. The goal of developing your suppliers is based on the need to provide high quality to the customer. Because variability is anathema to quality, the supplier's processes must be consistent with those of the customer. In the late 1980s, U.S. automakers developed certification programs for suppliers. The General Motors program was called "Targets for Excellence," and Ford used a program called "Q1." With the increase in popularity of ISO 9000, suppliers asked auto companies to adopt a single standard for certifying suppliers. The result, called **QS 9000**, provided a common standard for DaimlerChrysler, General Motors, and Ford. This standard has gone through an update and is being supplanted by ISO/TS 16949.

ISO/TS 16949

The ISO/TS 16949 standard applies only to automotive companies. ISO/TS 16949 is an International Standards Organization (ISO) Technical Specification that aligns existing automotive quality system requirements within the global automotive industry. ISO/TS 16949 specifies the quality system requirements for the design/development, production, and, where relevant, installation and servicing of automotive-related products.

ISO/TS 16949 was written by the International Automotive Task Force (IATF). The IATF consists of an international group of vehicle makers including Ford, General Motors, and DaimlerChrysler, as well as several automotive trade associations. Representatives and subcommittees of TC 176 also helped to prepare ISO/TS 16949. We will discuss ISO/TS 16949 in more depth below.

ISO/TS 16949 is founded on the model in Figure 9-3. This model shows that ISO/TS 16949 is closely aligned with ISO 9000:2000 in that it is founded upon a systems view of automotive production. This system for continual improvement involves management responsibility; resource management; product realization; and measurement, analysis, and improvement.

The sections of ISO/TS 16949[4] are shown in Table 9-3. We will discuss each of the major sections in more detail.

Quality Management System

For the quality management system, suppliers must recognize key processes and document these processes. They must establish sequences and linkages for these processes. The organization must determine how effective their operations are; make resources and information available in sufficient quality and quantity to run the business; monitor, measure, and analyze the business to ensure effective operations; and take actions to ensure the planned results are attained and continual improvements are being made.

---

[4]The facts for this section are drawn from Technical Specification ISO/TS 16949, International Standards Organization, 2006.

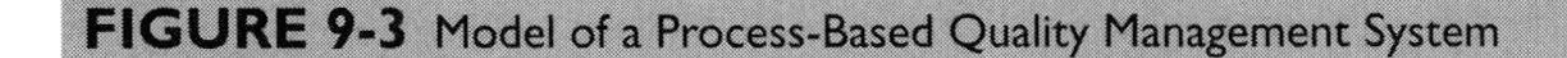
FIGURE 9-3 Model of a Process-Based Quality Management System

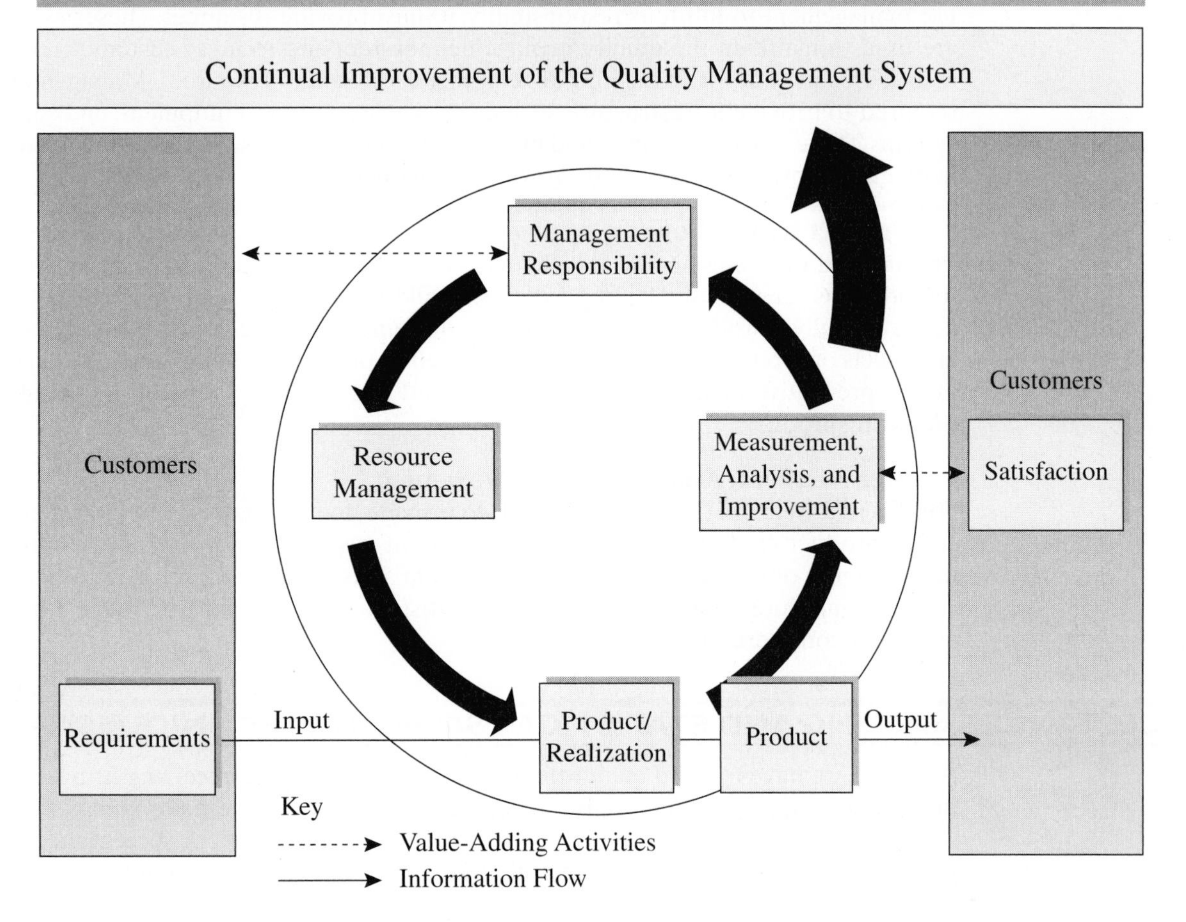

Management Responsibility

For this section of ISO/TS 16949, the extent to which management is committed to the development and implementation of quality management and continuous improvement is documented. Management is responsible for developing policy, communicating with the organization relative to customer service, establishing quality objectives, conducting managerial reviews, and providing resources. For example, managers with responsibility and authority for corrective actions will need to be informed when products do not meet specifications and see that corrective action is taken to ameliorate the problems.

TABLE 9-3 ISO/TS 16949 Sections

0. Introduction	6. Resource Management
1. Scope	7. Product realization
2. Normative reference	8. Measurement, analysis, and improvement
3. Terms and definitions	Annex—Control plan
4. Quality Management System	Bibliography
5. Management responsibility	

Resource Management

For management to fulfill its responsibility, it must provide resources. These resources are used to maintain the quality management system and to meet customer requirements. This includes training and development for human resources. Management is required to provide infrastructure such as bricks and mortar, equipment, and support systems. These must be planned and implemented properly. A safe, clean, and adequate work environment is established for worker satisfaction.

Product Realization

Product and processes should be adequately planned, including quality objectives for the products. Customer-related processes should be designed in a way that customer needs are fully considered and regulatory requirements are met. This section considers all aspects of product and process design as well as purchasing, suppliers, control plans, setups, preventive maintenance, traceability, and many other aspects of designing and producing products.

Measurement, Analysis, and Improvement

For this requirement, the company needs to provide documentation that it can demonstrate product conformity, quality management system conformity, and continual improvement of the quality management system. This includes aspects such as statistical tools, measurement systems, customer satisfaction measurement, internal audits, and other considerations.

ACCEPTANCE SAMPLING AND STATISTICAL SAMPLING TECHNIQUES

Although we have focused on developing standards so that the receiving firm has confidence in the quality of materials received from the supplier, there are times when the receiving firm must inspect incoming materials from its suppliers. **Acceptance sampling** is the technique used to verify that incoming goods from a supplier adhere to quality standards. Acceptance sampling inspection can range from 100% of the delivery to a relatively few items from which the receiving firm draws inferences about the whole shipment.

QUALITY HIGHLIGHT 9-2

INTEGRATING FORWARD ALONG THE SUPPLY CHAIN: 3M DENTAL PRODUCTS DIVISION[c]

A customer satisfaction rating of "good" is no longer good enough for 3M Dental Products Division (DPD), a Minnesota-based supplier of products used around the world. The 700-employee division of 3M has determined that only by striving to earn grades of "excellent" in all product and service areas can it set clear goals for performance improvement, continue to increase sales, and boost productivity at industry-leading rates.

Pursuit of excellence explains why 3M DPD's customer surveys no longer combine "good" and "excellent" responses in a single category, why it has developed a comprehensive network of customer "listening posts," and why it has built an information

[c]Adapted with permission from the Malcolm Baldrige National Quality Award Profiles of Winner, 2006.

system that tracks the purchasing decisions of dentists. It also accounts for how 3M DPD sets its priorities—by concentrating people and resources on opportunities most likely to improve products and services beyond customer expectations.

The division's careful reading of customer requirements drives a finely tuned innovation process that delivers a steady stream of new or improved products. Products introduced within the last five years now account for 45% of total annual sales, up from 12% in 1992.

3M DPD manufactures and markets more than 1,300 dental products, including restorative materials, crown and bridge materials, dental adhesives, and infection-control products. Most of its 700 employees are based at its St. Paul, Minnesota, headquarters and at its manufacturing and distribution facility in Irvine, California.

In the United States, where it has a leading share of the market, 3M DPD competes with more than 100 manufacturers of dental products. Sales and distribution to U.S. dentists are carried out through a network of independent distributors. In foreign markets, the division uses 3M subsidiaries for sales, marketing, and customer support. Sales of 3M DPD products outside the United States account for 65% of the division's total sales.

3M DPD aims "to become THE supplier of choice" of dental professionals worldwide. Setting a clear course to achieve this aim is the objective of the division's systematic strategic planning process, cited as an industry best practice by *Fortune* magazine. Led by a steering committee of top executives and senior managers, the process is designed to build consensus on what needs to be improved and how it will be accomplished. More than 20% of employees participate. The result is a 10-year vision, a detailed 5-year strategic plan, and a 1-year operating plan.

For each priority improvement, the steering committee negotiates with the appropriate department or functional unit to establish the anticipated business impact, determine resource allocations, and set metrics and target values for assessing progress.

The Employee Contribution and Development Plan is the division's chief personnel appraisal tool. It sets individual goals in the areas of business results, team effectiveness, and employee development and is used to determine performance ratings and to guide promotion decisions.

3M DPD's measurement system—the Business Performance Management Matrix—provides an easy-to-grasp framework for aggregating performance measures and for directly linking these measures to key business drivers and goals.

Most dentists in the United States and Europe—the division's largest markets—already use 3M DPD products. Future growth will depend largely on expanding existing customers' options—and spending—for 3M dental products. To do this, the division must have a thorough knowledge of customer requirements.

The division has graded the dentist market, resulting in five groups that reflect differences in satisfaction, purchasing behavior, referrals, repurchases, and number of 3M DPD products used. In-house and third-party surveys, focus groups, and hands-on evaluations are among the wide variety of methods that the division uses to listen to and learn from dentists in each segment. In addition, virtually all customer contacts—from visits by field representatives to calls to the technical hotline—provide additional information that also is entered into the division's customer information system. This extensive database provides the information necessary to determine whether specific products and services are meeting key customer satisfaction goals and to spot opportunities for new products.

Insights into changing customer requirements—combined with knowledge of technological, societal, and environmental trends—are the starting point for product and process innovations. Dentists, distributors, and major suppliers are involved in the division's systematic approach to translating key customer requirements into design requirements, prototypes, and—ultimately—reliable, quality products.

Continually raising the bar for performance improvement, 3M DPD is realizing benefits in nearly all facets of its business. Over the last 10 years the division has doubled global sales and market share.

Is Acceptance Sampling Needed?

Acceptance sampling is controversial. Some critics of the technique believe that the assumption in acceptance sampling that a percentage will be defective or less than perfect (called *acceptable quality level*, or AQL) is counter to Deming's concepts of continual improvement. However, there is still need for acceptance sampling in many different circumstances. See a situation where acceptance sampling and testing may be needed in A Closer Look at Quality 9-2. Following are some examples of when acceptance sampling might be needed:

- When dealing with unproven suppliers
- During start-ups and when building new products
- When products can be damaged in shipment
- When dealing with extremely sensitive products that can be damaged easily
- When products can spoil during shipment
- When problems with a certain supplier have been noticed in the production process that bring the supplier's performance into question

A CLOSER LOOK AT QUALITY 9-2

FOR RFID TO TAKE HOLD, RELIABILITY NEEDS TO IMPROVE

Radio frequency identification (RFID) tags are an important technology for supply chain management. For example, consider the supply chain of a hospital where the patient flows through the process. An elderly woman dozes quietly in her hospital bed, tucked under layers of blankets. Her doctor stops by on his rounds and clicks on a wireless tablet PC, which is equipped with an RFID reader. This device transmits a signal to an RFID tag in the woman's hospital bracelet, although it's hidden by the bedding. The tag transmits information to the PC, which is integrated with the hospital's information management system. On the display, the doctor sees the patient's name, her previously administered medications, plus recently collected vital information, such as temperature, heart rate, and blood pressure. Without disturbing the patient or having to check back at the nursing station, the doctor has received the accurate, up-to-date information needed to monitor the patient's progress.[d]

The patient may not be real but the application is. The hospital, Jacobi Medical Center in the Bronx, benefits tremendously from an RFID system. According to CIO Daniel Morreale, "We get 100% accuracy in identifying patients and an overall savings of clinician time because doctors and nurses get the patient information they need at the bedside." This is just one example of RFID usage in business. Businesses with logistics operations have been quick to embrace RFID. Large organizations such as Wal-Mart and the U.S. Department of Defense are mandating the use of RFID in their supply chain activities. However, there are concerns about RFID. They need to be nearly 100% reliable. The current standard is that pallet labels need to have a 100% read rate. That read rate is also nearly the expectation for individual items. Cost is also a consideration with one major retailer stating that they will adopt RFID as soon as the technology cost drops below 1 cent per tag.

It is expected that "RFID will go through a process similar to what happened 20 years ago with bar-codes," says Dan Mullen, president of AIM Global. "Initially, people thought they couldn't afford the technology. As it became more widespread, the payback grew. Now companies couldn't do without it. The emergence of RFID in the retail supply chain will drive further adoption. As companies implement the technology deeper within their own operations, the return on investment will grow and applications will expand." However, they need to be reliable.

[d]Adapted from "RFID After Compliance: Integration and Payback," *Business Week*, November 20, 2004, pp. 91–98.

Acceptance Sampling Fundamentals

We define *acceptance sampling* as a statistical quality control technique used in deciding to accept or reject a shipment of input or output. When compared with statistical quality control, acceptance sampling is defined by its occurrence after production has been completed. Acceptance sampling inspection can occur at the beginning of the process, such as when receiving components, parts, or raw materials from a supplier. Or it can occur at the end of production, as in the case of final inspection. Again, we focus here on inspection of incoming materials. One interesting application of acceptance sampling occurs with seed producers. Because of the biological nature of seed, a supplier can place the seed on a truck in perfect condition, and the seed can arrive in an unsuitable state. Therefore, large bulk purchasers of seed and other agricultural products are avid users of acceptance sampling techniques.

It should be noted that there are different methods for developing sampling plans. We discuss how to develop sampling plans using Dodge and Romig tables and OC curves.

Producer's and Consumer's Risk

Producer's risk is the risk associated with rejecting a shipment of materials that has good quality. Think of it this way. You are the producer of a product that has high quality. However, your customer has concluded that your product has poor quality and returns the product to you. In this case, you have been judged inaccurately. **Consumer's risk** is the exact opposite. As a consumer, you receive a shipment of poor-quality product and believe that it has good quality. Therefore, you pay for the product, use it in your production process, and suffer the consequences.

Producer's risk is denoted by *alpha* (α) and is called a *type I error*. Consumer's risk is denoted by *beta* (β) and is referred to as a *type II error*. Table 9-4 contrasts alpha and beta risk. The goal of acceptance sampling is to reduce producer's risk to low levels while maintaining consumer's risk at acceptable levels.

Acceptable Quality Level

The **acceptable quality level (AQL)** is "the maximum percentage or proportion of nonconforming items or number of nonconformities in a lot or batch that can be considered satisfactory as a process average." The AQL concept has been troublesome to many who consider it to condone an acceptance of less-than-perfect quality. To statisticians, AQL simply denotes an economic decision that is associated with producer risk.

Lot Tolerance Percent Defective (LTPD)

Lot tolerance percent defective (LTPD) is the level of poor quality that is included in a lot of goods. The difference between AQL and LTPD is sometimes confusing to students. Lots of AQL or better usually should have an alpha (that is, for example, a 5%) or less chance of rejection. AQL relates to type I error or producer's risk. Producer's risk is the probability that good products will be rejected by the consumer. For example, if a consumer concludes that an automobile is a lemon even though it is a great vehicle, a type I error has occurred. Lots of LTPD or worse should have a beta (that is,

TABLE 9-4 Alpha and Beta Risk

		State of Nature	
		Product Is Good	***Product Is Defective***
Outcome	Consumer accepts product	OK	Consumer's risk β
	Consumer rejects product	Producer's risk α	OK

say a 10%) or less chance of acceptance. LTPD relates to type II error or consumer's risk. Conversely, consumer's risk is the chance that an automobile that seems perfect is really a lemon. There is, theoretically, only one combination of sample size (n) and acceptance number (c) that meets both conditions simultaneously. In practice, we will be unable to meet both conditions precisely and must choose a combination of n and c that approximates both conditions simultaneously.

n and *c*

For the most part, the assignment of AQL, LTPD, alpha (α), and beta (β) is a management decision. Once these values have been determined, n and c can be determined; these values in turn define the **sampling plan**. The bottom line in acceptance sampling is that acceptance sampling plans are designed to give us two things: n and c, where

n = the sample size of a particular sampling plan

c = a number that, if exceeded by the number of defectives in the sample, causes rejection of the lot (acceptance number)

The average sampling plan can be stated in simple terms: $n = 20$ and $c = 2$. This clearly communicates the bounds of the sampling plan. That is, take a sample of 20 items, and if more than 2 are defective, reject the lot of materials. It is important to remember that you should always randomize when selecting products from a supplier to be inspected.

OC Curves

The **operating characteristic (OC) curve** provides an assessment of the probabilities of acceptance for a shipment, given the existing quality of the shipment. An OC curve is constructed to show the probability of accepting individual lots when the percent defective of the various individual lots is known or assumed to be at a given level. These curves also can be used to develop sampling plans.

Figure 9-4 shows an OC curve for an optimal sampling plan. This *ideal* sampling plan shows that given a probability of acceptance of 98%, this sampling plan should be

FIGURE 9-4 Perfect Operating Characteristic (OC) Curve

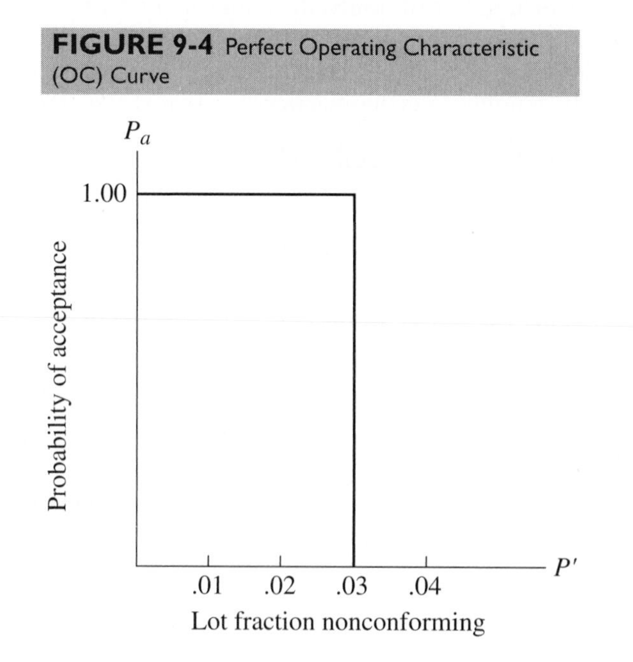

used. In this example, the probability of accepting a lot of goods with less than 3% defective is 100%. This means that the customer has considered the tradeoffs and has determined that a lot of materials with only 3% defective is acceptable given the circumstances.

However, OC curves never appear like the *ideal* case. Figure 9-5 shows OC curves for a sample size of $n = 100$ and $c = [0, 1, 2, 3]$. Notice that as c gets smaller, the OC curve gets steeper. Generally speaking, this means that higher values of c lead to higher probabilities of accepting bad shipments (consumer's risk). Also, higher values of n affect the OC curves in such a way that we have greater confidence that we have accepted a good shipment.

Figure 9-6 shows an OC curve for $n = 50$ and $c = 2$ (for a lot of size 200; $N = 200$). This shows that the probability of accepting a lot with 2% defective product is 92%. Also, the chance of accepting a lot of 12% defective product is 6%. This means that if a lot has 2% defective, and this sampling plan is used (i.e., $n = 50$, $c = 2$), then the producer's risk is 8%. Notice that if the lot really has much poorer quality, say, 8% defective, the chance of acceptance is quite low, about 24%. This means that there is a consumer's risk of 24% with the poor quality lot.

Example 9.1: OC Curves

Problem: Let's suppose that we use a sampling plan with $N = 200$, $n = 50$, and $c = 2$. In the past, *good-quality* shipments from suppliers had about 1% defective, and *poor-quality* shipments had about 6% defective. We desire to know producer's and consumer's risk. Using Figure 9-5, compute these probabilities.

FIGURE 9-5 Various OC Curves

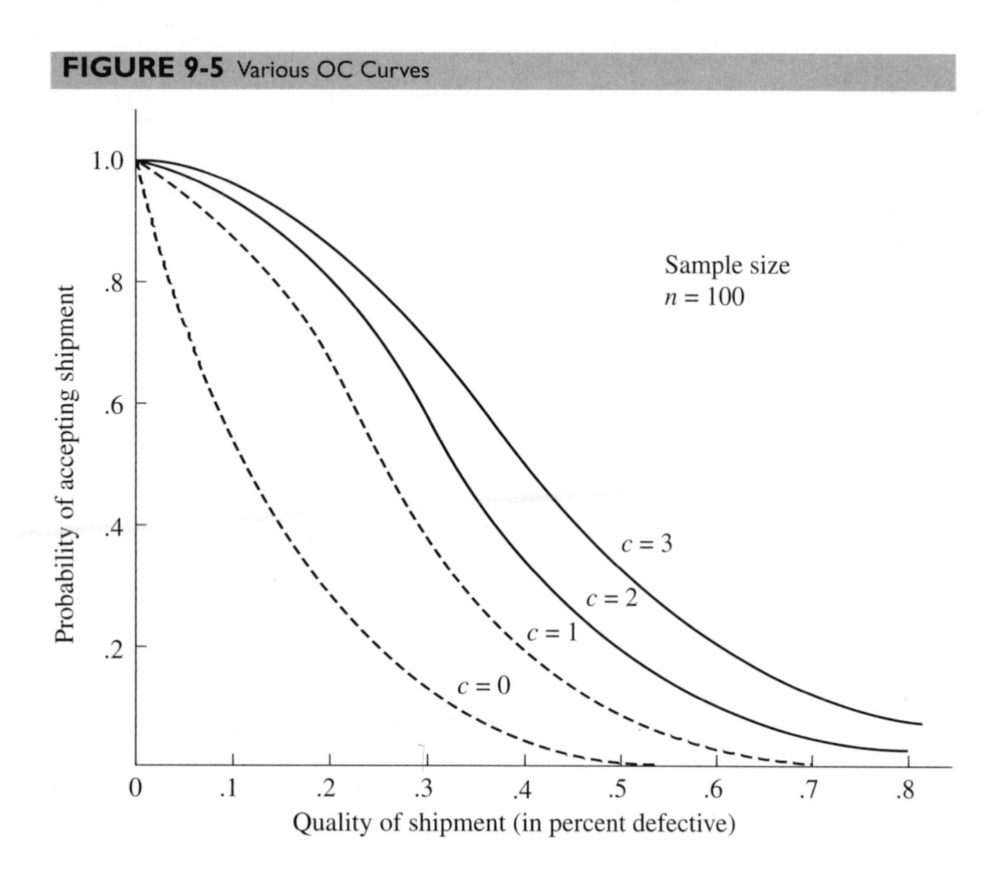

FIGURE 9-6 OC Curve for the Sampling Plan $N = 200, n = 50, c = 2$

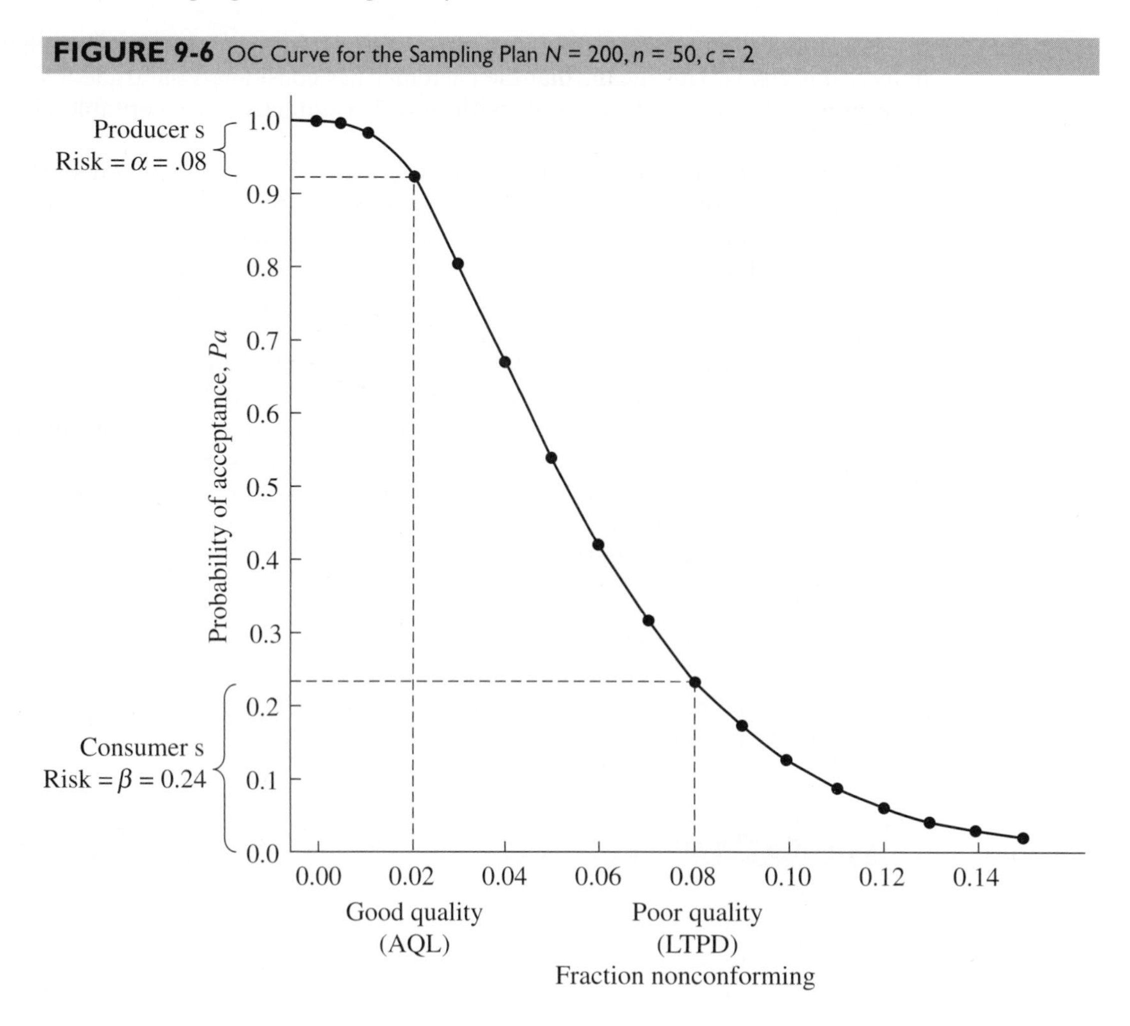

Solution: Using Figure 9-6, it appears that the probability of acceptance of a good lot is 99%. This translates to a producer's risk of 1%. Using the same figure, it appears that the probability of accepting a poor lot is 42%. This is consumer's risk.

Building an OC Curve

There are two ways to construct OC curves. The first uses the binomial distribution, and the second, the Poisson distribution. For simplicity, we will use the Poisson distribution. Using a sample size n and average percent defective p, we can develop an OC curve using a Poisson approximation of a binomial distribution.

Rather than computing the distributions from formulas, we rely on tables that have been developed for simplicity in computing *single-sampling plans*. Single sampling means that we use only one value of c to make a decision. Multiple-sampling plans are discussed later. The Dodge-Romig table[5] in Figure 9-7 shows a table that was developed to simplify such calculations. While holding c and n constant, we can vary the percent defective p to calculate various points on the OC curve.

---

[5]Dodge, H., and Romig, H., *Sampling Inspection Tables* (New York: John Wiley & Sons, Inc., 1959).

FIGURE 9-7 Probability Curves for Poisson Distribution

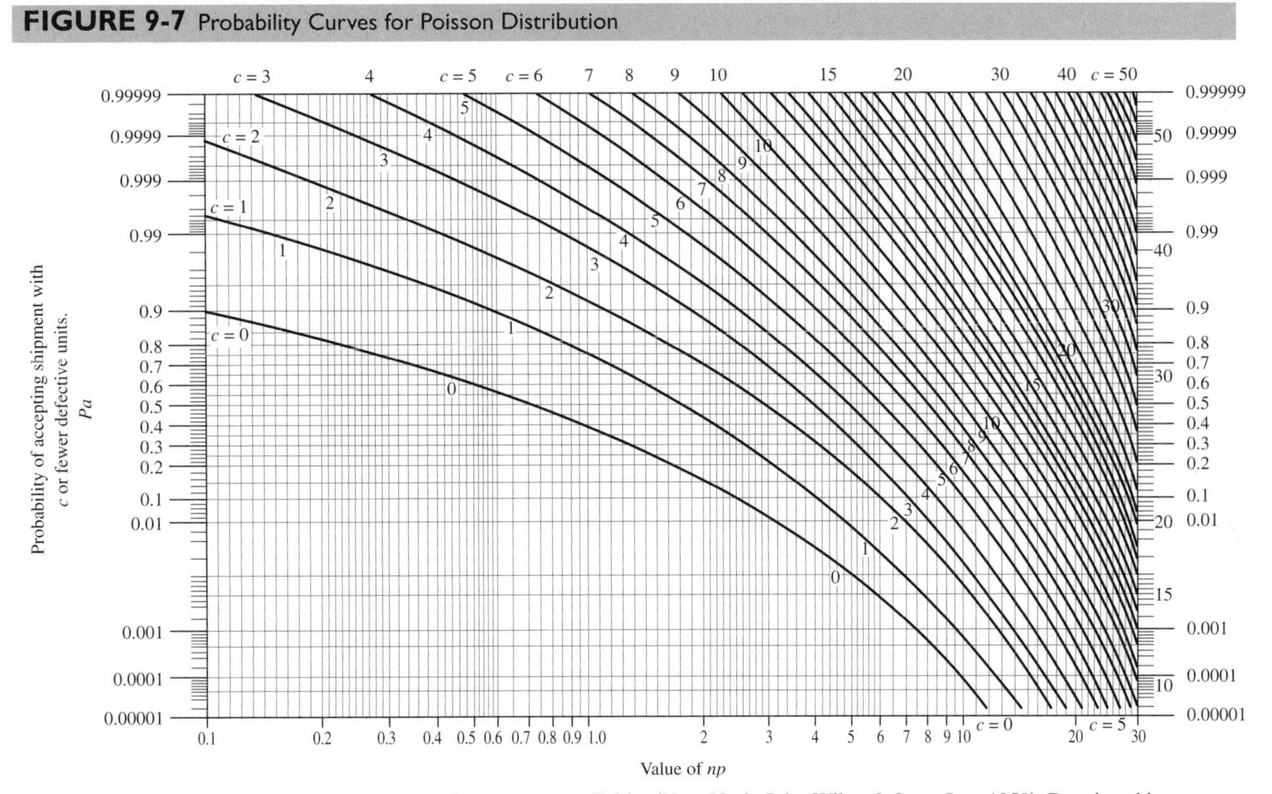

SOURCE: H. F. Dodge and H. G. Romig, *Sampling Inspection Tables* (New York: John Wiley & Sons, Inc., 1959). Reprinted by permission of John Wiley & Sons, Inc.

Steps in using Figure 9-7 to develop an OC curve:

1. Select values of P for percent defective in a shipment.
2. Multiply these values of np (where n is your sample size).
3. Using these values of np, go to Table 9-7 and find Pa (probability of acceptance).
4. Draw the OC curve.

Example 9-2: Developing Single-Sampling Plans Using OC Curves and the Dodge-Romig Table

Problem: We would like to develop an OC curve for a good single-sampling plan using the Dodge-Romig table provided in Figure 9-7. Our sample size n is 60 and maximum c is 2.

Solution: Table 9-5 shows the resulting values needed to construct the OC curve. The values of P_a are drawn from the Dodge-Romig graph (Fig. 9-7). Figure 9-8 shows the resulting OC curve.

Estimating AQL and LTPD

OC curves can be used to estimate both AQLs and LTPDs. MIL STD 105 (also known as ANSI/ASQC Z1.4) is a military standard for sampling procedures and tables for attributes that state that AQLs and LTPDs of sampling plans should always be documented. Figure 9-9 shows an OC curve for a single-sampling plan with $n = 50$ and $c = 1$. The AQL is generally stated with a 95% chance of acceptance ($\alpha = 0.05$). In Figure 9-9,

TABLE 9-5 OC Curve Probabilities

P	*n*	*np*	P_a	*p*	*n*	*np*	P_a
.005	60	0.30 = (.005 × 60)	.996	.055	60	3.30	.460
.015	60	0.90	.940	.065	60	3.90	.340
.025	60	1.50	.840	.075	60	4.50	.200
.035	60	2.10	.680	.085	60	5.10	.160
.045	60	2.70	.550	.095	60	5.70	.120
.050	60	3.00	.500	.105	60	6.30	.080

FIGURE 9-8 Resulting OC Curve for $n = 60, c = 2$

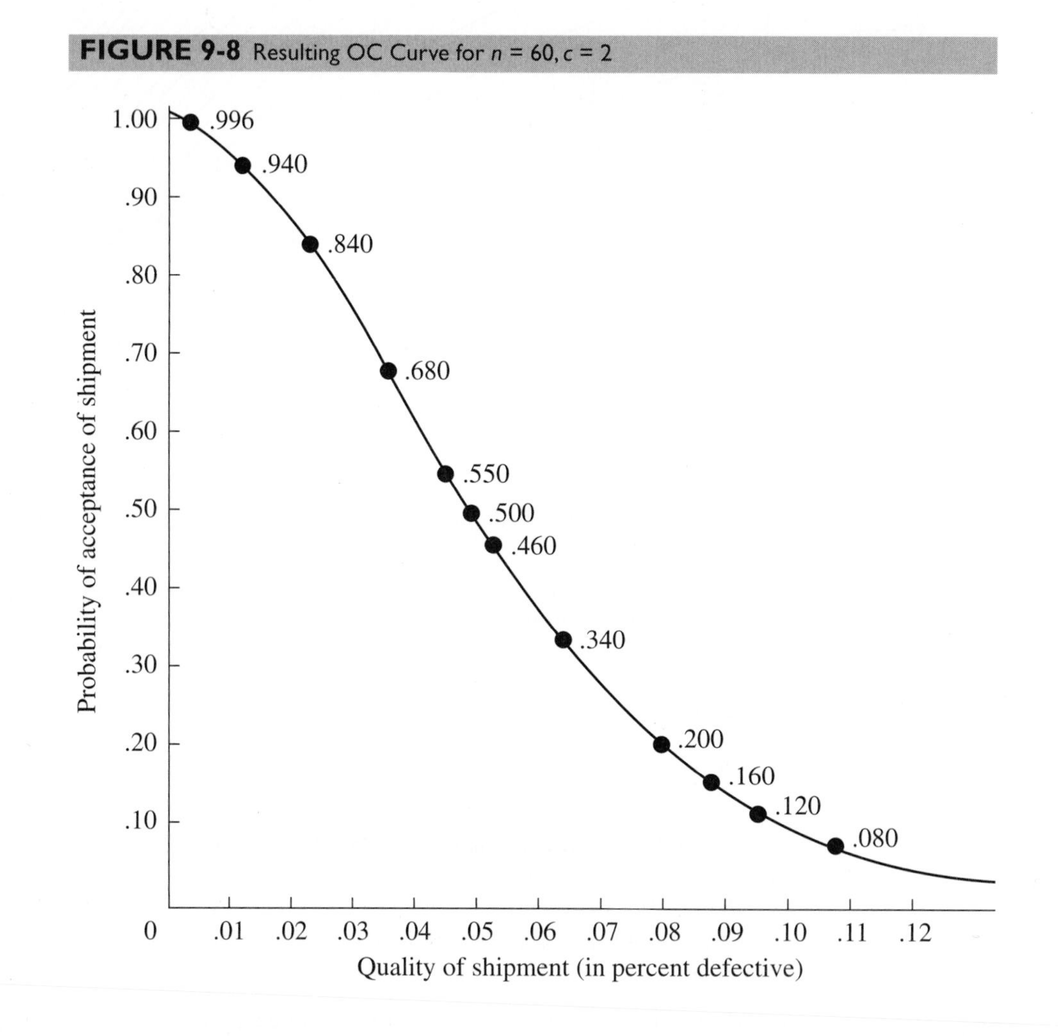

the AQL value is 72%. The LTPD is generally stated at a 10% chance of acceptance. In this case, the LTPD value is 7.6%. At the LTPD, 90% of the lots are rejected.[6]

More Complex Sampling Plans

So far we have focused on sampling plans for single samples. This is not the limit of sampling plans. More complex sampling plans are referred to as multiple-sampling

[6]Taylor, W., "Selecting Statistically Valid Sampling Plans," *Quality Engineering* 10, 2 (1997):365–370.

FIGURE 9-9 OC Curve of Single Sampling Plan n = 50 and c = 1

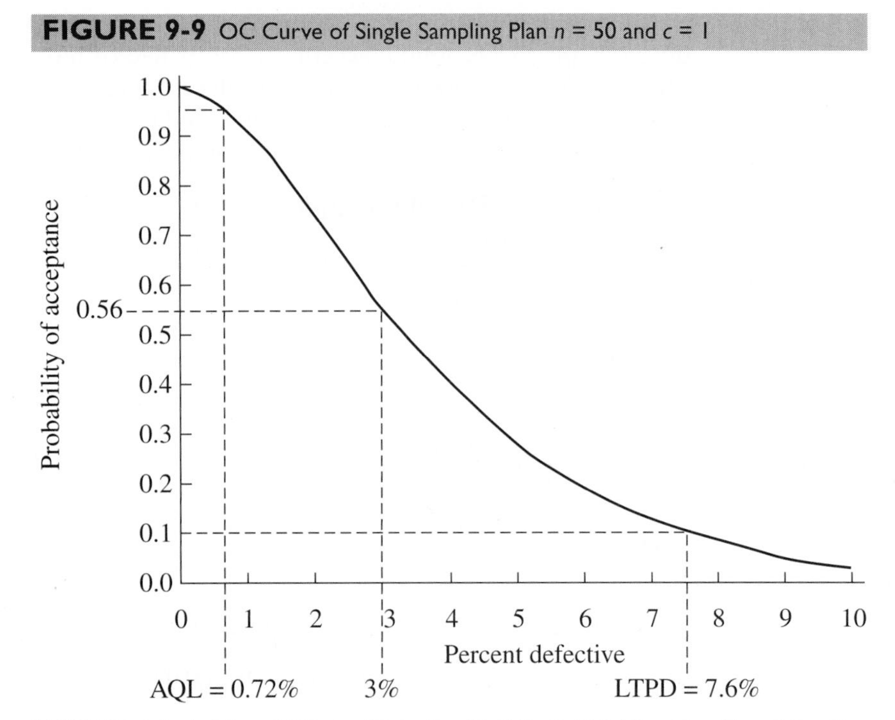

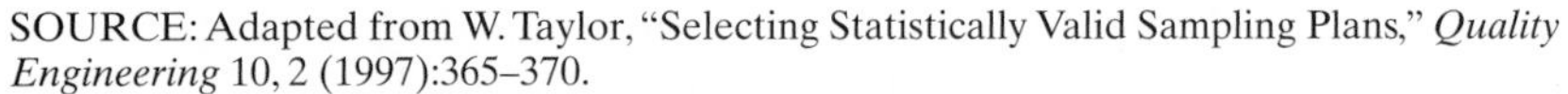
SOURCE: Adapted from W. Taylor, "Selecting Statistically Valid Sampling Plans," *Quality Engineering* 10, 2 (1997):365–370.

plans or sequential-sampling plans. With these sampling plans, the acceptance sampling rules might occur as follows:

n_1 = sample size for sample 1
n_2 = sample size for sample 2
n_n = sample size for sample n
c_1 = acceptance number for sample 1
c_2 = acceptance number for sample 2
c_n = acceptance number for sample n
r_1 = rejection number for sample 1
r_2 = rejection number for sample 2
r_n = rejection number for sample n

Multiple-sampling plans have advantages over single-sampling plans. The sample size used in multiple-sampling plans will be smaller, on average, with the same amount of protection as a single-sampling plan. The decision to use multiple-sampling plans usually will be made in the first phase of the sample. Example 9.3 provides an illustration of a double sampling plan.

Example 9.3: Demonstrating a Double-Sampling Plan

Problem: An engineer adopted a double-sampling plan with the following values: $n_1 = 40$, $c_1 = 1$, $n_2 = 60$, $c_2 = 5$. r_1 and r_2 are 5 and 6, respectively. Please explain to management how this double-sampling plan works.

Solution: Forty items from the incoming lot are drawn at random. If one or zero defective pieces are found, accept the lot. If five or more defective pieces are found, reject the lot. If between two and four pieces inclusive are found defective, perform a second sample with 60 pieces. If the combined number of defective pieces in both samples is less than or equal to five, the lot can be accepted.

Developing Double-Sampling Plans

Although OC curves can be used to develop double-sampling plans, the calculations to develop the OC curves are much more complex than for single-sample plans. For this reason, we will use a standard approach that is used by many practitioners. The main limitation of this plan is that sample sizes must be specified, as well as AQL, LTPD, producer's risk, and consumer's risk.

The double-sampling plan we will develop has the following parameters:

$$\begin{aligned} \text{AQL} &= 0.020 \\ \text{LTPD} &= 0.040 \\ \text{Producer's risk} &= 0.05 \\ \text{Consumer's risk} &= 0.10 \\ n_1 &= .5n_2 \end{aligned}$$

First, we start by computing the ratio of LTPD to AQL:

$$\text{LTPD/AQL} = 0.040/0.020 = 2 \qquad \textbf{(9.1)}$$

The parameters for the appropriate double-sampling plan are found in Tables 9-6 and 9-7. Table 9-6 is used when $n_1 = n_2$, and Table 9-7 is used when $n_1 = 0.5n_2$. For this example, we will use Table 9-7 because n_2 is twice the size of n_1.

TABLE 9-6 Values for Constructing a Double-Sampling Plan Having a Specified p_1' and p_2' ($n_1 = n_2$, $\alpha = 0.05, \beta = 0.10$)

Plan Number	$R = p_2'/p_1'$	*Acceptance Numbers* c_1	c_2	*Approximate Values of* $p'n_1$ $P = 0.95$	0.50	0.10	*Approximate (ASN)/n_1 for 0.95 Point*[a]
1	11.90	0	1	0.21	1.00	2.50	1.170
2	7.54	1	2	0.52	1.82	3.92	1.081
3	6.79	0	2	0.43	1.42	2.96	1.340
4	5.39	1	3	0.76	2.11	4.11	1.169
5	4.65	2	4	1.16	2.90	5.39	1.105
6	4.25	1	4	1.04	2.50	4.42	1.274
7	3.88	2	5	1.43	3.20	5.55	1.170
8	3.63	3	6	1.87	3.98	6.78	1.117
9	3.38	2	6	1.72	3.56	5.82	1.248
10	3.21	3	7	2.15	4.27	6.91	1.173
11	3.09	4	8	2.62	5.02	8.10	1.124
12	2.85	4	9	2.90	5.33	8.26	1.167
13	2.60	5	11	3.68	6.40	9.56	1.166
14	2.44	5	12	4.00	6.73	9.77	1.215
15	2.32	5	13	4.35	7.06	10.08	1.271
16	2.22	5	14	4.70	7.52	10.45	1.331
17	2.12	5	16	5.39	8.40	11.41	1.452

[a] *ASN is without curtailment on the second sample.*

SOURCE: Chemical Corps Engineering Agency, *Manual No. 2: Master Sampling Plans for Single, Duplicate, Double and Multiple Sampling* (Edgewood Arsenal, MD: Army Chemical Center, 1953).

TABLE 9-7 Values for Constructing a Double-Sampling Plan Having a Specified p'_1 and p'_2 ($n_1 = .5n_2$, $\alpha = 0.05, \beta = 0.10$)

Plan Number	$R = p'_2/p'_1$	*Acceptance Numbers* c_1	c_2	*Approximate Values of* $p'n_1$ $P = 0.95$	*0.50*	*0.10*	*Approximate (ASN)/*n_1 *for 0.95 Point*[a]
1	14.50	0	1	0.16	0.84	2.32	1.273
2	8.07	0	2	0.30	1.07	2.42	1.511
3	6.48	1	3	0.60	1.80	3.89	1.238
4	5.39	0	3	0.49	1.35	2.64	1.771
5	5.09	1	4	0.77	1.97	3.92	1.359
6	4.31	0	4	0.68	1.64	2.93	1.985
7	4.19	1	5	0.96	2.18	4.02	1.498
8	3.60	1	6	1.16	2.44	4.17	1.646
9	3.26	2	8	1.68	3.28	5.47	1.476
10	2.96	3	10	2.27	4.13	6.72	1.388
11	2.77	3	11	2.46	4.36	6.82	1.468
12	2.62	4	13	3.07	5.21	8.05	1.394
13	2.46	4	14	3.29	5.40	8.11	1.472
14	2.21	3	15	3.41	5.40	7.55	1.888
15	1.97	4	20	4.75	7.02	9.35	2.029
16	1.74	6	30	7.45	10.31	12.96	2.230

[a]*ASN is without curtailment on the second sample.*
SOURCE: Chemical Corps Engineering Agency, *Manual No. 2: Master Sampling Plans for Single Duplicate, Double, and Multiple Sampling* (Edgewood Arsenal, MD: Army Chemical Center, 1953).

Because LTPD/AQL was 2, in Table 9-7 we find that we will use double-sampling plan number 15 (because 1.97 is closest to our value of 2). We see immediately that c_1 and c_2 are 4 and 20, respectively. Next, we compute the sample size by using either column 5 or 6 from the table. If we want to use a probability of acceptance of 0.95, we use column 5. For sample plan 15, the value of $p'n_1$ is 4.75. Then we compute n_1 as

$$n_1 = p'n_{0.95}/\text{AQL} = 4.75/0.020 = 237.5 \cong 238 \quad \textbf{(9.2)}$$

As we can see, $n_1 = 238$ and $n_2 = 475$ (remember that n_2 is twice the size of n_1). These are rather large sample plans. We may want to reconsider our parameters to reduce these sample sizes.

We also can use this approach to determine a double-sampling plan if we want to maintain consumer's risk at a low level. To do this, we use column 7 of Table 9-7. If we wish to hold that value of consumer's risk to 0.10, we can use the column 7 value of plan number 15 of 9.35. Using Formula 9.2

$n_1 = p'n_{0.95}/\text{AQL}$
$9.35/0.020 = 467.5$ or 468 (rounding)

$n_2 = 935$ (remember that n_2 is twice the size of n_1)
c_1 and c_2 are 4 and 20, respectively

Again, because this sampling plan has large sample sizes, we may be wise to experiment with other parameters to find alternative sampling plans that meet targets for acceptance.

Acceptance Sampling in Continuous Production

The single- and double-sampling plans we have discussed so far are called *lot-by-lot sampling plans*. As we perform separate samples, we receive additional lots of materials. Sometimes it is not feasible to collect products into lots because they are produced in a continuous manner. In these cases, acceptance sampling procedures for continuous production are used. These procedures typically involve alternating between 100% inspection and sampling inspection. Although we will not develop a methodology for continuous acceptance sampling here, MIL STD 1235C (1988), named "Single and Multiple-Level Continuous Sampling Procedures and Table for Inspection by Attributes," is available from the Department of Defense. Any of the military standards that we have mentioned in this chapter can be ordered from the Department of Defense directly by accessing the DOD SSP on the Internet (Department of Defense Single Stock Point for Specifications and Standards). The URL is *http://dodssp.daps.dla.mil/*.

SUMMARY

In this chapter we have focused on the front end of the supply chain. Supplier development holds the greatest quality-related benefit for manufacturers and services firms. After all, if you purchase components or supplies from another firm, to ensure high quality, you need a process for making sure that you can believe in the products and services you are receiving.

The ISO standard TS 16949 was presented here as a good example of how automakers are developing their suppliers. Supplier development efforts have resulted in greatly improved product quality for many firms. Although this is one representative example, several firms are using approaches that are extensions of ISO 9000:2000. For newer suppliers, or where the situation dictates, we learned about acceptance sampling.

As firms learn more about developing suppliers, new practices will emerge. The eventual goal of supplier development is that the supplier resembles a de facto subsidiary of the customer. This closer relationship between supplier and customer holds great potential for improved quality in products and services.

KEY TERMS

- Acceptable quality level (AQL)
- Acceptance sampling
- Chain of customers
- Consumer's risk
- Dual sourcing
- Electronic data interchange (EDI)
- Hidden factory
- ISO/TS 16949
- Lot tolerance percent defective (LTPD)
- Operating characteristic (OC) curve
- Producer's risk
- QS 9000
- Sampling plan
- Single sourcing
- Sole-source filters
- Supplier audit
- Supplier certification of qualification programs
- Supplier development
- Supplier development programs
- Supplier evaluation
- Supplier partnering
- Supply chain
- Value chain
- Value-chain activities
- Value system
- Vendor-managed inventory

DISCUSSION QUESTIONS

1. What is the supply value chain? How does the supply value chain help organizations manage their supply chains?
2. Describe the concept of the hidden factory. How can a realization that the hidden factory exists help managers?

3. Think about a job that you have had or a volunteer organization with which you have been associated. Did this organization have a hidden factory? If so, describe at least two activities that you would associate with the organization's hidden factory.
4. Describe the concept of chain of customers. How does this concept benefit the ultimate consumer of a product or service?
5. What is the purpose of single sourcing? How can single sourcing help firms meet their quality objectives?
6. What are two of the most commonly used sole-source filters? Do you believe that these sources are appropriate? Please explain your answer.
7. What is the purpose of a supplier certification of qualification program? What is the difference between a supplier certification program and a supplier development program?
8. How does electronic data interchange facilitate the supply-chain process?
9. How does the contingency perspective apply to supplier partnering?
10. What is a supplier audit? How does a supplier audit differ from a supplier certification program?
11. Describe the concept that is referred to as supplier partnering. How does this concept differ from the traditional form of supplier–customer relationship?
12. ISO/TS 16949 is a concept that was developed by DaimlerChrysler, Ford, and General Motors. Explain what ISO/TS 16949 is and how it has helped its developers.
13. Is ISO/TS 16949 a good thing for auto suppliers as well as auto manufacturers? Explain your answer.
14. How does ISO/TS 16949 differ from ISO 9000:2000? Is the distinction between the two quality concepts important for the adopters of ISO/TS 16949?
15. What is acceptance sampling? When is acceptance sampling needed?
16. Define producer's risk and consumer's risk.
17. Define the concept of acceptable quality level. Why has this concept been troublesome to many people?
18. What is the term that is used to designate the level of poor quality that is included in a lot of goods? Please describe the role of this term in the quality management process.
19. What is an operating characteristic curve? What is the function of this curve in the quality management process?
20. Discuss the ways that firms develop their suppliers. Is supplier development an important process for many firms? Explain your answer.

PROBLEMS

1. Using Figure 9-5, with a sample size of $n = 100$ and an acceptance number $c = 1$, if a good shipment has no more than 0.05 defective, what is the probability of acceptance? What type of risk is this?
2. Using Figure 9-5, with a sample size of $n = 100$ and an acceptance number of $c = 2$, if a bad shipment has 40% defective, what is the probability of acceptance? What type of risk is this?
3. Using Figure 9-5, with a sample size of $n = 100$ and an acceptance number of $c = 3$, if a good shipment has no more than .02 defective, what is the probability of acceptance? What type of risk is this?
4. Using Figure 9-5, with a sample size of $n = 100$ and an acceptance number of $c = 1$, if a good shipment has no more than .30 defective, what is the probability of acceptance. What type of risk is this?
5. Develop an OC curve using Figure 9-7, a sample size of 100, and the following p values: .01, .02, .03, .05, .07, .09, .11, .13, .15, .17, .19, .21. The maximum acceptance number is $c = 2$.
6. From the OC curve developed in Problem 5, if a good shipment is 0.02 defective, what is the probability of a type I (producer's) error?
7. From the OC curve developed in Problem 5, if a bad shipment is defined as having at least 10% defective, what is your estimate of type II (consumer's) risk?

8. Develop an OC curve using Figure 9-7, a sample size of 100, and the following p-values: .01, .02, .03, .05, .07, .09, .11, .13, .15. The maximum acceptance number is $c = 4$.
9. From the OC curve developed in Problem 8, if a good shipment is .04 defective, what is the probability of a type I (producer's) error?
10. From the OC curve developed in Problem 8, if a bad shipment is defined as having at least 15% defective, what is your estimate of type II (consumer's risk) error?
11. Develop an OC curve using Figure 9-7, a sample size of 100, and the following p values: 0.01, 0.02, 0.03, 0.05, 0.07, 0.09, 0.11, 0.13, 0.15, 0.17, 0.19, 0.21. The maximum acceptance number is $c = 3$.
12. From the OC curve developed in Problem 11, if a good shipment has 5% defective, what is the probability of type I error? How is type I error defined?
13. From the OC curve in Problem 11, if a bad shipment has more than 20% defective, what is the probability of accepting a bad shipment?
14. Interpret the following sampling plan in plain English:

$n_1 = 50$
$c_1 = 2$
$n_2 = 100$
$c_2 = 5$
$r_1 = 4$
$r_2 = 6$

15. Interpret the following sampling plan in plain English:

$n_1 = 125$
$c_1 = 3$
$n_2 = 150$
$c_2 = 6$
$n_3 = 200$
$c_3 = 12$

16. We wish to develop a double sampling plan where $n_1 = n_2$ (see Table 9-6). Here are the needed parameters:

AQL = .010
LTPD = .030
Producer's risk = .05
Consumer's risk = .10

17. Rework Problem 16 where n_2 is twice the size of n_1 (see Table 9-7).
18. We wish to develop a double-sampling plan where $n_1 = n_2$ (Table 9-6). Here are the needed parameters:

AQL = .020
LTPD = .080
Producer's risk = .05
Consumer's risk = .10

19. Rework Problem 18 where n_2 is twice the size of n_1 (Table 9-7).
20. Your boss wants you to develop a double sampling plan where $n_1 = n_2$ (using Table 9-6). Here are some parameters for your use:

AQL = .015
LTPD = .040
Producer's risk = .05
Consumer's risk = .10

21. Rework Problem 20 where n_2 is twice the size of n_1 (Table 9-7).

Cases

Case 9-1 SBC: Setting High Standards for Suppliers and Rewarding Supplier Performance

SBC: ***www.sbc.com***

If you live in California or frequently call someone in the California area, you have invariably done business with SBC (formerly Pacific Bell). SBC provides telephone service to the majority of California's residents. Along with telephone service, the company provides a full array of wireless communications products for individuals and businesses.

In an effort to increase quality and decrease costs, SBC has been working hard to find new ways to manage its supply chain. The result has been the development of a comprehensive program that sets high standards for suppliers and rewards exemplary supplier performance. The program begins with training. All of SBC's procurement managers are required to participate in an Applied Total Quality program. The program consists of six 30-hour courses that teach TQM and supplier management. Suppliers also are encouraged to participate in the program at no cost.

The requirements that SBC places on its suppliers are demanding but are communicated clearly to the suppliers in advance. For its top suppliers, a contract is negotiated on a yearly basis that defines the objectives of the relationship between SBC and the supplier for the next year. Senior managers from SBC and the supplier meet twice a year to discuss the performance of the relationship and iron out any problems that have occurred. SBC's minor suppliers receive a one-page Supplier Quality Report every month. SBC and the company agree on criteria for performance in advance (e.g., on-time delivery, invoicing accuracy), and the supplier receives a score each month from SBC based on its record. Both SBC and its suppliers take these reports very seriously. If a supplier receives a poor score, it typically calls SBC to provide an explanation or ask for suggestions for improvement.

Although SBC's approach to supply management sounds rigid, the company works hard to develop lasting positive relationships with its suppliers. The company maintains a steady flow of communications with its suppliers to work through any problem that might arise. For example, the company has a toll-free 800 number that suppliers and potential suppliers can use to familiarize themselves with SBC requirements. SBC also asks its suppliers to tell it how it is doing, although this program has been only partially successful. According to the former executive director of contracting and supplier management at SBC, "Suppliers are always leery of telling customers about their problems."[7] As a result of supplier reluctance in this area, SBC is thinking about making the supplier feedback reports a requirement for certification.

To its credit, SBC goes to great lengths to reward supplier performance. Goals are in place:

1. To communicate to suppliers that their internal quality processes and performance results are critical.
2. To demonstrate that quality pervades all aspects of the business relationship.
3. To share expectations and information to build partnerships toward world-class performance.
4. To recognize a supplier's overall level of quality. There are three levels of recognition including the Gold Award, the Silver Award, and the Bronze Award. The Gold Award is awarded to the company's top suppliers. To win this award, a supplier must "delight" SBC by providing superior products/services and customer service for more than a year.

[7]George, S., and Weimerskirch, A., *Total Quality Management* (New York: John Wiley & Sons, Inc., 1994), p. 190.

As the telecommunications industry continues to become more competitive, SBC's efforts to maximize the performance of its supply chain will undoubtedly intensify. Setting high standards for suppliers and rewarding performance is the essence of SBC's philosophy of supply chain management. ■

DISCUSSION QUESTIONS

1. Do you believe that SBC's standards for its suppliers are too rigid? Why or why not?
2. Is SBC's "Supplier Quality Report" a good idea or are monthly reports too frequent and intrusive? Please explain your answer.
3. Do you believe suppliers should receive awards and designations of merit from the companies that purchase its products? Please explain your answer.

Case 9-2 Managing the Supply Chain at Honeywell

Honeywell: *www.honeywell.com*

If you own a home, work in an office building, or travel by air, the chances are excellent that a Honeywell product has affected your life. The Honeywell Corporation is the world's leading maker of control systems and related components for buildings, homes, industry, space, and aviation. The most recognizable Honeywell product is the automatic thermostat control for home heating that can be found in millions of U.S. homes. Less recognizable Honeywell products are used in commercial and military aircraft, industrial applications, home security systems, and the U.S. space program.

Like many companies, Honeywell is in a competitive environment that demands high-quality products at affordable prices. To meet these challenges, Honeywell has worked hard on its supply-chain management. In an effort to maximize the value that suppliers make to its business units, Honeywell has developed a distinctive approach to supplier management. The distinctive approach contains the following key components:

The company is shifting from a focus on price to a focus on total cost. As a result, the company does not always buy from the lowest price suppliers. Instead, it buys from the supplier that it believes will provide it the lowest cost in the "long term."

The company is shifting toward longer-term relationships with its suppliers and is treating its suppliers as if they are extensions of its own businesses rather than vendors.

The company is asking its suppliers to provide business solutions, rather than simply dropping products off at the loading dock.

Rather than treating suppliers in an adversarial manner, Honeywell is treating its suppliers like business partners and is looking for opportunities to partner with suppliers in a wide variety of areas.

The motivation for these criteria is not grounded in better public relations or altruism, but in the belief that prudent supply chain management saves money and creates a competitive advantage. The company's willingness to focus on issues other than price is central to the success of this philosophy. Commenting on this issue, a supply-chain manager at Honeywell remarked, "We have to reengineer our mindset to focus on improving the benefits we can receive from the suppliers versus just concentrating on prices. Many business organizations hold price so sacred that they miss the free things that suppliers will do for you."[8] Good suppliers want to be closely scrutinized by their buyers, because it gives the suppliers a chance to showcase their on-time performance and other positive attributes.

The consistent theme that is reflected in all aspects of Honeywell's approach to supply-chain management

[8]How Honeywell Works to Gain Greater Value from Their Suppliers," *Supplier Selection & Management Report*, The Institute of Management and Administration, *http://www.ioma.com* (2000).

is an effort to maximize the value of its supplier relationships. The company asks a lot of its suppliers, from on-time deliveries to providing input on the design of Honeywell products. In exchange, Honeywell gives its preferred suppliers millions of dollars a year in business and conducts itself in a responsible manner by making payments on time and treating its suppliers with respect. This overall approach to supplier management creates what Honeywell believes is a win-win partnership between itself and its suppliers. The more Honeywell's suppliers contribute to the profitability of its businesses, the more business Honeywell will give to its suppliers.

Although Honeywell is fully vested in its distinctive approach to supply-chain management, the company is also firmly committed to cost containment. Fortunately, supply-chain management and cost containment are not at odds at Honeywell. The firm is very careful about the suppliers it chooses and works hard to find the right mix of price and performance. Honeywell involves its suppliers in every aspect of improving the efficiency of the supply chain. This creates a healthy working relationship between Honeywell and its channel partners.

Honeywell is firmly committed to prudent supply-chain management. The company has quantified the results of its initiatives in several areas. For example, in terms of treating suppliers like business partners rather than arm's-length vendors, the company has learned that "partners" rather than "vendors" produce more cost savings, have better on-time delivery, supply better quality products, and provide more suggestions for supply chain improvements. These results, along with positive results in other areas of the company's approach to supply-chain management, have affirmed to the company the value of its approach to supply-chain management. ■

DISCUSSION QUESTIONS

1. Why does Honeywell spend so much time dealing with supply-chain management issues? Wouldn't the company be better off focusing on its own manufacturing operations? Explain your answer.
2. What aspect or aspects of Honeywell's approach to supply-chain management appear to be particularly prudent? Can you suggest any additions to Honeywell's supply-chain management program?
3. Do you believe that Honeywell's approach to supply-chain management creates a win-win situation between the company and its suppliers? Why or why not?

PART THREE

Implementing Quality

Everyone will tell you that the conceptual stuff is nice, but how do you actually implement quality throughout your supply chain? Once you understand and have planned and designed where you want to go, the tools of quality provide an important basis for improvement. More than 90% of successful quality implementations involve the basic tools of quality. These should be used and commonly practiced in most companies. Chapter 10 introduces the seven managerial tools for quality improvement. These tools are behavioral and help managers to optimize the use of team processes.

Quality improvement is often manifest in the performance of a long string of team projects. Chapter 11 discusses the behavioral side of managing teams and the technical side of planning and controlling projects. It also provides tools for overseeing and controlling multiple projects simultaneously. Many instructors may want to teach this chapter earlier in the text as team projects are important components of effective quality management classes.

In Chapters 12 and 13 you will learn about an efficient and effective method for teaching statistical quality and $\bar{x}$, R, p, np, u, and c charts. Many of the behavioral implications of statistical process control are also discussed.

Chapter 14 introduces Six Sigma. This is a widely adopted method for improving the designs of products and processes to assure quality.

CHAPTER 10

The Tools of Quality

When we introduce a particular method of doing a job, it is natural to consider whether the method is appropriate or not. The decision is usually based on past results and experience, or perhaps on conventional methods. Procedures will be most effective if a proper evaluation is made, and on-the-job data are essential for making a proper evaluation.

—KAORU ISHIKAWA,
Quality Tools Inventor[1]

Quality improvement in manufacturing or services, to be effective, should address the needs of the system as a whole. In this book we have attempted to address quality management from an integrative perspective. This perspective has encompassed the many functional areas of business, including supply chain management, marketing, accounting, human resources, operations, engineering, and strategy. None of these fields of endeavor operate in a vacuum. They are all interrelated and interdependent.

IMPROVING THE SYSTEM

To be successful, a business or organization must balance the needs of these different functional areas around a coherent business vision and strategy. The objective of the system is to satisfy the customer. Customer satisfaction means higher customer retention, which leads to improved profitability.

A quality system (Figure 10-1) uses the business model with a focus on the customer and includes the dynamics of continual improvement, change, planning, and renewal. Continual improvement is necessary for a company to learn to grow. Companies that are unable to adapt find themselves with stagnant cultures and labor forces. Many managers, on discovering that their organization has reached this point, believe they must resort to draconian measures such as layoffs and organizational reengineering to achieve change. If they had pursued continual improvement and learning in the first place, they might not have reached this juncture.

This quality system is not just a series of variables and relationships. It is an interconnected, interdisciplinary network of people, technology, procedures, markets, customers, facilities, legal requirements, reporting requirements, and assets that interact to achieve an end. The most important aspect of the system is the people. People are the engine of creativity and innovation. Technology is very good at performing rote tasks; however, technology in and of itself cannot innovate. Therefore, how we manage people may be the most important key in this system to unlock an organization's potential.

[1]Ishikawa, K., "Guide to Quality Control," Asia Productivity Organization, Tokyo, Japan, 1985.

FIGURE 10-1 Quality System Model

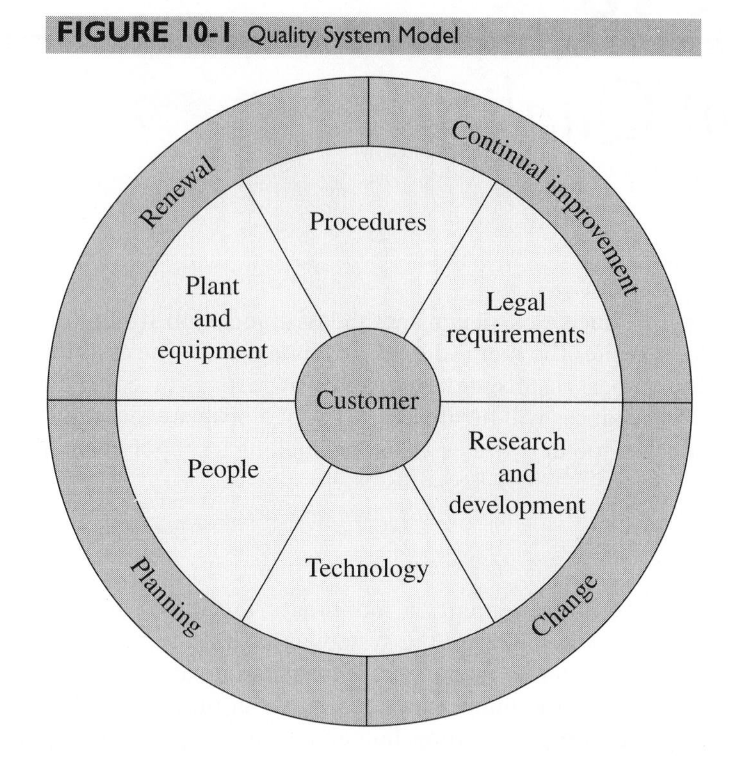

W. Edwards Deming was always adamant that we should continually and forever improve the system of production. The system includes people. In this chapter we present the tools that are commonly used to unlock this human potential for change and improvement.

In this chapter we introduce the **basic seven (B7) tools of quality** and the new seven (N7) tools (also referred to as the *managerial tools*). The seven basic tools are simple to use in continuous improvement efforts. The tools often are used by individuals and in teams, are useful at all levels of the organization, and can be applied by people of different educational levels. As you learn and apply the tools of quality, you too will appreciate their wide application and usefulness.

ISHIKAWA'S BASIC SEVEN TOOLS OF QUALITY

The basic seven tools of quality may be used in a logical order. Note that this is only a "typical" order of use for these tools. They can be used in any order. Figure 10-2 shows this order. The flowchart gives the team the big picture of the process to be improved. Process data are collected using a checksheet. The data are analyzed using either histograms, scatter plots, or control charts. The root causes of the problems associated with the process are identified using a cause-and-effect diagram. Finally, causes are prioritized using Pareto analysis. These tools are discussed in more depth on the following pages.

Process Maps

A **process map** is a picture of a process. The first step in many process improvement projects is to create a map of the process as it exists. This useful step also determines the parameters for process improvement. The concept is that we must know the process before we can improve it.

FIGURE 10-2 Logical Map of the Order for the Basic Seven (B7) Tools

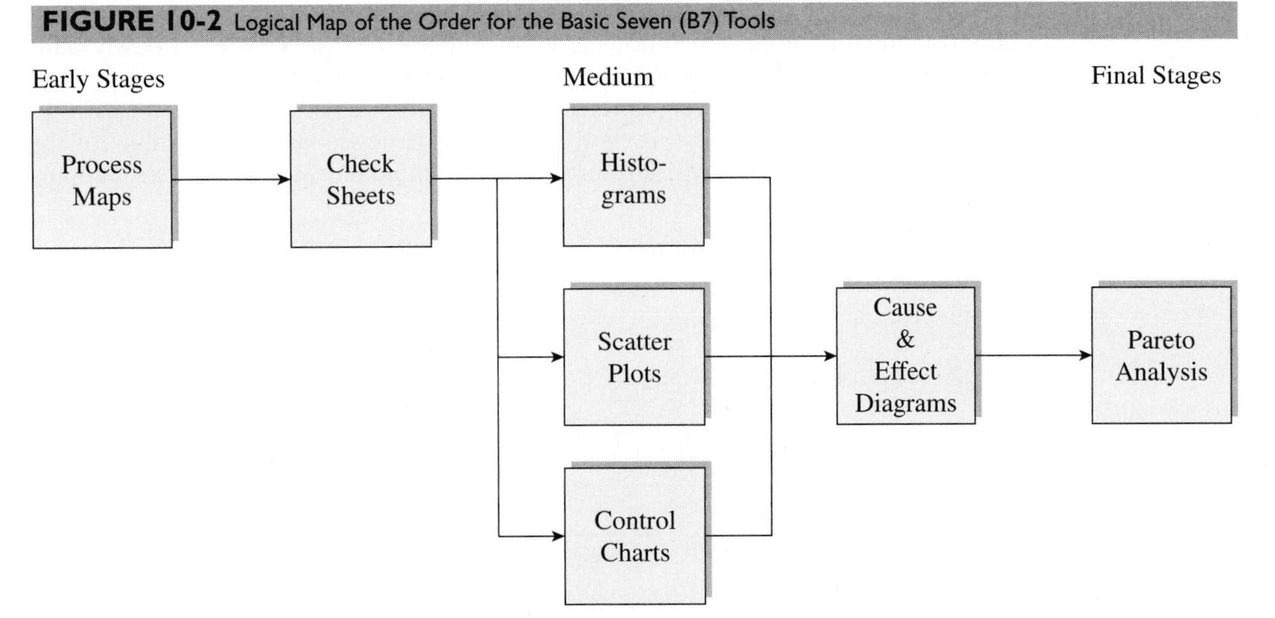

Adapted from *The Memory Jogger II*, published by GOAL/QPC, 2 Manor Parkway, Salem, New Hampshire, 2004.

The language of process maps can vary from the simple to the complex. A simple set of symbols is provided in Figure 10-3. The diamond indicates there is a decision to be made. Often these identify different paths of sequences in the process map. The parallelogram appears whenever materials, forms, or tools enter or leave the process. The rectangle is the processing symbol—the work that is actually performed. The start/stop symbol and the page connector are used for the convenience of the people using the process map. A few simple rules for process maps follow:

- Use these simple symbols to chart the process from the beginning, with all arcs in the process map leaving and entering a symbol. The arcs represent the progression from one step to the next. (See A Closer Look at Quality 10-1.)
- Develop a general process map and then fill it out by adding more detail or subflowchart each of the elements.
- Step through the process by interviewing those who perform it—as they do the work.
- Determine which steps add value and which don't in an effort to simplify the work.
- Before simplifying work, determine whether the work really needs to be done in the first place.

FIGURE 10-3 Basic Mapping Symbols

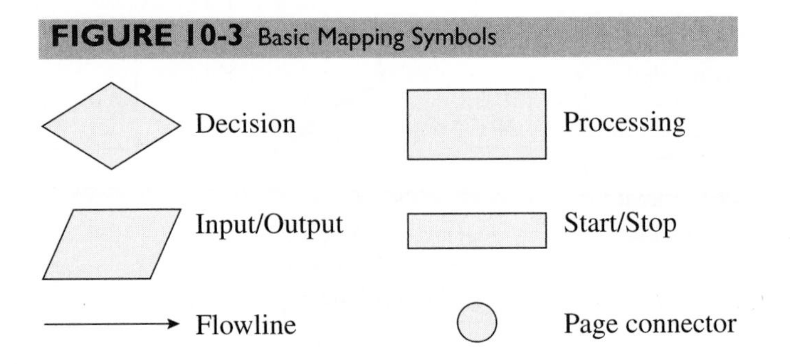

The process map in Figure 10-4 shows a simple process used in a city planning department to issue permits allowing applicants to take possession of newly built homes. Figure 10-4 shows the current process. In Figure 10-5 the process is simplified because the front desk is given more authority and training to process the forms without assigning them for analyst review. The analyst review does not add value for the organization or the customer. Therefore, it can be eliminated. Steps in process mapping include

1. Settle on a standard set of process mapping symbols to be used.
2. Clearly communicate the purpose of the process map to all the individuals involved in the exercise.

FIGURE 10-4 Process Mapping Home Occupation Process—Current

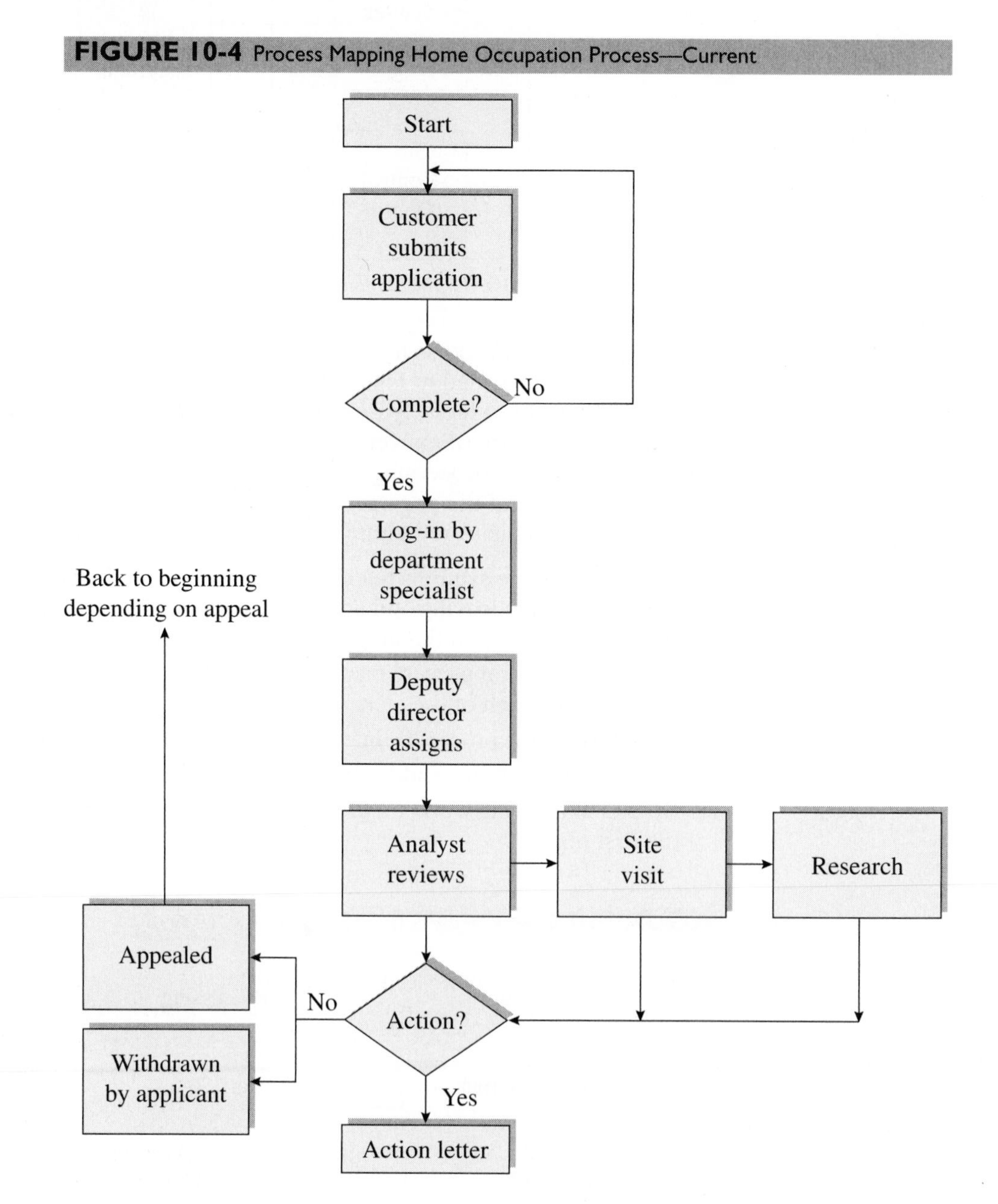

FIGURE 10-5 Process Mapping: Home Occupation Process—Proposed

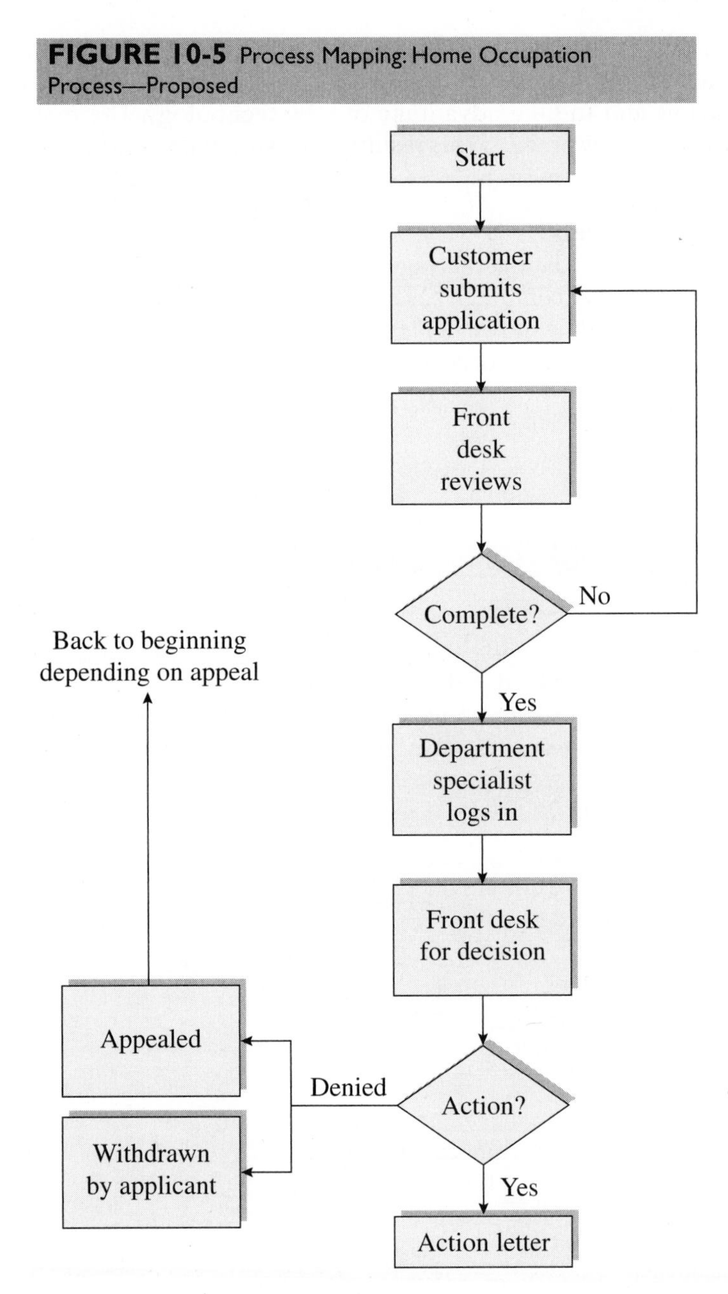

3. Observe the work being performed by shadowing the workers performing the work.
4. Develop a map of the process.
5. Review the process map with the employees to make needed changes and adjustments to the process map. (Note that it is often helpful to chart processes from the customer's point of view in addition to the worker's point of view.)

Example 10.1: Process Maps

Problem: The Well Construction Unit of a state Department of Water Resources entered into a multiyear project to update its database management system. As part of the process, the Well Construction staff was asked to document its current process flows.

Solution: The resulting process map is shown in Figure 10-6. Through a brainstorming process, the Well Construction team was asked to rethink its processes to simplify the workflow and to take advantage of new technology. The team worked together to develop the new process. This resulted in a streamlined flow that required less time for drillers to receive permits.

Check Sheets

Check sheets are data-gathering tools that can be used in forming histograms; they can be either tabular, computer based, or schematic. An example of a tabular check sheet for a Pareto chart is shown in Figure 10-11. This provides a chart for copier operators to mark each time a delay occurs in setting up new jobs.

FIGURE 10-6 Process Map with Responsibility of Existing Process

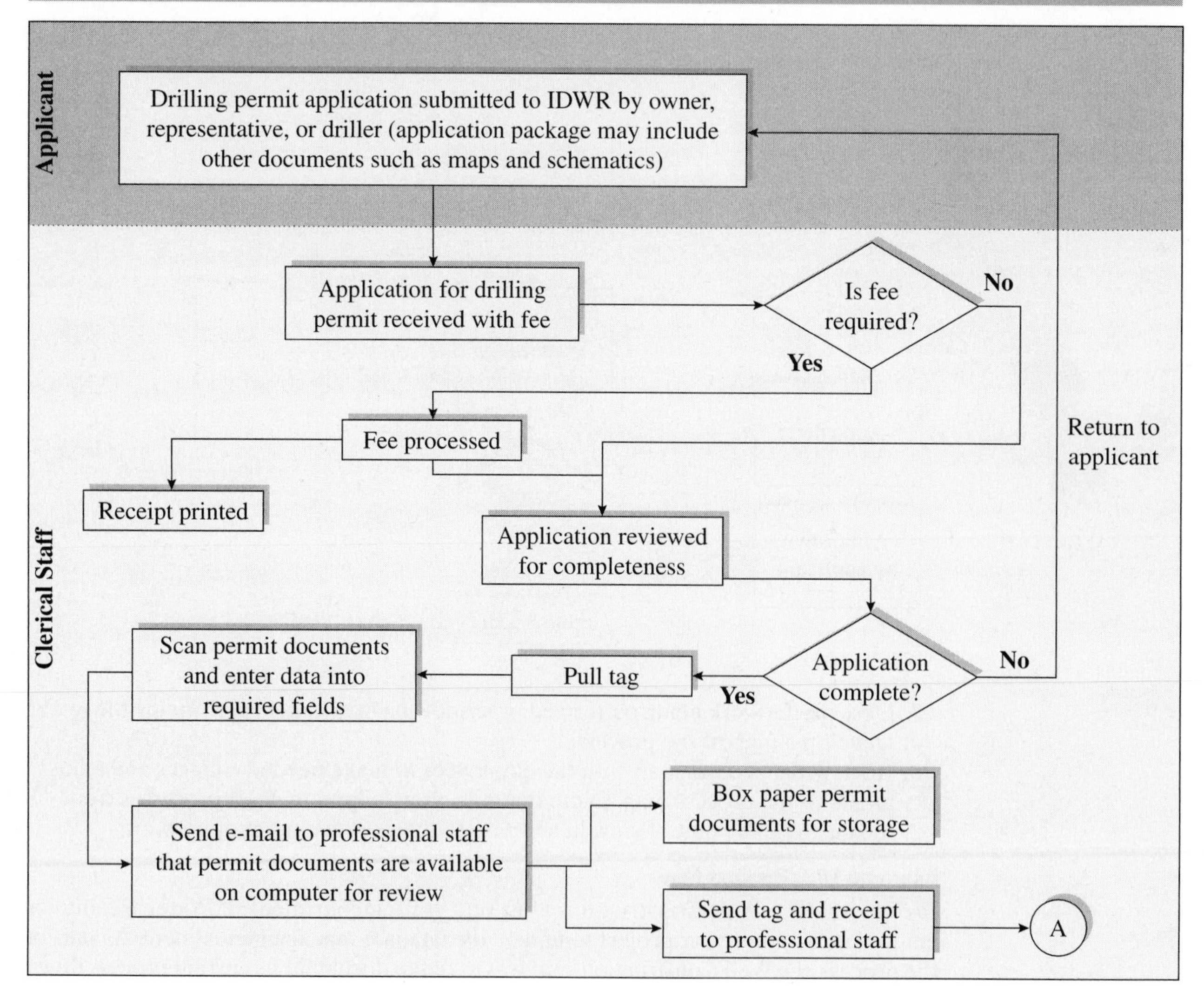

FIGURE 10-6 (Continued)

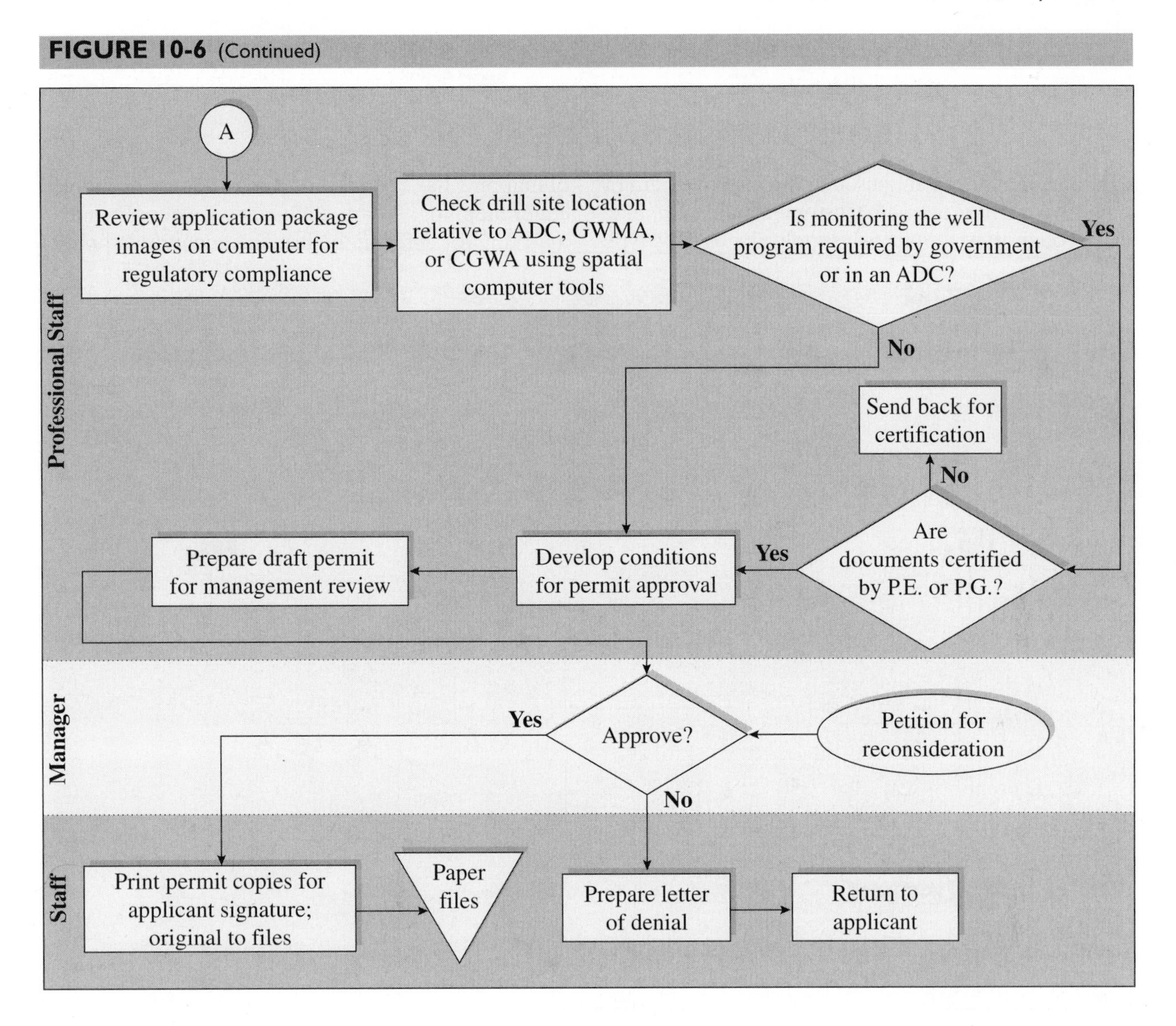

Figures 10-9 and 10-10 present two examples of schematic check sheets from Ishikawa. The first is for operators to mark bubbles where they occur in the finish of automobile windows. The second is for marking defects in a radiator. From these data, the types of defects are charted on a Pareto chart, cost analysis is performed, and the data are used to prioritize design improvements to the products. Setting up a check sheet involves the following steps:

1. Identify common defects occurring in the process.
2. Draw a table with common defects in the left column and time period across the tops of the columns (see Figure 10-8) to track the defects.
3. The user of the check sheet then places checkmarks on the sheet whenever the defect is encountered.

A CLOSER LOOK AT QUALITY 10-1

EXTENDED PROCESS MAPPING OF SUPPLY CHAINS

Process maps are being used in the improvement of supply chain processes. Customers and suppliers can collaborate to improve supply chains. This type of mapping has been referred to as extended supply chain mapping. Figure 10-7a shows a supply chain map for Mare Technologies. This includes supplier

FIGURE 10-7a Mare Technologies Current-State Extended Value Stream Map

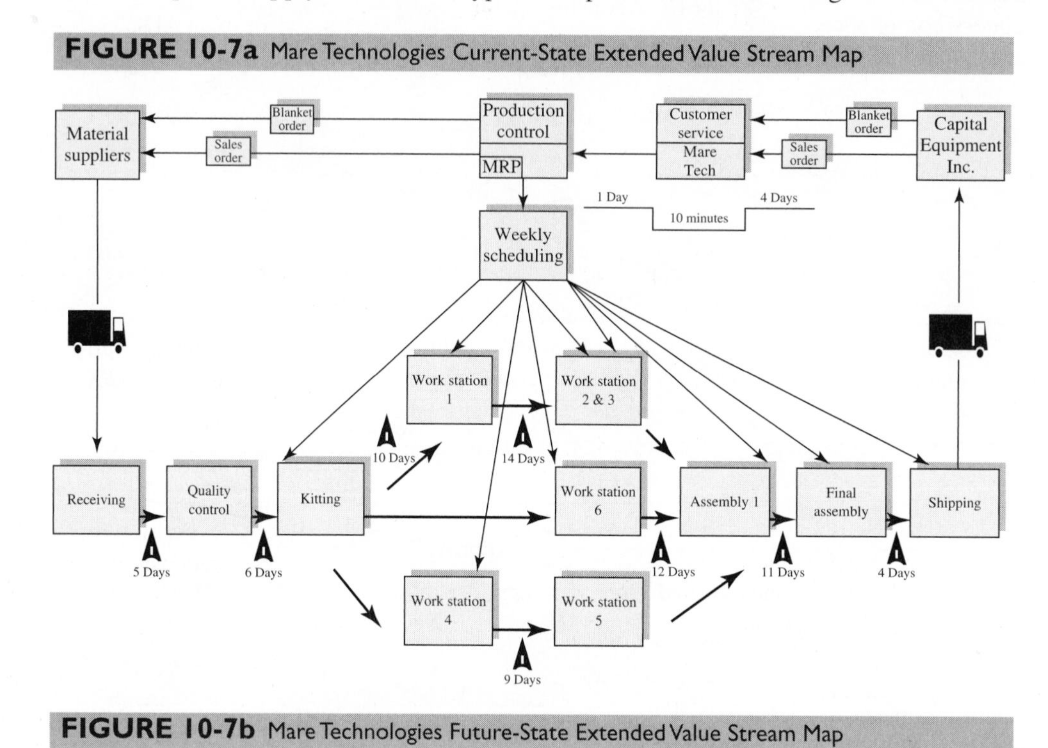

FIGURE 10-7b Mare Technologies Future-State Extended Value Stream Map

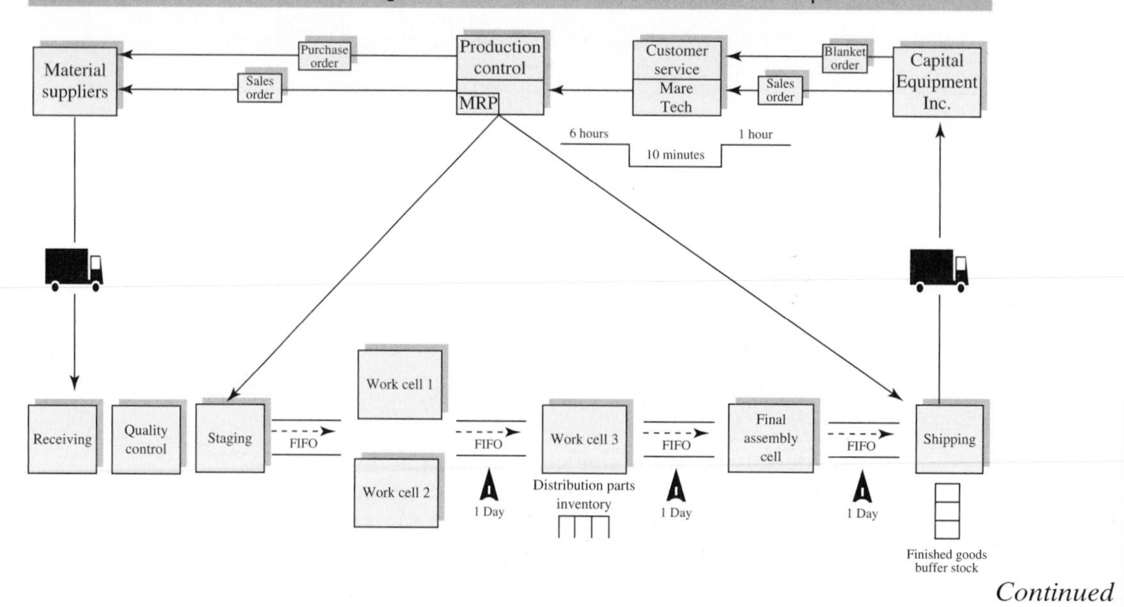

Continued

processes, receiving, internal processes, shipping, and customer service processes.

Figure 10-7b shows a map of the improved process. Some comparisons of the existing and improved processes are as follows:

Results Metrics	Current State	Improved State	% Improvement
Lead time (days)	55	42	24
WIP (days)	11	1	91
Flexibility	Limited	6.25% increase per week	400/year
Unit price	$6,440	$5,860	9

Source: Horton, P., and D. DelMonico, "Charting A New Course," *APICS: The Performance Advantage*, Oct 2004, 43–46.

Example 10.2: Check sheets

Problem: A copying company desires to set up a check sheet so that it can keep track of the sources of errors. Following are the major error types with frequencies.

Solution: Figure 10-11 shows a check sheet for these data. The check sheet will be kept to monitor how well workers are adhering to the new procedures.

Type of Problem	Frequency	Percentage
Setup routines are not standardized	315	52.1%
Equipment needed for setup is missing	124	20.5
Internal and external setup tasks are not separated	87	14.4
Extensive machine resetting and paper change is needed	56	9.2
Other	23	3.8
Total	605	100%

FIGURE 10-8 Check Sheet

Problem	M	Tu	W	T	F	Total
Totals						

FIGURE 10-9 Check Sheet

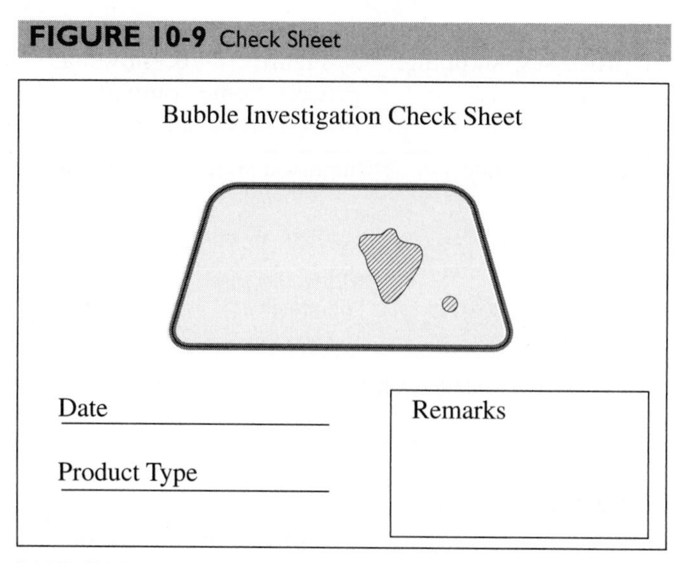

SOURCE: K. Ishikawa, "Guide to Quality Control," (Tokyo: Asia Productivity Organization, 1985).

Histograms

As shown in Figures 10-12 and 10-13, **frequency charts** and **histograms** are simply graphical representations of data in a bar format. The frequency chart in Figure 10-12 shows the number of occurrences of orders for maintenance on four production lines and the hours used to fill these maintenance orders. Note that a frequency chart is used for categorical data, while histograms are used for continuous numerical data. Histograms are also used to observe the shape of data (see Figure 10-13). For example, how are the data in an interval scale distributed? There are several rules for developing histograms:

- The width of the histogram bars must be consistent (i.e., class widths are the same where each bar contains a single class).
- The classes must be *mutually exclusive* and *all-inclusive* (or *collective exhaustive*).
- A good rule of thumb for the number of classes is given by the model

$$2^k \geq n \quad \textbf{(10.1)}$$

where n is the number of raw data values and k is the number of classes. Solving this equation for k, we obtain

$$k \geq \log n/\log 2 \quad \textbf{(10.2)}$$

Using this formula, we find

Number of Observations	Number of Classes
9 to 16	4
17 to 32	5
33 to 64	6
65 to 128	7
129 to 256	8

FIGURE 10-10 Radiator Check Sheet

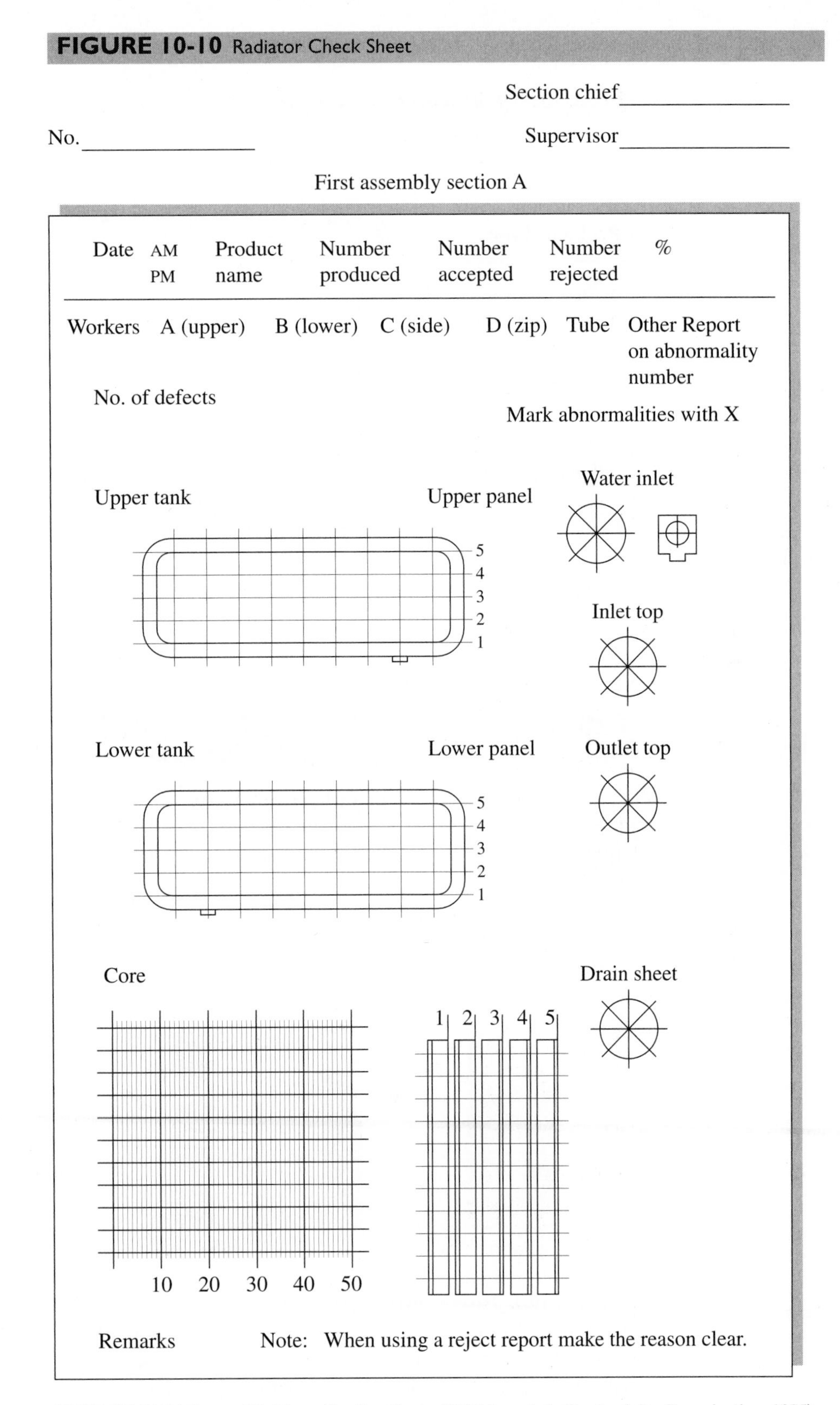

Section chief ______________

No. ______________ Supervisor ______________

First assembly section A

Date	AM / PM	Product name	Number produced	Number accepted	Number rejected	%

Workers	A (upper)	B (lower)	C (side)	D (zip)	Tube	Other Report on abnormality number
No. of defects						

Mark abnormalities with X

Upper tank — Upper panel — Water inlet

5 4 3 2 1

Inlet top

Lower tank — Lower panel — Outlet top

5 4 3 2 1

Core — Drain sheet

1 2 3 4 5

10 20 30 40 50

Remarks — Note: When using a reject report make the reason clear.

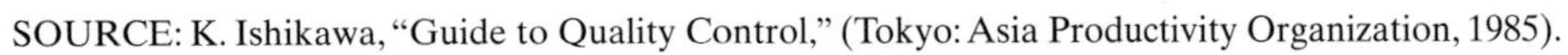

SOURCE: K. Ishikawa, "Guide to Quality Control," (Tokyo: Asia Productivity Organization, 1985).

FIGURE 10-11 Copier Problem Check Sheet

Problem Type	Monday	Tuesday	Wednesday	Thursday	Friday	Total
Setup routines not standardized						
Missing equipment for setup						
Failure to separate internal and external tasks						
Extensive machine resetting and paper change						
Other						

FIGURE 10-12 Frequency Chart of Number of Maintenance Occurrences and Service Hours for Four Production Lines

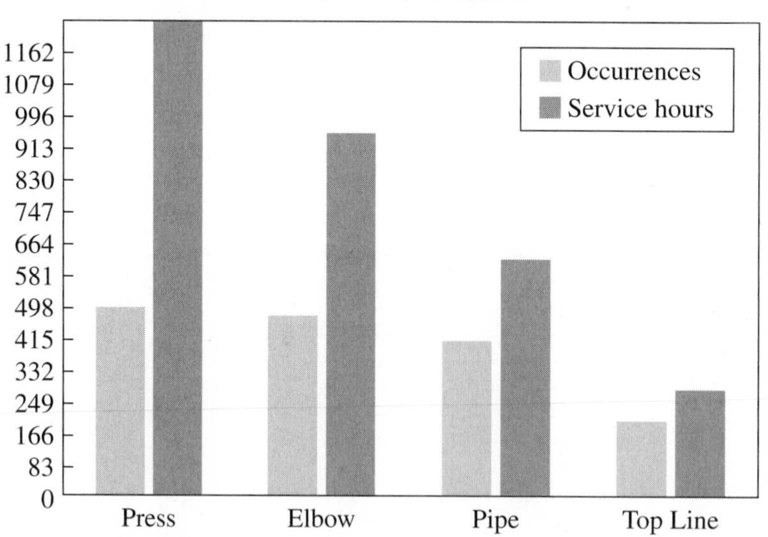

Example 10.3: Histograms

Problem: The Big City Cafeteria wants to determine the distribution of its sales during lunch time. On a given day the manager randomly selects 40 sales from the sales register receipt. The following table shows the sales:

Develop a histogram of the sales.

Solution: It is helpful to compute the mean, standard deviation, maximum value, and minimum value when developing a histogram because the histogram is often used to determine if the data are normally distributed. Following are these statistics from the previously given data:

Mean = 4.20

Maximum value = 8.95

Minimum value = .79

Difference = 8.16

Sum = 168

Using Formula (10.2)

$$k \geq \log 40/\log 2$$
$$k \geq 5.32$$

The number of classes is 6. Therefore,

Classes = 6

Class width = 8.16/6 = 1.36 – 1.40

Classes = 0.76–2.15; 2.16–3.55; 3.56–4.95; 4.96–6.35; 6.36–7.75; 7.76–9.15

The histogram is displayed in Figure 10-13. Thus the manager finds that sales occur in a skewed distribution with a mean of $4.20.

FIGURE 10-13 Histogram

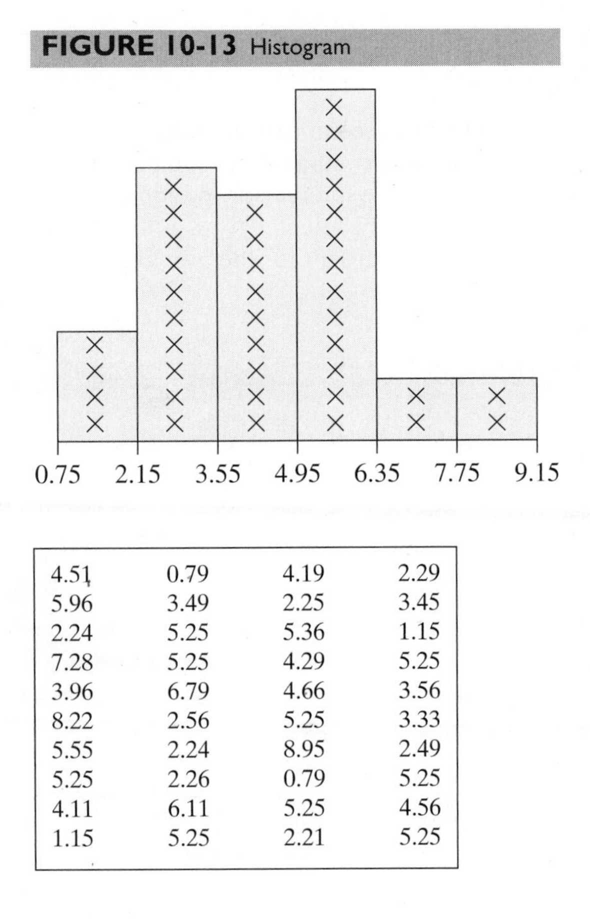

4.51	0.79	4.19	2.29
5.96	3.49	2.25	3.45
2.24	5.25	5.36	1.15
7.28	5.25	4.29	5.25
3.96	6.79	4.66	3.56
8.22	2.56	5.25	3.33
5.55	2.24	8.95	2.49
5.25	2.26	0.79	5.25
4.11	6.11	5.25	4.56
1.15	5.25	2.21	5.25

Scatter Diagrams

The **scatter diagram** or **scatter plot** is used to examine the relationships between variables. These relationships are sometimes used to identify indicator variables in organizations. For example, in a hospital, the postoperative infection rate has been found to be associated with many different factors such as the sterile procedure used by the doctors and nurses, cleanliness of the operating rooms, and sterile procedures in handling the utensils used in surgery. Therefore, the postoperative infection rate is an important variable for hospital quality measurement.

It is quite easy to develop scatter plots using the charting facilities in spreadsheet packages such as Excel. Figure 10-14 shows a scatter plot of the relationship between conformance data and prevention and appraisal quality-related costs in a real firm. Note that the figure shows the unexpected outcome of higher quality costs with higher levels of conformance. Later analysis showed that this firm was trying to "inspect in" quality, meaning that it was throwing a lot of in-process work away as a result of more rigorous inspection. Use the following steps in setting up a scatter plot:

1. Determine your x (independent) and y (dependent) variables.
2. Gather process data relating to the variables identified in step 1.
3. Plot the data on a two-dimensional plane.
4. Observe the plotted data to see whether there is a relationship between the variables. (Note that it is helpful to plot the data in Excel or another spreadsheet and to perform a correlation test to determine whether the variables have a statistically significant relationship.)

Example 10.4: Scatter Diagrams

Problem: Healthy People, Inc., a company specializing in home health care solutions for U.S. consumers, was a growing company. The company wished to study the relationship between absenteeism and the number of overtime hours worked by employees. Thirty employees were randomly selected, and numbers of overtime hours were graphed against numbers of days absent for the previous year (see Figure 10-15).

FIGURE 10-14 Prevention Costs and Conformance

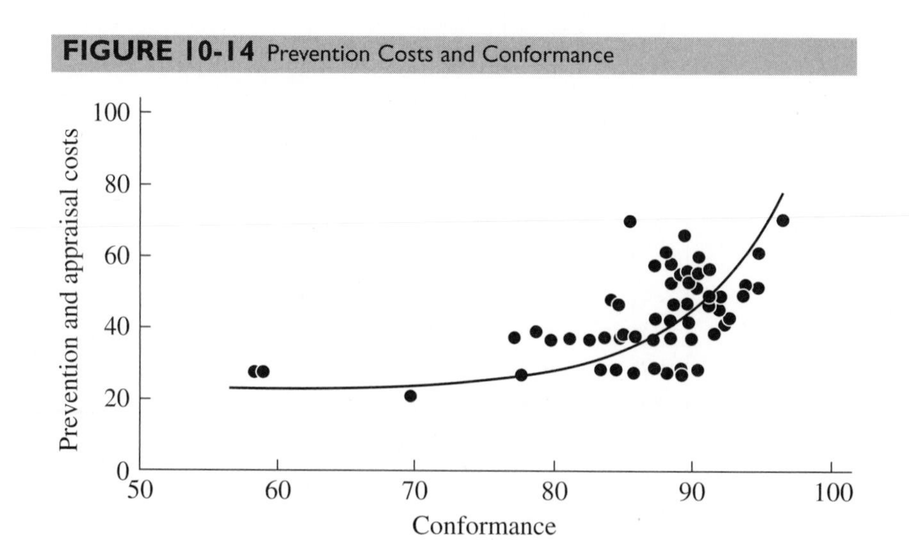

Employee	Hours of Overtime	Days Absent
1	243	3
2	126	2
3	86	0
4	424	6
5	236	3
6	128	0
7	0	0
8	126	2
9	324	3
10	118	0
11	62	0
12	128	3
13	460	6
14	135	1
15	118	1
16	260	2
17	0	1
18	126	1
19	234	2
20	246	3
21	120	1
22	80	0
23	112	1
24	237	3
25	129	2
26	24	1
27	36	0
28	128	2
29	246	3
30	326	6

FIGURE 10-15 Scatter Plot of Overtime Hours versus Days Absent

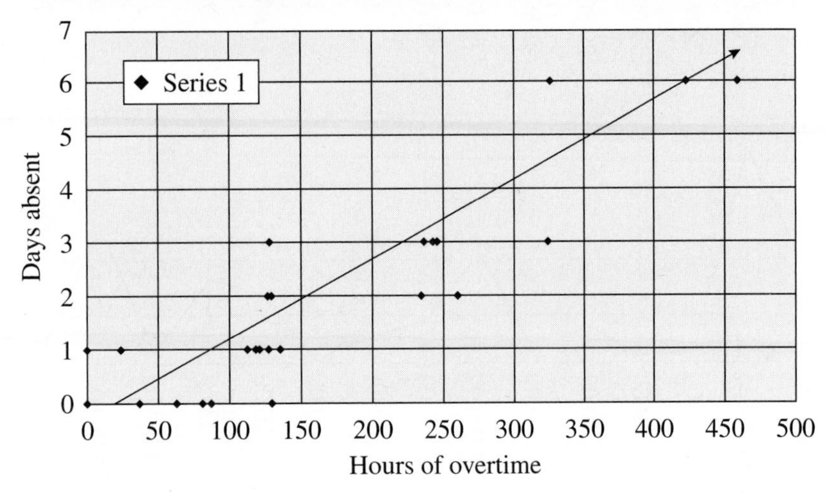

This analysis showed that there appeared to be a positive relationship between number of days absent and hours of overtime. Subsequent analysis showed that, in fact, these variables were significantly related. This led management to recalculate the actual cost of overtime.

Control Charts

Control charts are used to determine whether a process will produce a product or service with consistent measurable properties. Because control charts are discussed in Chapters 12 and 13, they will not be presented in detail here. Figure 10-16 illustrates two control charts usually used together.

Cause-and-Effect (Ishikawa) Diagrams

Often workers spend too much time focusing improvement efforts on the symptoms of problems rather than on the causes. The Ishikawa **cause-and-effect** or **fishbone** or **Ishikawa diagram** is a good tool to help us move to lower levels of abstraction in solving problems. The diagram looks like the skeleton of a fish, with the problem being the head of the fish, major causes being the "ribs" of the fish, and subcauses forming smaller "bones" off the ribs. A facilitator or designated team member draws the diagram after questioning why certain situations occur. It has been said that for each circumstance, the facilitator should ask "Why?" up to five times. This is sometimes

FIGURE 10-16 Partial $\overline{X}$ and R Charts for a Process

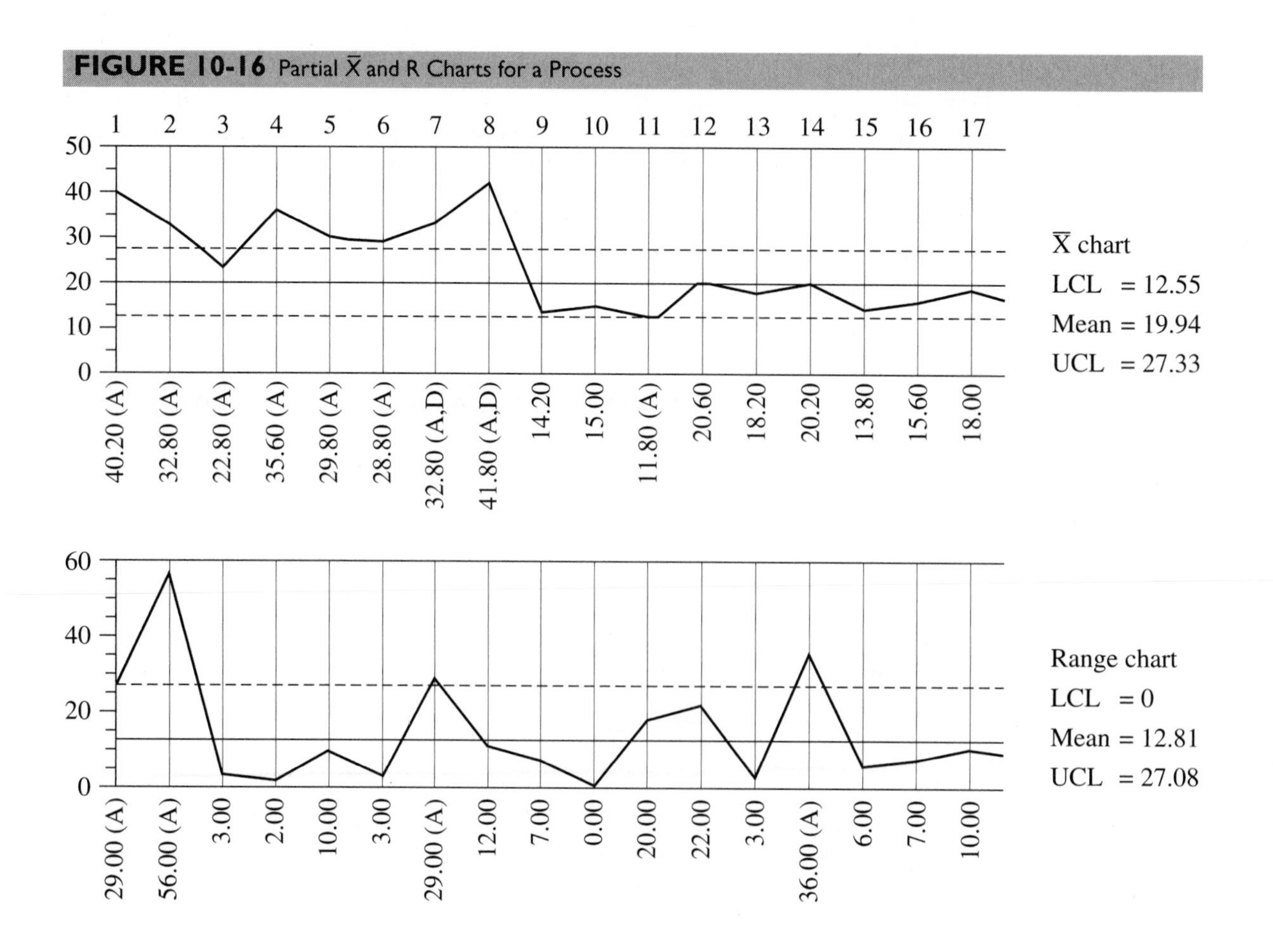

referred to as the "five whys." Fishbone (cause-and-effect) diagrams are created during brainstorming sessions with a facilitator by following these steps:

1. State the problem clearly in the head of the fish.
2. Draw the backbone and ribs. Ask the participants in the brainstorming session to identify major causes of the problem labeled in the head of the diagram. If participants have trouble identifying major problem categories, it may be helpful to use materials, machines, people, and methods as possible bones.
3. Continue to fill out the fishbone diagram, asking "Why?" about each problem or cause of a problem until the fish is filled out. Usually it takes no more than five levels of questioning to get to root causes—hence the "five whys."
4. View the diagram and identify core causes.
5. Set goals to address the core causes.

Figure 10-17 shows an Ishikawa diagram that was prepared for a wood mill that was experiencing problems with wobbling blades in its saws. The symptom of the problem was the wobbly blades. The major causes were associated with machines, materials, people, and methods. Concerning people, it was found that workers were not properly trained. For machines it was found that the blades were being incorrectly set up off-center.

Example 10.5: Ishikawa Diagrams

Problem: A team of employees from the Adjudication Team at a Department of Water Resources was assigned to improve its process. Adjudication is a process of going through the courts to settle legal disputes, in this case concerning water rights. Prior to brainstorming improvements for the process, the employees were asked to brainstorm some of the causes of problems with the existing system. A fishbone (Ishikawa) diagram was used to help to identify causes of problems they were experiencing.

FIGURE 10-17 Cause-and-Effect Diagram: Wobbling Saw Blade Example

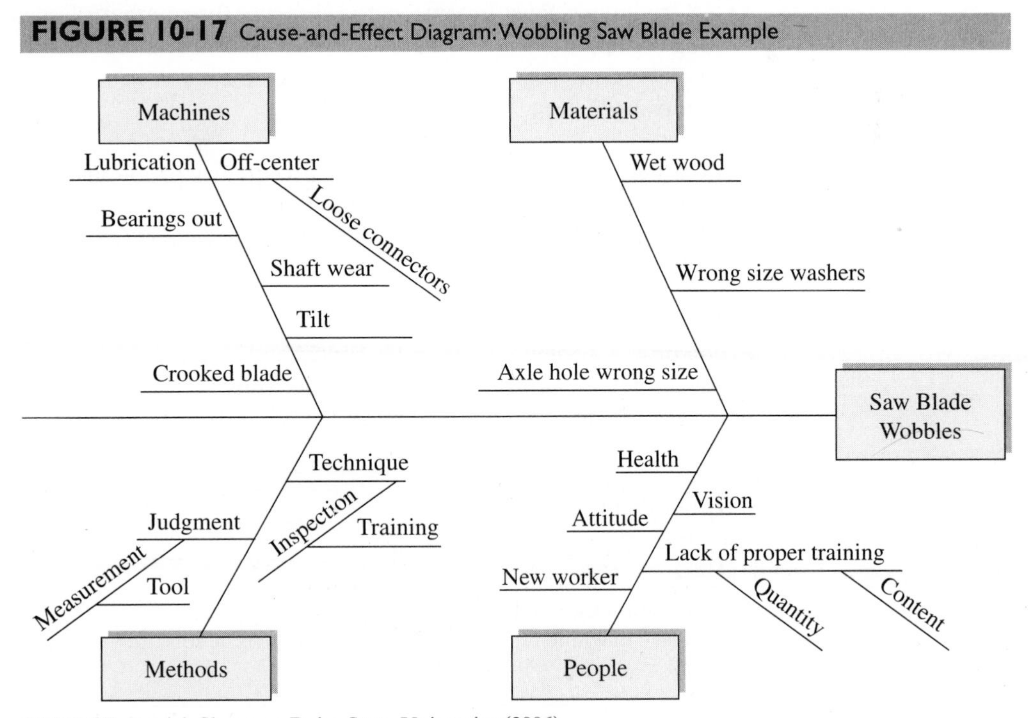

SOURCE: Patrick Shannon, Boise State University (2006).

Solution: Figure 10-18 shows the resulting fishbone diagram. The fishbone diagram shows that three major areas of concern are contractors, region office–state office communication, and database management. The facilitator used the "five whys" to get team members to reach lower levels of abstraction. After reaching these lower levels of abstraction, participants were asked to identify what they felt were major causes of the problems. This fishbone diagram was later complemented with further brainstorming for issues relating to the adjudication process.

Pareto Charts

Pareto charts are used to identify and prioritize problems to be solved. These are actually frequency charts that are aided by the 80/20 rule adapted by Joseph Juran from Vilfredo Pareto, the Italian economist. As you may remember, the 80/20 rule states that roughly 80% of the problems are created by roughly 20% of the causes. This means that there are a *vital few* causes that create most of the problems. This rule can be applied in many ways, and 80% and 20% are only estimates; the actual percentages may vary.

In a positive sense, a store manager could understand that 20% of the stock in a store holds 80% of the value of the store inventory. Twenty percent of the customers might provide 80% of the revenue. In a grocery store, a small number of quality problems created 80% of the complaints. The good news is that by focusing on the *vital few*, inventory can be controlled, satisfaction of the most important customers can be increased, or 80% of the complaints can be eliminated. There are some rules for constructing Pareto charts:

- Information must be selected based on types or classifications of defects that occur as a result of a process. An example of this might be the different types of defects that occur in a semiconductor.
- Data must be collected and classified into categories.
- A frequency chart is constructed showing the number of occurrences in descending order.

The steps used in Pareto analysis include

1. Gathering categorical data relating to quality problems.
2. Drawing a frequency chart of the data.
3. Focusing on the tallest bars in the frequency chart first when solving the problem.

Example 10.6: Pareto Charts

Problem: A copying company is concerned because it is taking too long for operators to set up new printing jobs. They decide to use Pareto analysis to find out why setup times are taking so long. The data gathered reflect the following major causes:

Type of Problem	Frequency (Number of Times)
Equipment needed for setup is missing	124
Internal and external setup tasks are not separated	87
Setup routines are not standardized	315
Extensive machine resetting and paper change is needed	56
Other	23

FIGURE 10-18 Adjudication Fishbone Diagram

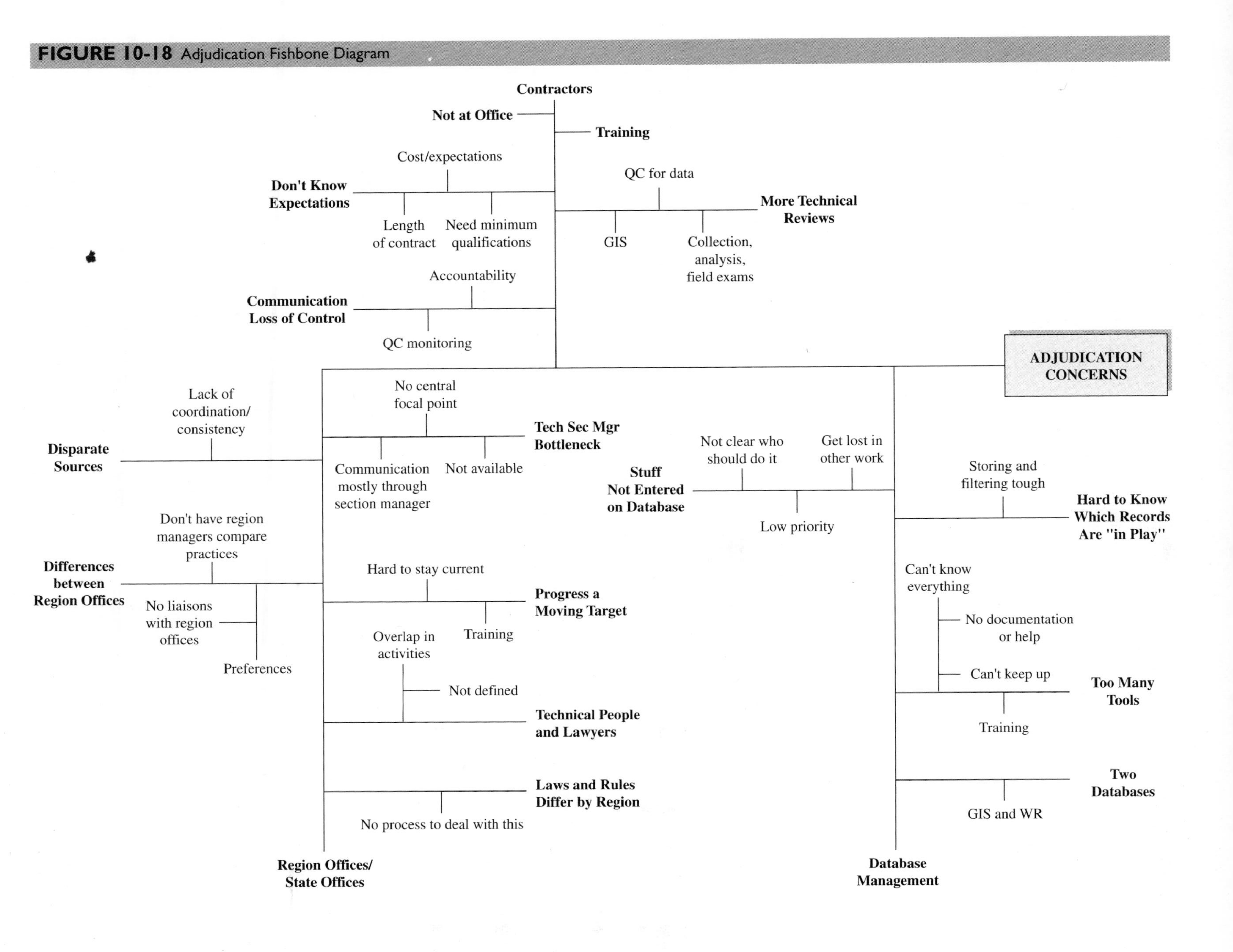

Solution: First, order the problems by frequency, and compute the percentage of problems related to each cause.

Type of Problem	Frequency	Percentage
Setup routines are not standardized	315	52.1%
Equipment needed for setup is missing	124	20.5
Internal and external setup tasks are not separated	87	14.4
Extensive machine resetting and paper change is needed	56	9.2
Other	23	3.8
Total	605	100%

Next, draw a frequency chart of the results (Figure 10-19). This Pareto chart shows that nonstandardized procedures for setting up copying jobs is the most frequently occurring problem causing slow setups. Therefore, the company can institute a training program to routinize its setups. This will result in a significant reduction in setup slowdowns.

Two points should be made. We also could analyze these data from a number of different perspectives, such as average time per type of delay or cost per type of delay. Also, this chart shows graphically that the law of diminishing marginal returns does have a place in quality thinking. As the group addresses each problem, the savings from correcting the problems decreases. There is no guarantee, however, that addressing the fourth problem will take any less effort than the first.

THE SEVEN NEW TOOLS FOR IMPROVEMENT

In addition to the basic seven tools of quality there is another set of tools that focuses on group processes and decision making. These are the new tools for management. The **new seven (N7) tools** were developed as a result of a research effort by a committee of the Japanese Society for QC Technique Development. They are shown in Figures 10-20 and 10-21 and are discussed in the following pages.

FIGURE 10-19 Pareto Analysis

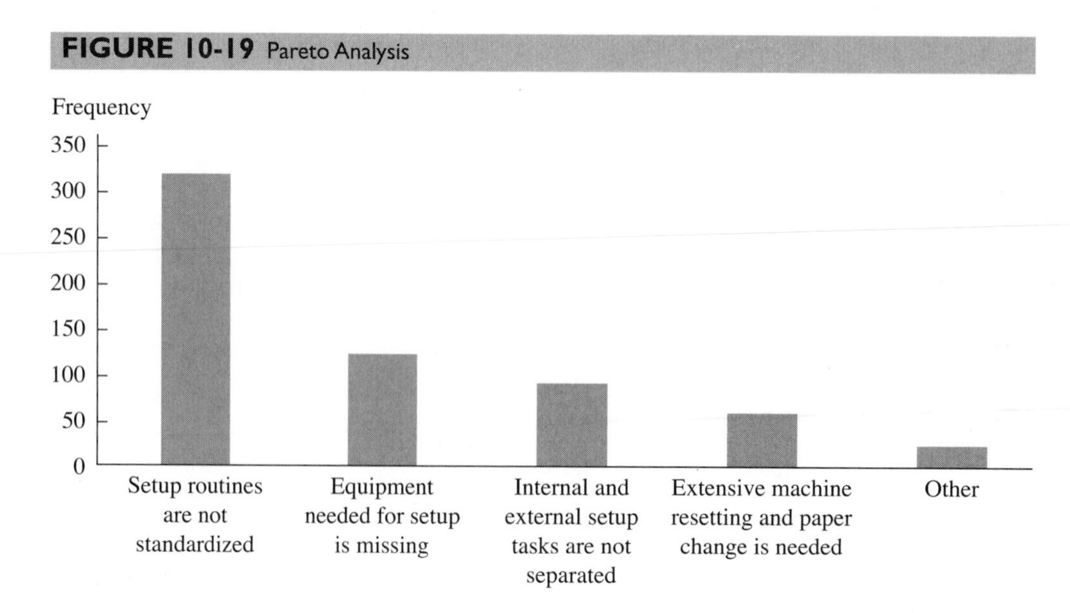

FIGURE 10-20 Three New Tools for Management

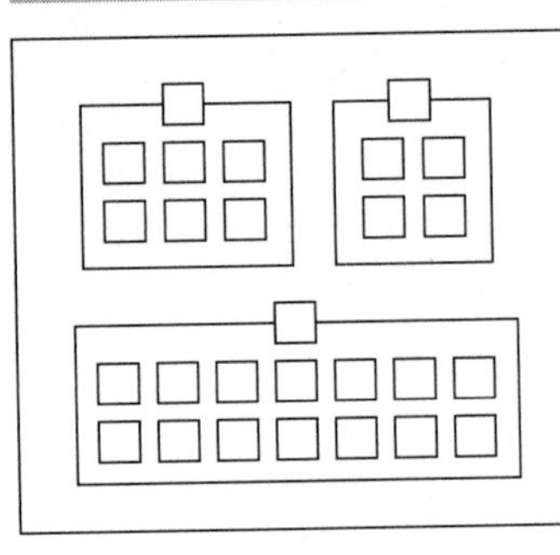

Affinity diagram

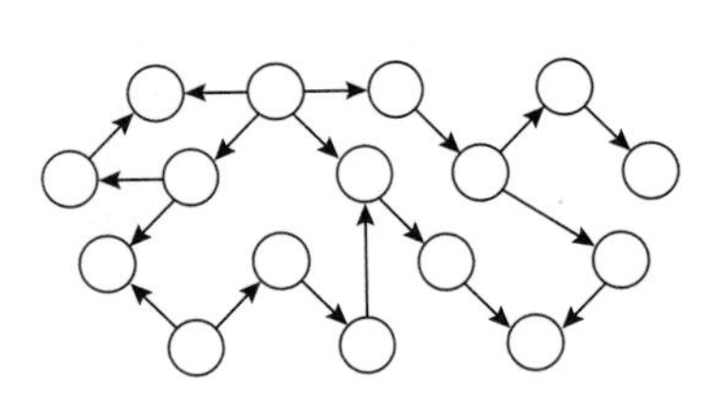

Interrelationship digraph

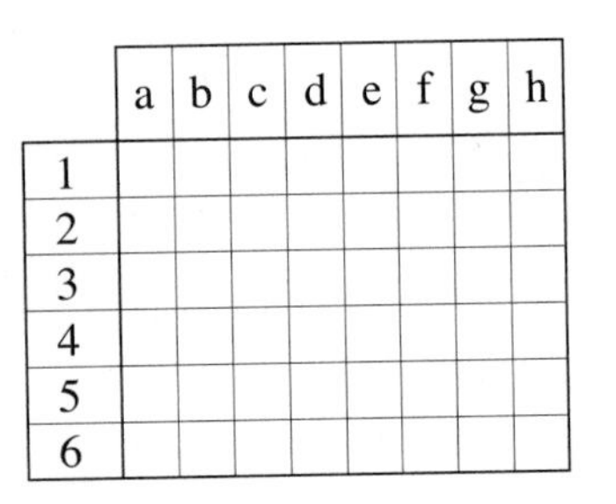

Matrix diagram

SOURCE: M. Brassard, "The Memory Jogger II," GOAL/QPC, Boston, 2004. Reprinted from "The Memory Jogger Plus+" with permission of GOAL/QPC, 12B Manor Parkway, Salem, NH 03079, *www.goalqpc.com*.

FIGURE 10-21 Four Other New Tools for Management

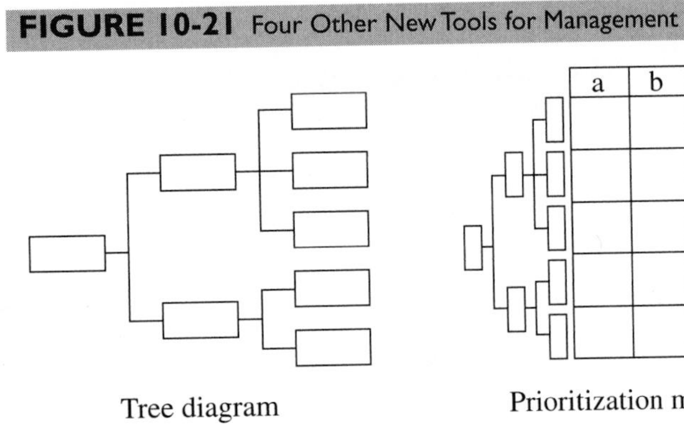

Tree diagram

Prioritization matrices

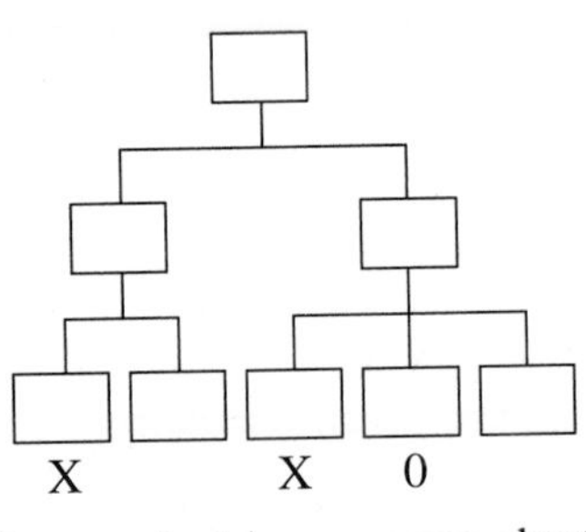

Process decision program chart

Activity network diagram

SOURCE: M. Brassard, "The Memory Jogger II," GOAL/QPC, Boston, 2004. Reprinted from "The Memory Jogger Plus+" with permission of GOAL/QPC, 12B Manor Parkway, Salem, NH 03079, *www.goalqpc.com*

GOAL/QPC, the consulting firm, is a major force for disseminating information about the N7 tools. GOAL/QPC recommends that the N7 tools be used in a "cycle of activity,"[2] wherein one tool provides inputs to another tool. One possible cycle is shown in Figure 10-22, where the affinity diagram or interrelation digraph are being used as inputs to the tree diagram, and so forth. Let us discuss each of these tools and the purposes they serve.

The Affinity Diagram

When we are solving a problem, it is often useful to first surface all the issues associated with the problem. A tool to do this is the **affinity diagram**. The affinity diagram helps a group converge on a set number of themes or ideas that can be addressed later. An affinity diagram creates a hierarchy of ideas on a large surface, as shown in Figure 10-23. The steps used in establishing an affinity diagram are as follows:

1. Identify the problem to be stated. Create a clear, concise statement of the issue that is understood by everyone.
2. Give the team members a supply of note cards and a pen. Ask them to write down issues that relate to the problem. There should only be one idea per card. Ask them to use at least four or five words to clearly explain their thinking.
3. Allow only about 10 minutes for this writing activity.
4. Place the written cards on a flat surface.
5. Lay out the finished cards so that all participants can see and have access to all the cards.
6. Let everyone on the team move the cards into groups with a similar theme. Do this work silently because it does not help to discuss your thinking. Work and move quickly.

FIGURE 10-22 Seven Management and Planning Tools Typical Flow

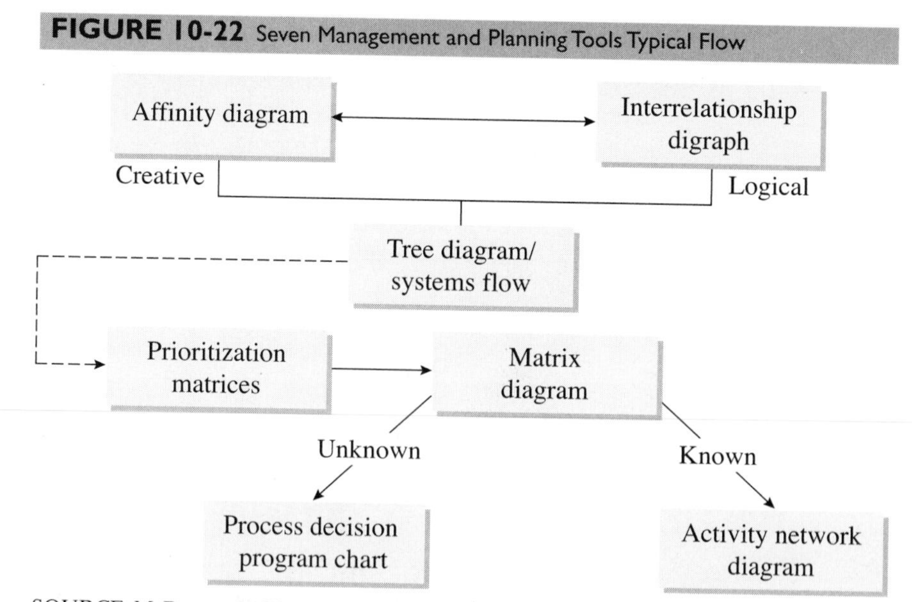

SOURCE: M. Brassard, "The Memory Jogger II," Goal/QPC, Boston, 2004.

[2]M. Brassard, "The Memory Jogger II," GOAL/QPC, Boston, 2004. Reprinted from "The Memory Jogger Plus+" with permission of GOAL/QPC, 12B Manor Parkway, Salem, NH 03079, *www.goalqpc.com*.

FIGURE 10-23 Affinity Diagram

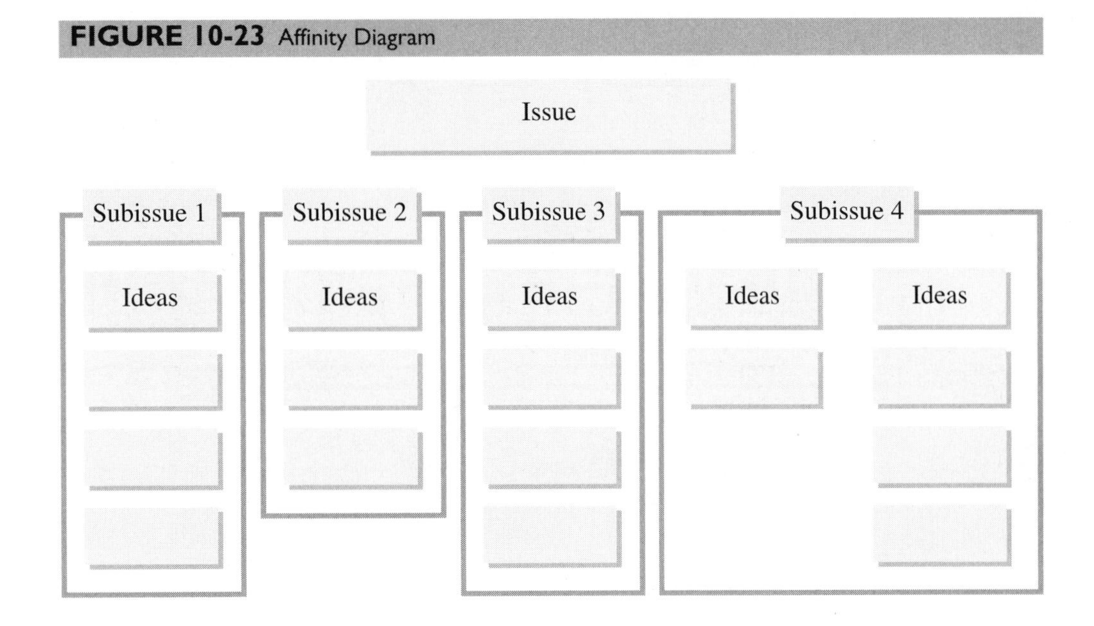

7. If you disagree with someone else's placement of a note card, say nothing but move it.
8. You reach consensus when all the cards are in groups, and the team members have stopped moving the cards. Once consensus has been reached concerning placement of the cards, you can create header cards.
9. Draw a finished affinity diagram and provide a working copy for all participants.

As illustrated in Figure 10-23, you should have a table with an issue statement, subissue header cards, and note cards with ideas. This will provide the basis for further discussion and brainstorming.

Zoo personnel at a zoological park used an affinity diagram to help develop a mission statement. The problem was stated as "Issues surrounding the mission of the Metropolitan City Zoo." The managers and zoo workers filled about 80 sticky notes with issues concerning the zoo's mission. Next, the team members placed the sticky notes into groups and ultimately defined a mission with six major elements. This provided a foundation for a final mission statement.

Example 10.7: Affinity Diagrams

Problem: The sales team at HealthPeople Corporation, a supplier of medical information, decided to develop a sales reference tool (SRT) as a means of improving its training processes for new sales people in the field. It was decided that this SRT would be available on the company intranet. A team of experienced salespeople was assembled who cataloged all the current sales material in many different locations. These materials were then reviewed by the team. Prior to performing preliminary design work for the SRT, the team members had to identify issues relating to the implementation of the SRT. The results are shown in Figure 10-24.

Solution: This analysis helped team members identify key issues in the design and implementation of the SRT. It was discovered that they needed to focus on eight issues

FIGURE 10-24 Affinity Diagram: Issues with Implementing the Sales Reference Tool

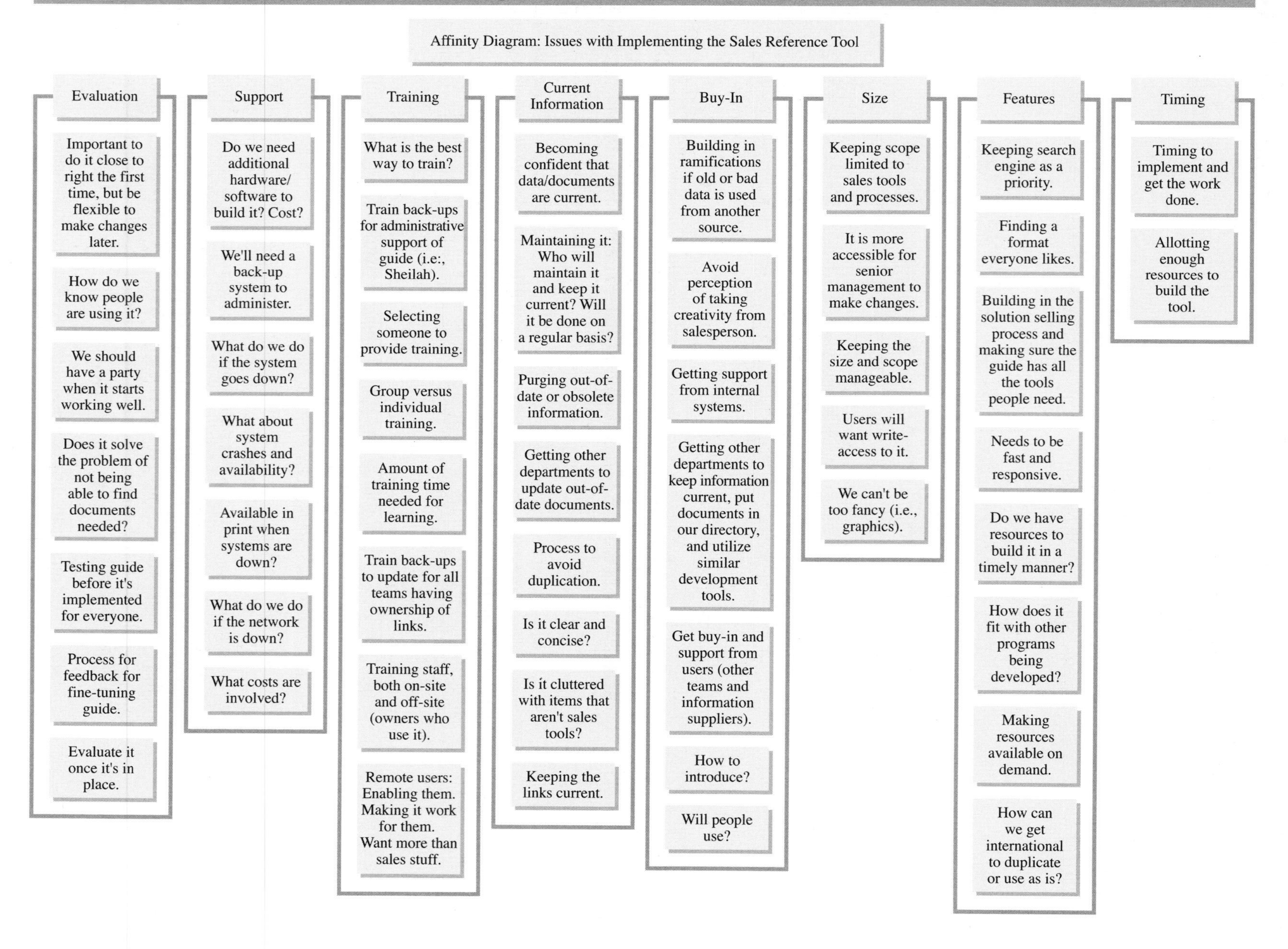

in implementation: evaluation, support, training, current information, buy-in, size, features, and timing. The notes underneath the headers present some of the issues identified by the participants.

The Interrelationship Digraph

After completing the affinity diagram, it might be useful to understand the causal relationships between the different issues that surfaced. Also, it is helpful to identify the most important issues to be focused on in pursuing the solution to a problem. A finished **interrelationship digraph** is shown in Figure 10-25. This interrelationship digraph shows the relationships between different issues. We will address how to develop this digraph, but you should notice that the shaded boxes are major issues that need to be addressed in developing improvement strategies. The steps to complete the interrelationship digraph are as follows:

1. Construct an affinity diagram to identify the issues relating to a problem. After you have done this, place the cards with related issues in columns with gaps between the cards. It is helpful to use sticky notes on a large piece of flipchart paper.
2. Create the digraph by examining the cards one by one asking, "What other issues on this digraph are caused or influenced by this issue?" As team members identify issues that are related, draw a one-way arrow from the first issue (the cause) to the second issue (the one influenced by the cause). Do this until all the issues have been discussed.

FIGURE 10-25 Interrelationship Digraph

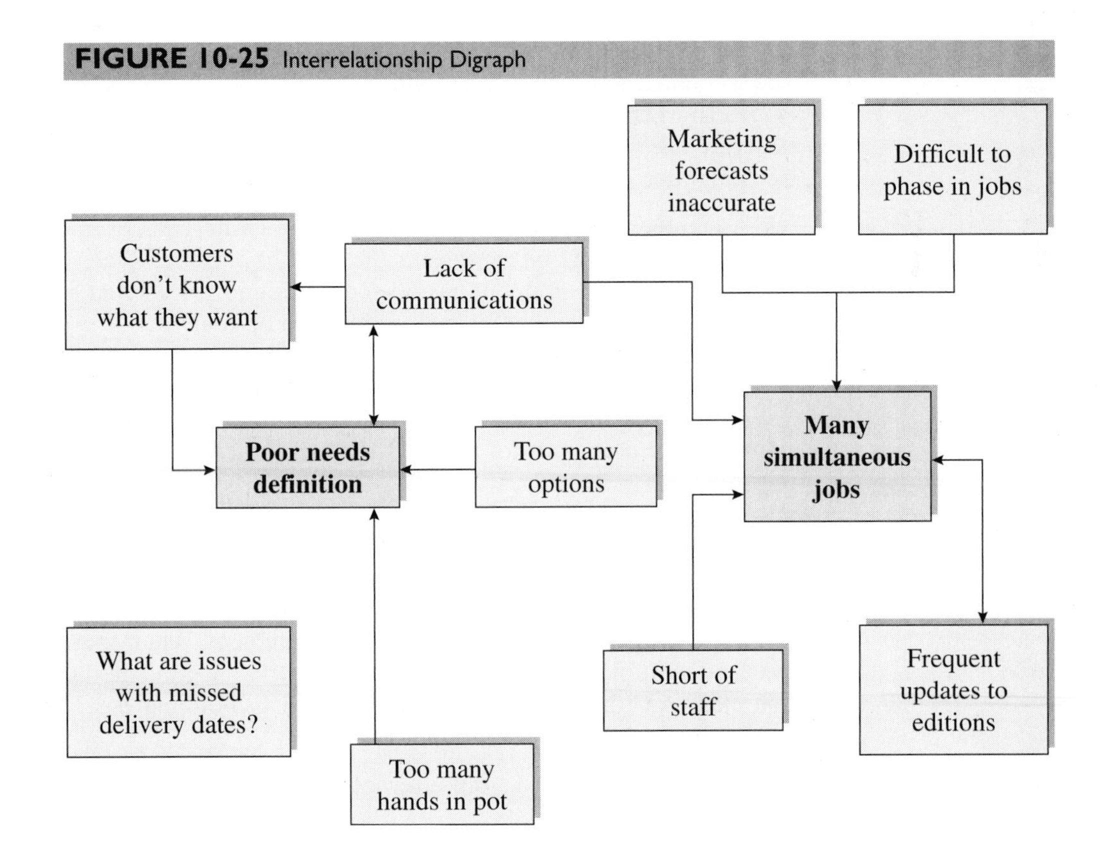

3. After reviewing the arrows and making needed revisions, count the numbers of arrows pointing to each note, and write the numbers on the notes.
4. Identify the cards with the most arrows as the "key factors." Experience has shown that there should not be more than 5 to 10 key factors, depending on the issue being discussed. Some cards may have several arrows, but for one reason or another they are not really key factors; these can be dropped from consideration at this point. Boxes with the most outgoing arrows tend to be root causes; those with incoming arrows tend to be performance indicators.
5. Draw a double box around the key factors and brainstorm ways to address these issues.

Example 10.8: Interrelationship Digraphs

Problem: For the issues relating to sales reference tools in Figure 10-24, team members were interested in knowing what issues had the greatest effects on other issues. This would help them to know where to focus their efforts in coming weeks.

Solution: The cards from the affinity diagram in Example 10.7 were used to identify the relationships between the different issues. For presentation purposes, we only used the cards from the first four columns in the affinity diagram (these were evaluation, support, training, and current information). The relationships were outlined using sticky notes and markers on a large piece of paper. The results, shown in Figure 10-26, reveal that the need for a backup system, training, and keeping the links current were key issues in developing the SRT. The team paid special attention to these aspects of the project. On a larger project, they might have established subteams to monitor these aspects of the project.

Tree Diagrams

The **tree diagram** is useful to identify the steps needed to address the given problem. Figure 10-27 shows a tree diagram. A tree diagram is very similar to a work breakdown structure used in planning projects. The following steps should be used to complete a tree diagram.

1. Assemble the header cards from the affinity diagram. From these cards, choose the header card that represents the most important issue.
2. Once the goal statement has been determined, ask, "What are the steps required to resolve or achieve this major objective or goal?"
3. Once the major tasks have been identified, move to the next level under each task, and ask for the second level tasks, "What are the steps required to resolve or achieve this objective or goal?"
4. Continue doing this for successive levels until you have exhausted your ideas for steps.

Prioritization Grid

A **prioritization grid** is used to make decisions based on multiple criteria. For example, in choosing a technology, we might have a variety of options. Also, the decision criteria vary as to how to choose possible desired outcomes. When there are multiple alternatives and multiple criteria, a prioritization matrix is a good method to inform your decision making without resorting to more sophisticated analysis. Following are the steps required to make a prioritization grid:

1. Determine your goal, your alternatives, and the criteria by which a decision is to be made.

FIGURE 10-26 Actual Interrelationship Digraph

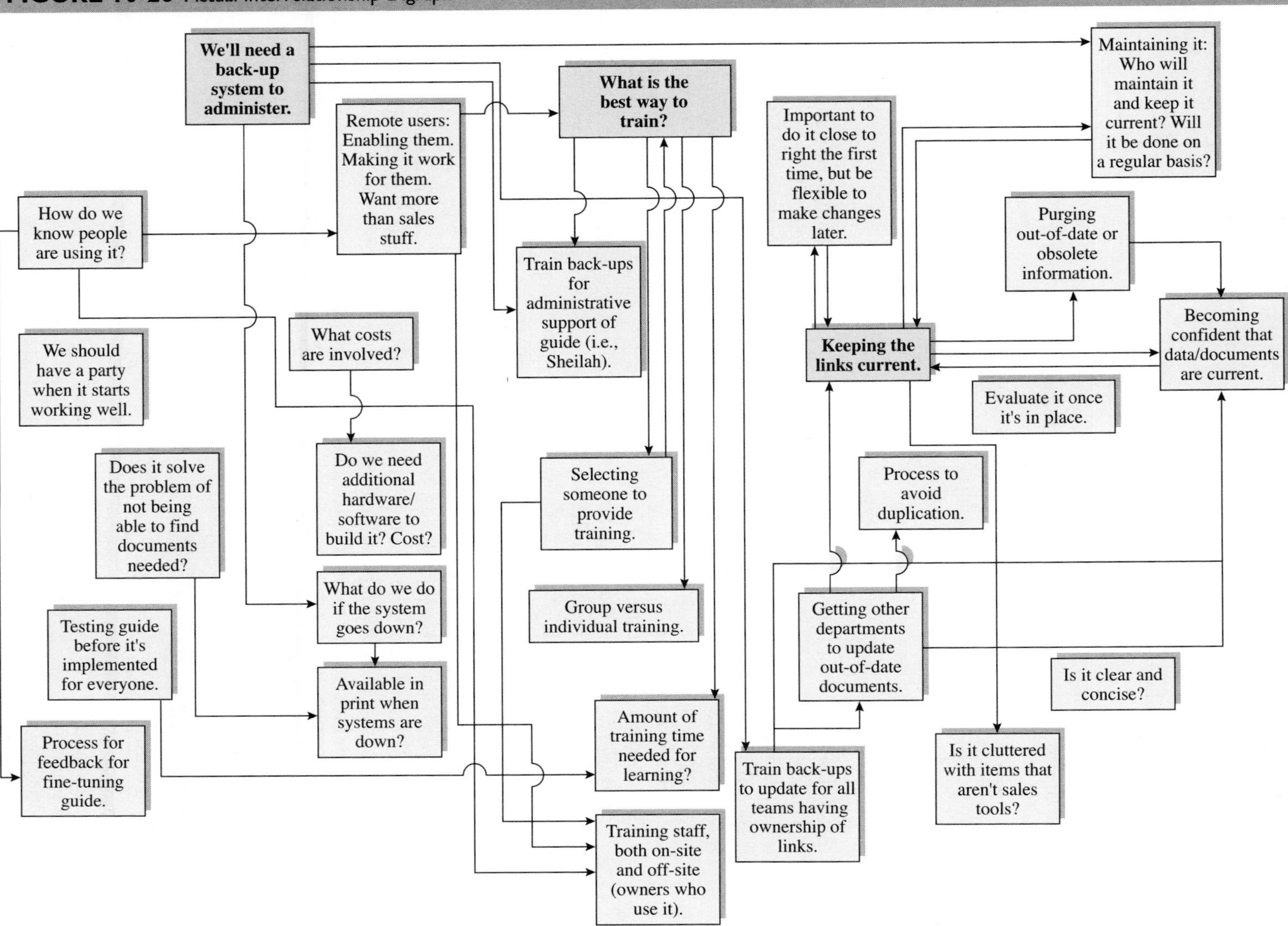

FIGURE 10-27 Tree Diagram

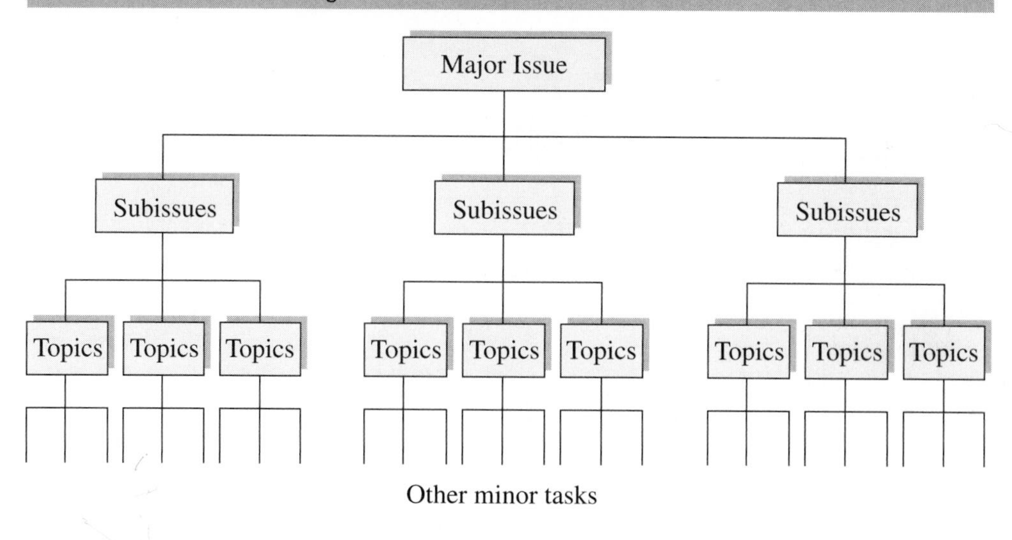

2. Place the selection criteria in order from most important to least important.
3. Apply a percentage weight to each of the criteria for each option. Apply a weight to each of the criteria such that all the weights add up to 1 (for example, A = .40, B = .30, C = .25 D = .05).
4. Add the individual rating for each criterion to come to an overall ranking. Divide by the number of options to find an average ranking.
5. Rank each option with respect to the criteria. Average the rankings, and apply a completed ranking.
6. Multiply the criteria weight by its associated criterion rank for each criterion in the matrix. Notice that in this case a ranking of 4 is best and 1 is worst. The result in each cell of the matrix is called an *importance score*.
7. Add the importance scores for each alternative.
8. Rank the alternatives according to importance.

Example 10.9: Prioritization Grid

A company had to choose between five possible machines for a service process with five criteria. The criteria were ease of use, necessary maintenance, cost of the machine, expected life of the machine, and reputation for the quality of the machine (see Table 10-1).

The three team members provided subjective importance ratings for each of the different decision criteria. These are in Table 10-2.

The team members then provided ratings for each of the different machines as they related to each criterion (see Table 10-3).

TABLE 10-1 Decision Criteria

Alternatives	*Criteria*
Machine A	Ease of use
Machine B	Maintenance
Machine C	Cost
Machine D	Expected life
Machine E	Reputation

TABLE 10-2 Importance Ratings

Criteria	*Person 1*	*Person 2*	*Person 3*	*Average Score*	*Final Criteria Ranking*
Ease of use	0.4	0.2	0.5	0.366	1
Maintenance	0.3	0.3	0.3	0.300	2
Cost	0.2	0.2	0.1	0.166	3
Expected life	0.05	0.1	0.05	0.066	5
Reputation	0.05	0.2	0.05	0.100	4
	1	1	1		

TABLE 10-3 Final Rankings

Ease of Use

Alternatives	*Person 1*	*Person 2*	*Person 3*	*Sum of Scores*	*Final Ease of Use Ranking*
Machine A	1	1	1	3	1
Machine B	2	3	2	7	2
Machine C	4	4	4	12	4
Machine D	5	5	5	15	5
Machine E	3	2	3	8	3

Maintenance

Alternatives	*Person 1*	*Person 2*	*Person 3*	*Sum of Scores*	*Final Maintenance Ranking*
Machine A	2	2	1	5	1
Machine B	1	3	2	6	2
Machine C	5	5	4	14	5
Machine D	4	4	5	13	4
Machine E	3	1	3	7	3

Cost

Alternatives	*Person 1*	*Person 2*	*Person 3*	*Sum of Scores*	*Final Cost Ranking*
Machine A	4	4	5	13	5
Machine B	5	3	4	12	4
Machine C	1	1	2	4	1
Machine D	2	2	1	5	2
Machine E	3	5	3	11	3

Expected Life

Alternatives	*Person 1*	*Person 2*	*Person 3*	*Sum of Scores*	*Final Expected Life Ranking*
Machine A	1	2	1	4	1
Machine B	2	3	2	7	2
Machine C	3	4	5	12	4
Machine D	4	5	4	13	5
Machine E	5	1	3	9	3

Reputation

Alternatives	*Person 1*	*Person 2*	*Person 3*	*Sum of Scores*	*Final Reputation Ranking*
Machine A	4	4	5	13	5
Machine B	5	3	4	12	4
Machine C	1	1	2	4	1
Machine D	2	2	1	5	2
Machine E	3	5	3	11	3

The final rankings were computed by multiplying the various rankings by their importance. It looks like alternative A is the best choice. Note that the lowest score is the best (see Table 10-4).

Matrix Diagram

The **matrix diagram** is similar in concept to quality function deployment in its use of symbols, its layout, and its application. Because the matrix diagram is one of the N7 tools, we mention it here. However, the prior presentation of QFD is much more complete, so we will keep this short. Like the other N7 tools, the matrix diagram is a brainstorming tool that can be used in a group to show the relationships between ideas or issues. Matrix diagrams are simple to use and can be used in two, three, or four dimensions. For our purposes, we provide an example of a simple two-dimensional matrix in Figure 10-28. The steps are as follows:

1. Determine the number of issues or dimensions to be used in the matrix.
2. Choose the appropriate matrix.
3. Place the appropriate symbols in the matrix:

Figure 10-29 shows a *responsibility* matrix diagram. The legend at the bottom of the figure shows the extent of responsibility among the different people. No example is provided here because we discussed QFD earlier in Chapter 7. This grid in Figure 10-29 gives a simplified version of the QFD demonstrated earlier.

TABLE 10-4 Combining Rankings

Final Criteria Ranking	*Final Ease of Use Ranking*	*Final Maintenance Ranking*	*Final Cost Ranking*
1	1	1	5
2	2	2	4
3	4	5	1
5	5	4	2
4	3	3	3

Final Expected Life Ranking	*Final Reputation Ranking*
1	5
2	4
4	1
5	2
3	3

Scores

Machine A: 1(1) + 2(1) + 3(5) + 5(1) + 4(5) = 43
Machine B: 1(2) + 2(2) + 3(4) + 5(2) + 4(4) = 44
Machine C: 1(4) + 2(5) + 3(1) + 5(4) + 4(1) = 41
Machine D: 1(5) + 2(4) + 3(2) + 5(5) + 4(2) = 52
Machine E: 1(3) + 2(3) + 3(3) + 5(3) + 4(3) = 45

Final Rankings

1	C
2	A
3	B
4	E
5	D

Machine A is the best choice.

FIGURE 10-28 Matrix Diagram

Reducing the Number of Billing Errors

Resources needed							**Total**
Capital investment	○					◎	12
Staff time	◎	◎	○	◎	◎	△	40
Training time		○		○	○	◎	18
Space						△	1
Equipment availability	○						3
Options	Improve database	Improve definition	Specify costs	Lessen options	Revise pricing	Simplify process	

◎ High (9)
○ Medium (3)
△ Low (1)

FIGURE 10-29 Responsibility Matrix Diagram

Reducing the Number of Billing Errors

People needed							**Total**
Layout engineers		◎				◎	18
Software designers			○	◎	◎	△	22
Engineers	◎	○		○		◎	24
Human resources			○			△	4
Systems	○				○		6
Options	Improve database	Improve definition	Specify costs	Lessen options	Revise pricing	Simplify process	

◎ High (9) [Prime responsibility]
○ Medium (3) [Secondary responsibility]
△ Low (1) [Kept informed]

Process Decision Program Chart

A **process decision program chart** is a tool to help brainstorm possible contingencies or problems associated with the implementation of some program or improvement. Figure 10-30 shows such a chart in tree form (the outline form is not presented here). The steps are as follows:

1. In developing the tree diagram, place the first-level boxes in sequential order. (These are the boxes in the first column in Figure 10-30.)
2. Moving to the second level, list implementation details at a fairly high level. Try to be all-inclusive at a macro level.
3. At the third level, ask the questions, "What unexpected things could happen in this implementation?" or "What could go awry at this stage?"
4. At the fourth level, brainstorm possible countermeasures to the problems identified at the third level.
5. Evaluate the countermeasures for feasibility, and mark those that are feasible with an **O** and those that are not feasible with an **X**.

Activity Network Diagram

The **activity network diagram** is also known as a *PERT* (program evaluation and review technique) *diagram* or *critical-path* (longest path in time from beginning to end) *diagram* and is used in controlling projects. Figure 10-31 shows an activity network PERT diagram with its nodes and times. The nodes are circles and the times are given in days. Activity network diagrams are discussed in depth in Chapter 11.

Reflections on the Managerial N7 Tools

As you can see, the N7 tools are useful for managing long projects that involve teams. With the B7 and N7 tools, you have a reasonably good set of skills that will help in managing many projects. They have been used successfully in many different settings and for many different purposes.

The power of these tools is that with the **Plan–Do–Check–Act (PDCA) cycle**, they give companies a simple, easy to understand methodology for solving unstructured problems. They are especially useful when used in teams. Many of these tools are also fun to use. By using them effectively, managers can reduce unproductive meeting time to a minimum and make good, fact-based decisions.

OTHER TOOLS FOR PERFORMANCE MEASUREMENT

There are other tools used in communicating performance to employees. The justification for these tools is to present data in an economical and understandable way. We will present three such tools.

Spider Charts

Spider charts are graphs that present multiple metrics simultaneously in a two-dimensional plane. Figure 10-32 shows a spider chart. In this case, we show six different metrics (A–F) and report goals and results. A quick review of the figure shows that the firm has not met performance goals on metrics A, B, D, and E. The firm has met the goal relative to metric C and has exceeded the goal on metric F.

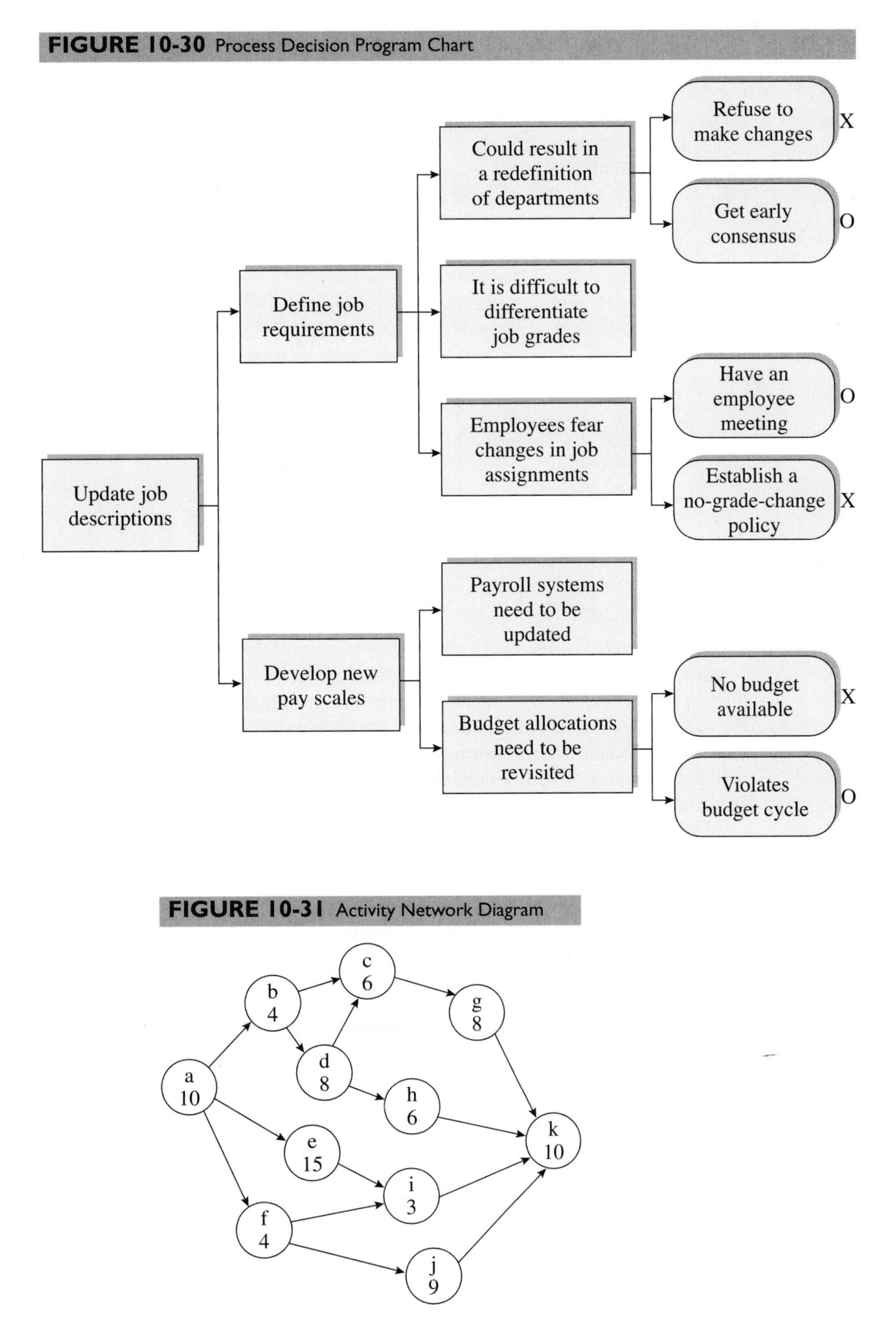

FIGURE 10-30 Process Decision Program Chart

FIGURE 10-31 Activity Network Diagram

FIGURE 10-32 Spider Chart Example

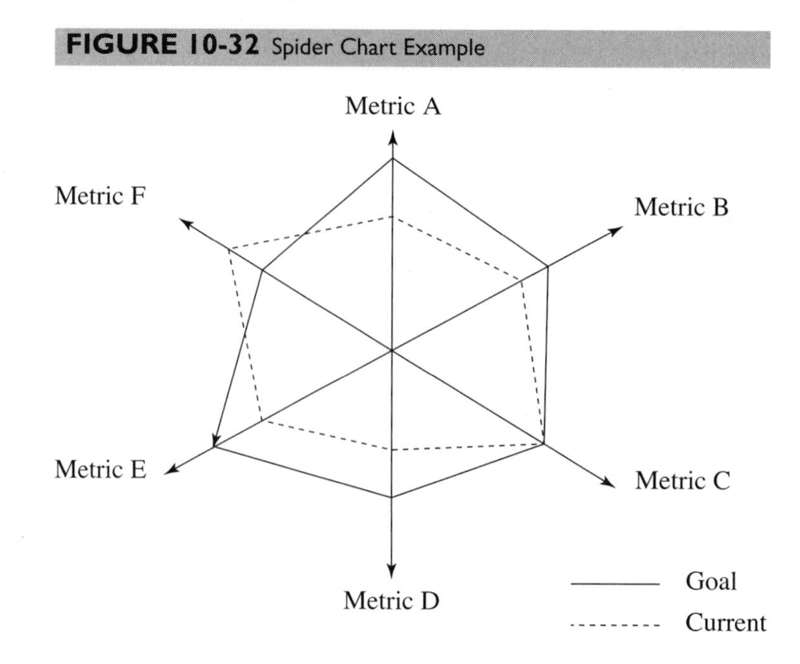

Other values that might be included on spider charts besides current performance and results are baseline performance measures and benchmark values. At times, this information can be found in QFD matrices.

Balanced Scorecards

A very important tool for measuring performance is a **balanced scorecard**. Balanced scorecards are usually spreadsheets that are communicated to management on a regular basis—weekly, monthly, quarterly, and annually. The usefulness of the balanced scorecard comes from integrating financial measures of business success, such as key metrics, along with nonfinancial, operational information about the business, such as customer satisfaction and process performance measures.

Figure 10-33 shows a very simplified layout for a balanced scorecard. Notice that this balanced scorecard combines financial, customer, process, and employee information into a single form. Often these forms are color coded to show if goals are being met or if performance is unsatisfactory. Scorecards, if used effectively, can be used to monitor and drive improvements in performance.

FIGURE 10-33 Balanced Scorecard Example

Strategic Theme:	Objectives	Measurement	Target	Initiative
Financial Performance	Profitability More Customers Less Investment	Market Value Truckload Rev. EVA Charge	Increase Market Share by 5% Increase Truck Revenue by 10%	Promote Delivery Service
Customer Satisfaction	Orders Delivered On Time Lowest Prices	Number of Orders Delivered on Date Promised	Exceed Customer Expectations 95%	Establish Specific Delivery Routes by Customer
Process Improvement	Efficient Staging and Loading of Customer Orders	Number of Orders Prestaged on Time for Loading	85% by June 20XX	Optimize Order Prestaging Process
Employee Satisfaction	Improved Communication Channels	Percent of Staff Trained in Teamwork	90% by June 20XX	Teamwork and Communication Skills Training

FIGURE 10-34 Dashboard Example

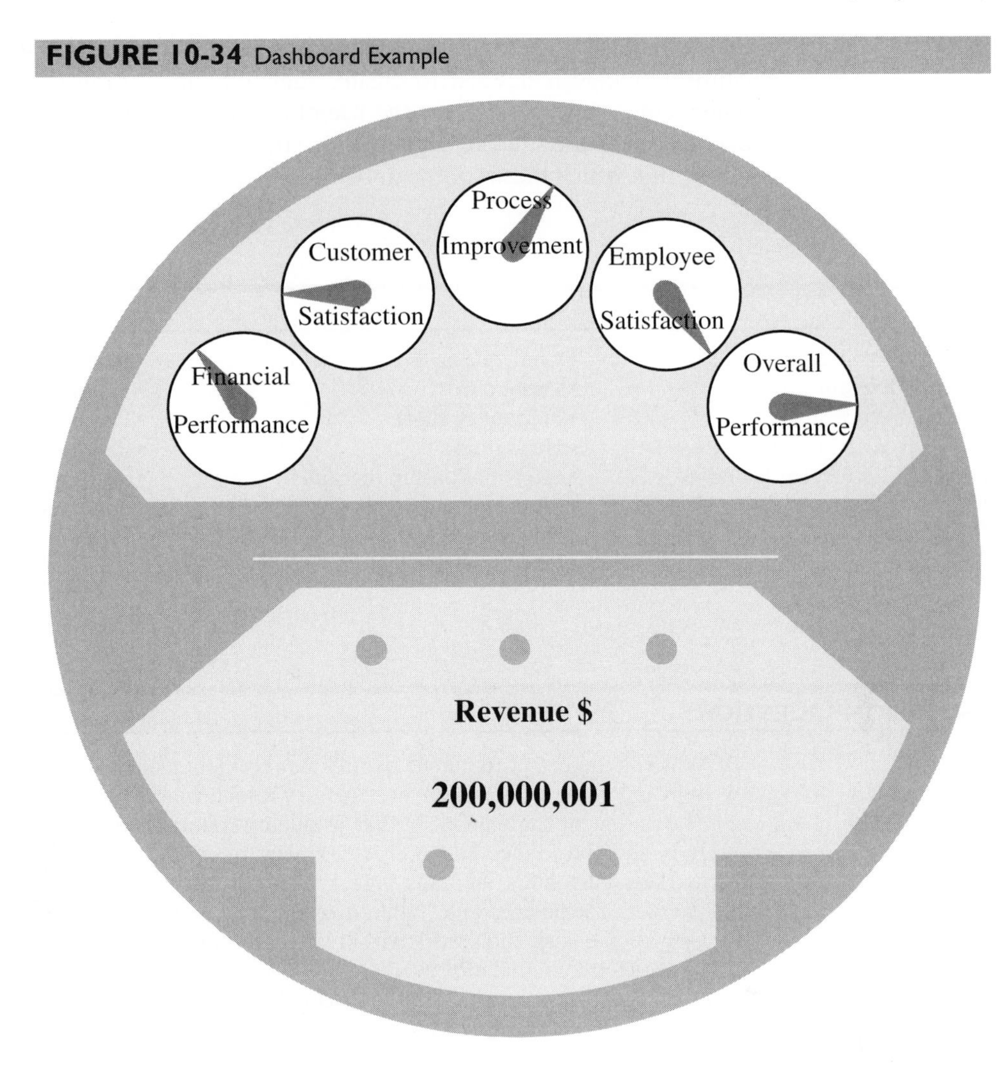

Dashboards

Dashboards look like electric meters or car dashboards. Figure 10-34 shows a dashboard that looks like an electric meter. Each of the "gauges" on the dashboard shows a different metric.

Notice in this case that the gauges in Figure 10-34 match the metrics reported in our balanced scoreboard discussed in Figure 10-33. This dashboard quickly communicates whether or not performance objectives have or have not been met. Again, the focus is on easy, clear communication.

SUMMARY

In this chapter we have briefly introduced the basic tools of quality. These tools should be adapted to the specific needs of your company. You also do not need always to use all the tools in a single project.

The B7 and N7 are useful in their simplicity and power. It is an easy undertaking to train employees to use these tools. However, many companies provide the training and then wonder why the employees don't use them. The reason is that along with the tools, cultural change is needed to ensure that implementation can be successful. Also, if management doesn't support the use of the tools in all possible situations, they do not become inculcated in the organization. In the following chapters, we discuss the context within which these tools can be successful.

KEY TERMS

- Activity network diagram
- Affinity diagram
- Balanced scorecard
- Basic seven (B7) tools
- Cause-and-effect (or fishbone or Ishikawa) diagram
- Check sheets
- Control charts
- Dashboards
- Frequency chart
- Histogram
- Interrelationship digraph
- Matrix diagram
- New seven (N7) tools
- Pareto charts
- Plan–do–check–act (PDCA) cycle
- Prioritization grid
- Process decision program chart
- Process map
- Scatter diagram or scatter plot
- Spider charts
- Tree diagram

DISCUSSION QUESTIONS

1. Why is it important to pursue quality management from a systems perspective?
2. Why is continual improvement necessary for a business organization?
3. The statement has been made that "A quality system is not just a series of boxes and arrows. It is an interconnected, interdisciplinary network of people, technology, procedures, markets, customers, facilities, legal requirements, reporting requirements, and assets that interact to achieve an end." What does this statement mean to you?
4. How do the basic tools work within W. E. Deming's plan–do–check–act (PDCA) cycle as a process for continual improvement?
5. What are the seven basic tools of quality? Who developed these tools?
6. Describe the purpose of a histogram or frequency chart.
7. Describe the purpose of a Pareto chart. Describe an instance (other than the one in the book) in which a Pareto chart could be effectively used.
8. What are the three basic rules for constructing Pareto charts?
9. What is the purpose of a cause-and-effect (Ishikawa) diagram?
10. Describe the purpose of a check sheet. Describe an instance (other than the ones in the book) in which a check sheet could be effectively used.
11. Describe the purpose of a scatter diagram.
12. Describe the purpose of a flowchart. What are three of the rules for designing and using flowcharts?
13. What is the purpose of a control chart?
14. Which of the seven (Ishikawa) tools of quality described in the chapter have been the most helpful to you in your experiences? Please make your answer as substantive as possible.
15. What is the purpose of an affinity diagram?
16. Describe the purpose of an interrelationship digraph?
17. Describe the purpose of tree diagrams. Describe an instance in which a tree diagram could be used.
18. Describe the purpose of a prioritization grid.
19. Describe the purpose of a matrix diagram. In what ways is the matrix diagram a brainstorming tool?
20. What is the purpose of a process decision program chart?

PROBLEMS

1. Develop a process map of washing a car. Include a high level of detail in your map. Make six recommendations for improvements to your process.
2. Take the process map from Problem 1 and develop it into an extended process map. Make five recommendations for simplifying the extended process as it exists.
3. Develop a process map for making chocolate chip cookies. Include a high level of detail if you need to. You may need to consult a cookbook. Make three recommendations for improvements to your process. Discuss these with the class.
4. Take the process map from Problem 3 and develop an extended process map. Make recommendations for three improvements to the extended process. Be sure to include all suppliers and logistics associated with the customers.
5. Develop a check sheet for defects in a flat-screen computer monitor.
6. Develop a check sheet for defects in a quality management class exam. Identify how you would use the check sheet to improve performance on future exams.
7. Develop a histogram for the following data:

Employee	Hours of Overtime	Days Absent
1	243	3
2	126	2
3	86	0
4	424	6
5	236	3
6	128	0
7	0	0
8	126	2
9	324	3
10	118	0
11	62	0
12	128	3
13	460	6
14	135	1
15	118	1
16	260	2
17	0	1
18	126	1
19	234	2
20	246	3
21	120	1
22	80	0
23	112	1
24	237	3
25	129	2
26	24	1
27	36	0
28	128	2
29	246	3
30	326	6

Develop two separate histograms for hours of overtime and days absent. How do the data appear to be distributed?

8. Using the data in Problem 7, develop a scatter plot of hours of overtime versus days absent. Do the data overtime hours and days absent appear to be correlated?

9. Develop a histogram using the following data:

   4.7, 4.7, 5.0, 5.6, 5.6, 5.6, 5.9, 5.9, 5.9, 5.9, 6.2, 6.2, 6.2, 6.2, 6.2, 6.2, 6.5, 6.5, 6.5, 6.5, 6.5, 6.5, 6.8, 6.8, 9.8, 9.8, 9.8, 9.8, 6.8, 6.8, 6.8, 6.8, 6.8, 6.8, 6.8, 6.8, 6.8, 6.8, 6.8, 7.1, 7.1, 7.1, 7.1, 7.1, 7.1, 7.1, 7.1, 7.1, 7.1, 7.1, 7.1, 7.1, 7.4, 7.4, 7.4, 7.4, 7.4, 7.4, 7.4, 7.4, 7.7, 7.7, 7.7, 7.7, 7.7, 7.7, 7.7, 7.7, 7.7, 7.7, 7.7, 7.7, 7.7, 7.7, 7.7, 7.7, 7.7, 8.0, 8.0, 8.0, 8.0, 8.0, 8.0, 8.3, 8.3, 8.3, 8.3, 8.3, 8.3, 8.6, 8.6, 8.6, 8.6, 8.6, 8.6, 8.9, 8.9, 8.9, 9.2, 9.2, 9.8.

   Do the data appear to be normally distributed?

10. If you have 60 data points, use the log formula to determine how many classes you should use in your histogram.
11. Use the logarithmic formula in the chapter to determine how many classes you should use for the following numbers of data:

    a. 35

    b. 200

    c. 600

12. Think about the following questions, and develop fishbone diagrams for each of them:

    a. What is the major service problem in your university?

    b. What is the major thing that interferes with your study?

    c. What is the major problem with your school newspaper?

    d. What is the major social problem in society?

13. For the following data, develop a Pareto analysis. The letters A, B, C, D, E, and F are problems that occur in a process. Which cause should you focus on first?

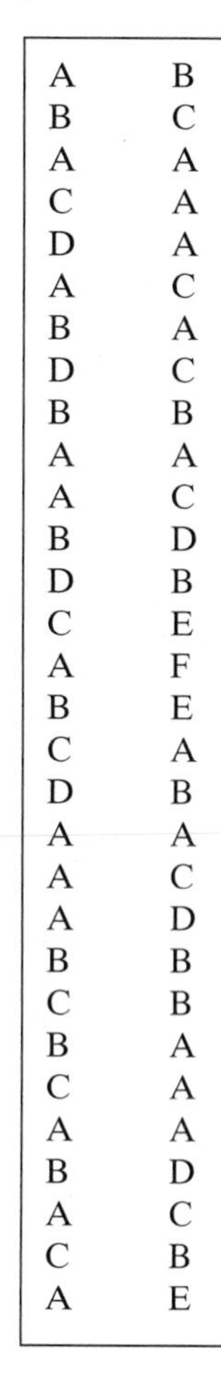

A	B
B	C
A	A
C	A
D	A
A	C
B	A
D	C
B	B
A	A
A	C
B	D
D	B
C	E
A	F
B	E
C	A
D	B
A	A
A	C
A	D
B	B
C	B
B	A
C	A
A	A
B	D
A	C
C	B
A	E

14. For the following data, develop a Pareto chart. The letters V, X, Y, and Z are problems that occur in a process. Which cause should you address first if the fixes for the problems are $1.00, $1.20, $.90, and $2.00 successively?

VXZZZYYXXYVVYYXXZZYZZYXXVYVYVYVYXXXYVYZVYXXXYVYVYV
XXXXVVVVYYYYVYVYVYXYXYXVXVZZZXYXYXYVYYXZZZ

15. Either alone or as a team, draw a fishbone diagram about the following topics:
    a. What are the causes of poor grades?
    b. Why do college students drink too much?
    c. Why do I not have enough money?
    d. What are the causes of poor response times on the Internet?

16. *Quality Is Personal* by Roberts and Sergesketter is a popular book. In it they recommend using the basic tools of quality in our personal lives. Develop a check sheet to keep track of personal defects you have in your life (such as sleeping too late, being too grumpy, and so on). Use this check sheet for two weeks to track these personal defects. After two weeks, perform a Pareto analysis to determine where you have the greatest need for improvement. Next, use the fishbone diagram to identify the underlying causes of the personal defects. After making changes, use a control chart to track your defects.

17. For the following data, draw a scatter diagram to see if time lost because of injuries and overtime hours are related. What do you conclude?

Plant	Lost Time Days	Overtime Hours
A	5	254
B	3	114
C	6	350
D	4	219
E	10	496
F	5	218
G	7	279

18. Develop a process map for the registration process at your university. Analyze the number of value-added and nonvalue-added steps.
19. Develop a process map for receiving financial aid at your university. Analyze the process and develop a proposal for improvement.
20. With a team of students, develop an affinity diagram relating to the following statement, "What are issues relating to finding a job in the current economy?" What did you find?
21. Form a team and develop an affinity diagram for the following problem statement: What are issues associated with completing a university or college degree in a reasonable amount of time?
22. Using the affinity diagram from Problem 21, develop an interrelationship digraph. What do you generalize from the interrelationship digraph?
23. Deveop a tree diagram for building a home.
24. Develop a tree diagram for writing a major research paper. Now take the steps you identified for this project and develop an activity network diagram.
25. Using the following criteria and rankings for each criterion, compute final ratings for each alternative.

Options	Customer Acceptance (Most Important)	Cost	Reliability	Strength (Least Important)
A				
Percentage weight	(.4)	(.3)	(.2)	(.1)
Rank	1	2	3	4
B				
Percentage weight	(.3)	(.4)	(.1)	(.2)
Rank	3	4	2	1
C				
Percentage weight	(.25)	(.25)	(.25)	(.25)
Rank	1	3	2	4
D				
Percentage weight	(.25)	(.25)	(.25)	(.25)
Rank	1	3	2	4
Sum of weights				
Average weight				
Criterion ranking				

26. Complete the analysis for the following prioritization matrix information by finding the final ranking:

Decision Alternatives
Machine 1
Machine 2
Machine 3
Machine 4

Decision Criteria	Importance
A	.3
B	.2
C	.5

Final Rankings

Criterion A	Ranking
Machine 1	1
Machine 2	2
Machine 3	3
Machine 4	4

Criterion B	Ranking
Machine 1	3
Machine 2	2
Machine 3	4
Machine 4	1

Criterion C	Ranking
Machine 1	4
Machine 2	2
Machine 3	1
Machine 4	3

27. Develop a process decision chart for completing your college degree.

Cases

Case 10-1 Corporate Universities: Teaching the Tools of Quality

Motorola University: ***www.motorola.com/motorolauniversity***
General Motors University: ***www.gm.com/company/careers/life/lif_gmu.html***
Sears University: ***www.aboutsears.com***

Although most of us are familiar with major public universities like Penn State, Colorado-Boulder, Georgia, and Ohio State, we are typically unfamiliar with corporate universities such as Motorola University, Intel University, and the AT&T Learning Center. This is because corporate universities are a fairly new concept, and they are created to serve the needs of a particular company's employees and other stakeholders.

The term *corporate university* has been adopted by firms that have significantly upgraded their training and development activities by creating learning centers within their corporations. These learning centers are typically designed to prioritize a firm's training initiatives, and to quickly share with a firm's employees the skills, techniques, and best practices that are necessary to remain competitive. For example, when a new quality tool or technique is developed, it is often the responsibility of a firm's corporate university to develop a plan to equip the firm's employees with the skills necessary to quickly incorporate the new tool or techniques into their work areas.

Following are brief descriptions of three corporate universities. After reading these descriptions, ask yourself the following rhetorical question, "Are these corporations well-equipped to teach their employees the tools of quality?"

MOTOROLA UNIVERSITY

Motorola University began in 1981 as the Motorola Training and Education Center. Initially, the purpose of the university was to help Motorola strengthen its training efforts and build a quality-focused corporate culture. Through the years, the university has grown in both size and stature and now has a staff of more than 400 employees and seven facilities across the world. The stated objectives of the university are as follows:

- To provide training and education to all Motorola employees.
- To prepare Motorola employees to be best-in-class in their industries.
- To serve as a catalyst for change and continuous improvement to position Motorola Corporation for the future.
- To provide added value to Motorola in the marketing and distribution of products throughout the world.

To accomplish these objectives, Motorola University does many things. For example, each of the company's employees is required to take a minimum of 40 hours a year of job-relevant training and education. The university also provides its employees consulting services in a number of areas, including benchmarking, cycle time reduction, quality improvement processes, and statistical tools and problem-solving techniques.

One unique aspect of Motorola University is that it reaches beyond the Motorola Corporation. The university provides training and certification programs for Motorola suppliers, and also provides consulting services and training for other corporations on a fee basis.

GENERAL MOTORS UNIVERSITY

General Motors University was founded in 1994 with the initial goal of providing focus to the company's many training and development activities. Although GM was actively involved in training prior to the establishment of the university, many of the courses offered to its employees were redundant and not connected to the company's strategic priorities. The company was also unable to quickly share with its employees new management techniques and best practice ideas.

General Motors University has brought focus to the company's training activities and has developed a standard curriculum to ensure that all of GM's employees understand the company's core values of continuous improvement, teamwork, integrity, and innovation. One of the more successful initiatives that the university has undertaken is using the case method in training GM employees. Cases are taught in areas such as leadership, quality, and union management. Often, the cases facilitate cross-functional training needs. For example, a case study designed as an ethics lesson may also involve issues pertaining to quality, integrity, and leadership.

SEARS UNIVERSITY

Sears University was established in 1994 with the ambitious goal of becoming an integral part of the company's turnaround efforts. The university was opened with the idea of offering a wide selection of formal training and self-study courses for Sears's employees. In its first year of operation, approximately 10,000 of the company's employees participated in formal programs ranging from one day to one week in duration. Another 4,000 employees completed self-study courses each month.

In addition to offering training programs in areas such as merchandising, operations, customer service, and human resources management, Sears University also provides the company's employees programs designed to help them function as change agents and strategic leaders within the corporation. For example, participants in financial management training programs use computer-based simulations to model the impact of various financial strategies on business unit performance. Particular attention is paid to trying to help employees see the company's operations from the customer's perspective. The courses are taught by seasoned line managers along with professional facilitators and Sears University personnel. ■

DISCUSSION QUESTIONS

1. Are corporate universities a good idea? If so, why?
2. How can a corporate university do a better job of teaching a firm's employees the "tools of quality" than traditional training programs?
3. Select one of the three corporate universities in this case study and visit its Web site. How does the company's corporate university facilitate the company's overall quality-related goals and initiatives?

Case 10-2 Lanier: Achieving Maximum Performance by Supporting Quality Products with Quality Services

Lanier home page ***www.lanier.com***

Lanier, a wholly owned subsidiary of Harris Corporation, is the largest independent distributor of office equipment in the world. The company, which is headquartered in Atlanta, Georgia, has more than 1,600 sales and services centers in more than 100 countries worldwide. Lanier's product mix includes copy machines, fax machines, voicemail, dictation/transcription systems, presentation systems, and other related office products.

Throughout most of its corporate history, Lanier has been a sales driven organization. The company was founded in 1934 by Tommy Lanier and his two brothers as the distributors of "Ediphone" dictation machines in the southeastern portion of the United States. In 1955 Lanier entered the copier business as an independent distributor of 3M "Thermofax" dry process copiers. Through the years, the company's product line has broadened, and it has experienced consistent growth and profitability. Lanier also has earned for itself an excellent reputation in the office products industry.

Rather than manufacturing its own products, Lanier's business strategy has been to partner with companies like 3M, Toshiba, and Canon to develop a cohesive line of high-quality office equipment. The biggest challenge for Lanier has been to differentiate itself from its competitors. Although the company sells good quality products, its products are similar to the products sold by other office equipment vendors. To find a point of differentiation from its competitors, in the early 1990s Lanier decided to shift from a sales driven company to a company focused on customer satisfaction. The company realized that to make this shift successfully, it had to build a corporate culture that supported its products with quality customer service.

Throughout the early and mid-1990s, Lanier worked hard to develop quality services to complement its products. To accomplish this, the company developed several specific quality-related programs. These programs included the following:

- Customer Vision
- Performance Promise
- 100 Percent Sold
- Lanier Team Management Process

The premise behind the Customer Vision program was to encourage each employee to see the company's business through the customers' eyes and respond appropriately. The Performance Promise program was designed to offer the industry's best performance pledge by guaranteeing total product satisfaction (or replacement at no charge); and by providing a 24-hour, toll-free helpline; free loaners when repairs are necessary; and a 10-year guarantee on the availability of service, parts, and supplies for all Lanier products. The 100 Percent Sold program challenged the company's employees with the goal of having every Lanier customer buy all of their office products from Lanier. Finally, the Lanier Team Management Process was a quality program that stressed a never ending process of continuous improvement in quality, reliability, and performance in all things Lanier did at all levels within the company.

In addition to specific programs to support service and product quality, the company also started to see itself as a consulting organization rather than a sales organization. By giving potential customers good advice before the sale, the company found that it could create a seamless stream of Lanier involvement in satisfying a customer's office product needs. The stream includes presale advice, the actual sale, and after-sale service. Lanier also has coupled its new initiatives with extensive training for its employees and incentive programs tied to the company's quality-related goals.

Lanier has been successful in complementing its quality products with quality customer service. As evidence of this, the company received several prestigious awards from its customers including DuPont's "Partners in Excellence Award," Pacific Bell's "Quality Partner Award," and Chevron's "Alliance Supplier Award." ■

DISCUSSION QUESTIONS

1. Why was it important for Lanier to develop specific programs, such as Customer Vision and Performance Promise, to facilitate its dual emphasis on quality products and quality services?
2. What steps has Lanier taken to reinforce the importance of quality services to its employees?
3. Do you believe that Lanier continued to be successful? Why or why not?

CHAPTER 11

Managing Quality Improvement Teams and Projects

Teamwork is sorely needed throughout the company. Teamwork requires one to compensate with his strength someone else's weakness, for everyone to sharpen each other's wits with questions.
—W. EDWARDS DEMING

In their classic article on quality and participation, Robert Cole[1] and his coauthors explained the need for employee participation as a key element in managing changing organizations in an increasingly complex world. As this chapter's opening quote shows, W. Edwards Deming also argued for employee participation and teamwork.

Why do you suppose so many influential voices call for participation and teamwork to manage businesses today? There are several reasons. One of the biggest is *complexity* in the workplace. Given the large volumes of data available to managers, is it any surprise that unilateral decision making is a thing of the past? Also, business is transforming itself from a "command and control" environment to one of *collaboration*. Such collaboration is needed as complexity drives workers from routine work to **knowledge work**, or work that involves the development and transmission of knowledge and information. Knowledge work implies a greater amount of ambiguity, searching, researching, and learning in the job environment.

Figure 11-1 shows the difference between routine work and knowledge-based work. Knowledge work is effective when workers are given a certain amount of autonomy and decision-making authority. Companies such as 3M encourage their employees to become more entrepreneurial in their approach to work. This regularly results in new products and markets for the company.

As more collaborative practices are being adopted in business, teamwork is the natural result. For our purposes, a **team** is defined as a finite number of individuals who are united in a common purpose. This chapter discusses an approach to improving quality that uses teams and collaboration as a means for improvement. Often these team approaches are used in conjunction with the tools of quality as discussed in Chapter 10.

Joseph Juran has long emphasized the importance of teams and projects in the improvement of quality. He stated that the improvement of quality should be approached on a "project-by-project basis, and in no other way." Teams are a fundamental part of projects. Philip Crosby also supports the use of teams in improving quality. Although there are notable failures of teams, on the whole, this is such a widely practiced approach to quality improvement that we discuss teams in some depth.

[1]Cole, R., Bacdayan, P., and White, J., "Quality, Participation, and Competitiveness," *California Management Review* 35, 3 (1992):68–81.

FIGURE 11-1 Differences between Routine and Knowledge Work

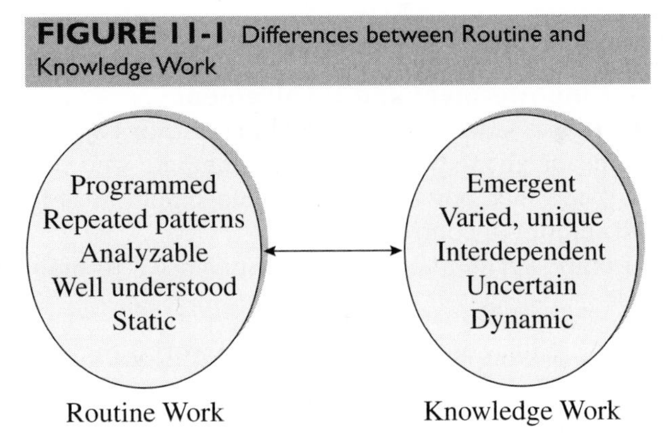

SOURCE: Adapted from S. Mohrman, S. Cohen, and A. Mohrman, *Designing Team-based Organizations* (San Francisco, CA: Jossey-Bass, 1995).

This chapter focuses on two interrelated topics: *leading teams* and *managing projects*. The first half of the chapter focuses on the behavioral aspects of making teams work. The second half of the chapter is about managing and controlling projects. We treat these topics together in the same chapter because they are interrelated.

Why Employees Enjoy Teams

We should note that well-led teams often lead to improved employee morale. Employees like teams for many reasons. In a study of project managers who were involved in project teams, five motivators emerged[2]:

Mutuality. The need for mutual support and encouragement between line management and project managers as well as personal loyalty of project managers to their teams and organizations.

Recognition for personal achievement. The opportunity for personal development as well as recognition for personal achievement through rewards, incentives, or status.

Belonging. The individual's need for supportive, cohesive, and friendly team relations. This implies clear communications both within the team and within the larger organization, as well as clear information and project goals.

Bounded power. The need for authority and control over project resources and people, personal accountability and challenge, individuals' abilities to influence decisions that affect the project, and opportunities for personal growth and development.

Creative autonomy. The need for individuals to have opportunities to use their creativity and potential during the course of a project and to enjoy good working conditions.

These motivators provide strong reasons for individuals to be involved in teams and show why teamwork is often correlated with positive attitudes.

---

[2]Tampoe, M., and Thurloway, L., "Project Management: The Use and Abuse of Techniques and Teams," *International Journal of Project Management* 11, 4 (November 1993):245–250.

LEADING TEAMS FOR QUALITY IMPROVEMENT

Employee Empowerment and Involvement

When we begin to use teams, decision-making authority is given to team members. **Empowerment** means giving power to team members who perhaps had little control over their jobs. When such power is given, management must follow through and give up a reasonable amount of control.

Implicit in empowerment is a series of promises to employees. These implicit promises include

Video Clip: Teams at the Ritz

You will have greater control over your own work.

You will not be penalized for making decisions that don't pan out.

Management is changing and becoming more contemporary.

Management is committed to quality improvement over the long haul.

Management will concede more control over company systems to you.

Management values your ideas and opinions and will give them serious consideration.

Management trusts you and is worthy of trust in return.

You will be rewarded for making decisions that benefit the company.

Labor is capable of decision making concerning its own jobs and company processes.

This approach to managing labor is an important factor in improving employee morale. It means a lot to employees to be told that their thoughts and ideas are valued. The Baldrige criteria encourage employee participation, adding that

> A company's success depends increasingly on the knowledge, skills, and motivation of its workforce. Employee success depends increasingly on having opportunities to learn and to practice new skills. Companies need to invest in the development of the workforce through education, training, and opportunities for continuing growth. Opportunities might include classroom and on-the-job training, job rotation, and pay for demonstrated knowledge in skills. On-the-job training offers a cost-effective way to train and to better link training to work processes. Workforce education and training programs may need to utilize advanced technologies, such as computer-based learning and satellite broadcasts. Increasingly, training, development, and work units need to be tailored to a diverse workforce and to more flexible, high-performance work practices.
>
> Major challenges in the area of workforce development include: (1) integration of human resources practices—selection, performance, recognition, training, career advancement, and the alignment of human resource management with strategic change processes. Addressing these challenges requires use of employee-related data on knowledge, skills, satisfaction, motivation, safety, and well-being. Such data need to be tied to indicators of company or unit performance such as customer satisfaction, customer retention, and productivity. Through this approach, human resource management may be better integrated and aligned with business directions.[3]

[3]2002 Criteria for Performance Excellence, Malcolm Baldrige National Quality Award, NIST, Gaithersburg, MD, pp. 8–9.

A number of preconditions are necessary for empowerment. These are

- ***Clear authority and accountability.*** Employees must know what is expected of them and be given authority over their own work.
- ***Participation in planning at all levels.*** Employees should be involved in planning related to their jobs. They should be provided with planning tools.
- ***Adequate communication and information for decision making.*** If employees are to make decisions related to their jobs, they need the right managerial information.
- ***Responsibility with authority.*** Employees should be given a definition of power that focuses on getting things done rather than exerting influence over people.[4]

Of course, granting authority to employees doesn't guarantee that people will work well together or necessarily achieve all the lofty goals that are espoused in this approach. Many issues surround empowerment and teamwork that must be addressed (see A Closer Look at Quality 11-1). These issues range from operations and behavior to organizational design. For example, if the existing culture does not reward this type of activity, it is doubtful that participatory approaches will work until the cultural issues are resolved. However, using teams can lead to cultural changes that facilitate improvement. This chapter focuses on the issues related to managing projects and teams to help make the transition succeed.

From a behavioral perspective, empowerment is a way to enhance organizational learning. **Organizational learning** leads to change in organizational behavior in a way that improves performance.[5] This type of learning takes place through a network of interrelated components. These components include teamwork, strategies, structures, cultures, systems, and their interactions. Corporate learning relies on an open culture where no one feels threatened to expose opinions or beliefs—a culture where individuals can engage in learning, questioning, and not remain constrained by "taboos" or existing norms. This strategy includes continuous improvement projects as a governing principle for all team members.[6]

Flattening Hierarchies for Improved Effectiveness

Along with the emphasis on teamwork and empowerment, there has been a move toward flattening hierarchies in organizations. Led by consultants such as Tom Peters and others, top managers have eliminated layers of bureaucratic managers in order to improve communication and simplify work. Having many layers of management can have the effect of increasing the time required to perform work. For example, it has been reported that in the 1980s, one of the largest automobile manufacturers in the United States required six months to determine its standard colors for office phones. Probably this decision required many, many meetings and proper authorization. However, such decisions are minor when compared with competitive decisions that need to be made. The time required to make this decision was excessive.

---

[4]Adapted from class notes from Dr. Catherine Beise, Kennesaw State College, Kennesaw, GA (Summer 1998).
[5]Swieringa, J., and Wierdsma, A., *Becoming a Learning Organization* (Reading, MA: Addison-Wesley, 1992).
[6]Ayas, K., "Integrating Corporate Learning with Project Management," *International Journal of Production Economics* 51, 1–2 (1997):59–67.

A CLOSER LOOK AT QUALITY 11-1

EMPOWERMENT IN ACTION

www.engelhard.com, www. srcreman.com

Empowerment has been used in a number of companies very successfully. These include Springfield Remanufacturing of Springfield, Missouri. This company, founded by Jack Stack, has successfully applied empowerment principles among its employees. Stack uses surprising candor with his employees, sharing all financial information with them each month. Stack tells how a janitor persuaded him to expand into a new product line. The company experienced difficult times. In order to avoid bankruptcy, Stack persuaded his employees to invest in the company. Since that time, monthly bank reports that include cash flow projections have been distributed to all employees. During this period, the company experienced 15% annual growth. The company uses the following credo:

> As employees of SRC, we are more than just operators—we're owners. This gives us the incentive to be disciplined and determined to make SRC the best remanufacturer of engines and components. Each department has a healthy competitive spirit, but we're all in the game together. Our sole objective is to produce the best remanufactured engines and components that money can buy. As an employee-owned corporation, we at Springfield Remanufacturing Corporation can account our success to the commitment of each SRC employee. Our philosophy is simply, "providing the customer with the highest quality product possible at a competitive price."[a]

Engelhard Corporation of Huntsville, Alabama was losing money, morale was low, employee turnover was 150%, and one in five products was returned to be scrapped or reworked. After empowerment, employees took responsibility for completing work. Productivity increased by 324% and employee turnover reduced to 2% per year. In the process of empowering employees, time clocks were eliminated, performance was emphasized, and employees received pay increases. Safety was emphasized with the hiring of a full-time nurse, and vacation and pension benefits were improved. The company focused on training managers, and supervisors were given time to join the process. In order to accept change, workers were given counseling and coaching to speak up in meetings and express themselves openly. According to their human resources manager, "when workers feel listened to, they put more energy into their work."[b]

[a]Lee, C., "Open-Book Management," *Training*, 31, 7 (1994):21.
[b]"Engelhard Corporation," *Plastics World*, 55, 6 (1997):G–62.

Too many layers of management also can impede creativity, stifle initiative, and make empowerment impossible. With fewer layers of management, companies tend to rely more on teams. When Lee Iacocca took the reins at Chrysler Corp., one of his first acts was to eliminate several levels of management. Iacocca credits this move with making other needed changes easier within the organization.

Team Leader Roles and Responsibilities

Quality professionals are unanimous—to be successful in achieving teamwork and participation, strong leadership both at the company level and within the team is essential. However, what is not always clear is what it means to be an effective team leader. We know that leaders are responsible for setting team direction and seeking future opportunities for the team. Leaders reinforce values and provide a system for achieving desired goals. Leaders establish expectations for high levels of performance, customer focus, and continuous learning. Leaders are responsible for communicating effectively, for evaluating organizational performance, and for providing feedback concerning such performance.

An important aspect of leadership is the organization's preparedness to follow the leadership. The best general is probably not going to be successful if the troops are not well trained or prepared. Hersey and Blanchard[7] propose a theory called a **situational leadership model** that clarifies the interrelation between employee preparedness and effectiveness of leadership. According to Hersey and Blanchard, situational leadership is based on interplay among the following:

The amount of guidance and direction a leader gives (task behavior)

The amount of socioeconomic support a leader provides (relationship behavior)

The readiness level that followers exhibit in performing a specific task, function, or objective

Therefore, if team members are trained and prepared so that they are "task ready," leadership will be more effective. **Readiness**, in this context, is the "extent to which a follower has the ability and willingness to accomplish a specific task." Readiness is a function of two variables. These are ability and technical skills (job maturity) and self-confidence in one's abilities (psychological maturity). Therefore, effective leadership helps employees become competent and instills confidence in employees that they can do the job.

Figure 11-2 shows the Hersey and Blanchard model of situational leadership with four different styles of leadership. As the model shows, different contingencies drive different approaches to leading. According to Hersey and Blanchard, the best approach to leading (i.e., telling, selling, participating, or delegating) depends on the readiness of employees to perform tasks and functions or accomplish objectives.

As it relates to quality management, leadership is especially difficult. Leaders are told that they should empower employees. To many leaders this implies a laissez-faire or a hands-off approach to management. In other words, many leaders feel that they are to provide resources but that they should not be involved in overly controlling employee behavior. Although the literature contains examples of companies that have

FIGURE 11-2 Hersey and Blanchard's Situational Leadership Model

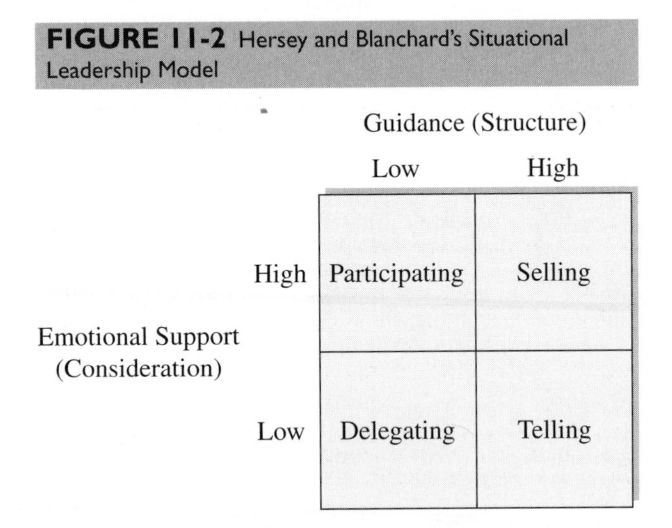

[7]Hersey, P., and Blanchard, K., *Management of Organizational Behavior: Utilizing Human Resources* (Englewood Cliffs, NJ: Prentice Hall, 1996).

been successful in delegating authority to this extent, quality management is not a vehicle by which leaders abdicate their responsibility.

In most organizations, employees want leaders who provide clear direction, necessary information, feedback on performance, insight, and ideas. Skilled team leaders need to demonstrate this ability to lead. The single most important attribute of companies with failed quality management programs is a lack of leadership. A close second is poor communication, which is related to leadership. Effective leaders are people who are able to provide visions, ideas, and motivation to others to achieve the greater good.

Team Roles and Responsibilities

Besides team leaders, there are a variety of roles that individuals occupy in teams. Meredith Belbin provides a widely adopted typology of team roles. Table 11-1 contains names and profiles for each of these roles. Belbin notes that each of these roles may be more relevant at different stages during a project. Also, these roles are not mutually exclusive. This means that one person can fulfill different roles on a team.

Also, team roles can be defined functionally. Often teams require different functional talents such as management, human resources, engineering, operations, accounting, marketing, management information systems, and others. In these cases, the managers overseeing the project help to identify the talents needed and then search for the team members to provide these talents.

TABLE 11-1 Belbin's Team Roles

Key Stages of Team Activity	*Team Roles Relevant to Particular Stages*
1. Identifying needs	Key figures at this stage are individuals with a strong goal awareness. Shapers and coordinators make their mark here.
2. Finding ideas	Once an objective is set, the means of achieving it are required. Here plant and resource investigators play a crucial role.
3. Formulating plans	Two activities help ideas turn into plans. One weighing up the options, the second making good use of all relevant experience and knowledge to ensure a good decision. Monitor evaluators make especially good long-term planners and specialists play a key role at this stage.
4. Making contacts	People must be persuaded that an improvement is possible. Champions of the plans and cheerleaders must be found. This is an activity where resource investigators are in their element. However, to appease disturbed groups, a team worker is required.
5. Establishing the organization	Plans need turning into procedures, methods, and working practices to become routines. Implementers are the people required here. These routines, however, need people to make them work. Coordinators are good at getting people to fit the system.
6. Following through	Too many assumptions are made that all will work out well in the end. Good follow-through benefits from the attentions of completers. Implementers, too, pull their weight in this area, for they pride themselves on being efficient in anything they undertake.

SOURCE: Adapted from R. M. Belbin, *Management Teams* (Boston: Butterworth-Heinemann, 2004).

FIGURE 11-3 Stages of a Team's Development

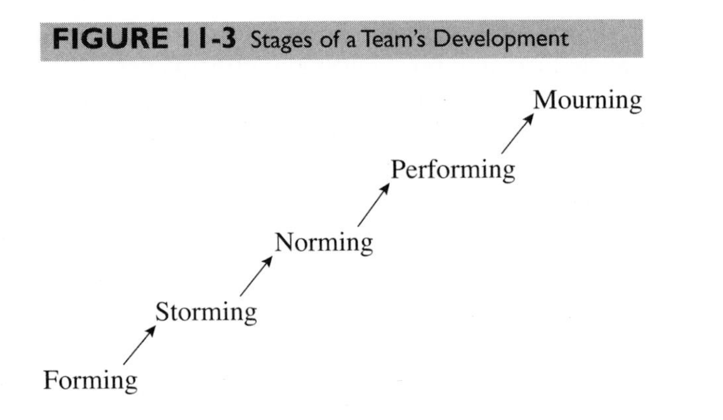

Team Formation and Evolution

The way a team is formed depends—to an extent—on the objectives or goals of the team. Regardless of the type of team your firm employs, teams experience different stages of development (see Figure 11-3).[8] These stages include the following. **Forming**, where the team is composed, and the objective for the team is set; **storming**, where the team members begin to get to know each other, and agreements have not yet been made that facilitate smooth interaction between team members; **norming**, where the team becomes a cohesive unit, and interdependence, trust, and cooperation develop; and **performing**, where a mutually supportive, steady state is achieved. And in successful projects, the final stage is **mourning**, where team members regret the ending of the project and the breaking up of the team.

Team Rules

During the norming stage, teams develop ground rules. Such ground rules can forestall conflict. Common ground rules for teams in projects are shown in Table 11-2. It is often useful to establish ground rules first in order for a team to be functional. If a team is

TABLE 11-2 Ground Rules for Effective Teams

1. Test assumptions and inferences.
2. Share all relevant information.
3. Focus on interests, not positions.
4. Be specific—use examples.
5. Agree on what important words mean.
6. Explain the reasons behind one's statements, questions, and actions.
7. Disagree openly with any member of the group.
8. Make statements, then invite questions and comments.
9. Jointly design ways to test disagreements and solutions.
10. Discuss undiscussable issues.
11. Keep the discussion focused.
12. Do not take cheap shots or otherwise distract the group.
13. All members are expected to participate in all phases of the process.
14. Exchange relevant information with nongroup members.
15. Make decisions by consensus.
16. Do self-critiques.

SOURCE: R. Schwarz, "Ground Rules for Effective Groups," *Popular Government* 54, 4 (1989):25–30. Reprinted by permission of the Institute of Government, the University of North Carolina at Chapel Hill.

---

[8]Sommerville, J., and Dalzie, A., "Project Teambuilding: The Applicability of Belbin's Team-Role Self-Perception Inventory," *International Journal of Project Management* 16, 3 (1998):165–171.

functional, individual participation enhances the group's effectiveness. If the team is dysfunctional, such participation reduces the effectiveness of the group. Acts of commission include talking behind the backs of other team members or otherwise acting out one's feelings. There are also acts of omission in such passive-aggressive behavior as forgetting to attend meetings or withholding information. Counteractive behavior improves the group's effectiveness by negating dysfunctional behavior. Counteractive behavior can be enacted either by the team, the facilitator, the team manager, or even the offending individual.

TYPES OF TEAMS

At this point we will pause to define the various types of teams that are used in improving quality. Continuous process improvement often requires small teams that are segmented by work areas. Projects with multiple departments in a company require cross-functional teams. Large projects require teams with large budgets and multiple members. Smaller projects, such as "formulating a preventive maintenance plan for oiling the metal lathes," probably will require a much smaller team. The literature is full of different types of teams and approaches to teamwork. Table 11-3 contains a list of a few of the major types of teams found in the literature. In the following sections we list and define a number of team types.

Process Improvement Teams

Process improvement teams are teams that work to improve processes and customer service. These teams may work under the direction of management or may be self-directed. In either case, process improvement teams are involved in some or all of the following activities: identifying opportunities for improvement, prioritizing opportunities, selecting projects, gathering data, analyzing data, making recommendations, implementing change, and conducting postimplementation reviews. Many process improvement teams are an outgrowth of quality-related training. These teams use the basic tools and the plan–do–check–act cycle to effect change relating to processes.

Cross-Functional Teams

Cross-functional teams enlist people from a variety of functional groups within the firm. In the real world, problems often cut across functional borders. As a result, problem-solving teams are needed that include people from a variety of functions. These cross-functional teams often work on higher-level strategic issues that involve multiple functions. Such teams often work on macrolevel, quality-related problems such as communication or redesigning company-wide processes.

Tiger Teams

A **tiger team** is a high-powered team assigned to work on a specific problem for a limited amount of time. These teams are often used in reengineering efforts or in projects

TABLE 11-3 Types of Teams

Team Type	*Scope*
Process improvement team	Local or single department
Cross-functional team	Multiple departments
Tiger team	Organization-wide
Natural work group	Customer- or region-centered
Self-directed work team	Narrow or broad

where a specific problem needs to be solved in a very short period of time. The work is very intense and has only a limited duration.

Natural Work Groups

Natural work groups are teams organized around a common product, customer, or service. Many times these teams are cross-functional and include marketers, researchers, engineers, and producers. The objective of natural work groups includes tasks such as increasing responsiveness to customers and market demand. In order to implement natural work groups, a great deal of effort is typically expended relating to organizational redesign and systems redesign. Commonly cited outcomes of natural work groups are improved job design and improved work life for employees. The key, elemental impact of natural work groups is to improve service by focusing work units in an organization on the customer. A by-product is improved communication with customers. Often a natural work group will be established for a specific customer.

Self-Directed Work Teams

A **self-directed work team** is a team chartered to work on projects identified by team members themselves. There is little managerial oversight except to establish the teams and fund their activities.

Self-directed teams are identified as either *little s* or *big S* teams. Little s self-directed work teams are made up of employees empowered to identify opportunities for improvement, select improvement projects, and complete implementation. Big S self-directed teams are involved in managing the different functions of the company without a traditional management structure. These types of teams contain totally self-directed employees who make decisions concerning benefits, finances, pay, processes, customers, and all the other aspects of running the business. Often big S self-directed work teams hold partial ownership of the companies they work for so that they participate in the benefits of their teamwork.

Technology and Teams

New tools for teamwork are constantly emerging. The model for team effectiveness in Figure 11-4 shows how teams use information systems tools and demonstrates that organizational integration (or how well the cross-functional units in the company work together) and simultaneity methods (such as concurrent design teams) are important components of team effectiveness. Also, team effectiveness is a precursor to project task performance. The figure shows different variables such as organizational integration methods, organizational redefinition, and simultaneous methods that affect the outcome of the project. Integrated IS tools involve integrated information systems such as CAD (computer-aided design)/CAM (computer-assisted manufacturing) and CIM (computer-integrated manufacturing). These aid in achieving improvement in efficiency and effectiveness. Process technology is used in helping to improve task performance. Process standardization methods such as the tools of quality and customer input methods complete the model.[9]

This model amplifies that more and more, team effectiveness is assisted by integrated tools and technologies, and the impact of technology should increase. As software becomes cheaper and easier to use, more tools will be used by everyone involved with the project.

---

[9]Moffat, L., "Tools and Teams: Competing Models of Integrated Product Development Project Performance," *Journal of Engineering and Technology Management* 15 (1998):55–85.

FIGURE 11-4 A Team Effectiveness Model

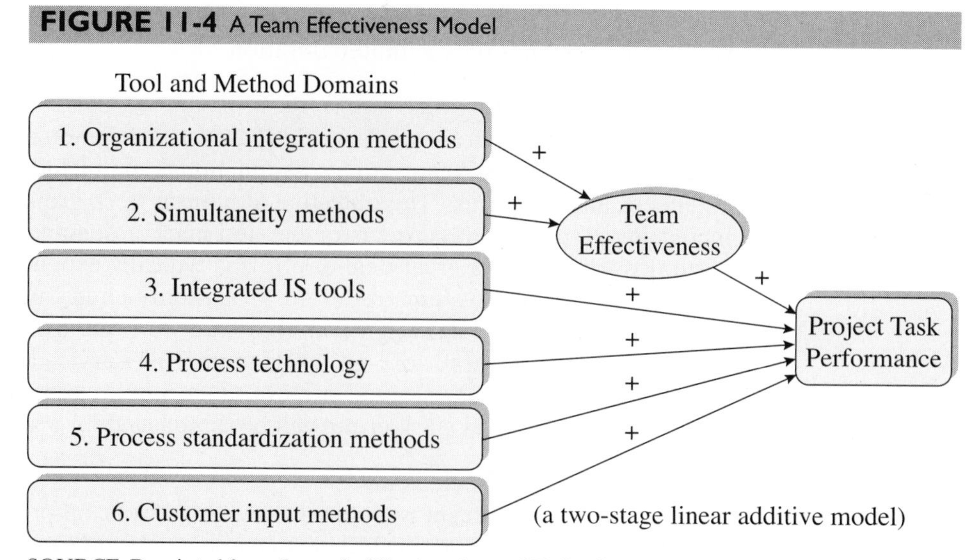

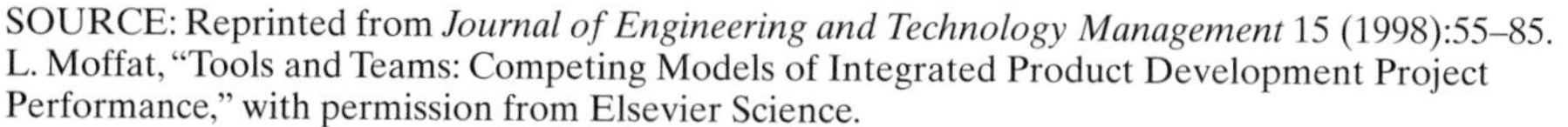
SOURCE: Reprinted from *Journal of Engineering and Technology Management* 15 (1998):55–85. L. Moffat, "Tools and Teams: Competing Models of Integrated Product Development Project Performance," with permission from Elsevier Science.

Virtual Teams

The term **virtual teams** is emerging as more companies become "virtual organizations," loosely knit consortia that produce products and services. Virtual teams are teams that rarely or never physically meet, except in electronic meetings using group decision software. Among virtual organizations, projects often cross organizational boundaries. Today, Internet and intranet-based applications called **teamware** are emerging that allow us to access the World Wide Web and build a team, share ideas, hold virtual meetings, brainstorm, keep schedules, and archive past results with people in far-flung locations around the world. Hectic schedules and the difficulty in finding convenient times to meet to solve problems will make teams of this type more important in the future (see A Closer Look at Quality 11-2).

IMPLEMENTING TEAMS

The teams in our examples have something in common. The performance of the team is essential to their individual success, and in some cases, even lives hang in the balance. If the NASCAR team performs ineffectively, the driver loses. If the Massachusetts General hospital team is ineffective, people die. If the SEALs don't function properly, lives are lost, and the mission fails. How do we engender this sense of urgency in quality improvement teams? How do we create a momentum or team ethic that will help us beat the odds and be successful? Accomplishing this often requires facilitation and team building. **Facilitation** is helping or aiding teams by maintaining a process orientation and focusing the group. **Team building** is accomplished by following a process that identifies roles for team members and then helps them to become competent in achieving those roles.

The role of the facilitator is very important in managing teams, particularly when team members have little experience with teamwork. The role of the **facilitator** is to make it easy for the group to know where it is going, know why it wants to get there, know how to get there, and know what it is going to do next.[10]

[10]Anson, R., "Facilitation Skills for Focused Meetings," Working paper, 1998.

A CLOSER LOOK AT QUALITY 11-2

LESSONS FROM EFFECTIVE TEAMS OUTSIDE THE BUSINESS WORLD[c]

Many teams outside business can serve as examples to business. Obviously, sports teams such as the Los Angeles Lakers and the Boston Celtics of the 1980s, and the Chicago Bulls of the 1990s provide excellent examples of effective teams. In this Closer Look at Quality we consider several outstanding teams that display many of the attributes that are needed in quality improvement teams.

NAVY SEALS

Navy SEALs are an elite team of individuals who perform very dangerous missions for the Navy. Even though the first four weeks of Navy SEAL training are grueling, they pale in comparison with the fifth week, known as hell week. During hell week, recruits swim many, many miles in cold water in the Pacific Ocean, they row rubber boats for hours on end, they run obstacle courses over and over, they perform grueling calisthenics using 300-pound logs, and they sustain personal insults from trainers. During the entire hell week, recruits will sleep for, perhaps, four hours.

About 30% of the recruits drop out during the five-day hell week experience. Commitment is needed because hell week is followed by months of rigorous underwater training, weapons training, explosives training, parachute training, and a six-month probationary period. The ultimate success rate for recruits is about 30%.

Teamwork is essential because SEALs never operate on their own, and team members identify totally with the group. Navy SEALs will never leave the battlefield if a fallen SEAL remains. The SEALs teach us about the necessity of training and cohesion among team members.

MASSACHUSETTS GENERAL HOSPITAL EMERGENCY ROOM

Every day about 200 patients arrive in the emergency room at Massachusetts General Hospital—one of the finest hospitals in the world. Working as a team in an emergency-trauma unit, doctors, nurses, and aides have to deal with great amounts of stress, fatigue, and emotional baggage.

Seamless role playing is important as gurneys are rolled into the trauma center. The triage nurse determines the severity of an injury. Next, the team of doctors, nurses, and aids leaps into action, performing the tasks they have performed many times before. Each person performs his or her task—assessing the patient, starting an IV, scoping the wounds, or hooking up heart monitors. At Mass General, "nobody bosses anybody around; if someone has a thought that's useful, we are open to suggestions." As with any effective team, the members understand and flawlessly execute their roles under stressful conditions.

THE CHILDRESS NASCAR TEAM

Richard Childress, a former racer himself, has taken part in several Winston Cup championships for the top prize in NASCAR racing since founding his Childress Racing maintenance teams. How did he do it?

Childress has a keen eye for young mechanical talent that fits into the team concept he has built. "I'd rather train them our way than try to break old habits," says Childress.

Almost everyone starts at the bottom—cleaning up after the other mechanics and running for parts and coffee. On the road, they fill water bottles and stack tires. During this time, they are judged less for their actual performance than for their ability to work with the team. Says one team member, "Attitude is more important than expertise—you've got to be able to have people who won't let you down." The proof of this message is in the results. A team of 17 people, 7 working on the car itself and 10 behind the wall, can change 4 tires, pump 21 gallons of gas, clean the windshield, and give the driver a drink in less than 18 seconds. The Childress team excels because of its focus on detail and maintenance of team member roles.

[c]Labich, K., "Elite Teams Get the Job Done," *Fortune*, 133, 3 (1996):90.

A facilitator focuses the group on the process it must follow. Successful facilitation does not mean that the group always achieves its desired results. The facilitator is responsible for ensuring that the team follows a meaningful and effective process to achieve its objectives.

How is this accomplished? The facilitator should plan how the group will work through a task, help the group stay on track and be productive, draw out quiet members, discourage monopolizers, help develop clear and shared understanding, watch body language and verbal cues, and help the group to achieve closure. Again, facilitators must remain neutral on content. Facilitators cannot take sides or positions on important areas of disagreement. However, facilitators should help key members reach points of agreement.

Meeting Management

Effective **meeting management** is an important skill for a facilitator of quality improvement teams. Often quality improvement involves a series of meetings of team members who meet to brainstorm, perform root-cause analysis, and carry out other activities. Tools for successful meeting management include an agenda, predetermined objectives for the meeting, a process for running the meeting, processes for voting, and development of an action plan. Using these tools requires outstanding communication skills as well as human relations skills. The steps required for planning a meeting are

1. Defining an agenda
2. Developing meeting objectives
3. Designing the agenda activity outline
4. Using process techniques

Structured processes, a set of rules for managing meetings, work well in conducting meetings. It is paradoxical that structured processes are inhibiting, time-consuming, and unnatural—which is why they work. Why do we use processes in meetings? The answers are clear. We wish meetings to stay focused, to involve deeper exploration, to separate creative from evaluative activities, to provide equal opportunity for contribution, to encourage reflection, to provide objective ground rules to reduce defensiveness, and to separate the person from the idea.

Figure 11-5 identifies some meeting structured process techniques. Tools such as flipcharts, sticky dots, whiteboards, and sticky notes are used commonly in structured process activities. The focus of team meetings moves from clarifying to generating ideas, to evaluating ideas, and to action planning. Some of the techniques, such as silent voting and idea writing, help team members reach consensus rapidly.

Another useful meeting management tool that was pioneered by Hewlett Packard is the "**parking lot**." The parking lot (see Figure 11-6) is a flipchart or whiteboard where topics that are off the subject are *parked* with the agreement that these topics will be candidates for the next meeting's agenda. At the end of the meeting, the group agrees on the agenda for the following meeting, and the parking lot is erased.

Conflict Resolution in Teams

As people work closely together in teams, conflicts arise. Conflict resolution is a hugely important topic for team leaders and members. Conflicts are endemic to all kinds of team projects. Using team processes, assumptions are questioned, change is brainstormed, and cultures are challenged. This type of creative activity results in possible

FIGURE 11-5 Structured Process Activities

By Activity Type	*Approaches*	*Tools*
• Clarity techniques	• Individual	• Verbal
—Lasso	• Subgroups ("buzz groups")	• Flipcharts
• Generate techniques	• Full group	• Whiteboard
—Structured brainstorming		• Paper pads
—Round-robin contribution		• Sticky notes
—Silent writing		• Sticky dots
—Sticky notes recording		• Computer
—Brain writing		
• Evaluate techniques		
—Reduce list		
—Pros and cons		
—Force fields		
—Silent voting		
—Sticky dots		
—Idea writing		
• Action planning		

SOURCE: R. Anson, "Facilitation Skills for Focused Meetings," Working Paper (Boise State University, 1998).

FIGURE 11-6 Parking Lot

conflict. It is claimed that team leaders and project managers spend more than 20% of their time resolving conflict.[11] If this is true, then conflict resolution resounds as one of the very important underdiscussed topics in team building.

There are many sources of conflict. Some conflicts are internal, such as personality conflicts or rivalries; some are external, such as disagreements over reward systems,

[11]Thomas, K., and Schmidt, W., "A Survey of Managerial Interests with Respect to Conflict," *Academy of Management Journal* 10 (1976):315–318.

FIGURE 11-7 Modes of Conflict Behavior

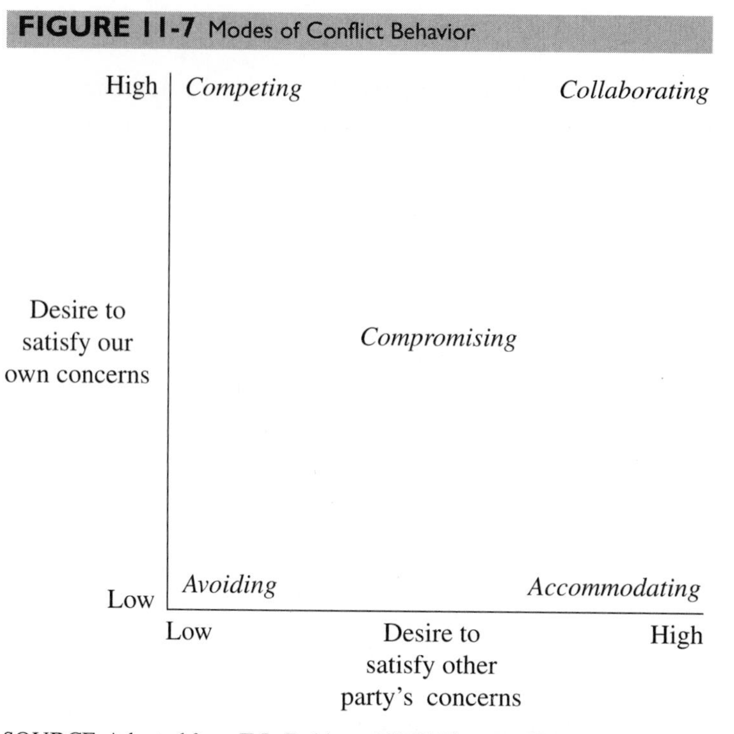

SOURCE: Adapted from T. L. Ruble and K. W. Thomas, "Support for a Two-Dimensional Model of Conflict Behavior," *Organizational Behavior and Human Decision Processes* 16 (1976):221–237, with permission of Elsevier Science.

scarce resources, lines of authority, or functional differentiation (see Fig. 11-7). Teams bring together individuals from a variety of cultures, backgrounds, and functional areas of expertise. Being on a team can create confusion for individuals and insecurity as members are taken out of their comfort zones. It is interesting to note that these are also some of the reasons teams are successful. Some organizational causes of differences are more insidious: faulty attribution, faulty communication, or grudges and prejudice.

Four recognizable stages occur in the conflict resolution process[12]:

Frustration. People are at odds, and competition or aggression ensues.

Conceptualization and orientation. Opponents identify the issues that need to be resolved.

Interaction. Team members discuss and air the problems.

Outcome. The problem is resolved.

One of the things a leader must be able to do is manage conflict in the organization. In order to foster a well-run workplace, leaders must be able to resolve conflict effectively in organizations. Leaders resolve conflict in a variety of ways:

Passive conflict resolution. Some managers and leaders ignore conflict. This is probably the most common approach to working out conflict. There may be positive reasons for this approach. The leader may prefer that subordinates

[12]Pinto, J., and Kharbanda, O., "Project Management and Conflict Resolution," *Project Management Journal* 26, 4 (1995):45–54.

work things out themselves. Or the conflict may be minor and will take care of itself over time. Leaders may feel that some issues are small enough not to merit micromanagement.

Win–win. Leaders might seek solutions to problems that satisfy both sides of a conflict by providing win-win scenarios. One form of this is called *balancing demands* for the participants. This happens when the manager determines what each person in the conflict wants as an outcome and looks for solutions that can satisfy the needs of both parties.

Structured problem solving. Conflicts can be resolved in a fact-based manner by gathering data regarding the problem and having the data analyzed by a disinterested observer to add weight to the claims of one of the conflicting parties.

Confronting conflict. At times it is best to confront the conflict and use active listening techniques to help subordinates resolve conflicts. This provides a means for coming to a solution of the conflict.

Choosing a winner. In some cases, where the differences between the parties in the conflict are great, the leader may choose a winner of the conflict and develop a plan of action for conflict resolution between the parties.

Selecting a better alternative. Sometimes there is an alternative neither of the parties to the conflict has considered. The leader then asks the conflicting parties to pursue an alternative plan of action.

Preventing conflict. Skilled leaders use different techniques to create an environment that is relatively free of conflict. These approaches are more strategic in nature and involve organizational design fundamentals. As shown in Table 11-4, these organizational design fundamentals are useful in reducing

TABLE 11-4 Constructive Conflict Resolution Components

1. *Goal structure*—Goals should be well defined and operational and should reflect each unit's contribution to the total organization. Managers, in turn, should convey to their subordinates as precisely as possible the feeling that their own unit is dependent on the work of other units.
2. *Reward systems*—Each unit's contribution to the effectiveness of the total organization should be assessed carefully and rewarded accordingly. Where high levels of interdependence exist, reward systems should be designed specifically to reflect interdependence. Such a reward system will encourage cooperation among organization units.
3. *Contact and communication*—Frequent contact and communication between organization units need to be encouraged. Individuals should be rotated through related organization units in order to have them gain experience, understanding, and empathy for the work done and the problems encountered in other units.
4. *Coordination*—Liaison roles should be established where potential communication and coordination problems exist (e.g., between R&D and manufacturing at the point where a new product moves from advanced development into the pilot stage of production). The liaison role can be used to facilitate necessary interaction, thus reducing the time and information content lost when using formal channels. In addition, the liaison becomes better acquainted with the work of the different units and can provide continuous updating to each of the other units.
5. *Competitive systems*—Competition, where it does exist, should be examined carefully. Although competition can facilitate productivity, it can also produce conflict whenever organization units are interdependent. In such situations, competition need not be eliminated, but its benefits should be evaluated against its potential for causing conflict. Clearly, organization units should not be forced into win-lose situations.

SOURCE: D. Greene, E. Adams, and R. Ebert, *Management for Effective Performance* (Englewood Cliffs, NJ: Prentice Hall, 1985), p. 454. Reprinted by permission of Pearson Education, Inc., Upper Saddle River, NJ.

conflict in organizations. By carefully defining goals, rewards, communication systems, coordination, and the nature of competition in a firm, conflict can be reduced or eliminated. Conflicts often are the results of the reward systems in the firm. A systems approach will focus attention on organizational design rather than individual interactions.

Generally speaking, these conflict-resolution approaches involve three alternatives: *avoidance*, *defusion*, or *confrontation*. Avoidance involves letting things work themselves out without involving a leader. Defusion means smoothing ruffled feelings while getting the team project back on track. Confrontation involves injecting the leader into the conflict to find a solution.

Saving Quality Teams from Failure: Diagnosing Problems and Intervening before It Is Too Late

In the prior section we talked about handling conflict. Sometimes quality improvement teams embark on improvement projects and, for whatever reason, things begin to fall apart, and the team risks failure. It is usually easy to see when a quality improvement team project is going awry. There are cost overruns. Deadlines and schedules are not met. Frustration abounds. People are trying to deflect responsibility or bailing out. At such times it is critical to be able to diagnose and intervene to save the project.

Figure 11-8 shows the diagnosis-intervention cycle that must be undertaken by the facilitator, team leader, or the team members themselves. The figure shows the cycle that is followed to diagnose team failure and to intervene before the team fails. This requires observing the behavior of team members—you may see team members exhibiting nonverbal body language during team meetings. As a facilitator, you may have team members contacting you outside of team meetings to let you know what is "really happening." Drawing inferences about the meaning of the behavior means that, as a facilitator or team leader, you need to determine and understand the root causes of the behavior you are observing. Is fear present? Are people feeling external pressure? Do certain team members behave in a way that is dysfunctional?

FIGURE 11-8 Diagnosis-Intervention Cycle

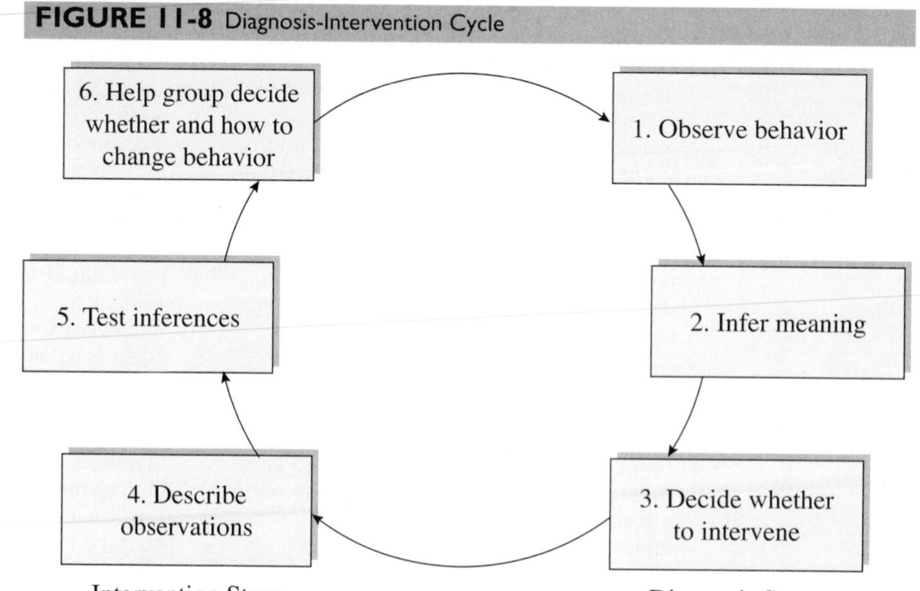

SOURCE: Adapted from R. Schwarz, *The Skilled Facilitator* (San Francisco: Jossey-Bass, 1994).

Next, you need to decide whether to intervene to improve the behavior. This is an important decision that requires insight. Is it better to intervene, or will the problems solve themselves over time? Describing your observations to the team is a means of "putting the issues on the table" so that they can be discussed. When you test inferences about observations, you open up a group discussion to let others understand why they are exhibiting difficult behavior. Helping the group decide whether and how to change the problematic behavior is the remedial step in which the leader or facilitator helps to resolve the problem.[13]

Recognition and action are keys to saving teams that are troubled. Eliyahu Goldratt,[14] in his book *Critical Chain*, a thought-provoking text on project management, demonstrates that people usually blame others when projects fail, or they blame uncertainty. The problem is that uncertainty is common in all projects. Skilled facilitators and team leaders recognize this fact and use effective communication to keep projects on track.

MANAGING AND CONTROLLING PROJECTS

Up to this point we have discussed team building from an organizational behavior perspective. Our discussion has focused on topics such as leadership, team roles, forming teams, facilitating teams, resolving conflict, and intervening to save teams. At this point, we shift our discussion to the activities associated with controlling projects.

Too often companies attempt to implement teams and projects in a poorly planned manner. The ultimate result is a number of operating teams with no clear methodology for coordinating the different activities of the teams. In the following pages we discuss how to coordinate and manage projects. From this perspective, we adopt an organizational theory-based approach to project management. The question is, How can we manage quality improvement projects in a way that they will be coherent, thoughtful, and in alignment with organizational objectives?

We introduce the tools used in controlling projects in order of sequence. Because some of these tools were introduced in Chapter 10, we now consider them in the context of their implementation.

Qualifying Projects

As will be discussed in more detail in Chapter 14, an important function of management is to select and qualify projects. To qualify a project means to determine the worthiness of a project on different dimensions. Commonly used methods for qualifying projects are **cost benefits analysis (CBA)** using **payback period** calculation. Both of these involve identifying direct and indirect project costs and expected returns for projects. Often, these analyses are used for comparing and selecting projects. For these calculations, use the following formulas:

$$C_t = \Sigma(C_d + C_i) \qquad \textbf{(11.1)}$$

where:

C_t = total project costs
C_d = direct project costs
C_i = indirect project costs

[13]Schwarz, R., *The Skilled Facilitator* (San Francisco: Jossey-Bass, 1994).
[14]Goldratt, E., *Critical Chain* (Great Barrington, MA: North River Press, 1997).

$$PP = C_t/B_a \tag{11.2}$$

where:

PP = payback period
C_t = total project costs
B_a = annualized benefits

When evaluating projects, you need to understand the differences between **soft costs** and **hard costs**. Soft costs are costs that are not easily recovered in project savings. This is usually because the benefits of the project add to organizational slack without resulting in actual dollar savings. An example of soft cost savings is the reduction of task time in a noncritical process task. This means that the process will have more slack. However, it does not mean that cycle time is reduced for production of a particular product—thereby resulting in no actual increase in productive capacity. Hard costs are just that—the reduction of rent, equipment costs, or labor direct costs—hard savings. It is best to justify savings based on hard costs that accrue to the bottom line.

Example 11.1: Cost Benefit Analysis in Action

A Six Sigma master black belt has asked you to help to analyze a possible project. This project involves implementing a computer-based sales system to improve supply chain performance. Some of the direct and indirect costs are as follows:

Direct Costs

- 20 networkable PCs—$1,500/each
- A server—$2,000
- Peripherals—$2,000
- Network installation—$5,000
- Sales system software—$10,000

Indirect Costs

- Training—$10,000
- Lost time—30 days × $120/day
- Sales-related losses during implementation—$25,000

Annualized Benefits

- Increased sales capacity—$200,000
- Improved customer retention—$500,000
- Improved follow-up sales opportunities—$100,000

Total costs = $49,000 + $38,600 = $87,600
Benefits = $800,000 per year
Payback period = $87,600/$800,000 = .11 years

Project Charters

Project charters are simple tools to help teams identify objectives, participants, and expected benefits from projects. The charter includes spaces for signatures to identify reporting relationships for planning purposes. In some firms, these signatures are for approvals and in other firms they are for information only.

Figure 11-9 shows a charter for an actual project. Notice that the prior identification of benefits helps team members and management identify measures for postimplementation review to see whether the project reaped the expected benefits.

FIGURE 11-9 An Actual Project Charter

Date: June 9, 2006

Project Charter for the Document Checklist Team

Team members: Jody, Hollie, Elaine

Objective(s): Create document checklists to accompany texts, covers, and fulfillment.

Projected benefits: Help the Roles and Responsibilities team; clarify status of jobs; enable backup people to take over when needed; reduce time to complete a project; reduce stress; reduce errors, omissions, and rework.

Approvals:

G. Andrea ________ *Responsible Manager*
Jim ________ *Department Head*
St ________ *Leadership Council*

Completion Date: 9/14/08

FORCE-FIELD ANALYSIS

A useful tool for planning projects is **force-field analysis**. This tool is designed to identify and quantify all of the forces for or against organizational change. This allows you to identify possible project assassins and to fortify support for a project. To perform force-field analysis, perform the following steps:

1. List all forces for change in the first column and all forces against change in the third column.
2. Assign a score for each force, where 1 = weak and 5 = very strong.
3. Sum the forces for and against the change and draw a diagram showing the forces.

Example 11.2: Force Fields in Action

A project was performed to determine the feasibility of implementing a new customer relationship management system (CRMS). A group of experts within the firm identified major forces for and against implementing the new CRMS. These forces and scores are shown in Figure 11-10.

Solution: The scores show that there are significant forces that mitigate against change. Work is needed to develop more forces in favor of change.

FIGURE 11-10 Force-Field Analysis

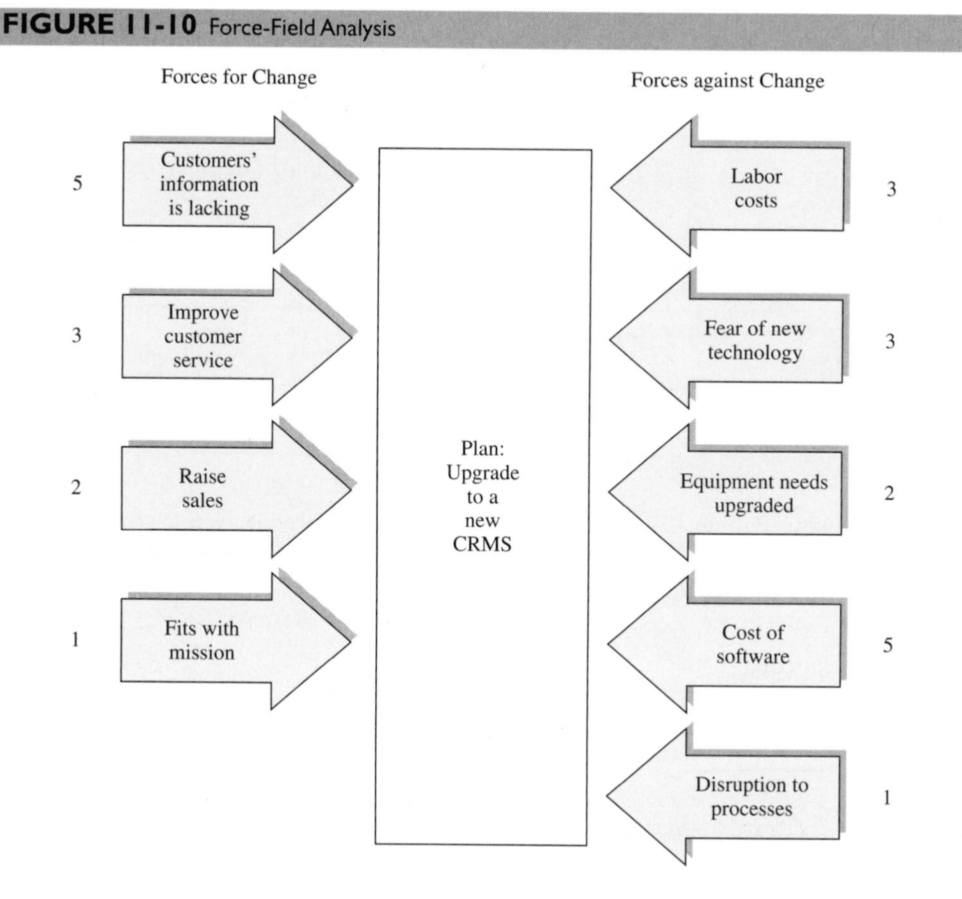

Work Breakdown Structure (WBS)

After chartering the project, the next step is to begin planning the project. The work breakdown structure (WBS) that was introduced in Chapter 10 is an excellent tool for determining the tasks to complete a project. Figure 11-11 shows an outline of a WBS for a project. Team members identify the major tasks required to perform the project. The tasks for an actual project—the Documentation Checklist Project—are listed in Table 11-5 with their appropriate start times, number of workdays, and predecessor tasks. After the tasks were identified, the question was asked, "What subtasks will be needed to complete the identified major tasks?" Next, the same questions were asked for the subtasks until we had a complete tree of major tasks, subtasks, and minor tasks.

It is important to note that with large projects it is often best to develop a separate WBS for each of the major tasks because separate individuals or groups may be involved in different tasks. Therefore, the marketing department can develop a WBS for the marketing portion of the project, and operations can develop a WBS for the operations portion. In the end, these WBSs must be combined into an overall WBS for the project.

Identifying Precedence Relationships

The WBS tasks are placed on individual note cards, and precedence relationships are identified for all the tasks. The note cards are placed in order on a large sheet of paper, and arrows are drawn between the tasks.

FIGURE 11-11 Work Breakdown Structure

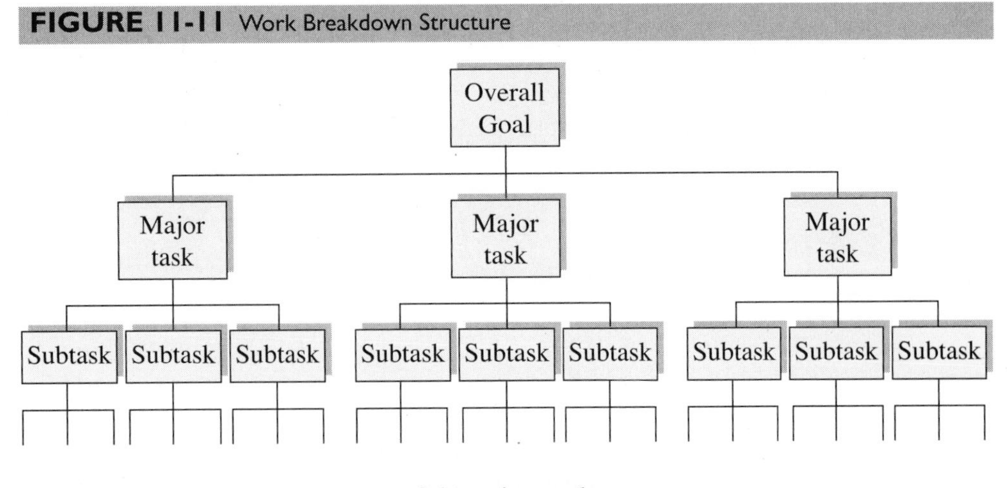

TABLE 11-5 Tasks for the Document Checklist Project

Tasks	*Workdays*	*Start Date*	*End Date*	*Preceding Tasks*
1. Use flowcharts to list steps	5	Fri 6/9/06	Thurs 6/15/06	
2. Preliminary design of form	5	Fri 6/16/06	Thurs 6/22/06	1
3. Check against Roles team and Andrea	10	Fri 6/16/06	Thurs 6/29/06	
4. Develop procedures for using form	10	Fri 6/16/06	Thurs 6/29/06	
5. Preliminary draft of form completed	5	Fri 6/30/06	Thurs 7/6/06	4,2,3
6. Test run form and procedures	40	Fri 7/7/06	Thurs 8/31/06	5
7. Revise and finalize form and procedures	10	Fri 9/1/06	Thurs 9/14/06	6

Identifying Outcome Measures

During the planning stage of the project, the team should refer to the project charter and identify measures against which team performance can be judged. This oft-overlooked step in planning is useful because some needed measures may not exist in current cost accounting systems. It is important to have preliminary and post hoc data so that baselining or intervention analysis can be performed effectively. This involves gathering preliminary data and postimplementation data. Then statistical tests are used to test for improvement.[15]

Identifying Task Times

Now we identify completion times for all the tasks. To do this, identify three time estimates for each task:

a = the optimistic completion time
m = the most likely completion time
b = the pessimistic completion time

Next, a weighted average of the tasks is calculated using the following formula:

$$\text{Expected time} = (a + 4m + b)/6 \qquad \textbf{(11.3)}$$

[15]For a good example of project-based intervention analysis, see Foster, S. T., Jr., and Franz, C. R., "Assessing Process Reengineering Impacts through Baselining," *Benchmarking for Quality Management and Technology* 2, 3 (1995):4–19.

The task variance is computed as

$$\sigma_t^2 = \left[(b-a)/6\right]^2 \tag{11.4}$$

The project variance is computed as

$$\sigma_T^2 = \sum_{t=l}^{n} \sigma_t^2 \tag{11.5}$$

Finally, the project standard deviation is

$$\sigma_T = \sqrt{\sigma_T^2} \tag{11.6}$$

Example 11.3: Calculating Task Times

The following tasks were identified for a major project using a WBS. Team members also identified optimistic, most likely, and pessimistic task completion times for each task. They have asked you to compute *task times* and *variance* for this project (solution in italics).

Task	a	m	b	Expected Times	Variances		Predecessors
A	1	3	5	*3 ((1 + (4)3 + 5)/6) = 3*	*.444*	*((5 − 1)/6)² = .444*	—
B	5	6	7	*6*	*.111*		—
C	6	9	12	*9*	*1.000*		—
D	3	4	5	*4*	*.111*		A
E	3	4	23	*7*	*11.111*		A
F	8	10	12	*10*	*.444*		A
G	12	17	22	*17*	*2.778*		A
H	5	9	13	*9*	*1.778*		B, D
I	13	15	17	*15*	*.444*		B, D
J	10	11	12	*11*	*.111*		B, D
K	1	8	15	*8*	*5.444*		C, E
L	2	3	16	*5*	*5.444*		F, H, K
M	8	10	12	*10*	*.444*		F, H, K
N	4	7	11	*7.17*	*1.361*		G, I, L
O	3	10	11	*9*	*1.778*		J
P	3	4	5	*4*	*.111*		M, N, O

Activity Network Diagrams

The following steps are used to develop an activity network diagram (PERT chart):

1. Using the inputs from a tree diagram listing tasks to be performed in the project, list all the tasks.
2. Determine task times.
3. Determine the precedence relationships between the tasks; that is, indicate which tasks depend on the completion of other tasks in the process.
4. Draw the network diagram.
5. Compute early start and early finish times by working from left to right in the network. These are the earliest times that individual tasks can be started and finished.

6. Compute late start and late finish times by moving from right to left in the network. These times are the latest times that tasks can possibly be started or finished.
7. Compute slack times and determine the critical path. The critical path links activities with zero slack.

$$\text{Slack time} = \text{late start} - \text{early start} \tag{11.7}$$

Example 11.4: Activity Network Diagram

A company developing a new advertising brochure identified the steps in the project. These were placed in the tree diagram shown in Figure 11-12. Table 11-6 lists all the tasks with their brainstormed times and predecessors. Figure 11-13 shows the network with tasks. This also shows times and precedence relationships using activity-on-node (AON).

In the next step we compute the early start and early finish times. This shows that the project is expected to be completed in 14 weeks (see Figure 11-14). Next, we compute the late start and late finish times by working from right to left. Notice that the late finish time for task A is 14 and the late start is 10 (see Figure 11-15). Task B then has a late finish of week 10 and an early start of week 9.

The slack times for this project are computed by taking the late start times minus the late finish time. From this diagram it appears that initially there is no slack for activities M, N, O, I, E, B, and A and a lot of slack for all other activities.

FIGURE 11-12 Tree Diagram of Tasks

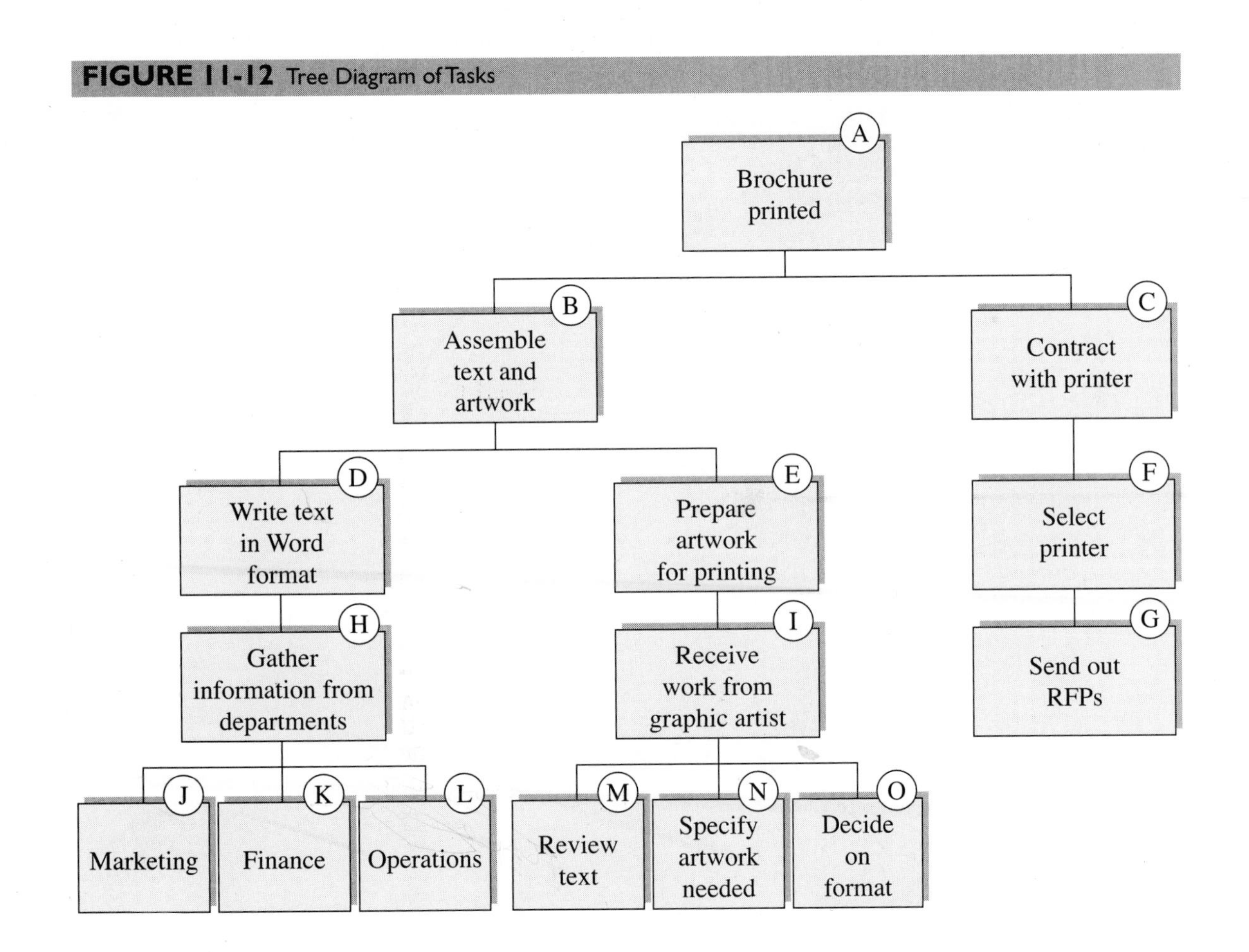

TABLE 11-6 Tasks

Task	*Task IDs*	*Predecessors*	*Expected Time (Weeks)*
Brochure printed	A	B, C	4
Assemble text and artwork	B	E, D	1
Contract with printer	C	F	2
Write text in Word format	D	H	2
Prepare artwork for printing	E	I	4
Select printer	F	G	2
Send out RFPs	G	—	3
Gather information from departments	H	J, K, L	1
Receive work from graphic artist	I	M, N, O	4
Marketing	J	—	1
Finance	K	—	1
Operations	L	—	1
Review text	M	—	1
Specify artwork needed	N	—	1
Decide on format	O	—	1

Arrow Gantt Charts

Most college texts covering project management treat PERT (program evaluation and review technique) charts separately from Gantt charts. However, with many new software packages, these differences are becoming inconsequential. In this chapter we demonstrate both methods. Since software packages vary, it is best that you understand both methods.

FIGURE 11-13 AON Network

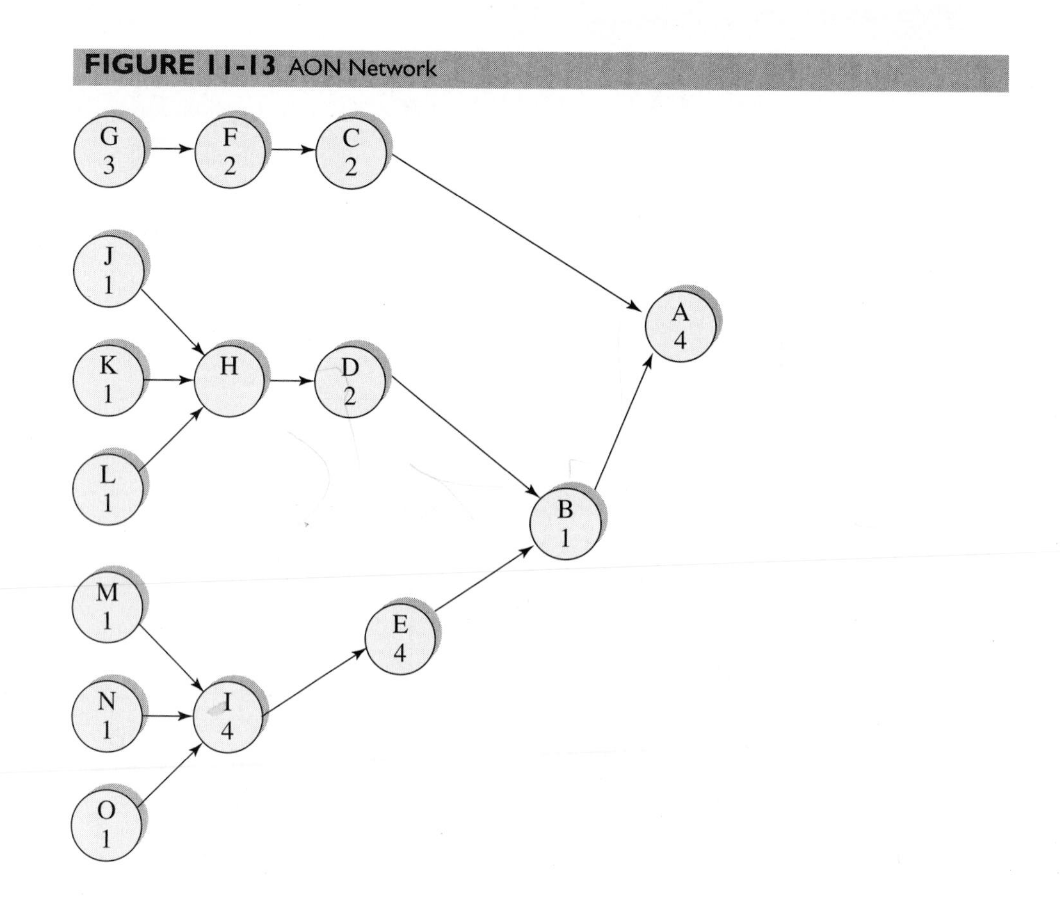

FIGURE 11-14 AON with Early Times

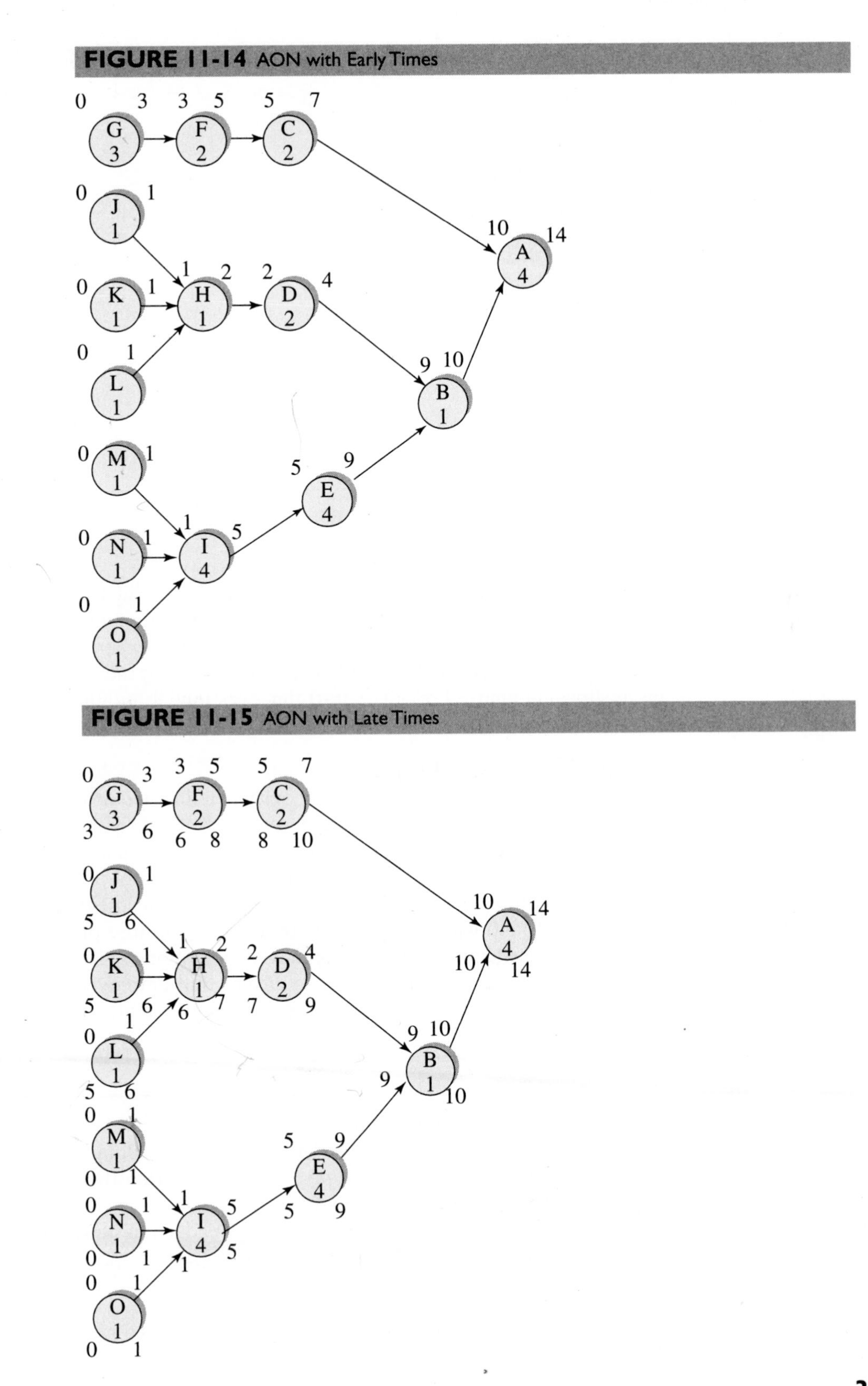

FIGURE 11-15 AON with Late Times

FIGURE 11-16 Project PERT Chart Showing Tasks Needed to Perform the Document Checklist Project

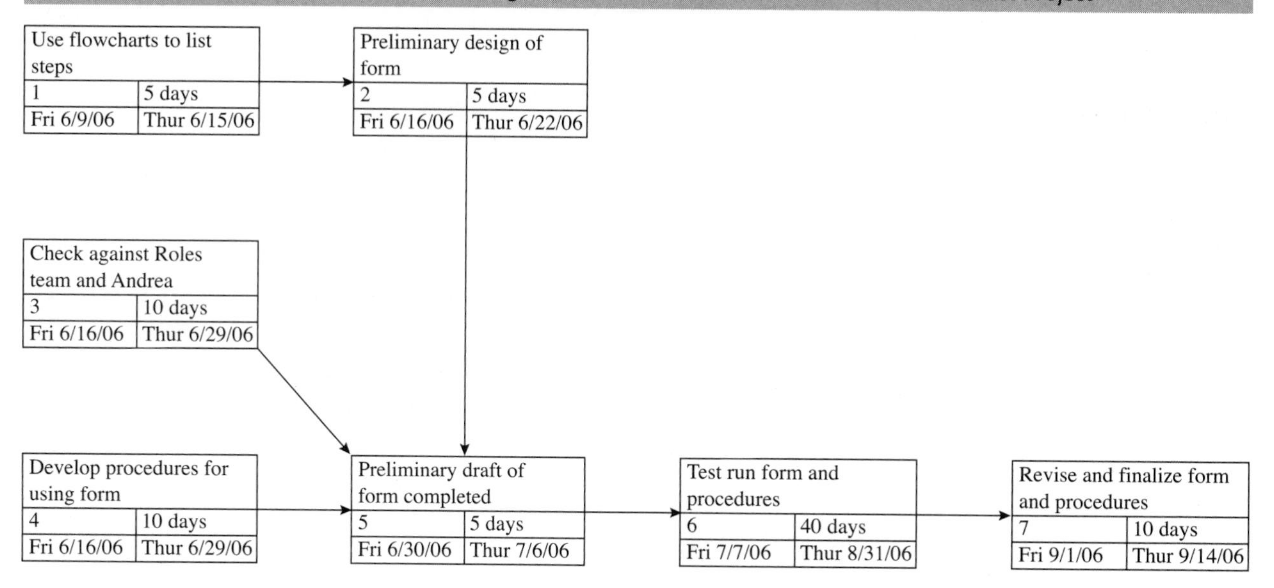

Continuing our example, the already-identified tasks have been entered in a planning software package called Microsoft Project. This is a commonly used and inexpensive project control software package that requires little expertise to run.

Using the data given in Table 11-5, the PERT chart for the document checklist project is as shown in Figure 11-16. The start time was June 9, and the expected completion date was September 7.

Managing Multiple Projects

At times, several projects occur simultaneously in a firm. Coordination can be difficult. Implementations have to be coordinated. Also, individuals within the company should not be involved in too many teams in order for them to remain effective. The multiple project control form[16] in Figure 11-17 is a management tool to aid company-wide coordination of multiple projects. By using this form, management can keep track of the multiple projects and which employees are involved in those projects. Notice that the form identifies participants with varying levels of responsibility as well as the project managers. Also, management's duty is to ensure that employees have ample time to work on projects so that their successful completion is facilitated.

SUMMARY

In this chapter we have explored the use of teams and collaboration as a means of improvement. We focused on the behavioral aspects of building and leading effective teams. As companies focus on knowledge work and flatten their hierarchies, we have seen a movement toward more teamwork.

Team leaders are responsible for providing leadership and training so that team members have the skills to perform required tasks. Leaders also need to be sensitive to

[16]Van Der Merwe, A. P., "Multi-project Management Organizational Structure and Control," *International Journal of Project Management* 15, 4 (1997):223–233.

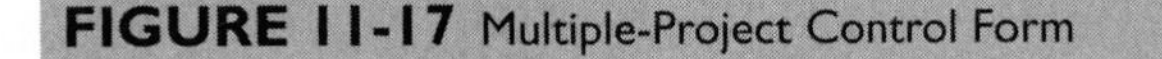

FIGURE 11-17 Multiple-Project Control Form

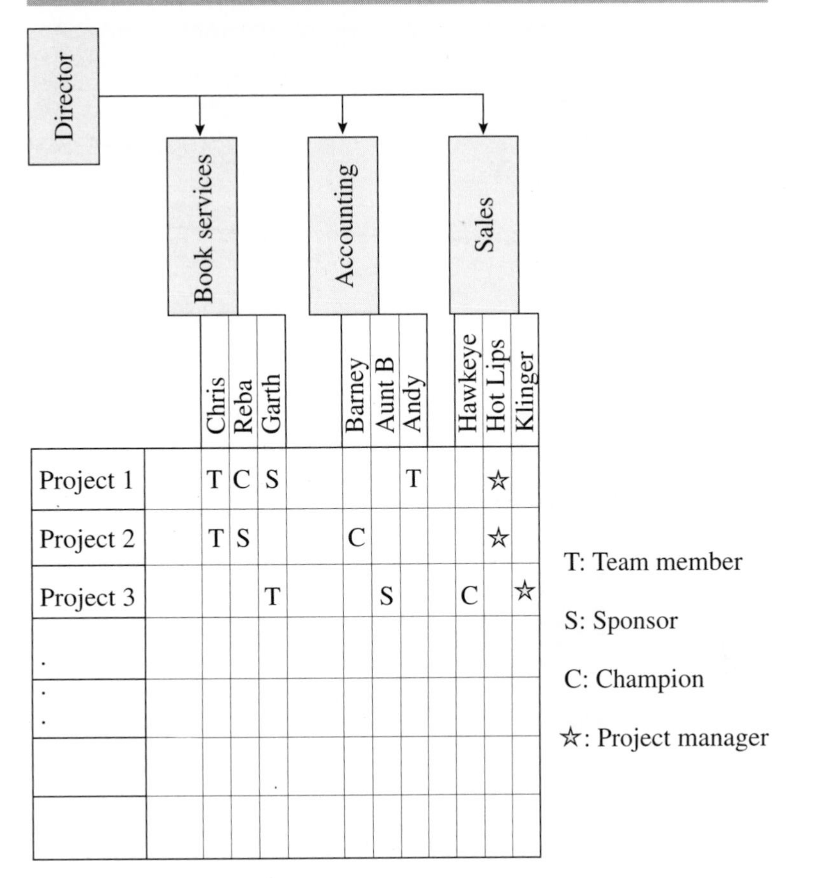

	Chris	Reba	Garth	Barney	Aunt B	Andy	Hawkeye	Hot Lips	Klinger
Project 1	T	C	S			T		☆	
Project 2	T	S		C				☆	
Project 3			T		S		C		☆

the amount of social support team members need to be effective. Part of this is to identify and assign team roles and to establish team norms and rules.

Once the teams are established, they evolve through the stages of forming, storming, norming, performing, and, with successful projects, mourning. Team members should be aware of these stages so that they can successfully complete each stage.

The type of team should be considered as teams are formed. Teams might be process improvement teams, cross-functional teams, tiger teams, natural work groups, self-directed teams, or virtual teams. The type of team chosen depends on the team objective.

Team leaders may use facilitators and meeting management techniques to help the team achieve success. When teams struggle, conflict resolution might be needed to overcome problems. Before problems threaten the success of a team, the process-intervention cycle provides a basis for recognizing and remediating problems in time.

Finally, we presented some project planning fundamentals that can be combined with the new seven tools discussed in Chapter 10. These planning tools, such as charters, work breakdown structures, models for determining task times, and PERT/CPM, are useful for controlling team projects for quality improvement. Note that many projects are controlled using software such as Primavera or MS Project. Figure 11-18 shows an example of a simple project in MS Project.

Teams can either be a drag on an organization's resources or they can provide the basis for improved competitiveness. There is some ambiguity in the quality literature as

FIGURE 11-18 MS Project Plan

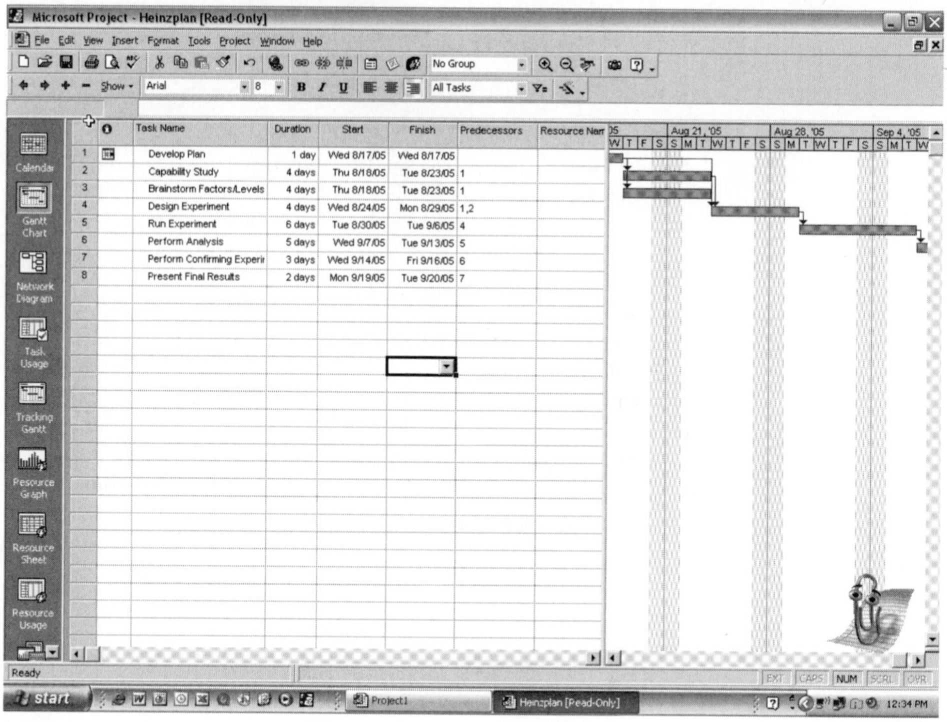

	Task Name	Duration	Start	Finish	Predecessors	Resource Names
1	Develop Plan	1 day	Wed 8/17/05	Wed 8/17/05		
2	Capability Study	4 days	Thu 8/18/05	Tue 8/23/05	1	
3	Brainstorm Factors/Levels	4 days	Thu 8/18/05	Tue 8/23/05	1	
4	Design Experiment	4 days	Wed 8/24/05	Mon 8/29/05	1,2	
5	Run Experiment	6 days	Tue 8/30/05	Tue 9/6/05	4	
6	Perform Analysis	5 days	Wed 9/7/05	Tue 9/13/05	5	
7	Perform Confirming Experi	3 days	Wed 9/14/05	Fri 9/16/05	6	
8	Present Final Results	2 days	Mon 9/19/05	Tue 9/20/05	7	

to whether managers should aggressively manage these teams. In truth, there is probably a balance that needs to be established between too tight or too loose control. Achieving that balance requires insight and skill. The concepts and tools given here will help you to achieve this balance. Remember that many firms cannot absorb too much change. Therefore, the rate of change needs to be managed and planned.

KEY TERMS

- Cost benefits analysis (CBA)
- Cross-functional teams
- Empowerment
- Facilitation
- Facilitator
- Force-field analysis
- Forming
- Hard costs
- Knowledge work
- Meeting management
- Mourning
- Natural work groups
- Norming
- Organizational learning
- Parking lot
- Payback period
- Performing
- Process improvement teams
- Project charters
- Readiness
- Self-directed work team
- Situational leadership model
- Soft costs
- Storming
- Team
- Team building
- Teamware
- Tiger team
- Virtual teams

DISCUSSION QUESTIONS

1. Why has increased complexity in the workplace resulted in an increased emphasis on employee participation and teamwork?

2. What is the difference between routine work and knowledge work? Provide examples of each type of work.
3. What is meant by employee empowerment? What is the relationship between employee empowerment and teamwork?
4. In what ways can employee empowerment facilitate and contribute to organizational learning?
5. What are the major disadvantages of multiple layers of management in a business organization? Is teamwork typically implemented in organizations that have multiple layers of management or organizations that have fairly flat hierarchies? Explain your answer.
6. Describe the principal attributes of Hersey and Blanchard's situational leadership model. Is this model of leadership appropriate for a quality-minded company?
7. What does readiness refer to in a leadership context? What are the two variables that an employee's level of readiness is a function of?
8. In general, do employees enjoy working in teams? Why or why not?
9. Briefly describe the five stages of the life cycle of a team. Is it important for a team to pass through each of these stages? Why or why not?
10. What is the purpose of a process improvement team? Provide an example of a process improvement team for a business organization.
11. What is meant by a self-directed work team? In your judgment, are self-directed work teams a good idea? Explain your answer.
12. Describe the role of the facilitator of a team. Is the role of the facilitator important, or is it primarily ceremonial in nature?
13. What is meant by team building? Is team building a concept from which all teams can benefit?
14. In a meeting management context, to what does *parking lot* refer?
15. What are the primary sources of conflict in work teams? What are some of the methods for resolving team conflict?
16. Describe what is meant by virtual teams. Provide an example of a virtual team in a business organization.
17. How does a project charter help a team identify issues relevant to team success?
18. Describe the critical path method for organizing work projects. When is the use of this method appropriate?
19. Describe the difference between critical tasks and a critical path.
20. Describe an experience you have had working in a team (in a work setting or in a college class). Describe how your experience either confirms or refutes one of the principles of teamwork described in this chapter.

PROBLEMS

1. Perform a cost benefit analysis with payback period for a project where indirect costs are $50,000, direct costs are $25,000, and annualized benefits are $60,000.
2. Perform a cost benefit analysis using the following data:

Direct Costs

- 10 laptops—$2,000/each
- A server—$2,000
- Network installation—$15,000
- Software—$20,000

Indirect Costs

- Training—$15,000

Annualized Benefits

- Increased capacity—$100,000

3. Following are activities, precedence relationships, and task times for a number of tasks.

Task	Task Time (days)	Preceding Task
A	5	None
B	2	A
C	4	A
D	5	B
E	6	C
F	7	D
G	4	E
H	5	D
I	1	C
J	4	I, F
K	3	G, J
L	7	H, K
M	4	L

a. Construct a precedence diagram.
b. Compute early and late times.
c. Find the critical path.
d. Which tasks could the project manager delegate?
e. Compute the slack for each task.
f. Which tasks would you shorten first to shorten the project completion time?
g. What is the expected project completion time?
h. What is the probability of completing the project in the time identified in G?

4. Following are activities, precedence relationships, and task times for a number of tasks.

Task	Task Time (days)	Preceding Task
A	70	None
B	25	A, C
C	89	None
D	50	A
E	46	B, D
F	72	C
G	46	E
H	25	G
I	30	F
J	14	I
K	36	H, J
L	77	J

a. Construct a precedence diagram.
b. Compute early and late times.
c. Find the critical path.
d. Which tasks could the project manager delegate?
e. Compute the slack for each task.
f. Which tasks would you shorten first to shorten the project completion time?
g. What is the expected project completion time?

5. For the following data:

Task	Predecessor	Time Estimates (Days)		
		t_o	t_m	t_p
A	—	3	5	13
B	A	2	5	8
C	A	1	4	6
D	A	4	6	10
E	B	2	8	11
F	B	5	9	16
G	C	4	12	20
H	C	6	9	13
I	D	3	7	14
J	D	8	14	22
K	F, G	9	12	20
L	H, I	6	11	15
M	E	4	7	12
N	J	3	8	16
O	M, K, L, N	5	8	10

a. Draw an AON diagram.
b. Compute the expected completion time for each task.
c. Identify the critical path and the expected completion time.
d. Compute slack for all tasks.
e. Your project manager wishes for you to compute a completion time (in days) that gives you a 95% chance of success. He will use this time estimate in negotiating a completion date for the project.

6. For the following data:

Task	Predecessor	Time Estimates (Days)		
		t_o	t_m	t_p
A	—	1	2	9
B	—	2	5	8
C	—	1	3	5
D	A	4	10	25
E	A	3	7	12
F	B	10	15	25
G	C	5	9	14
H	D, E	2	3	7
I	D, E, F	1	4	6
J	D, E, F, G	2	5	10
K	H, I, J	3	3	3

a. Draw a PERT diagram.
b. Compute expected completion times for all tasks.
c. Find the critical path.
d. What is the completion time that gives you a 75% chance of success?

7. (Team project) Develop a project plan for buying a house. Use the following steps:
   a. Use a work breakdown structure (tree diagram—Chapter 10) to identify tasks for completing the project. Be complete (at least 50 tasks), and use sticky notes.
   b. Laying out your sticky notes, identify the precedence relationships for each task.
   c. Brainstorm optimistic time, most likely time, and pessimistic time for each task.
   d. Compute expected completion time for each task.
   e. Draw your network diagram, compute early and late times, and find the critical path.
   f. Compute the 90% completion time for your project.
8. (Team project) Complete the steps in Problem 7 for the following project: completing a university degree program (choose any major).
9. (Team project) Develop a Gantt chart for welcoming a new baby into the world (conception has already occurred). Remember you only have 9 months to complete the project.
10. Complete a cost benefit analysis for a college degree at your university. Be sure to include all direct and indirect costs. Take into account the time value of money for future income.
11. Complete a cost benefit analysis for a marriage. Be sure to include both the benefits and costs in your model. Model the marriage given two scenarios—a strong marriage and a poor marriage. (Hint: You should find that the benefits for a strong marriage approach infinity and the benefits of a poor marriage approach zero.)
12. Meet with a favorite professor to perform a force-field analysis relating to adding a new course to your major in your university. Would you recommend this project?

Cases

Case 11-1 Whole Foods: Using Teamwork as a Recipe for Success

Whole Foods Inc.: ***www.wholefoods.com***

Whole Foods, Inc., is the nation's number one chain of natural foods supermarkets, operating more than 140 stores under the names of Whole Foods Market, Bread & Circus, Bread of Life, Fresh Fields, Merchant of Vino, and Wellspring Grocery. The stores are much different from the small "health food" stores that sprang up in the United States during the 1980s and 1990s. They are complete supermarkets with an emphasis on organically grown produce, fresh-baked bread, wholesome deli foods, and other health food products. Conspicuously absent at Whole Foods stores are soft drinks in plastic containers, coupon dispensers for laundry detergent, salted potato chips, sugared cereals, and other high-sugar or high-fat products.

Now you know what the customer sees—a company that is passionate about health food and the people who buy health food products. But there is more to the Whole Foods story, which is the part that the customer doesn't see. In the midst of the aging supermarket industry, Whole Foods has created a new approach to managing its employees—an approach based on teamwork and employee empowerment. Here is how it works.

Each of Whole Foods' 143 stores is an autonomous profit center composed of an average of 10 self-managed teams. A separate team operates each of the departments of the store such as produce, canned goods, the bakery, and so on. Each team has a team leader and specific team goals. The teams function as autonomous units and meet monthly to share information, exchange stories, solve problems, and talk about how to improve performance. The team concept is present throughout the organization. The team leaders in each store are a team, store leaders in each geographic region are a team, and the leaders of each of the company's seven regions are a team.

Why teams? There are two primary benefits that Whole Foods believes results from its emphasis on teamwork. First is to promote cooperation among the firm's employees. The teamwork approach facilitates a strong sense of community, which engenders pride and discipline in the work ethic of the employees. An example of this is Whole Foods' hiring practices. The teams, rather than the store managers, have the power to approve new hires for full-time jobs. The store leaders do the initial screening, but it takes a two-thirds vote of the team, after what is usually a 30-day trial period, for the candidate to become a full-time employee. This type of exclusivity helps a team bond, which facilitates a cooperative atmosphere. Another example of how teamwork promotes cooperation among employees is evident in Whole Foods' team meetings. Each team holds a team meeting at least once a month. There is no rank at the team meetings. Everyone is afforded an equal opportunity to contribute to the discussion.

The second benefit that Whole Foods realizes from its emphasis on teamwork is an increased competitive spirit among its employees. The individual teams, stores, and regions of the company compete against each other in terms of quality, service, and profitability. The results of the competitions determine employee bonuses, recognition, and promotions. To facilitate competition, the company is extraordinarily open in terms of team performance measures. For example, at a Bread & Circus store in Wellesley, Massachusetts, a sheet posted next to the time clock lists the previous day's sales broken down by team. A separate sheet lists the sales numbers for the same day the previous year. This information is used by the teams to determine "what it will take" to be the top team for the store during a particular week. This type of competition also exists at the store level. Near the same time clock, once a week a fax is posted that lists the sales of every store in the Northeast region broken down by team with comparisons to the same week the previous year. There is one note of caution that Whole Foods has learned through these experiences. Sometimes competition between teams can become too intensive. As a result, the company has had to "tone down" the intensity of the competition between teams and stores on occasion.

The overall results of Whole Foods' management practices have been encouraging. The company has grown from one store in 1980 to 143 stores today. The grocery store industry is intensely competitive and Whole Foods' decision to use teamwork as a "recipe for success" represents a novel and innovative approach to management. ■

DISCUSSION QUESTIONS

1. Do you believe that Whole Foods' emphasis on teamwork is appropriate for the grocery store industry? Why or why not?
2. What is your opinion of Whole Foods' practice of sharing team performance data with all company employees? Do you believe that this practice risks creating "too competitive" a spirit among the firm's teams and employees? Explain your answer.
3. Would you enjoy working on a team at Whole Foods? Why or why not?

Case 11-2 The Boeing 777: Designing and Building an Airplane Using Virtual Teams

Boeing: ***www.boeing.com***

In the early 1990s the worldwide commercial aircraft industry was undergoing significant changes. The demand for large jumbo jets like the Boeing 747 was receding, and the airline industry was asking the airplane manufacturers for smaller jets that were more efficient and could fly longer routes without refueling. The airline industry also wanted the new jets quickly and wanted them in service in record time.

Boeing's answer to these challenges was the Boeing 777. At the outset, the 777 project was designed with two objectives in mind. First, to produce the most reliable and efficient two-engine aircraft ever made. Second, to design and build the airplane quickly and efficiently. To achieve these simultaneous objectives. Boeing knew that it had to set aside its traditional design process. There simply was not enough time for the design of the plane to move through each functional area in the firm in a sequential manner. Instead, Boeing designed the 777 largely by using virtual teams. Virtual teams are groups of geographically and/or organizationally dispersed coworkers who use information technology (such as the Internet), intranet videoconferencing, and other products to communicate and work with one another. The advantage of virtual teams is that they are flexible, and employees can work with each other, regardless of physical location, time zone, or organizational affiliation. Virtual teams are also fairly cost efficient, because teammates do not meet face to face and can avoid the downtime associated with traditional meetings and travel.

In the spirit of this concept, to design the 777, Boeing created 238 virtual design teams to work on separate components of the airplane simultaneously. The design teams consisted of Boeing employees and, in some cases, customers and suppliers. To facilitate this process, Boeing created a sophisticated and robust computer network. As a result of the computer power available, the teams were able to produce "paperless" prototypes of the components of the airplane, permitting the engineers to simulate the assembly of the 777 on computer, rather than building actual mockups. The system worked so well that only a nose mockup (to check critical wiring) was built before assembly of the first plane. The design teams also worked in a true "virtual" fashion. The 238 teams used a network of 1,700 individual computer systems, which had links to Japan, Boeing facilities in Philadelphia, Wichita, and Seattle, and other Boeing and non-Boeing locations. The teams were not limited to managers and engineers. Boeing included a wide array of people in the design process, from customers and operators, down to line mechanics. To facilitate the teamwork aspect of the process, Boeing also greatly reduced the secrecy of the design of the 777. Within reasonable limits, the teams were able to share information with each other and work toward the common goal of designing a high-quality airplane in an expedient and cost-efficient manner.

In the end, the 777 project was highly successful. The aircraft was developed in just more than two years, which is a record in the commercial aircraft industry. The Boeing 777 teams also achieved impressive results in terms of quality and efficiency. In comparison to the design of other Boeing aircraft, the 777 teams reduced changes in design, errors, and rework by 50%. The 777 family of planes fly up to 550 passengers for a range that exceeds the range of the Boeing 747. In addition to its favorable passenger capacity and fuel efficiency attributes, the 777 also was designed with pilot and group crew efficiencies in mind.

As a tribute to the members of the 238 virtual teams that designed the 777, Boeing named the first 777 placed in service *Working Together*, a name that appeared prominently on the fuselage of the plane. The high-profile nature of the 777 project also attracted the attention of other companies benchmarking against Boeing's virtual teams concept. For example, one high-profile project emulating the design process of the 777 is the International Space Station.

The use of virtual teams is an exciting development for increasing the quality, speed, and efficiency of the design of products and services. For Boeing, the result was not simply the acquisition of a new plane, it was the acquisition of a new way of building planes—a new way of building planes that Boeing is confident will become the industry standard. ■

DISCUSSION QUESTIONS

1. In your opinion, did the use of virtual teams play a key role in the successful development of the Boeing 777? Explain your answer.
2. Why do you believe that virtual teams speed up the product design process?
3. Other than the design of an airplane, what other types of design projects might benefit from using a virtual team concept? Why?

CHAPTER 12

Statistically Based Quality Improvement for Variables

> Data are required to obtain the average dimensions and the degree of dispersion (in a process) so that we can determine . . . whether the production process used for manufacturing the lot was suitable, or if some action must be taken. In other words, action can be taken on a process on the basis of data gained from the samples.
>
> —KAORU ISHIKAWA

As we view the world about us, statistics are everywhere. We hear statistics about politics, health, inflation, or the economy on the radio or TV on a daily basis. Yet many people view the topic of statistics with fear, loathing, and trembling. The purpose of statistics is clear. Statistics are a group of tools that allow us to analyze data, make summaries, draw inferences, and generalize from data.

Statistics are very important in the field of quality. In fact, during the first half century of the quality movement, nearly all the work done in the field of quality related to statistics. This work resulted in a body of tools that are used worldwide in thousands of organizations.

This chapter focuses on the use of statistical tools, not as control mechanisms, but as the foundation for continual improvement. We present many statistical techniques and different types of control charts. These tools represent powerful techniques for monitoring and improving processes. We also discuss the behavioral aspects of statistical process improvement. It is important to recognize that it is not enough to learn the different charts and statistical techniques. We also must know how to apply these techniques in a way that will document and motivate continual improvement in organizations.

These techniques can be enjoyable to use, and we present them in a way that is intuitive and easy to understand. Where possible, we develop shortcuts and simple statistical techniques instead of more complex models. The primary goal is that these tools be used.

STATISTICAL FUNDAMENTALS

What Is Statistical Thinking?

Statistical thinking is a decision-making skill demonstrated by the ability to draw conclusions based on data. We make a lot of decisions based on intuition and gut feelings. Often we choose friends, homes, and even spouses based on feelings. Therefore, intuition and feelings are very important in making good decisions in certain circumstances.

However, intuitive decisions are sometimes biased and wrong-headed. Consider the case of government. Many times it is the most vocal groups that seem to control political agendas. It is difficult for mayors, governors, or presidents to determine exactly what the voting public wants on any issue. As a result, decisions are sometimes made that satisfy the few but irritate the many. Statistical thinking is based on these three concepts:

- All work occurs in a system of interconnected processes.
- All processes have variation (the amount of variation tends to be underestimated).
- Understanding variation and reducing variation are important keys to success.

In business, decisions need to be made based on data. If you want to know how to satisfy your customers, you need to gather data about the customers to understand their preferences. It is one thing to watch a production process humming along. It is a completely different thing to gather data about the process and make adjustments to the process based on data. Statistical thinking guides us to make decisions based on the analysis of data (see Quality Highlight 12-1).

Why Do Statistics Sometimes Fail in the Workplace?

Before beginning a discussion of statistical quality improvement, we must remember that many times statistical tools do not achieve the desired results. Why is this so? Many firms fail to implement quality control in a substantive way. That is, they prefer form over substance. We provide several reasons as a guide. You can use this guide to assess whether your organization will be successful in using statistics to improve. Reasons for failure of statistical tools include the following:

Video Clip: Behavioral Aspects of SPC

- Lack of knowledge about the tools; therefore, tools are misapplied.
- General disdain for all things mathematical creates a natural barrier to the use of statistics. When was the last time you heard someone proclaim a love for statistics?
- Cultural barriers in a company make the use of statistics for continual improvement difficult.
- Statistical specialists have trouble communicating with managerial generalists.
- Statistics generally are poorly taught, emphasizing mathematical development rather than application.
- People have a poor understanding of the scientific method.
- Organizations lack patience in collecting data. All decisions have to be made "yesterday."
- Statistics are viewed as something to buttress an already-held opinion rather than a method for informing and improving decision making.
- People fear using statistics because they fear they may violate critical statistical assumptions. Time-ordered data are messy and require advanced statistical techniques to be used effectively.
- Most people don't understand random variation, resulting in too much process tampering.
- Statistical tools often are reactive and focus on *effects* rather than *causes*.
- Another reason people make mistakes with statistics is founded in the notions of type I and type II errors. In the study of quality, we call type I error **producer's risk** and type II error **consumer's risk**. In this context, producer's risk is the probability that a good product will be rejected. Consumer's risk is the probability that a

QUALITY HIGHLIGHT 12-1

STATISTICAL TOOLS IN ACTION

Statistical tools have long been staples of the quality manager. Around the world, many firms have adopted statistical tools with good results. One of the appealing features of statistical tools is that they can be adapted and used in a wide variety of situations. For example, Ore-Ida Corporation, a subsidiary of Heinz Corporation and a nationally known producer of consumer food products, uses statistics to ensure that its food meets weight and measure requirements. One of the products that Ore-Ida produces is called a Pita Pocket Sandwich. The problem with the Pita sandwich was that if the sandwiches were too large, they would not fit into the formed plastic Pita-holding package, and excess costs would be incurred. If the sandwiches were too small, customers would perceive them as having less value. As a result, Ore-Ida used statistical process control, experimental design, and process capability studies to ensure that the sandwiches met requirements.

Statistical process control is not always immediately successful. Simplot Corporation, a competitor of Ore-Ida's, attempted statistical process control and other tools of quality in its Caldwell, Idaho, facility during the 1980s. According to Bob Romero, manager of total quality management services, the company had to do an educational assessment of its employees, which resulted in a picture that was anything but flattering. Many of Simplot's employees had marginal literacy skills. As a result, the company undertook a lengthy program of training and education in literacy, after which new standards were created for employees that included overall standards for literacy and the ability to use word-processing software and spreadsheets. After completing this program, management again implemented statistical process control and other quality management tools. This time they were successful in improving processes and reducing costs.

Jaco Manufacturing Company, a producer of industrial components, tube fittings, and injection-molding machines, implemented statistical process control, process capability studies, and quality management tools as a means for improving customer service. G. K. Products, Inc., of Ann Arbor, Michigan—a Jaco customer—asked Jaco to reduce its costs by improving its inspection of plastic float bodies. These float bodies are used in gas tanks so that in the case of rollover the flow of fuel to the car's engine will shut off. Benefits that were achieved through this program included a 14% reduction in cycle time, a decrease in scrap, thousands of dollars in cost savings, improved morale, and improved customer satisfaction.

As we learn from these examples and countless others, companies around the world either now have implemented statistical quality tools or are in the process of adopting these tools. All processes exhibit variability. This fact alone makes statistical process tools invaluable to manufacturing and services companies alike.

nonconforming product will be available for sale. Consumer's risk happens when statistical quality analysis fails to result in the scrapping or reworking of a defective product. When either type I or type II errors occur, erroneous decisions are made relative to products. These erroneous decisions can result in high costs or lost future sales. Given these problems, we adopt the approach that statistics should be used, they should be used correctly, and they should be taught correctly.

What Do We Mean by the Term *Statistical Quality Control?*

The age of control-oriented management is over. We're not sure when control ended, but we think it was sometime in the early 1980s in the United States. Now the focus is on continual improvement—not on process or organizational control. In its pure form, the term *control* is fairly benign, implying receiving feedback for improvement. However, over time, control came to mean centralized quality control functions, centralization of authority, and lack of empowerment. As a result, we prefer not to use the term *quality control* here. This is a departure from historical quality thought. We use the term *control* sparsely in this chapter, such as in *control limits*. It is difficult to completely eliminate a term that has been around for such a long time.

Understanding Process Variation

All processes exhibit variation. There is some variation that we can control and other variation that we cannot control. If there is too much variation, process parts will not fit correctly, products will not function properly, and a firm will gain a reputation for poor quality.

Two types of process variation commonly occur. These are random and nonrandom variation. Random variation is uncontrollable, and nonrandom variation has a cause that can be identified. The statistical tools discussed later in this chapter are useful for determining whether variation is random.

Random variation is centered around a mean and occurs with a consistent amount of dispersion. This type of variation cannot be controlled. Hence we refer to it as "uncontrolled variation." The amount of random variation in a process may be either large or small. When the variation is large, processes may not meet specifications on a consistent basis.

The statistical tools discussed in this chapter are *not* designed to detect random variation. Figure 12-1 shows normal distributions resulting from a variety of samples taken from the same population over time. We find a consistency in the amount of dispersion and the mean of the process. The fact that not all observations within the distributions fall exactly on the target line shows that there is variation. However, the consistency of the variation shows that only random causes of variation are present within the process. This means that in the future when we gather samples from the process, we can expect that the distributions associated with such samples also will take the same form.

FIGURE 12-1 Random Variation

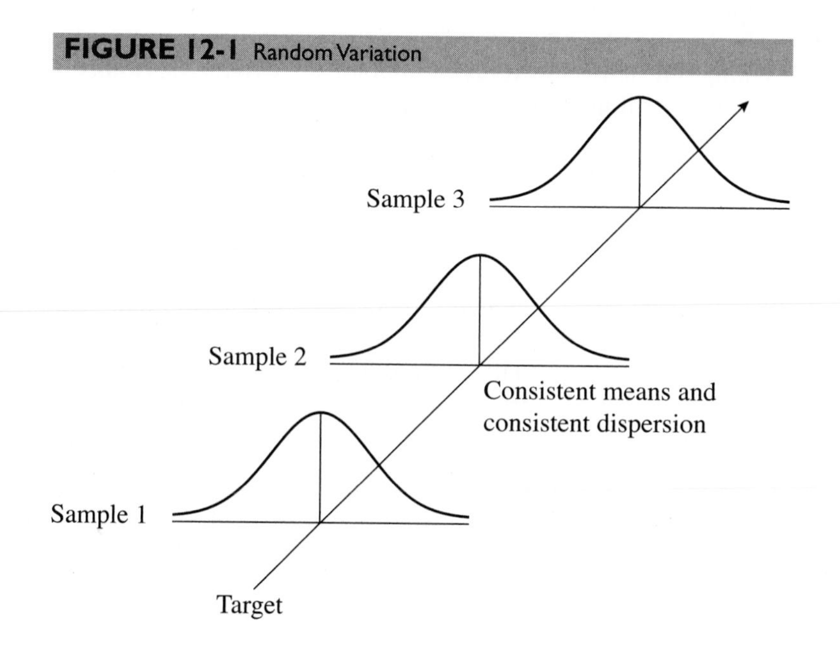

Nonrandom (or "special cause") **variation** results from some event. The event may be a shift in a process mean or some unexpected occurrence. For example, we might receive flawed materials from a supplier. There might be a change in work shift. Joe might come to work under the influence of drugs and make errors. The machine may break or not function properly. Figure 12-2 shows distributions resulting from a number of samples taken from the same population over time where nonrandom variation is exhibited. Notice that from one sample to the next, the dispersion and average of the process are changing. When we compare this figure to random variation, it is clear that nonrandom variation results in a process that is not repeatable.

Process Stability

Process **stability** means that the variation we observe in the process is random variation (common cause) and not nonrandom variation (special or assignable causes). To determine process stability, we use process charts. **Process charts** are graphs designed to signal process workers when nonrandom variation is occurring in a process.

Sampling Methods

To ensure that processes are stable, data are gathered in samples. Process control requires that data be gathered in samples. For the most part, sampling methods have been preferred to the alternative of 100% inspection. The reasons for sampling are well established. Samples are cheaper, take less time, are less intrusive, and allow the user to frame the sample. In cases where quality testing is destructive, 100% inspection would be impossible and would literally drive the company out of business. In some processes, chemicals are used in testing or pull tests are applied to cables. These destructive tests ruin the sample but are useful to show that a good product is being made.

However, recent experience has shown that 100% inspection can be effective in certain instances. One hundred percent samples are also known as *screening samples, sorting samples, rectifying samples, or detailing samples*. They have been most common

FIGURE 12-2 Nonrandom Variation

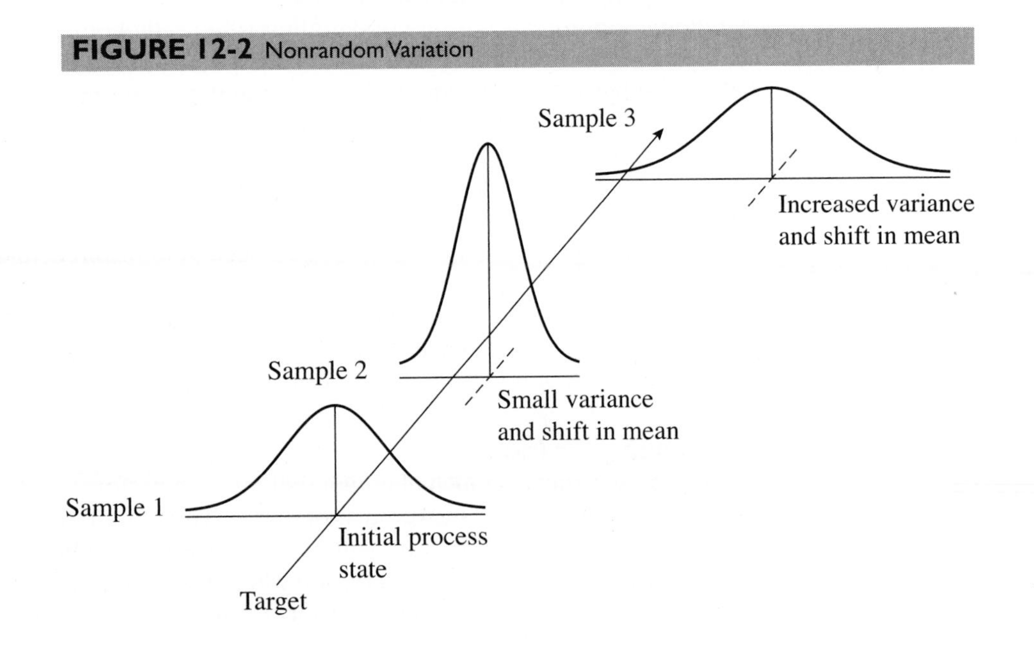

in acceptance sampling (Chapter 9) where a lot of material has been rejected in the past and materials must be sorted to keep good materials and return defective materials for a refund.

Another example of 100% inspection is used when performing in-process inspection. Many companies have asked their employees to inspect their own work as the work is being performed. This can result in 100% inspection at every stage of the process! We should clarify that in-process inspection also can be performed on a sampling basis. Because sampling is so important, let's look at some different types of samples.

Random Samples

Randomization is useful because it ensures independence among observations. To *randomize* means to sample in such a way that every piece of product has an equal chance of being selected for inspection. This means that if 1,000 products are produced in a single day, each product has a 1/1,000 chance of being selected for inspection on that day. Random samples are often the preferred form of sampling and yet often the most difficult to achieve. This is especially true in process industries where multiple products are made by the same machines, workers, and processes in sequence. In this case, there is not independence among observations because the process results in ordered products that can be subject to machine drift (going out of adjustment slowly over time).

Systematic Samples

Systematic samples have some of the benefits of random samples without the difficulty of randomizing. Samples can be systematic according to *time* or according to *sequence*. If a sample is systematic according to time, a product is inspected at regular intervals of time, say, every 15 minutes. If a systematic sample is performed according to sequence, one product is inspected every tenth iteration. For example, every tenth product coming off the line is sampled.

Sampling by Rational Subgroups

A *rational subgroup* is a group of data that is logically homogeneous; variation within these data can provide a yardstick for computing limits on the standard variation between subgroups. For example, in a hospital it may not make sense to combine measurements such as body temperature or medication levels taken in the morning with measurements taken in the evening. Morning measurements occur before medications are given and before the first meal of the day and constitute a rational subgroup. Evening measurements, another such subgroup, are taken after treatment has been provided during the day, medications have been administered, and patients have been nourished. If data are gathered that combine these two subgroups (e.g., the night and day measurements), then differences between morning results and evening results will not be detected. If variation among different subgroups is not accounted for, then an unwanted source of nonrandom variation is being introduced.

Planning for Inspection

As you can see, there is much planning that must be performed in developing sampling plans. Questions must be answered about what type of sampling plan will be used, who will perform the inspection, who will use in-process inspection, sample size, what the critical attributes to be inspected are, and where inspection should be performed. There are rules for inspection that help to prioritize where inspection should be performed.

Many firms compute the *ratio between the cost of inspection and the cost of failure* resulting from a particular step in the process, in order to prioritize where inspection should occur first.

PROCESS CONTROL CHARTS

Now that we have learned about variation, it is time to learn about the tools used to understand random and nonrandom variation. Statistical process control charts are tools for monitoring process variation. Figure 12-3 shows a control chart. It has an upper limit, a center line, and a lower limit. Several different types of control charts will be discussed later. In this chapter we introduce different statistical charts one by one. However, you should know that there is a generalized process for implementing all types of process charts that we introduce first. This is a useful approach to learning control charts because the process for establishing different types of control charts is the same. Although the process for establishing different control charts is the same, the formulas used to compute the upper limit, center line, and lower limit are different. You will learn the formulas later.

Variables and Attributes Control Charts

To select the proper process chart, we must differentiate between variables and attributes. As we already stated, a **variable** is a continuous measurement such as weight, height, or volume. An **attribute** is an either-or situation. Here are examples of attributes: The motor is either starting or not starting, or either the inspector is bald or he is not. We will discuss measurements in this chapter and attributes in Chapter 13. While discussing attributes, we will also introduce reliability theory.

Table 12-1 shows the most common types of variable and attribute charts. The variables charts are X, x-bar ($\bar{x}$), R, MR, and s charts. The attributes charts are p, np, c, and u charts. In the following pages we introduce and develop these basic charts. We begin by introducing the generic process for developing all charts and then discuss the charts individually. There are four central requirements for properly using process charts:

1. You must understand this generic process for implementing process charts.
2. You must know how to interpret process charts.
3. You need to know when different process charts are used.
4. You need to know how to compute limits for the different type of process charts.

We treat each of these topics separately.

FIGURE 12-3 Control Chart

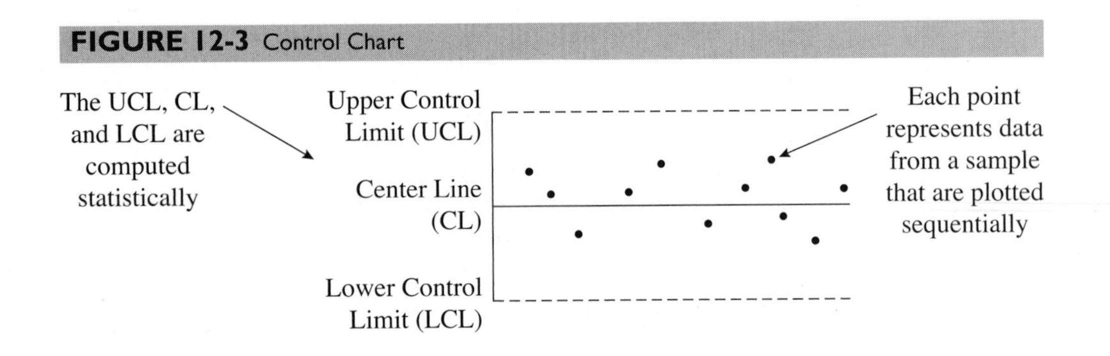

TABLE 12-1 Variables and Attributes

Variables	*Attributes*
X (process population average)	p (proportion defective)
$\bar{x}$ (mean or average)	np (number defective or number nonconforming)
R (range)	c (number nonconforming in a consistent sample space)
MR (moving range)	
s (standard deviation)	u (number defects per unit)

A Generalized Procedure for Developing Process Charts

The process for developing a process chart is the same for almost all charts. The only differences are in the actual statistical computations. Following are steps used in developing process control charts:

1. Identify *critical operations* in the process where inspection might be needed. These are operations in which, if the operation is performed improperly, the product will be negatively affected.
2. Identify *critical product characteristics*. These are the aspects of the product that will result in either good or poor functioning of the product.
3. Determine whether the critical product characteristic is a variable or an attribute.
4. Select the appropriate *process control chart* from among the many types of control charts. This decision process and the types of charts available are discussed later.
5. Establish the *control limits* and use the chart to *continually monitor and improve*.
6. *Update the limits* when changes have been made to the process.

Understanding Control Charts

Before showing how to establish control charts, we need to understand what control charts are and how they work. We use the chart to illustrate the fact that the process chart is nothing more than an application of hypothesis testing where the null hypothesis is that the process is stable. An $\bar{x}$ chart is a variables chart that monitors average measurements. For example, suppose that you were a producer of $8\frac{1}{2}$-inch × 11-inch notebook paper. Because the length of paper is measured in inches, a variables chart such as the $\bar{x}$ chart is appropriate. If the length of the paper is a key critical characteristic, we might inspect a sample of sheets to see whether the sheets are indeed 11 inches long.

To demonstrate how a control chart works, we could use a hypothesis test instead of a control chart to determine whether the paper is really 11 inches long. Therefore, the null hypothesis is

$$H_0: \mu = 11 \text{ inches}$$

The alternative hypothesis is

$$H_1: \mu \neq 11 \text{ inches}$$

To perform the hypothesis test, we establish a distribution with the following 95% ($Z = 1.96$) rejection limits. If the standard error of the sample distribution ($n = 10$) is .001 inch, the rejection limits are $11 \pm 1.96(.001) = \{11.00196, 10.99804\}$. Figure 12-4 shows the distribution with its rejection regions.

FIGURE 12-4 Hypothesis Testing

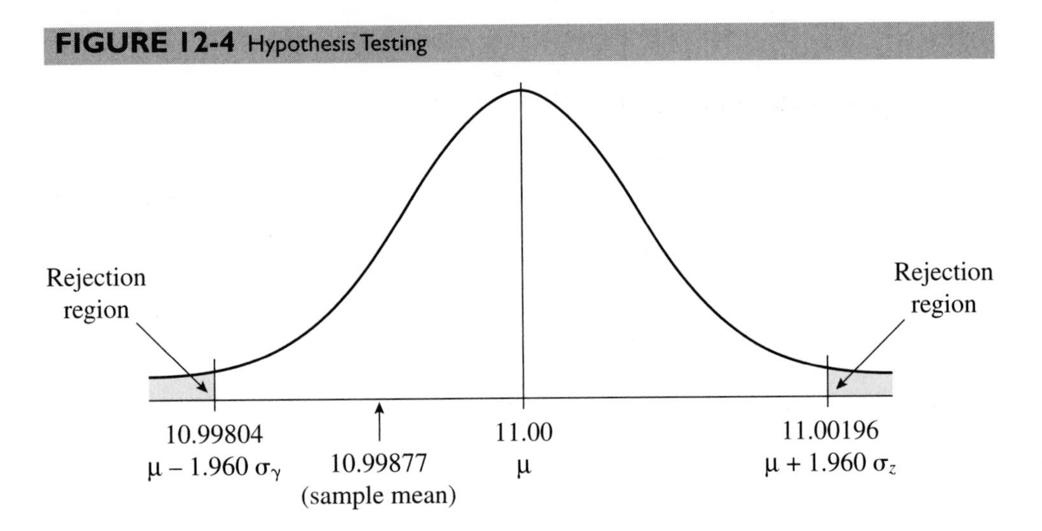

Next, to test this hypothesis, we draw a sample of $n = 10$ sheets of paper and measure the sheets. The measurements are shown in the following table:

Sheet Number	Measurement
1	11.0001
2	10.9999
3	10.9998
4	11.0002
5	11.0004
6	11.0020
7	10.9980
8	10.9999
9	10.9870
10	11.0004
Sum	109.9877
Sample mean	10.99877

Because 10.99877 does not fall within either of the rejection regions shown in Figure 12-4, we fail to reject the null hypothesis and conclude that the sheets do not differ significantly from an average of 11 inches.

This is a basic hypothesis test. Now we use a process control chart to monitor this paper production process. With process charts, we place the distribution on its side, as shown in Figure 12-5. We draw a center line and upper and lower rejection lines, which we call *control limits*. We then plot the sample average (10.99877) on the control chart. Because the point falls between the control limits, we conclude that the process is in control. This means that the variation in the process is random (common).

Notice that the preceding example was based on a sample of $n = 10$. This means that the distribution we drew was a sampling distribution (not a population distribution). Therefore, we can invoke the *central limit theorem*. The central limit theorem states that when we plot the sample means, the sampling distribution approximates a normal distribution.

FIGURE 12-5 Process Control Chart

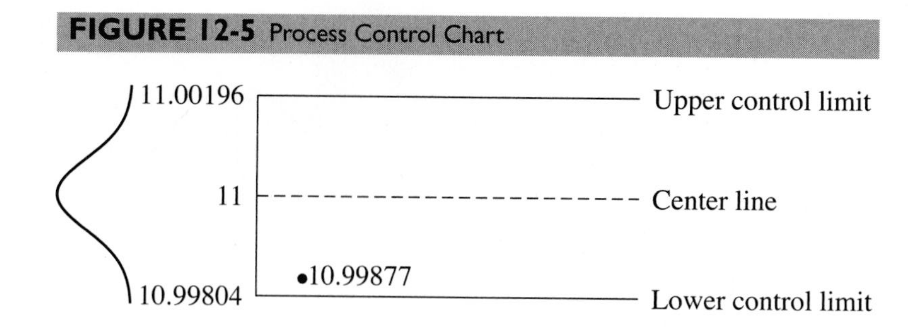

$\bar{x}$ and *R* Charts

Now that we have developed a process chart, we need to understand the different types of charts. First we discuss two charts that go hand in hand—$\bar{x}$ and R charts. We developed an $\bar{x}$ chart in the prior example. When we are interested in monitoring a measurement

FIGURE 12-6 $\bar{x}$ and R Charts

Variables Control Chart ($\bar{x}$ and R)

Part number	Chart number

Part name (product)	Operation (process)	Specification limits

Operator	machine	Gauge	Unit of measure	Zero equals

Date																									
Time																									
Sample measurements 1																									
2																									
3																									
4																									
5																									
Sum																									
Average, $\bar{x}$																									
Range, R																									
Notes																									
	1	2	3	4	5	6	7	8	9	10	11	12	13	14	15	16	17	18	19	20	21	22	23	24	25
Averages																									
Ranges																									

for a particular product in a process, there are two primary variables of interest: the mean of the process and the dispersion of the process. The $\bar{x}$ chart aids us in monitoring the process mean or average. The R chart is used in monitoring process dispersion.

The **$\bar{x}$ chart** is a process chart used to monitor the average of the characteristic being measured. To set up an $\bar{x}$ chart, select samples from the process for the characteristic being measured. Then form the samples into rational subgroups. Next, find the average value of each sample by dividing the sums of the measurements by the sample size, and plot the value on the process control $\bar{x}$ chart.

The ***R* chart** is used to monitor the dispersion of the process. It is used in conjunction with the $\bar{x}$ chart when the process characteristic is a variable. To develop an R chart, collect samples from the process and organize them into subgroups, usually of three to six items. Next, compute the range R by taking the difference of the high value in the subgroup minus the low value. Then plot the R values on the R charts.

A standard process chart form is shown in Figure 12-6. This form has spaces for measurements and totals. In the example in Figure 12-7, our control chart form is filled

FIGURE 12-7 Completed $\bar{x}$ and R Chart

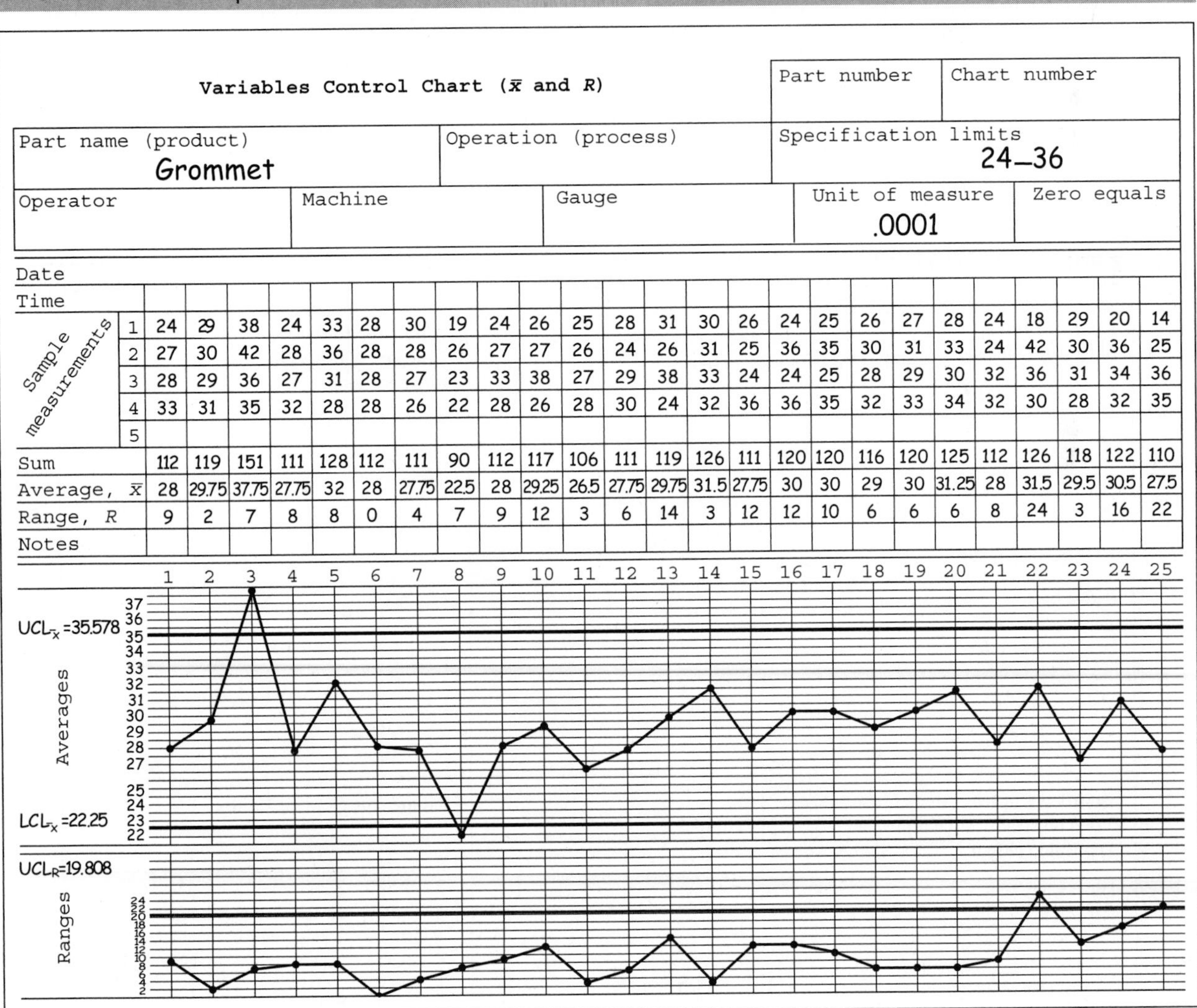

Variables Control Chart ($\bar{x}$ and R)

Part number | Chart number

Part name (product): Grommet | Operation (process) | Specification limits: 24–36

Operator | Machine | Gauge | Unit of measure: .0001 | Zero equals

	1	2	3	4	5	6	7	8	9	10	11	12	13	14	15	16	17	18	19	20	21	22	23	24	25
Date																									
Time																									
Sample measurements 1	24	29	38	24	33	28	30	19	24	26	25	28	31	30	26	24	25	26	27	28	24	18	29	20	14
2	27	30	42	28	36	28	28	26	27	27	26	24	26	31	25	36	35	30	31	33	24	42	30	36	25
3	28	29	36	27	31	28	27	23	33	38	27	29	38	33	24	24	25	28	29	30	32	36	31	34	36
4	33	31	35	32	28	28	26	22	28	26	28	30	24	32	36	36	35	32	33	34	32	30	28	32	35
5																									
Sum	112	119	151	111	128	112	111	90	112	117	106	111	119	126	111	120	120	116	120	125	112	126	118	122	110
Average, $\bar{x}$	28	29.75	37.75	27.75	32	28	27.75	22.5	28	29.25	26.5	27.75	29.75	31.5	27.75	30	30	29	30	31.25	28	31.5	29.5	30.5	27.5
Range, R	9	2	7	8	8	0	4	7	9	12	3	6	14	3	12	12	10	6	6	6	8	24	3	16	22
Notes																									

out with measurements from a process. Notice that there are $k = 25$ samples of size $n = 4$. For each of the samples, totals, ranges, and averages are computed. The range is the difference between the largest measurement and the smallest measurement in a particular sample.

Now that we have measurements, we need to compute a center line and control limits for our $\bar{x}$ and R charts. The center line is the process average. The upper and lower control limits are usually located three standard deviations from the center line. The formulas for computing these lines are given in Figure 12-8. Figure 12-9 shows the completed formulas for the example in Figure 12-7. Notice that the A_2 and D_4 table values come from the Factor for Control Limits table in the lower right corner of Figure 12-9. These table values provide estimates for the three standard deviation limits for the $\bar{x}$ and R charts [when combined with R-bar (the mean of the R values)]. You also may notice that no formula is given for the lower limit of the R chart. This is because the lower limit of R is zero for sample sizes less than or equal to six. For sample sizes greater than six, D_3 values must be used from Table A-1 in the appendix (the formula for the lower control limit is shown in Table A-1). Notice that we have superimposed the control limits computed in Figure 12-9 on the charts in Figure 12-7.

Interpreting Control Charts

Before introducing other types of process charts, we discuss the interpretation of the charts. Figure 12-10 shows several different signals for concern that are sent by a control chart, as in the second and third boxes. When a point is found to be outside of the

FIGURE 12-8 $\bar{x}$ and R Chart Calculation Work Sheet

Control Limits
Subgroups included ______ ______

$\bar{R} = \frac{\Sigma R}{k} = \_\_\_\_ = \_\_\_$

$\bar{\bar{X}} = \frac{\Sigma \bar{x}}{k} = \_\_\_\_ = \_\_\_$

or

$\bar{x}$ (Midspec or std) =

$A_2\bar{R} = \quad \times \quad = \_\_\_$

$UCL_{\bar{X}} = \bar{\bar{X}} + A_2\bar{R} =$

$LCL_{\bar{X}} = \bar{\bar{X}} - A_2\bar{R} =$

$UCL_{\bar{R}} = D_4\bar{R} = \quad \times \quad =$

Limits for Individuals
Compare with specification or tolerance limits

$\bar{\bar{X}}$ =

$\frac{3}{d_2}\bar{R} = \quad \times \quad = \_\_\_$

$UL_x = \bar{x} + \frac{3}{d_2}\bar{R} =$

$LL_x = \bar{x} - \frac{3}{d_2}\bar{R} =$

US =

LS = ___

US – LS =

$6\sigma = \frac{6}{d_2}\bar{R} =$

Modified Control Limits for Averages
Based on specification limits and process capability
Applicable only if US – LS > 6σ

US = LS =

$A_m\bar{R} = \quad \times \quad = \_\_\_$ $A_m\bar{R} = \_\_\_$

$URL_{\bar{X}} = US - A_m\bar{R} =$ $LRL_{\bar{X}} = LS + A_m\bar{R} =$

Factors for Control Limits

n	A_2	D_4	d_2	$\frac{3}{d_2}$	A_m
2	1.880	3.268	1.128	2.659	0.779
3	1.023	2.574	1.693	1.722	0.749
4	0.729	2.282	2.059	1.457	0.728
5	0.577	2.114	2.326	1.290	0.713
6	0.483	2.004	2.534	1.184	0.701

FIGURE 12-9 Calculation Work Sheet for Figure 12-7 Data

Control Limits

Subgroups included ______ ______

$\overline{R} = \frac{\Sigma R}{k} = \frac{217}{25} = 8.68$ ______

$\overline{\overline{X}} = \frac{\Sigma \overline{x}}{k} = \frac{731.25}{25} = 29.25$ ______

or

$\overline{x}$ (Midspec or std) =

$A_2\overline{R} = .729 \times 8.68 = 6.328$

$UCL_{\overline{X}} = \overline{\overline{X}} + A_2\overline{R}$ 29.25+6.328 = 35.578

$LCL_{\overline{X}} = \overline{\overline{X}} - A_2\overline{R}$ 29.25-6.328 = 22.922

$UCL_{\overline{R}} = D_4\overline{R} = 2.282 \times 8.68 = 19.808$

Limits for Individuals

Compare with specification or tolerance limits

$\overline{\overline{X}}$ =

$\frac{3}{d_2}\overline{R}$ = × = ______

$UL_x = \overline{x} + \frac{3}{d_2}\overline{R}$ =

$LL_x = \overline{x} - \frac{3}{d_2}\overline{R}$ =

US =

LS = ______

US – LS =

$6\sigma = \frac{6}{d_2}\overline{R}$ =

Modified Control Limits for Averages

Based on specification limits and process capability

Applicable only if US – LS > 6σ

US = LS =

$A_m\overline{R}$ = × = ______ $A_m\overline{R}$ = ______

$URL_{\overline{X}} = US - A_m\overline{R}$ = $LRL_{\overline{X}} = LS + A_m\overline{R}$ =

Factors for Control Limits

n	A_2	D_4	d_2	$\frac{3}{d_2}$	A_m
2	1.880	3.268	1.128	2.659	0.779
3	1.023	2.574	1.693	1.722	0.749
4	0.729	2.282	2.059	1.457	0.728
5	0.577	2.114	2.326	1.290	0.713
6	0.483	2.004	2.534	1.184	0.701

control limits, we call this an "out-of-control situation." When a process is out of control, variation is probably no longer random. If there are three standard deviation limits, the chance of a sample average or range being out of control when the process is stable is less than 1%. Because this probability is so small, we conclude that this was a nonrandom event and search for an assignable cause of variability.

Figure 12-10 presents examples of where nonrandom situations occur. You need not only have an out-of-control situation to signal that a process is no longer random. Two points in succession farther than two standard deviations from the mean likely will be a nonrandom event because the chances of this happening at random are very low. Five points in succession (either all above or below the center line) is called a *process run*. This means that the process has shifted. Seven points that are all either increasing or decreasing result in *process drift*. Process drift usually means that either materials or machines are drifting out of alignment. An example might be a saw blade that is wearing out rapidly in a furniture factory. Large jumps of more than three or four standard deviations result in *erratic behavior*. In all these cases, process charts help us to understand when the process is or is not in control.

If a process loses control and becomes nonrandom, the process should be stopped immediately. In many modern process industries where just-in-time is used, this will result in the stoppage of several workstations. The team of workers who are to address the problem should use a structured problem-solving process using brainstorming and cause-and-effect tools such as those discussed in Chapter 10 to identify the root cause of the out-of-control situation. Typically, the cause is somewhere in the interaction among processes, materials, machinery, or labor. Once the assignable cause of variation

FIGURE 12-10 Control Chart Evidence for Investigation

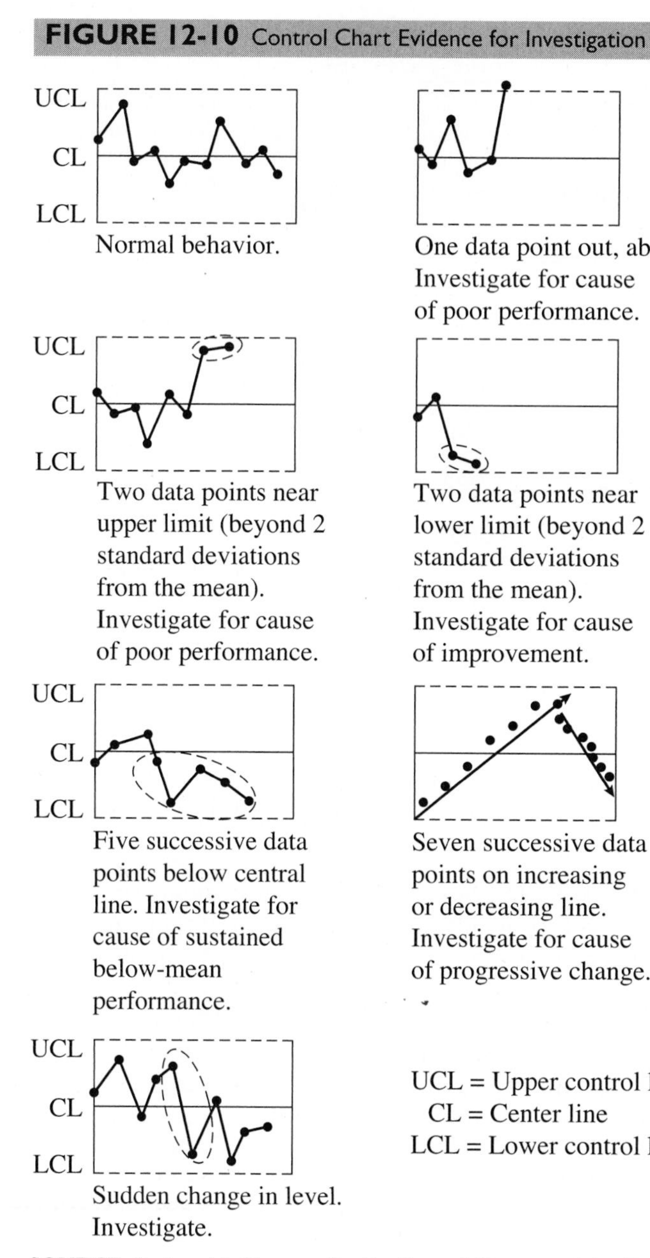

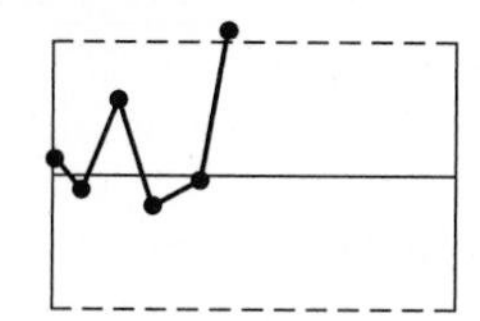

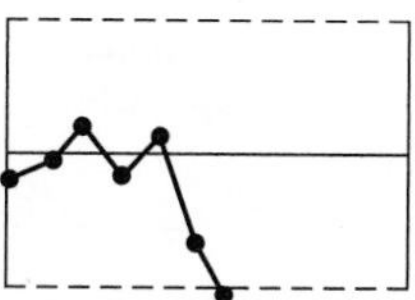

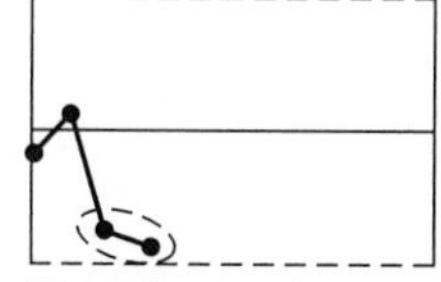

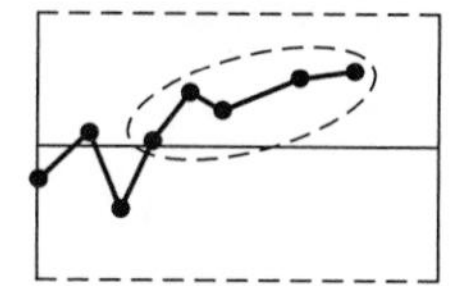

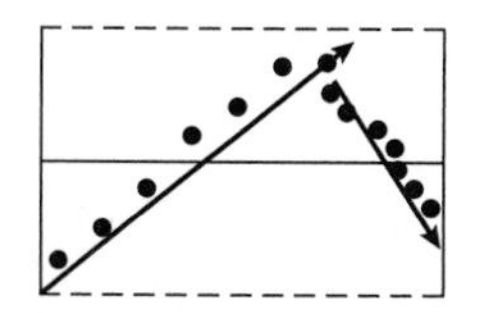

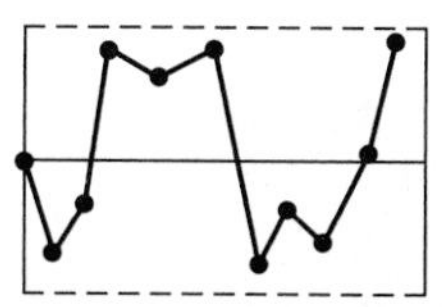

Normal behavior.

One data point out, above. Investigate for cause of poor performance.

One data point out, below. Investigate for cause of improvement.

Two data points near upper limit (beyond 2 standard deviations from the mean). Investigate for cause of poor performance.

Two data points near lower limit (beyond 2 standard deviations from the mean). Investigate for cause of improvement.

Five successive data points above central line. Investigate for cause of sustained poor performance.

Five successive data points below central line. Investigate for cause of sustained below-mean performance.

Seven successive data points on increasing or decreasing line. Investigate for cause of progressive change.

Erratic behavior. Investigate.

Sudden change in level. Investigate.

UCL = Upper control limit
CL = Center line
LCL = Lower control limit

SOURCE: Bertrand L. Hansen, *Quality Control: Theory and Applications*, © 1963, p. 65. Reprinted by permission of Prentice-Hall, Inc., Englewood Cliffs, NJ.

has been discovered, corrective action can be taken to eliminate the cause. The process is then restarted, and people return to work.

The cause of the problem should be documented and discussed later during the weekly departmental meeting. All workers should know why a problem in the process occurred. They should understand the causes and the corrective actions that were taken to solve the problem.

Production companies that embark on this level of delegation of authority and development of employees find the transition difficult because the process is often stopped and work is interrupted. However, as time passes, the processes become more stable as causes of errors are detected and eliminated. One manufacturer regularly produced poor-quality material that needed to be scrapped. As a result, it had increased its master production schedules by 20% to cover up this problem. The company decided instead to embark on a lot-size reduction program coupled with giving the workers line-stop authority. During the first shift, the company reduced the number of scrapped pieces from an average of more than 1,000 to 6! At first, production suffered. However, within two weeks of implementation, output volume had increased by more than 30%. This was the result of less rework, scrap, and other problems because of poor quality. It is interesting to note that staff and machinery were not changed during this period. At first, management thought its workers were unmotivated, resulting in the poor work. It wasn't the people; it was the process and the management.

Example 12.1: Using $\bar{x}$ and *R* Charts

Problem: The Sampson Company produces high-tech radar that is used in top-secret weapons by the Secret Service and the Green Berets. They have had trouble with a particular round component. The target diameter for this component is 6 centimeters. Samples of size four were taken during four successive days. The results are in the following table.

Active Model: Example 12.1

Excel File: Example 12.1

Day	x				Means	Ranges
1	6	6	5	7	6	2
2	8	6	6	7	6.75	2
3	7	6	6	6	6.25	1
4	6	7	5	4	5.5	3

Solution: The grand mean is 6.125. $\bar{R}$ is 2.

Develop a process chart to determine whether the process is stable. Because these are measurements, use $\bar{x}$ and R charts. Using the calculation work sheet, Figure 12-11 shows the values for the process control limits.

The $\bar{x}$ control chart for this problem is shown with the appropriate limits. The R chart is also in control. The sample averages were placed on the control chart, and the process was found to be historically in control. Because the averages and ranges fall within the control limits and no other signals of nonrandom activity are present, we conclude that the process is random. Note that this example is very simple. Generally, you use 15 to 20 subgroups to establish control charts.

Using Excel to Draw $\bar{x}$ and *R* Charts

The problem in Example 12.1 can be solved easily using Excel. While there are more elegant ways to develop control charts in Excel,[1] we will demonstrate a simple "brute force" method for creating $\bar{x}$ and R charts in Excel.

[1]There are several software packages and Excel add-ins that create control charts. A good place to find free Excel control chart templates is *www.freequality.org* or on your CD-ROM.

FIGURE 12-11 Calculation Work Sheet and $\bar{x}$ Chart

Control Limits

Subgroups included ________ ________

$\bar{R} = \frac{\Sigma R}{k} = \frac{8}{4} = 2$ ___

$\bar{\bar{X}} = \frac{\Sigma \bar{x}}{k} = \frac{24.5}{4} = 6.125$ ___

or

$\bar{x}$ (Midspec or std) =

$A_2\bar{R} = .729 \times 2 = 1.458$

$UCL_{\bar{X}} = \bar{\bar{X}} + A_2\bar{R} = 7.583$

$LCL_{\bar{X}} = \bar{\bar{X}} - A_2\bar{R} = 4.667$

$UCL_{\bar{R}} = D_4\bar{R} = 2.282 \times 2 = 4.564$

Limits for Individuals

Compare with specification or tolerance limits

$\bar{\bar{X}}$ =

$\frac{3}{d_2}\bar{R} = \quad \times \quad =$ ___

$UL_x = \bar{x} + \frac{3}{d_2}\bar{R}$ =

$LL_x = \bar{x} - \frac{3}{d_2}\bar{R}$ =

US =

LS = ___

US − LS =

$6\sigma = \frac{6}{d_2}\bar{R}$ =

Modified Control Limits for Averages

Based on specification limits and process capability

Applicable only if US − LS > 6σ

US = LS =

$A_m\bar{R} = \quad \times \quad =$ ___ $A_m\bar{R}$ = ___

$URL_{\bar{X}} = US - A_m\bar{R}$ = $LRL_{\bar{X}} = LS + A_m\bar{R}$ =

Factors for Control Limits

n	A_2	D_4	d_2	$\frac{3}{d_2}$	A_m
2	1.880	3.268	1.128	2.659	0.779
3	1.023	2.574	1.693	1.722	0.749
4	(0.729)	(2.282)	2.059	1.457	0.728
5	0.577	2.114	2.326	1.290	0.713
6	0.483	2.004	2.534	1.184	0.701

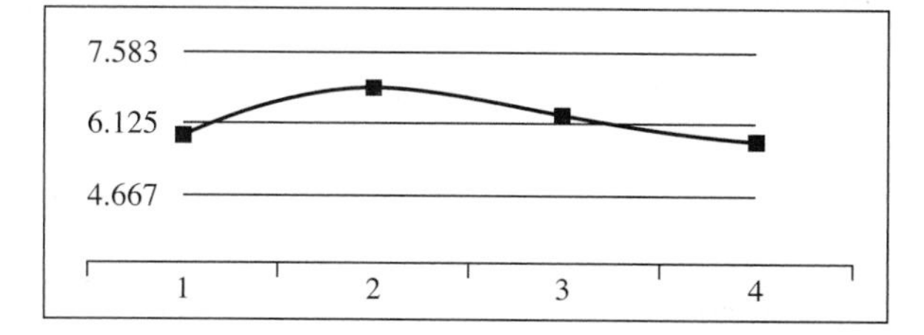

As you can see in Figure 12-12, we place the data in rows. From this we compute averages ($\bar{x}$s), R's, and $\bar{R}$. Using these data, the center line (CL), the upper limits (UCL), and lower limits (LCL) are computed. Figure 12-12 provides all the needed equations. Try doing this for yourself. Note that we have now included access to Quik Sigma, a quality control software on the CD accompanying this text.

X and Moving Range (*MR*) Charts for Population Data

At times it may not be possible to draw samples. This may occur because a process is so slow that only one or two units per day are produced. If you have a variable measurement that you want to monitor, the X and MR charts might just be the thing for you.

There are important caveats associated with the X and MR charts. Because you will not be sampling, the central limit theorem does not apply. This may result in the data being nonnormally distributed and an increase in the likelihood that you will draw an erroneous conclusion using a process chart. Therefore, it is best to first make sure that the data are normally distributed.

FIGURE 12-12 Example 12.1 Using Excel

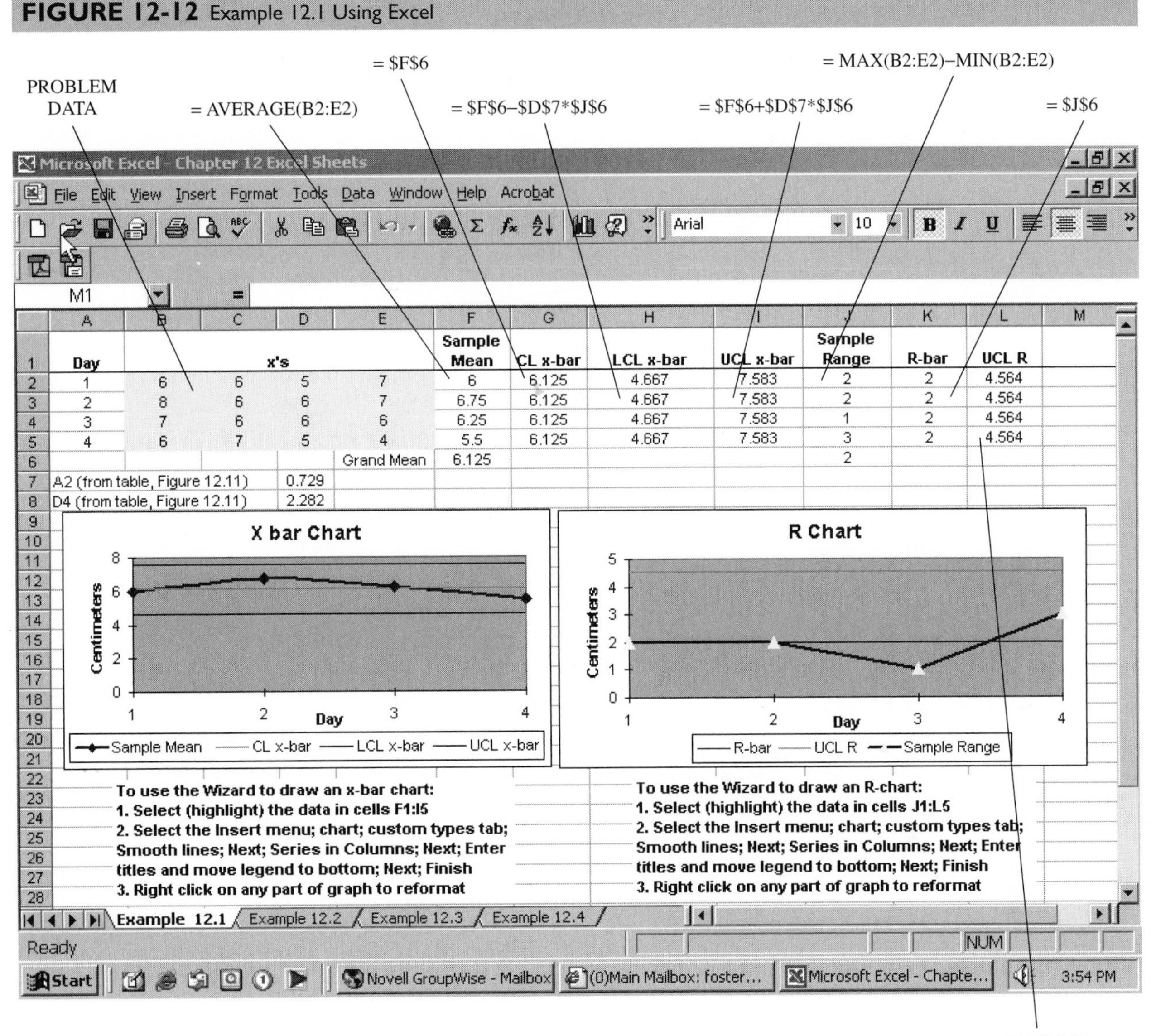

	Day	x's				Sample Mean	CL x-bar	LCL x-bar	UCL x-bar	Sample Range	R-bar	UCL R
2	1	6	6	5	7	6	6.125	4.667	7.583	2	2	4.564
3	2	8	6	6	7	6.75	6.125	4.667	7.583	2	2	4.564
4	3	7	6	6	6	6.25	6.125	4.667	7.583	1	2	4.564
5	4	6	7	5	4	5.5	6.125	4.667	7.583	3	2	4.564
6					Grand Mean	6.125				2		

If data are not normally distributed, other charts are available. A g chart is used when data are geometrically distributed, and h charts are useful when data are hypergeometrically distributed. Figure 12-13 presents geometric and hypergeometric distributions. If you develop a histogram of your data and it appears like either of these distributions, you may want to use either an h or a g chart instead of an X chart.

In statistics, an X is an individual observation from a population. Therefore, the ***X* chart** reflects a population distribution. We call the three standard deviation limits in an X chart the *natural variation* in a process. This natural variation can be compared with specification limits. So, strictly speaking, *X chart limits are not control limits; they are natural limits.*

FIGURE 12-13 *h* and *g* Distributions

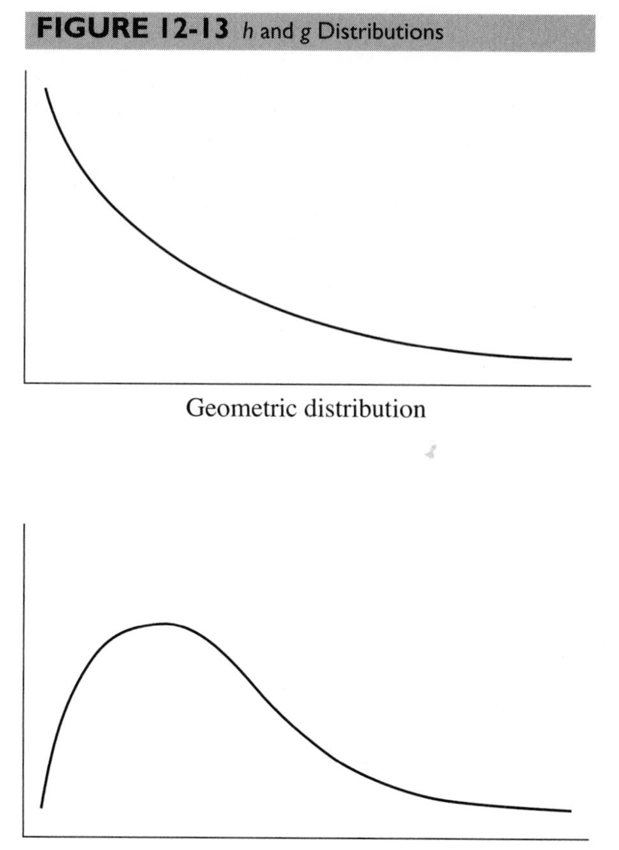

Geometric distribution

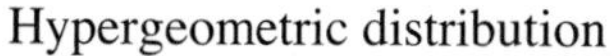

Hypergeometric distribution

The formula for the center line and the natural limits for an X chart is as follows:

$$\bar{x} \pm E_2(\overline{MR}) \tag{12.1}$$

where

$$\bar{x} = \frac{\Sigma X}{k}$$

and

X = a population value
k = the number of values used to compute $\bar{x}$
$E_2 = 2.66$ ($n = 2$) (see Table A-1 in the Appendix)

The formula for the MR chart is similar to that for the R chart (where $n = 2$), only, the ranges are computed as the differences from one sample to the next [$n = 2$; UCL = $D_4(\overline{MR})$; LCL = 0].

Active Model: Example 12.2

Example 12.2: *X* and *MR* Charts in Action

Problem: The Adam Everett Trucking Company of Columbia, Missouri, hauls corn from local fields to the Lee Sang Processing Plant in Lincoln, Nebraska. The trucks generally take 6.5 hours to make the daily trip. However, recently there seems to be more variability in the arrival times.

Excel File: Example 12.2

Mr. Everett, the owner, suspects that one of his drivers, Paul, may be visiting his girlfriend Janice en route in Kansas City. The driver claims that this is not the case and that the increase is simply random variation because of variability in traffic flows. The drivers keep written logs of departure and arrival times. Mr. Everett has listed these times in the following table. You are chosen as the analyst to investigate this situation. What do you think?

Date	Travel Times (Hours)	Moving Range
1	6.4	—
2	6.2	0.2
3	5.8	0.4
4	7.3	1.5
5	8.6	1.3
6	6.0	2.6
7	6.5	0.5
8	6.3	0.2
9	7.2	0.9
10	7.3	0.1
11	7.5	0.2
12	7.2	0.3
13	8.0	0.8
14	7.8	0.2
15	8.2	0.4
16	7.0	1.2
17	7.8	0.8
	$\bar{x} = 7.1235$	$\overline{MR} = .725$

Solution: You decide to develop an X and MR process chart to test the hypothesis concerning the change. The calculation work sheet yields the results shown in Figure 12-14. You conclude that, in fact, a run (from point 9 to point 15) indicates that trip times may be increasing. However, this does not imply that the girlfriend is the cause. Further investigation may be needed. (Note that $E_2 = 2.66$ and $D_4 = 3.268$).

Using Excel to Draw X and MR Charts

The problem in Example 12.2 is now solved using Excel. Again, we use the "brute force" method for creating X and MR charts in Excel. The process is very similar to what we did before. Notice that E_2 and D_4 are both constants.

Interpreting the charts, there is a run on the X chart and an out-of-control point on the sixth (fifth observation in the graph because there was no moving range for the first). This was because of the jump from 8.6 hours down to 6 hours. It might be that our hero thought he should be on better behavior after the long day on the fifth (see Figure 12-14).

Median Charts

While $\bar{x}$ charts generally are preferred for variables data, sometimes it is too time-consuming or inconvenient to compute subgroup averages. Also, there may be concerns about the accuracy of computed means. In these cases, a **median chart** may be used (a.k.a. an $\tilde{x}$ chart). The main limitation is that you will use an odd sample size to avoid calculating the median. Generally, sample sizes are 3, 5, or 7. Like the $\bar{x}$ chart, small sample sizes generally are used, although the larger the sample size, the better is the sensitivity of the chart as a tool to detect nonrandom (special cause) events (this is also true for $\bar{x}$ charts).

FIGURE 12-14 Example 12.2 Using Excel

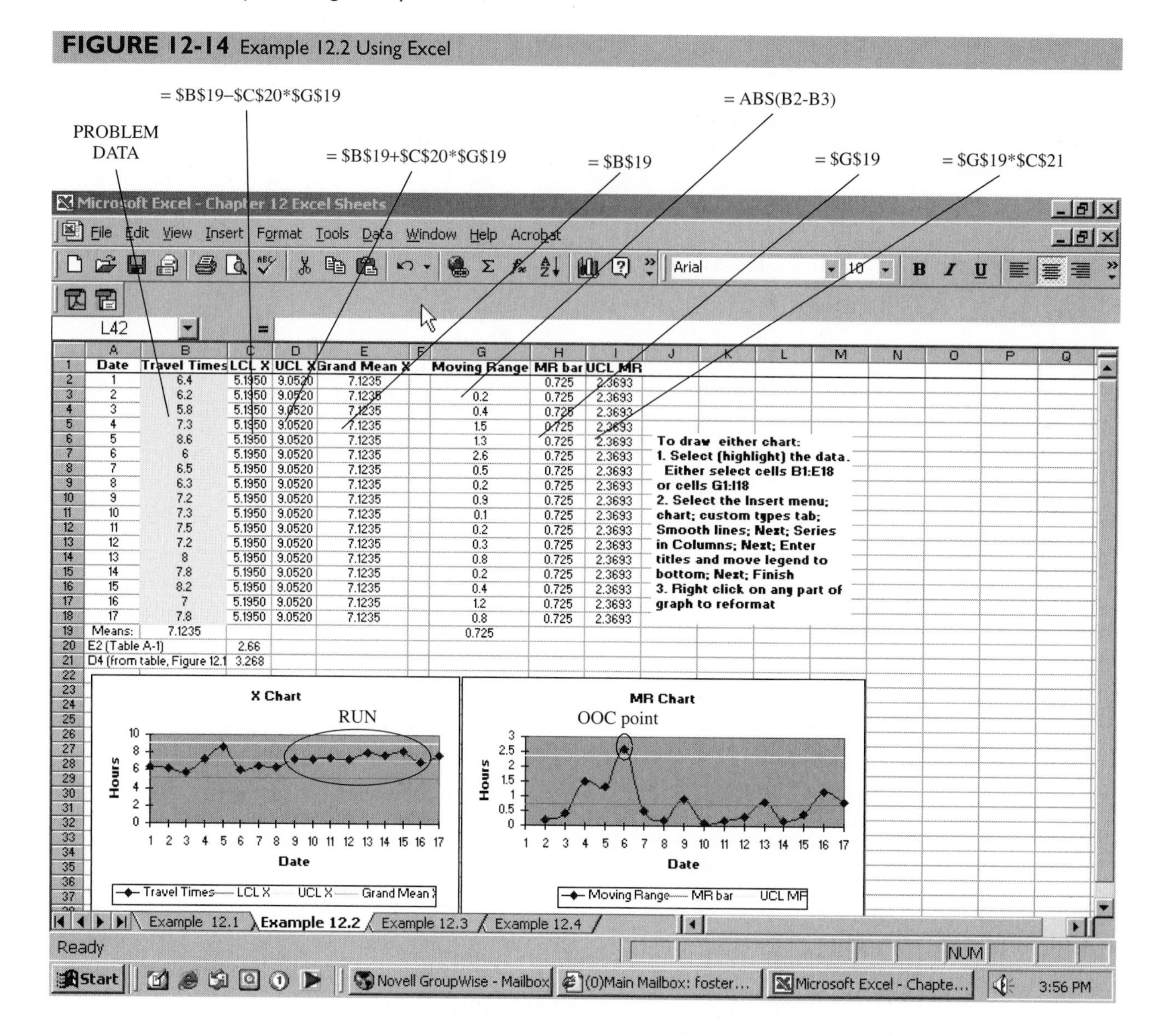

To prepare median charts, determine your subgroup size and how often you will sample. The rule of thumb to establish a median control chart is to use 20 to 25 subgroups and a total of at least 100 individual measurements.

Equations for computing the control limits are

$$\text{Mean of medians} = \text{sum of the medians/number of medians} = \bar{\tilde{x}} \qquad \textbf{(12.2)}$$

$$\text{LCL}_x = \bar{x} - \tilde{A}_2\bar{R} \qquad \textbf{(12.3)}$$

$$\text{UCL}_x = \bar{x} + \tilde{A}_2\bar{R} \qquad \textbf{(12.4)}$$

$\tilde{A}_2$ values are found in Table 12-2. Median charts are usually used with R charts.

TABLE 12-2 Median Chart Values

n	$\tilde{A}_2$	D_4
3	1.187	2.575
5	0.691	2.115
7	0.508	1.924
9	0.412	1.816

Active Model: Example 12.3

Excel File: Example 12.3

Example 12.3: Median Charts in Action

Problem: The Jeffrey Luftig food company has gathered the following data with weights of its new health food product. Since the published weight on the package is 6 ounces, Mr. Luftig wants to know if the company is complying with weight requirements. Twenty samples of size 5 were drawn.

Solution: The data are given below. Twenty samples of size 5 were drawn. Results show that the process is not in control with an average of 6.23. The median process chart (see Figure 12-15) does show that some product is being made that is below 6 ounces. Also, points 4, 7, and 10 are out of control.

Sample	Observation 1	Observation 2	Observation 3	Observation 4	Observation 5
1	6.2	6.1	6.3	6.5	6.4
2	6.2	6.2	6.2	6.3	6.4
3	6.3	5.9	6.2	6.4	6.3
4	5.3	5.1	5.3	5.1	5.3
5	6.1	6.6	6.3	6.2	6.4
6	6.2	6.2	6.2	6.2	6.2
7	5.8	5.7	5.9	7.2	5.2
8	6.3	5.9	6.2	6.4	6.3
9	6.3	5.9	6.2	6.4	6.3
10	7.4	7.4	7.1	7.3	7.1
11	6.2	6.3	6.2	6.3	6.2
12	6.4	6.3	6.2	6.1	6.1
13	6.3	6.4	6.2	6.3	6.1
14	6.1	6.1	6.1	6.1	6.1
15	6.3	6.4	6.1	6.3	6.1
16	6.4	6.2	6.4	6.2	6.2
17	6.2	6.4	6.3	6.4	6.2
18	6.1	6.2	6.3	6.4	6.5
19	6.2	6.1	6.1	6.1	6.1
20	6.4	6.3	6.2	6.5	6.3

Using Excel to Draw Median Charts

Figure 12-15 shows the results for Example 12.3. Again, the columns are ordered such that the data can be grouped properly and drawn using the Chart Wizard. Excel makes creation of the chart quick and easy. You will have the best results if you start with Example 12.1 and work all the Excel examples. By now you should have the hang of it. A good shortcut is to highlight the data in columns B through F prior to invoking the Chart Wizard. Have fun!

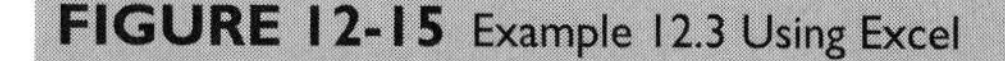

FIGURE 12-15 Example 12.3 Using Excel

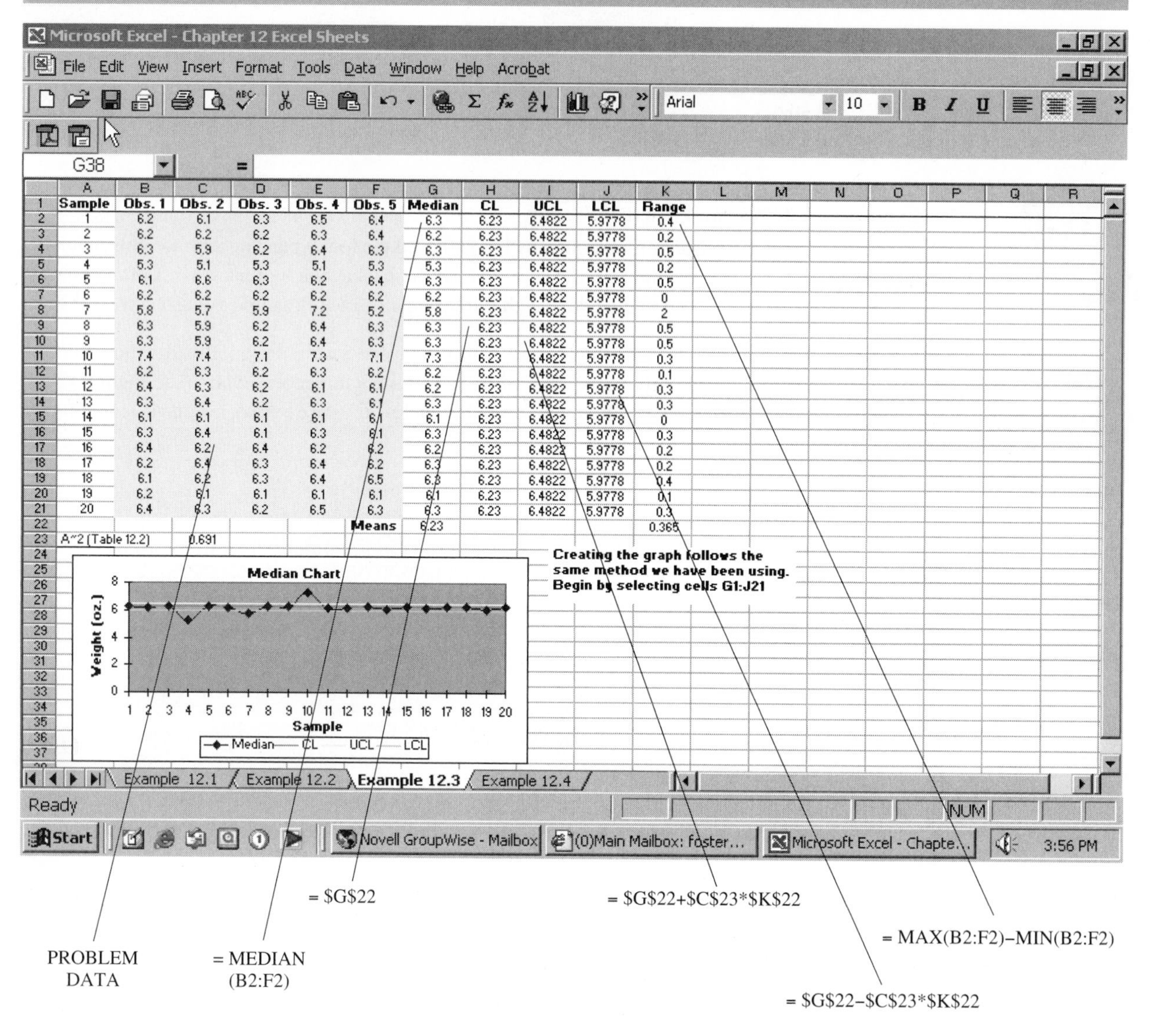

	A	B	C	D	E	F	G	H	I	J	K
1	Sample	Obs. 1	Obs. 2	Obs. 3	Obs. 4	Obs. 5	Median	CL	UCL	LCL	Range
2	1	6.2	6.1	6.3	6.5	6.4	6.3	6.23	6.4822	5.9778	0.4
3	2	6.2	6.2	6.2	6.3	6.4	6.2	6.23	6.4822	5.9778	0.2
4	3	6.3	5.9	6.2	6.4	6.3	6.3	6.23	6.4822	5.9778	0.5
5	4	5.3	5.1	5.3	5.1	5.3	5.3	6.23	6.4822	5.9778	0.2
6	5	6.1	6.6	6.3	6.2	6.4	6.3	6.23	6.4822	5.9778	0.5
7	6	6.2	6.2	6.2	6.2	6.2	6.2	6.23	6.4822	5.9778	0
8	7	5.8	5.7	5.9	7.2	5.2	5.8	6.23	6.4822	5.9778	2
9	8	6.3	5.9	6.2	6.4	6.3	6.3	6.23	6.4822	5.9778	0.5
10	9	6.3	5.9	6.2	6.4	6.3	6.3	6.23	6.4822	5.9778	0.5
11	10	7.4	7.4	7.1	7.3	7.1	7.3	6.23	6.4822	5.9778	0.3
12	11	6.2	6.3	6.2	6.3	6.2	6.2	6.23	6.4822	5.9778	0.1
13	12	6.4	6.3	6.2	6.1	6.1	6.2	6.23	6.4822	5.9778	0.3
14	13	6.3	6.4	6.2	6.3	6.1	6.3	6.23	6.4822	5.9778	0.3
15	14	6.1	6.1	6.1	6.1	6.1	6.1	6.23	6.4822	5.9778	0
16	15	6.3	6.4	6.1	6.3	6.1	6.3	6.23	6.4822	5.9778	0.3
17	16	6.4	6.2	6.4	6.2	6.2	6.2	6.23	6.4822	5.9778	0.2
18	17	6.2	6.4	6.3	6.4	6.2	6.3	6.23	6.4822	5.9778	0.2
19	18	6.1	6.2	6.3	6.4	6.5	6.3	6.23	6.4822	5.9778	0.4
20	19	6.2	6.1	6.1	6.1	6.1	6.1	6.23	6.4822	5.9778	0.1
21	20	6.4	6.3	6.2	6.5	6.3	6.3	6.23	6.4822	5.9778	0.3
22						Means	6.23				0.365
23	A^2 (Table 12.2)		0.691								

$\bar{x}$ and s Charts

When you are particularly concerned about the dispersion of the process, it might be that the R chart is not sufficiently precise. In this case, the $\bar{x}$ chart is recommended in concert with the **s chart** or **standard deviation chart**. The standard deviation chart is often used where variation in a process is small. For example, s charts are often used in monitoring the production of silicon chips for computers.

Unfortunately, when using the s chart, since we do not compute ranges, new formulas are used to compute the $\bar{x}$ limits. We will introduce the formulas for the $\bar{x}$ and s charts because of their importance for high-tech production.

The control limits for the s chart are computed using the formulas:

$$UCL_s = B_4 \times \bar{s} \quad (12.5)$$
$$LCL_s = B_3 \times \bar{s} \quad (12.6)$$

where

B_3 and B_4 come from Table 12-3;

and

$$\bar{s} = \Sigma s_i / k \quad (12.7)$$

where
s_i is the standard deviation for sample i
k is the number of samples.

Note that it is easy to find the sample standard deviation in Excel. If you don't have Excel, use the usual formulas for computing the sample standard deviations. We will show you how to do this in Excel in Example 12.4.

After computing the limits, plot your sample means to see if the process is in control. If the s chart is not in control, determine the cause for the out-of-control point, eliminate the cause, and then recompute your control limits by throwing out the out-of-control data point(s). Do not eliminate samples with out-of-control points if a cause cannot be identified.

When your s chart is in statistical control, use the following formula to estimate the process standard deviation:

$$\sigma_{est} = \bar{s} \times \sqrt{(1 - C_4^2)} / C_4 \quad (12.8)$$

where C_4 can be found in Table 12-3.

Formulas for the $\bar{x}$ chart can now be created using the following formulas:

$$UCL_{\bar{x}} = \bar{\bar{x}} + A_3(\bar{s}) \quad (12.9)$$

$$LCL_{\bar{x}} = \bar{\bar{x}} - A_3(\bar{s}) \quad (12.10)$$

where A_3 can be found in Table 12-3 and $\bar{\bar{x}}$ is the grand mean.

TABLE 12-3 Values for $\bar{x}$ and s Charts

n	B_3	B_4	C_4	A_3
2	0	3.267	0.7979	2.659
3	0	2.568	0.8862	1.954
4	0	2.266	0.9213	1.628
5	0	2.089	0.9400	1.427
6	0.030	1.970	0.9515	1.287
7	0.118	1.882	0.9594	1.182
8	0.185	1.815	0.9650	1.099
9	0.239	1.761	0.9693	1.032

FIGURE 12-16 Example 12.4 Using Excel

	Sample	Obs. 1	Obs. 2	Obs. 3	Std. Dev.	CLs	UCLs	x-bar	CLx-bar	UCLx-bar	LCLx-bar
2	1	2.0000	1.9998	2.0002	0.0002	0.0002	0.000617	2.0000	2.0000	2.0005	1.9996
3	2	1.9998	2.0003	2.0002	0.0003	0.0002	0.000617	2.0001	2.0000	2.0005	1.9996
4	3	1.9998	2.0001	2.0005	0.0004	0.0002	0.000617	2.0001	2.0000	2.0005	1.9996
5	4	1.9997	2.0000	2.0004	0.0004	0.0002	0.000617	2.0000	2.0000	2.0005	1.9996
6	5	2.0003	2.0003	2.0002	0.0001	0.0002	0.000617	2.0003	2.0000	2.0005	1.9996
7	6	2.0004	2.0003	2.0000	0.0002	0.0002	0.000617	2.0002	2.0000	2.0005	1.9996
8	7	1.9998	1.9998	1.9998	0.0000	0.0002	0.000617	1.9998	2.0000	2.0005	1.9996
9	8	2.0000	2.0001	2.0001	0.0001	0.0002	0.000617	2.0001	2.0000	2.0005	1.9996
10	9	2.0005	2.0000	1.9999	0.0003	0.0002	0.000617	2.0001	2.0000	2.0005	1.9996
11	10	1.9995	1.9998	2.0001	0.0003	0.0002	0.000617	1.9998	2.0000	2.0005	1.9996
12	11	2.0002	1.9999	2.0001	0.0002	0.0002	0.000617	2.0001	2.0000	2.0005	1.9996
13	12	2.0002	1.9998	2.0005	0.0004	0.0002	0.000617	2.0002	2.0000	2.0005	1.9996
14	13	2.0000	2.0001	1.9998	0.0002	0.0002	0.000617	2.0000	2.0000	2.0005	1.9996
15	14	2.0000	2.0002	2.0004	0.0002	0.0002	0.000617	2.0002	2.0000	2.0005	1.9996
16	15	1.9994	2.0001	1.9996	0.0004	0.0002	0.000617	1.9997	2.0000	2.0005	1.9996
17	16	1.9999	2.0003	1.9993	0.0005	0.0002	0.000617	1.9998	2.0000	2.0005	1.9996
18	17	2.0002	1.9998	2.0004	0.0003	0.0002	0.000617	2.0001	2.0000	2.0005	1.9996
19	18	2.0000	2.0001	2.0001	0.0001	0.0002	0.000617	2.0001	2.0000	2.0005	1.9996
20	19	1.9997	1.9994	1.9998	0.0002	0.0002	0.000617	1.9996	2.0000	2.0005	1.9996
21	20	2.0003	2.0007	1.9999	0.0004	0.0002	0.000617	2.0003	2.0000	2.0005	1.9996
22				means:	0.0002402			2.0000			
23	Normal Variate		2.568								
24	Normal variate		1.954								

Example 12.4: s and $\bar{x}$ Charts in Action

Problem: Twenty samples were taken for a milled rod. The diameters are needed to determine if the process is in control. Since these milled rods must be measured within 1/10,000 of an inch, it is determined that the process dispersion is important. Therefore, you need to use an s and $\bar{x}$ chart to monitor the process. The data are found in Figure 12-16. We have 20 samples with $n = 3$.

Active Model: Example 12.4

Solution: The control charts in Figure 12-16 show that the process is in control. There is no need for corrective action. The solution method is demonstrated in the next section.

Using Excel to Draw s and $\bar{x}$ Charts

Figure 12-16 shows the solution method for Example 12.4. Using the preceding formulas, we computed the CL, UCL, and LCL for each chart. Notice that the LCL for the s chart is zero. Also, notice that we have taken some shortcuts here compared with some of the other charts we have drawn.

TABLE 12-4 Summary of Variables Chart Formulas

Chart	*LCL*	*CL*	*UCL*	
$\bar{x}$	$\bar{\bar{x}} - A_2\bar{R}$	$\bar{\bar{x}}$	$\bar{\bar{x}} + A_2\bar{R}$	
R	$D_3\bar{R}$	$\bar{R}$	$D_4\bar{R}$	
$\bar{x}$ (with s)	$\bar{\bar{x}} - A_3\bar{s}$	$\bar{\bar{x}}$	$\bar{\bar{x}} + A_3\bar{s}$	(Appendix Table A-3)
X	$\bar{x} - E_2(\overline{MR})$	$\bar{x}$	$\bar{x} + E_2(\overline{MR})$	
s	$B_3\bar{s}$	$\bar{s}$	$B_4\bar{s}$	(Appendix Table A-3)
Median	$\bar{\tilde{x}} - \tilde{A}_2\bar{R}$	$\bar{\tilde{x}}$	$\bar{\tilde{x}} + \tilde{A}_2\bar{R}$	

Excel File: Example 12.4

Other Control Charts

Table 12-4 shows all the formulas for the process charts we have discussed in this chapter. These are the major charts that are used the vast majority of times. Some other charts that are used more rarely should be mentioned.

Moving Average Chart

The moving average chart is an interesting chart that is used for monitoring variables and measurement on a continuous scale. This chart uses past information to predict what the next process outcome will be. Using this chart, we can adjust a process in anticipation of its going out of control.

Cusum Chart

The cumulative sum, or cusum, chart is used to identify slight but sustained shifts in a universe where there is no independence between observations. A cusum chart looks very different from a Shewhart process chart as shown in Figure 12-17.

FIGURE 12-17 Cusum Chart

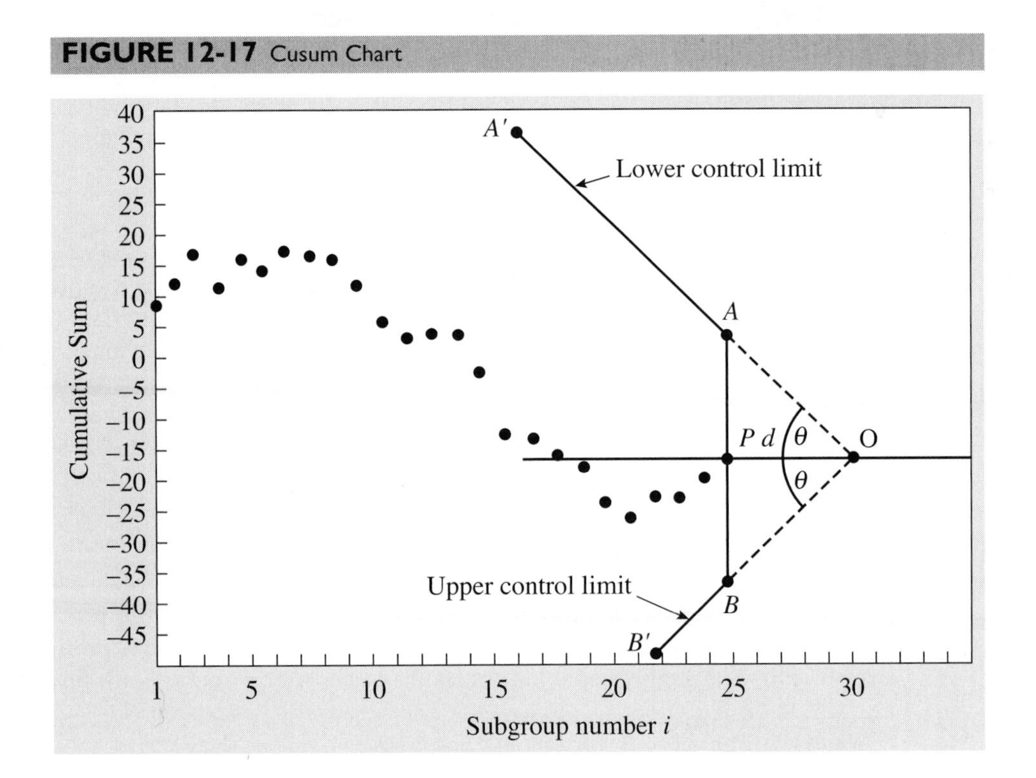

SOME CONTROL CHART CONCEPTS FOR VARIABLES

Choosing the Correct Variables Control Chart

Now that we have developed control charts, we are in a position to briefly discuss some important control chart concepts before moving to process capability. The first concept has to do with choosing the correct chart. Obviously, it is key to choose the correct control chart. Figure 12-18 shows a decision tree for the basic control charts. This flow chart helps to show when certain charts should be selected for use.

Corrective Action

When a process is out of control, corrective action is needed. Corrective action steps are similar to continuous improvement processes. They are

1. Carefully identify the quality problem.
2. Form the correct team to evaluate and solve the problem.
3. Use structured brainstorming along with fishbone diagrams or affinity diagrams to identify causes of problems.
4. Brainstorm to identify potential solutions to problems.
5. Eliminate the cause.
6. Restart the process.
7. Document the problem, root causes, and solutions.
8. Communicate the results of the process to all personnel so that this process becomes reinforced and ingrained in the organization.

How Do We Use Control Charts to Continuously Improve?

One of the goals of the control chart user is to reduce variation. Over time, as processes are improved, control limits are recomputed to show improvements in stability. As upper and lower control limits get closer and closer together, the process is improving. There are two key concepts here:

- The focus of control charts should be on continuous improvement.
- Control chart limits should be updated only when there is a change to the process. Otherwise any changes are unexpected.[2]

Tampering with the Process

One of the cardinal rules of process charts is that you should never tamper with the process. You might wonder, Why don't we make adjustments to the process any time the process deviates from the target? The reason is that random effects are just that—random. This means that these effects cannot be controlled. If we make adjustments to a random process, we actually inject nonrandom activity into the process. Figure 12-19 shows a random process. Suppose that we had decided to adjust the process after the fourth observation. We would have shifted the process—signaled by out-of-control observations during samples 12 and 19.

PROCESS CAPABILITY FOR VARIABLES

Once a process is stable, the next emphasis is to ensure that the process is capable. Process **capability** refers to the ability of a process to produce a product that meets specification. A highly capable process produces high volumes with few or no defects

[2]Wheeler, D., "When Do I Recalculate My Limits?," *Quality Progress* (May 1996):79–80.

FIGURE 12-18 Process for Selecting the Right Chart

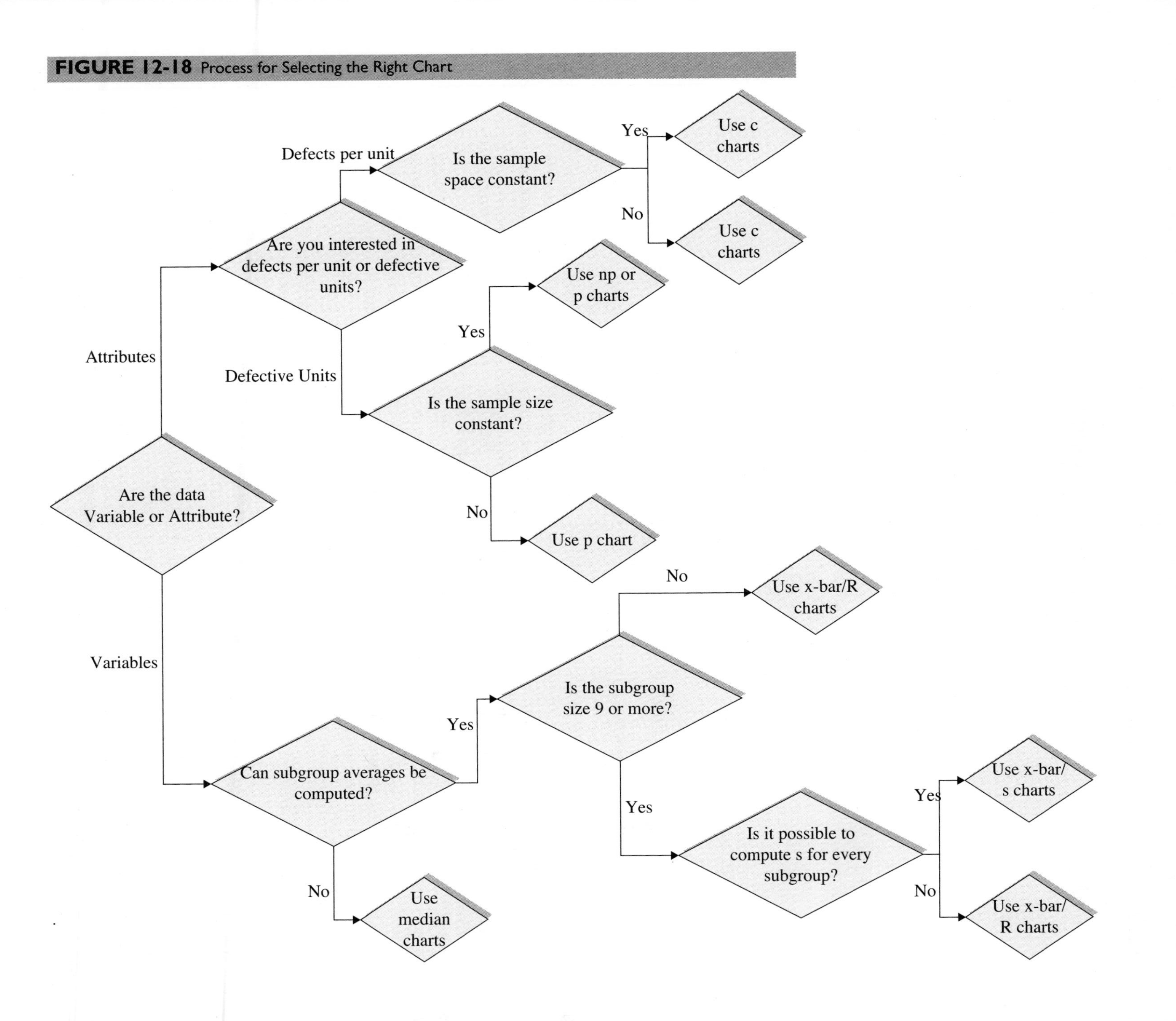

Video Clip: Process Capability

FIGURE 12-19 The Effects of Tampering

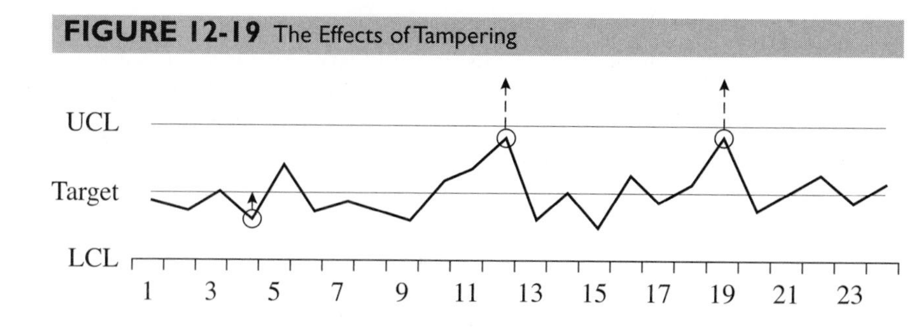

and is the result of optimizing the interactions between people, machines, raw materials, procedures, and measurement systems. World-class levels of process capability are measured by parts per million (ppm) defect levels, which means that for every million pieces produced, only a small number (less than 100) are defective. A Closer Look at Quality 12-1 looks at the need for capability in software.

A CLOSER LOOK AT QUALITY 12-1

A JUSTIFICATION FOR MEETING STANDARDS IN SOFTWARE QUALITY

In this chapter, we discuss the importance of meeting standards and having controlled processes. We see all around us the results of poor quality and defects. Among the areas where this is important is software quality. Consider the following examples.

In the mid 1980s, poor software design in a radiation machine, known as Therac-25, contributed to the deaths of three cancer patients.[a] The Therac-25 was built by Atomic Energy of Canada Ltd., which is a Crown corporation of the government of Canada. In 1988, the company incorporated and sold its radiation-systems assets under the Theratronics brand. According to Nancy Leveson, now a professor at MIT, the design flaws included the inability of the software to handle some of the data it was given and the delivery of hard-to-decipher user messages.

During Operation Desert Storm, an Iraqi Scud missile hit a U.S. Army barracks in Saudi Arabia, killing 28 Americans. The approach of the Scud should have been noticed by a Patriot missile battery. A subsequent government investigation found a flaw in the Patriot's weapons-control software, however, that prevented the system from properly tracking the incoming missile.

During Operation Iraqi Freedom, the Patriot missile system mistakenly downed a British Tornado fighter and, according to the *LA Times*, an American F/A-18c Hornet. Reports show that investigators were looking at a glitch in the missile's radar system that made it incapable of properly distinguishing between a friendly aircraft and an enemy missile.

In 2002, the Food and Drug Administration (FDA), which oversees medical-device software, said of 3,140 medical-device recalls, 242 were attributed to software failures. The FDA also says the number of software-related recalls may be underreported as it is often hard to determine the exact cause of a problem in the immediate aftermath of an accident.

It is expected that these types of losses are likely to mount as complex software programs are tied across networks. Imagine all of the various pieces of corporate data that come together in systems for CRMS, SCMs, or ERPs. "Software is the most complicated thing that the human mind can come up with and build," says Gary McGraw, the chief technology officer at Citigal, a consulting firm specializing in improving software quality. Tools introduced in this chapter will be key in detecting if future software is functioning properly.

[a]Gage, D., and J. McCormick, "Why Software Quality Matters," *Baseline* 28 (March 2004):34–59.

Six-Sigma programs such as those pioneered by Motorola Corporation result in highly capable processes. Six Sigma is a design program that emphasizes engineering parts so that they are highly capable. As shown in Figure 12-20, these processes are characterized by specifications that are ±6 standard deviations from the process mean. This means that even large shifts in the process mean and dispersion will not result in defective products being built. If a process average is on the center line, a six-sigma process will result in an average of only 3.4 opportunities for defects per million units produced. The Taguchi method is a valuable tool for achieving Six-Sigma quality by helping to develop robust designs that are insensitive to variation.

Population versus Sampling Distributions

To understand process capability, we must first understand the differences between population and sampling distributions. *Population distributions* are distributions with all individual responses from an entire population. A *population* is defined as a collection of all the items or observations of interest to a decision maker. A **sample** is a subset of the population. *Sampling distributions* are distributions that reflect the distribution of sample means. We can demonstrate the difference between a sample and a population. Suppose that you want to understand whether a product conforms to specifications. Over a month's time, a firm produces 10,000 units of product to stock. Because the product is fragile, it is not feasible to inspect all 10,000 units and risk damaging some of the product in the inspection process. Therefore, 500 units are randomly selected from the 10,000 to inspect. In this example, the population size m is 10,000 and the sample size n is 500.

We now demonstrate the difference between a sampling distribution and a population distribution. Understanding the differences between sampling and population distributions is important: Population distributions have much more dispersion than sampling distributions. Consider a class of 40 students where the tallest student is 6 feet 4 inches and the shortest is 5 feet in height. As shown in Figure 12-21, student height for this population is normally distributed, with a mean of 5 feet 8 inches and a distribution ranging from 5 feet to 6 feet 4 inches.

Now suppose that you draw samples of size five from the population (with replacement). Notice in Figure 12-21 that the mean of the sample is still 5 feet 8 inches, but the distribution ranges only from 5 feet 4 inches to 6 feet. This is so because it is difficult to randomly obtain a sample average that is more than 6 feet or less than 5 feet 4 inches. As a result, we see that sampling distributions have much less dispersion than population distributions.

FIGURE 12-20 Six-Sigma Quality

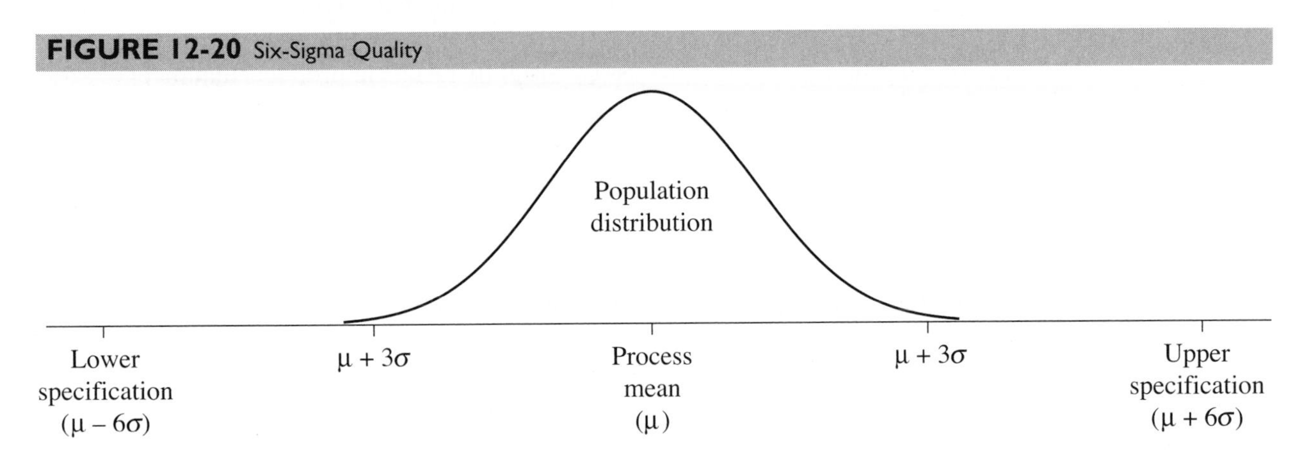

FIGURE 12-21 Population and Sampling Distributions for Class Heights

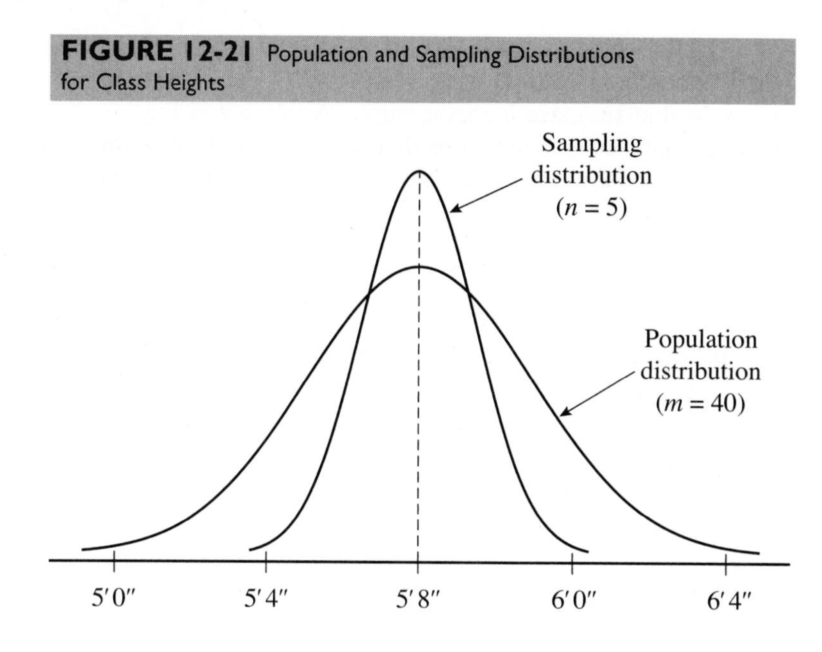

In the context of quality, specifications and capability are associated with population distributions. However, sample-based process charts and stability are computed statistically and reflect sampling distributions. Therefore, *quality practitioners should not compare process chart limits with product specifications*. To compare process charts limits with specification limits is not so much like comparing apples to oranges as it is comparing apples to watermelons. We later show that process chart limits are statistically computed from sample data. Specification (or tolerance) limits are set by design engineers who establish limits based on the design requirements for a product. These design requirements might have to do with making parts fit together properly or with the properties of certain materials used in making products.

Capability Studies

Now that we have defined process capability, we can discuss how to determine whether a process is capable. That is, we want to know if individual products meet specifications. There are two purposes for performing process capability studies:

1. To determine whether a process consistently results in products that meet specifications
2. To determine whether a process is in need of monitoring through the use of permanent process charts

Process capability studies help process managers understand whether the range over which natural variation of a process occurs is the result of the system of common (or random) causes. There are five steps in performing process capability studies:

1. Select a critical operation. These may be bottlenecks, costly steps of the process, or places in the process where problems have occurred in the past.
2. Take k samples of size n, where x is an individual observation.

- Where $19 < k < 26$
- If x is an attribute, $n > 50$ (as in the case of a binomial)
- Or if x is a measurement, $1 < n < 11$
  (*Note*: Small sample sizes can lead to erroneous conclusions.)

3. Use a trial control chart to see whether the process is stable.
4. Compare process natural tolerance limits with specification limits. Note that natural tolerance limits are three standard deviation limits for the population distribution. This can be compared with the specification limits.
5. Compute capability indexes: To compute capability indexes, you compute an upper capability index (Cpu), a lower capability index (Cpl), and a capability index **(Cpk)**. The formulas used to compute these are

$$\text{Cpu} = (\text{USL} - \mu)/3\hat{\sigma} \tag{12.11}$$

$$\text{Cpl} = (\mu - \text{LSL})/3\hat{\sigma} \tag{12.12}$$

$$\text{Cpk} = \min\{\text{Cpu}, \text{Cpl}\} \tag{12.13}$$

where

USL = upper specification limit
LSL = lower specification limit
μ = computed population process mean
$\hat{\sigma}$ = Estimated process standard deviation $= \hat{\sigma} = \overline{R}/d_2$ **(12.14)**

Make a decision concerning whether the process is capable. Although different firms use different benchmarks, the generally accepted benchmarks for process capability are 1.25, 1.33, and 2.0. **We will say that processes that achieve capability indexes (Cpk) of 1.25 are capable, 1.33 are highly capable, and 2.0 are world-class capable (Six Sigma)**.

Example 12.5: Process Capability

Problem: For an overhead projector, the thickness of a component is specified to be between 30 and 40 millimeters. Thirty samples of components yielded a grand mean $(\bar{x})$ of 34 millimeters with a standard deviation $(\hat{\sigma})$ of 3.5. Calculate the process capability index by following the steps previously outlined. If the process is not highly capable, what proportion of product will not conform?

Solution:

$$\text{Cpu} = (40 - 34)/(3)(3.5) = .57$$

$$\text{Cpl} = (34 - 30)/(3)(3.5) = .38$$

$$\text{Cpk} = .38$$

The process capability in this case is poor. To compute the proportion of nonconforming product being produced, we use a Z table (Appendix A-2) with a standardized distribution. The formula is

$$Z = (x - \mu)/\hat{\sigma} \tag{12.15}$$

Thus, for the lower end of the distribution

$$Z = (30 - 34)/3.5 = -1.14$$

and for the upper end of the distribution

$$Z = (40 - 34)/3.5 = 1.71$$

Using a Z table (as shown in Figure 12-22, using Table A-2 from the Appendix), the probability of producing bad product is .1271 + .0436 = .1707. This means that, on average, more than 17% of the product produced does not meet specification. This is unacceptable in almost any circumstance.

Ppk

If your data are not arranged in subgroups and you only have population data to compute your capability, use Ppk to compute your capability. **Ppk** stands for *population capability index*. Rather than using the within-groups variation to estimate the sigma that you used in Cpk, you use the population standard deviation to compute your capability. Otherwise, the computations are the same as Cpk. Here are the formulas:

$$\text{Ppk} = \min\{\text{Ppu}, \text{Ppl}\} \quad \textbf{(12.16)}$$

$$\text{Ppu} = (\text{USL} - \mu)/3\sigma \quad \textbf{(12.17)}$$

$$\text{Ppl} = (\mu - \text{LSL})/3\sigma \quad \textbf{(12.18)}$$

$$\sigma = \sqrt{\Sigma(x_i - \bar{x})^2/(n-1)} \quad \textbf{(12.19)}$$

where

USL = upper specification limit
LSL = lower specification limit
μ = population mean
σ = population process standard deviation

Interpretation for Ppk is the same as for Cpk. The only difference is the use of population parameters in computing the indexes.

FIGURE 12-22 Proportion of Product Nonconforming for Example 12.5

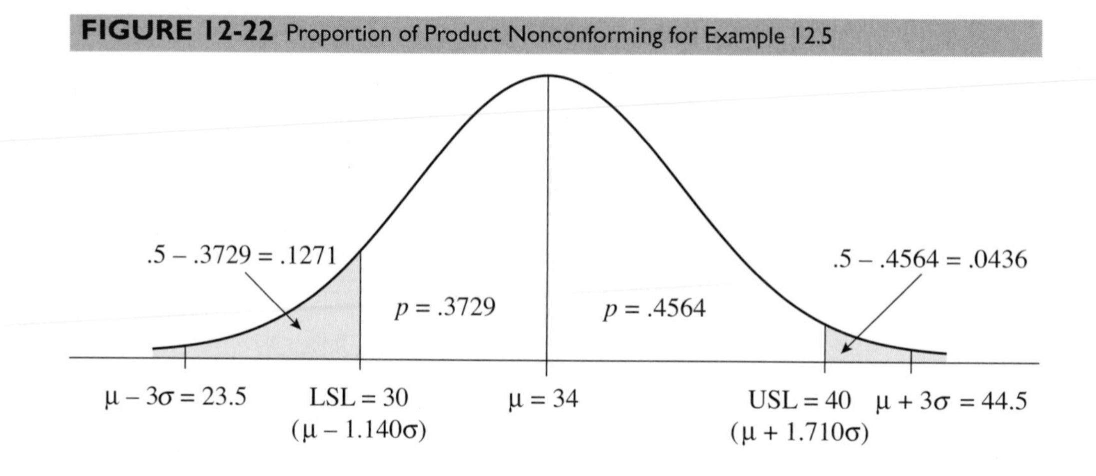

Example 12.6: Population Process Capability

Problem: The upper and lower specification limits (tolerances) for a metal plate are 3 millimeters ±0.002 millimeters. A sample of 100 plates yielded a mean $\bar{x}$ of 3.001 millimeters. We know that the population standard deviation is .0006. Compute the Ppk for this product.

Solution:

$$\text{Ppu} = (3.002 - 3.001)/.0006 = 1.67$$
$$\text{Ppl} = (3.001 - 2.999)/.0006 = 3.33$$
$$\text{Cpk} = 1.67$$

Therefore, the process is highly capable.

The Difference between Capability and Stability

Once again, *a process is capable if individual products consistently meet specification. A process is stable if only common variation is present in the process.* This is an important distinction. It is possible to have a process that is stable but not capable. This would happen where random variation was very high. It is probably not so common that an incapable process would be stable.

OTHER STATISTICAL TECHNIQUES IN QUALITY MANAGEMENT

Throughout this chapter we have focused on hypothesis testing and process charts. In Chapter 14 we discuss experimental design and off-line experimentation. Correlation and regression also can be useful tools for improving quality, particularly in services.

Although it is almost never appropriate to use regression on process data used in developing control charts, there are other types of data that can be correlated and regressed to understand the customer. For example, Figure 12.23 shows where conformance rates and quality costs were correlated in one company. As conformance

FIGURE 12-23 Plot of Prevention and Appraisal Costs with Conformance

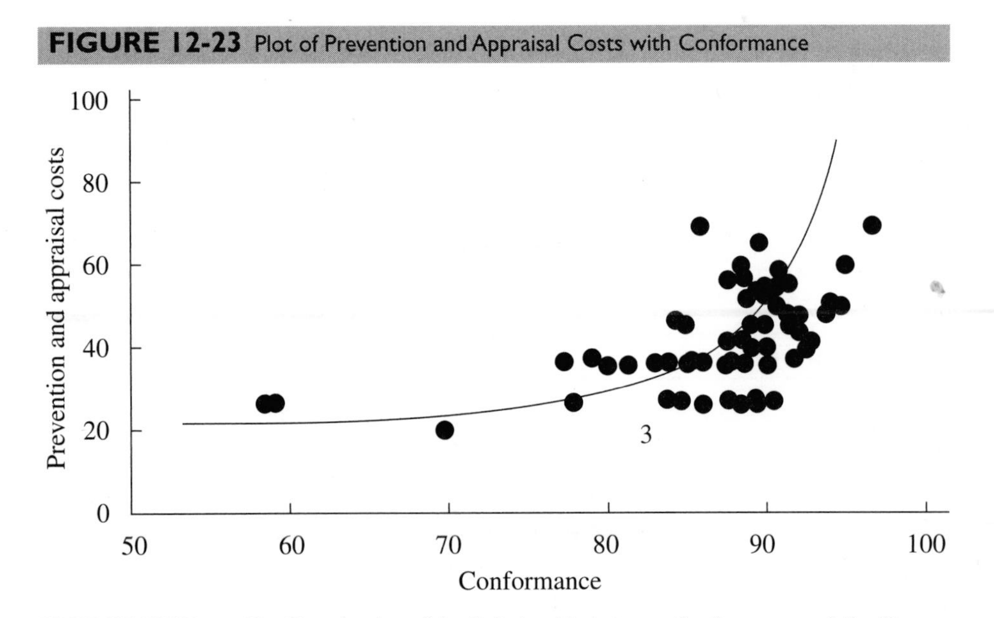

SOURCE: S.T. Foster, "An Examination of the Relationship between Conformance and Quality-Related Costs," *International Journal of Quality and Reliability Management* 13, 4, (1996):50–63.

TABLE 12-5 Relationship between Conformance and PA Costs

Model	R^2	p
First order	0.4002	0.0001
quadratic	0.4675	0.0001

SOURCE: S. T. Foster, "An Examination of the Relationship between Conformance and Quality-Related Costs," *International Journal of Quality and Reliability Management* 13, 4 (1996):50–63.

increased, costs increased as well. Table 12-5 shows that these variables were significantly and positively related. The R^2 values show the strength of the relationships between the variables for linear and nonlinear (quadratic) models.[3]

Such correlation is called *interlinking*.[4] Interlinking is useful in helping to identify causal relationships between variables.

SUMMARY

In this chapter we have introduced the basic process charts and the fundamentals of statistical quality improvement. The process for developing process charts is the same regardless of chart. Therefore, the things that are required are:

You need to know the generic process for developing charts.

You need to be able to interpret charts.

You need to be able to choose which chart to use.

You need the formulas to derive the charts.

You need to understand the purposes and assumptions underlying the charts.

We have given you all these things for variables in this short chapter. You have everything you need to get started. Have fun and enjoy yourself. Remember, the purpose of process charts is to help you to continually improve.

KEY TERMS

- Attribute
- Capability
- Consumer's risk
- Cpk
- Median chart
- Nonrandom variation
- Ppk
- Process charts
- Producer's risk
- *R* chart
- Random variation
- Sample
- *s* chart or standard deviation chart
- Stability
- Statistical thinking
- Variable
- *X* chart
- $\bar{x}$ *chart*

DISCUSSION QUESTIONS

1. The chapter states that the era of control has ended. Do you agree? Why or why not?
2. The concept of statistical thinking is an important theme in this chapter. What are some examples of statistical thinking?
3. Sometimes you do well on exams. Sometimes you have bad days. What are the assignable causes when you do poorly?

[3]To learn more about this, see Foster, S.T., "An Examination of the Relationship between Conformance and Quality-Related Costs," *International Journal of Quality and Reliability Management* 13, 4 (1996):50–63.
[4]Collier, D., *The Service/Quality Solution* (Homewood, IL: Irwin, 1994).

4. What is the relationship between statistical quality improvement and Deming's 14 points?
5. What are some applications of process charts in services? Could demerits (points off for mistakes) be charted? How?
6. What is random variation? Is it always uncontrollable?
7. When would you choose an *np* chart over a *p* chart? An *X* chart over an $\bar{x}$ chart? An *s* chart over an *R* chart?
8. Design a control chart to monitor the gas mileage in your car. Collect the data over time. What did you find?
9. What does "out-of-control" mean? Is it the same as a "bad hair day?"
10. Design a control chart to monitor the amounts of the most recently written 50 checks from your checkbook. What did you find?

PROBLEMS

1. Return to the chart in Figure 12-7. Is this process stable? Explain.
2. Return to the data in Figure 12-7. Is this process capable? Compute both Cpk and Ppk.
3. For the following product characteristics, choose where to inspect first:

Characteristic	Cost of Inspection	Cost of Failure
A	$2.50	$20
B	$2.00	$19
C	$4.00	$37
D	$3.00	$38

4. For the following product characteristics, choose where to inspect first:

Characteristic	Cost of Inspection	Cost of Failure
A	$35	$200
B	$37	$225
C	$38	$175
D	$40	$182

5. Interpret the following charts to determine if the processes are stable.

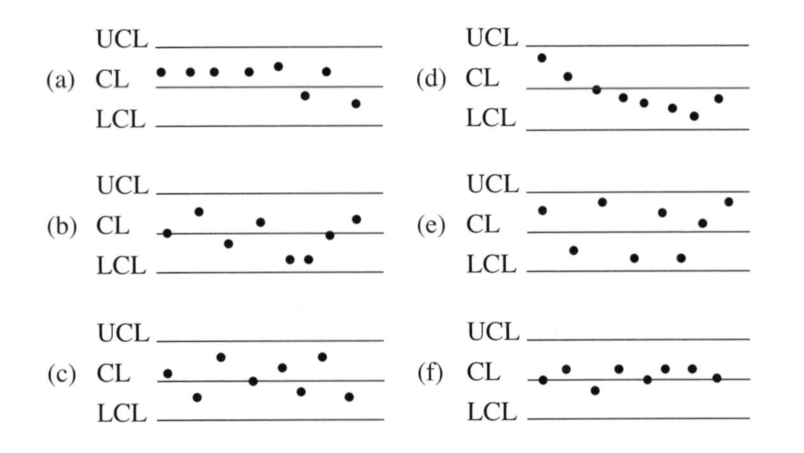

6. Interpret the following charts to determine if the processes are stable.

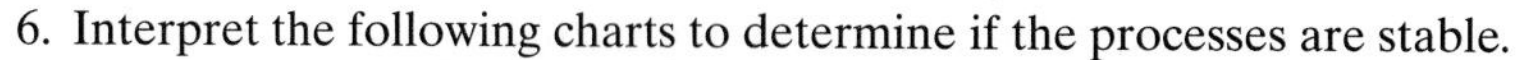

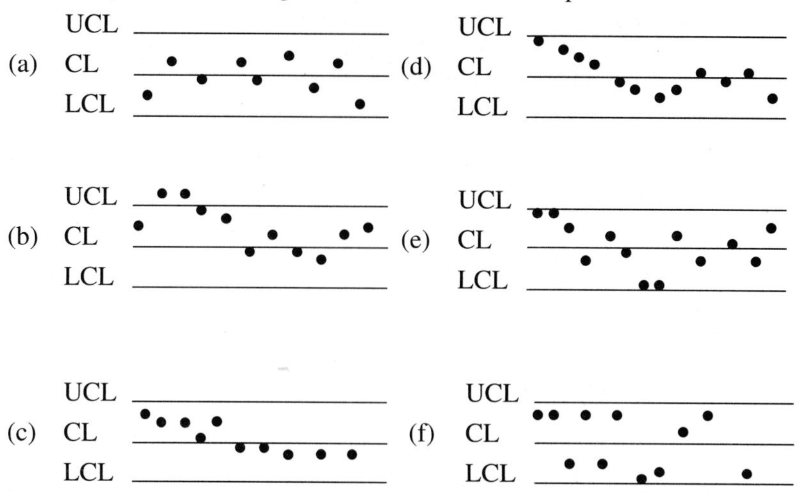

7. Tolerances for a new assembly call for weights between 32 and 33 pounds. The assembly is made using a process that has a mean of 32.6 pounds with a population standard deviation of .22 pounds. The process population is normally distributed.
   a. Is the process capable?
   b. If not, what proportion will meet tolerances?
   c. Within what values will 99.5% of sample means of this process fall if the sample size is constant at 10 and the process is stable?

8. Specifications for a part are 62″ +/− .01.″ The part is constructed from a process with a mean of 62.01″ and a population standard deviation of .033. The process is normally distributed.
   a. Is the process capable?
   b. What proportion will meet specifications?
   c. Within what values will 95% of sample means of the process fall if the sample size is constant at 5 and the process is stable?

9. Tolerances for a bicycle derailleur are 6 cm +/− .001 cm. The current process produces derailleurs with a mean of 6.0001 with a population standard deviation of .0004. The process population is normally distributed.
   a. Is the process capable?
   b. If not, what proportion will meet specs?
   c. Within what values will 75% of sample means of this process fall if the sample size is 6 and the process is stable?

10. A services process is monitored using $\bar{x}$ and R charts. Eight samples of $n = 10$ observations have been gathered with the following results:

Sample	Mean	Range
1	4.2	.43
2	4.4	.52
3	3.6	.53
4	3.8	.20
5	4.9	.36
6	3.0	.42
7	4.2	.35
8	3.2	.42

a. Using the data in the table, compute the center line, the upper control limit, and the lower control limit for the $\bar{x}$ and R charts.
b. Is the process in control? Please interpret the charts.
c. If the next sample results in the following values (2.5, 5.5, 4.6, 3.2, 4.6, 3.2, 4.0, 4.0, 3.6, 4.2), will the process be in control?

11. A production process for the JMF Semicon is monitored using x-bar and R charts. Ten samples of $n = 15$ observations have been gathered with the following results:

Sample	Mean	Range
1	251	29
2	258	45
3	233	36
4	275	25
5	234	35
6	289	20
7	256	3
8	265	19
9	246	14
10	323	46

a. Develop a control chart and plot the means.
b. Is the process in control? Explain.

12. *Experiment*: Randomly select the heights of at least 15 of the students in your class.
a. Develop a control chart and plot the heights on the chart.
b. Which chart should you use?
c. Is this process in control?

13. A finishing process packages assemblies into boxes. You have noticed variability in the boxes and desire to improve the process to fix the problem because some products fit too tightly into the boxes and others fit too loosely. Following are width measurements for the boxes.

Sample							
1	2	3	4	5	6	7	8
68.51	68.94	68.66	68.49	68.64	68.34	68.99	68.92
68.46	68.20	68.44	68.94	68.63	68.42	68.94	68.91
68.54	68.54	68.55	68.56	68.62	68.99	68.95	68.97
68.34	68.56	68.77	68.62	68.32	68.02	68.95	68.93
68.46	68.70	68.70	68.69	68.34	68.03	68.94	68.96
68.46	68.70	68.64	68.56	68.24	68.47	68.97	68.95

Using $\bar{x}$ and R charts, plot and interpret the process.

14. For the data in Problem 13, if the mean specification is 68.5 ± .25 and the estimated process standard deviation is .10, is the process capable? Compute Cpu, Cpl, and Cpk.
15. For the data in Problem 13, treat the data as if they were population data, and find the limits for an X chart. Is the process in control? Compare your answer with the answers to Problem 8. *Hint*: Use the formula $CL_x = \bar{x} \pm (3/d_2)\bar{R}$ (Figure 12-8).

16. A Rochester, New York firm produces grommets that have to fit into a slot in an assembly. Following are dimensions of grommets (in millimeters):

Sample	x				
1	46	33	54	46	64
2	52	45	54	75	64
3	34	64	36	46	63
4	34	45	47	37	62
5	46	64	75	55	16

a. Use x-bar and R charts to determine if the process is in control.

17. Using the data from Problem 13, compute the limits for x-bar and s charts. Is the process still in control?
18. Using the data from Problem 16, compute the limits for x-bar and s charts. Is the process still in control?
19. Use a median chart to determine if the process for the following data is centered.

Sample	Observation 1	Observation 2	Observation 3	Observation 4	Observation 5
1	8.06	7.93	8.19	8.45	8.32
2	8.06	8.06	8.06	8.19	8.32
3	8.19	7.67	8.06	8.32	8.19
4	6.89	6.63	6.89	6.63	6.89
5	7.93	8.58	8.19	8.06	8.32
6	8.06	8.06	8.06	8.06	8.06
7	7.54	7.41	7.67	9.36	6.76
8	8.19	7.67	8.06	8.32	8.19
9	8.19	7.67	8.06	8.32	8.19
10	9.62	9.62	9.23	9.49	9.23
11	8.06	8.19	8.06	8.19	8.06
12	8.32	8.19	8.06	7.93	7.93
13	8.19	8.32	8.06	8.19	7.93
14	7.93	7.93	7.93	7.93	7.93
15	8.19	8.32	7.93	8.19	7.93
16	8.32	8.06	8.32	8.06	8.06
17	8.06	8.32	8.19	8.32	8.06
18	7.93	8.06	8.19	8.32	8.45
19	8.06	7.93	7.93	7.93	7.93
20	8.32	8.19	8.06	8.45	8.19

20. Use an $\bar{x}$ chart to determine if the data in Problem 19 are in control. Do you get the same answer?
21. The following data are for a component used in the space shuttle. Since the process dispersion is closely monitored, use an $\bar{x}$ and s chart to see if the process is in control.

Sample	Observation 1	Observation 2	Observation 3
1	4.8000	4.7995	4.8005
2	4.7995	4.8007	4.8005
3	4.7995	4.8002	4.8012

4	4.7993	4.8000	4.8010
5	4.8007	4.8007	4.8005
6	4.8010	4.8007	4.8000
7	4.7995	4.7995	4.7995
8	4.8000	4.8002	4.8002
9	4.8012	4.8000	4.7998
10	4.7988	4.7995	4.8002
11	4.8005	4.7998	4.8002
12	4.8005	4.7995	4.8012
13	4.8000	4.8002	4.7995
14	4.8000	4.8005	4.8010
15	4.7986	4.8002	4.7990
16	4.7998	4.8007	4.7983
17	4.8005	4.7995	4.8010
18	4.8000	4.8002	4.8002
19	4.7993	4.7986	4.7995
20	4.8007	4.8017	4.7998

22. Develop an R chart for the data in Problem 21. Do you get the same answer?
23. Using the data from Problem 21, compute limits for a median chart. Is the process in control?
24. Design a control plan for exam scores for your quality management class. Describe how you would gather data, what type of chart is needed, how to gather data, how to interpret the data, how to identify causes, and remedial action to be taken when out-of-control situations occur.
25. For the sampling plan from Problem 24, how would you measure process capability?
26. For the data in Problem 16, if the process target is 50.25 with spec limits +/−5, describe statistically the problems that would occur if you used your spec limits on a control chart where $n = 5$. Discuss type I and type II error.

Case

Case 12-1 Ore-Ida Fries

www.heinz.com

One of the new innovations in the frozen french fry industry is the upright bag. When new equipment was introduced to produce the bags, the Heinz Frozen Food Corporation facility in Ontario, Oregon, was selected to produce the new bag type.

When the new bags were produced, there were problems with consistency. It was unclear whether the problem was with the machinery or the "film" (the material used in the bags). One of the key measurements was the distance from the UPC (universal product code) and a black mark on the bag. A number of rolls of film were randomly selected, and this measurement was taken. The result of this actual study was the data on the next page.

We need to know if the film is consistent. Please take the data on the next page and use control charts to determine if the measurements are consistent. Please report your results to management. ■

Sample	Millimeters from code to UPC Box				
1	7	7	8	6	7
2	6	5	6	5	7
3	7	7	8	6	8
4	6	8	8	7	7
5	6	7	6	6	7
6	6	6	5	6	5
7	5	6	4	4	4
8	4	5	5	5	6
9	5	6	5	5	5
10	5	5	5	5	5
11	6	6	7	7	7
12	7	7	6	7	7
13	6	7	7	7	7
14	6	7	7	7	7
15	6	6	6	6	6
16	6	6	6	6	6
17	6	7	7	6	7
18	6	7	6	7	7
19	6	6	6	6	6
20	5	6	5	6	6
21	9	12	10	10	10
22	10	10	9	10	10
23	10	10	10	9	10
24	10	10	10	10	10
25	10	10	10	10	10
26	10	10	10	11	10
27	11	12	10	11	11
28	11	12	10	11	12
29	10	11	11	11	11
30	10	11	12	10	10
31	10	11	11	11	11
32	11	11	11	12	12
33	11	11	0	0	5
34	6	4	4	5	7
35	7	6	6	0	1
36	6	7	6	7	6
37	6	6	5	6	7
38	10	9	10	10	9
39	10	9	8	8	11
40	10	10	10	10	10

Sample	Millimeters from code to UPC Box				
41	10	10	10	11	10
42	11	11	11	10	10
43	10	10	10	10	10
44	11	10	10	10	10
45	10	10	10	10	10
46	10	10	10	10	10
47	10	10	10	10	10
48	10	10	10	11	12
49	10	11	10	11	11
50	12	12	11	11	11
51	12	11	11	10	10
52	12	12	11	11	10
53	10	11	11	11	11
54	11	10	11	12	11
55	11	10	12	11	11
56	11	11	12	11	11
57	10	10	12	12	11
58	10	11	11	11	11
59	11	11	16	16	17
60	18	17	17	16	16
61	18	17	16	16	16
62	17	17	17	17	16
63	16	16	16	15	16
64	16	17	18	16	16
65	16	17	17	17	16
66	16	17	17	17	17
67	15	15	17	16	17
68	16	15	16	17	17
69	16	16	16	18	16
70	16	15	17	16	16
71	16	15	16	15	16
72	16	16	16	16	16
73	15	15	15	16	16
74	16	15	16	16	16
75	16	16	16	16	15
76	16	16	15	16	17
77	16	16	16	16	16
78	17	16	15	16	16
79	17	17	17	16	16
80	16	16	16	16	16

CHAPTER 13

Statistically Based Quality Improvement for Attributes

To be or not to be, that is the question.
—William Shakespeare

This quote from *Hamlet* might seem to you like an odd one for a quality management book. However, the quote gets at the core of what is an attribute. In supply chain quality, we are usually asking ourselves, "Is it a defect or not?" Is the piece defective, or is it not? Defined by Webster, an *attribute* is a "peculiar and essential characteristic." It is something that either does or does not exist.

As shown in Table 13-1, there are five types of attributes. **Structural attributes** have to do with physical characteristics of a particular product or service. For example, an automobile might have electric windows. Services have structural attributes as well, such as a balcony in a hotel room.

Sensory attributes relate to senses of touch, smell, taste, and sound. For products, these attributes relate to form design or packaging design to create products that are pleasing to customers. In services such as restaurants and hotels, atmosphere is very important to the customer experience.

Performance attributes relate to whether or not a particular product or service performs as it is supposed to. For example, does the lawn mower engine start? Does the stereo system meet a certain threshold for low distortion?

Temporal attributes relate to time. Were delivery schedules met? This often has to do with the reliability of delivery.

More and more, **ethical attributes** are important to firms. Do they report properly? Is their accounting transparent? Is the service provider empathetic? Is the teacher kind or not?

As you can see, all these types of attributes have to do with some state of being. These are often binary in nature. Either the flaw exists, or it does not. From a quality management point of view, we understand that there is a difference between attributes that are desired by customers and attributes that are monitored in the production processes. *Customer-based attributes* are more associated with customer satisfaction.

TABLE 13-1 Types of Attributes
Structural attributes
Sensory attributes
Performance attributes
Temporal attributes
Ethical attributes

Production-related attributes are more internal and engineering-oriented. Both types of attributes are important. In this chapter we will focus more on process-oriented attributes.

In this chapter we introduce quality control for attributes. This includes four types of control charts. After introducing these control charts, we will then present reliability models. We do this because reliability has to do with whether or not products fail—where failure is an attribute.

GENERIC PROCESS FOR DEVELOPING ATTRIBUTES CHARTS

In Chapter 12 we discussed the use of control charts. To reduce redundancy, we will not repeat control chart basics. Attribute charts are developed and interpreted the same way as variables charts. The only difference is the statistic of interest. In Chapter 12 we dealt with measurements. In this chapter we deal with states of being.

To reiterate, the generic process for developing process charts consists of the following six steps:

1. Identify *critical operations* in the process where inspection might be needed. These are operations that will have a negative effect on the product if performed improperly.
2. Identify *critical product characteristics*. These are the attributes of the product that will result in either good or poor form, fit, or function for the product.
3. Determine whether the critical product characteristic is a variable or an attribute.
4. Select the appropriate *process chart* from the many types of charts. The decision process and the types of charts are defined and discussed in Chapters 12 and 13.
5. Establish the control limits and use the chart to continually monitor and improve.
6. Update the limits when changes have been made to the process.

UNDERSTANDING ATTRIBUTES CHARTS

Attributes charts deal with binomial and Poisson processes that are not measurements. We will now think in terms of defects and defectives rather than diameters and widths. A *defect* is an irregularity or problem with a larger unit. The larger unit may contain many defects. For example, a piece of glass may contain several bubbles or scratches. Some of these may be detectable only with a magnifying glass. Defects are countable, such as six flaws within a particular glass pane. A *defective* is a unit that, as a whole, is not acceptable or does not meet performance requirements. For example, a letter with a wrong address is defective. Also, a letter could have several defects, but the entire unit is labeled defective if there is a single defect. Defectives are monitored using p and np charts. Defects are monitored using c and u charts. We will now introduce these charts.

p Charts for Proportion Defective

The ***p* chart** is a process chart that is used to graph the proportion of items in a sample that are defective (nonconforming to specification). p charts are effectively used to determine when there has been a shift in the proportion defective for a particular product or service. Typical applications of the p chart include things like late deliveries, incomplete orders, calls not getting dial tones, accounting transaction errors, clerical errors on written forms, or parts that do not mate properly.

The subgroup size on a p chart is typically between 50 and 100 units. The subgroups may be of different sizes for a p chart. However, it is best to hold subgroup sizes constant. Usually at least 25 subgroups are used to establish a p chart. The formulas for the p chart are as follows:

$$\text{Control limits for } p = \bar{p} \pm 3\sqrt{[(\bar{p})(1-\bar{p})/n]} \qquad \textbf{(13.1)}$$

where

p = the proportion defective
$\bar{p}$ = the average proportion defective
n = the sample size

Active Model: Example 13.1

Excel File: Example 13.1

Example 13.1: p Charts in Action

Problem: A city police department was concerned that the number of convictions was decreasing relative to the number of arrests. The suggestion was raised that the district attorney's office was becoming less effective in prosecuting criminals. You are asked to perform an analysis of the situation. The data for the previous 27 weeks are provided in the following table:

Sample	Number of Cases Reviewed	Number of Convictions	Proportion
1	100	60	.60
2	95	65	.68
3	110	68	.62
4	142	62	.44
5	100	56	.56
6	98	58	.59
7	76	30	.39
8	125	68	.54
9	100	54	.54
10	125	62	.50
11	111	70	.63
12	116	58	.50
13	92	30	.33
14	98	68	.69
15	162	54	.33
16	87	62	.71
17	105	70	.67
18	110	58	.53
19	98	30	.31
20	96	68	.71
21	100	54	.54
22	100	62	.62
23	97	70	.72
24	122	58	.48
25	125	30	.24
26	110	68	.62
27	100	54	.54
		$\bar{p}$ = 14.63/27 = .54	Sum = 14.63

Solution: Notice that in this problem, the sample size is not constant. When this happens, you have at least two options:

1. Compute the control limits using an average sample size (this is easier to understand).

2. Compute the control limits using the different sample sizes (this is statistically more correct).

Based on this analysis, a p chart for this process was established. As seen in Figure 13-1, the p charts that were computed using the two methods show that although it is not clear that the number of prosecutions is declining on a consistent basis, the process is becoming much more erratic, resulting in 2 months of poor performance (periods 19 and 25). Investigations should be undertaken to identify assignable causes of variation.

Using Excel to Draw p Charts

Figure 13-2 shows the Excel® solution for Example 13.1. As in Chapter 12, we generated this graph using the "brute force" method. Review Example 12.1 (Figure 12-12) for an in-depth discussion of how to generate the charts. Again, the chart shows that the process is erratic and out of control. This chart was drawn using the average sample size to compute the limits.

FIGURE 13-1 p Charts for Example 13.1

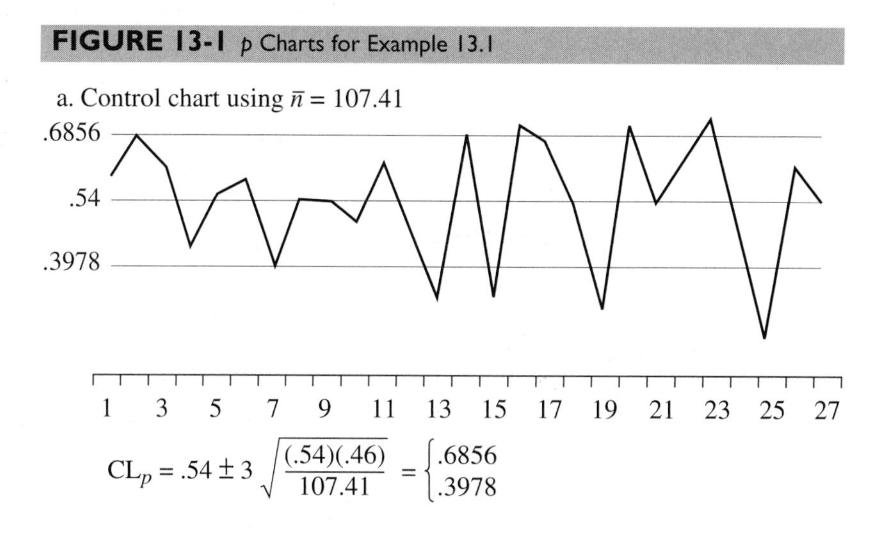

$$CL_p = .54 \pm 3\sqrt{\frac{(.54)(.46)}{107.41}} = \begin{cases} .6856 \\ .3978 \end{cases}$$

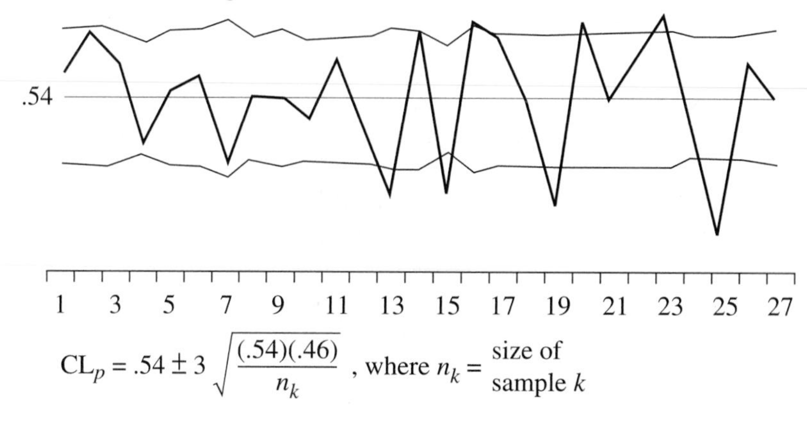

$$CL_p = .54 \pm 3\sqrt{\frac{(.54)(.46)}{n_k}}, \text{ where } n_k = \text{size of sample } k$$

FIGURE 13-2 Example 13.1 Using Excel

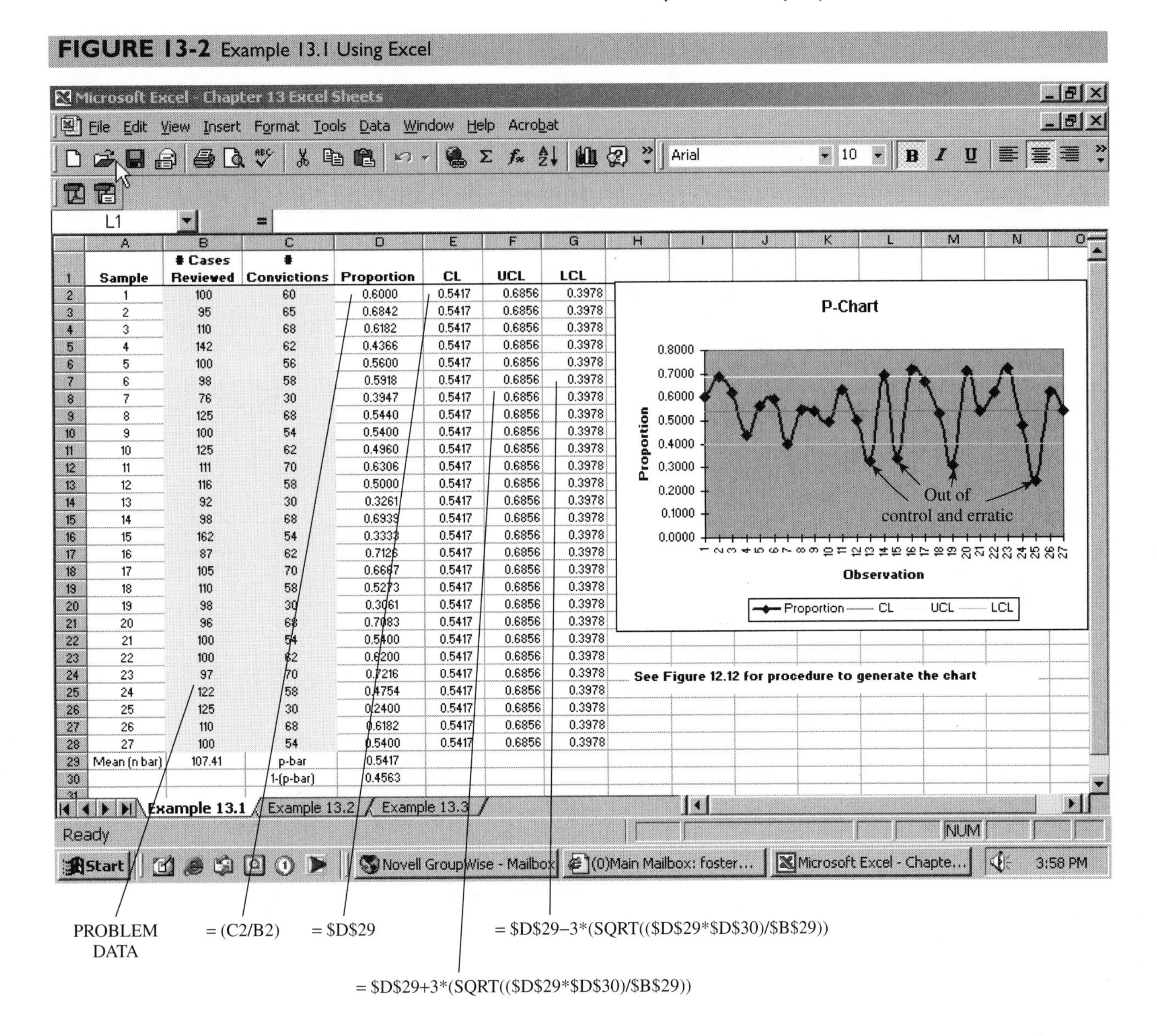

Sample	# Cases Reviewed	# Convictions	Proportion	CL	UCL	LCL
1	100	60	0.6000	0.5417	0.6856	0.3978
2	95	65	0.6842	0.5417	0.6856	0.3978
3	110	68	0.6182	0.5417	0.6856	0.3978
4	142	62	0.4366	0.5417	0.6856	0.3978
5	100	56	0.5600	0.5417	0.6856	0.3978
6	98	58	0.5918	0.5417	0.6856	0.3978
7	76	30	0.3947	0.5417	0.6856	0.3978
8	125	68	0.5440	0.5417	0.6856	0.3978
9	100	54	0.5400	0.5417	0.6856	0.3978
10	125	62	0.4960	0.5417	0.6856	0.3978
11	111	70	0.6306	0.5417	0.6856	0.3978
12	116	58	0.5000	0.5417	0.6856	0.3978
13	92	30	0.3261	0.5417	0.6856	0.3978
14	98	68	0.6939	0.5417	0.6856	0.3978
15	162	54	0.3333	0.5417	0.6856	0.3978
16	87	62	0.7126	0.5417	0.6856	0.3978
17	105	70	0.6667	0.5417	0.6856	0.3978
18	110	58	0.5273	0.5417	0.6856	0.3978
19	98	30	0.3061	0.5417	0.6856	0.3978
20	96	68	0.7083	0.5417	0.6856	0.3978
21	100	54	0.5400	0.5417	0.6856	0.3978
22	100	62	0.6200	0.5417	0.6856	0.3978
23	97	70	0.7216	0.5417	0.6856	0.3978
24	122	58	0.4754	0.5417	0.6856	0.3978
25	125	30	0.2400	0.5417	0.6856	0.3978
26	110	68	0.6182	0.5417	0.6856	0.3978
27	100	54	0.5400	0.5417	0.6856	0.3978
Mean (n bar)	107.41	p-bar	0.5417			
		1-(p-bar)	0.4563			

np Charts

The ***np* chart** is a graph of the number of defectives (or nonconforming units) in a subgroup. The *np* chart requires that the sample size of each subgroup be the same each time a sample is drawn. When subgroup sizes are equal, either the *p* or *np* chart can be used. They are essentially the same chart. Some people find the *np* chart easier to use because it reflects integer numbers rather than proportions. The uses for the *np* chart are essentially the same as the uses for the *p* chart.

Subgroup sizes for the *np* chart are normally between 50 and 100. Usually, at least 20 subgroups are used in developing the *np* chart. Again, *subgroup sizes must be equal.* To compute the control limits on an *np* chart, the following formula is used:

$$\mathrm{CL}_{np} = n(\bar{p}) \pm 3s_{np} \tag{13.2}$$

where

n = the sample size
$\bar{p}$ = is the average proportion defective

$$s_{np} = \text{standard error of } np = \sqrt{n\bar{p}(1 - \frac{n\bar{p}}{n})}$$

Active Model: Example 13.2

Example 13.2: *np* Charts in Action

Problem: Within the J. Kim Insurance Company of Boston, Massachusetts, management felt that too many of its policies were rated incorrectly. Management directed that policy applications be reviewed for the past 24 months on a sampling basis. One hundred policies from each month were selected for review. As an analyst, you are asked to review the policies for correct rating. If any problem is found with the rating of a policy, it is said to be defective.

Excel File: Example 13.2

Month	Number of Policies Reviewed	Number of Policies with Rating Errors	p
1	100	11	.11
2	100	10	.10
3	100	12	.12
4	100	6	.06
5	100	14	.14
6	100	8	.08
7	100	10	.10
8	100	9	.09
9	100	12	.12
10	100	2	.02
11	100	14	.14
12	100	18	.18
13	100	7	.07
14	100	13	.13
15	100	14	.14
16	100	12	.12
17	100	11	.11
18	100	8	.08
19	100	9	.09
20	100	17	.17
21	100	18	.18
22	100	20	.20
23	100	25	.25
24	100	28	.28
			Mean = .13

Solution: The results of the control chart are shown in Figure 13-3. The chart shows that rating errors are increasing. Assignable causes should be identified through investigation.

Using Excel to Draw *np* Charts

Figure 13-4 contains the Excel spreadsheet for Example 13.2. As you can see, the process has drifted out of control.

FIGURE 13-3 *np* Chart for Example 13.2

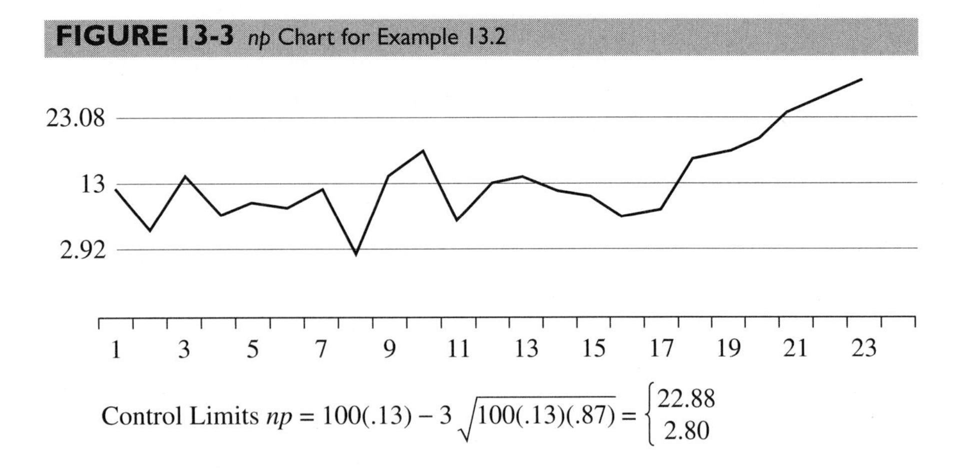

$$\text{Control Limits } np = 100(.13) - 3\sqrt{100(.13)(.87)} = \begin{cases} 22.88 \\ 2.80 \end{cases}$$

FIGURE 13-4 Example 13.2 Using Excel

Microsoft Excel - Chapter 13 Excel Sheets

A1 = Month

	A	B	C	D	E
1	Month	Rating Errors	CL	UCL	LCL
2	1	11	12.8333	22.8672	2.7995
3	2	10	12.8333	22.8672	2.7995
4	3	12	12.8333	22.8672	2.7995
5	4	6	12.8333	22.8672	2.7995
6	5	14	12.8333	22.8672	2.7995
7	6	8	12.8333	22.8672	2.7995
8	7	10	12.8333	22.8672	2.7995
9	8	9	12.8333	22.8672	2.7995
10	9	12	12.8333	22.8672	2.7995
11	10	2	12.8333	22.8672	2.7995
12	11	14	12.8333	22.8672	2.7995
13	12	18	12.8333	22.8672	2.7995
14	13	7	12.8333	22.8672	2.7995
15	14	13	12.8333	22.8672	2.7995
16	15	14	12.8333	22.8672	2.7995
17	16	12	12.8333	22.8672	2.7995
18	17	11	12.8333	22.8672	2.7995
19	18	8	12.8333	22.8672	2.7995
20	19	9	12.8333	22.8672	2.7995
21	20	17	12.8333	22.8672	2.7995
22	21	18	12.8333	22.8672	2.7995
23	22	20	12.8333	22.8672	2.7995
24	23	25	12.8333	22.8672	2.7995
25	24	28	12.8333	22.8672	2.7995
26	Mean	12.833			
27	Sample size, n	100			
28	p bar	0.128			
29	1-p bar	0.872			
30	Std err of np, S_{np}	3.345			
31					

PROBLEM DATA

= \$B\$26

= \$B\$26+3*\$B\$30

= \$B\$26–3*\$B\$30

c and *u* Charts

The ***c* chart** is a graph of the number of defects (nonconformities) per unit. The units must be of the same sample space; this includes size, height, length, volume, and so on. This means that the "area of opportunity" for finding defects must be the same for each unit. Several individual units can comprise the sample, but they will be grouped as if they are one unit of a larger size. If multiple units are used, the same number of units must be in each subgroup. The control limits for the *c* chart are computed based on the Poisson distribution.

Like other process charts, the *c* chart is used to detect nonrandom events in the life of a production process. Typical applications of the *c* chart include number of flaws in an auto finish (for a particular model), number of flaws in a standard typed letter, number of data errors in a standard form, and number of incorrect responses on a standardized test.

Again, the *c* chart is used when you are always inspecting the same size sample space. When the sample space is varied, such as in the inspection for flaws of different models of cars within a model family, a *u* chart is used.

The ***u* chart** is a graph of the *average number* of defects per unit. This is contrasted with the *c* chart, which shows the *actual number* of defects per standardized unit. The *u* chart allows for the units sampled to be different sizes, areas, heights and so on and allows for different numbers of units in each sample space. The uses for the *u* chart are the same as the *c* chart.

The formulas for the *c* and *u* charts are

$$CL_c = \bar{c} \pm 3\sqrt{\bar{c}} \tag{13.3}$$

$$CL_u = \bar{u} \pm 3\sqrt{\frac{\bar{u}}{n}} \tag{13.4}$$

where

n = average sample size
$\bar{c}$ = process average number of nonconformities
$\bar{u}$ = process average number of nonconformities per unit

As the following example shows, the limits for the *u* chart are more conservative than are the limits for the *c* chart.

Active Model: Example 13.3

Example 13.3: c and *u* Charts in Action

Problem: The J. Grout Window Company makes colored-glass objects for home decoration. J. Grout, the owner, has been concerned about scratches in the finish of recently made product. The company makes two products. These are the Demi-Glass, which comes in one standard configuration, and the Streakless-Glass, which comes in three similar models. Using high-power magnifying glasses, the company examined 25 each of both the Demi (one style only) and the Streakless (randomly selected in all three styles). As an analyst, you are asked to evaluate the process by determining whether the processes are stable. Note that, on average, the Streakless are 1.5 times the size of the Demis.

Excel File: Example 13.3

Item Number	Demi Defects	Streakless Defects
1	5	6
2	4	4

3	6	7
4	3	9
5	9	5
6	4	8
7	5	7
8	4	4
9	3	5
10	7	4
11	9	5
12	12	4
13	3	5
14	6	6
15	2	4
16	8	8
17	5	5
18	7	7
19	12	10
20	4	5
21	6	4
22	8	7
23	5	5
24	7	6
	Sum $c = 144$	Sum $u = 140$
	$\bar{c} = 6$	$\bar{u} = 5.83$

Solution: As shown in Figures 13-5 and 13-6, the process for Demis appears to be in control. However, the process for Streakless shows a run of five points below the mean. An assignable cause should be sought.

Using Excel to Draw *c* and *u* Charts

Figure 13-7 shows the Excel® *c* and *u* charts from Example 13.3. As you can see, these charts are drawn using the typical method. Formulas are provided in the figure. Note that since the lower control limits were computed to be negative numbers, zero is used as the lower limit for both charts.

FIGURE 13-5 c Chart for Demis

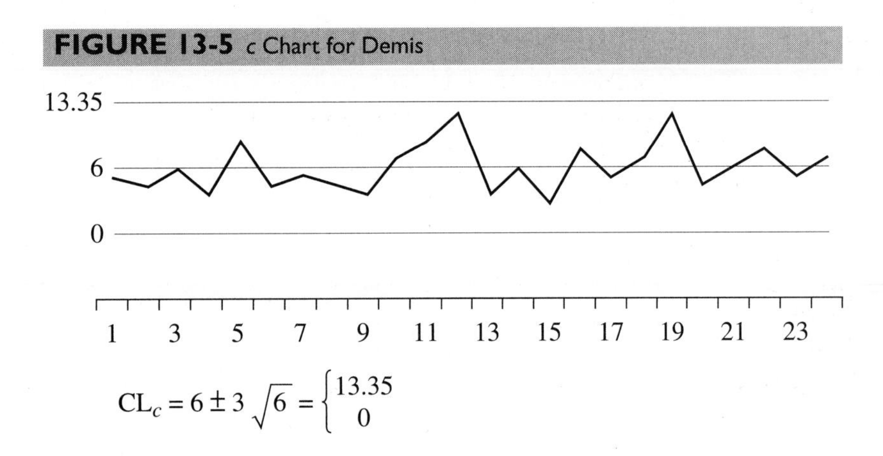

$$CL_c = 6 \pm 3\sqrt{6} = \begin{cases} 13.35 \\ 0 \end{cases}$$

FIGURE 13-6 *u* Chart for Streakless

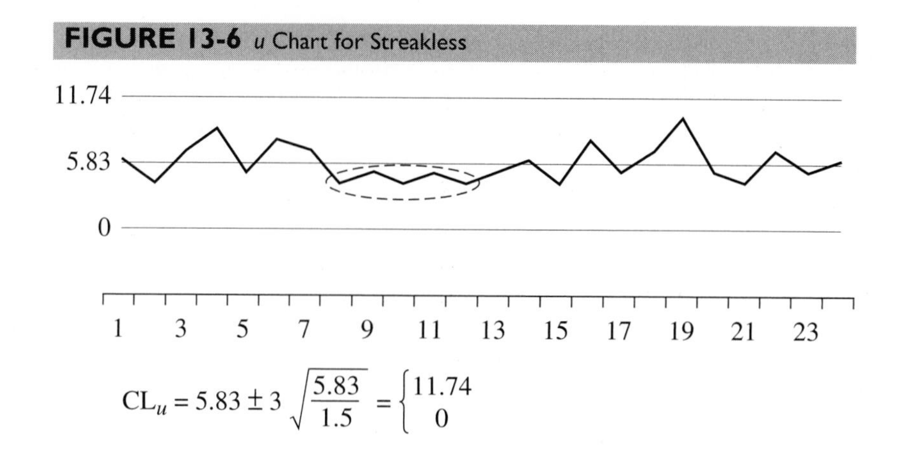

$$CL_u = 5.83 \pm 3\sqrt{\frac{5.83}{1.5}} = \begin{cases} 11.74 \\ 0 \end{cases}$$

FIGURE 13-7 Example 13.3 Using Excel

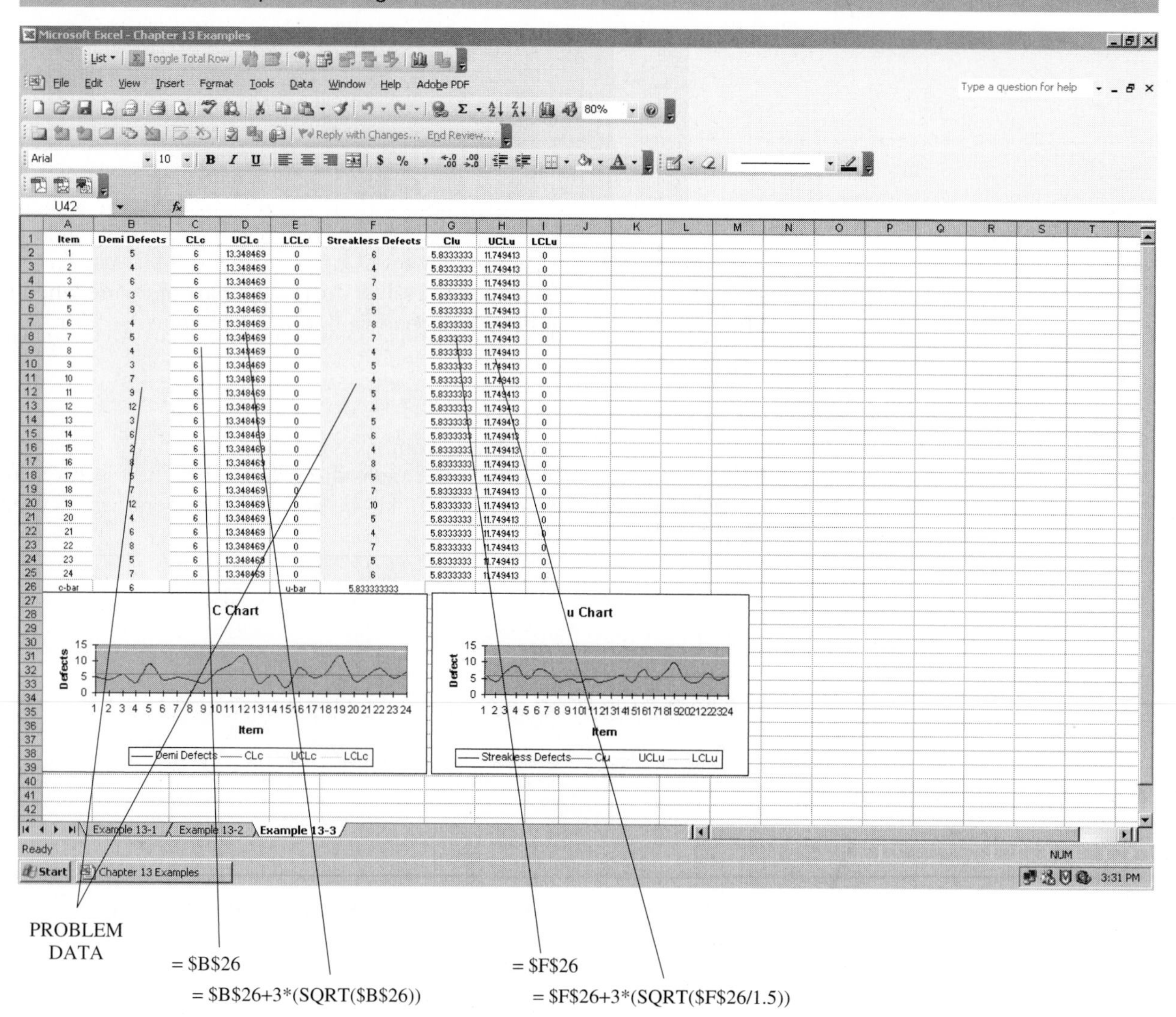

	A	B	C	D	E	F	G	H	I
1	Item	Demi Defects	CLc	UCLc	LCLc	Streakless Defects	Clu	UCLu	LCLu
2	1	5	6	13.348469	0	6	5.8333333	11.749413	0
3	2	4	6	13.348469	0	4	5.8333333	11.749413	0
4	3	6	6	13.348469	0	7	5.8333333	11.749413	0
5	4	3	6	13.348469	0	9	5.8333333	11.749413	0
6	5	9	6	13.348469	0	5	5.8333333	11.749413	0
7	6	4	6	13.348469	0	8	5.8333333	11.749413	0
8	7	5	6	13.348469	0	7	5.8333333	11.749413	0
9	8	4	6	13.348469	0	4	5.8333333	11.749413	0
10	9	3	6	13.348469	0	5	5.8333333	11.749413	0
11	10	7	6	13.348469	0	4	5.8333333	11.749413	0
12	11	9	6	13.348469	0	5	5.8333333	11.749413	0
13	12	12	6	13.348469	0	4	5.8333333	11.749413	0
14	13	3	6	13.348469	0	5	5.8333333	11.749413	0
15	14	6	6	13.348469	0	6	5.8333333	11.749413	0
16	15	2	6	13.348469	0	4	5.8333333	11.749413	0
17	16	8	6	13.348469	0	8	5.8333333	11.749413	0
18	17	5	6	13.348469	0	5	5.8333333	11.749413	0
19	18	7	6	13.348469	0	7	5.8333333	11.749413	0
20	19	12	6	13.348469	0	10	5.8333333	11.749413	0
21	20	4	6	13.348469	0	5	5.8333333	11.749413	0
22	21	6	6	13.348469	0	4	5.8333333	11.749413	0
23	22	8	6	13.348469	0	7	5.8333333	11.749413	0
24	23	5	6	13.348469	0	5	5.8333333	11.749413	0
25	24	7	6	13.348469	0	6	5.8333333	11.749413	0
26	c-bar	6			u-bar	5.833333333			

ATTRIBUTES CHARTS SUMMARY

Table 13-2 shows all the formulas for the process charts we have discussed in this chapter. These are the major attributes charts that are used the vast majority of times.

CHOOSING THE RIGHT ATTRIBUTES CHART

Figure 13-8 provides a flowchart for choosing the correct attributes chart. The key questions are whether you are interested in defects or defectives and whether your sample sizes are constant. As we said before, for c and u charts, we are more interested in whether or not the sample space is constant. This is the standardized chart used by auto companies such as Ford in selecting charts.

RELIABILITY MODELS

Although there are several reliability models, we will only discuss some of the simpler ones. The first model is graphic (see Fig. 13-9) and is called the **bathtub-shaped hazard function**. The vertical axis on the bathtub function is failure rate. The horizontal axis is time. This model shows us that products are more likely to fail either very early in their life or late in their useful life. Consider this function when you purchase major appliances. It is now very common for appliance vendors to offer service contracts at an additional cost. However, notice from the bathtub function that products likely will fail either very early in their life or after their expected useful life. Because most major appliances include a 1-year warranty that covers all the labor and parts needed to repair the appliance and most appliances are made to last several years, by purchasing a service contract, you are really insuring the product during the part of its life when it is least likely to fail. This appears to be a very good deal for the appliance vendor.

Series Reliability

Components in a system are in series if the performance of the entire system depends on all the components functioning properly. The components need not be physically wired sequentially for the system to be in series. However, all parts must function for the system to function. Figure 13-10 shows n components in a series. System reliability for the series is expressed as[1]

$$R_s = P(x_1 x_2 \cdots x_n) \quad \textbf{(13.5)}$$

$$= P(x_1)P(x_2|x_1)P(x_3|x_1x_2) \cdots P(x_n|x_1x_2 \cdots x_{n-1}) \quad \textbf{(13.6)}$$

TABLE 13-2 Summary of Chart Formulas

Chart	*LCL*	*CL*	*UCL*
p	$\bar{p} - 3\sqrt{\bar{p}(1-\bar{p})/n}$	$\bar{p}$	$\bar{p} + 3\sqrt{\bar{p}(1-\bar{p})/n}$
np	$n\bar{p} - 3\sqrt{n\bar{p}(1-\bar{p})}$	$n\bar{p}$	$n\bar{p} + 3\sqrt{n\bar{p}(1-\bar{p})}$
c	$\bar{c} - 3\sqrt{\bar{c}}$	$\bar{c}$	$\bar{c} + 3\sqrt{\bar{c}}$
u	$\bar{u} - 3\sqrt{\bar{u}/n}$	$\bar{u}$	$\bar{u} + 3\sqrt{\bar{u}/n}$

[1]Ramakumar, R., *Engineering Reliability* (Englewood Cliffs, NJ: Prentice Hall, 1993).

FIGURE 13-8 Process for Selecting the Right Chart

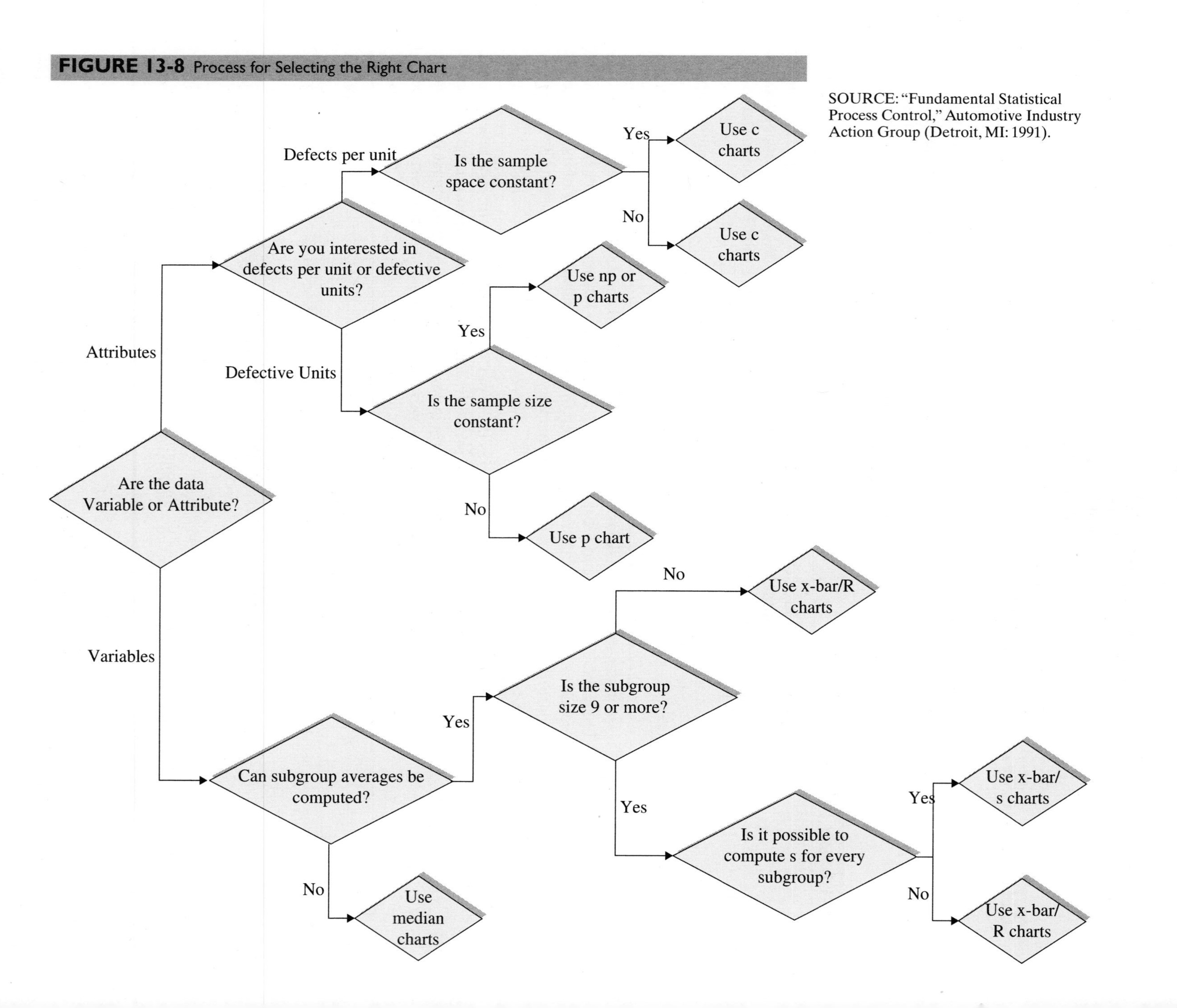

SOURCE: "Fundamental Statistical Process Control," Automotive Industry Action Group (Detroit, MI: 1991).

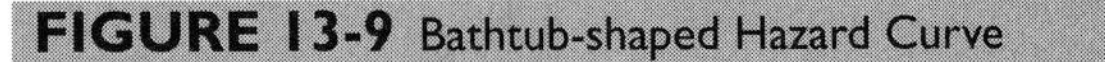
FIGURE 13-9 Bathtub-shaped Hazard Curve

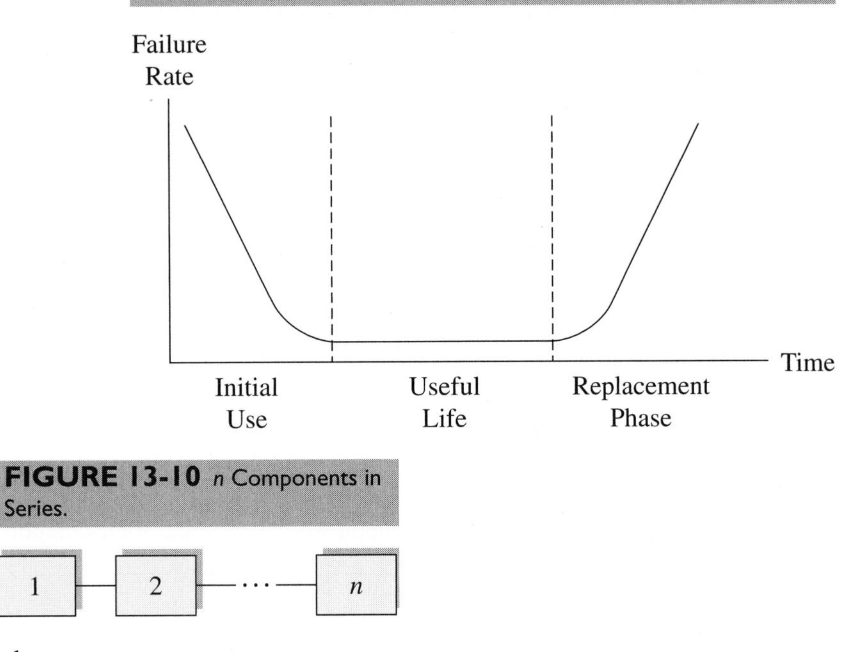

FIGURE 13-10 *n* Components in Series.

where

R_s = system reliability

$P(x) = 1 -$ probability of failure for component x_i

By the same token, system unreliability can be modeled as

$$Q_s = 1 - R_s \tag{13.7}$$

where Q_s = system unreliability.

These reliability models assume independence between failure events. This means that the failure of one component does not influence another component to fail. Following is an example of simple reliability for one component: Imagine a component with 99% reliability over 5 years. This component will have a 99% chance of lasting 5 years—only a 1% chance of failure. This probably sounds pretty good to you. However, consider a television set made up of 700 components with each component having a 99% reliability. If this is a series system, where the failure of any one component will cause the entire system to fail, the overall reliability will be $.99^{700} = .00088$ or .088% reliable. In other words, the television has less than a 1% chance of surviving 5 years—not so good. To compare, think of an automobile with 17,000 parts or a space shuttle with millions of components. This gives you an understanding of how difficult it is to make a product that will last.

Now let's suppose that we wanted this television with 700 components to have 90% overall reliability. If we want $0.90 = R^{700}$, we find that $R = (0.90)^{1/700} = .99985$, which is the required component reliability.

Parallel Reliability

As we have already seen, a high-reliability system often requires extremely high component reliability. At times when such high reliability is an impossibility, an alternative is to use a backup system. Other words for a backup system are *redundant* or *parallel systems*. If a set of components is in parallel, as opposed to being in series, the system can function if a given component in the system fails. System reliability is then expressed as

$$R_p = P(x_1 + x_2 + \cdots + x_n) = 1 - P(\bar{x}_1\bar{x}_2 \cdots \bar{x}_n) \quad \textbf{(13.8)}$$

Given that component failures are independent of one another, redundant reliability is modeled as

$$R_p = 1 - P(\bar{x}_1)P(\bar{x}_2|\bar{x}_1)P(\bar{x}_3|\bar{x}_1\bar{x}_2) \cdots P(\bar{x}_n|\bar{x}_1\bar{x}_2 \cdots \bar{x}_{n-1}) \quad \textbf{(13.9)}$$

Therefore, system unreliability is modeled as

$$Q_p = \prod_{i=1}^{n} Q_i \quad \textbf{(13.10)}$$

Example 13.4: Parallel and Series Reliability

At times, systems have some components in series and some components in parallel (or redundancy). Figure 13-11 has one such system.

Overall reliability for this system is

$$R = .98 \times .99 \times (1 - (.1 \times .1)) \times .97 = .932$$

To continue the example, it is interesting to compare the overall reliability of this system without component C_2. This equals

$$R = .98 \times .99 \times .90 \times .97 = .847$$

Thus the overall improvement in system reliability by adding the additional component is

$$D = .932 - .847 = .085$$

This is an 8.5% improvement in system reliability resulting from the additional component.

A space shuttle would be unreliable using a series system, given the millions of components contained in the shuttle. Therefore, many redundant systems are used in the shuttle. This choice only improves the likelihood that the shuttle will function properly; however, it is not a guarantee. For example, the O-rings that exploded on the

FIGURE 13-11 Series and Redundant Reliability

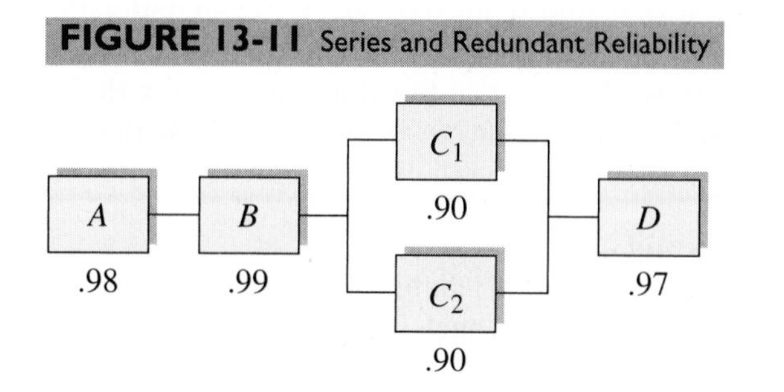

space shuttle *Challenger* included a second set of redundant O-rings. It has been suggested that the engineers working for NASA made the mistake of concluding that failure of the redundant O-rings would be independent of the failure of the original O-rings. However, they were not independent, and both sets of O-rings failed. This demonstrates the importance of the independence assumption. These formulas assume that the failure of one component is independent of another component. Independence is modeled using conditional probability notation as

$$\text{If } P(x_1|x_2) = P(x_1), \text{ then events } x_1 \text{ and } x_2 \text{ are independent} \quad \textbf{(13.11)}$$

where $P(x)$ is the probability of some event (x).

Measuring Reliability

Let's discuss some basic reliability functions. Failure rate is measured using the following equation:

$$\text{Failure rate} = \lambda = \text{number of failures}/(\text{units tested} \times \text{number of hours tested}) \quad \textbf{(13.12)}$$

Note that some care should be exercised in using this function because there is no distinction as to whether the hours of testing are continuous or performed at separate times. For example, there is no difference in testing hours if five units are tested for 100 hours each or if one unit is tested for 500 hours.

Example 13.5: Reliability Measurement—Failure Rates

Problem: Suppose that we tested 25 ski exercise machines under strenuous conditions for 100 hours per machine. Of the machines tested, 3 experienced malfunctions during the test. What is the failure rate for the exercise machines?

Solution: Failure rate = $3/(25 \times 100) = .0012$ failures per operating hour

MTTF

Since reliability can be defined as the probability that a product will not fail over its defined product life, if λ is the product failure rate, the function representing failure can be modeled using the following exponential function:

$$R(T) = 1 - F(T) = e^{-\lambda T} \quad \textbf{(13.13)}$$

where

$R(T)$ = reliability of the product
$F(T)$ = unreliability of a product
λ = failure rate
T = product's useful life expressed as a function of time

Using another useful function, $1/\lambda$ is called the **mean time to failure (MTTF)**. This is the average time before the product will fail.

Example 13.6: MTTF

Problem: Suppose that a product is designed to operate for 100 hours continuously with a 1% chance of failure. Find the number of failures per hour incurred by this product and the MTTF.

Solution:

$$0.99 = e^{-\lambda(100)}$$
$$\ln 0.99 = -100\lambda$$
$$\lambda = -(\ln 0.99)/100$$
$$= .01005/100$$
$$= .0001005$$

$$\text{MTTF} = 1/.0001005 = 9950.25 = 1/\lambda \quad \textbf{(13.14)}$$

This means that the failure rate for this product is .0001005 and that the average time before failing is 9950.25 hours.

Another function of interest is the **mean time between failures (MTBF)**. This tells us the average time from one failure to the next when a product can be repaired. The formula is

MTBF = total operating hours/number of failures

This formula is important in scheduling service calls. Consider the plight of Otis Elevator company in trying to determine the number of service representatives needed in New York City. If they know their MTBF, the number of elevators in service, and the hours those elevators are in use, they can calculate how many service reps are needed. A Closer Look at Quality 13-1 looks at failures and reliability from a more macroperspective. What do you think? Is quality getting worse?

A CLOSER LOOK AT QUALITY 13-1

IS QUALITY ON THE DECLINE?

Has quality been on the decline in the past decade? Some say, "Yes!" Admittedly, some of the evidence is anecdotal. But its sources are diverse: repair shop techs, blog gripes, current and former quality management consultants, and others. Lamp housings crack, VCRs rewind slowly and haltingly, cell phone batteries fall out, shirt buttons crumble, washing machines falter, and TVs render flesh tones in rainbow hues.

Overall, automobile performance is improved, but many components are still very fragile. For instance, the dashboard "idiot light" on many cars signals a mysterious computer-detected malfunction somewhere in the engine (often in sensors tied to a catalytic converter). It typically requires a repair shop visit to diagnose and shut off the light. But then, a day or two after the ostensible repair, the light reactivates itself, like the villain in a Halloween movie who cannot be escaped.

The American Society for Quality and the University of Michigan have cosponsored the American Customer Satisfaction Index (ACSI) based on customer surveys. They show recent declines in customer perceptions of quality for many sectors of the economy. Some big-name companies showing declines include Hewlett-Packard and General Electric.

Many product manufacturers are lowering engineering standards to shave costs. "One thing that often goes wrong with VCRs is the loading mechanism," says Tod Marks, an author for *Consumer Reports* magazine. "That used to be metal, attached with screws. Now it's a piece of extruded plastic fused to the chassis."

There is some evidence that warranty costs are increasing—although these have only been reported since 2003. "It's happening on so many dimensions," says Greg Brue, president of Albuquerque-based Six Sigma Consultants. "Companies are going to shorter and shorter warranties, and dealing with more and more repairs, and responding with rebates and price promotions instead of improving their products—and they feel like they are getting away with it."[a]

The bottom line is that we need to be demanding of those who provide products.

[a]This is adapted from C-net news.com, March 30, 2005—from Booz Allen and Hamilton.

Example 13.7: MTBF

Problem: A product has been operated for 10,000 hours and has experienced four failures. What is the MTBF?

Solution:

$$\text{MTBF} = 10{,}000/4 = 2500 \text{ hours between failures}$$

The failure rate is then calculated as $\lambda = 1/2500 = .0004$ failures per hour.

System Availability

It should be remembered that mean time between failures is a useful measure for many products. In Example 13.7, 2500 hours between failures may not make sense because many products are never used that many hours. To clarify, MTBF generally is used as an average over several products.

A useful measure for maintainability of a product is system availability, which considers both MTBF and a new statistic, **mean time to repair (MTTR)**. **System availability (SA)** gives us the "uptime" of a product or system. Here is the formula:

$$\text{SA} = \text{MTBF}/(\text{MTBF} + \text{MTTR}) \tag{13.15}$$

Example 13.8: System Availability

Problem: Jane Bell has to decide between one of three suppliers for a server for a network. Other factors equal, she is going to base her decision on system availability. Given the following data, which supplier should she choose?

Supplier	MTBF (h)	MTTR (h)
A	67	4
B	45	2
C	36	1

Solution: Using Formula 13.15, we find the following solutions:

$$SA_A = 67/(67 + 4) = .944$$
$$SA_B = 45/(45 + 2) = .957$$
$$SA_C = 36/(36 + 1) = .973$$

Choose supplier C. As you can see, service does matter.

SUMMARY

In Chapters 12 and 13 we introduced the basics of quality control. Remember that the object of using process charts is to continually improve your processes. Monitoring processes is not enough. As we make changes and improvements to the processes, our attributes charts will improve—there will be fewer defects and defectives. As this improvement takes place, the control limits constrict and draw closer to zero (for attributes charts). This is the goal. If you are not seeing this type of improvement in your processes, you should work more to improve processes.

We also introduced a number of reliability models. These included series reliability, parallel reliability, and reliability functions. Reliable processes are cost-effective and productive.

In the last few chapters we have introduced several quality tools and models. In the next chapter we will apply these in a Six-Sigma setting.

KEY TERMS

- Bathtub-shaped hazard function
- *c* chart
- Ethical attributes
- Mean time between failures (MTBF)
- Mean time to failure (MTTF)
- Mean time to repair (MTTR)
- *np* chart
- Parallel reliability
- *p* chart
- Performance attributes
- Sensory attributes
- Series reliability
- Structural attributes
- System availability
- Temporal attributes
- *u* chart

DISCUSSION QUESTIONS

1. What are key attributes for a high-quality university?
2. What are some attributes that you can identify for an automobile tire?
3. What are some attributes for a university financial aid process?
4. What are some personal attributes that you could monitor using control charts? Which control chart would you use?
5. What are examples of structural attributes?
6. What are some examples of sensory attributes?
7. What are some examples of performance attributes?
8. What are some examples of temporal attributes?
9. What are some examples of ethical attributes?
10. What ethical attributes might you use to determine where you should go to work after graduation?

PROBLEMS

1. Suppose you want to inspect a lot of 10,000 products to see whether or not they meet requirements. Design a sampling plan used to test these products.
2. Suppose a product is made of 100 components, each with a 97% reliability. What is the overall reliability for the product?
3. Suppose a product is made of 1,000 components, each with .999 reliability. What is the unreliability of this product? Is this acceptable? Why or why not?
4. A product consists of 45 components. Each component has an average reliability of .97. What is the overall reliability for this product?
5. A radio is made up of 125 components. What would have to be the average reliability for each component for the radio to have a reliability of 98% over its useful life?
6. List five products with low reliability. List five that have high reliability. What are the elemental design differences between these products? In other words, what are the factors that make some products reliable and others unreliable?
7. An assembly consists of 240 components. Your customer has stated that your overall reliability must be at least 99%. What needs to be the average reliability factor for each component?
8. A product is made up of six components. They are wired in series with reliabilities of .95, .98, .94, .96, .98, and .97. What is the overall reliability for this product?
9. Suppose that redundant components are introduced for the two components in Problem 8 with the lowest reliability. What is now the overall reliability for this product?

10. Suppose that redundant components are introduced for all of the components in Problem 8. What is now the overall reliability for the product?
11. A product is made up of components A, B, C, and D. These components are wired in series. Their reliability factors are .98, .999, .97, and .989, respectively. Compute the overall reliability for this product.
12. A product is made up of components A, B, C, D, E, F, G, H, I, and J. Components A, B, C, and F have a 1/10,000 chance of failure during useful life. D, E, G, and H have a 3/10,000 chance of failure. Components I and J have a 5/10,000 chance of failure. What is the overall reliability for this product?
13. For the product in Problem 12, if parallel components are provided for components I and J, what is the overall reliability for the product?
14. A product is made up of 20 components in a series. Ten of the components have a 1/10,000 chance of failure. Five have a 3/10,000 chance of failure. Four have a 4/10,000 chance of failure. One component has a 1/100 chance of failure. What is the overall reliability for this product?
15. For the product in Problem 14, if parallel components are used for any component with worse than a 1/1,000 chance of failure, what is the overall reliability? How many components will the new design have? What will be the average component reliability for the redesigned product?
16. An inspector visually inspects 200 sheets of paper for aesthetics. Using trained judgment, the inspector will either accept or reject sheets based on whether they are flawless. Following are the results of recent inspections:

Sample	1	2	3	4	5	6	7	8	9	10
Defectives	10	15	12	14	26	3	10	14	12	11

a. Given these results, using a p chart, determine if the process is stable.
b. What would need to be done to improve the process?

17. Using the data in Problem 16, compute the limits for an np chart.
18. Suppose a company makes the following product with the following numbers of defects. Construct a p chart to see if the process is in control. $n = 100$

Sample	Defectives
1	67
2	28
3	45
4	32
5	30
6	48
7	32
8	24
9	25
10	27
11	28
12	29
13	65
14	66
15	69
16	70
17	26

Continued

18	13
19	45
20	46
21	47
22	48
23	28
24	29
25	75

19. Using the data from Example 13.3, evaluate the Demis using a u chart and evaluate the Streakless using a c chart. Assume that the Demis are twice the size as the Streakless on average.
20. Politicians closely monitor their popularity based on approval ratings. For the previous 16 weeks, Governor Johnny's approval ratings have been (in percentages):

Month	1	2	3	4	5	6	7	8	9	10	11	12	13	14	15	16
Approval %	65	62	59	64	61	60	58	52	51	53	54	52	62	65	66	67

a. Prepare a report for the governor outlining the results of your analysis. Use control charts to analyze the data ($n = 200$).
b. What action would you propose to the governor based on your analysis?

21. Construct and interpret a c chart using the following data:

Sample	Defects
1	6
2	5
3	7
4	6
5	8
6	5
7	6
8	7
9	6
10	8
11	7
12	6
13	7
14	8
15	7
16	6
17	5
18	2
19	1
20	0
21	12
22	4
23	6
24	7
25	8
26	3

27	2
28	3
29	2
30	3

22. Construct and interpret a u chart using the following data. Note that the average size is two times the original product.

Sample	Defects
1	4
2	7
3	6
4	7
5	4
6	5
7	7
8	4
9	5
10	7
11	5
12	3
13	5
14	6
15	3
16	7
17	6
18	8
19	4
20	5
21	6
22	7
23	3
24	2
25	3
26	2
27	3

23. Dellana company tested 50 products for 75 hours each. In this time, they experienced four breakdowns. Compute the number of failures per hour. What is the mean time between failures?
24. The Collier Company tested 200 products for 100 hours each. In this time, they experienced 12 breakdowns. Compute the number of failures per hour. What is the MTBF?
25. Crager company tested 100 products for 50 hours each. During the test, three breakdowns occurred. Compute the number of failures per hour and MTBF.
26. Suppose a product is designed to function for 10,000 hours with a 3% chance of failure. Find the average number of failures per hour and the MTTF.
27. Suppose a product is designed to function for 100,000 hours with a 1% chance of failure. Suppose that there are six of these in use at a facility. Find the average number of failures per hour and the MTTF.

28. Suppose that there are 42 pumps used in a refinery. These pumps are continuously being used with a 2% chance of failure over 50,000 hours. If repair time is 10 hours to install a new rebuilt pump, how many pumps should be kept on hand to keep the chance of a plant shutdown to less than 1%. (Hint: Treat this problem as a traditional safety stock problem and use a z table.)
29. Suppose that a product is designed to work for 1,000 hours with a 2% chance of failure. Find the average number of failures per hour and the MTTF.
30. A product has been used for 5,000 hours with 1 failure. Find the mean time between failures (MTBF) and λ.
31. You are to decide between three potential suppliers for an assembly for a product you are designing. After performing life testing on several assemblies, you find the following:

Supplier	MTBF (h)	MTTR (h)
A	45	2
B	100	6
C	150	9

Based on system availability, which supplier should you choose?

32. You are to choose a supplier of a copier based on reliability and service. After gathering data about the alternatives, here is what you found. What do you recommend?

Supplier	MTBF (hr)	MTTR (hr)
1	45	2
2	90	2
3	120	6
4	200	6

Case

Case 13-1 Decision Sciences Institute National Conference

During a recent Decisions Sciences Institute National Conference, the author of this text served as a track chair for the Manufacturing Track. Papers were submitted by 174 authors to this track of the conference with the hopes that their papers would be accepted, and they would be given the opportunity to present their research to colleagues from around the world at the conference in Las Vegas.

The 174 papers were sent to reviewers with the results shown in Table 13-3. Following is a key to understanding the codes:

r = reject
t = table topic (a lower level of acceptance)
f = full session presentation
fy = full session and recommended for best paper contest

One of the difficulties of statistical analysis is to figure out how to analyze, interpret, and present the results of your analysis. Take these raw data and develop research questions. Next, using the statistical tools from this chapter, analyze the data. Finally, put the data into a form that will be useful for decision makers. Good luck. ■

TABLE 13-3 DSI Review Results

Paper Number	*Reviewer 1*	*Reviewer 2*	*Reviewer 3*	*Paper Number*	*Reviewer 1*	*Reviewer 2*	*Reviewer 3*
1	r	f	t	55	f		f
2				56	f		f
3	f	f	f	57	f		f
4				58	t	f	f
5	f	f	f	59	t	f	f
6	r		r	60	fy	f	f
7	t		t	61	f	f	f
8	f		f	62	fy	fy	f
9	t	t	t	63	f		f
10	fy		f	64	t	f	t
11	fy	fy	f	65	f	r	f
12	f	f	f	66	fy	f	t
13	f		f	67	r	f	t
14	f		f	68	f	f	f
15	r		r	69	f		f
16	f		f	70	f		f
17	r	f	t	71		t	Workshop
18	t	f	f	72	t	t	t
19	f		f	73	f		t
20	t		t	74	t	f	t
21	f		f	75	t		f
22	r		r	76	f		f
23	t		t	77		r	
24	f	t	f	78	r	f	r
25	fy		f	79	f	f	f
26	t		t	80	f		f
27	f	f	f	81	fy		f
28				82	f		f
29				83			
30	fy	fy	f	84	t		t
31	fy	fy	f	85	f	f	f
32	t		t	86	f		f
33	f		f	87	fy	f	f
34	f	f	f	88	fy		f
35	f	f	f	89	fy	fy	f
36	f	f	f	90	f	f	f
37	fy	fy	f	91	r		r
38	fy		f	92	t		t
39	f	fy	f	93	f	f	f
40	f	t	f	94	t	t	t
41	f	fy	f	95	t		t
42				96	f	f	f
43	f		f	97	f		f
44	f		f	98	f	f	f
45	f	f	f	99	t	f	f
46	r	f	t	100	r	t	t
47				101	t	f	f
48	r		r	102	f	f	f
49	f	t	f	103	f		f
50	fy		f	104	f	r	t
51	f	f	f	105	f	t	f
52	f	f	f	106	t	t	t
53	t	f	f	107	f	t	f
54	f	t	f	108	f		f

(continued)

TABLE 13-3 Continued

Paper Number	*Reviewer 1*	*Reviewer 2*	*Reviewer 3*	*Paper Number*	*Reviewer 1*	*Reviewer 2*	*Reviewer 3*
109	f		f	142	fy	f	f
110	fy	f	f	143	f		f
111	f	f	f	144	r	f	t
112	t	f	f	145	f	t	f
113	r	f	t	146	f	t	f
114	f	f	f	147	f		f
115	t	t	t	148			
116	t	f	f	149	fy	f	f
117	t	f	f	150	fy	f	f
118	t	f	f	151	f	f	f
119	t		t	152	t		t
120				153	f		f
121	f		f	154	t	f	f
122	t		t	155	t	r	t
123	f	f	f	156	t	f	f
124				157			
125	f	t	f	158	f	f	f
126				159			
127	f	f	f	160	fy	f	f
128				161	t	f	f
129	f	f	f	162	f	f	f
130	fy		f	163			
131	f	r	t	164	f	f	f
132	t	t	t	165	f	f	f
133	fy		f	166	f		f
134	f	fy	f	167	r	f	t
135	r	r	r	168	f	f	f
136	f		f	169	f		f
137	f	f	f	170	r	t	t
138	f		f	171	f	f	f
139	t		t	172	t		t
140	t	f	f	173	f		f
141	r		r	174	f		f

NOTE: If a space is blank, the reviewer failed to return a review. This should be considered a missing value.

CHAPTER 14

Six-Sigma Management and Tools

I look at Six Sigma as a foundation on which you can build more innovation.
—JEFFREY R. IMMELT, CHAIRMAN GENERAL ELECTRIC

As you can see by the preceding quote, Immelt and General Electric (GE) placed a lot of importance on Six Sigma as a method for improvement. They initiated the Six-Sigma program in 1995 with a goal of being a Six-Sigma company by 2000.

As is evidenced by a quick Internet search, **Six Sigma** is a very popular approach to improving quality. There are several distinctions about Six Sigma that differentiate it from traditional continuous improvement. First, Six Sigma represents a well-thought-out packaging of quality tools and philosophies in an honest effort to provide rigor and repeatability to quality improvement efforts.

Second, Six Sigma is much more cost-reduction-oriented than traditional continuous improvement. It is this second aspect of Six Sigma that has made it so popular with CEOs. In fact, many quality practitioners are uncomfortable with the focus on results, stating that this approach violates several of Deming's points, especially in setting targets and goals for cost reduction. On the other hand, proponents of Six Sigma state that this focus on profits is one of the strengths of a Six-Sigma approach.

The third fundamental nuance of Six Sigma is the way it is organized. Six Sigma is a bonanza for consultants and providers of training because it is organized around creating champions, black belts, green belts, and in some situations, yellow belts. Later in the chapter we will discuss how Six-Sigma efforts are normally organized.

For this text, we approach Six Sigma from a contingency perspective. This is simply one of the more popular current approaches to quality improvement. If after studying this chapter you feel that the Six-Sigma approach will be helpful to your company, then you can strongly support implementation.

WHAT IS SIX SIGMA?

The *sigma* in Six Sigma refers to the Greek symbol σ, which designates a standard deviation in statistics. The *six* refers to the number of standard deviations from a specification limit to the mean of a highly capable process.

Video Clip: Six Sigma at Kurt Manufacturing

There are two key versions of Six Sigma. From one perspective, Six Sigma is a program begun at Motorola in 1982. That year, Motorola's CEO requested that costs be cut in half. He then repeated the same request the following year. These efforts pointed out that Motorola needed to improve its product designs and analytical techniques to achieve these goals. Motorola emphasized designing products to achieve Six Sigma. Figure 14-1 shows what this means. In the figure, distribution *a* shows a typical product design with 3-standard-deviation specifications (or tolerances). If this is the case, about 0.5% of products will not meet specification. As shown in part *b* of the figure, if the

FIGURE 14-1 Six-Sigma Variation

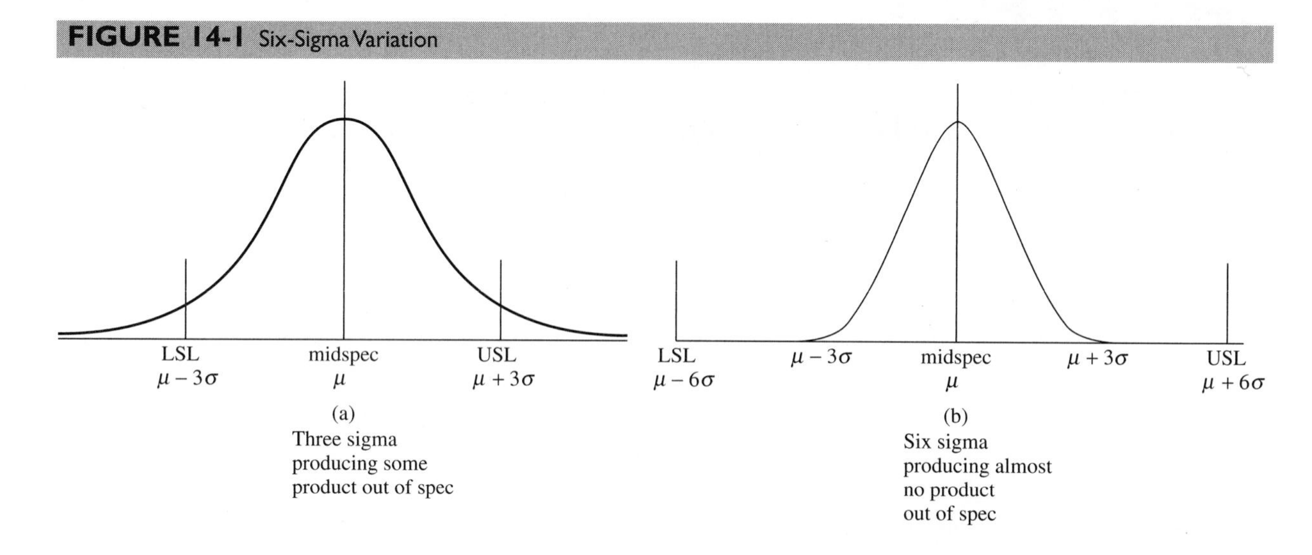

tolerances are 6 standard deviations, the probability of producing a bad part is very low. Notice how in part *b* the mean or dispersion of the process could change significantly, and the product still would meet specs. Table 14-1 shows the number of defective parts per million (ppm) that are produced between one and Six-Sigma levels. Using this definition, Six Sigma translates into more robust designs, radically lower defect levels, and lowered costs of poor quality (COPQ).

From the early days of improving the robustness of design at Motorola, Six Sigma has morphed into an organization-wide program for improvement involving hierarchical training, organizational learning, and pay for learning. As you will see in this chapter, none of the analytical tools used in Six-Sigma efforts are new. What is new is how they are packaged and deployed within a company.

Some argue that Six Sigma is an advanced quality improvement approach designed to help to tackle the most difficult quality problems. As you can see in the pyramid in Figure 14-2, the basic tools of quality can be used to handle 90% of quality problems. Most of the next 10% requires advanced training and analytical techniques. Beyond that, there are a few problems that require expertise that may not be found within the company. Thus you can see that care should be taken in determining what projects should be undertaken by Six-Sigma specialists.

At the core of Six Sigma is the following equation:

$$Y = f(X) \tag{14.1}$$

TABLE 14-1 Sigma Levels and ppm Defects

Sigma Level	*Long-Term ppm* Defects*
1	691,462
2	308,538
3	66,807
4	6,210
5	233
6	3.4

*ppm = parts per million.

FIGURE 14-2 Six-Sigma Effectiveness

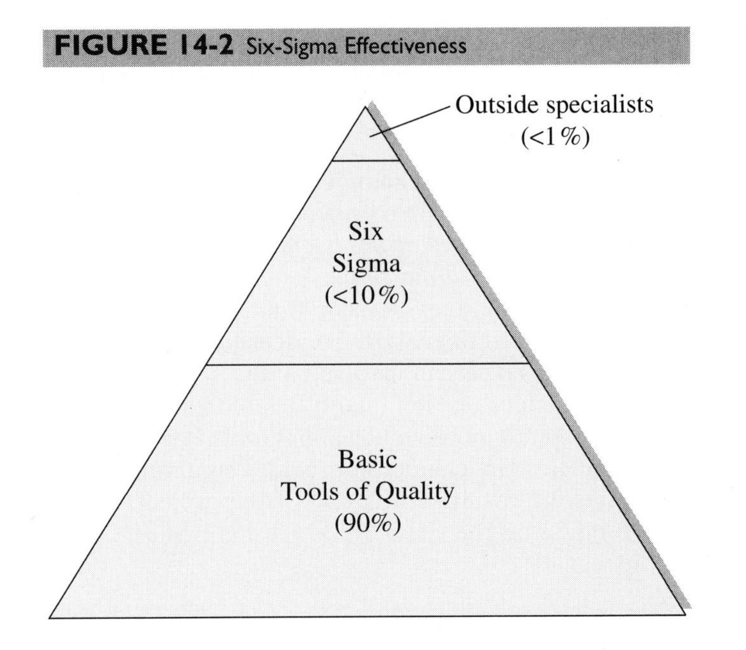

Strictly speaking, this means that Y (the dependent variable) is a function of X (an independent variable). To Six-Sigma practitioners, this means that an output is a function of inputs and processes, where

Y = output (key business objectives and measures)
f = function (interrelationships to be controlled and managed)
X = controllable and noncontrollable variables that affect Y

For example, the profitability of a company (Y) is affected by several variables (Xs), including customer retention, inventory turnovers, rolled throughput yield, production costs, and many others. If our objective is to improve profits, we focus on these variables on a project-by-project basis and improve our performance. In this scenario, the job of management is to identify and prioritize projects to achieve the goal of higher profits. The job of employees is to obtain the training and expertise required to meet these objectives.

As you can see, Six Sigma started as a single firm's approach to reducing costs and improving quality. Currently, it is much more. It involves planning, organization, training, human resources planning, and pay for knowledge. This requires both organizational and individual cooperation to achieve a goal. At GE, management made it clear that participation in Six Sigma was a prerequisite for advancement within the company.

ORGANIZING SIX SIGMA

Probably you have heard about Six Sigma black belts. This is the designation for a person who has completed rigorous (and costly) black-belt training and has completed between one and two Six Sigma projects (depending on the company providing the training and the certificate) with demonstrated results. The cost of training generally runs between $10,000 and $20,000 for a single black belt. Expected returns from Six

Sigma projects can run into the hundreds of thousand dollars. While these payoffs are attractive to management, they do provide quite a bit of pressure for the organization to achieve outstanding results from their Six Sigma efforts.

Below we list some of the key players in Six Sigma efforts:

Champion. The job of the **champion** is to work with black belts and potential black belts to identify possible projects. They get information from a variety of sources such as the *voice of the business* (VOB), the *voice of the customer* (VOC), and the *voice of the employee* (VOE) for potential project ideas. As shown in Figure 14-3, they act as a funnel for project ideas and use Pareto analysis (Chapter 10) to analyze the ideas to determine where the best return on investment lies. This can involve COPQ analysis and regression studies to determine the main causes of quality-related losses. From the champion's perspective, Six Sigma is not so much about tools. It is about managing the process for improvement. The champion provides continuing support for the project and validates the results at the end of the project. In a small company, the champion might be the CEO. In larger companies, they may be senior vice presidents.

FIGURE 14-3 Champion Decision Making

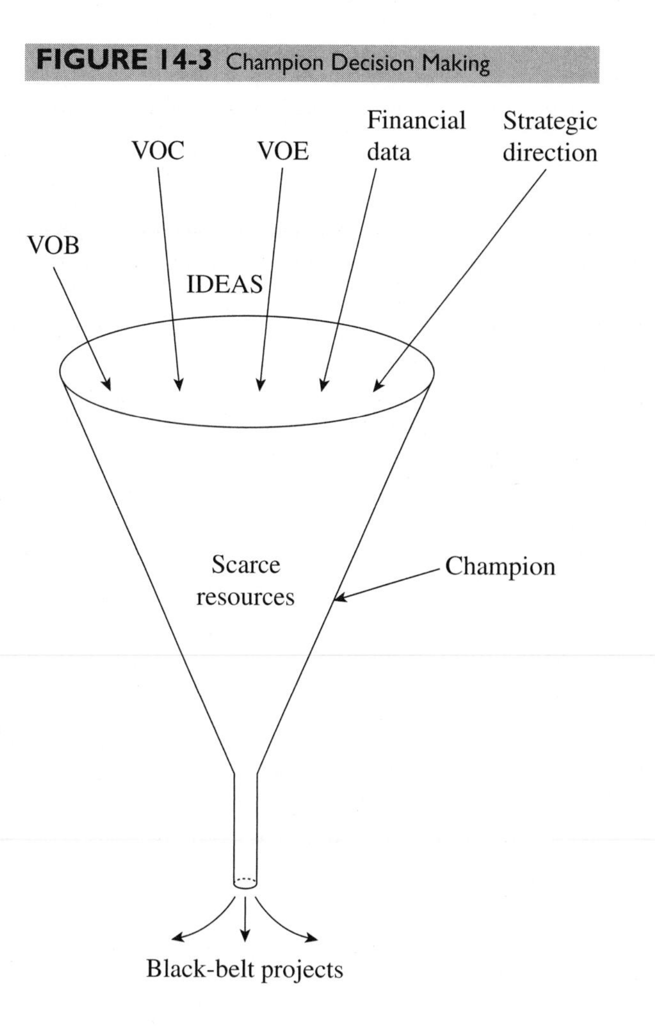

Master black belts. In some firms, experienced black belts are designated **master black belts**. In these cases, master black belts serve as mentors and trainers for new black belts. This brings the training in-house and can reduce costs.

Black Belt. The **black belt** is the key to Six Sigma. These are specially trained individuals. The training usually lasts about 4 months. After completing training, these individuals are committed full time to completing cost-reduction projects. At GE, black belts were expected to complete two more projects after their certification. Each project lasts from 2 months to a year depending on the project scope. The black belt is a specialist. Within 9 months of beginning Six Sigma within its appliance division, GE had advertised, created, and filled 115 black belt positions.[1] It has been suggested that small to midsized companies may only need between one and five black belts at one time. Individuals usually spend about 2 years as a black belt and are then moved into management jobs. The black-belt designation is also very valuable for finding new positions in the job marketplace. A review of the *Wall Street Journal* will show that black belts are in high demand and draw attractive salaries.

Green Belts. **Green belts** are trained in basic quality tools and work in teams to improve quality. Green belts are assigned part time to work on process and design improvement. In some cases, the results of green-belt activities are the same as black belts. In other organizations, green belts are involved in less critical projects. In a small company of 100 employees, there might be 1 black belt and 60 green belts. Some companies also have **yellow belts**, who are employees who are familiar with improvement processes.

DMAIC OVERVIEW

Table 14-2 shows the steps in the **DMAIC process**. DMAIC stands for *define, measure, analyze, improve, and control*. This is very similar to the PDCA cycle proposed by Shewhart and Deming. We will discuss each of these steps in two parts. Here we will define each of the steps. Then we will fit quality tools to each of these steps. Figure 14-4 shows an overview of the tools used at each stage of the DMAIC process. We will discuss these in much more depth over the next several pages. In some cases, where we have already presented a tool, such as the basic seven (B7) tools, we will mention the tool. However, it is up to you to refer to the other chapter where the tool is defined and explained. A Closer Look at Quality 14-1 shows some results of DMAIC processes in different companies.

TABLE 14-2 The Six-Sigma Process—DMAIC

DMAIC	**Define**	Define the project goals and customer (internal and external) deliverables
	Measure	Measure the process to determine current performance
	Analyze	Analyze and determine the root cause(s) of the defects
	Improve	Improve the process by eliminating defects
	Control	Control future process performance

SOURCE: *www.Freequality.org* (2006).

[1]Hendricks, C. A., and Kelbaugh, R. L., "Implementing Six Sigma at GE," *Journal for Quality and Participation* 21, 4 (1998):48–53.

FIGURE 14-4 Overview of the Six-Sigma Process

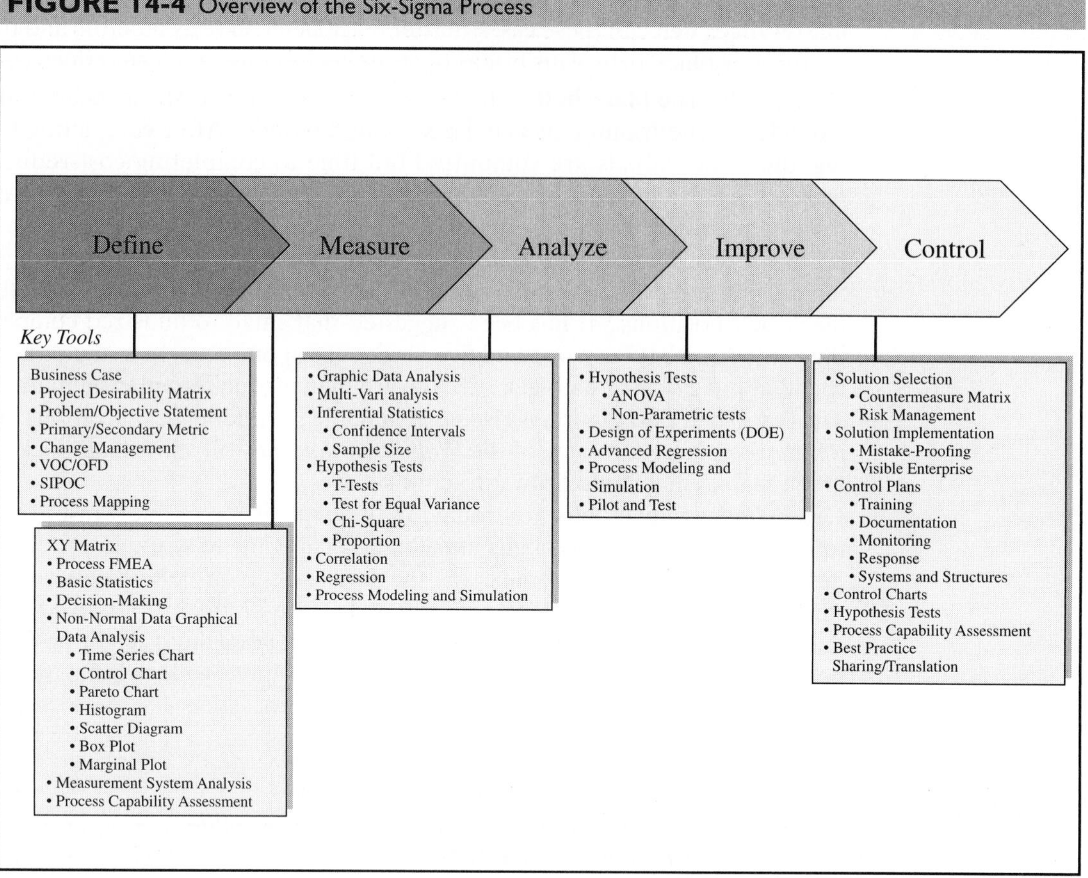

A CLOSER LOOK AT QUALITY 14-1

DMAIC IN ACTION

While Kevin Colby was working on a Six-Sigma project at the Truck Components Automated Products Division of Eaton Corporation, the company was examining cost savings opportunities. The division produced transmissions that included speed sensors, which measure shaft speeds and work in conjunction with the gears produced by the Cleveland, Ohio-based company. The gears with holes caused signal fluctuations that affected the sensors. Two electronic control units (ECUs) with different circuit speeds were manufactured to allow the sensor to work with both types of gears.

Engineers within the division's design group who were involved with the gear project realized that they could simultaneously have an impact on two divisions. Jerry Ganski, principal engineer who led the effort to eliminate the second ECU, said, "We realized that removal of the holes in the gears would allow the Automated Products Division to eliminate the special ECU we had to manufacture to deal with the holes. We now use a common ECU for all our platforms and thus save the money it took to build, stock, and handle two ECU styles where the only difference between them was the speed sensor circuit. The savings is estimated at approximately 12 percent." Based on the improvements realized from these three projects, Eaton is investigating other gear-related projects for potential improvement opportunities.

DEFINE PHASE

In the **define phase**, projects are identified and selected. Project selection is performed under the direction and with the participation of the champion. Also involved in selection are master black belts and black or green belts. We will discuss this in four phases:

1. Developing the business case
2. Project evaluation
3. Pareto analysis
4. Project definition

Developing the Business Case

Business case development involves

- Identifying a group of possible projects
- Writing the business case
- Stratifying the business case into problem statement and objective statements

Following is a sample business case. As you can see, the **business case** is a short statement outlining the objectives, measurables, and justification for the project.

> ***Business Case:*** *During the four-week period from January 1, 2006, to February 1, 2006, the throughput yield for plant number 3 in region 4 was at 57% of capacity, resulting in an annualized COPQ of $5.6 million. This gap of rolled throughput yield mandates a business objective to improve throughput by 50% from 57% of design capacity to 85% by February 1, 2007, representing $3 million in savings. This project will increase the throughput for plant 3 in region 4 to meet the year 2007 corporate goal of increasing sales in region 4 by $10 million.*

The mnemonic device **RUMBA** is used to check the efficacy of a business case. Evaluating your business case, is it

- ***R***ealistic—Are the goals attainable, is the time line feasible?
- ***U***nderstandable—Do I understand the case?
- ***M***easurable—Do we show the measures?
- ***B***elievable—That is a lot of money. Can it be done?
- ***A***ctionable—Can it be implemented?

If the business case meets all these requirements, it probably will be a good project.

Project Evaluation

There are several methods for evaluating a project. Here we will demonstrate a **project risk assessment** for a potential Six-Sigma project in Example 14.1.

Example 14.1: Project Risk Assessment

Problem: Figure 14-5 shows a sample risk assessment for your candidate project. Using management input, you determine a rating of yes, uncertain, or no for each of the questions. Each item is weighted on a scale of 1 to 10 for importance, where a yes is 0 points, uncertain is 3 points, and no is valued 5 points.

Solution: The 0-, 3-, and 5-point scale values are multiplied by their related weights. Notice that the weights sum to 200 points and that the sum of the weighted scaled values is 390 points. Since the possible total points is $200 \times 5 = 1000$ points, dividing 390 by 1000 gives 39%. Therefore, 39% is our risk factor. We will use this later to determine the attractiveness of the potential project.

Figure 14-6 shows the Six-Sigma project return analysis. For this analysis, the potential project is evaluated in three dimensions—growth, urgency, and impact. We will demonstrate this in Example 14.2.

Example 14.2: Project Return Assessment

Problem: For our project we have performed a project return assessment (see Fig. 14-6). As is shown, using the return scales for growth, urgency, and impact each time the project rates a 2.

Solution: Totaling the score, 6 out of a possible 15 points yields a score of 40%. Combining our scores for both risk and return into a project risk and return matrix (Fig. 14-7), we see that this project is classified as low-hanging fruit. This means that this project is worthwhile if it can be completed quickly.

Pareto Analysis

Part of the responsibility of the champion is to perform a cost of poor quality (COPQ) analysis. This is based on the PAF categorization of costs (see Chap. 4). Performing a study of internal and external failure costs will help to determine where the most benefit can be found. Figure 14-8 shows a two-level Pareto analysis of COPQ. The first-level analysis shows $5.6 million quality costs in plant A. This is the plant with the highest losses. When we study causes of poor quality in plant A, it becomes clear that operation *P* accounts for about 62% of the $5.6 million loss (in the second-level analysis). For this reason, this project holds great promise for breakthrough improvement.

Problem Definition

Once the risk analysis and Pareto analysis have been completed for the project, a project definition consists of a problem statement, project goals/objectives, primary metrics, secondary metrics, and team member identification. Figure 14-9 shows an example of problem definition.

MEASURE PHASE

The **measure phase** involves two major steps:

1. Selecting process outcomes
2. Verifying measurements

We will discuss these separately.

FIGURE 14-5 Risk Assessment

Six-Sigma Project Risk Worksheet

Project Name: Plant 3 Throughput
Belt: Foster
Sponsor(s): Shannon
Mentor(s):
Date: Sept. 12, 2007

Before worksheet can be completed, the following questions must be answered

1. *Is the defect/key characteristic known?*
2. *Is the defect/key characteristic measurable?*
3. *Is the solution to the problem unknown?*

For questions 1 through 3...
If you answered Yes proceed with answering each criteria question below
If you answered No see your mentor your project may be better completed by means other than Six Sigma

Risk Value: 39%

Category	Criteria	Rating		Weight	Total
Define Opportunity	(Six-Sigma Risk Rating Scale: Yes = 0; Uncertain = 3; No = 5)	Scale values			
	1 Are we currently measuring the defect(s)/key characteristic(s)?	Yes	0	10	0
	2 Is historical data currently available?	Yes	0	10	0
	3 Is it easy to acquire additional data?	Uncertain	3	10	30
	4 Are the specifications for the process or product defined?	No	5	5	25
	5 Do you know how the specifications were defined?	No	5	5	25
	6 Is the defect measured where it occurs in the process?	No	5	10	50
	7 Is the defect frequency continuous? ("No" for sporadic or cyclical)	No	5	5	25
Customer Focus					
	8 Has/have the customer(s) been identified?	Yes	0	5	0
	9 Have you verified what is important to the customer?	Yes	0	10	0
	10 Is this defect/key characteristic important to the customer?	Yes	0	10	0
	11 Will the customer see the result of eliminating/reducing the defect?	Uncertain	3	10	30
Company Benefit/ Leveraging					
	12 Does the defect relate to the mission, a business driver, or a reliability measure?	Yes	0	10	0
	13 Does the defect impact operations?	Yes	0	10	0
	14 Can the results of the project be applied to other processes or products?	Yes	0	5	0
	15 Can the impact be quantified in dollars?	Uncertain	3	10	30
Project Leadership/ Global Bounding					
	16 Are all managers, at all levels, in agreement that your project is important?	No	5	5	25
	17 Is your team the only effort currently pursuing this defect?	Yes	0	10	0
	18 Can adequate visibility for the problem and solution be created?	Uncertain	3	10	30
	19 Is the team comprised of representatives from only one location/business function?	Uncertain	3	10	30
	20 Are appropriate resources available to participate on the team?	Uncertain	3	10	30
	21 Can the project be bounded to an effective size?	Yes	0	10	0
	22 Can the project be completed on schedule within 4-6 months?	Uncertain	3	5	15
	23 Do we know the boundaries of the process(es)?	Uncertain	3	5	15
	24 Can process changes be implemented within the project schedule?	Uncertain	3	10	30
		Totals		200	390
		390/(200 × 5) = 39%			

FIGURE 14-6 Project Return Analysis

Six-Sigma Project Return

Before this worksheet can be completed, the following questions must be answered: *Does the defect / key characteristic for this project relate to one of the company's business drivers, or a reliability or service measure?*

If you answered *Yes* proceed with the worksheet
If you answered *No* see your sponsor or mentor this project may not provide appropriate returns

Project:	Plant 3 Throughput	Growth Score:	2/5
Belt:	Foster	Urgency Score:	2/5
Sponsor(s):	Shannon	Impact Score:	2/5
Date:	Sept. 12, 2007	Return Value:	6/15 = 40%

Growth = Competitive Advantage
Choose the single best answer

Return Scale	
0	☐ The project does not result in incremental sales with paying customer(s).
1	☐ The project does not create incremental sales with paying customer(s), but does improve the competitive position of the company by improving operating efficiencies that bear on competitive performance.
2	☑ The project does not create incremental sales with paying customer(s), but does improve the competitive position of the company by improving operating efficiencies in a key strategic area.
3	☐ The project provides some degree of incremental sales with paying customer(s) and moderately improves the competitive position of the company.
4	☐ The project provides a moderate degree of incremental sales with paying customer(s) and substantially improves the competitive position of the company by providing a higher level of service.
5	☐ The project provides a high degree of incremental sales with paying customer(s) and greatly improves the competitive position of the company by providing a level of service unmatched by competitors.

Urgency = Competitive Response
Choose the single best answer

Return Scale	
0	☐ The project can be postponed for at least 12 months without affecting competitive position, or existing processes or procedures can produce substantially the same result and will not affect competitive position.
1	☐ The postponement of the project does not affect competitive position, and minimal incremental operating costs are expected to be incurred to produce substantially the same result.
2	☑ The postponement of the project does not affect competitive position; however, operating costs may escalate to produce substantially the same result.
3	☐ If the project is postponed for now, the company remains capable of responding to the needed change without affecting its competitive position. However, it is expected the company will be substantially hindered in responding rapidly and effectively to future changes in the competitive environment.
4	☐ The postponement of the project may result in further competitive disadvantage to the company, or in a loss of competitive opportunity; or existing successful activities in the company may be curtailed because of the lack of the proposed system.
5	☐ The postponement of the project will result in further competitive disadvantage to the company, or in a loss of competitive opportunity; or existing successful activities in the company must be curtailed because of the lack of having a solution to this problem.

Impact = Annual Expected Financial Savings
Choose the single best answer

Return Scale	
0	☐ Less than $20,000*
1	☐ $20,000-$99,999*
2	☑ $100,000-$200,000
3	☐ $200,001-$300,000
4	☐ $300,001-$400,000
5	☐ Over $400,000

(*Projects are expected to have, on average, an annual expected savings of $100,000 or more. If the project does not meet this criteria, it should have significantly high scores in the Growth and Urgency categories)

FIGURE 14-7 Project Risk & Return Matrix

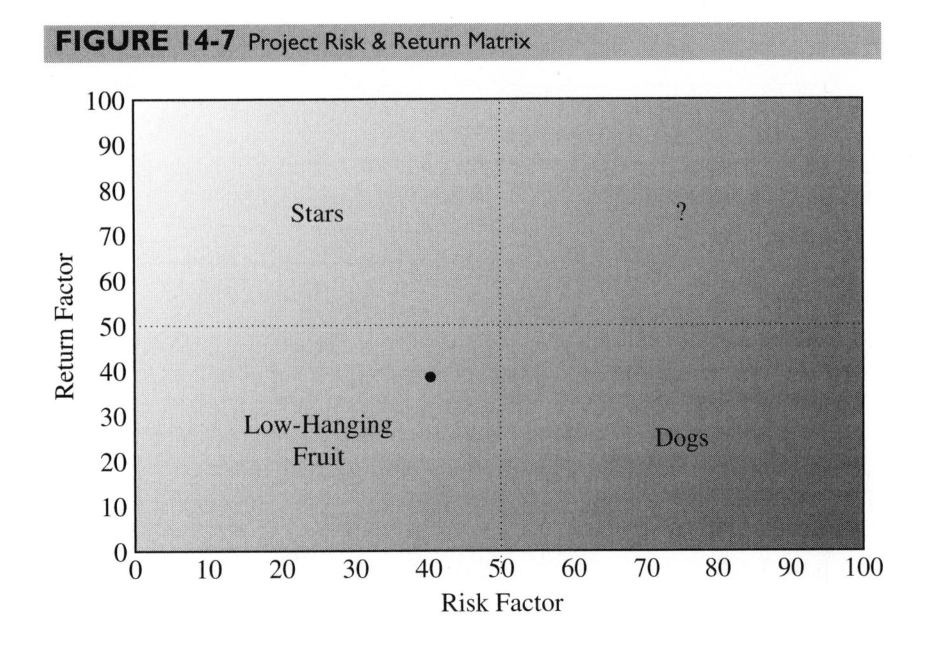

Selecting Process Outcomes

Table 14-3 shows the tools often used in the measure phase. To define process outcomes, you first need to understand the process. This involves process mapping. The process map uses the same approach defined in Chapter 10. A *process map* is a flowchart with responsibility. In Figure 14-10 there is a high-level process map showing champion responsibilities in the Six-Sigma process. Notice that any of the individual steps could be broken out into lower-level process maps. The goal with a process map is to identify non-value-added activities. Two important measures that are monitored are **defects per unit (DPU)** and **defects per million opportunities (DPMO)**.

The ***XY* matrix** is used to identify inputs (*X*s) and outputs (*Y*s) from a project you have mapped and are desiring to pursue. Figure 14-11 shows an *XY* matrix for a potential project. The inputs include dimensions, standard operating procedures (SOPs), and other inputs along the left-hand column. Output variables include key dimensions, sizes, flashing, and the presence of all needed welds. Each of the outputs is provided an importance weight (1-10). The relationship between each of the *X*'s and *Y*'s is placed in the matrix (1-10 scale). These are multiplied horizontally, with ranks and scores being computed by multiplying each matrix cell by its weight and summing the products horizontally. As you can see, the most important aspects of the process are SOPs, weld schedules, and daily tip dressing.

Example 14.3

Problem: Figure 14-12 shows a matrix for a services process. On the left, inputs *A* through *G* are listed with their associated correlations with five different outputs.

Solution: Figure 14-13 shows the solution for Example 14.3. As you can see, inputs *F* and *D* should be studied especially closely.

FMEA

Failure modes and effects analysis was introduced in Chapter 7. FMEA is used to identify ways a process or product can fail to meet critical customer requirements.

FIGURE 14-8 Pareto Analysis

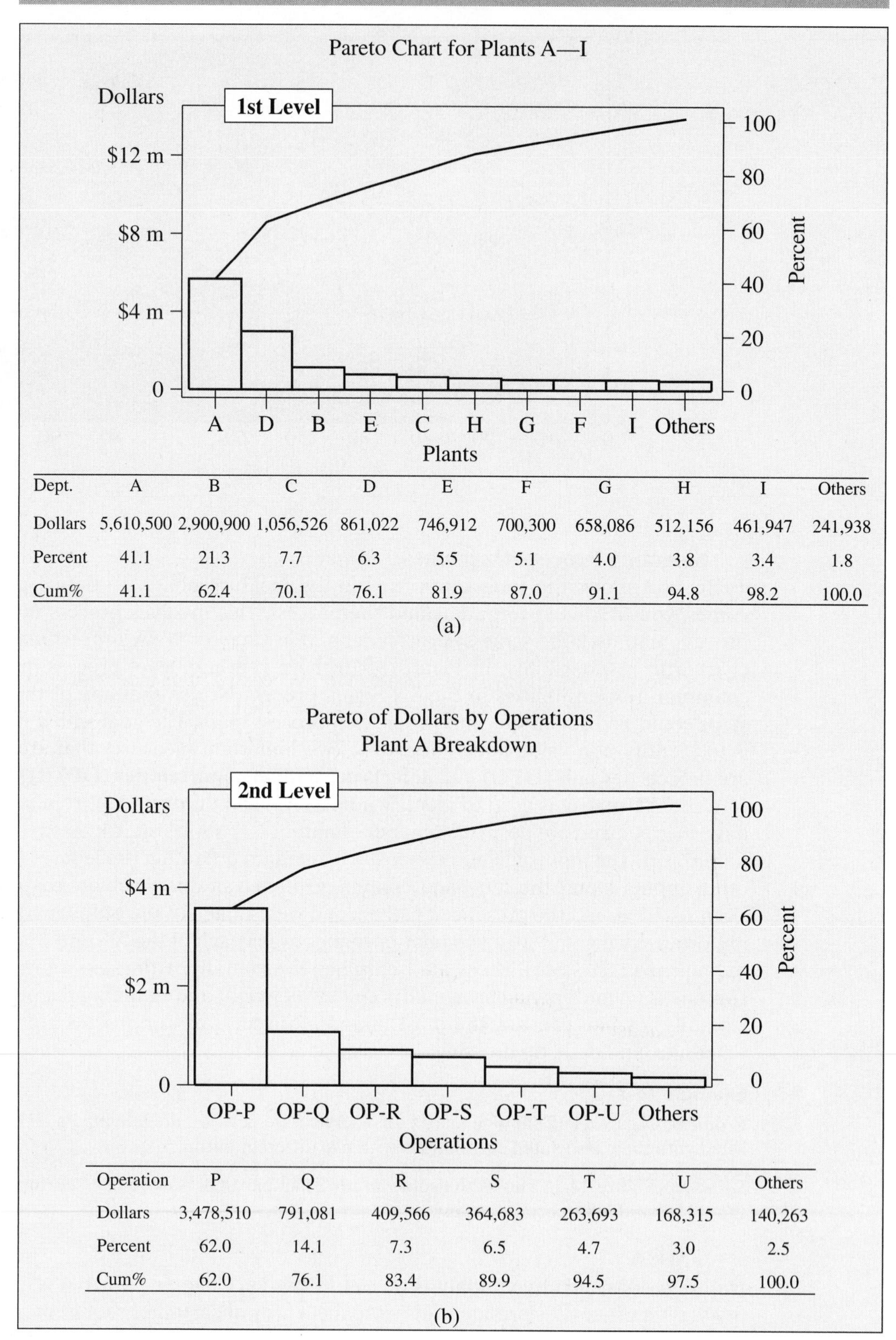

Dept.	A	B	C	D	E	F	G	H	I	Others
Dollars	5,610,500	2,900,900	1,056,526	861,022	746,912	700,300	658,086	512,156	461,947	241,938
Percent	41.1	21.3	7.7	6.3	5.5	5.1	4.0	3.8	3.4	1.8
Cum%	41.1	62.4	70.1	76.1	81.9	87.0	91.1	94.8	98.2	100.0

(a)

Operation	P	Q	R	S	T	U	Others
Dollars	3,478,510	791,081	409,566	364,683	263,693	168,315	140,263
Percent	62.0	14.1	7.3	6.5	4.7	3.0	2.5
Cum%	62.0	76.1	83.4	89.9	94.5	97.5	100.0

(b)

FIGURE 14-9 Problem Definition

Problem Statement:	In 2006, plant A lost $5.6 million on COPQ. Of this, almost $3.5 million occurred in operation P (see Fig. 14-8). This has resulted in a loss of profitability for the firm.		
Project Goals/Objective:	Reduce COPQ for operation P by 30% by year-end.		
Primary Metrics:	COPQ	Secondary Metrics:	Downtime for process
	Rework (% of sales)		Plant sales
	Scrap (% of sales)		Labor productivity
Team Members:	Bill Sawaya	Stan Fawcett	
	Cynthia Wallin	Scott Sampson	

Verifying Measurements

When measuring critical characteristics of processes, it is necessary to use gauges, calipers, and other tools. While these tools are often very accurate, there can be problems with variation in measurements. As a result, **measurement system analysis (MSA)** is used to determine if measurements are consistent. Another approach for verifying measurements is to perform product and process capability analysis. This was discussed in Chapter 13. We will discuss gauge R&R in this chapter.

Gauge R&R

The most commonly used MSA is **Gauge Repeatability and Reproducibility Analysis (gauge R&R, sometimes referred to as gage R&R)**. Gauge R&R is used to determine the accuracy and precision of your measurements. If your measurements are imprecise, there will be a large amount of variation as a result of measurement error. Obviously, you do not want to draw incorrect conclusions as a result of measurement error. Problems in measurement can result for a variety of reasons:

- The measurement gauges are faulty.
- Operators are using gauges improperly.

TABLE 14-3 Measure-Phase Tools

Measure-Phase Tools
Process map
XY matrix
FMEA
Gauge R&R
Capability assessment

FIGURE 14-10 High-Level Process Map

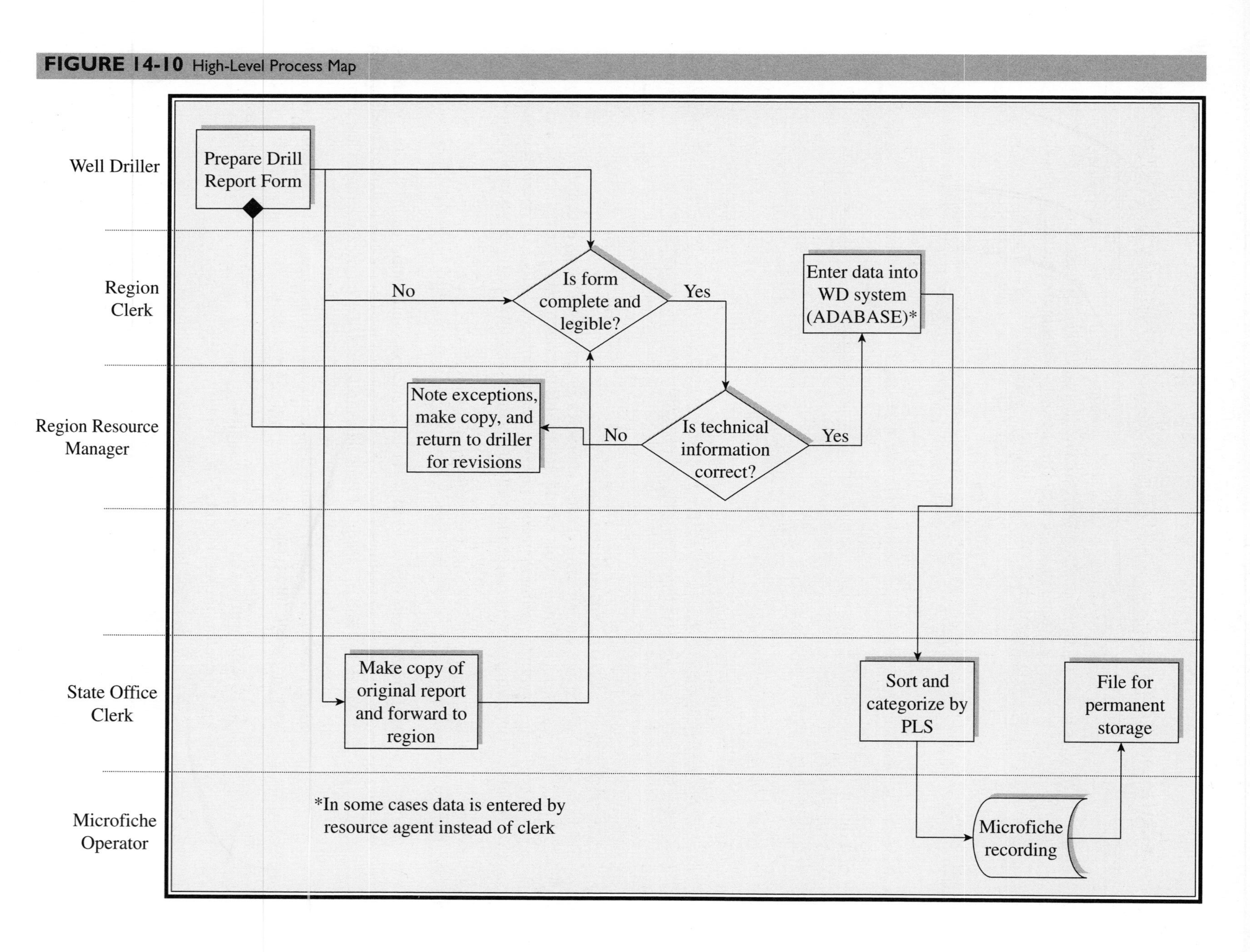

FIGURE 14-11 *XY* Matrix

XY Matrix Project: Welding Operation

Date: Oct. 14, 2007

	Output Variables	Width Dimension	Length Dimension	Assembly Size	Flash Free	All Welds Present	Product	Rank
	Rank	5	7	9	8	10		
	Inputs							
1	Width Grommet 1	10					50	8
2	Length Grommet 1		7				49	10
3	Subassembly Width	7					35	12
4	Subassembly Length		10				70	1
5	All SOPs	9	9		9	6	240	1*
6	Weld Schedule			10		10	190	2*
7	Air Pressure			5	5	5	135	4
8	Line Voltage			5	5	5	135	4
9	H_20 Circulation			5	5	5	135	4
10	Width Bracket 2	10					50	8
11	Length Bracket 2		7				49	10
12	Daily Tip Dressing			10	8		154	3*
13								
14								
15								

Importance weights

$(9\times5)+(9\times7)+(9\times8)+(6\times10) = 240$

Relationship weights

- Training in measurement procedures is lacking.
- The gauge is calibrated incorrectly.

Statistical experiments using analysis of variance (ANOVA) are useful in performing gauge R&R. Two-way ANOVA is used to determine whether variation comes from the part being measured, differences in operator measurements, or the measurement instrument.

Example 14.4: Gauge R&R In Action

Problem: Table 14-4 contains measurement data for a particular operation with three operators taking two measurements per part.[2] In other words, there are 3 operators and 20 parts. Each operator measures each part twice using a gauge and logs these measurements. For example, operator 1 measures part 1 twice, with resulting measurements of 21 and 20 millimeters. The same operator

[2]Montgomery, Douglas, *Design and Analysis of Experiments* (New York: Wiley, 1997), p. 473.

FIGURE 14-12 *XY Matrix for Example 14.3*

Project: Service Process
Date: Oct. 14, 2007

	Output Variables	Vendor Selection	Component Selection	Internal Lead Time	Cost	Ship To Location					
	Rank	7	10	8	9	5					
	Inputs										
1	*A*		10	4	5						
2	*B*	9		7	2						
3	*C*	8	4		10						
4	*D*	10	9	8							
5	*E*	4		3		10					
6	*F*	4	7	8	5	10					
7	*G*	3	7		10						
8											
9											

FIGURE 14-13 Solution to Example 14.3

Project: Service Process
Date: Oct. 14, 2007

	Output Variables	Vendor Selection	Component Selection	Internal Lead Time	Cost	Shift To Location				Total	Rank	
	Rank	7	10	8	9	5						
	Inputs											
1	*A*		10	4	5					177	5	
2	*B*	9		7	2					137	6	
3	*C*	8	4		10					186	3	
4	*D*	10	9	8						224	2	*
5	*E*	4		3		10				102	7	
6	*F*	4	7	8	5	10				257	1	*
7	*G*	3	7		10					181	4	
8												
9												

TABLE 14-4 Gauge R&R Data

Part Number	*Operator 1*	*Operator 2*	*Operator 3*
1	21	20	19
	20	20	21
2	24	24	23
	23	24	24
3	20	19	20
	21	21	22
4	27	28	27
	27	26	28
5	19	19	18
	18	18	21
6	23	24	23
	21	21	22
7	22	22	22
	21	24	20
8	19	18	19
	17	20	18
9	24	25	24
	23	23	24
10	25	26	24
	23	25	25
11	21	20	21
	20	20	20
12	18	17	18
	19	19	19
13	23	25	25
	25	25	25
14	24	23	24
	24	25	25
15	29	30	31
	30	28	30
16	26	25	25
	26	26	27
17	20	19	20
	20	20	20
18	19	19	21
	21	19	23
19	25	25	25
	26	24	25
20	19	18	19
	19	17	17

measures part 2 and gets the results of 24 and 23 millimeters. As we can see, there is measurement variation. However, we do not know whether the variation comes from the gauge, the part, or the operators.

Solution: Figure 14-14 shows the ANOVA table for these data. From the ANOVA table we can see that parts contribute most of the variation (sample row in the ANOVA table, $p = .000$), operator effect is insignificant (columns row in the ANOVA table, $p = .275$), and part-operator interactions are insignificant. As a result, we conclude that the gauge measurement is repeatable and reproducible.

FIGURE 14-14 Gauge R&R in Excel: Data Screen

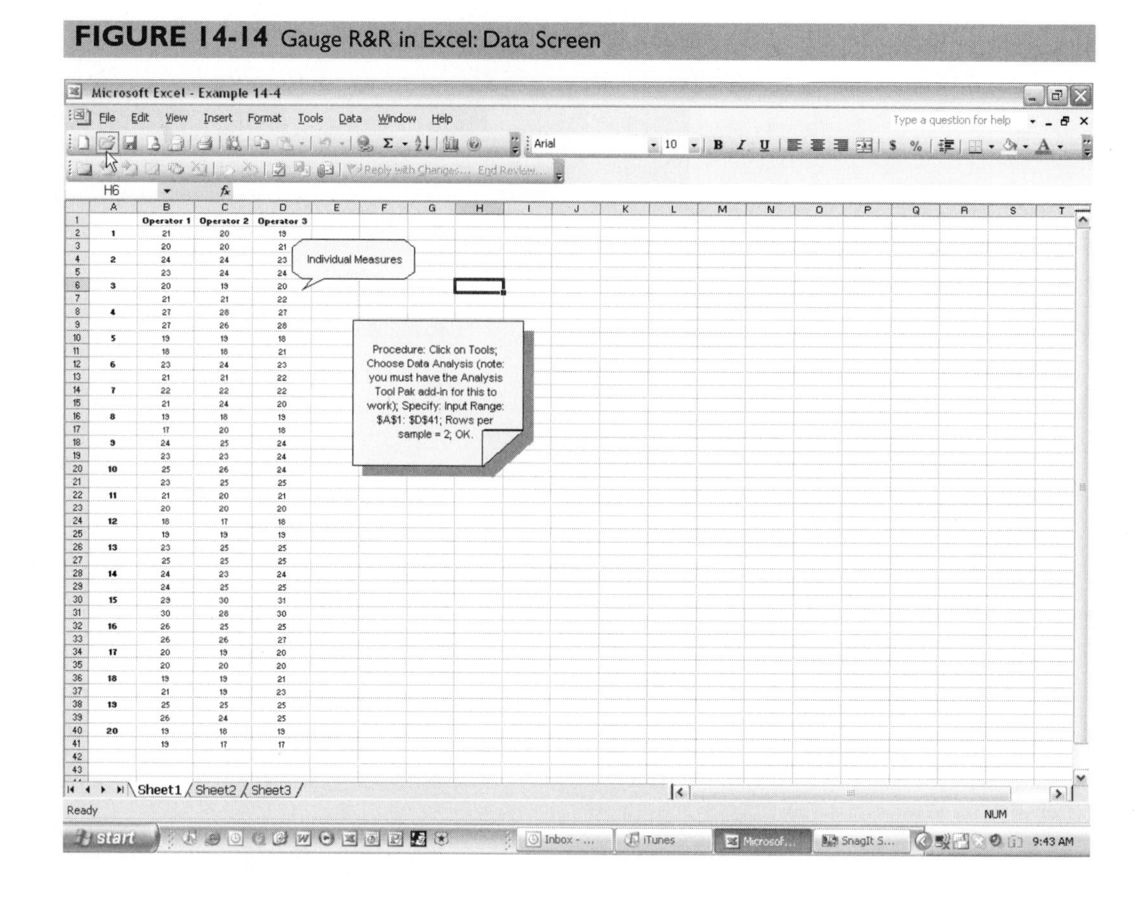

	A	B	C	D
1		Operator 1	Operator 2	Operator 3
2	1	21	20	19
3		20	20	21
4	2	24	24	23
5		23	24	24
6	3	20	19	20
7		21	21	22
8	4	27	28	27
9		27	26	28
10	5	19	19	18
11		18	18	21
12	6	23	24	23
13		21	21	22
14	7	22	22	22
15		21	24	20
16	8	19	18	19
17		17	20	18
18	9	24	25	24
19		23	23	24
20	10	25	26	24
21		23	25	25
22	11	21	20	21
23		20	20	20
24	12	18	17	18
25		19	19	19
26	13	23	25	25
27		25	25	25
28	14	24	23	24
29		24	25	25
30	15	29	30	31
31		30	28	30
32	16	26	25	25
33		26	26	27
34	17	20	19	20
35		20	20	20
36	18	19	19	21
37		21	19	23
38	19	25	25	25
39		26	24	25
40	20	19	18	19
41		19	17	17

Using Excel to Perform Gauge R&R Analysis

The data need to be entered into an Excel spreadsheet exactly as shown in Figure 14-14. The labels for the operators are inserted on each row. The part numbers are listed only once in column A as shown. And the data are entered as shown. This problem is set up for only two measurements per part. Follow the steps outlined in the figure. *Note*: Use two-factor analysis of variance with replication.

The results for Example 14.4 are shown in Figure 14-14. The *P*-value column in the ANOVA table shows what variables significantly contribute to variation.

ANALYZE PHASE

The **analyze phase** involves gathering and analyzing data relative to a particular black-belt project. Following are the analyze-phase steps:

1. Define your performance objectives.
2. Identify independent variables (*X*s).
3. Analyze sources of variability.

We will discuss each of these steps separately. Since the tools used in this analysis were already discussed in Chapters 10, 12, and 13, we will only refer to these tools. You may need to refer to the other chapters to refresh your memory about these tools.

Defining Objectives

When defining performance objectives, you are attempting to determine what characteristics of the process need to be changed to achieve improvement. First, capability analysis is reviewed to determine where the processes are incapable. These areas are prioritized in order of importance. As is shown in Figure 14-15, capability analysis demonstrates whether certain quantitative parameters or discrete events are meeting specification. It helps to determine whether these parameters or events are centered on the mean and whether or not they meet specification. If they are not centered on the desired mean, the process mean needs to be adjusted. If there is too much variability, then the variability is reduced. We will discuss means for reducing variation in the improve phase.

FIGURE 14-15 Capability Results

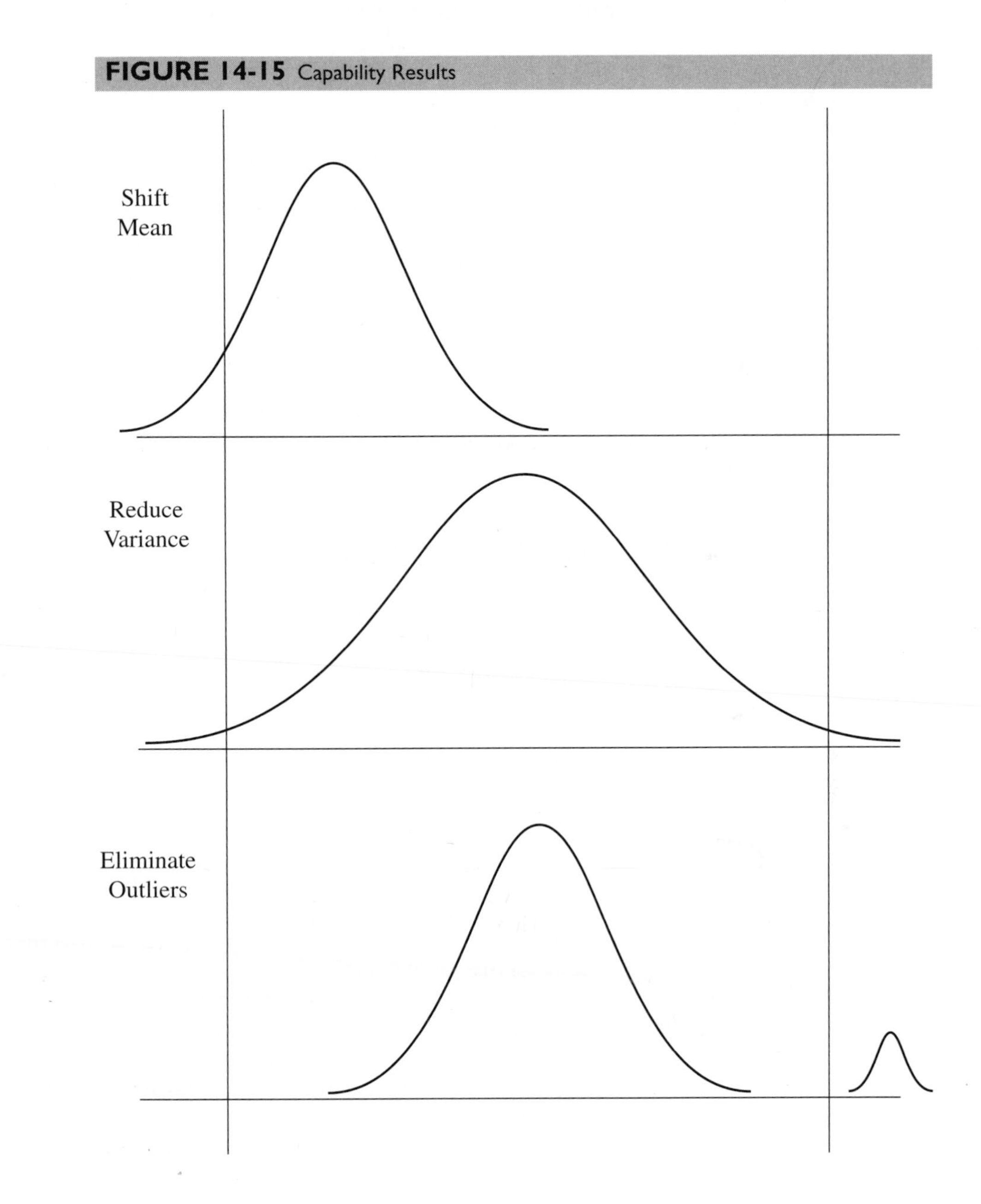

Identifying *X*s

This step involves identifying the independent variables where data will be gathered. These are variables that contribute significantly to process or product variation. Process maps, *XY* matrices, brainstorming, and FMEAs are the primary tools that are used in identifying *X*s.

Analyzing Sources of Variation

The goal of this step is to use visual and statistical tools to better understand the relationships between dependent and independent variables (*X*s and *Y*s) for use in future experimentation. A number of tools are used in this analysis. They include histograms, box plots, scatter plots, regression analysis, and hypothesis tests.

IMPROVE PHASE

The **improve phase** of the DMAIC process involves off-line experimentation. Off-line experimentation involves studying the variables we have identified and using ANOVA to determine whether these independent variables significantly affect variation in our dependent variables. We will introduce an important method for performing off-line experiments, the *Taguchi method*.

CONTROL PHASE

The **control phase** involves managing the improved processes using process charts These topics were covered in Chapters 12 and 13.

TAGUCHI DESIGN OF EXPERIMENTS

Many different factors, inputs, or variables need to be considered when making a product. For example, suppose that you wanted to bake a cake. How much flour should you use? How many eggs? How long should it bake? At what temperature should you set the oven? Probably you would find a recipe to follow. What if there were no recipes, and you were the pioneer trying to invent the best combinations of inputs to bake a cake? Likely you would have to resort to trial and error. However, there is a better way to design an experiment to find out the best combination of variables to make your product (cake).

The **Taguchi method** is a standardized approach for determining the best combination of inputs to produce a product or service. This is accomplished through **design of experiments (DOE)** for determining parameters. DOE is an important tool in the arsenal of tools available to the design and process engineer. It provides a method for quantitatively identifying just the right ingredients that go together to make a high-quality product or service. In this section we discuss first the Taguchi definitions, stages, and behavioral issues that form Taguchi's approach to design of experiments. The purpose here is to introduce concepts and processes relating to the Taguchi method from a managerial perspective. The more technical engineering explanation is available in a variety of engineering books. Taguchi approaches design from four perspectives: robust design, concept design, parameter design, and tolerance design. These are defined in the following paragraphs.

Robust Design

The Taguchi concept of **robust design** states that products and services should be designed so that they are inherently defect-free and insensitive to random variation. The concept is not necessarily new. The notion that products and services should be designed to be of high quality or that processes should be defect-free is as old as mass assembly. However, Taguchi has provided new approaches for creating robust designs through a three-step method of concept design, parameter design, and tolerance design.

Concept design is the process of examining competing technologies to produce a product. Concept design includes process technology choices and process design choices. Appropriate choices in these areas can reduce production costs and result in higher quality products. In a *copying* store, concept design includes layout choices and choices of technology. Each candidate copying machine is tested separately to determine its suitability for the job. In financial services companies, this step likely will involve user groups, MIS staff, and systems analysts in defining processes and choices of equipment and technology.

Parameter design refers to the selection of **control factors** and the determination of **optimal levels** for each of the factors. Control factors are those variables in a process that management can manipulate. For example, the type and amount of training provided to customer service representatives is controlled by management. If it is determined that the amount of training received by customer service representatives determines the quality of service provided the customer, then training is identified as a control factor. Control factors do not affect production costs. Optimal levels are the targets or measurements for performance. For example, a sheet of paper is 8.5 inches wide. This would be the target. The goal is to find the most efficient process and service design. Parameter design involves selecting the best level for performance. For example, in baking cookies, what is the best temperature and time for baking? These parameters can be determined through experimentation.

Tolerance design deals with developing specification limits. Tolerance design occurs after parameter design has been used to reduce variation and the resulting improvement has been insufficient. This often results in an increase in production costs. For example, in tolerance design, engineers selectively tighten specified tolerances and require the use of higher-grade materials in production.

Of these four design considerations, the Taguchi method focuses primarily on parameter design. Getting back to our cake-baking example, using the Taguchi method, we could identify the best amounts of heat, baking time, flour, eggs, and other ingredients to make the best tasting cake. These ingredients are called *parameters*. Their best amounts are referred to as *levels* or *settings*.

BACKGROUND OF THE TAGUCHI METHOD

The Taguchi method was first introduced by Dr. Genichi Taguchi to AT&T Bell Laboratories in 1980. Thanks to its wide acceptance and utilization, the Taguchi method for improving quality is now commonly viewed as comparable in importance to statistical process control (SPC), the Deming approach, and the Japanese concept of total quality control. From a historical perspective Taguchi's method is a continuation of the work in quality improvement that began with Shewhart's work in statistical quality control (SQC) and Deming's work in improving Japanese quality. The Taguchi method provides

1. A basis for determining the functional relationship between controllable product or service design factors and the outcomes of a process.

2. A method for adjusting the mean of a process by optimizing controllable variables.
3. A procedure for examining the relationship between random noise in the process and product or service variability.

Among the unique aspects of the Taguchi method are the Taguchi definition of quality, the quality loss function (QLF), and the concept of robust design. These are discussed briefly in the following paragraphs.

Taguchi Definition of Quality

The traditional definition of quality was conformance to specification. However, Taguchi diverges from the traditional view of conformance quality. In Taguchi terms, **ideal quality** refers to a reference point or target value for determining the quality level of a product or service. This reference point is expressed as a target value. Ideal quality is delivered if a product or a tangible service performs its intended function throughout its projected life under reasonable operating conditions without harmful side effects. In services, because production and consumption of the service often occur simultaneously, ideal quality is a function of customer perceptions and satisfaction. Taguchi measures service quality in terms of loss to society if the service is not performed as expected.

Quality Loss Function

In Figure 14-16 a measurement is taken of the critical product characteristic. This is shown in the figure as A. If A is within the specification limits, the traditional conclusion was that it wasn't a problem. However, point A is closer to being out of specification than to being at the target measurement. This means that over time it might cause a problem. Taguchi calls this potential for problem a potential **loss to society**. In a "hard core" manufacturing operation, Taguchi identifies these losses to society not only in terms of rejection, scrap, or rework but also in terms of pollution that is added to the environment, products that wear out too quickly, or other negative effects that occur. Loss to society is the cost of a deviation from a target value.

To quantify loss to society, Taguchi used the concept of a *quadratic loss function*. Figure 14-17 shows that any variation from the target of six (where $T = 6$) results in some loss to the company. The **quality loss function (QLF)** focuses on the economic and societal penalties incurred as a result of purchasing a nonconforming product. Losses may include maintenance costs, failure costs, ill effects to the environment such

FIGURE 14-16 Classical QC-Step Function

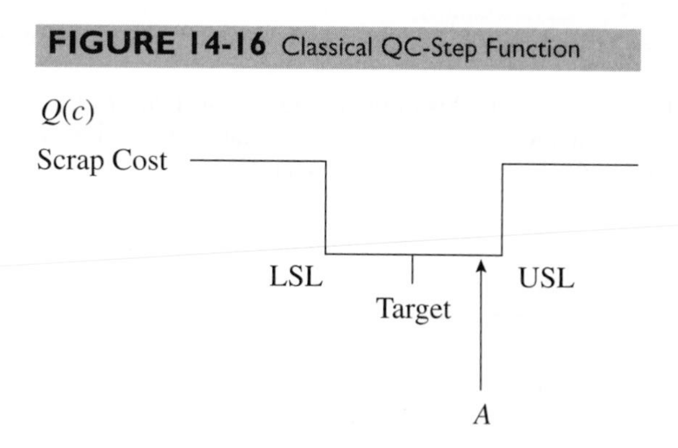

FIGURE 14-17 Taguchi Quadratic Loss Function

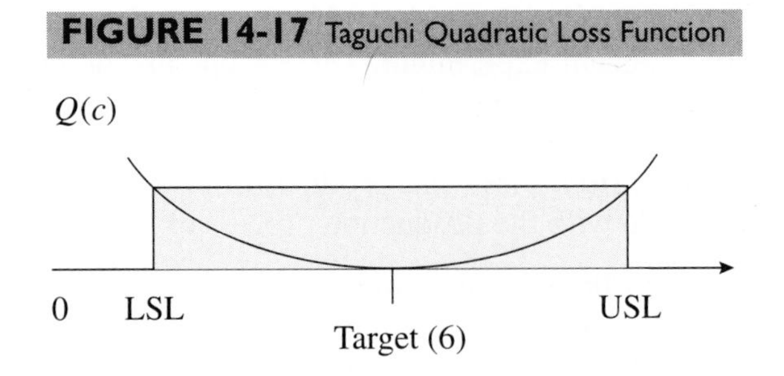

as pollution, or excessive costs of operating the product. The QLF is determined by first computing the constant

$$K = C/T^2 \tag{14.2}$$

where

K = a constant
C = the unit repair cost
T = a tolerance interval (the allowable variation in a parameter)

After computing the constant, next compute

$$L = K * V^2 \text{ (\$/unit)}$$

where
L = the economic penalty incurred by the customer as a result of the product quality deviation
V^2 = the mean squared deviation from the desired target value

The application of this concept is demonstrated in the followed example.

Example 14.5: Application of the QLF

Problem: Suppose the cost to repair a radiator on an automobile is \$200. Compute the QLF for losses incurred as a result of a deviation from a target setting where a tolerance of 6 ± 0.5 mm is required and the mean squared deviation from the target is $(1/6)^2$.

Solution: $K = 200/.5^2 = 800$, and

$$L = K * V^2 = 800*(1/6)^2 = \$22.22/\text{unit}$$

Therefore, the loss caused by deviation from the target standard is \$22.22 per unit. When we can compare the costs of other defectives, we can establish priorities for implementing product design improvement.

The QLF deviated from the historical concept of statistical based control charting and the establishment of specification limits in that any deviation from the target or mean specification is expressed in terms of an economic loss to the customer.

THE TAGUCHI PROCESS

An outline is presented here of the steps in the Taguchi process. Although the Taguchi process is viewed as fairly technical and statistical, a major component involves behavioral steps such as teamwork and brainstorming. We will now examine the steps in the

Taguchi process. As shown in Figure 14-18, a series of six steps is followed in the initial phase of the Taguchi experiment. These steps are described in the following paragraphs.

1. *Problem identification.* First, the production problem must be identified. The problem may have to do with the production process or the service itself.

2. *Brainstorming session.* Second, a brainstorming session to identify variables that have a critical effect on service or product quality takes place. At a minimum, the brainstorming session is attended by the project facilitator/leader and workers taking part in the process to be changed. Managers attending the brainstorming session should be careful that their attendance does not stifle frank discussion among the session participants. In services and manufacturing environments, managerial practices are often critical variables impeding ideal, quality results. When appropriate, technical staff members such as computer programmers or systems analysts also attend. The role of the facilitator is to initiate the brainstorming session, to maintain a nonjudgmental environment conducive to discussion, and to document the discussion for future use.

The critical variables identified in this session are referred to by Taguchi as *factors*. These may be identified as either **control factors** (variables that are under the control of management) or **noise factors** (uncontrollable variation). Examples of control factors within a production process might be procedures, amount of lighting, or ambient temperature setting. Noise-factor examples include uncontrollable variation in temperature, variations in human performance, or environmental variables that cannot be controlled.

Once these factors have been identified, different levels or settings of the control factors are defined. For example, three or four possible levels of ambient temperature settings may be identified for the production of silicon wafers. At least three levels should be used for each factor in order to identify functional forms (such as interactions) of the effects more clearly. Possible interactions between factors should be identified during the brainstorming session. Noise factors can be measured at the time of the experiment and included in the analysis.

FIGURE 14-18 Taguchi Process

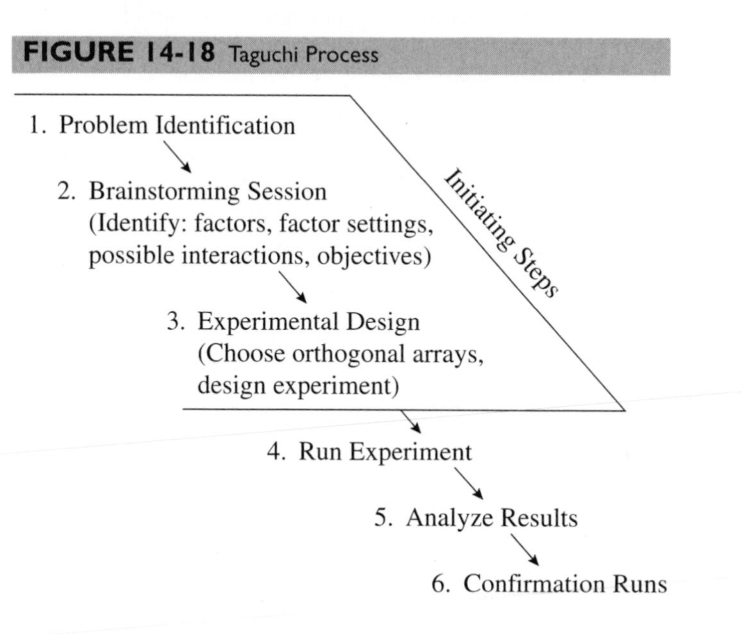

Once the decision variables are established, objectives of the experiments should be defined. Examples of objectives are the less the better, nominal is best, or the more the better. These objectives are defined as follows:

- *The less the better*. This desired level of defectives or errors is as close to zero as possible. For example, in a cake-baking process, we may want zero "fallen" cakes. In an egg-packaging process, we desire the lowest number possible of broken eggs.
- *Nominal is best*. This desired outcome usually relates to a measurement. For example, we may desire to have all boards exactly $\frac{3}{4}$ of an inch thick or all sheets of paper $8\frac{1}{2}$ inches wide.
- *The more the better*. This desired outcome is the opposite of less the better. We may desire the maximum number of computer chips per lot without defects. We may want maximum weight gain from a nutrient we give to a farm animal.

3. ***Experimental design.*** Using the factors, factor levels, and objectives from the brainstorming session, the experiment is designed. The Taguchi method uses **off-line experimentation** as a means of improving quality. This contrasts with traditional on-line (in-process) quality measurement. For this reason, the experimental design is a key consideration in conducting a Taguchi experiment. As with any experiment, care should be taken in selecting an appropriate number of trials and with the conditions for each trial, such as means of performing measurements, maintaining continuity with objectives, and reducing random noise by providing sufficient controls. The number of **replications** to be used in the experiment should be established beforehand.

4. ***Experimentation.*** There are different Taguchi analysis approaches that use quantitatively rigorous techniques such as analysis of variance (ANOVA), signal-to-noise (S/N) ratios, and response charts. These approaches, although not always theoretically sound,[3] are useful in engineering related projects involving engineered specifications, torques, and tolerances.

For services, the approach advocated by Ross[4] may be the most useful. This methodology is more intuitively understandable for management and provides essentially the same results as ANOVA and S/N ratios. What is compromised with the Ross methodology is the additional information provided by the more quantitative results of ANOVA and S/N ratios. The experimental steps used in this methodology are

a. Choose the appropriate orthogonal array for the experiment. (**Orthogonal arrays** are tools to maintain independence between the successive trials of a Taguchi experiment.) The appropriate orthogonal array is determined by the number of factors and levels chosen from the brainstorming session. A number of standard orthogonal arrays can be found in any of the books mentioned in the footnotes.
b. Run the experiment for the appropriate number of replications and record the results.
c. Compute average performance levels for each of the factors and levels.
d. Plot the average responses on a response chart showing the best outcomes in accordance with the objective of the experiment.

---

[3]Box, G. E. P., and Bisgaard, S., "The Scientific Context of Quality Improvement," *Quality Progress* 20, 6 (June 1987):54–61.
[4]Ross, P., *Taguchi Techniques for Quality Engineering* (New York: McGraw-Hill, 1988).

5. ***Analysis.*** Experimentation is used to identify the factors that result in closest-to-target performance. In essence, the best levels for all factors are determined. If interactions between factors are evident, two alternatives are possible. Either ignore the interactions (there is inherent risk to this approach) or, provided the cost is not prohibitive, run a full factorial experiment to detect interactions. The full factorial experiment tests all possible interactions among variables.

6. ***Confirming experiment.*** Once the optimal levels for each of the factors have been determined, a confirming experiment with factors set at the optimal levels should be conducted to validate the earlier results. If earlier results are not validated, the experiment may have somehow been significantly flawed. If results vary from those expected, interactions also may be present, and the experiment should, therefore, be repeated.

Example 14.6: The Taguchi Method in Action

Problem: Below is a standard Taguchi problem. In this experiment it was determined that there were three important factors (*X*s) in producing a wood product. These factors were

- *A*: Pressure applied in treatment
- *B*: Drying temperature
- *C*: Process time

For each of these factors, two levels were established for each setting:

- *A*: 250 psi, 300 psi
- *B*: 150 degrees, 180 degrees
- *C*: 3 hours, 4 hours

We need to determine the best levels for each of the settings. The objective is more is better.

Solution: When performing a Taguchi experiment, you need to use an orthogonal array. Since we have three levels with two factors, a full-factorial experiment would require $2^3 = 8$ trials. Since the Taguchi method is much more economical, the L_4 (2^3) orthogonal array (the right array for the job) is used to perform this experiment. The array looks like this:

Trials	Factors A	Factors B	Factors C	Responses
1	1	1	1	25
2	1	2	2	30
3	2	1	2	28
4	2	2	1	36

Interpreting the orthogonal array, you see that there are three factors and that only four trials are needed to run the experiment. Following the first line, this means that for the first trial we use settings A_1, B_1 and C_1. This means that we use pressure applied at 250 psi, drying temperature at 150 degrees, and a process time of 3 hours. We then measure the outcome, and the responsiveness of the wood product is rated at 25. Similarly, for trial 2, we set the pressure at 250 psi, the temperature at 180, and the time at 4 hours. As you can now see, the 1s and 2s in the orthogonal array correspond to levels for each of the factors to be used during each trial of the experiment. Remember that a trial is an iteration or "run" of the experiment.

Determining the best levels for each factor requires computing averages and analysis of variance. We will demonstrate this in Excel.

Using Excel to Solve Taguchi Experiments

Figure 14-19 shows the results of our Taguchi experiment. To compute our mean scores, we averaged the responses for each factor and level. That is, in trials 1 and 2, factor A was set at level 1. Therefore, the average response when factor A was at level 1 was $(25 + 30)/2 = 27.5$. However, when factor A was set at level 2, the mean response was 32. Since our objective is more is better, the higher response is preferred. Similarly, B_2 and C_1 are preferred settings. This means that the best settings are as follows:

- Pressure applied in treatment = 300 psi
- Drying temperature = 180 degrees
- Process time = 3 hours.

DESIGN FOR SIX SIGMA

Design for Six Sigma (DFSS) is used in designing new products and services with high performance as measured by customer-based critical-to-quality metrics. Instead of the DMAIC methodology, DFSS requires the **DMADV process** (design, measure, analyze, design, verify). Another method for DFSS is **IDOV** (identify, design, optimize, verify). IDOV is focused on final engineering design optimization. These methods are customer-focused, encompassing the entire business-to-market process, and pertain to both services and products. Whereas DMAIC pertains to improving existing processes and products, DMADV pertains to developing new processes and products.

LENSING SIX SIGMA FROM A CONTINGENCY PERSPECTIVE

As we stated early in the chapter, Six Sigma is a very popular approach for improving the robustness of designs and processes. As you have seen, this approach is very technical and requires special expertise in the form of black-belt specialists. This method can be very useful for companies that need to improve their cost and efficiency through quality efforts.

As with any quality improvement approach, when people implement Six Sigma without thoroughly understanding their processes, sometimes processes fail. Some reasons for Six-Sigma failures include

- Lack of leadership by champions
- Misunderstood roles and responsibilities
- Lack of appropriate culture for improvement
- Resistance to change and the Six-Sigma structure
- Faulty strategies for deployment
- Lack of data

As with any quality improvement approach, there need to be a culture, leadership, and commitment in place to make the effort successful. Also, key to Six-Sigma success is the availability of data for projects. Companies where good process data are not available will struggle getting outstanding results from their Six-Sigma efforts.

FIGURE 14-19 Example 14.6 Results

Microsoft Excel - Example 14.6

Orthogonal Array and Responses

	A	B	C	D	E
1	Trial	Fac A	Fac B	Fac C	Responses
2	1	1	1	1	25
3	2	1	2	2	30
4	3	2	1	2	28
5	4	2	2	1	36
6	Mean Scores				
7	A1	27.5	= (E2+E3)/2		
8	A2	32	= (E4+E5)/2		
9	B1	26.5	= (E2+E4)/2		
10	B2	33	= (E3+E5)/2		
11	C1	30.5	= (E2+E5)/2		
12	C2	29	= (E3+E4)/2		

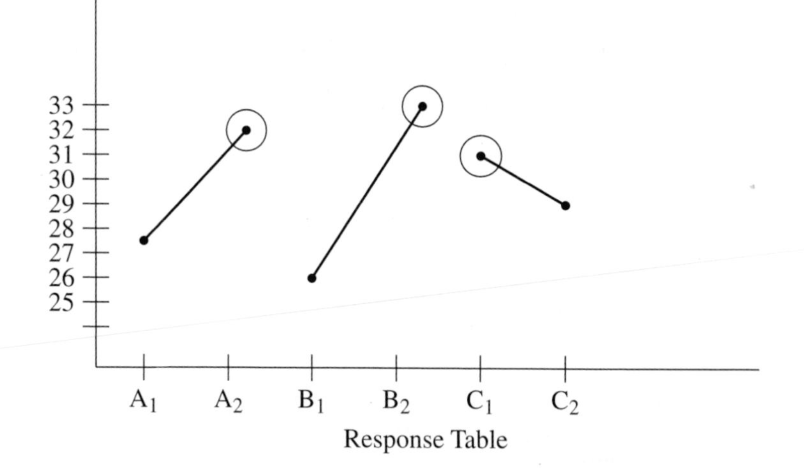

Response Table

SUMMARY

In this chapter we have discussed Six Sigma. We emphasized both managerial and technical requirements for Six Sigma. The process for Six Sigma is define, measure, analyze, improve, and control.

Many companies have reported outstanding results with Six Sigma. There are also many failures. Keys to Six-Sigma success are skilled management, leadership, and long-term commitment.

KEY TERMS

- Analyze phase
- Black belt
- Business case
- Champion
- Concept design
- Control factors
- Control phase
- Defects per million opportunities (DPMO)
- Defects per unit (DPU)
- Define phase
- Design for Six Sigma (DFSS)
- Design of experiments (DOE)
- DMADV process
- DMAIC process
- Gauge repeatability and reproducibility analysis (gage R&R)
- Green belts
- Ideal quality
- Improve phase
- IDOV
- Loss to society
- Master black belts
- Measure phase
- Measurement system analysis (MSA)
- Noise factors
- Off-line experimentation
- Optimal levels
- Orthogonal arrays
- Parameter design
- Project risk assessment
- Quality loss function (QLF)
- Replications
- Robust design
- RUMBA
- Six Sigma
- Taguchi method
- Tolerance design
- *XY* matrix
- $Y = f(X)$
- Yellow belts

DISCUSSION QUESTIONS

1. Can you think of an example where the Taguchi quality loss function (QLF) would work in real life? Discuss how it might work.
2. How does the Taguchi concept of ideal quality compare to other definitions of quality discussed in Chapter 1?
3. How could the Taguchi method be used to design a course in quality management? Identify all the variables, measures, and objectives.
4. Why are behavioral processes such as brainstorming important for the Taguchi method?
5. How would you cost-benefit a Taguchi experiment? What might be some of the quantifiable parameters you would use in evaluating the worth of a Taguchi experiment.
6. The chapter cites different services implementations of the Taguchi method. Do you think the Taguchi method is useful for services? Why or why not? Why do you think the technique has not been widely adopted in services?
7. Where do you think that Six Sigma can be used effectively?
8. How will risk assessments vary from industry to industry?
9. What industries would be the best candidates for the Six-Sigma approach? Why?
10. What is different between Six Sigma and traditional quality improvement?

PROBLEMS

1. Part of a Six-Sigma project is to identify Xs and Ys. What are the Xs you can identify for student satisfaction with a quality management course? Use the Y (dependent variable) of student satisfaction with a quality management class.

2. Part of a Six-Sigma project is to identify *X*s and *Y*s. Identify *X*s and *Y*s for an athletic director of a major university (insert the name of your university here) who is interested in increasing attendance at football games.
3. Part of a Six-Sigma project is to identify *X*s and *Y*s. Identify these variables for the owner of a copy shop who is interested in reducing mistakes in orders.
4. Develop a Problem Definition (see Figure 14-9) for the project in Problem 3.
5. Complete the *XY* matrix for the following data:

XY Matrix						
	Outputs					
	A	B	C	D	Total	Rank
Ranks:	4	6	5	9		
Inputs						
1	10					
2			5			
3		7				
4				8		
5		3	5			

Which inputs are the most important?

6. Complete the *XY* matrix for the following data:

XY Matrix							
	Outputs						
	A	B	C	D	E	Total	Rank
Ranks:	3	8	9	6	7		
Inputs							
1	4	3	5	1	2		
2	5	3	6		2		
3		7	9		2		
4	3	5	4	8	2		
5	1	3	5	1	2		

Which inputs are the most important?

7. Find the QLF for the following information:

   $C = 300$
   $T = .25$
   $V = 1/3$

8. Compute the QLF for the following information:

   $C = 250$
   $T = .40$
   $V = 1/6$

9. It costs \$50 to repair a component in a VCR. Compute the QLF for losses incurred as a result of a deviation from a target setting with a nominal tolerance of 10 ± 0.25 mm required. The mean squared deviation is 1/2.

10. It costs $350 to repair a refrigerator compressor. Compute the QLF for losses incurred as a result of a deviation from a target setting with a nominal tolerance of 60 amps, where a 2-amp variation is acceptable. The mean squared deviation is 1/5.
11. For a component, the following measurements were taken:

2.04	2.05	2.03	2.04
1.96	1.97	1.95	1.96
2.03	2.04	2.02	2.03
2.02	2.03	2.01	2.02
1.99	2.00	1.98	1.99

If the nominal target value is 2 ± .05, compute the QLF for this component where the repair cost is $200.

12. Below are answers to the worksheets in Figures 14-5 and 14-6. Using these responses, develop a risk assessment for this project. Produce a risk and return matrix to determine if this project is worth pursuing. Use the weights in Figure 14-5.

    1. Yes
    2. No
    3. No
    4. Uncertain
    5. Yes
    6. No
    7. No
    8. No
    9. Uncertain
    10. Yes
    11. No
    12. Uncertain
    13. Yes
    14. Yes
    15. Yes
    16. No
    17. No
    18. No
    19. Yes
    20. Yes

    21–24. All Uncertain

    Return scale A: 1 (growth)
    Return scale B: 4 (urgency)
    Return scale C: 3 (impact)

13. Below are answers to the worksheets in Figures 14-5 and 14-6. Using these responses, develop a risk assessment for this project. Produce a risk and return matrix to determine if this project is worth pursuing. Use the weights in Figure 14-5.

    1. No
    2. Yes
    3. No
    4. Yes
    5. Uncertain
    6. Yes
    7. No
    8. No

9. Uncertain
10. No
11. Yes
12. Yes
13. Yes
14. Yes
15. Yes
16. Yes
17. Uncertain
18. Yes
19. Yes
20. Yes

21–24. All Uncertain

Return scale A: 3 (growth)
Return scale B: 4 (urgency)
Return scale C: 5 (impact)

14. Using the following data, perform a gauge R&R analysis where there are two replications for each part number.

Part Number	Operator 1	Operator 2	Operator 3
1	28.35	27.20	26.03
	27.00	27.20	28.77
2	32.40	32.64	31.51
	31.05	32.64	32.88
3	27.00	25.84	27.40
	28.35	28.56	30.14
4	36.45	38.08	36.99
	36.45	35.36	38.36
5	25.65	25.84	24.66
	24.30	24.48	28.77
6	31.05	32.64	31.51
	28.35	28.56	30.14
7	29.70	29.92	30.14
	28.35	32.64	27.40
8	25.65	24.48	26.03
	22.95	27.20	24.66
9	32.40	34.00	32.88
	31.05	31.28	32.88
10	33.75	35.36	32.88
	31.05	34.00	34.25
11	28.35	27.20	28.77
	27.00	27.20	27.40
12	24.30	23.12	24.66
	25.65	25.84	26.03
13	31.05	34.00	34.25
	33.75	34.00	34.25
14	32.40	31.28	32.88
	32.40	34.00	34.25
15	39.15	40.80	42.47
	40.50	38.08	41.10
16	35.10	34.00	34.25

	35.10	35.36	36.99
17	27.00	25.84	27.40
	27.00	27.20	27.40
18	25.65	25.84	28.77
	28.35	25.84	31.51
19	33.75	34.00	34.25
	35.10	32.64	34.25
20	25.65	24.48	26.03
	25.65	23.12	23.29

15. Perform a gauge R&R analysis for the following data where there are three replications. What are your findings?

Part Number	Operator 1	Operator 2	Operator 3
1	21	20	19
	20	20	21
	21	20	21
2	24	24	23
	23	24	24
	22	23	24
3	20	19	20
	21	21	22
	22	22	22
4	27	28	27
	27	26	28
	26	25	26
5	19	19	18
	18	18	21
	19	18	16
6	23	24	23
	21	21	22
	22	22	21
7	22	22	22
	21	24	20
	24	24	23
8	19	18	19
	17	20	18
	19	18	19
9	24	25	24
	23	23	24
	24	25	24
10	25	26	24
	23	25	25
	25	24	26
11	21	20	21
	20	20	20
	26	26	25
12	18	17	18
	19	19	19
	18	19	20

(continued)

13	23	25	25
	25	25	25
	24	24	24
14	24	23	24
	24	25	25
	23	23	24
15	29	30	31
	30	28	30
	25	26	31
16	26	25	25
	26	26	27
	26	25	28
17	20	19	20
	20	20	20
	20	20	20
18	19	19	21
	21	19	23
	22	22	22
19	27	26	25
	25	25	25
	26	24	25
20	19	18	19
	19	17	17
	18	19	18

Case

Case 14-1 The Neiman-Marcus Cookie

One of the fun urban legends to pop up on the Internet has been the Neiman-Marcus Cookie. This work is in the public domain and you might have seen it before. We present the legend here and propose that you develop a Taguchi experiment based on this recipe. It is written in the first person by an anonymous author.

My daughter and I had just finished a salad at Neiman-Marcus in Dallas and decided to have a small dessert. Because our family are such cookie lovers, we decided to try the "Neiman-Marcus Cookie." It was so excellent that I asked if they would give me the recipe, and they said with a small frown, "I'm afraid not." Well, I said, would you let me buy the recipe? With a cute smile she said, "Yes." I asked how much, and she responded, "Two-fifty." I said with approval, just add it to my tab.

Thirty days later, I received my Visa statement from Neiman-Marcus, and it was $285. I looked again and I remembered I had only spent $9.95 for two salads and about $20 for a scarf. As I glanced at the bottom of the statement, it said, "Cookie Recipe—$250." Boy, was I upset!! I called Neiman's accounting department and told them the waitress said it was "two-fifty," and I did not realize she meant $250 for a cookie recipe.

I asked them to take back the recipe and reduce my bill, and they said they were sorry, but because all the recipes were this expensive so that not just anyone could duplicate any of the bakery recipes, . . . the bill would stand.

I waited, thinking of how I could get even or even try and get any of my money back.

I just said, "Okay, you folks got my $250, and now I'm going to have $250 worth of fun." I told her that I was going to see to it that every cookie lover will have a $250 cookie recipe from Neiman-Marcus for nothing. She replied, "I wish you wouldn't do this." I said, "I'm sorry, but this is the only way I feel I can get even," and I will.

So here it is, and please pass it to someone else or run a few copies. . . . I paid for it; now you can have it for free.

The Neiman-Marcus Cookie (recipe may be halved)

5 cups blended oatmeal
2 cups brown sugar
2 cups sugar
2 cups butter
2 tsp. soda
1 8-oz. Hershey bar, grated
2 tsp. baking powder
2 tsp. vanilla
4 cups flour
24 oz. chocolate chips
1 tsp. salt
4 eggs
3 cups chopped nuts

Measure oatmeal and blend in a blender to a fine powder. In a separate bowl, cream the butter and both sugars. Add eggs and vanilla, mix together with oatmeal, flour, salt, baking powder, and baking soda. Add chocolate chips, Hershey bar, and nuts. Roll into balls and place two inches apart on a cookie sheet. Bake for 10 minutes at 375 degrees. Makes 112 cookies.

Using this recipe and these production procedures, develop a Taguchi experiment to find the optimal process for making chocolate chip cookies. Be careful in identifying control factors, noise factors, objectives, and design the whole experiment. ■

PART FOUR

Forever Improving the Quality System

This section is for the organizations that have gone through the hard work outlined in the first 14 chapters of this book and are trying to get to the next level—managing the growth of the individuals within the organization.

This section takes a hard look at team facilitation and training. As a manager, the future will demand that you can effectively assess, develop, and train your employees. This certainly is a key aspect of quality management. Many of the quality concepts that you have learned in this text need to be taught to your fellow employees. Managing organizational learning is key to your success as a manager—regardless of whether you specialize in quality. It is interesting that this concept is largely overlooked in the quality literature. Yet everyone in quality is involved in training.

Finally, Chapter 16 provides a means for outstanding companies to assess themselves and to determine where they need to improve. If you are the benchmark or standard, you often have to look within to see where to improve. Here you will find a method for accomplishing this important task.

CHAPTER 15

Managing Learning for Quality Improvement

I believe that the quality movement as we have known it up to now in the U.S. is in fact the first wave in building learning organizations—organizations that continually expand their ability to shape their future.

—PETER SENGE
Massachusetts Institute of Technology[1]

One major objective of quality management is to enhance organizational learning. At all levels of the supply chain, this is certainly a laudable objective. However, as with other aspects of quality management, the devil is in the details. To create a learning environment, many firms turn to training. Indeed, the old quality control departments that once focused on inspection have now turned their attention to training employees. However, many firms find that once they have trained employees, there is still much work to do. Training often does very little to create cultural change, is usually inadequately planned, and is often poorly implemented. The result is much activity with little in the way of results. With this chapter we hope to provide insight that is necessary to design and implement effective training programs.

EFFECTIVE STORY TELLING

Analogies, metaphors, and stories often go a long way toward helping workers understand quality principles. One excellent example is Brian Joiner's whack-a-mole story. We repeat part of the story here to amplify this point.[2]

> A beautifully simple example of the link between quality and productivity comes from Tim Fuller, who managed a large department that assembled electronic devices.
>
> The diagram [see Fig. 15-1] shows how the assembly operation was supposed to work: get a kit of parts, put them together, then move the assembled product to the stock area. (The real assembly had 100 parts, not 3 as shown, but the diagram gives an idea of what was involved.)
>
> Tim knew that each of the 100 parts had a 98% chance of being there when needed. That sounds pretty good, doesn't it? Well, as Tim found out, at least 75% of the time one or more parts would be missing; in other words, 75% of

[1]Senge, P., "Building Learning Organizations," *Journal for Quality and Participation* (March 1992):30–38.
[2]Joiner, B., *Fourth Generation Management* (New York: McGraw Hill, 1994), pp. 17–20.

FIGURE 15-1 The Way It Was Supposed to Work

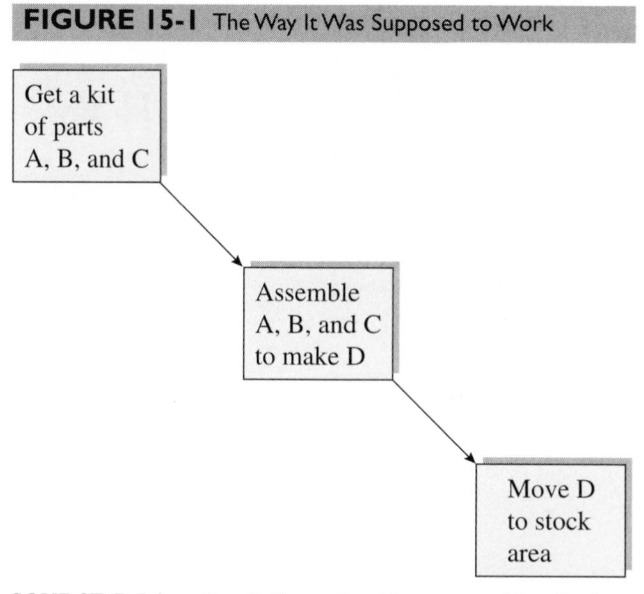

SOURCE: B. Joiner, *Fourth Generation Management* (New York: McGraw Hill, 1994). Reproduced with permission of the McGraw Hill Companies.

the kits were incomplete when the employees began the assembly operation. So the work actually went like this [see Fig. 15-2].

Employees would use what parts they had, complete a partial assembly, log it into the computer, and then store it on a shelf. When a part arrived, they'd go to the computer, find all the partial assemblies that were missing that component, retrieve them from the shelves, and complete the assemblies—or return them to the shelves if other parts were still missing.

In the diagram, steps above the dashed line are what Tim called the real work. They are the only steps that are necessary when everything works perfectly. Tim came to see all the steps below the dashed line as complexity, work that would never have to happen if complete kits were always available. That led Tim to a radical idea: He asked his employees to pretend there was a curtain between the beginning of their assembly operation and everything that came before it. They were not to take any kits past that curtain, into the assembly process, until they were complete.

Referring to the work performed below the line: What do you call this work? Some people call it busy work, fire-fighting. I call it **whack-a-mole:** In many carnival arcades, you'll find a game where you take a big mallet and whack the moles as they pop out of their holes. Every time you whack one down, another one pops up somewhere else. The more moles you hit, the better your score. Isn't that what most of us do all day? Take care of one mole only to have another pop up somewhere else. Isn't that whack-a-mole? The difference of course, is that when you go to a carnival, you have to pay them to play whack-a-mole. When you come to work, your company pays you to play!

The whack-a-mole story is an excellent parable for actual work life. The experienced quality professional will see many issues that can be discussed (e.g., reliability theory, plan-do-check-act, motivation, variation, process improvement, flowcharting,

FIGURE 15-2 The Way It Really Worked

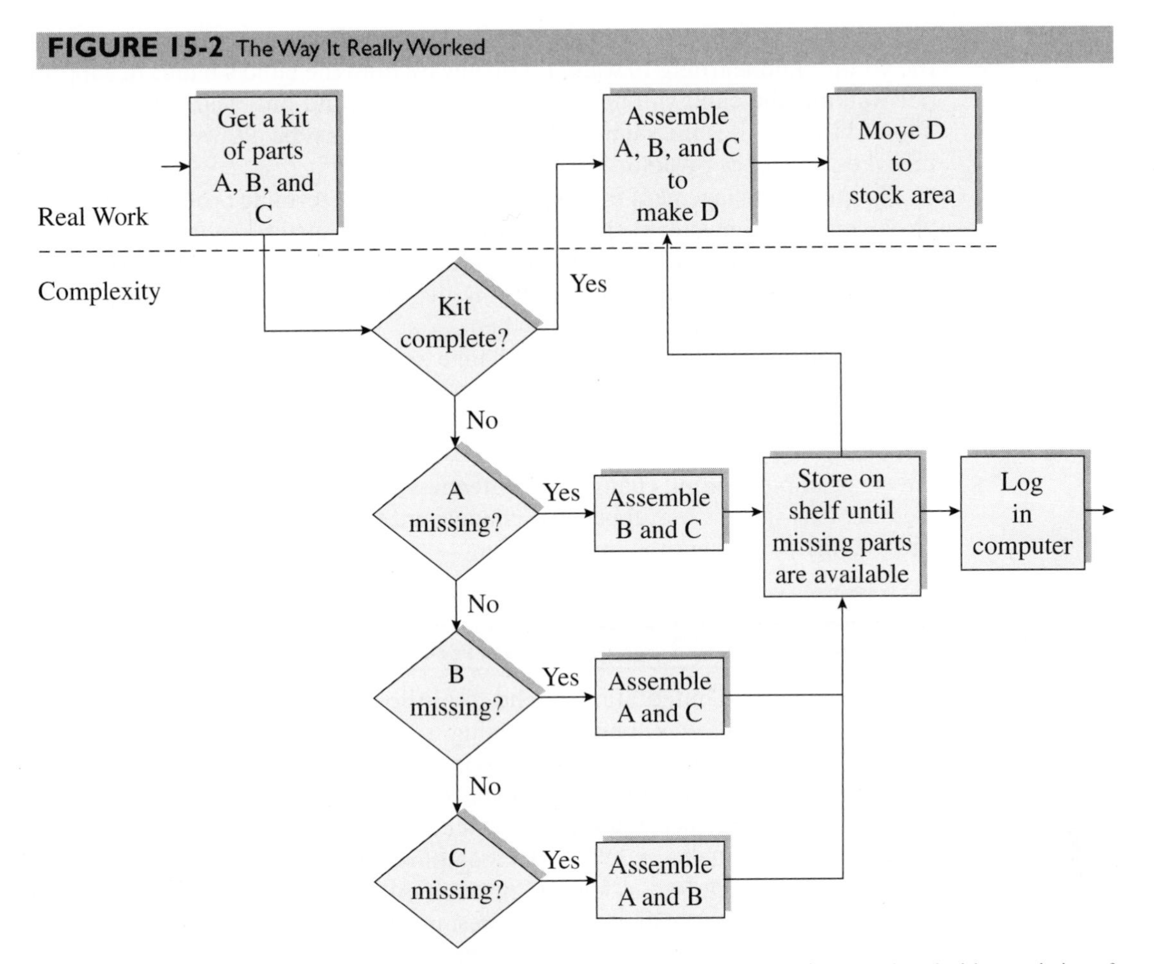

SOURCE: B. Joiner, *Fourth Generation Management* (New York: McGraw Hill, 1994). Reproduced with permission of the McGraw Hill Companies.

random variability, teamwork, and many other topics). Not everyone will understand the statistical topic of variation and the problems caused by variation. However, Joiner's whack-a-mole story makes this subject understandable.

At another level, the Joiner story teaches us a more important lesson. Organizational learning occurs when ideas are put into action. At first, the firm had a process that was very familiar; within the context of the existing process, *people learned how to get around the process*. At the same time, greater organizational learning could not take place until the assumptions of the original system were questioned. This learning is akin to the scientific method that involves forming and testing hypotheses.

Real organizational learning took place only after Tim questioned fundamental assumptions and tested his theory by implementing and testing a process change. The educator John Dewey understood learning and isolated it into four fundamental stages:

Discover: The formation of new insights.

Invent: Creating new options.

Produce: Creating new actions.

Observe: Seeing the consequences of the actions, leading to new discoveries and continuing the cycle.

This cycle must be allowed to run its course for real learning to take place. Consider the young child learning to walk. If you always hold the child's hands to support his or her walking, the child will never learn to walk. A child only learns when he or she is allowed to fall—it is the same with employees. We must be allowed to take risks, make mistakes, and learn by doing.

In quality management training, studying or brainstorming problems in the work environment is one means to create a business case for quality improvement. Identifying the costs of quality also can create this understanding. This is the process of *discovery*.

Next, new options are created during the *inventing* stage. As participants become more creative, new ideas open up to them.

Producing means implementing change and testing ideas. Through this type of interaction, people begin to understand how variables in the work environment interact to create process outcomes such as defects.

Observing is the final stage or the postimplementation review. By observing the effects of implementing change, the outcomes can be observed and the change documented. This stage is useful because change and learning are frozen through finalizing and making permanent the improvement.

INDIVIDUAL LEARNING AND ORGANIZATIONAL LEARNING

Companies in the United States spend too little time on training. Statistics show that less than 1% of salary is spent on training in the United States.[3] Employee development, of which training is a major component, is an important ingredient in retaining employees. As long as employment levels remain high, employee satisfaction and retention will become increasingly important issues. According to the Center for Workplace Development, about 70% of what employees learn about their jobs is informally learned.[4] Only about 30% is learned through training.

It is not clear, however, what is the best means for training. Although several training approaches are discussed in this chapter, a few concepts emerge from experience and the literature. Most of what we learn in the organization we learn informally. Most companies provide new-employee orientation, but they offer very little in the way of training about how the job is really performed in a formal sense. Many times workers are given a day or two of on-the-job training and are allowed to move at their own rate. This happens even with many professional and managerial jobs. In this environment, people are left to figure out for themselves what they should do, and this is the origin of many of the problems associated with confusing and nonrepeatable work methods in firms today.

Organizational learning is equal to the sum of the change in knowledge among its employees. In-house training provides greater potential for long-term benefits to an organization than off-site training and is more likely to be tailored to the needs of the organization. Also, training that is relevant to the immediate work environment is more likely to be valued and rated highly by the trainees. This suggests that organizations need to develop frameworks so that individual learning and experience can be transferred among employees. For an example of one approach to learning and training, see Quality Highlight 15-1.

---

[3]June Poll Results: Readers Views: Training Needed," *Nation's Business* (August 1998): 66.
[4]Stack, J., "The Training Myth," *Inc.* (August 1998):41–42.

QUALITY HIGHLIGHT 15-1

MOTOROLA'S QUALITY UNIVERSITY[a]

Motorola's system for training and educating employees is known worldwide and is reflective of the key value "Constant Respect for People." As a result, Motorola employees believe that training and education are necessary tools for helping the company respond to today's changing competitive environment. Those who receive training from Motorola's quality university are prepared to implement best practices in customer satisfaction, quality improvement, and cycle-time reduction. This level of training and education also leads to improved momentum for individual development throughout a career.

Motorola uses a three-pronged approach to training and education. First, each business organization has its own training organization. Because there are differences in processes, services, and approaches from one business sector to another, these training units focus on the unique training requirements of each business sector. Second, the Motorola human resources department provides training on a sector or geographic basis and focuses on the unique training requirements of its customers. Third, the Motorola university provides training in education for employees. Motorola University is part of corporate human resources and reports through the executive vice president of human resources.

As a result of this three-pronged orientation, education and training at Motorola are both centralized and decentralized. This allows Motorola to apply strategic initiatives and plans and to support the organization as a whole and within particular sectors. The mission of Motorola University is to

- Be an "agent of change."
- Provide training, education, and development to every Motorola associate worldwide.
- Be part of the value-added chain of doing business with Motorola.
- Be the protector and conveyor of all the ethics, values, and history of Motorola to all its associates.

To accomplish this, Motorola University designs, develops, and delivers training in education to support various corporate initiatives at Motorola. In this way, Motorola University provides associates worldwide with needed training and provides a basis for creating a unified culture and language of customer satisfaction and quality improvement.

[a]Adapted from *http://mu.motorola.com*, 2006.

First-rate universities know and appreciate these facts. They provide resources so that individual initiative can be turned into research and teaching results. They also provide colloquia where faculty can share ideas and research. The University of Chicago Graduate School of Business has such a colloquium for faculty that is legendary. Faculty present ideas that are vigorously attacked by other faculty. Such an environment guarantees an intellectually stimulating environment.

Effectively Planning Quality Training

Like other quality efforts, training is a planned process. Prior to beginning training, firms should embark on a **training needs assessment**. Needs assessment includes assessing the educational acumen of employees, assessing organizational needs, and identifying gaps to be addressed through training. Such planning should be part of normal, annual strategic planning processes. The portion of the human resources strategic plan that identifies training needs and how the organization will respond to those needs is called the *strategic training plan*. The strategic training plan should be in alignment with overall company objectives and an integral part of the process through which strategic objectives are achieved.

Training needs assessment helps the organization become aware of the factors to consider when developing a training program. The assessment consists of two phases. The first phase is an *employee assessment* to provide an objective basis to determine and prioritize program goals by directly involving the employees through personal interviews, surveys, and focus groups. The second part is an *environmental assessment* to develop an inventory of available resources to meet training needs and to determine the characteristics of existing resources, company needs, and employee characteristics. Focus groups, personal interviews, and questionnaires provide information both to determine training needs and to ascertain whether resources exist in-house to provide such training.

Training needs assessment lets employees know that the organization is planning its training and thus improves employee acceptance of training programs. This benefits the organization by providing company feedback about current management and the current status of employee knowledge. Management may receive information concerning the current status of its employees' knowledge. In addition, serious problems may be uncovered that must be addressed through training. Either way, better information leads to better understanding of company training needs.

Needs assessment can be provided from within the firm or from a variety of external organizations. Baldrige examiners have found that many successful firms enter into strategic alliances with local universities to provide needs assessment and training. Outside firms can provide assessment testing, and employees often are more honest with a third party than with an immediate supervisor. Every firm should begin quality initiatives with training. Ultimately, the best training should be performed in-house. Even if consultants begin the training, the goal should be for the company to perform its own training over the long term. If the consultants are unhappy with this long-term objective, perhaps they are the wrong consultants.

The goal of internal assessment is to provide insights that are used to develop the strategic training plan. By performing assessments, the organization indicates that it is serious about addressing customer needs and cares about the development of employees. Some firms are afraid of losing employees as a result of developing them and providing them with new know-how. However, the alternative is to populate the firm with employees who are ill-prepared to meet consumer needs.

Employees appreciate the fact that assessment involves them in developing future plans related to training. Many benefits accrue to employees as a result of performing training assessment. Employees begin to realize that they can make choices concerning their careers through the needs assessment program, express concerns about the present status of their job knowledge (without having the information show up on their performance appraisal), facilitate the company in achieving its objectives and goals, and evaluate their own knowledge and learning.

Although many benefits accrue to the employees, there are several benefits for the organization as well. Reflecting on the contingency approach to quality, effective training programs likely will be tailored to the specific needs of any firm or organization. A needs assessment provides a systematic, repeatable approach for customizing training programs. As a result, organizations can know what their training needs are and develop a base for sequencing and phasing in new knowledge. Firms also develop an understanding as to why they are providing particular training. This approach facilitates budgeting. Other benefits to the firm include

Knowing what training is being planned and why
Justifying costs in relation to training benefits

Evaluating training based on measurable, written objectives

Encouraging continuous employee participation and enthusiasm for training

Knowing what resources already exist to meet training objectives and what resources must be purchased

Discovering what type of employee development is desired and for which employees

Confirming how much time and money are needed to achieve the objectives

The bottom line of needs assessment is to determine organizational needs, employee needs, and organizational resources to provide needed training.

A MODEL TO GUIDE TRAINING IN AN ORGANIZATION

Imagine that you are asked to develop a quality management training program for a large organization. Where do you start? How do you do it?

As with anything else, there is a process for developing training. This five-phase process involves needs analysis, definition of instructional objectives, training (curriculum) design, implementation, and evaluation (see Fig. 15-3). Although these steps are covered in more depth later in this chapter, it is useful to briefly define them at this point.

Training needs analysis begins with identifying organizational needs in terms of capabilities, **task needs assessment** in terms of skill sets that are needed within the firm, and **individual needs analysis** to determine how employee skills fit with company needs. A *gap assessment* shows what skills are lacking in the organization and leads to a definition of instructional objectives for the organization. Once the instructional objectives have been defined, a determination is made as to whether the training resources are available in-house. If the skill sets are not available, outside resources are sought.

Training program design includes the specifics of tailoring a course or set of courses to the needs of the company. Support is necessary to provide training in terms of facilities, equipment, and infrastructure.

Implementation occurs when the training materials are actually delivered to the students. Evaluation is an important step in training because the feedback can be used to adjust future training.

Hierarchical Approaches to Training Design

Training needs are assessed on a hierarchical basis. In most firms, training and development needs are different for executives, managers, supervisors, and other employees. In Figure 15-4, different needs are shown by level in the organization. Relating to quality, executives need to understand general quality principles, leadership skills, and strategic quality planning. Managers need to understand general quality principles such as Deming's 14 points. Supervisors should understand facilitation and the management of worker teams, and employees need to understand how to implement quality improvement with basic tools.

Figure 15-5 shows an outline of a hierarchical training model for the U.S. Army. This governmental plan is based on organizational, occupational, and individual levels of needs. Training plans are developed to fit each of these levels of needs.

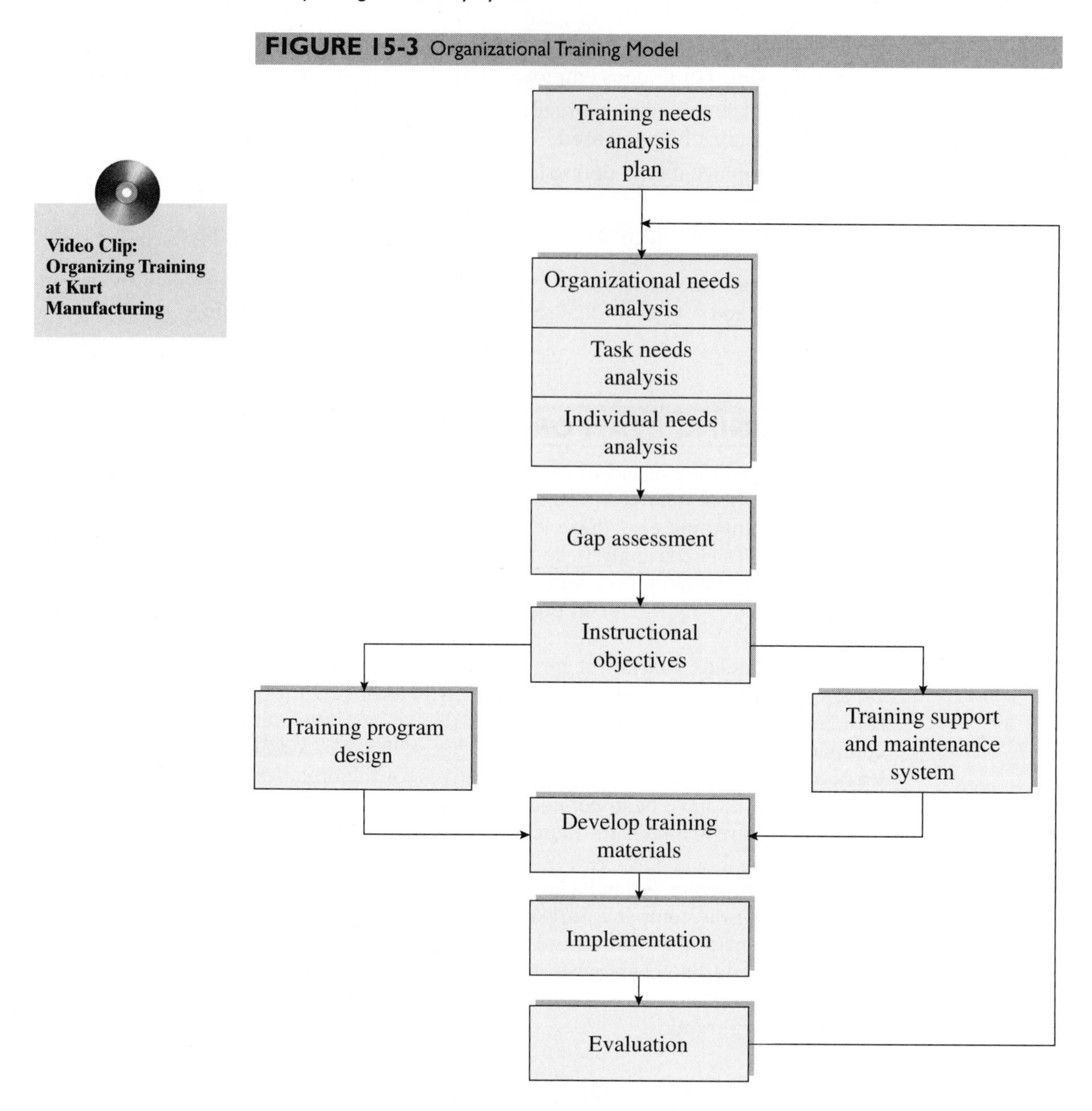

FIGURE 15-3 Organizational Training Model

Example 15.1: An Example of a Hierarchical Approach to Quality Management Training

This example shows how a quality training program delivered to Qwest Corporation was designed. The hierarchy had three levels with a focus on executives, managers and supervisors, and employees. Notice that the needs that were addressed at each level were similar to those identified by the U.S. Army in Figure 15-5.

Table 15-1 shows the topics that were delivered to executives in 2 three-hour sessions. These sessions focused on organizational objectives and management philosophies contained in quality improvement (the topics discussed in the first three chapters of this text). The executives would not be directly involved in quality improvement processes. However, their support and resources were needed to implement organizational change.

FIGURE 15-4 Hierarchical Training Needs

Level	*Needs*
Executives	General principles Strategic quality planning Needed resources
Managers and Supervisors	General principles Facilitator Team management
Employees	Quality tools General principles Working in teams

FIGURE 15-5 Hierarchical Planning in the Military

Training Needs Assessment

Organizational *Chief of Staff, Army* *Commanders* *Commands/Activities*	Supports mission, goals, objectives Ensures budget/program resources Focus on organizational performance Leader development common core New employee orientation
Occupational *Functional Chiefs/* *Personnel Proponents*	Short- and long-term skills requirements to support organization Promotes development of career paths Developmental assignments Functional training Education
Individual *(Supervisors/* *Managers)*	Focus on personal development requirements Individual needs in terms of organizational goals Promotes development of employee in career path Communication Skills Filing systems PC training

SOURCE: U.S. Army, *http://www.cpol.army.mil/permiss/710.html* (1998).

TABLE 15-1 Executive-Level Training Topics

Module 1 Executives	Quality management philosophy What is quality? The three spheres of quality Deming, Juran, and Crosby The importance of leadership Shifting paradigms Services gap model ISO 9000:2000 overview Self-assessment Overview of the improvement process

The training module contained in Table 15-2 shows the training that was provided for managers and supervisors at Qwest. This training consisted of the management tools and techniques needed for supervising quality management projects. Notice that the focus here was on developing managers who could provide leadership and guidance in specific projects. Although the specific tools of improvement were not given in extensive detail, the managers and supervisors were familiarized with these tools so that they could guide improvement.

Table 15-3 shows the training that was provided to the rank-and-file employees. This training included the overviews provided to executives and managers in an abbreviated format. Many sessions then focused on particular quality tools (see Chap. 10). The tools training was followed up with facilitated quality management projects that emphasized the implementation of the tools and philosophies.

Although this training was viewed as very effective by executives, managers, and employees, it presents only one possible design of a basic quality management program. An interesting project is to benchmark against other firms to see what is helpful and unhelpful in training employees with these tools and techniques.

Example 15.2: An Example of a Real-Life Training Session

To go beyond simply discussing training programs, we provide an overview of one module discussed in the previous example. The following pages contain an actual example of training

TABLE 15-2 Manager and Supervisor Training

Module 2 Managers and Supervisors	Quality management philosophy What is quality? The three spheres of quality Deming, Juran, and Crosby The importance of leadership Shifting paradigms Services gap model ISO 9000:2000 overview Self-assessment Overview of the improvement process Kaizen—People-based improvement Process improvement Process management Supervising teams Cross-functional teams Kaizen approach to problem solving PDCA cycle Overview of tools

TABLE 15-3 Employee-Level Training

Module 3 *Employees*	Overview of quality management philosophy What is quality? Overview of the improvement process
	Kaizen—People-based improvement Process improvement Process management Supervising teams Cross-functional teams Kaizen approach to problem solving PDCA cycle Training on tools
	Flowcharting Histograms Pareto charts Brainstorming/nominal group techniques Fishbone diagrams Root cause analysis Affinity diagrams Cycle time reduction Waste removal Statistical process control Sample survey procedures Quality team projects (experiential learning)

provided to a state Department of Water Resources in preparation for flowcharting processes prior to process improvement and migration to a state-of-the-art enterprise database system. This single full-day session introduced the techniques of developing process flow diagrams and process flowcharts.

A good rule of thumb for successful training sessions is a ratio of no more than one-third lecture and no less than two-thirds in-class exercises and reports. This ratio allows participants to stay active and to learn by doing. It is difficult for trainers to understand that the less they talk and the more they facilitate, the more satisfied their audience will be. Most businesspeople are bright and require very little actual lecturing.

Figure 15-6 shows the cover slide for the presentation. Note that these slides were all included in a notebook that was provided during the training. In addition, a handbook (not shown here) was developed in concert with the training that provided a ready reference for use after completion of the training by employees. Weekly meetings were also provided where flowcharts were reviewed and approved.

Figure 15-7 is the overhead that provides a brief overview of the session. Notice that this provides the key objectives of the four-hour session. The bullet items are few and basic.

The importance of people as a resource should not be overlooked. Training provides a good opportunity for people to become acquainted with others from other work groups. Brief introductions were performed where people gave their names, their divisions within the organization, and an interesting fact about themselves (see Fig. 15-8).

Figures 15-9, 15-10, and 15-11 provide definitions of processes, flows, and business processes. Several minutes of discussion went into clarifying these terms, because, in preliminary interviews prior to the training, managers had stated that there was some ambiguity in people's minds as to the definition of a process.

In the next phase of the class session (see Figs. 15-12 and 15-13), the instructor provided classroom work about using Microsoft PowerPoint to draw flowcharts. Prior to the training, the

FIGURE 15-6 Introduction to Idaho Department of Water Resources Training Program

Process Flowcharts
Presented by
Dr. Tom Foster
Brigham Young University

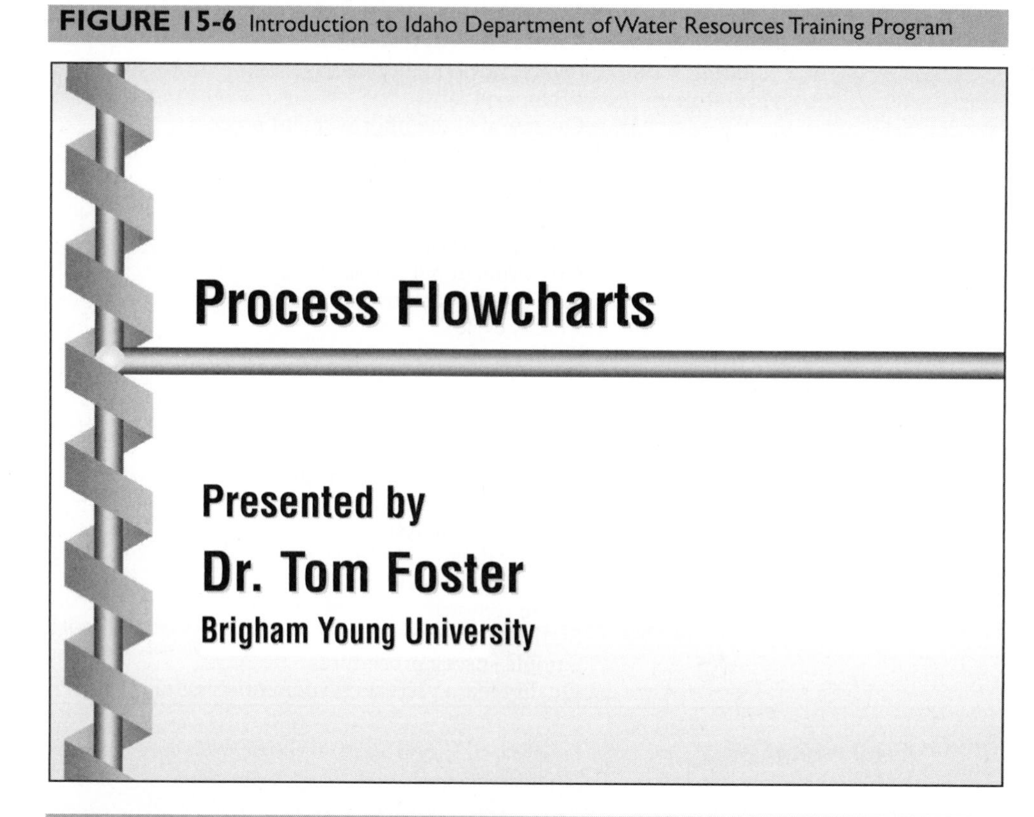

FIGURE 15-7 Key Objectives of Training Program

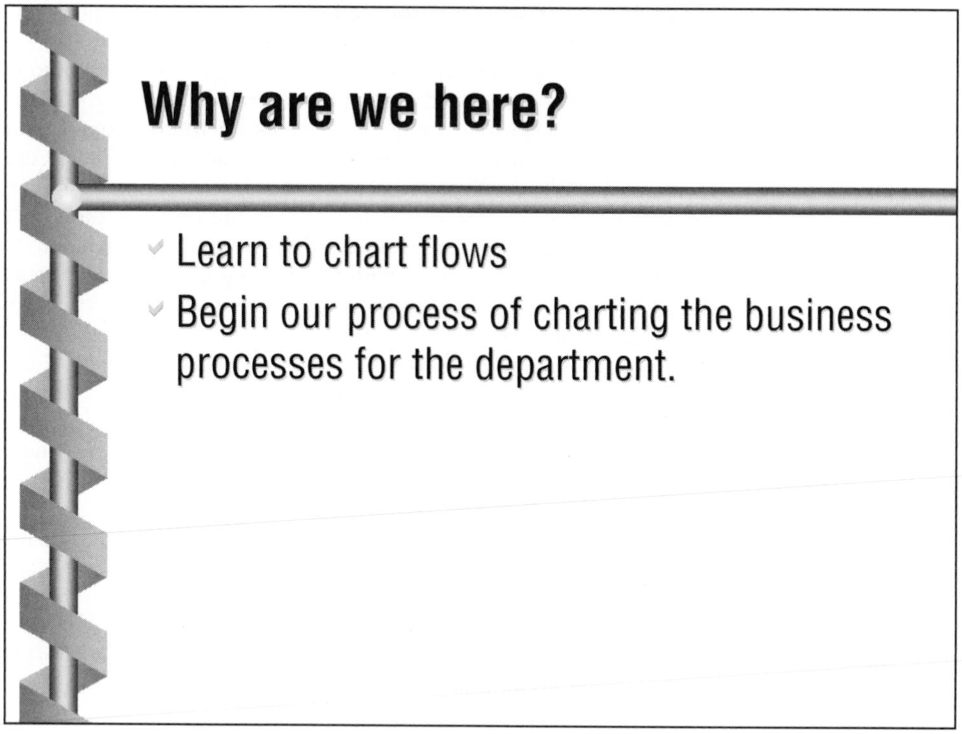

Why are we here?
Learn to chart flows
Begin our process of charting the business processes for the department.

FIGURE 15-8 People Are an Important Resource

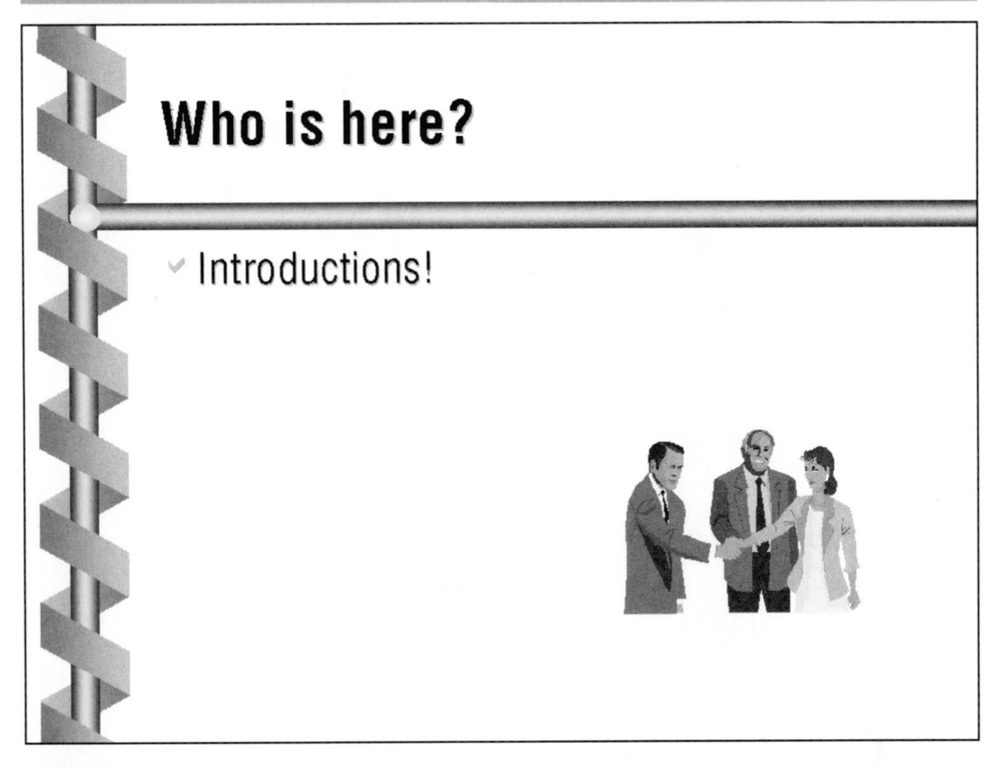

FIGURE 15-9 Definitions: Process

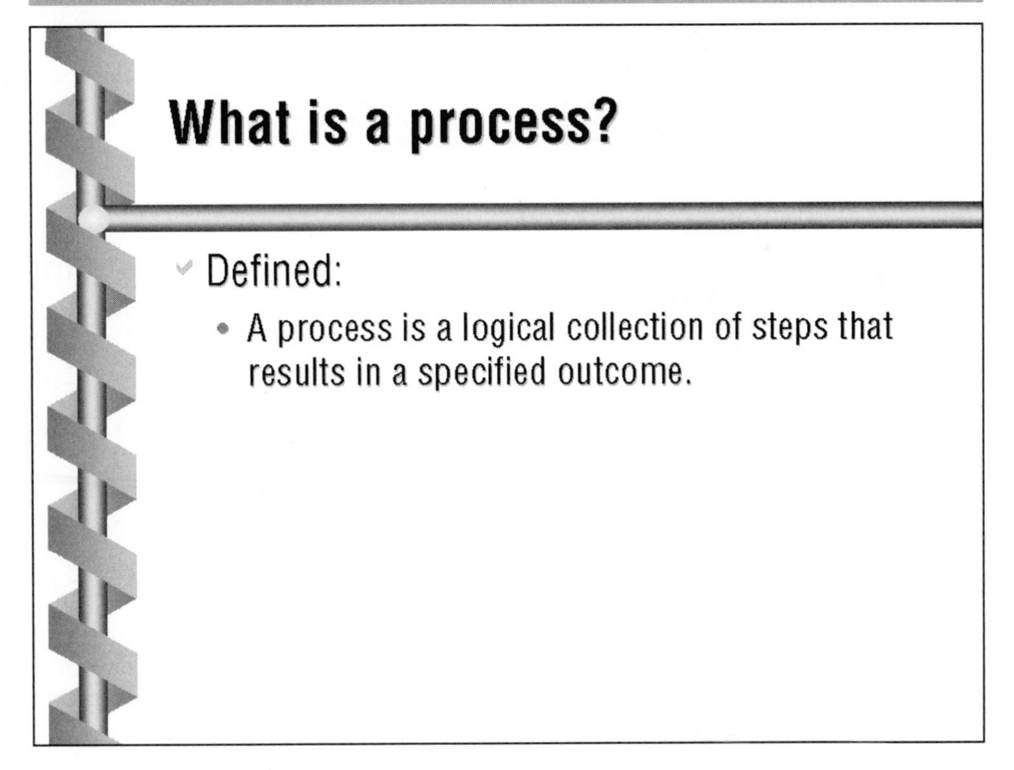

FIGURE 15-10 Definitions: Flow

What is a "flow?"
A flow indicates the *direction* of a process from one action to another.

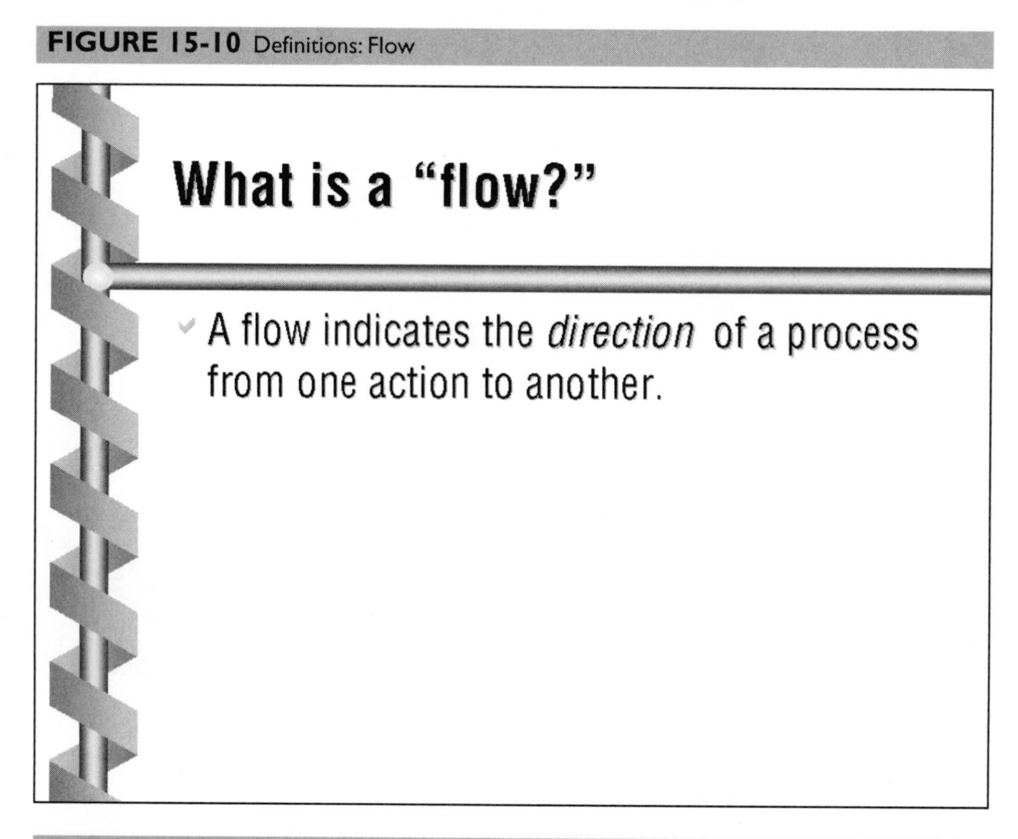

FIGURE 15-11 Definitions: Business Process

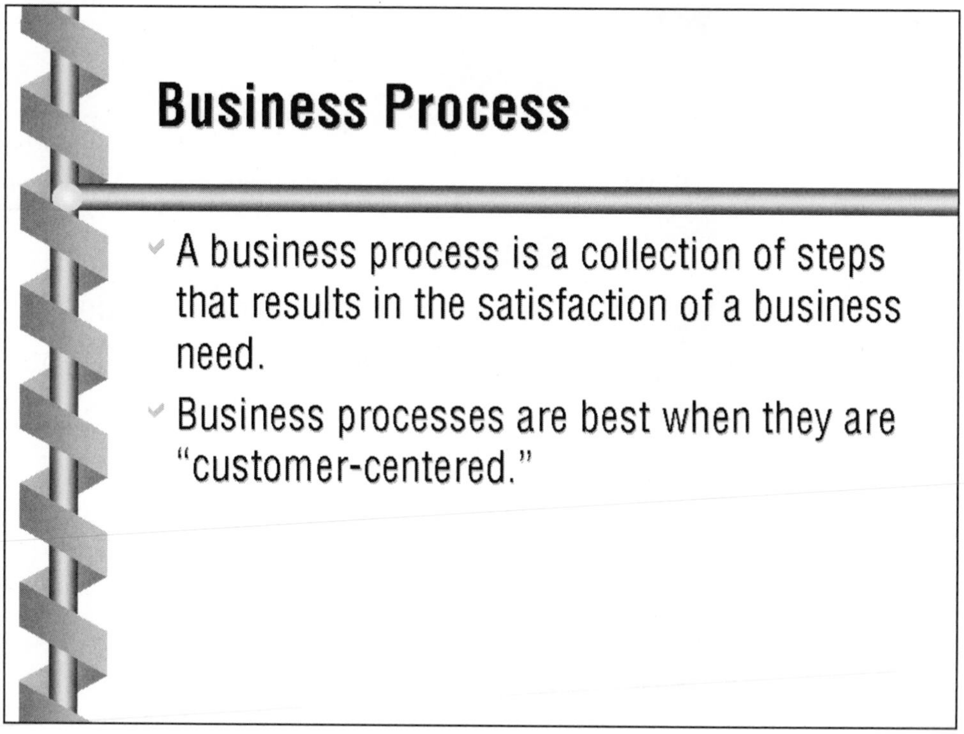

Business Process
A business process is a collection of steps that results in the satisfaction of a business need.
Business processes are best when they are "customer-centered."

FIGURE 15-12 Objectives of PowerPoint Flowcharts

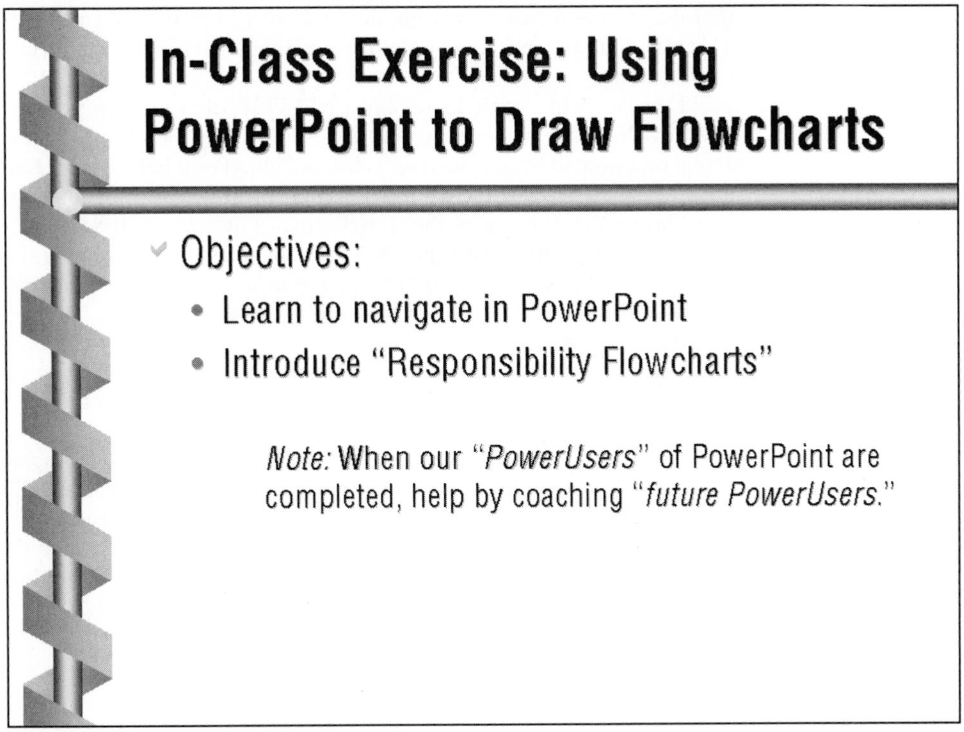

In-Class Exercise: Using PowerPoint to Draw Flowcharts

- Objectives:
  - Learn to navigate in PowerPoint
  - Introduce "Responsibility Flowcharts"

Note: When our "*PowerUsers*" of PowerPoint are completed, help by coaching "*future PowerUsers*."

FIGURE 15-13 Steps for Creating Flowcharts Using PowerPoint

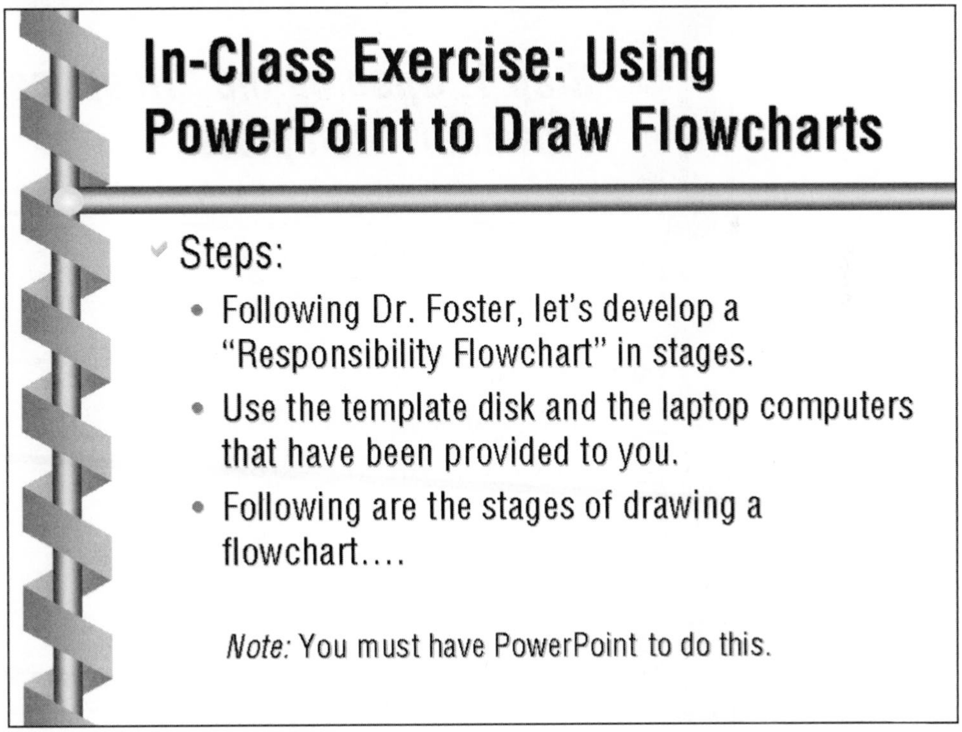

In-Class Exercise: Using PowerPoint to Draw Flowcharts

- Steps:
  - Following Dr. Foster, let's develop a "Responsibility Flowchart" in stages.
  - Use the template disk and the laptop computers that have been provided to you.
  - Following are the stages of drawing a flowchart....

Note: You must have PowerPoint to do this.

decision was made to use PowerPoint as a platform because all employees were familiar with MS Office. The flowcharting facility in PowerPoint, although not leading edge, was usable. Notice that the session focused on using the tool prior to providing specific training on charting flows. This was done to familiarize employees with a flowchart format prior to developing flowcharts from scratch.

Figures 15-14, 15-15, 15-16, and 15-17 show the four-step process for drawing a flowchart. The flowchart format here is called a "responsibility high-level process map." A flowchart of an actual process had been placed on a disk in PowerPoint beforehand. Each employee was provided a laptop PC to practice the exercise on their own. The exercise lasted about 45 minutes.

During the drawing of the flowcharts in PowerPoint, the instructor roamed the classroom and helped individuals with specific problems in using the software. This took the instructor out of the *lecture mode* and placed him in the *coaching mode*. After the session, time was taken to list what had been learned on a flipchart and to answer questions.

After a break, the participants returned to learn the specifics of flowcharting processes. Figures 15-18, 15-19, and 15-20 show the overheads used in training employees on the basics of flowcharting. The flowcharting steps were discussed as well as the meanings of the flowcharting symbols.

Figures 15-21 and 15-22 outline the class exercise. Notice again the ratio of lecture to actual in-class activities. During this exercise, employees were asked to form teams and flowchart a process from a dialogue prepared by the instructor. The dialogue outlined a services process flow in a company. The class members flowcharted the process using a process flow work sheet and developed a flowchart with their teams. These flowcharts were presented to the class.

FIGURE 15-14 Creating a Flowchart: Step 1

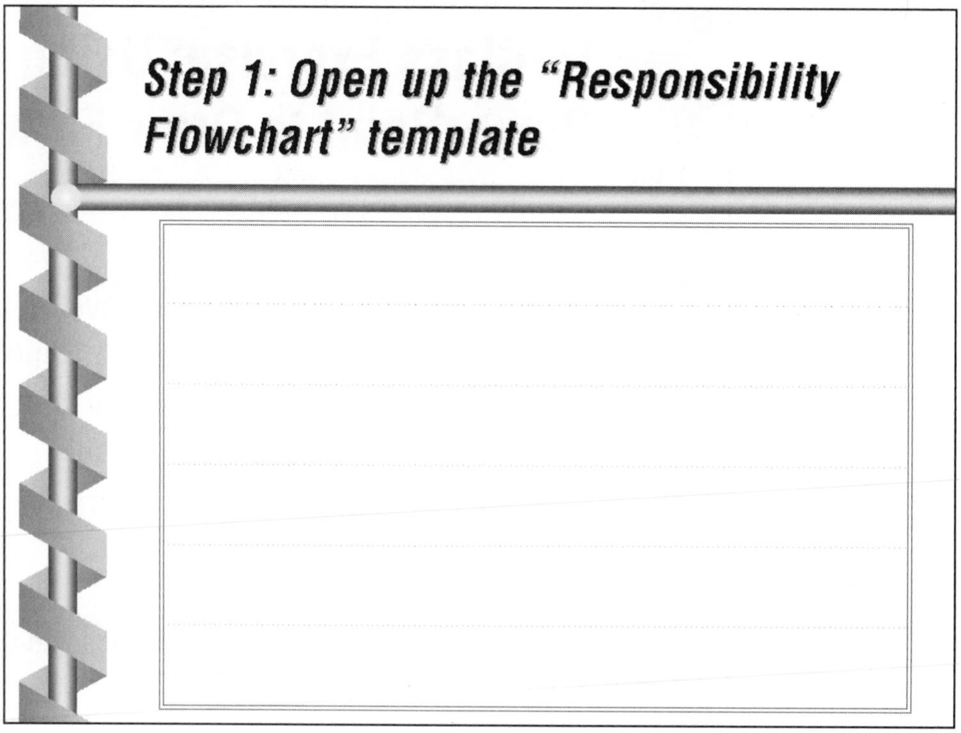

FIGURE 15-15 Creating a Flowchart: Step 2

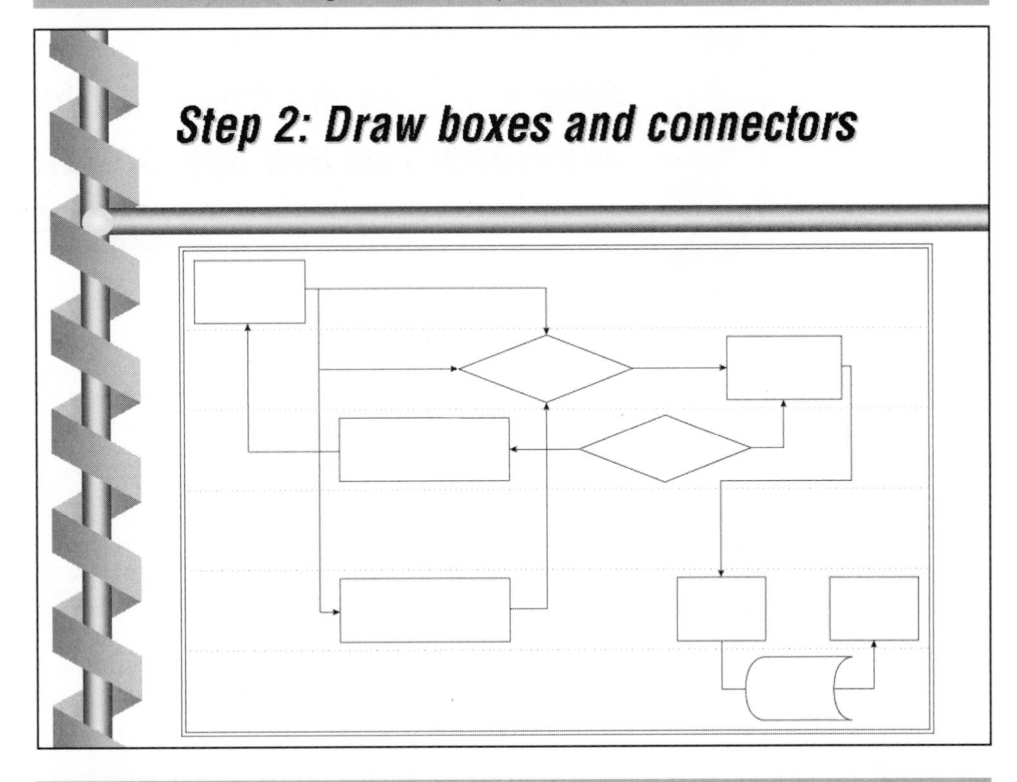

FIGURE 15-16 Creating a Flowchart: Step 3

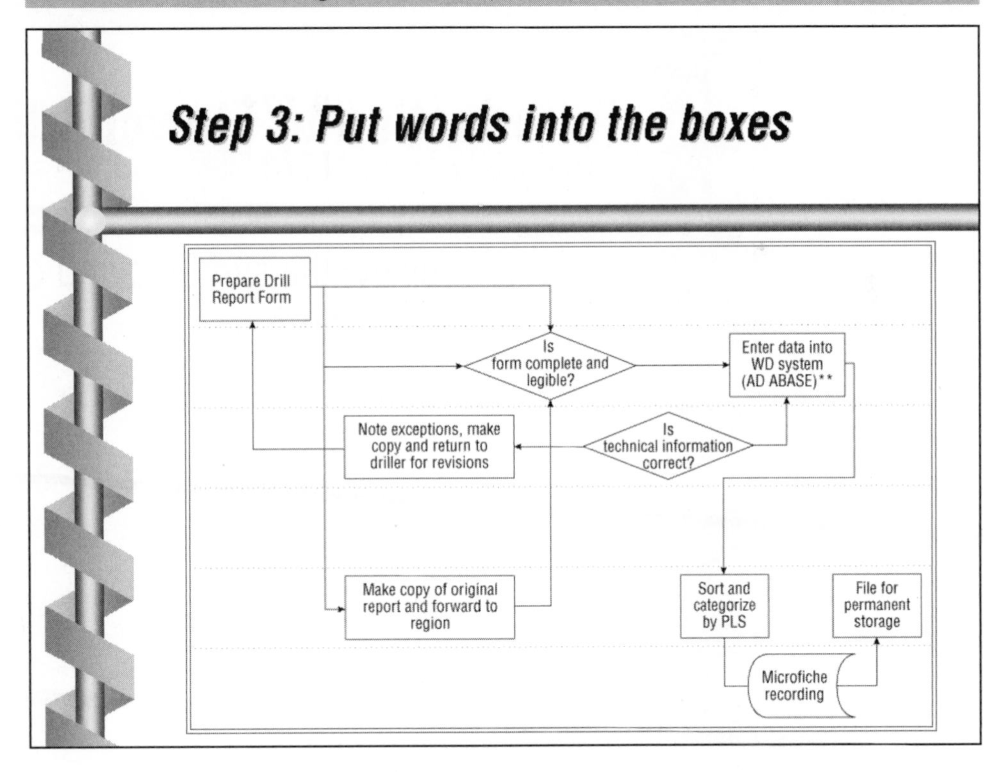

FIGURE 15-17 Creating a Flowchart: Step 4

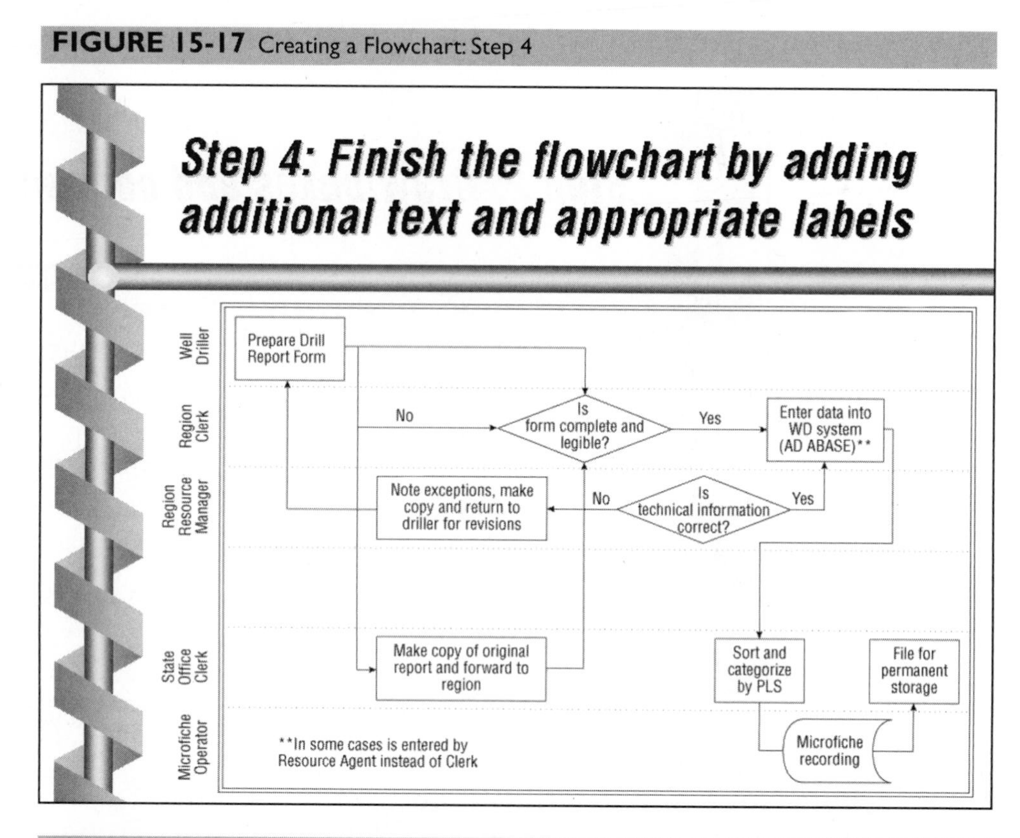

FIGURE 15-18 Flowcharting Specifics: Introduction

- We will review the symbols and discuss the steps of flowcharting...

FIGURE 15-19 Flowcharting Specifics: The Steps

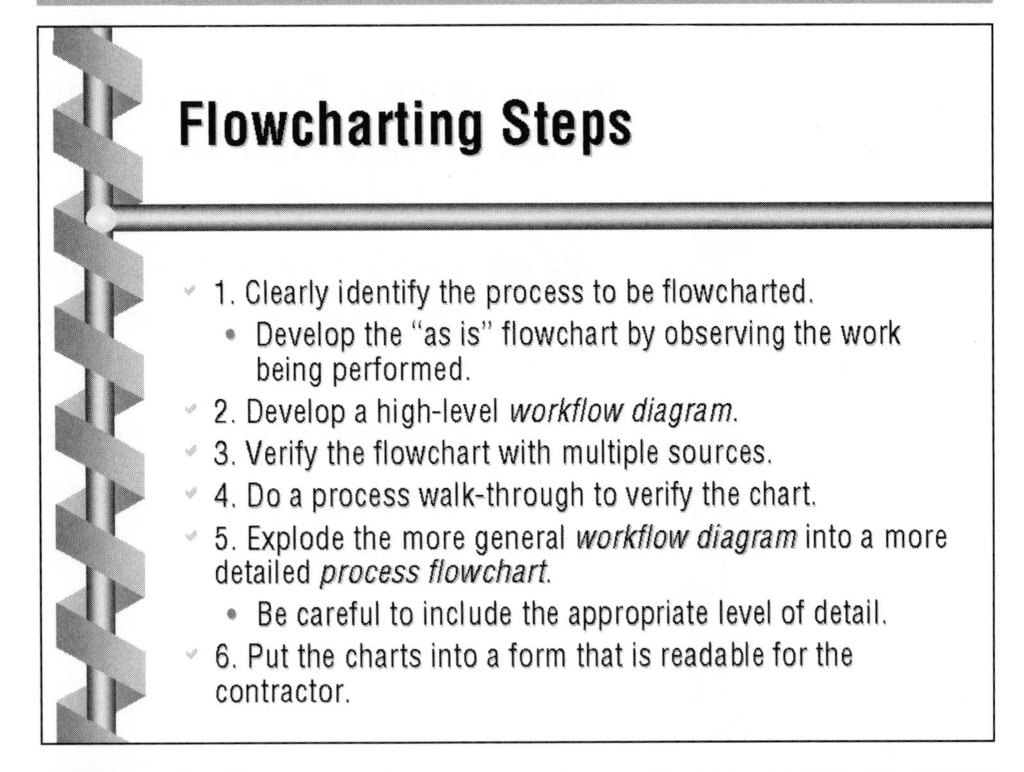

FIGURE 15-20 Flowcharting Specifics: Symbols

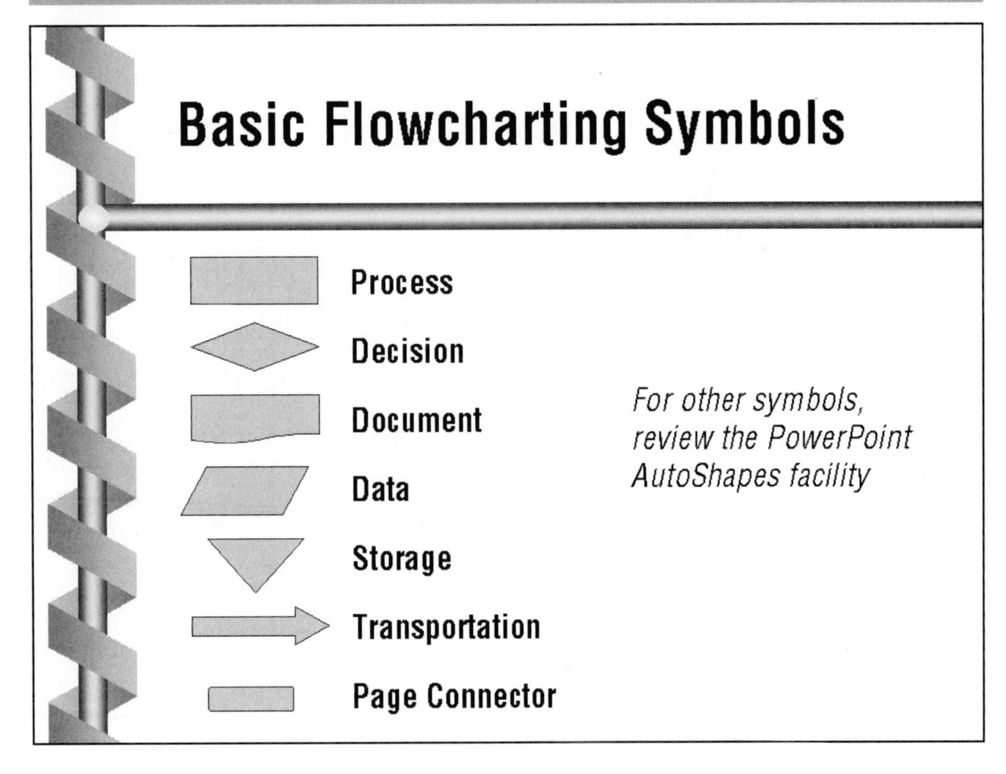

FIGURE 15-21 In-class Exercise

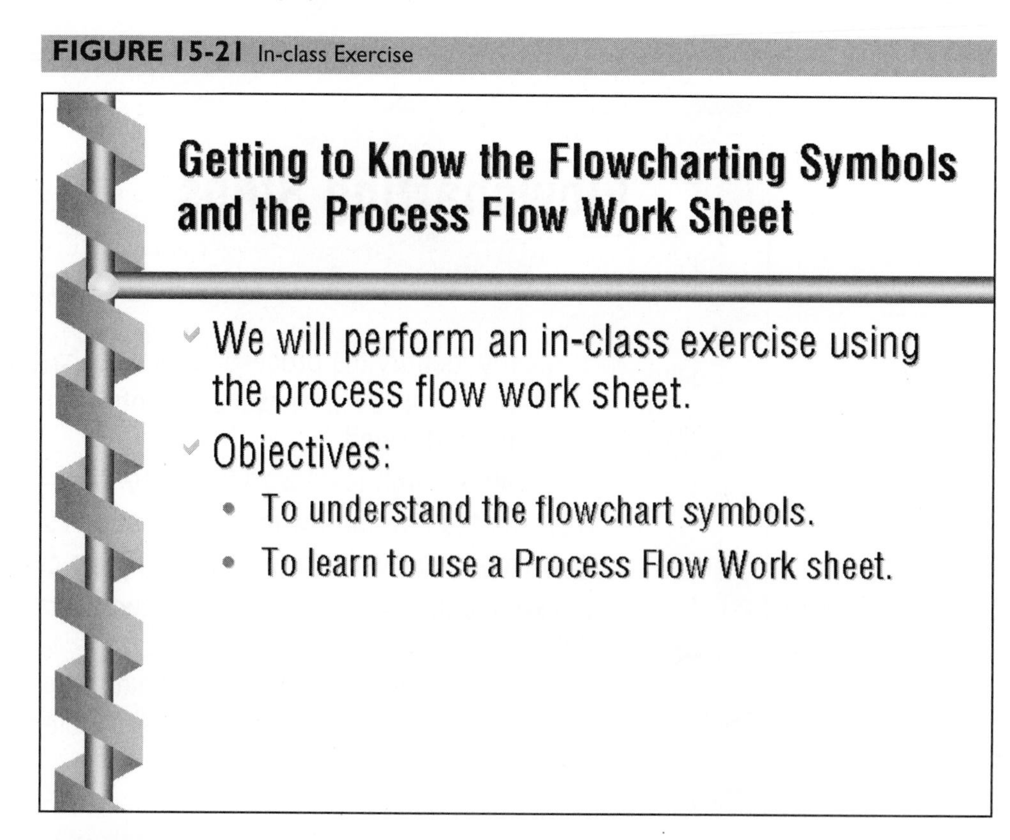

FIGURE 15-22 Flowcharting a Process Practice

- Process:
  - Obtain work sheets.
  - The instructor reads the paragraph.
  - The class documents the process using the work sheet.
  - When we have finished, we will go over a sample work sheet for this process.
    - Activity Time: 60 minutes

After the flowcharting practice, participants presented what they had done and were given a pre-prepared flowchart of the flow. The participants were then asked to follow their flowcharts and the instructor's answer while the instructor reread the process description. This allowed the participants to learn by seeing an example.

After listing what had been learned and entertaining questions, a similar process was followed to introduce workflow diagrams. Figures 15-23, 15-24, and 15-25 show the overheads used in this training.

Finally, another in-class exercise was used for each participant to actually begin to flowchart a process within the organization (see Figs. 15-26 and 15-27). Time limits were set for participants to have their "piece of a process" completed and drawn on PowerPoint. After doing their work, volunteers spent 45 minutes presenting what they had done by placing their disk in the instructor's laptop and projecting the flowcharts onto a screen. Class participants could critique the work, ask questions, discuss issues that were unclear, and cement learning.

During the final 10 minutes, a senior manager reviewed the project facing the class participants. He discussed timelines and expectations. Final questions were answered and the session evaluation was performed.

ADULT LEARNING

One of the difficulties encountered by trainers is understanding that adult training needs are different from youth educational needs. Therefore, course design for adult training is fundamentally different from course design for a college course or a high

FIGURE 15-23 Workflow Diagrams: Introduction

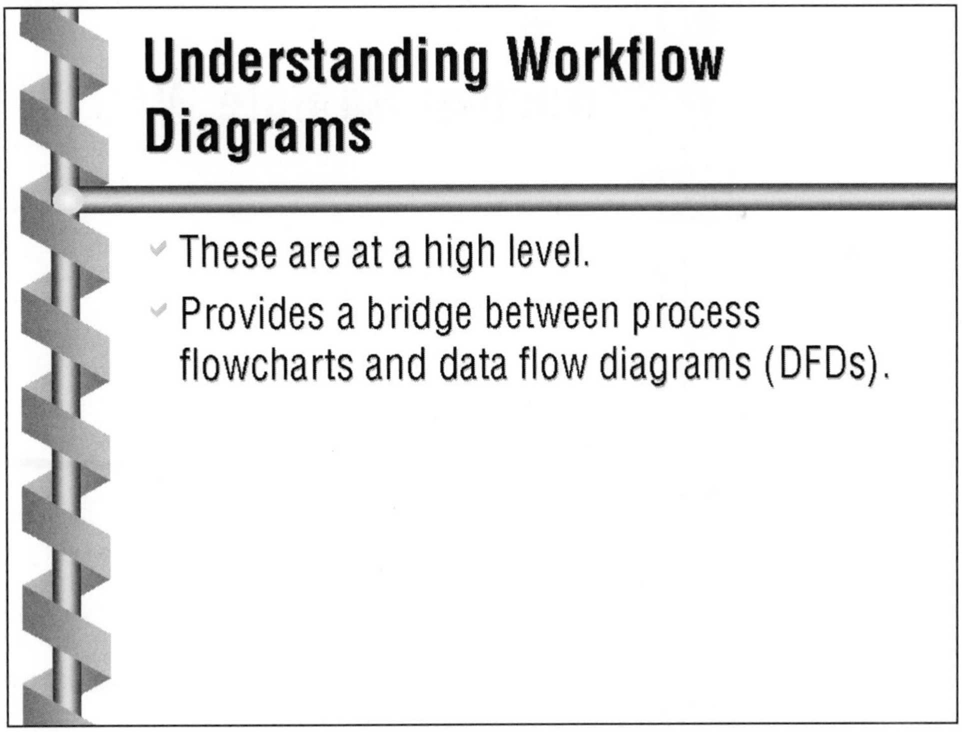

Understanding Workflow Diagrams

- These are at a high level.
- Provides a bridge between process flowcharts and data flow diagrams (DFDs).

FIGURE 15-24 Workflow Diagram

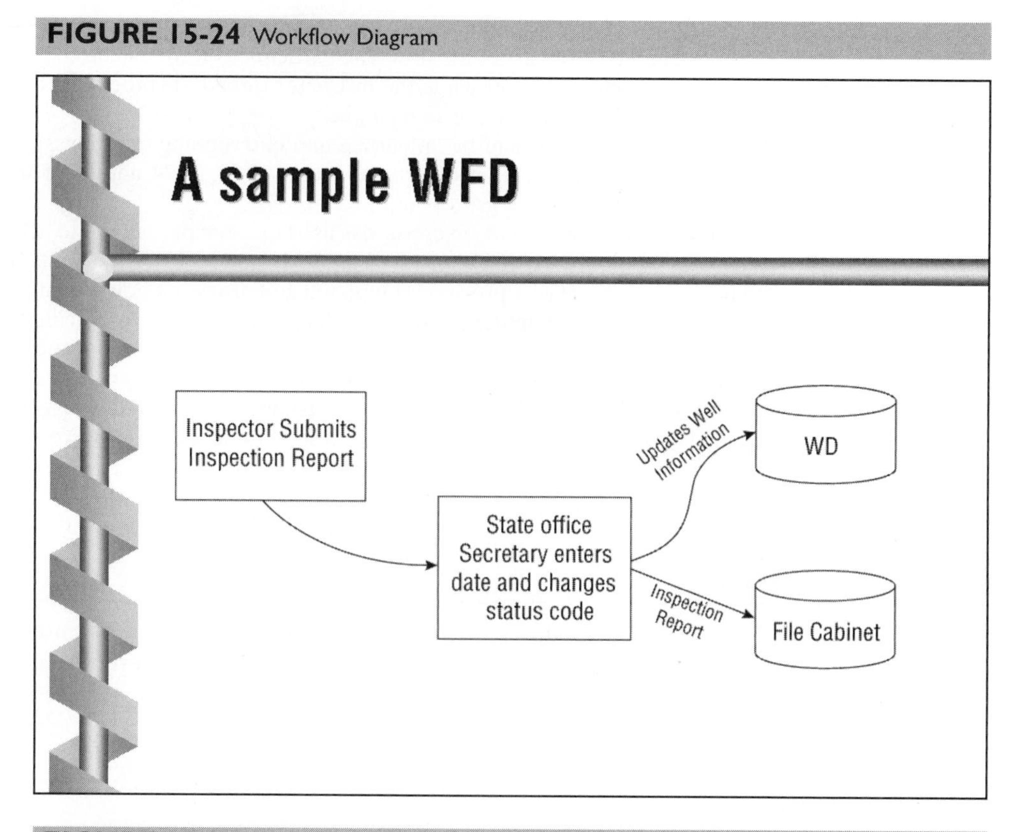

FIGURE 15-25 Workflow Diagram

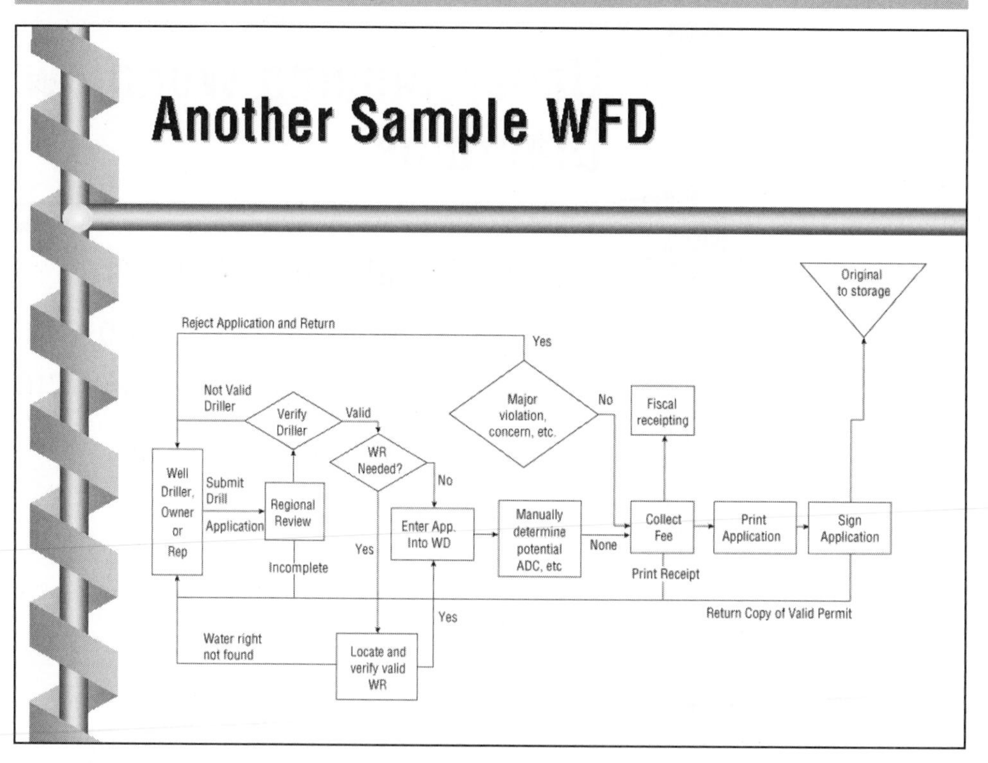

FIGURE 15-26 Beginning the Flowcharting Process

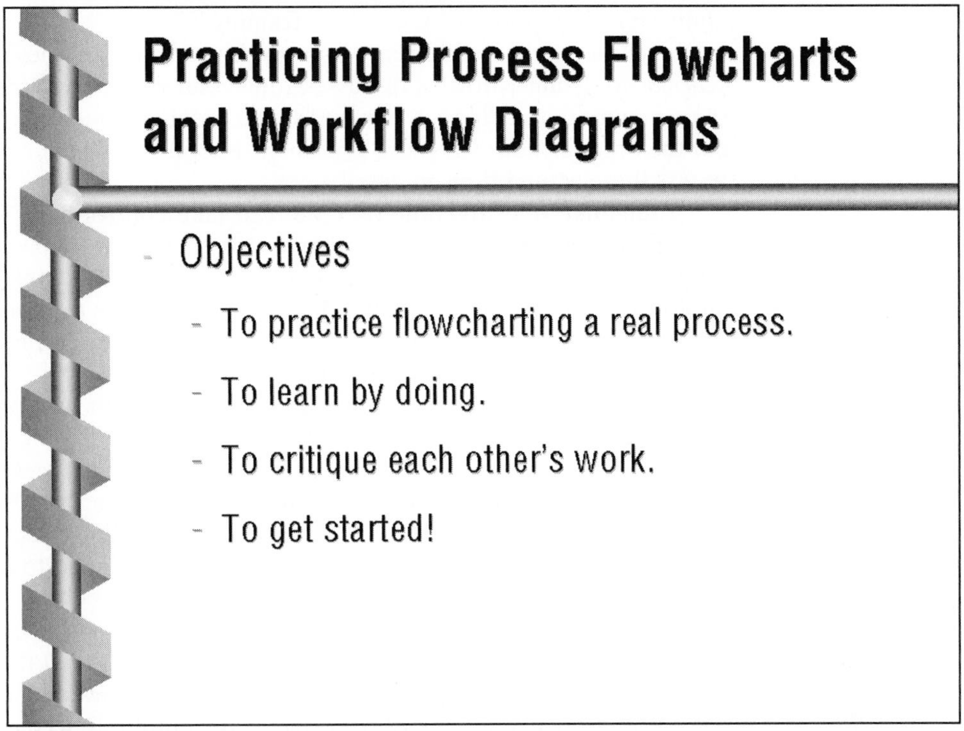

Practicing Process Flowcharts and Workflow Diagrams

- Objectives
  - To practice flowcharting a real process.
  - To learn by doing.
  - To critique each other's work.
  - To get started!

FIGURE 15-27 Developing a Workflow Diagram

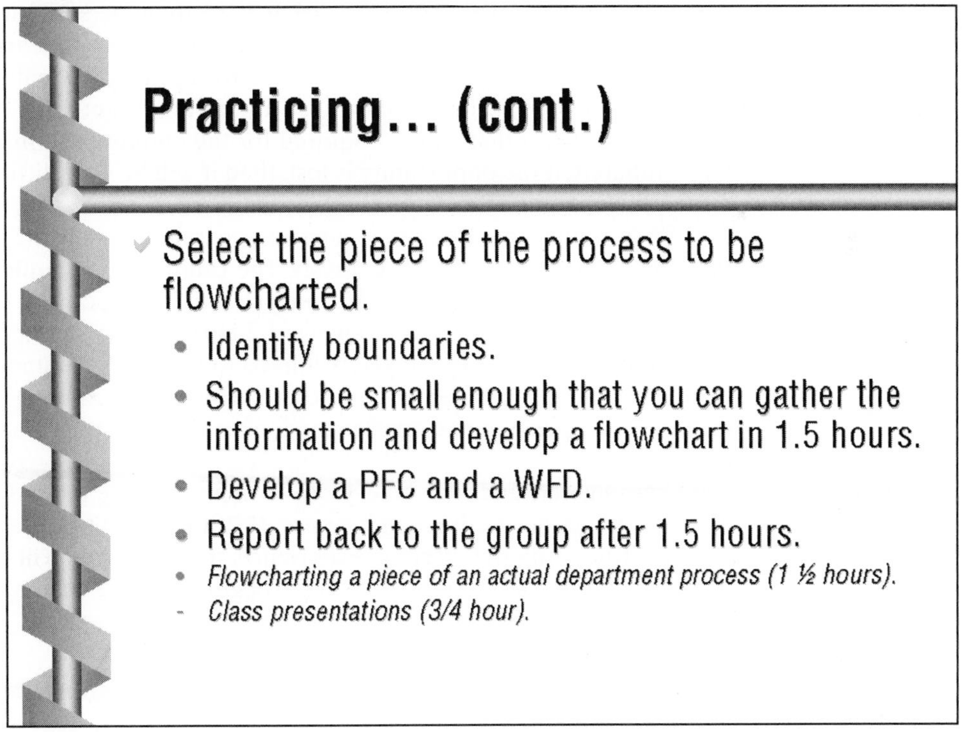

Practicing... (cont.)

- Select the piece of the process to be flowcharted.
  - Identify boundaries.
  - Should be small enough that you can gather the information and develop a flowchart in 1.5 hours.
  - Develop a PFC and a WFD.
  - Report back to the group after 1.5 hours.
  - *Flowcharting a piece of an actual department process (1 ½ hours).*
  - *Class presentations (3/4 hour).*

school course. Because trainers are faced with the challenge of teaching new and existing employees job skills and tasks they need to be effective, an understanding of how adults learn is important for effective training.

Teaching adults is different from teaching young people. In his book, *The Adult Learner*, Malcolm Knowles discusses four assumptions regarding the adult learner.[5] These assumptions help to design and shape training to meet the needs of the employee and the organization. They also give insight about how to motivate adult learners. We discuss these assumptions in detail.

Adults like **self-direction**; they want to take control or at least have some say in their training agenda or plan. In other words, they want to have choices and make decisions about training that affects them. Trainers should use this knowledge and solicit employee involvement during the design phase for any training program. Employees like to be asked about their needs, expectations, and preferences. This is not always possible, but it can lead to effective learning.

Experiential training techniques are useful for adults who prefer to learn by doing. Therefore, training should include plenty of hands-on practice. This follows the age-old, four-step training philosophy of *tell, show, do*, and *feedback*.

Adults will know when they have to learn something and when they will be motivated to learn. Therefore real-life events help in providing effective training. For example, a newly promoted supervisor will understand the need to learn supervisory skills and will be motivated to learn.

The final assumption of application is that people want to learn things they can use. Learning should have some application that is related to what the employee currently is doing or will be expected to do in the near future. If the training is conducted too far before application, it will be hard to convince the employee to participate and the retention of what is taught will be affected. It is also much more difficult when you have to convince employees to learn something simply for the sake of learning.

Adults are motivated to learn what is in their personal interest. As we already noted, a newly promoted supervisor will be very willing to learn and will seek out information. Adults can also be motivated to learn by appealing to *personal growth objectives*. This might be accomplished by putting together a **development plan** that identifies the skills that will be required for the employee to move up in the company. Unfortunately, if the opportunity is lost, then it will be very difficult for the manager or trainer to encourage the employee to learn.

When trainers are designing curricula or training plans they should consider carefully the assumptions noted previously. The training also should take into account the individual's personal needs. The learning should be an active, experiential process that keeps the learner involved and engaged. Learning also should be transferable to the job or the capacity the employee will have to fulfill. Understanding and retention are related to student involvement in the process. Some techniques that increase involvement include question and answer sessions, subgroups, case studies, debating, and role playing.

Learning can and will take place in many different ways or places and occurs whether the trainer is directing it or not. If trainers don't take an active role in designing and structuring the learning experience, the employee will do it informally, on his or her own. Innovative companies have begun to make training more effective by in-

[5]Knowles, M. S., *The Adult Learner: A Neglected Species* (Houston, TX: Gulf Publishing Co., 1984).

volving managers in the process and, more importantly, by holding them accountable for the effectiveness of training. To accomplish this, the role of who is responsible for training has broadened to include everyone in the company.

At the Chiat/Day Company,[6] for example, training is not a discrete event but rather an ongoing process in which the employee learns from an experienced team member. Bell Labs involved its engineers to help set up an expert model for training. This information was used to develop nine prioritized work strategies that are integrated into training sessions that occur during normal work days. The program resulted in an initial 10% increase in training effectiveness, which grew to 25% by the end of the first year. Motorola has taken learning to a higher level, guaranteeing at least 40 hours of training per year per employee. In addition, informal training on the job is considered as important as formal training. Informal training time includes at least four hours for every hour of formal training. The informal training doesn't just occur by chance, though, it is a structured process Motorola calls the "learning environment."[7]

TRAINING TOOLS

There are a variety of formal and informal tools available to the trainer. **On-the-job training** can be used as a part of a structured training program. Even though on-the-job training seems informal, it can be used to teach a single task or skill, multiple programs, or work processes, or it can occur in conjunction with other training.

The most common means of using on the job training is as a single training program that addresses a specific set of skills. Approaches may include the use of a supervisor or an experienced employee to teach a task or skill to a person. For example, a supervisor who is particularly effective at giving feedback to employees regarding performance may be asked to teach the skill to all supervisors.

Multiple programs is a package of specific skills in which an employee chooses to become qualified that can be addressed by developing a curriculum of several topics that need to be taught. The new employees progress through the program as they learn to do the jobs. The jobs may be set up from easiest to most difficult or by linking similar tasks. The training can be done during nonpeak hours so as not to create bottlenecks and to allow focused training.

On-the-job training can be used to help teach tasks within a work process as well. Employees can be taught to operate specific equipment or perform all the tasks involved in the process, effectively **cross-training** the entire group. By learning the entire set of process skills, employees will have a better appreciation for their jobs and the jobs of other employees.

Finally, on-the-job training can be used in combination with other types of training. This process can be used to support or enhance classroom or other types of training. For example, flight attendants might be taught skills such as serving and controlling passengers through role-playing. Then they may be assigned to flights with senior attendants. During these flights they will have an opportunity to observe, practice, and receive immediate feedback regarding their performance.

---

[6]The Price Waterhouse Change Integration Team, *The Paradox Principles* (Chicago: Irwin Professional Publishing, 1996).

[7]Stamps, D., "Learning Ecologies," *Training* (January 1998):32–38.

Five selection criteria can help determine that on-the-job training is appropriate as a training tool in a given situation. These include the nature of the task, available resources, constraints in the workplace, financial considerations, and individual differences. These variables need to be considered in conjunction with the tasks that employees need to learn (see Table 15-4).

Internet/Intranet Training

The Internet provides another approach for meeting the training needs of an organization. The Internet can be an easy, convenient, and affordable way to get information and training materials out to large groups. Once a company builds a Web site, it is relatively easy to keep it up to date and refreshed. Network training can be distributed to an unlimited number of people without the difficulty of having to distribute the information manually. Table 15-5 shows five ways in which the Internet can be used to deliver training information.

TABLE 15-4 Five Factors to Consider in Training Adults

Nature of the Task

Four issues are related to the nature of the task. These include immediacy, frequency, difficulty, and consequences.

1. Immediacy—On-the-job-training (OJT) is best used when the need is not immediate, or the logistics do not allow the people to be brought together away from the workplace.
2. Frequency—OJT is more practical for reoccurring or frequent tasks. If something is done only occasionally, this is not the best approach.
3. Difficulty—OJT is suitable for difficult tasks because it provides concrete information and feedback. It should not be used for high-risk tasks involving high speed or safety.
4. Consequences—Tasks that have a low consequence of error and difficulty are more suited to OJT.

Available Resources

Three types of resources need to be considered when implementing OJT: people, time, and equipment. OJT requires experienced people to provide the role models and actual training to the new people. There must also be time available to do the training. The training cannot be fit into the normal work day without providing for extra help. Finally, the equipment must be available to accomplish the training.

Constraints

Constraints to the training are typically found in the location and in distractions. The location should be suitable for the training without having negative consequences on the business. Likewise distractions should be at a minimum to facilitate effective training.

Financial Considerations

There should be a clear cost-benefit relationship for all training. OJT is most appropriate when the number of people needing training at a given time is low, even though the total number of people to be trained is high.

Individual Differences

Individual differences will affect the overall effectiveness of the training. Managers must consider two individual differences: trainee prerequisites and preferences. Prerequisites are factors in any type of training as are preferences. Before beginning the training it must be determined what skills or experience are necessary. When possible the training should also be set up to meet the preferences of the individual.

SOURCE: The Price Waterhouse Integration Team, *The Paradox Principles* (Chicago: Irwin Professional Publishing, 1996). Reproduced with permission of The McGraw-Hill Companies.

TABLE 15-5 Internet Training

1. General Communication—General communication is the simplest form of Internet-based training. Basically, the trainers and trainees use e-mail to distribute schedules, information, tests, and so on.
2. On-line Reference—The network can be used as an on-line library of company policies, information, and reference materials. Employees can access information on demand at their convenience.
3. Testing and Assessment—The network can be used for distributing, grading, and providing feedback on tests.
4. Distribution—The Internet can be used to distribute computer-based training modules, which can be downloaded as needed. This eliminates the time and the logistical problems of distribution and puts the materials where they are needed at any time.
5. Delivery of Multimedia—This technology can provide a multimedia training program.

SOURCE: Reprinted from *HR Focus*. November 1996, copyright © 1996, American Management Association International. Reprinted by permission of American Management Association International, New York. All rights reserved.

Computer-Based Training and CD-ROM

Computer-based training uses specialized software known as courseware that addresses specific topics. Companies may customize courses to meet specific needs. This training can be delivered quickly and cost effectively, reducing training time. The disadvantage of computer-based training is that there is no peer interaction or personal feedback.

Distance Learning

Distance learning incorporates technologies such as videoconferencing, Internet, and satellite delivery of courses. Although many companies cannot afford satellite systems, many can afford videoconferencing. Both technologies allow interaction and broad delivery to large groups. It is possible to use rental services such as those available in hotels and in service firms such as Kinko's. This kind of training still has the drawback of sometimes being a passive learning experience. With the use of interactive voice and video technologies, however, the learner's experience has improved.

Electronic Performance Support Systems (EPSS)

Electronic performance support systems (EPSS) are a type of just-in-time training. Many times these take the form of software "helps." Examples include "wizards" and technical queue cards. Basically, the training is made available when it is needed, as opposed to during a training session. This approach reduces or eliminates some training time, although many people find it difficult to understand the prompts provided in programs.

Multimedia

A multimedia approach can boost retention by 50% to 70% over traditional methods. Multimedia training is accomplished by incorporating several of the tools already mentioned into an interactive process focusing on several senses. Multimedia productions also can reduce training time by 40% to 60%.

Other Training Tools

Of course, training relies on many less formal tools. Competent instructors are required for training to be effective. Humor is a training tool that some trainers have learned to use very well. Incorporating humor into training allows the audience to relax and enjoy the material being presented. Many times, if material is funny, it is likely to be remembered. Finally, games are often used in training to reinforce concepts.

Whether material is presented electronically or by classroom means, great attention should be paid to workbooks, handouts, and other information that is provided to the participants. Like any other service provider, you will be judged by the tangibles that you bundle with the services you provide.

EVALUATING TRAINING

An important component of a training program is evaluation. Figure 15-28 shows a sample training evaluation form. These forms should measure the applicability of the training, the relevance to the job, the materials, and the performance of the instructor in presenting the materials.

You should realize that if you receive a poor evaluation, it might be because of random variation, or because of other factors that you cannot control. Consistently poor evaluations indicate that you should pay attention to how you teach, and you should receive training yourself to develop into a good instructor.

The Learning Effect

So far in this chapter we have discussed specific tools relating to training and learning. To gain perspective, we would like to step back and discuss the theory of learning and the effects that learning can have on the productivity and competitiveness of an organization. To do this, we discuss the **learning curve** and its relation to quality. In a nutshell, the learning curve can be interpreted as, "the more you do something, the better you become at doing it." The quote of Ralph Waldo Emerson comes to mind, "That which we persist in doing becomes easier, not that the thing has changed, but that our *power to do* has increased." An objective of training must be to increase employee's "power to do."

Improvement occurs as a job is repeated. If the improvement is repeatable and predictable after time, it is likely to be the result of learning. Progress depends on the continuation of learning in an organization. The learning effect can be described as an empirical relationship between output quantities and quantities of certain inputs where learning inducement improvement is present.[8] The philosophy of the learning curve is as follows:

- Where there is life, there can be learning.
- The more complex the life, the greater the rate of learning. Human-paced operations are more susceptible to learning or can give greater rates of progress than machine-paced operations.
- The rate of learning can be sufficiently regular to be predictive. Operations can develop trends that are characteristic of themselves. Projecting such established trends is more valid than assuming a level performance or no learning.[9]

The learning effect applies at the individual and organizational levels in an organization. At the individual level, the two elements to consider are the initial starting level of improvement and the rate of learning. At the organizational level, learning is a function of the interactions between the individual and the supply chain.

[8]Belkaoui, A., *The Learning Curve* (Westport, CT: Quorum Books, 1986).
[9]Ibid.

FIGURE 15-28 Sample Course Evaluation

Program Evaluation

Instructor Name________________________________ Date __________________________

Session____

Was the instructor effective in these areas?

	Effective				*Ineffective*
A. Knowledge and understanding of the subject matter	5	4	3	2	1
B. Ability to present ideas clearly	5	4	3	2	1
C. Ability to actively involve me	5	4	3	2	1
D. Use of relevant and practical ideas	5	4	3	2	1
E. Ability to respond to my questions and concerns	5	4	3	2	1
F. Openness to others' ideas & opinions	5	4	3	2	1

Session Content:

A. What did you learn from this session that was the most valuable to you?____________

____

B. What was covered from this session that was of least value to you?________________

____

C. The session materials were: ☐ Very appropriate ☐ Adequate ☐ Inadequate

If not adequate, comments?____

D. My expectations were: ☐ Exceeded ☐ Met ☐ Not met

If not met, comments?____

E. Were there ideas presented in this session that you will use on the job?

☐ Yes ☐ No ☐ Maybe

Comments:____

____

F. What one thing could we have done differently to improve this session?____________

____

____

G. What other session topics are you interested in?______________________________

____

Overall Rating:

Please check the statement that best describes your feelings about this session.

☐ Excellent ☐ Good ☐ Satisfactory ☐ Fair ☐ Poor

Please sign below if we may use your comments for promotional purposes.

Name: ________________________________ Company: ____________________

Signature: ______________________________ Date: _______________________

Please use space on back for additional comments

Quality-based learning occurs as people discover the causes of errors, defects, and poor customer service in a firm. Once these causes of errors are discovered, systems are put in place to ensure that the cause of the error never reoccurs. Such learning is termed **profound organizational learning** and is permanent. The changes that are made to address the cause should be documented and communicated so that the learning can be frozen.

The Relationships Between Organizational Learning and Quality Costs

As learning takes place, the various costs of quality are reduced. *Prevention costs* are reduced as use of preventive action becomes more efficient and focused. *Appraisal costs* decrease as the need for inspection decreases and firms apply resources to inspection in a more skilled manner. *External and internal failure costs* are decreased by the fewer number of errors and defects. As learning takes place, capacity increases because fewer resources are applied to rework and down time.

Pay-for-Learning Programs

Pay-for-learning programs compensate employees for knowledge and skills rather than for the job they actually perform. Other names for pay for learning include *pay for knowledge, skill-based compensation, knowledge-based pay*, and *pay for skill*. The basic premise of pay for knowledge is that employees with a wide inventory of skills will receive higher rates of compensation than others who do not possess the same levels of knowledge.

There are two basic forms of pay-for-learning schema, **knowledge-growth systems** and **multiple-skills systems**. Knowledge-growth systems increase employees' pay as they establish competence at different levels relating to job knowledge in a single job classification. These are sometimes called *technical skills ladders*. Multiple-skills systems are much more experimental and use training for job skills in a variety of job classifications. This promises the advantages of greater labor flexibility and job mobility for employees.

The benefits of skill-based pay include better employee development, increased cross-training, increased labor flexibility, lower levels of staff, improved problem solving, and increased job satisfaction. The disadvantages of pay-for-learning systems include increased costs in training, salaries, and administration of the program. As with any quality management approach, if there is a lack of focus in implementation, the program can result in increased salaries with negligible benefits.

SUMMARY

As more jobs are created in the information and services sectors of the economy, learning is becoming more critical to the competitiveness of firms. We have discussed and presented various approaches to training. Not all firms can start a Motorola University. However, all firms can train their employees effectively by following the training processes outlined in the chapter.

Reflecting on the contingency approach outlined in this book, it is likely that effective training programs can be provided through the different approaches. For example, an insurance company can use training booklets to prepare employees to understand insurance laws. It may send executives to training sessions to improve their knowledge and provide classes and electronic forms of training for employees within the company to satisfy a variety of needs.

KEY TERMS

- Computer-based training
- Cross-training
- Development plan
- Distance learning
- Experiential training techniques
- Individual needs analysis
- Knowledge-growth systems
- Learning curve
- Multiple-skills systems
- On-the-job training
- Organizational learning
- Pay-for-learning programs
- Profound organizational learning
- Self-direction
- Task needs assessment
- Training needs analysis
- Training needs assessment
- Training program design
- Whack-a-mole

DISCUSSION QUESTIONS

1. Describe the concept of organizational learning. Is organizational learning a useful concept for quality conscious firms?
2. How does the whack-a-mole analogy help us better understand the importance of organizational learning in a quality-focused organization?
3. What are the four stages of John Dewey's concept of learning? Why is it necessary to move through all four steps for meaningful learning to take place?
4. Why do you think that companies in the United States spend so little on training? Please make your answer as thoughtful as possible.
5. How has Motorola's system for training and educating employees helped the company focus on quality-related issues? What elements of the system could other firms successfully adopt for their own use?
6. Describe the purpose of a training needs analysis. How can a training needs analysis help a business become more of a "learning" organization?
7. What is a learning needs assessment? Should an organization perform a learning needs assessment on a periodic basis? Why?
8. "Training plays a key role in the development of a quality culture in any organization." Why do you think this is so? Explain your answer.
9. Describe the purpose of a strategic training plan. Should a strategic training plan be a part of a firm's overall strategic plan, or should it be conducted separately? Explain your answer.
10. What specific techniques can managers use to determine an organization's training needs?
11. Many successful firms enter into strategic alliances with local universities to provide needs assessment and training. Why would a successful firm need the help of a local university? If you were the manager of a business organization, would you consider developing a strategic alliance with a local university to help you with your training needs assessment?
12. Do employees at different hierarchical levels in an organization need the same type of training or different types of training? Explain your answer.
13. In what ways are adult training needs different from youth educational needs? How do these differences affect the design of a training program?
14. What does self-direction mean in a training context?
15. What are some of the important heuristics or rules of thumb to keep in mind when training adults?
16. Is on-the-job training a common tool for training employees or an unusual type of training? What is the purpose of on-the-job training?
17. What is meant by cross-training? Under what circumstances does cross-training make sense?
18. Describe the idea of computer-based training. Do you think that computer-based training will become more common or less common in the future? Explain your answer.
19. Describe the concept of distance learning. How can distance learning help facilitate a company's training needs?
20. What is an electronics performance support system (EPSS)? How does an EPSS help facilitate a company's training needs?

Cases

Case 15-1 British Petroleum: Achieving Organizational Learning Through the Creative Use of Video Technology

British Petroleum: *www.bp.com*

Imagine yourself on an offshore oil platform in the North Sea off the coast of Scotland. The wind is blowing, it is cold, and the sea is dark and rough. You have just experienced an equipment failure that a year ago would have forced you to shut down drilling operations and wait for a team of engineers to be flown in from shore. You always dreaded doing that, because it put you behind schedule and pulled a team of engineers away from their other responsibilities. But now you have an option short of shutting down and calling in the helicopters from shore. Your company, British Petroleum (BP), has equipped you with a communication tool called a Virtual Teamworking Kit. The kit allows you to call ashore, set up a videoconference with a team of BP engineers, show the engineers real-time video of your problem, and discuss with the engineers solutions to the problem without ever asking anyone to leave their original locations. With a little luck, the problem will be solved in an hour or two rather than two or three days.

This approach to using video technology to troubleshoot problems and share information among employees is a very serious initiative at BP for several reasons. Solving a problem on an oil rig through a Web-enabled videoconference rather than flying a group of engineers from Scotland to a location in the North Sea has obvious cost advantages. But beyond the cost savings, the use of the Internet serves a much larger potential purpose. BP has approximately 100,000 employees who work in 26 different countries. Many of these employees have similar jobs and are constantly challenged to troubleshoot similar problems. There is tremendous potential to save money, improve quality, and improve safety if employees can learn from each other, rather than repeating each other's mistakes. The trick is to get employees to talk to each other and exchange information in a meaningful way. Obviously, employees could talk to each other on the phone. But it is a well-documented fact that the majority of communication that takes place is nonverbal, even when people are face to face. This makes the telephone, in many respects, a poor communication medium. As a result, at BP and other companies, there is a growing awareness that employees share information in a more meaningful manner when they can "see" each other and see each other's work environment.

The Virtual Teamworking Kit is one of several tools that BP is using to make this happen. The hardware for the kit includes a personal computer with videoconferencing equipment, multimedia e-mail, shared applications, a scanner, and an electronic whiteboard. The kits were custom built for BP and are available to BP employees everywhere the company operates, no matter how remote the location. The idea behind the Virtual Teamworking Kit originated from BP's knowledge management team, which is a group of BP personnel charged with finding ways for employees to easily communicate and share knowledge with each other. BP estimates that knowledge sharing through the virtual teamwork technology saved the company at least $30 million in its first full year.

As a result of its positive experience with the expanded use of video technology, BP is working hard to help its employees share knowledge in other areas. For example, every employee at BP now has the authority and the capability to create a Web page for the company's internal intranet. As a result, many employees have created Web pages that provide an overview of their personal areas of expertise and examples of how they have solved problems. Other sites are dedicated to sharing technical information on specific aspects of BP operations. Employees can access this information and find specific examples of how to tackle tough problems through powerful search engines. Again, the purpose of this initiative is to improve communication among BP employees and to make it easier for BP employees to share information with each other.

The search for improved ways for employees to communicate with each other and share knowledge will continue at British Petroleum. Perhaps these efforts help explain why British Petroleum is one of the most profitable companies in the oil-producing industry. ■

DISCUSSION QUESTIONS

1. Do you believe that British Petroleum's investment in the Virtual Teamworking Kit has been money well spent? Explain your answer.
2. To what extent do you believe that it is important for the employees of a multinational firm to be able to communicate easily with one another? How does ease of communication influence organizational learning?
3. Suggest one additional way that a company could use either videoconferencing or the Internet to enhance organizational learning.

Case 15-2 At Globe Metallurgical, Training Is a Way of Life

Globe Metallurgical: ***www.globemetallurgical.com***

Globe Metallurgical, an Ohio-based manufacturer of metal products, is somewhat of a Cinderella story. In the early 1980s, the company was in a tailspin and was losing business at an alarming rate to higher-quality foreign competition. It was clear to the managers of the company that something had to be done and that business as usual with only minor improvements would not save the firm. Rather than give up, the company decided to take a stand. Globe Metallurgical set out to become the lowest-cost, highest-quality producer of ferroalloys and silicon metal in the United States.

Remarkably, the company was true to its word. Globe was the first small business awarded the Malcolm Baldrige National Quality Award. Since that time, Globe's share of the U.S. market for high-quality metal products has risen significantly, sales in Canada and Europe have increased, and the company has experienced consistent profitability. Several of the company's accomplishments have been truly remarkable. For example, Globe's products are so highly regarded in Europe that when European traders place an order for magnesium ferrosilicon alloy they often specify that the material must be "Globe quality," an informal standard that other companies must now meet.

In light of this remarkable turnaround, the obvious questions to ask are, "What did Globe do to reverse course so quickly?" and "What can other companies learn from Globe's experiences?" The driving force behind Globe's turnaround was the implementation of a company-wide, quality improvement program, nicknamed QEC (standing for quality, efficiency, and cost). A major component of the QEC program was an emphasis on employee training. In developing its approach to employee training, however, Globe did not copy the approach of another firm. Instead, the company developed an approach to training that matched its culture, and it was able to identify quickly the information and skills that its employees needed to be successful.

The first thing that Globe did was to expose all of its new employees to quality system training. In addition, employees were provided with a broad overview of the industries in which the company competes. Although this involved some classroom work, the majority of the training was done on the job. After the initial orientation period, each new employee worked with a production team for three months before going on his or her own.

After a new employee settles into a permanent job, training by no means stops. The company has learned, however, that there is a fine line between being proactive and reactive in terms of determining training needs. Because Globe is a small firm, managers often assess employee training needs through casual conversations with employees and informal feedback from production teams. The danger in this approach is that training can end up being reactive, and management could settle into a pattern of responding to problems rather than anticipating and preventing them. To guard against this, Globe systematically probes many sources of information in its firms to assess training needs and responds quickly when training needs are determined. By adopting this approach, training has become an everyday event at Globe rather than a semiannual or yearly activity.

One thing Globe does not do is hire outside consultants to conduct training programs. The company prefers to handle the majority of its own training needs. Globe does take advantage of training offered by customers. Globe's employees have received training from General Motors, Ford, Dow-Corning, General Electric, and other customers. Still, Globe maintains close control over the content and the process of this training.

Globe's overall QEC program, with its special emphasis on employee training, has reaped rich rewards. Since the QEC program was launched, the company has realized efficiency improvements of more than 50%. Customer complaints have decreased by 91%. Life has also improved for Globe's employees. The company's employees now have flexible job assignments, and their quality improvement efforts are recognized with personal letters from management and small gifts. More importantly, the accident rate in the plant has dropped, and the rate of employee absenteeism has steadily declined. ■

DISCUSSION QUESTIONS

1. In your opinion, did employee training play a key role in Globe's QEC program? Explain your answer.
2. Why do you believe that Globe is reluctant to hire outside consultants to conduct training sessions? Do you believe that the company's policy to provide training in-house is motivated by a desire to save money, or by some other reason?
3. What type of company would benefit by benchmarking against Globe's approach to employee training?

CHAPTER 16

Implementing and Validating the Quality System

The systems approach to quality begins with the basic principle of total quality control that customer satisfaction cannot be achieved by concentrating upon any one area of the plant or company alone—design engineering, reliability analysis, inspection, quality equipment, reject troubleshooting, operator education, or maintainability studies—important as each phase is in its own right. Its achievement depends, instead, both upon how well and how thoroughly these quality actions in the several areas of the business work individually and upon how well and how thoroughly they work together.
—ARMAND FEIGENBAUM[1]

A firm that skillfully implements the tools, philosophies, and techniques in this book will improve its quality management system. Such improvement should result in better performance. When you reach best-in-class or best-in-the-world status, how do you get beyond that point? If you have reached the limit of your knowledge and yet feel that more improvement can be made, how do you determine the next step? Answering these questions is the goal of this chapter—that is, to discuss how to discover the next step that will form the foundation for continually improving once you have reached quality maturity.

Remember that world-class competitors did not achieve that status by merely adopting tools or techniques. They got there because they were very good at performing the fundamentals. They understand their customers, products, employees, competitors, markets, and technologies. World-class performers have a good understanding of where they are and where they want to be. They are excellent at being creative and devising processes for getting to the future.

This type of progression is like life. Successful people are good at progressing from one plateau to another in a relatively short period of time. Rarely do people become successful overnight; it is more a process of continual growth and improvement. The same holds true for firms. They have to grope, feel, make mistakes, and stumble on the way to the next level.

In this chapter, we present processes for assessing where you are and identifying how you can move to the next level. These include defining, understanding, and assessing the quality system in your organization.

[1]Feigenbaum, A., *Total Quality Control* (New York: McGraw-Hill, 1983).

BUILDING BLOCKS FOR THE SYSTEM OF QUALITY IMPROVEMENT

A quality system depends on the interactions of many different variables, as evidenced in the quality system model shown in Figure 16-1. As we have demonstrated throughout this book, quality improvement is not a stand-alone discipline. Quality improvement requires the interactions on a contingency basis of many different disciplines to create products, services, processes, and systems that effectively serve customers. Another factor driving the integrative approach is the increasing complexity of work. Complexity requires people from different disciplines to settle on standardized methods for serving customers that result in work simplification. In the following section we discuss the parts that go together to create a quality system.

People

The model in Figure in 16-1 is built on a base of people. People represent the core of a firm's capabilities because they provide the intellect, empathy, and ability that is required to provide outstanding customer service. Therefore, systems must be in place to develop, train, care for, and motivate people to serve the customer properly. The management of a small metal fabrication plant was concerned that a particular production line was exhibiting particularly bad quality. The managers discussed all they had tried to do to improve quality on this line. They stated that they had even fired all

FIGURE 16-1 Quality System Model

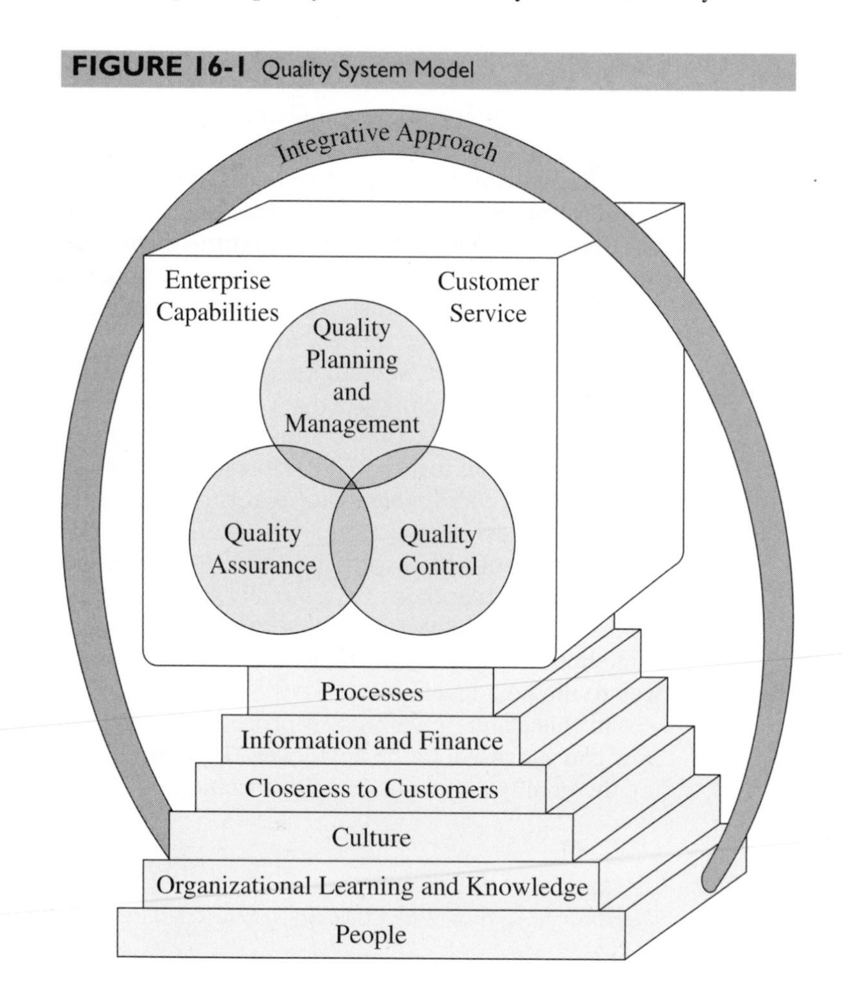

the employees who previously worked on the line and replaced them with new employees. However, they were disconcerted that the quality problems persisted. The managers were then told that they were primarily responsible for the poor quality of the production. The employees had changed; however, the management had remained constant. To their credit, they peacefully responded, "we suspected we were the problem." At this point, they began to change the culture they had created to include improved input from employees.

For a quality system to function effectively, employees must understand that they are integral to that system. They must be made to feel important and necessary for the continued survival and growth of the company. This does not mean that management is unnecessary. Remember that employee empowerment does not mean abdicating authority. Rather, it means to share in decision making in areas that affect employees.

Another people-related issue involves corporate restructuring and reengineering. Often, these efforts result in mass layoffs and much turmoil among employees. Although companies must sometimes necessarily take these measures to respond to economic realities such as downturns in sales or the national economy, layoffs should never be associated with quality improvement efforts.

Also, layoffs should occur only as a "last resort." Quality improvement efforts, when performed properly, are associated with improved morale and improved confidence among employees that the company is becoming more competitive and the employees' personal security is improving. Many managers forget that human beings desire security. At the same time, some managers believe that insecurity results in better job performance. However, the opposite is most likely true. At times of high insecurity, the best employees leave and the unmarketable employees remain. As a result, the ability for organizational learning to take place decreases, and even the ability of the firm to attract top talent in the future is impaired.

Recently, a major military contractor announced the layoffs of several hundred engineers in a large metropolitan city. At the same time, the company was recruiting engineers from a nearby engineering school at a major university. Needless to say, the best engineering graduates were disinterested in interviewing with this firm. They knew the obvious. This firm did not care about employees' personal needs over the long term. Therefore, they would not interview nor accept employment with such a firm.

There is evidence that some employees will forego higher wages in order to achieve higher job security. A good example is the tenured professor who turns down employment that would pay much better in the private sector in order to enjoy the benefits of tenure. Without tenure, there would be upward pressure on faculty salaries.

Organizational Learning and Knowledge

The second building block of a quality system is organizational learning and knowledge. In Chapter 15 we discussed various approaches to improving learning and knowledge in organizations. Knowledge is the capital that fuels outstanding quality results. Business leaders report that life-long learning is a key attribute for employees. These same CEOs and presidents of companies should also carefully consider the organizational and cultural aspects of their companies that inhibit and fail to reward life-long learning.

A visitor to a manufacturer of components for automobiles heard from the employees that they felt as if they were hired from the "neck down" to do a job and did not feel that management valued their knowledge or opinions. One of the by-products of a quality improvement effort is the realization by managers that rank-and-file employees are primary sources for in-depth knowledge relating to processes and that organizational learning is the sum of the learning of individual employees.

Learning needs assessment, training design, and delivery of training are important for competitiveness. Outstanding customer service results from providing employees with outstanding knowledge and training. Organizational learning is required for consistency in operations, approaches, and customer contact. After developing a new database system to tie together customer information with production planning, financial planning, and other functions, a service firm recently found the database system was not yielding the expected integration and the desired results. On studying the problem, management found that employees were inputting customer-related information on the system using inconsistent methods. Therefore, customer information was found in inconsistent places in the system. A cause-and-effect study revealed that employees had received minimal and ineffective training on the new system during implementation. Such costly lack of attention to employee needs resulted in greater cost and poor use of information resources.

Culture

Culture refers to the norms and beliefs that lead to decision-making patterns and actions in an organization. Experienced quality practitioners and consultants claim that some organizations have cultures conducive to quality improvement, and others make quality improvement very difficult. Some of the key aspects of culture include attitudes toward change; presence or absence of fear; degree of openness, fairness, and trust; and employee behavior at all levels. Companies that play the so-called blame game end up with cultures where trust is absent and employees act in ways that appear self-defensive. Companies that work in an environment of fear find that distrust takes root between labor and management, between midmanagers and vice presidents, and between departments. Such distrust is the opposite of the open and trusting culture that is required for a company to be able to respond rapidly to changing customer needs. When distrust is present, control systems are put into place that result in lost time, capacity, and flexibility. (Traditional cost accounting systems have been criticized for these reasons.)

Companies that respond quickly to customers' needs have cultures where decision making is open, information is available to everyone, and risks are rewarded. Some companies have cultures that are very control oriented, and a large number of younger companies foster quite different environments, including "granola cultures," where managers and employees view the company as a means of achieving happiness for all employees.

Closeness to Customers

Closeness to customers describes the firm's understanding of the customers, their needs, and their wants. Notice that the building block of "closeness to customers" is built on people, organizational learning and knowledge, and culture. All these things result in a supportive environment in which employees can be close to customers. Companies that value knowledge will gather data about customers and will study and understand changing customer needs. Companies that are close to their customers retain those customers. And, as we have seen, customer retention is closely related to profitability.

As we discussed in Chapter 6, for customer closeness to be achieved, systems must be put in place for gathering data about customers, analyzing the data, and implementing change systems based on the analysis. Not only is an understanding of customers required, but also an understanding of the competitor's customers. When you understand the customers of your competitors, you'll be able to put into place marketing methods that will attract those customers away from your competitors.

Customer closeness engenders loyalty. People who drive Harley-Davidsons and Mercedes identify with those products. As a result, these customers return to buy new products, replacement components, and accessories and to get service. Brand loyalty may become less important as time passes and consumers move to an ethos where product choices are valued over brands. This is particularly true with e-commerce. In this environment, personal relationships between vendors, suppliers, and customers become important differentiators. Consider the example of a doctor. Once patients find a doctor they believe is competent and trustworthy, they will remain with that doctor for many years. How does one achieve such loyalty? The answers are different for different markets. However, close, empathic service appears to be one way to achieve closeness to customers.

Customer closeness is especially key in services. The level of customer contact achieved in services requires an understanding not only of customers' current wants but also of their emerging wants. Sensitivity to changing needs, moods, stages in life, and environmental conditions are possible only when supported by knowledgeable employees and a culture that values people. This means that a company should be focused on whom it chooses to serve. An understanding of the types of customers that a firm will serve and a focus on key capabilities are important in determining to which customers a firm should become close.

Information and Finance

Information systems provide the core of the support system for satisfying the customer. Well-designed information systems, such as those we've described at Ritz-Carlton hotels (see Quality Highlight 8-1), become the institutional memory for customer needs. If information systems are not well designed and information is difficult or slow to obtain, customers will go elsewhere. Banks that do not provide ready account information over the Internet or by phone will find that customers go elsewhere for their banking.

Electronic data interchange (EDI) is of increasing importance for satisfying customer needs. These systems allow customers and suppliers to tie their systems together to enhance planning, purchasing, and coordination. The objective of a quality information system is to gather information relating to the key variables that affect customer services and product quality. The scope of the information system includes the entire organization, linkages to customers, and linkages to suppliers. Better information leads to better customer service. The actions required to make the information system effective are problem identification, analysis, and corrective action.

Finances are listed in the building block in Figure 16-1 with information because financial resources are needed to provide the infrastructure and services that customers want. It takes money for doctors to decorate their offices in a way that is pleasing to patients, to have state-of-the-art technology that will result in state-of-the-art care, and to provide the equipment needed to achieve customer satisfaction. It is interesting to note that most firms that are world-class examples of quality are also financially successful. Research has shown that high quality can improve bottom-line results and enhance financial stability. Therefore, finance is a two-way street. Companies that are financially successful are able to invest in systems that will satisfy customers; and high-quality processes, products, and services lead to financial success.

The Three Spheres of Quality

The building blocks that we have discussed provide the foundation to the quality system that supports the three key spheres of quality—quality planning and management, quality assurance, and quality control. These spheres were discussed in Chapter 1. Notice that they are closely related to enterprise capabilities and customer service. **Enterprise capabilities** are those capabilities firms have that make them unique and attractive to customers. Steven Appleton, the CEO of Micron Corporation, the worldwide cost leader in D-RAM computer chip production, was asked what his firm's core competency was. Appleton answered, "Micron's ability to improve productivity at a rate of 14% to 20% per year, year after year." Such capability is only possible if quality assurance, quality control, and quality management systems are effective. All Micron's systems and culture are focused on achieving this goal. *Customer service* is both a goal and an outcome of systems. Processes, procedures, training, and enterprise capabilities must be focused on providing good customer service.

The Integrative Approach

The integrative approach outlined in this book provides the glue binding together the systems that result in high-quality products and services. Because quality is not under the purview of any specific functional group and is the responsibility of everyone, cross-functional approaches are required to achieve the desired results. The integrative systems view recognizes that all the building blocks must be in place in all the functional areas and throughout all levels of the organization for quality improvement to be both horizontally and vertically deployed in world-class companies.

Successful companies that provide high-quality products and services are increasingly able to take the focus away from narrow functional orientations and turn it to broad, customer-centered systems. In such organizations, all the foundation blocks we have identified in Figure 16-1 are in place to support the customer-focused environment.

Alignment Between the Quality System and Strategy

The design of the quality system must have focus. A strategic framework is necessary to achieve this focus, which means that the design of the quality system must be in alignment with the strategic objectives and plans of the organization.

Hoshin planning or policy deployment, discussed in Chapter 4, provides a framework for achieving alignment through the catchball process. Through this process, strategic objectives are translated into specific projects and plans. Project teams are empowered to carry out the strategic objectives. Strategic objectives could be to develop and design a quality management system, a quality assurance system, or a quality control system. See Quality Highlight 16-1 to see how Ford is linking quality and strategy.

QUALITY HIGHLIGHT 16-1

BACK TO BASICS AT FORD[a]

It was an interesting day when Nick Scheele, COO of Ford Motor Company announced that the company would move forward by "going back to the basics." What does this mean? Table 16-1 shows the different phases Ford has gone through—from mass production to the new era. By making the decision to focus more on the basics of running the company, Ford saw a 27% decrease in warranty spending and saved more than $2 billion with Six Sigma.

According to Scheele, "Going back to basics means building quality products on time and at the

[a]Smith, L., "Back To the Future at Ford," *Quality Progress*, 38, 3 (March 2005): 50–56.

Continued

TABLE 16-1 Ford's Timeline

	1977 to 1980: Mass production	*1981 to 1993: Competitive quality*	*1994 to 1998: Global economy*	*1999 to 2001: Niche markets/ acquisitions*	*Back to basics vision*
Management	• Mass production in a captured market, with Big Three competition. • Reduce short-term cost.	Emphasis on: • People (employee involvement). • Teamwork. • Processes. • Systems thinking. • Cost.	• Cycle plan. • Worldwide centers of excellence. • Common world vehicles and processes. • Customer satisfaction emphasis via added vehicle features.	• Emphasis on developing markets and acquisitions. • Push for youth and diversity in management (outside hires). • A, B, C ranking of people. • Bias for actions that provide rapid returns.	Emphasis on: • People and niche knowledge. • Teamwork. • Processes. • Systems thinking • Cost/profit. • Eco-effective design.
Manufacturing	• Reliance on manufacturing experience. • Inspection department to find and contain defective product.	• Quality is job number one. • Q1: basic quality system. • Variability reduction using statistical process control (SPC) and design of experiments (DoE). • Process improvement. • Regular senior management quality meetings.	• Plant vehicle teams established for find and fix problem solving. • Focus on top 25 issues. • Advanced product quality planning. ISO 9000 and lean emphasis.	• Strategy of "seek contain repair" for product issues and improvement. • Implementation of lean and flexible manufacturing systems. • Short-term cost (not profit) focus. • Six Sigma used for find and fix.	• Q1: basic quality system. • Variability reduction SPC and DOE. • Process improvement. • Lean/flexible manufacturing. • Six Sigma problem solving • Regular senior management quality meetings.
Engineering	• Reliance on engineers with great experience. • Find and fix waranty. • Push on failure mode and effects analysis (FMEA) and basic reliability tools. • Design standards and verification manuals.	• Increased emphasis on "prevent." • Training and books available on statistics, learning organization, SPC, quality function deployment (QFD), DoE and Taguchi from world experts. • Increased interaction with customers, manufacturing and suppliers.	• Reorganized into platform teams; engineers rapidly rotate jobs. • Quality training centralized, but no longer taught by subject matter experts. • QFD replaced with marketing reports activity.	• Program content costly and complex. • Emphasis on analytical models. • Heavy cost reduction focus and a cutting of programs. • Quality training available on the Web. • Push on FMEA and basic reliability disciplines.	• Q1 program for engineering. • A disciplined system to implement "prevent" quality methods tied to reward and recognition (design for Six Sigma). • Training and software on powerful, cutting edge methods of TRIZ and axiomatic design.
Suppliers	• Multiple suppliers for each part. • Most business sourced to the lowest bidder.	• Strive to reduce supply base and establish col-laborative partnerships. • Increased interaction with Ford engineers in design, quality and cost. • Transfer of engineering competence to full-service suppliers (FSSs) begins.	• Supplier technical assistance reorganized under purchasing. • FSSs operate independently. • Supplier technical assistance (STA) staff greatly reduced; suppliers self-certify. • Cost reduction emphasis.	• Visteon becomes a supplier. • Further reductions in supply base emphasize lowest bidder. • Increased STA staff to deal with program issues.	• Longer term collaborative partnerships operating in a lean value stream.

Adapted from: Smith, L., "Back to the Future of Ford," *Quality Progress* 38, 3 (March 2005).

right price—with a value proposition that is absolutely compelling." It blends a driving vision with a mission and guiding principles that lead to quality by valuing people, teams and processes.

According to Ford's CEO, Jim Padilla, the following are vital priorities:

- Improve quality
- Improve quality
- Deliver exciting products
- Achieve competitive cost and revenue
- Build relationships.

Only time will tell if Ford can execute this plan. This focus does show that it is not all about new efforts and programs. It is about understanding your business and establishing a strategy that works for competitiveness.

INTERNAL VALIDATION: DOCUMENTING AND ASSESSING THE QUALITY SYSTEM

Once a quality system has been established, the system must be allowed to operate effectively over time. Of course, the system is in constant change, as continuous improvement efforts occur throughout the organization. At times, managers in firms have difficulty finding new ideas for improvement. The firm is operating well, is competitive, and is profitable. Benchmarking can be used to observe the practices of others and achieve even higher levels of performance.

Once this higher performance has been achieved, however, what happens? What is next when your firm is the benchmark or at the top of its industry? Where do you go from there? This is a nice problem to have.

When you have achieved benchmark or role model status, the ideas and initiatives for improvement must come from within. In this case, self-assessment is a good tool to spur improvement. The goal of self-assessment is to observe current practices, to assess those practices, and to identify gaps in deployment. Once gaps have been identified, new levels of performance can be achieved by filling gaps. It is interesting that Solectron Corp., a California company, is the only two-time Baldrige award winner as of this writing. Solectron management has stated that the Baldrige self-assessment process has resulted in continual improvement over several years.

Figure 16-2 shows a generic process for performing self-assessment in a four-stage process of surveying, categorizing, investigating, and evaluating. In the following pages we discuss various types of audits. The generic process is reflected in each of the audit types. **Surveying** is the means of generating lists of strengths and weakness for the organization. Some attempt should be made to prioritize these strengths and weaknesses, as many bear no relationship to company competitiveness.

Stage 2 is **categorizing** the strengths and weaknesses. Strengths can be categorized as strategic resources or strategic capabilities that help define the path for improvement the firm should undertake. The next stage (3) is **investigating** the sources of competitive advantage. This investigation should answer questions concerning how the firm markets itself, differentiates itself from competitors, wins orders in the marketplace, and achieves competitive advantage over competitors. Companies who are at the later stages of competitive advantage in manufacturing or services should identify internal operational efficiencies that result in competitive advantages. These sources of competitive advantage then are related to the resources and capabilities defined in stage 2. This is done to determine how the competitive factors actually add value for the customers. If they are not found to add value, adjustments should be made to improve operational focus or become more customer oriented.

FIGURE 16-2 Internal Environmental Analysis Process

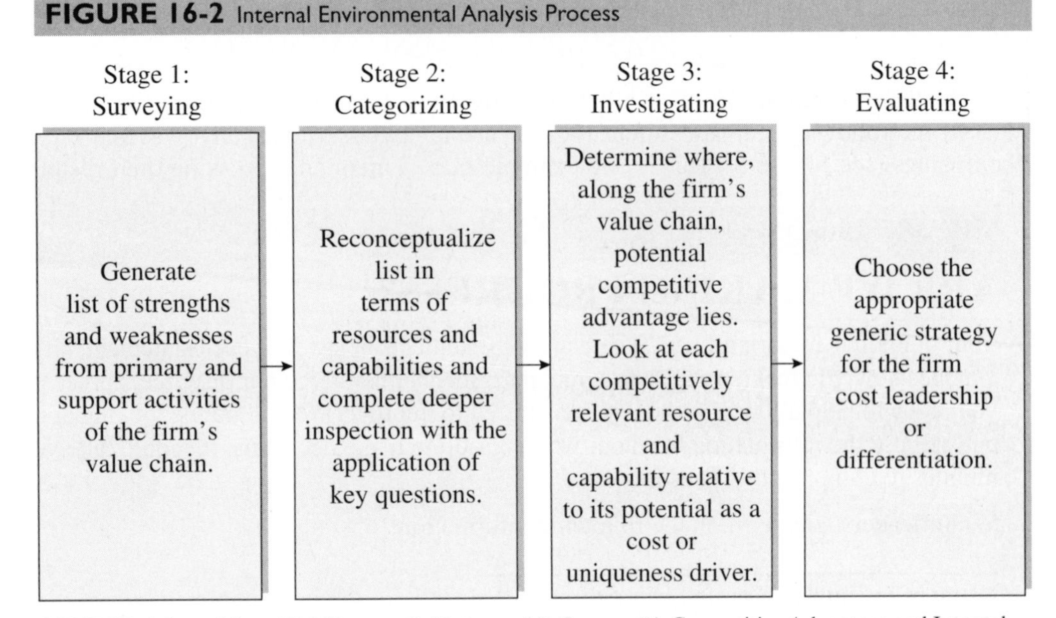

SOURCE: Adapted from W. J. Duncan, P. Ginter, and L. Swayne, "A Competitive Advantage and Internal Organization Assessment," *Academy of Management Executive* 12, 3 (1998):6–16.

Stage 4 relates to **evaluation** of competitive advantage to assess how relevant resources and capabilities are in terms of generic strategies. Remember from Chapter 3 that the generic strategies are cost leadership, differentiation, and focus. Again, the primary question is the extent to which there is alignment between resource allocation in the firm and primary marketing and operational objectives. A Closer Look at Quality 16-1 shows a simple self-assessment tool.

QUALITY AUDITS

The model in Figure 16-2 is a generic internal assessment model that can be used in many firms. We now discuss specific internal assessment models such as the Baldrige criteria and presidential audits. Note that ISO 9000:2000 also uses repetitive audits to ensure compliance and to maintain registration. However, the approach to auditing used in ISO 9000:2000 has not yet achieved general acceptance as a useful model for internal assessment. The approaches discussed here involve helping competitive organizations to become even more competitive. There are other ways to improve that we will discuss. In the abstract, these techniques involve audit processes. We start with internal assessment models based on auditing.

There are a number of different approaches to auditing company systems. We should first define what we mean by audits. These are not the compliance audits that internal auditing departments perform to ensure that generally accepted accounting principles are being followed. These are not audits to ensure that companies and employees are acting honestly. These are audits that can be defined as *an internal assessment tool to identify areas for improvement.*

Quality audits study ways to improve customer service and ascertain whether current customer service processes are being performed. The audit process is based on a

A CLOSER LOOK AT QUALITY 16-1

A SIMPLE SELF-ASSESSMENT TOOL

The Malcolm Baldrige office at NIST recently released the following self-assessment tool for use in companies (see Fig. 16-3). This is one example of a tool that can be used in improving quality. A good class exercise involves teams performing this assessment and reporting their results to the company.

FIGURE 16-3 A Simple Self-Assessment Tool

ARE WE MAKING PROGRESS?

Your opinion is important to us. There are 40 statements below. For each statement, check the box that best matches how you feel (strongly disagree, disagree, neither agree nor disagree, agree, strongly agree. How you feel will help us decide where we most need to improve. We will not be looking at individual responses but will use the information from our whole group to make decisions. It should take you about 10 to 15 minutes to complete this questionnaire.

Senior leaders, please fill in the following information:

____

Name of organization or unit being discussed

CATEGORY 1: LEADERSHIP	Strongly Disagree	Disagree	Neither Agree nor Disagree	Agree	Strongly Agree
1a I know my organization's mission (what it is trying to accomplish).	❑	❑	❑	❑	❑
1b My senior (top) leaders use our organization's values to guide us.	❑	❑	❑	❑	❑
1c My senior leaders create a work environment that helps me do my job.	❑	❑	❑	❑	❑
1d My organization's leaders share information about the organization.	❑	❑	❑	❑	❑
1e My senior leaders encourage learning that will help me advance in my career.	❑	❑	❑	❑	❑
1f My organization lets me know what it thinks is most important.	❑	❑	❑	❑	❑
1g My organization asks what I think.	❑	❑	❑	❑	❑
CATEGORY 2: STRATEGIC PLANNING					
2a As it plans for the future, my organization asks for my ideas.	❑	❑	❑	❑	❑
2b I know the parts of my organization's plans that will affect me and my work.	❑	❑	❑	❑	❑
2c I know how to tell if we are making progress on my work group's part of the plan.	❑	❑	❑	❑	❑

FIGURE 16-3 (continued)

CATEGORY 3: CUSTOMER AND MARKET FOCUS

Note: Your customers are the people who use the products of your work.

		Strongly Disagree	Disagree	Neither Agree nor Disagree	Agree	Strongly Agree
3a	I know who my most important customers are.	❑	❑	❑	❑	❑
3b	I keep in touch with my customers.	❑	❑	❑	❑	❑
3c	My customers tell me what they need and want.	❑	❑	❑	❑	❑
3d	I ask if my customers are satisfied or dissatisfied with my work.	❑	❑	❑	❑	❑
3e	I am allowed to make decisions to solve problems for my customers.	❑	❑	❑	❑	❑

CATEGORY 4: MEASUREMENT, ANALYSIS, AND KNOWLEDGE MANAGEMENT

		Strongly Disagree	Disagree	Neither Agree nor Disagree	Agree	Strongly Agree
4a	I know how to measure the quality of my work.	❑	❑	❑	❑	❑
4b	I know how to analyze (review) the quality of my work to see if changes are needed.	❑	❑	❑	❑	❑
4c	I use these analyses for making decisions about my work.	❑	❑	❑	❑	❑
4d	I know how the measures I use in my work fit into the organization's overall measures of improvement.	❑	❑	❑	❑	❑
4e	I get all the important information I need to do my work.	❑	❑	❑	❑	❑
4f	I get the information I need to know about how my organization is doing.	❑	❑	❑	❑	❑

CATEGORY 5: HUMAN RESOURCE FOCUS

		Strongly Disagree	Disagree	Neither Agree nor Disagree	Agree	Strongly Agree
5a	I can make changes that will improve my work.	❑	❑	❑	❑	❑
5b	The people I work with cooperate and work as a team.	❑	❑	❑	❑	❑
5c	My boss encourages me to develop my job skills so I can advance in my career.	❑	❑	❑	❑	❑
5d	I am recognized for my work.	❑	❑	❑	❑	❑
5e	I have a safe workplace.	❑	❑	❑	❑	❑
5f	My boss and my organization care about me.	❑	❑	❑	❑	❑

FIGURE 16-3 (continued)

CATEGORY 6: PROCESS MANAGEMENT	Strongly Disagree	Disagree	Neither Agree nor Disagree	Agree	Strongly Agree
6a I can get everything I need to do my job.	❑	❑	❑	❑	❑
6b I collect information (data) about the quality of my work.	❑	❑	❑	❑	❑
6c We have good processes for doing our work.	❑	❑	❑	❑	❑
6d I have control over my work processes.	❑	❑	❑	❑	❑

CATEGORY 7: BUSINESS RESULTS					
7a My customers are satisfied with my work.	❑	❑	❑	❑	❑
7b My work products meet all requirements.	❑	❑	❑	❑	❑
7c I know how well my organization is doing financially.	❑	❑	❑	❑	❑
7d My organization uses my time and talents well.	❑	❑	❑	❑	❑
7e My organization removes things that get in the way of progress.	❑	❑	❑	❑	❑
7f My organization obeys laws and regulations.	❑	❑	❑	❑	❑
7g My organization has high standards and ethics.	❑	❑	❑	❑	❑
7h My organization helps me help my community.	❑	❑	❑	❑	❑
7i I am satisfied with my job.	❑	❑	❑	❑	❑

Would you like to give more information about any of your responses? Please include the number of the statement (for example, 2a or 7d) you are discussing.

____

____

____

SOURCE: NIST, Gaithersburg, MD, 2005. Used with permission.

FIGURE 16-4 Generic Auditing Steps

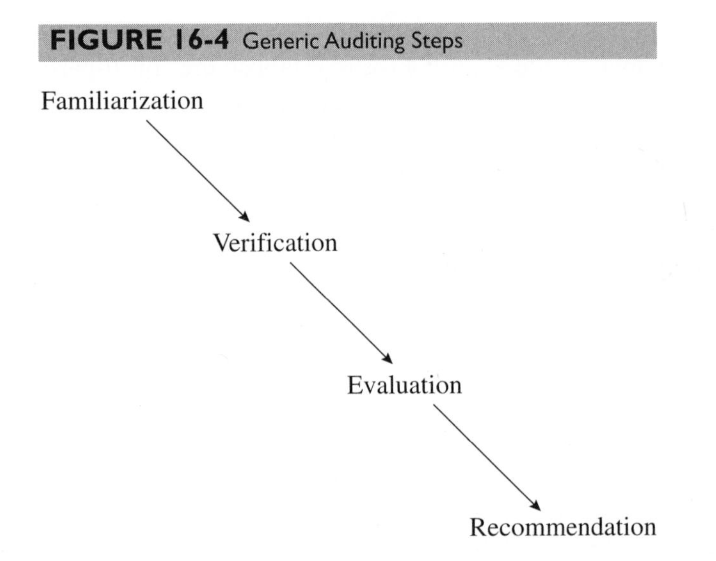

framework of standards, concepts, procedures, and reporting practices. Steps for auditing include familiarization, verification, evaluation, and recommendation (see Fig. 16-4). This disciplined process relies heavily on evidence, analysis, convention, and informed professional judgment.

Although intuition from past experience has an important role in auditing, intuition alone does not constitute a sufficient conceptual basis for a proper audit. In-depth knowledge of auditing standards, proper establishment of audit objectives, carefully planned audit procedures, controlled execution of audit steps, and disciplined adherence to proper auditing methods in arriving at conclusions are all requisite for accurate audits. An audit must be both planned and executed in a highly objective and unbiased manner.

Investigations designed to select only the "right" evidence and ignore the "other" cannot be termed audits. An audit must be designed and conducted to gather all the facts relevant to the matter in question, weighing the "good" against the "bad." Once these objectives are achieved, the auditor then renders professional judgment.

There are three main ingredients to an audit. These are auditing principles, auditing standards, and auditing procedures. An *auditing principle* is a fundamental truth, primary law, or doctrine. Although an auditing principle is not a primary truth or law in the philosophical sense, it does constitute a rule that is derived from reasoning and experience. Auditing principles guide the auditor. Therefore, auditing principles are basic truths and doctrines that indicate the objective of auditing. They also suggest the manner in which the objectives are achieved. In other words, auditing principles constitute the basis for the application of audit procedures in a logical manner, which in turn will fulfill the objectives of the audit. A principle may result from accepted practice or it may develop as a result of general acceptance of a consistently applied procedure.

An *auditing standard* is a measurement of performance or a criterion establishing professional authority and consent. Therefore, a principle is a primary law, whereas a standard is a performance-measuring device. Some examples of general standards include the following:

- The audit must be performed by a person having adequate technical training and proficiency as an auditor.

- In all matters relating to the assignment, independence in mental attitude is to be maintained by the auditor so as to establish the "third-party viewpoint."
- Professional care is to be exercised in the performance of the examination and in the preparation of the report.

The *auditing procedure* establishes the courses of action available to the auditor to judge the adherence to the standards and the validity of the application of the principles. Although auditing concepts and methods are not as rigid as methods in the physical sciences, there is a similarity. For example, in auditing there is evidential support for every conclusion; otherwise, there could not be an audit opinion. For auditing purposes, an "audit-scientific" method should be developed and followed, because any scientific method is concerned with searching for the truth based on evidence. In a scientific method, unsupported ideas and prejudices must be excluded from consideration because of lack of evidential support.

Quality Audit Process

Although there are many types of quality-related audit approaches, the basic quality audit follows these steps:

Preparation: Develop lists of questions, gather materials, form a list of candidates for the audit team, establish schedules, and perform the other activities required for beginning an audit.

Audit team selection: Select the right members for the team. Technical and managerial expertise is very important.

Develop checklists: The checklists contain the questions to be studied in the audit. Checklists also identify who will perform audits in various departments.

The opening meeting: A meeting between the auditing team and the management of the area being audited is called in which ground rules for the audit are established. An especially important agreement is that all pertinent information will be made available to the auditing team.

Implementing the audit: The audit is conducted and pertinent information gathered.

Analysis: The data are analyzed and preliminary results are developed.

The exit meeting: Preliminary results are shared with management at the exit meeting.

Reporting and corrective action: The final report is provided to management and plans are made for taking corrective action. Management implements the corrective action.

Follow-up: A post-implementation review is performed to ensure that the corrective actions were taken and the desired results obtained.

Closure: The audit is closed for the current audit cycle.

Types of Audits

In the following pages we discuss the different approaches to quality auditing that are used most often. Remember the focus is on studying current systems to understand how they can be improved.

Operational Audits

Quality audits are based on the practice of operational auditing. **Operational auditing** is the term that was first used by nontraditional internal auditors many years ago to describe the work they were doing. This work includes a specific objective of improving the operations that are being audited. Much of the application of operational auditing has been in industrial companies. However, operational auditing is being used now frequently for services and government.

In a broad sense, the approach and the state of mind of the auditor characterizes operational auditing—not the methods. Operational studies are seldom made as special and distinct audits. Instead, operational audits represent the application of the talents, background, and techniques of the auditors to the operating controls that exist in the business.

The general objective of the operational audit is to "assist all members of management in the effective discharge of their responsibilities by furnishing them with objective analyses, appraisals, recommendations, and pertinent comments concerning the activities reviewed." Operational audits may follow organizational or functional lines, but the majority are organization-wide because such an audit presents a complete appraisal of the internal operations of an organizational unit. However, the auditor must always have in mind the division of functional responsibilities among organizational units.

Performance Audits

As shown in Table 16-2, there are several major types of performance audits. **Supplier audits** are conducted by purchasers of their suppliers. In the automotive industry, for example, these are conducted using the TS/ISO 16949 standards. Toyota Motor Company in the United States spends months developing its suppliers using its own framework. Periodic audits are then performed to ensure that the supplier is maintaining standards and to improve the performance of suppliers. This approach is referred to as *supplier development*.

Certification audits are used to maintain a certification such as ISO 9000:2000, ISO 14000, QS 9000, or other standards. These audits focus on the documentation of systems and adherence to those standards.

Award audits such as the Baldrige, state quality awards, customer awards, and other prizes involve site visits to externally validate the claims made by applicants in their applications. These visits exist to clarify and verify the information that has been provided by the applicant.

Consultant audits are studies performed by consultants to determine the maturity of a company in the quality pursuit and to help identify areas to be addressed in future quality plans. These are sometimes called *quality maturity studies*.

Presidential audits are audits performed by a team led by the president of the company. These audits are usually operational and quality-related in focus. This gets the president of the firm actively involved in the quality audits and design of the quality system. During a presidential audit, the questions in Table 16-3 are asked. Question-and-answer

TABLE 16-2 Types of Performance Audits

Supplier audits
Certification audits
Award audits
Consultant audits
Presidential audits
Qualitative audits

TABLE 16-3 Questions to Be Answered in a Presidential Audit

1. Under what policies and objectives has the unit proceeded with quality control?
2. What kinds of results have been obtained, and by means of what procedures? (The report must not consist merely of the results; rather, the unit must show the process through which the results were obtained. The unit should report its efforts as QC stories.)
3. What kinds of problems still exist today?
4. Under what policies and objectives does the unit expect to proceed with quality control in the future?
5. What suggestions does the unit want to give to the president and to the headquarters staff?

SOURCE: K. Ishikawa, "The Quality Control Audit," *Quality Progress* (January 1987):39–41.

sessions are held in the morning with the people who work in the areas being audited. Visits are conducted in the afternoon to areas such as research and development, design, purchasing, manufacturing, quality control, marketing, training, and administrative services. In these audits, an overall assessment of the organization is performed, and this information is inserted into the following year's strategic plans.

Kaoru Ishikawa was an advocate of presidential audits. Table 16-4 shows expected benefits of presidential audits as he defined them. See A Closer Look at Quality 16-2.

Qualitative and Quantitative Elements in Audits

Some audits have more of a quantitative flavor when focusing on issues such as conformance and quality control. These audits may involve reviewing quality control check sheets to model trends in defect rates. Cost-of-quality audits may be conducted to determine whether quality costs are lessening and the programs are working.

TABLE 16-4 Benefits of Presidential Audits

1. First of all, such an audit is good for the president. The audit depends on him or her, so he or she is forced to study about quality control. He or she can also observe the actual operations of and facts about factories and other units, which deepens his or her understanding of the company. Knowing everything through paperwork and data is not enough. The president may have an idea of how a particular unit operates and can conceptualize its position in the company, but nothing can replace actual knowledge obtained through first-hand experience.
2. The president can discover the true state of his company. Normally, the truth is not reported to the president. Bad news is suppressed and only good news is reported. If subordinates write candid reports, they risk being scolded. So I advise presidents who are about to begin their own presidential audit, "Never get angry when something bad is reported to you. As long as it is true, never lose your temper. Instead, let your employees report on things that are not going well. Let them give you a candid report of what troubles them. Discuss these problems and try to find solutions together in a spirit of cooperation. After all, the audit by the president is conducted for this very purpose."
3. There will be an improvement in the human relationship between the president and subordinates. The president usually does not have a chance to meet section chiefs, staff members, and supervisors face to face. The audit provides an opportunity to meet, to talk, and to listen. The people involved will develop a feeling for one another and their relations will improve. After the audit, why not have a drink and dinner together?
4. For the people whose QC activities are audited, it is also a significant occasion. There are always ups and downs in human activities. There are times when a person can devote full energy to work, and there are times when a person only goes through the motions. The presidential audit is an occasion for challenging employees and stimulating vigorous activities in total quality control. It also ensures continuation of QC circle activities.

SOURCE: K. Ishikawa, "The Quality Control Audit," *Quality Progress* (January 1987):39–41.

A CLOSER LOOK AT QUALITY 16-2

QUALITY AUDITS IN ACTION

Komatsu Ltd.: www.komatsu.com
Boise, Idaho: www.cityofBoise.org

Komatsu, Ltd., is a large producer of construction equipment. Established in 1921, the company boasts more than $7 billion a year in sales. Komatsu has extensive experience with quality, having begun implementing quality control in 1961. At that time, Komatsu had very poor quality and began to improve its quality in response to a desire to partner with another firm in a joint venture.

Company managers came to realize that if they were going to improve in stature, they needed to improve quality. Thus, they turned to Kaoru Ishikawa, the late professor at Tokyo University. Ishikawa's first efforts centered around helping the company to "correctly understand the problems." This questioning led to the performance of quality audits.

Ryochi Kawai, Komatsu's former chairman of the board, reflected on his experiences with quality audits:

> When I was in the head office and wanted to know about a plant, I would call the plant manager to obtain the status report, and I could understand the plant's condition. But when I listened to the exchanges with Ishikawa, I was quite surprised to find that I was able to see through the detail. This did not mean, however, that the report by the plant manager was false. What really happened was that he reported what I liked to hear. Now I had two routes for understanding the true state of the plant. As a result, I now could decide correct policies much faster.[b]

FIGURE 16-5 Boise City Quality Model

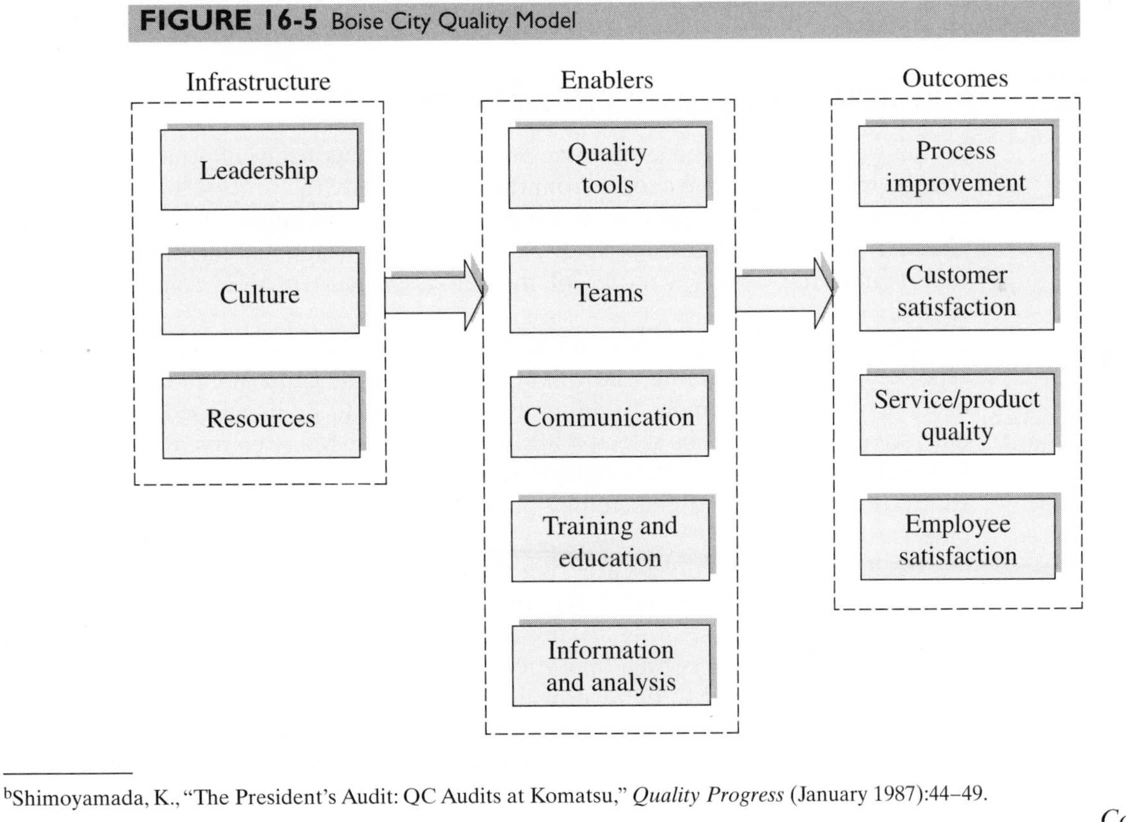

[b]Shimoyamada, K., "The President's Audit: QC Audits at Komatsu," *Quality Progress* (January 1987):44–49.

Continued

This approach to quality auditing laid the foundation for Komatsu to achieve world-class status in quality products.

Can these approaches also work in government? The answer is, "Yes." The city of Boise, Idaho, underwent a **quality maturity analysis (QMA)**. Boise had performed team and quality tools training for a number of years. The QMA allowed a disinterested third party to take a step back and assess the city's status in terms of improving service to Boise customers. The mayor of Boise commissioned the study. This support was key for the success of the study. With the input of city management, a model for improvement in the city was developed (see Fig. 16-5). Based on this model, a survey was performed asking questions relating to each of the variables in the model. In addition, focus groups with rank-and-file employees were formed and interviews were conducted with city department heads. The study was used to develop strategic planning processes to provide a basis for future improvement.

Qualitative audits compare current practice against structural measures. *Structural measures* are the documents relating to processes. These include standards, contracts, procedures, regulatory requirements, and hierarchical standards. In qualitative audits, studies are performed to see that procedures are being followed. For example, as quality improvement becomes more behavioral and emphasizes decision making, it will be important to know whether employees are using structured processes for decision making. Hence qualitative audits should become more important as time passes.

EXTERNALLY VALIDATING THE QUALITY SYSTEM

So far we have discussed examining company systems using internal validation. At times it is helpful to have people from the outside study the process and identify areas for improvement. The process is termed **external validation**. As we discussed in Chapter 3, the strength of the Baldrige criteria is their applicability as an assessment tool. Figure 16-6 shows a model of the self-assessment process using the Baldrige criteria.

The first step is to plan and define the purpose of the self-assessment. Important aspects include determining who will be involved in the gathering of data and writing of the evaluation document. Training materials have to be developed and inside and outside readers need to be selected for the application. Many companies hire previous Baldrige examiners to provide outside reading and training support. The National Institute of Standards and Technology publishes a list of Baldrige examiners for such purposes.

Collecting data involves gathering the elements needed to write the application in each of the seven Baldrige categories. Typically, teams are formed from each of the functional areas relating to the Baldrige categories (i.e., top management, strategic planning, marketing, operations, accounting, and so on). Firms usually assign a single individual to be responsible for compiling the final report and coordinating printing and terminology. The findings are bound into evaluation documents. Next, internal and external examiners are assigned to read the application and evaluate the application.

FIGURE 16-6 Self-Assessment Process

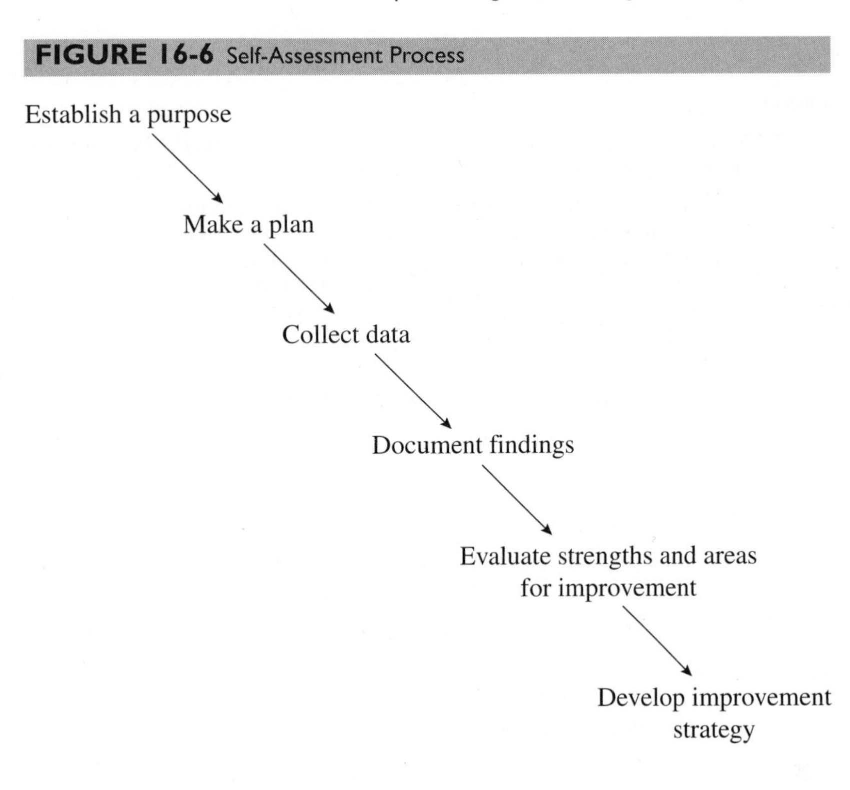

Video Clip: Baldrige Assessment at the Ritz

After evaluating the reports, the examiners write responses to the evaluation document outlining areas that should be improved. Figures 16-7 through 16-9 show examples of the type of feedback received. The first feedback is the overview. Next, feedback is provided for each of the Baldrige categories. We are showing only sample feedback for category 1 to a fictitious firm named Landmark Dining, Inc. The most useful information for improvement comes from the areas for improvement items.

Notice that the feedback reports provide very specific areas for improvement. From the various areas for improvement, management can now make a list of potential improvements in the various criteria areas—remember the criteria addresses company-wide improvement. After the list of improvements has been created, projects can be chartered for different functional areas and coordinated at the strategic level.

State Quality Awards

State quality awards can provide a low-cost means of using the Baldrige criteria in firms. There are currently 41 state quality awards. Some quality awards are struggling because of difficulty in finding funding for the awards. More than 1,000 companies annually apply for state quality awards, and most of the award programs are Baldrige based. American firms also can apply for the Japanese Deming Prize.

State quality awards allow companies to receive external reviews for a nominal cost. Because many award programs are struggling financially, private industry and the National Institute of Standards and Technology need to explore means to provide better support for state awards.

FIGURE 16-7 Example of a Baldrige Summary

Scoring Summary

Landmark Dining Inc. (Landmark) scored in band 5 in the consensus review of written application for the Malcolm Baldrige National Quality Award.

An organization in band 5 typically demonstrates effective, systematic, well-deployed approaches responsive to the overall requirements of the items. Landmark demonstrates a fact-based, systematic evaluation and improvement process and organizational learning that result in improving the effectiveness and efficiency of key processes. Results address most key customer/stakeholder, market, and process requirements, and they demonstrate areas of strength against relevant comparisons and/or benchmarks. Improvement trends and/or good performance are reported for most areas of importance to the organization's key requirements.

a. The most important strengths or outstanding practices (of potential value to other organizations) are as follows:

- Landmark maintains a focus on the future through its systematic and well-deployed strategic planning process, which is integrated and aligned with data and information systems such as Voices and Foodtrak. These systems provide fact-based data and information to support short- and longer-term planning by using key input from customers, suppliers/partners, key stakeholders, and employees. The Strategic Planning Process results in the development of a Strategy Matrix that helps Landmark align its strategic objectives with its strategic challenges, action plans, and competitive success factors related to both current and future operations. The alignment and integration in the Strategic Planning Process may help the organization to remain agile when responding to its current operational challenges while positioning itself to best address strategic challenges.
- The organization clearly demonstrates its commitment to management by fact and continuous improvement with systematic approaches to data collection and analysis and process improvement. DINERS Teams use a systematic improvement process to address opportunities for improvement across Landmark. Data and information from the Voices and Foodtrak systems undergo multiple analyses to provide senior leaders, DINERS Teams, and employees at all levels actionable information on which to base their improvement recommendations. Landmark presents several examples of performance improvement resulting from its well-deployed and integrated approaches to process improvement, data collection, and analysis.
- Landmark capitalizes on its planning, process improvement, and data and information collection and analysis approaches by creating a team-based environment and an operating style empowered and enabled by data and information available through a number of avenues. Landmark ensures access to data and information and creates an environment of organizational learning through its Communication Process, Foodtrak Knowledge Management system, and a variety of two-way communication vehicles, including line-up meetings and periodic performance reviews. Employees have real-time access to key performance data and information, enabling them to make informed decisions in the course of their day-to-day work. Best practices are shared with all employees through line-up meetings, the Foodtrak Knowledge Management system, and storytelling designed to support Landmark's culture and individual and organizational learning.
- The organization demonstrates in its systematic approaches to employee learning, development, satisfaction, and well-being that it is committed to and values its employees. Landmark provides employees with a "cafeteria" plan that allows them to select benefits that best suit their individual needs. Through numerous education and training opportunities, employees acquire multiple skills that increase their capabilities and overall value to the organization. To capitalize on its well-trained and motivated workforce, Landmark empowers employee teams to schedule, manage, and improve their work processes. All employees complete an Individual Review and Development Plan (IRDP) that is aligned with the organization's direction and balances the needs of the individual and the organization. In addition, a systematic succession planning approach is integrated with the organization's employee development and training approaches to help ensure Landmark's long-term sustainability. These approaches support the organizational values of family culture with teamwork and employee development, as well as its strategic objective to be an employer of choice.
- Landmark is committed to being a valued citizen in the communities where it operates. It uses a systematic approach to identify key communities for organizational-level support as part of its Strategic Planning Process and ensures deployment of its approaches for community support by authorizing time off for employees to participate in key community activities. In addition, members of Landmark's Leadership Team serve in volunteer positions side-by-side with employees. Some senior leaders are board members in key community support organizations, including the Houston and Galveston Food Funds, historic preservation associations, and area Chambers of Commerce.

SOURCE: Landmark Dining, Inc. Feedback Report (Gaithersburg, MD: NIST, 2005).

FIGURE 16-8 Item 1.1 Example of a Feedback Report

Details of Strengths and Opportunities for Improvement

Category 1 Leadership

1.1 Senior Leadership

Strengths

- The Senior Leadership Team, including the Advisory Board, reviews Landmark's vision, mission, and values during the Strategic Planning Process (Figure 2.1-1) and incorporates them into the Strategy Matrix (Figure 2.2-3). The Strategy Matrix aligns the competitive success factors, values, strategic objectives, short- and longer-term plans, and related measures to gauge success, and it provides Landmark with a means to link its day-to-day operations with its values and performance expectations. The Strategy Matrix is reviewed with all employees, and their IRDPs are linked to it; portions of the matrix are shared with suppliers; and the vision, mission, and values are printed on menus for customers to see.
- Senior leaders' personal actions reflect a commitment to organizational values through communication, reinforcement, and role modeling of values and expectations. Examples include providing discounted health care options to all part-time employees to support Landmark's value of family culture with teamwork, spending 10% to 20% of their time working with employees in the restaurants or catering service each week, and leading a half-day employee orientation to discuss the organization's values and expectations.
- Senior leaders use the Communication Process, annual ethics training for all employees, and annual signing of the ethics statement by all employees to promote an environment that fosters and requires legal and ethical behavior. The organization further requires legal and ethical behavior by making compliance to its ethics policy a condition of employment for employees and a condition of contractual relationships with suppliers. Organizational learning is demonstrated by Landmark's recent refinement of its values to include ethics, honesty, and integrity.
- Sustainability is addressed through a three-tiered approach: (1) a vision and direction to provide a focus for employee decisions, (2) a process orientation, and (3) accountability for performance through the measurement system and review structure. Employees are encouraged to suggest innovative approaches and to identify improvement opportunities. Each senior leader further fosters sustainability through involvement in succession planning, which includes identifying talented employees, developing IRDPs, coaching and mentoring high-potential employees, and discussing future leadership issues during monthly executive reviews.
- Senior leaders use multiple methods to communicate with, motivate, and empower employees. These methods include a formal Communication Process (Figure 5.1-l) to determine key factors for communicating important information; daily line-up, weekly staff, and monthly all-employee and team leader meetings; feedback from and to senior leaders when they work in the restaurants; and public reward and recognition of employees. A team leader approach helps empower teams, which develop their own daily and weekly work schedules responsive to company and employee needs.
- Senior leaders create an environment that focuses on both accomplishing strategic objectives and on improving performance by integrating Landmark's vision, mission, and values with its strategic planning and deployment process, action plans, goals, and key measures on the Balanced Scorecard (Scorecard). An environment of improvement and innovation is supported through format and systematic assessment processes that include aligned and linked organization, department, and individual performance reviews; Landmark's DINERS Improvement Process; and annual Baldrige self-assessments. During reviews and meetings, employees' ideas and feedback are solicited, discussed, and recognized by senior leaders.

Opportunities for Improvement

- Although Landmark's suppliers and partners are asked to report ethics violations, a systematic process is not evident for monitoring and assessing Landmark's effectiveness in deploying ethical requirements to its suppliers. Given that 90% of all supplier costs are for products and services from an external purchasing consortium and related transactions may not be transparent to Landmark, the company may have difficulty ensuring these transactions are consistent with its values of ethics, honesty, and integrity.
- It is unclear how the Advisory Board members, as members of the Senior Leadership Team, are personally involved in succession planning and the development of future organizational leaders. This may be of particular importance since the Advisory Board includes external members of the local business community with key competencies that Landmark identifies as not being present elsewhere in Landmark's leadership.
- Although Landmark utilizes a Communication Process (Figure 5.1-1) that includes daily line-up meetings, other meetings, and communication logs, it is not clear how Landmark ensures that all employee groups, including part-time, on-call, catering, and HMR employees, are able to participate in the various communication methods.

SOURCE: Landmark Dining, Inc. Feedback Report (Gaithersburg, MD: NIST, 2005).

FIGURE 16-9 Item 1.2 Example of a Feedback Report

Governance and Social Responsibilities

Strengths

- Landmark uses an external Advisory Board, composed of prominent business leaders, that provides independent guidance and feedback on leadership and governance and takes an active leadership role in meetings and strategic planning activities. Two criteria are used to select Advisory Board members: (1) they must be comfortable with and supportive of the organization's value system, and (2) they must have specific skills that complement those of the existing Senior Leadership Team. Annual financial audits are conducted by external independent auditors; results are shared with the Advisory Board to help ensure fiscal accountability; and, while not required, Landmark is in the process of implementing compliance under the Sarbanes-Oxley Act.
- Senior leaders and managers use 360-degree reviews and track completion of IRDPs to improve their effectiveness as individual leaders and as a leadership team. The results of these assessments are discussed and appropriate actions planned during a special meeting prior to starting the Strategic Planning Process. Senior leaders and the Advisory Board also receive feedback from an external consultant who attends their meetings quarterly.
- Senior leaders hold key positions in local community committees and associations, including the Chamber of Commerce, the NRA, and the Galveston and Houston Health and Human Services advisory boards. Information they gain from these positions is used in Landmark's Strategic Planning Process to help anticipate and identify potential concerns with current and future products, services, and operations.
- Landmark has established procedures, training and certification, and measurement and reporting practices to help ensure compliance with multiple local, state, and federal regulatory requirements (e.g., food safety requirements, waste removal requirements, local zoning and building codes, licensing, employee safety requirements, and human resource requirements). Key compliance goals and measures are used to ensure that requirements are met. One result of these activities was the implementation of Hazard Analysis and Critical Control Point (HACCP) elements in anticipation of future Food and Drug Administration (FDA) food safety regulations and to ensure customer safety.
- Landmark helps promote ethical behavior by communicating in multiple ways that ethical behavior is an organizational value and a condition of employment. All employees receive ethics training and sign an ethics statement annually. The organization monitors ethical behavior by tracking code of conduct violations, the number of employees terminated due to ethical issues, and regulatory compliance measures, and it reviews the results of customer, supplier, and employee surveys. Senior leaders and/or the Advisory Board investigate potential breaches of ethical behavior, and appropriate actions are taken.
- The organization identifies key communities and areas for support during its Strategic Planning Process and reviews them annually. Landmark has selected its two primary communities of operation, Houston and Galveston, and has identified the key support areas for these communities (Figure 1.2.1). Senior leaders and employees participate in a number of community events, and senior leaders fill leadership positions in several key community support organizations.

Opportunities for Improvement

- Although Landmark appears to have a systematic approach to the selection and use of its Advisory Board, it is not clear how this approach ensures accountability for management's actions. Further, although Landmark shares its Strategy Matrix, key performance measures, and financial audit results with the Advisory Board and its employees, it is not clear how this approach provides internal controls on governance processes that would support transparency in operations or the selection and disclosure policies for the Advisory Board and the Senior Leadership Team that constitute Landmark's governance system. It also is unclear how Landmark's governance approach addresses fiscal accountability and risk in its supplier/vendor relationships to protect the interests of all key stakeholders.
- While Landmark contracts with a professor from the business department of a local university to attend quarterly board meetings and provide feedback to the senior leaders and the Advisory Board on their performance, a systematic process is not evident for using this feedback to improve the personal leadership effectiveness of senior leaders, the Advisory Board and its individual members, as well as the leadership system as a whole.
- Although Landmark addresses many of the concerns associated with safe food handling at its restaurants with employee training and certification, it is not clear how it addresses other potentially adverse impacts of its products and operations. For example, a systematic process is not evident for addressing the potential adverse impacts of serving beer, wine, and other alcoholic beverages—key requirements for several customer segments. It also is not clear how potential adverse impacts of its transportation operations (e.g., HMR deliveries to distributors, catering event deliveries) are addressed. Without adequately addressing the potential of adverse impacts from these products, services, and operations, Landmark may not be able to effectively identify potential risks.

FIGURE 16-9 (continued)

- Although perceptions of ethical behavior are tracked through surveys and Landmark collects data on code of conduct violations, a systematic process is not evident for monitoring the ethical behavior of Landmark's governance structure or interactions with customers and partners, including HMR distributors. Further, while some measures are provided, it is not clear how these measures enable Landmark to monitor and respond to several key ethical challenges associated with its operations and interactions with direct and/or HMR customers, such as the abuse of customer credit card information (identified as a key concern by Landmark) and legal/ethical issues associated with the sale of alcoholic beverages.

SOURCE: Landmark Dining, Inc. Feedback Report (Gaithersburg, MD: NIST, 2005).

Summary

In this chapter we have identified ways for outstanding firms to get better. Once your firm has achieved role model status, it is difficult to find ways to improve. However, the recognition that there are endless ways to improve should lead firms to audit and self-assess.

The audit processes and Baldrige self-assessment discussed here provide a means for top management to improve its leadership in the area of quality management. These models reinforce the importance of the role of top management leadership in enhancing the system for improvement.

Although the firm that is looking to improve on already-high standards need not be large or world-class, it should be moderately mature in its quality journey. It is unlikely that you would want to perform a full-blown Baldrige assessment if you are a novice to quality improvement. However, once you are at the point where you have measurable results, self-assessment can help to prioritize where to go next.

Companies also should recognize that these approaches are not short-term fixes. They require long-term commitment and support that is probably going to be more obvious in firms that have established quality programs.

Key Terms

- Award audits
- Categorizing
- Certification audits
- Consultant audits
- Enterprise capabilities
- Evaluation
- External validation
- Investigating
- Operational auditing
- Presidential audits
- Qualitative audits
- Quality maturity analysis (QMA)
- Supplier audits
- Surveying

Discussion Questions

1. The model in Figure 16-1 shows people as the basis of the quality system. Do you agree with this assertion? Why or why not?
2. There are regions where the three spheres of quality overlap. What are some of the overlaps between management, assurance, and control? Why are they important?
3. Review the concept of enterprise capabilities. Pick a firm and determine what you think is the enterprise capability for that firm.
4. Why is internal assessment a necessary tool for outstanding companies?
5. At what stage do you believe a company would be ready for internal assessment?

6. At what stage does a company become ready for Baldrige-based internal assessment?
7. Define the different types of audits. Pick a company and define which type would be best for it. Support your answer.
8. The Boise City Leadership model is an interesting model for a governmental entity. Is this model different from a model that would be used for a for-profit firm? Why or why not?
9. What are the *enablers* for quality improvement in a school? What are they in a firm where you have worked (see Fig. 16-4)?

PROBLEMS

1. Administer the survey instrument in A Closer Look at Quality 16-1 to a local business owner and report your findings.
2. Administer the survey instrument in A Closer Look at Quality 16-1 to the employees of the business from Problem 1 above. Compare the employees' and owner's perceptions to see where they differ significantly.

Case

Case 16-1 Setting Priorities Using The Baldrige Criteria

www.nist.gov

On the following pages are examiner evaluation notes for categories 2 through 6 for the fictional company discussed in the chapter (the entire report is not provided because of length consideration). Given this feedback, develop a plan for improvement for Landmark. Include prioritization of the different new projects. ■

DISCUSSION QUESTIONS

1. How did you select particular projects from the feedback report?
2. What are some of the weaknesses of this approach?
3. Why would it be good to use a mix of internal and external examiners as is recommended in the chapter?
4. How did you prioritize projects for improvement?
5. Was all of the feedback meaningful? What are some of the attributes of useful feedback?

Category 2 Strategic Planning

2.1 Strategy Development

Strengths

- Landmark conducts its Strategic Planning Process at its annual three-day retreat with key participants (the Senior Leadership Team, the Board of Directors, and key suppliers) and also involves partners, Advisory Board members, and other community representatives, as appropriate. Participation by a variety of external stakeholders helps the organization identify blind spots, as well as gain insight into various changes that may impact its future. As part of its Strategic Planning Process, Landmark has identified a short-term (one-year) planning horizon, and a longer-term

Category 2 Strategic Planning (continued)

planning horizon, which was established at five years to be responsive to the value of historic preservation. Planning horizons are addressed in the planning process through the development of interim milestones to track progress from short- to longer-term goals.

- The Strategic Planning Process is reviewed each year at the annual retreat, and it has evolved since 1990 to include key steps, as well as a more rigorous Strengths, Weaknesses, Opportunities, Threats, and Trends (SWOTT) analysis. In 2001, the Strategy Matrix was introduced to align Landmark's vision, mission, and values with its key strategic challenges, strategic objectives, action plans, and goals. Subsequent refinements include the integration of competitive success factors, key stakeholders, and the Approach-Deployment-Learning-Integration (ADLI) concept. This alignment and integration may help Landmark to maintain its focus on the future while addressing its key strategic challenges.
- Prior to the strategic planning retreat, each member of the Leadership Team collects and analyzes data on one or more key factors, which ensures that customer and market needs, financial risks, technology, human resource needs, regulatory and societal risks, and economic changes are integrated into the Strategic Planning Process. This information then is used in an environmental scan and SWOTT analysis to identify relevant opportunities, review progress, and ensure the availability of financial resources necessary to carry out the strategic plan. To ensure agility in the execution of the plan, any changes to the key factors or performance are presented at the scheduled executive reviews or at midyear. For example, as a result of an analysis and review of occupancy rates, funds were allocated to purchase new tables that are more easily configured for varying sizes of parties, which has increased the occupancy rate to 4% over the national average.
- Landmark has identified key strategic objectives and the key goals for achieving these objectives in its Strategy Matrix. Short- and long-term objectives are aligned with strategic goals, competitive success factors, the organization's values, and its strategic challenges. Landmark has identified its most important goals for 2005 as maintaining a 15% growth rate per year in new service results, increasing customer satisfaction to 96.5%, and increasing its occupancy rate to 85%.

Opportunities For Improvement

- Although Landmark's Leadership Team collects and analyzes data and information on a number of factors to support its Strategic Planning Process, a systematic process is not evident for analyzing its supply chain strengths and weaknesses to ensure that needs can be met for factors related to business continuity and growth. Because suppliers are an integral part of its operations, without a systematic process to capture and analyze such information, Landmark may not be able to effectively identify and address risks associated with its suppliers.
- Although Landmark's Strategy Matrix includes the strategic challenges aligned to its strategic objectives, it is not clear how its strategic objectives specifically address each of the challenges (e.g., challenges associated with the sophistication of the American palate, heightened interest in food safety, or intensified government impact through increased mandates and their associated cost impacts). In addition, it is not clear how the strategic objectives balance short- and longer-term challenges and opportunities and the needs of all key customer groups, such as families, business patrons, and tourists.

2.2 Strategy Deployment

Strengths

- During the Strategic Planning Process, the Leadership Team identifies the specific actions required to accomplish Landmark's strategic objectives, along with associated measures, and identifies who, what, when, and how the specific actions/tasks will be accomplished. A Strategy Matrix is developed to ensure that short- and long-term action plans are linked to the competitive success factors, strategic challenges, and values of the organization. The action plans are then deployed throughout Landmark and to suppliers and partners through the Communication Process. Specific shorter-term actions are further deployed through the development of action plans that support the organization-level direction at the department and employee levels, and employees' action plans are linked to their IRDPs. The DINERS Improvement Process and monthly performance reviews are used to formalize process

SOURCE: Landmark Dining, Inc. Feedback Report (Gaithersburg, MD: NIST, 2005).

Category 2 Strategic Planning (continued)

changes, ensure organizational learning, and ensure that key changes resulting from the accomplishment of strategic action plans are integrated and that performance is sustained.

- All measures in the Strategy Matrix are tracked through Foodtrak, and these measures are reviewed weekly and monthly by the Leadership Team. If there are emergencies or changes in the business climate, market conditions or customer requirements, or if performance projections are not being met, the DINERS Improvement Process is used to determine causes and recommend changes. The Strategy Matrix is then modified, appropriate measures are added to the Scorecard to track performance, employees are notified of changes during line-up or all-employee meetings, and managers and supervisors assist employees in modifying IRDPs, if necessary. These approaches allow Landmark to react quickly and with agility to changes as they occur.
- The key performance measures for tracking progress on action plans are identified in the Strategy Matrix. The Leadership Team evaluates action plans at weekly and monthly executive review meetings to ensure alignment of the action plan measurement system with organizational strategies and stakeholder needs.
- Landmark has identified performance projections for its 28 key short- and longer-term action plan measures in the Strategy Matrix. Its 2005 performance projections are better than or as good as its competitors' 2010 performance projections in most measures presented. Projected performance gaps are addressed using the DINERS Improvement Process.

Opportunities for Improvement

- Although the Leadership Team members take ownership of various action plans (their development and the alignment of their numbers), it is not clear how Landmark allocates resources, other than financial resources, to ensure the accomplishment of its action plans. Without a systematic process to allocate resources according to its priorities, Landmark may not be able to ensure achievement of all its action plans and, in turn, its strategic objectives.
- Although the organization identifies many of its human resource plans with respect to its short-term action plans, human resource plans that derive from longer-term key action plans are not provided. Without specific human resource action plans, it may be difficult for Landmark to accomplish longer-term strategic objectives that may be dependent on recruiting and retaining skilled and motivated employees.
- Although key longer-term competitors' performance projections for the year 2010 are shown in the Strategy Matrix, it is not clear how Landmark's short term projections compare to those of its competitors. This may limit Landmark's ability to gauge its progress toward being recognized as one of the cities' top 10 dining experiences.

Category 3 Customer and Market Focus

3.1 Customer and Market Knowledge

Strengths

- Landmark uses product, market, and pricing requirements identified by the restaurant industry to determine its customer and market segments. The company competes in the semicasual dining steak and seafood market, with $35–$50 dinner pricing. Customers within this market are segmented by customer type (e.g., family, business, tourist) and by type of service (dine-in, take-out, catering, and dinner delivery). Landmark uses market research to identify potential customers and customers of competitors for current as well as future products and services.
- To listen to and learn from its customers, Landmark uses its Voices system, which includes the "voices" of experience, the customer, the server, and the process. This systematic, integrated process is used to capture information before, during, and after dining experiences occurring with varying frequencies, and a 360-degree analysis is conducted to compare and validate data across the various voices. In addition, a Satisfaction and Importance Levels matrix is used to analyze the relative importance of various factors and their impact on customers' satisfaction. The matrix is used to determine priorities that will enhance customer loyalty and retention.
- Information from the Voices system, data from the Our Family program and Secret Diners program, and complaint data are aggregated through the Foodtrak system and used for multiple purposes,

Category 3 Customer and Market Focus (continued)

including as input into the Value Creation Processes. Based on the feedback, DINERS Teams may be chartered, resulting in menu adjustments, job redesign, and communication refinements. Senior leaders also use the information and knowledge gathered as input to the Strategic Planning Process.

- A variety of methods are available for customer groups to provide information. These communication methods include verbal responses; multiple-choice written responses; telephone, Web site, and written surveys; comment cards; focus groups; and comments via e-mail. Mechanisms are tailored according to the needs of various customer groups and markets, including frequency of contact.
- The Voices system was initially designed in 1997 and has been through numerous cycles of improvement. Landmark uses the DINERS Improvement Process, Baldrige self-assessment, and the Strategic Planning Process to refine its approaches to listening and learning to keep them current with business needs and directions, including changes in the marketplace. A recent example is the change from conducting an annual customer survey to conducting an ongoing survey to ensure agility in reacting to changing needs, such as dietary and palate preferences and the need for convenience.

Opportunities for Improvement

- Although the Voices system provides a systematic approach for listening and learning related to restaurant customers, it is not clear how or whether the system is used with customers of the catering or HMR services. This may be particularly important given Landmark's strategic challenge of continued expansion of its products and services.
- Although Landmark has identified multiple customer and market segments, it is not clear that there is a systematic process to use the unique requirements identified for each segment to better satisfy customer needs, determine needs for current or future products or services, or identify and prioritize opportunities for segments specific improvements.
- While Landmark does collect customer retention data on catering customers and Our Family program members, it is unclear whether it collects and analyzes this type of data on other dine-in customers or its takeout customers.

3.2 Customer Relationships and Satisfaction

Strengths

- To build relationships with customers, Landmark uses an outside advertising vendor to promote public awareness of its reputation among targeted customer segments through television, radio, magazine, and Web-based advertising; displays; and promotions. It also builds relationships at multiple points of contact with customers by identifying specific customer requirements for all aspects of its food and beverage preparation and service. Loyalty is developed and strengthened through the Our Family frequent-diner program, which personalizes service for program members and offers incentives (e.g., two-for-one meals and "treat a friend" coupons) to increase repeat business and positive referrals. The relationship-building process is reviewed annually through the DINERS Improvement Process.
- To enable all of its customer segments to seek information, conduct business, and voice complaints, Landmark provides multiple mechanisms, including personal contact, telephone, the Internet, fax, e-mail, surveys, and focus groups. The Voices system is used to identify customer contact requirements based on customer satisfaction ratings and comments related to the various contact methods. Customer contact standards are deployed throughout Landmark through the "Prospective Employee Guide," the Employee Handbook, reinforcement at daily line-up meetings, and automated reminders through the Foodtrak system.
- Landmark manages customer complaints by using its systematic Service Recovery Process, which enables identification and resolution of customer complaints on the spot or before the customer leaves the restaurant, thus minimizing customer dissatisfaction—and promoting repeat business due to the customers' perception of special treatment during the recovery. The Service Recovery Process is used in all stages of the customer experience, and all employees receive training on contact requirements and the Service Recovery Process. Complaints surfaced during this process and from all other sources are integrated, aggregated, and analyzed through Foodtrak to identify root causes and trends, to prevent reoccurrence, to improve other customer-related approaches, and to refine the Voices system and the

Category 3 Customer and Market Focus (continued)

customer contact and relationship-building process. Successful use of the Service Recovery Process is rewarded and celebrated at weekly staff meetings and in internal publications.

- Landmark reviews and improves its approaches for building customer relationships and determining customer satisfaction to keep them current with business needs and directions by using the DINERS Improvement Process and conducting at least one Baldrige self-assessment annually. Results and processes related to the Voices system, Our Family program, contact methods and standards, and the Service Recovery Process are systematically reviewed and evaluated by senior leaders during strategic planning to ensure alignment with strategic directions, and action plans are created to address necessary changes. External satisfaction surveys are reviewed by an academic expert to ensure their validity and reliability.
- Landmark uses a variety of methods to determine customer satisfaction and dissatisfaction before, during, and after the dining process. These mechanisms include internal and external customer surveys, point-of-service input, and focus groups. Surveys are available in Spanish, English, Braille, and TTY (text telephone) systems. DINERS Teams use the correlation between satisfaction and importance levels, along with complaint factor analysis, to capture actionable information to exceed customer expectations, secure future business, gain positive referrals, design new processes, and redesign/improve existing processes.
- Landmark uses the personal customer contact standards and the methods identified in the Voices system to ensure immediate follow-up on the performance of products and services. If there is negative feedback generated from these contacts, the Service Recovery Process is implemented to integrate immediate action with the feedback, and follow-up calls are made by shift managers to verify resolution of the complaint. Information is documented in the Foodtrak system to capture learning and facilitate aggregation with other data collected.
- Landmark obtains information regarding its customers' satisfaction relative to their satisfaction with competitors regarding food, service quality, timeliness, price, value, and facilities through the Secret Diners Association and external customer satisfaction surveys conducted by a third party. Additional information is gathered from local publications in news and trade journals, including reviews by food critics and industry benchmark information from the NRA. Results from internal customer satisfaction surveys also are used to analyze strengths and weaknesses of specific competitors identified by customers.

Opportunities for Improvement

- It is not evident that the approaches Landmark uses to build relationships and to increase loyalty and retention address takeout, catering, or HMR customers. For example, the Our Family program, a key mechanism to increase customer loyalty and retention, appears to focus only on dine-in customers. This gap may be important given that Landmark intends to develop and expand its newer business lines.
- It is not clear how access mechanisms and personal customer contact standards are deployed to HMR customers, who are directly served by Landmark's distributor partners. Contact requirements for access mechanisms other than personal contact, such as telephone, fax, and Web access, also are not described. Without a systematic approach to determine contact requirements for all customer segments and access mechanisms, it may be difficult for Landmark to ensure that the specific needs of all customers are being met.

Category 4 Measurement, Analysis, and Knowledge Management

4.1 Measurement, Analysis, and Review of Organizational Performance

Strengths

- Senior leaders use the annual Strategic Planning Process and the Strategy Matrix to systematically select and align measures for tracking organizational performance. From the matrix, senior leaders create an integrated Scorecard with key organizational measures that are reviewed at monthly Senior Leadership Team meetings to track performance and progress on strategic action plans. During the monthly review meetings, senior leaders also assess Landmark's external and internal environment and adjust the Strategy Matrix and Scorecard, as appropriate. Key Scorecard measures

Category 4 Measurement, Analysis, and Knowledge Management (continued)

are integrated from performance data that reside in the Foodtrak system. This system is linked to all sites and all operational functions and supports Landmark's value creation and support processes.

- Data used to track daily operations are selected and refined by DINERS Teams, which use a formal, systematic process that includes five selection criteria (including a direct relationship to the strategic plan) to align new and existing measures as business processes are refined. Significant operational measures are integrated through the Foodtrak system from linked supplemental databases that are used for data collection.
- During the Strategic Planning Process, Landmark selects the comparative data and information that are used to understand its competitive position, to help determine action plans and goals, to design processes, and to facilitate the DINERS Improvement Process. Also, comparative data are part of the organizational performance reviews in monthly executive meetings, daily lineup meetings, and all-employee meetings. Comparative data are collected from the NRA, Secret Diners Association, employee dining reports, the Chamber of Commerce, Staffing Solutions, local industry surveys, Landmark's financial auditor, and best-in-class sources, such as Baldrige Award recipients.
- Daily and weekly performance trends are analyzed quarterly to verify that key leading indicators are predictive of organizational performance. Landmark's performance measurement system is refined annually by senior leaders during the Strategic Planning Process. Employee feedback collected through the Foodtrak Knowledge Management system and Advisory Board feedback are used to evaluate the measures and their linkages to the Strategy Matrix. Performance measures are refined as needed during monthly executive reviews of leading and outcome measure analyses to address more frequent and unexpected changing business needs and directions. Real-time changes are made through Foodtrak for rapid deployment to all employees.
- To assess organizational capabilities and performance, all areas of the company use a systematic process of regularly scheduled, cascading performance review meetings with varying frequencies (Figure 4.1-2), ranging from annually (e.g., strategic planning) to daily (e.g., line-up meetings). DINERS Teams are created to address areas identified for improvement, and a number of analyses are performed in support of the various cascading organizational reviews. These include correlation analyses that are used by senior leaders and employees at all levels to assess organizational performance results relative to goals, strategic objectives, action plans, and competitive performance.
- Gaps in performance discovered as part of the analysis process are translated by senior leaders into priorities for improvement in various ways. These include refinements in key measures and goals, the development of action items or action plans and the deployment of DINERS Teams. Scorecard performance measures and any changes in priorities, directions, action plans, or allocation of resources resulting from senior leaders' performance reviews are systematically deployed to all employees. Deployment occurs via the Foodtrak system, by sharing the Strategy Matrix and Scorecard at all-employee meetings, and by modifying IRDPs, which are linked to the Scorecard and Strategy Matrix.

Opportunities for Improvement

- Although Landmark uses comparative data in its selection of organizational and operational measures, it is not clear what criteria are used to select the various measures available from multiple sources. In addition, it is not evident how Landmark ensures the effective use of comparative and competitive data and information in support of daily operational decision making and innovation for some of its divisions, such as catering and HMR Dinner Delivery.
- While Landmark provides an example of its ability to respond quickly to findings in organizational reviews, it is not clear how the organization uses its various performance reviews to assess its overall ability to rapidly respond to changing organizational needs and challenges.

4.2 Information and Knowledge Management

Strengths

- Landmark makes needed data and information available to its employees through its Foodtrak system. Employees can have immediate access to required data from the system through wired computers, wireless PDAs (personal digital assistants), touch pads, and touch terminals. For example, servers carry wireless POS (point of service)/PDA units that place orders, provide order status, and prompt staff when actions are required. The Foodtrak system is integrated with Landmark's public Web site, enabling its customers to access transactions and order-related information. It also enables

Category 4 Measurement, Analysis, and Knowledge Management (continued)

access to applicable data for Our Family program members and to appropriate suppliers for inventory management purposes,
- The Foodtrak system's information technology (IT) vendor provides technical support during operating hours and remotely monitors the system and software to ensure network security. The IT vendor also provides backup systems and operates databases to ensure data security. The LANs use secure encryption access codes, and WANs are electronically protected behind an access-restricted firewall. In addition, user feedback is captured in the Foodtrak Knowledge Management system, and suggested changes are reviewed prior to acceptance by a sampling of staff members.
- Landmark has a disaster recovery program to ensure continued availability of data and information in the event of an emergency. Program components include replacement for interface hardware at all locations to immediately replace breakdowns, battery-backed power supplies, daily data backups to on-and off-site locations, and contracted replacement of key system hardware components within 12 hours and of all customer contact systems within 24 hours. System performance is evaluated by the IT vendor with input from Foodtrak customers during the annual Improvement Day.
- Constant user feedback is solicited and monitored by technical staff from Landmark's contracted IT vendor. The vendor also uses its annual Foodtrak Improvement Day to keep this data and information availability system current with business needs and directions. In addition, user feedback is captured in the Foodtrak Knowledge Management system, where employees can provide input and questions about system capabilities, and DINERS Teams formally address improvement opportunities with the vendor.
- The Knowledge Management system within Foodtrak is used by Landmark on an ongoing basis to collect, organize, and share knowledge, including best practices, among key stakeholders. In addition, best practices are shared during team leader meetings and with all employees during line-up meetings. Vendors and suppliers are included in discussions when appropriate and are encouraged to enter into Foodtrak comments, suggestions, and ideas regarding their products.
- Landmark addresses data accuracy by using selection options, information scanning technologies, and forced-review elements in the design of its data entry processes. Electrical systems and manual backups are used to ensure reliability, and touchpads, computer access, and PDAs help ensure the timeliness of data and information. Security is ensured through the use of passwords and firewalls. To help ensure confidentiality, senior lenders must authorize access to protected electronic information, such as credit card data, customer profiles, and critical organizational data.

Opportunities For Improvement

- Although some product suppliers such as the restaurant purchasing consortium have access to online inventory data and some vendors have Web access to their performance data, it is not clear whether outside suppliers of support services (e.g., custodial services, human resource management, advertising, and marketing) or the HMR distributor partners have access to similar information.
- Although Landmark has several mechanisms for sharing best practices (e.g., team leader and line-up meetings), a systematic process is not evident for identifying best practices. In addition, a systematic process is not described for effectively implementing a best practice once it is identified.
- While Advisory Board members are required to sign nondisclosure agreements, it is not clear whether a similar approach is used for employees and vendors/suppliers to ensure confidentiality.

Category 5 Human Resource Focus

5.1 Work Systems

Strengths

- Landmark has organized employees in all business divisions, including Catering, Dinner Delivery Service, and Administration, into process teams that align with each of its key processes to ensure alignment with the strategic plan and to promote cooperation, empowerment, and innovation. Teams are responsible for scheduling and managing work to operate and improve their key processes. All team leaders meet monthly to assess performance, review customer feedback, identify opportunities for improvement, and share best practices. To keep current with business needs, promote agility, and encourage professional growth and development, cross-training is provided to all employees in at least two to three positions, and lateral service is emphasized.

Category 5 Human Resource Focus (continued)

- Landmark achieves effective communication and skill sharing across work units, jobs, and locations through its Communication Process, the Foodtrak Knowledge Management system, and other systems, such as a communication log (a benchmarking process of a Baldrige Award recipient), cross-training, and meetings (all-employee, monthly team leader, and shift meetings). Skill sharing also is facilitated through process improvements that are documented and included in Foodtrak to ensure standardized processes and procedures and by sharing best practices. These approaches may support Landmark's commitment to its success factors of superior service and operational excellence.
- Landmark's formal employee performance management system is its IRDP process, which is designed to provide two-way communication between employees and managers and includes a performance appraisal. IRDPs are aligned with organizational and department action plans and are reviewed quarterly during the first year and annually thereafter, with midyear check-ins to assess progress and identify barriers. Managers also participate in a biennial 360-Degree Feedback Process. Multiple reward and recognition mechanisms (e.g. dining certificates, birthday recognition, and on-the-spot awards such as gift cards and monetary bonuses) reinforce and support high performance and a focus on customer and business goals.
- Landmark uses a systematic Job Review Process to identify characteristics and skills needed by potential employees that are then documented in formal job descriptions. The job descriptions are based on process requirements and are systematically refined as part of the annual Strategic Planning Process or after major process changes. Job descriptions are updated and shared through IRDPs. Skills needed for newly created positions are identified by the hiring manager based on goals for the position. Functional flowcharts of key processes are validated and updated through weekly reviews after employees are hired.
- In addition to the Job Review Process, Landmark uses its systematic eight-step Recruiting and Hiring Process to recruit and hire employees; this process has been refined through input from employees and managers and annual DINERS Team reviews. Both processes are integrated with Landmark's Strategic Planning Process to ensure each addresses both short- and longer-term organizational needs and directions. Employees and managers are involved in interviews of potential employees, and a staffing agency is contracted to address targeted recruitment efforts to reduce diversity gaps.
- Landmark's formal succession planning initially identifies individuals to be developed for each leadership position and has been refined to include team leaders. The succession plan includes a career path, rotational assignments, training, development activities, and job shadowing of the future role. The plan is reviewed every six months by the Chief Executive Officer and the Business Excellence (BE) Director. In addition, as part of their IRDPs, all employees are asked to develop career goals that include developmental goals, action plans, and estimated timelines; those who express an interest in the industry are supported through special training. These processes support Landmark's commitment to its value of employee development.

Opportunities for Improvement

- Although Landmark's work systems approach includes employee representation on process teams, on DINERS Teams, and in leadership positions to promote cooperation and empowerment, it is not clear how this approach helps the organization to capitalize on the diverse cultures and ideas of its workforce. Without a systematic approach, Landmark may have difficulty addressing its strategic challenge of having available skilled and motivated employees to match the growth of the organization.
- Although Landmark has a process to recruit and hire new employees, a systematic approach to retain employees is not evident, As a result, the organization may be limiting its effectiveness in addressing its strategic challenge of having available skilled and motivated workers and in addressing its key success factors of providing superior service and operational excellence.

5.2 Employee Learning and Motivation

Strengths

- Landmark ensures that education and training efforts for employees align with organizational strategies and action plans identified by using the Strategic Planning Process and resulting strategic objectives to create IRDPs. This linkage and alignment also ensures that key requirements associated with Landmark's business needs and directions and the accomplishment of its action plans are addressed at the organizational level in its training and development approaches. Also, DINERS

Category 5 Human Resource Focus (continued)

Teams are used to identify improvement strategies that frequently are supported by training; for example, changes in the strategic plan have focused training for 2005 on ethics monitoring, the catering service and HMR delivery service, the Foodtrak system, and strategic planning.

- Landmark uses formal training methods to address needs associated with new employees, including a four-hour orientation provided by senior leaders, a virtual tour, and the Employee Handbook. Team leaders provide on-the-job training. New employees are assigned a coach/mentor for the first three months and job-shadow the coach for three to five days. Throughout the year, training is provided on CPR, safety, workplace violence, OSHA requirements, safe handling of equipment, and building security through line-ups, on-line modules, and all-employee meetings. Employees are required to pass the NRA course for food handling and food safety, and all managers complete the Food Service Manager's Certification.
- Landmark seeks and uses input from employees and their supervisors and managers on education, training, and development needs through employees' IRDPs, the Employee Satisfaction Survey, and informal feedback during line-up meetings. Organizational learning and knowledge assets are formally incorporated into education and training through coaching, sharing best practices at monthly all-employee meetings, using the Foodtrak Knowledge Management system as part of research for DINERS Teams, and providing outside trainers with key information from Foodtrak to incorporate into training,
- Landmark reinforces the use of new knowledge and skills through line-ups, coaching, and on-the-job training, which includes immediate reinforcement and ongoing oversight of a team leader to help reinforce the use of new skills. For external training, supervisors develop a plan for the employee's use of a new skill, and employees are expected to share their key learnings by entering lessons learned into the Knowledge Management System. Further reinforcement is provided by including training in employees' IRDPs and evaluating employees on their attainment of skills.
- Landmark evaluates the effectiveness of its training using formal end-of-class evaluations, feedback from annual employee surveys, and correlations of improvement activities associated with related training. Other indicators include the accomplishment of action plans and the percentage of goals attained in employee IRDPs.
- Employees are motivated to develop and utilize their full potential through their IRDPs, which are linked to organizational strategic and action plans, as well as each individual's career goals, and are systematically reviewed with the supervisor. Raises and promotions are tied to performance appraisal results but can be given any time at the manager's discretion. This process demonstrates Landmark's commitment to its value of employee development and its strategic objective of being the employer of choice.

Opportunities For Improvement

- Although employees develop IRDPs through the Strategic Planning Process, it is not clear how Landmark addresses training related to performance measurement or how it balances the individual training needs and those associated with career progression with short- and longer-term organizational objectives. Given the high turnover of staff within the industry, without a systematic process Landmark may not be effective in meeting the needs of its employee groups or in addressing its key success factors of providing superior service and operational excellence.
- Although employees and supervisors can provide input on training delivery approaches through their evaluation of current training and as part of the IRDP development process, it is not evident that there is a systematic process to seek and use input from employees and their supervisors and managers to determine appropriate delivery approaches prior to establishing the training. Also, it is not evident that there is a systematic process to provide input regarding informal training approaches, such as on-the-job training, which is most often used.
- Although Landmark generally has a "debrief period" for retiring and departing employees to train their replacements and document best practices, it is not clear if this approach to transferring knowledge is systematic and consistently deployed throughout Landmark. This may be particularly important given Landmark's stated desire to keep a core of employees who are the knowledge base of the company, given the strategic challenge of the availability of skilled and motivated employees, and the generally high industry turnover rates. Although there is an expectation that employees will share learning from external training by entering lessons learned into the Knowledge Management System, it is not evident that Landmark has a systematic approach to document employees' new knowledge and skills from internal training for long-term organizational use.

Category 5 Human Resource Focus (continued)

5.3 Employee Well-Being and Satisfaction

Strengths

- Landmark ensures and improves workplace health, safety, security, and ergonomics through a contractor who monitors OSHA compliance, provides health and safety training, and conducts regular inspections. Employees provide suggestions for improving workplace factors at line-up meetings and IRDP sessions. The BE Director manages the vendor relationship and evaluates vendor performance. Team leaders monitor measures, and DINERS Teams are created when opportunities for improvements are identified by vendor inspections or employee suggestions.
- Landmark has a formal Disaster Preparedness Plan to ensure workplace preparedness at its restaurant locations for general business disasters (e.g., fire) and natural disasters that are likely to occur in the area (e.g., hurricanes). The plan details actions employees should take and identifies ongoing activities to support disaster recovery, such as daily data system backup, off-site data storage, and a backup technology plan. The plan is reviewed and updated annually, and the information is available to all employees of the organization, with hard copies at each restaurant and in senior leaders' homes. An electronic version also is available in the Foodtrak Knowledge Management system. During orientation, new employees receive initial information on emergency procedures that is reviewed on an ongoing basis. Drills arc conducted monthly.
- Landmark uses its Employee Satisfaction Survey, IRDP Process, and the results of exit interviews to determine the key factors that affect employee well-being, satisfaction, and motivation. Key factors are segmented for hourly and salaried workers. The Employee Satisfaction Survey can be analyzed to identify the factors for various employee groups, and it asks employees to rank order various satisfaction factors by importance and their degree of satisfaction with these factors.
- To support its employees, Landmark uses a cafeteria-type health care plan with a dollar limit that allows employees to tailor their benefits to meet their own diverse needs. Examples of these benefits include a 401k plan, subsidized medical insurance, a child care subsidy, health club membership, subsidized transportation, paid time off for holidays, time off for community involvement/volunteer activities, in-restaurant dining discounts, and recognition for participating in improvement activities. Benefits are prorated for part-time employees, and on-call workers may purchase medical insurance at reduced rates. These benefits reflect Landmark's commitment to addressing its strategic objective of being the employer of choice.
- As its key tool for determining employee satisfaction. Landmark conducts an on-line, semiannual Employee Satisfaction Survey that is modeled after a national survey by the NRA and enables comparison to national results in the hospitality industry and best-in-class benchmarks. Response rates exceed 90% and are segmented by job, location, gender, age, and ethnicity. Landmark also monitors employee turnover, the rate of IRDP completion, absenteeism, sales per server, results from exit interviews, and work environment measures as other indicators of employee satisfaction and well-being. When declining results occur, DINERS Teams are created to conduct reviews and make improvements.
- Senior leaders review indicators of employee satisfaction and motivation, including Employee Satisfaction Survey results, and regularly conduct correlation analyses with Landmark's Voice of the Process and Voice of the Customer measures. These analyses are used to identify potential opportunities for improvement in the work environment that impact key business results. Results of these analyses are addressed by DINERS Teams.

Opportunities for Improvement

- While the organization addresses safety through the use of an outside firm and employees address safety needs and improvements at line-ups, it is not clear how Landmark addresses safety and other issues for employee segments, such as administrative office workers or outside contractors' employees (e.g., custodial and security personnel) who work on Landmark's premises.
- It is unclear how Landmark's workplace Disaster Preparedness Plan considers the needs of on-site contracted employees or its employees working off-site, such as delivery and catering personnel, to make them aware of the plan and their actions and responsibilities in the event of a disaster. Without this information, Landmark's Disaster Preparedness Plan may not effectively address the needs of all employee groups.

Category 6 Process Management

6.1 Value Creation Processes

Strengths

- Using information gathered through the Voices system, Landmark determines that its key value creation processes are those that add value to the dining experience according to the customer's perspective. This same system provides information during strategic planning to identify emerging processes, which are then added to the Strategy Matrix and undergo an annual review. Landmark has identified key processes for its restaurants, catering and HMR business lines, and various product/service segments.
- Value creation process requirements are determined by using Voices data from multiple stakeholders and are gathered before, during, and after the dining experience. Landmark's suppliers provide information through their participation in reviews and input to the Foodtrak Knowledge Management component. The Purchasing Consortium Manager and key suppliers participate in the monthly executive review meetings, where key metrics are discussed and opportunities for revisions are identified.
- Cross-functional and cross-restaurant DINERS Teams use a formal, systematic, nine-step method to design value creation and support processes. This process begins with stakeholder requirements and includes flowcharts, in-process metrics, targets from the Scorecard, a pilot phase, communication, training, and an annual evaluation. The Process Design Process also includes searches for new technology and a search of the Knowledge Management System for relevant information.
- Landmark has identified the key performance measures and indicators for its value creation processes. To ensure the day-to-day operation of its processes meets key requirements, Landmark uses on-line and hard copy documentation of its value creation processes, training and on-the-job reinforcement visual management and job aids, walk-throughs for certain events, and twice-daily line-ups, where key performance information is shared and reviewed. Customer input obtained through the Voices system is used in value creation process management, and supplier input is acquired through a variety of periodic meetings.
- Landmark uses its DINERS Improvement Process to annually review and improve its value creation processes and to keep them current with business needs and directions, Cross-functional employee teams are trained in the DINERS Improvement Process and related tools, and employees are trained to identify potential improvement opportunities that might necessitate a DINERS Team review at times other than the annual cycle. Process improvements are shared through monthly team leader process meetings and the Foodtrak Knowledge Management system. Process changes are documented and included in employee training guidelines within 10 days.

Opportunities for Improvement

- A systematic process is not evident for incorporating cycle time, cost control, productivity, and other effectiveness and efficiency factors into its value creation process design approach. Further, it is not clear how Landmark implements the processes, once designed, to ensure they perform as expected and meet design requirements. Without a systematic process to incorporate efficiency and effectiveness factors into process design, it may be difficult for Landmark to ensure that its value creation processes are achieving the desired performance.
- Although Landmark conducts "quick and economical" preaudits and daily observation of processes, it is not clear how these approaches enable Landmark to systematically minimize the cost of inspections and audits or to prevent rework or defects, as appropriate.

6.2 Support Processes and Operational Planning

Strengths

- Landmark determines its key support processes, as well as related key requirements, in-process measures, and outcome measures at the same time and in a similar fashion as its key value creation processes. The processes are identified either through Step 4 in the Process Design Process or through strategic planning. Many support process measures also are on the Scorecard, which is aligned with strategic objectives and action plans to help achieve business success.
- Key support process requirements are determined by process owners and suppliers based on information from the Voices system. Support process requirements include hiring of suitable employees;

Category 6 Process Management (continued)

an accurate, timely, and cost-efficient payroll; information system availability; and multiple requirements for suppliers.

- Landmark's team leaders and DINERS Teams use a nine-step approach to design Landmark's key support processes to meet all key requirements. The approach starts with the determination of the desired outcomes and incorporates new technology in Step 4 of (the design approach).
- Landmark has identified its key support processes, as well as associated measures and indicators used to control and operate the processes, The Foodtrak system provides prompts to guide and standardize support processes. The system is also used to communicate changes in processes to all employees.
- Support processes are improved using the DINERS Improvement Process and are reviewed annually by DINERS teams for needed improvements of approaches or measures. Improvements are shared departmentally and with internal customers, and they are documented in Foodtrak to ensure they are used for organizational learning and innovative approaches for other processes. Process changes are included in employee training guidelines within 10 days, and employees receive updated training.
- Landmark uses its annual Budget Process to ensure adequate resources are available to support its operations. The Budget Process follows the Strategic Planning Process, and departments present requirements for both current operations and for accomplishing their respective action plans. The Leadership Team then reviews all requests, prioritizes them based on operational and investment priorities related to the strategic plan, and allocates the required resources accordingly.
- Landmark has a Disaster Recovery Program designed to ensure operations can resume within a reasonable amount of time after an emergency. This program includes IT systems backup, employee safety procedures, and return-to-work instructions that focus on disasters likely to happen in its region (e.g., a hurricane).

Opportunities for Improvement

- While Landmark considers employees working in value creation processes to be internal customers, and while value creation processes results are shared with all employees, it is not clear how Landmark systematically uses input from its internal customers in the determination of key support process requirements. Without input from these key stakeholders, it may be difficult for Landmark to identify valid and important key support process requirements.
- It is not clear how Landmark systematically incorporates cycle time, cost control, productivity, and other effectiveness and efficiency factors in its support process design approach. Further, it is not clear how it implements the processes, once designed, to ensure they perform as expected and meet design requirements.

Appendix

TABLE A-1 Factors for Determining Control Limits for $\overline{X}$ and R Charts

Number of Observations in Subgroup n	*Factor for* $\bar{x}$ *Chart* A_2	*Factor for X Chart* E_2	*Factors for R Chart* *Lower Control Limit* D_3	*Upper Control Limit* D_4
2	1.88	2.66	0	3.27
3	1.02	1.77	0	2.57
4	0.73	1.46	0	2.28
5	0.58	1.29	0	2.11
6	0.48	1.18	0	2.00
7	0.42	1.11	0.08	1.92
8	0.37	1.05	0.14	1.86
9	0.34	1.01	0.18	1.82
10	0.31	0.98	0.22	1.78
11	0.29		0.26	1.74
12	0.27		0.28	1.72
13	0.25		0.31	1.69
14	0.24		0.33	1.67
15	0.22		0.35	1.65
16	0.21		0.36	1.64
17	0.20		0.38	1.62
18	0.19		0.39	1.61
19	0.19		0.40	1.60
20	0.18		0.41	1.59

$$\text{Upper control limit for } \bar{x} = UCL_{\bar{x}} = \bar{\bar{x}} + A_2\bar{R}$$
$$\text{Lower control limit for } \bar{x} = LCL_{\bar{x}} = \bar{\bar{x}} - A_2\bar{R}$$

(If aimed at or standard value $\bar{x}'$ is used rather than $\bar{\bar{x}}$ as the central line on the control chart, $\bar{x}'$ should be substituted for $\bar{\bar{x}}$ in the preceding formulas.)

$$\text{Upper control limit for } R = UCL_R = D_4\bar{R}$$
$$\text{Lower control limit for } R = LCL_R = D_3\bar{R}$$

All factors in Table A-1 are based on the normal distribution.

$$\text{Upper control limit for } X = UCL_X = \bar{\bar{x}} + E_2\overline{MR}$$
$$\text{Lower control limit for } X = LCL_X = \bar{\bar{x}} - E_2\overline{MR}$$
$$\text{Upper control limit for } MR = UCL_{MR} = D_4\overline{MR}$$
$$\text{Lower control limit for } MR = LCL_{MR} = D_3\overline{MR}$$

$$\text{Upper control limit for } \bar{x} = UCL_{\bar{x}} = \bar{\bar{x}} + A_3\bar{\sigma}$$
$$\text{Lower control limit for } \bar{x} = LCL_{\bar{x}} = \bar{\bar{x}} - A_3\bar{\sigma}$$

(If aimed-at or standard value $\bar{x}'$ is used rather than $\bar{\bar{x}}$ as the central line on the control chart, $\bar{x}'$ should be substituted for $\bar{\bar{x}}$ in the preceding formulas.)

$$\text{Upper control limit for } \sigma = UCL_\sigma = B_4\bar{\sigma}$$
$$\text{Lower control limit for } \sigma = LCL_\sigma = B_3\bar{\sigma}$$

All factors in this table are based on the normal distribution.
Median chart formulas:

$$CL_{\tilde{x}} = \bar{\tilde{x}} \pm \bar{\tilde{A}}\bar{R}$$

$$\text{Average} = \frac{\text{sum of medians}}{\text{number of medians}} = \bar{\tilde{x}}$$

TABLE A-2 Normal z Curve Areas

z	*.00*	*.01*	*.02*	*.03*	*.04*	*.05*	*.06*	*.07*	*.08*	*.09*
0.0	.00000	.00399	.00798	.01197	.01595	.01994	.02392	.02790	.03188	.03586
0.1	.03983	.04380	.04776	.05172	.05567	.05962	.06356	.06749	.07142	.07535
0.2	.07926	.08317	.08706	.09095	.09483	.09871	.10257	.10642	.11026	.11409
0.3	.11791	.12172	.12552	.12930	.13307	.13683	.14058	.14431	.14803	.15173
0.4	.15542	.15910	.16276	.16640	.17003	.17364	.17724	.18082	.18439	.18793
0.5	.19146	.19497	.19847	.20194	.20540	.20884	.21226	.21566	.21904	.22240
0.6	.22575	.22907	.23237	.23565	.23891	.24215	.24537	.24857	.25175	.25490
0.7	.25804	.26115	.26424	.26730	.27035	.27337	.27637	.27935	.28230	.28524
0.8	.28814	.29103	.29389	.29673	.29955	.30234	.30511	.30785	.31057	.31327
0.9	.31594	.31859	.32121	.32381	.32639	.32894	.33147	.33398	.33646	.33891
1.0	.34134	.34375	.34614	.34850	.35083	.35314	.35543	.35769	.35993	.36214
1.1	.36433	.36650	.36864	.37076	.37286	.37493	.37698	.37900	.38100	.38298
1.2	.38493	.38686	.38877	.39065	.39251	.39435	.39617	.39796	.39973	.40147
1.3	.40320	.40490	.40658	.40824	.40988	.41149	.41309	.41466	.41621	.41174
1.4	.41924	.42073	.42220	.42364	.42507	.42647	.42786	.42922	.43056	.43189
1.5	.43319	.43448	.43574	.43699	.43822	.43943	.44062	.44179	.44295	.44408
1.6	.44520	.44630	.44738	.44845	.44950	.45053	.45154	.45254	.45352	.45449
1.7	.45543	.45637	.45728	.45818	.45907	.45994	.46080	.46164	.46246	.46327
1.8	.46407	.46485	.46562	.46638	.46712	.46784	.46856	.46926	.46995	.47062
1.9	.47128	.47193	.47257	.47320	.47381	.47441	.47500	.47558	.47615	.47670
2.0	.47725	.47778	.47831	.47882	.47932	.47982	.48030	.48077	.48124	.48169
2.1	.48214	.48257	.48300	.48341	.48382	.48422	.48461	.48500	.48537	.48574
2.2	.48610	.48645	.48679	.48713	.48745	.48778	.48809	.48840	.48870	.48899
2.3	.48928	.48956	.48983	.49010	.49036	.49061	.49086	.49111	.49134	.49158
2.4	.49180	.49202	.49224	.49245	.49266	.49286	.49305	.49324	.49343	.49361
2.5	.49379	.49396	.49413	.49430	.49446	.49461	.49477	.49492	.49506	.49520
2.6	.49534	.49547	.49560	.49573	.49585	.49598	.49609	.49621	.49632	.49643
2.7	.49653	.49664	.49674	.49683	.49693	.49702	.49711	.49720	.49728	.49736
2.8	.49744	.49752	.49760	.49767	.49774	.49781	.49788	.49795	.49801	.49807
2.9	.49813	.49819	.49825	.49831	.49836	.49841	.49846	.49851	.49856	.49861
3.0	.49865	.49869	.49874	.49878	.49882	.49886	.49889	.49893	.49897	.49000
3.1	.49903	.49906	.49910	.49913	.49916	.49918	.49921	.49924	.49926	.49929

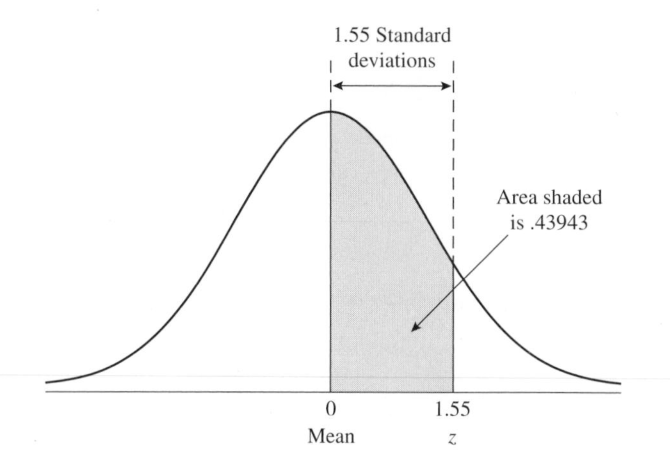

The numbers represent the proportion of the total area away from the mean, μ, to one side. For example, the area between the mean and a point that is 1.55 standard deviations to its right is .43943.

TABLE A-3 Factors for Determining the 3-Sigma Control Limits for $\bar{x}$ and **s** Charts

Number of Observations in Subgroup n	Factor for $\bar{x}$ Chart A_3	Factors for s Chart		
		Lower Control Limit B_3	Upper Control Limit B_4	Divisors for Estimate of σ C_4
2	2.659	0	3.27	0.7979
3	1.954	0	2.57	0.8862
4	1.628	0	2.27	0.9213
5	1.427	0	2.09	0.9400
6	1.287	0.03	1.97	0.9515
7	1.182	0.12	1.88	0.9594
8	1.099	0.19	1.81	0.9650
9	1.032	0.24	1.76	0.9693
10	0.975	0.28	1.72	0.9727
11	0.927	0.32	1.68	0.9754
12	0.886	0.35	1.65	0.9776
13	0.850	0.38	1.62	0.9794
14	0.817	0.41	1.59	0.9810
15	0.789	0.43	1.57	0.9823
16	0.763	0.45	1.55	0.9835
17	0.739	0.47	1.53	0.9845
18	0.718	0.48	1.52	0.9854
19	0.698	0.50	1.50	0.9862
20	0.680	0.51	1.49	0.9869
21	0.663	0.52	1.48	0.9876
22	0.647	0.53	1.47	0.9882
23	0.633	0.54	1.46	0.9887
24	0.619	0.55	1.45	0.9892
25	0.606	0.56	1.44	0.9896

TABLE A-4 Factors for Median Charts

n	$\bar{\tilde{A}}_2$	D_4
3	1.187	2.575
5	0.691	2.115
7	0.508	1.924
9	0.412	1.816

Glossary

acceptable quality level (AQL) The maximum percentage or proportion of nonconformities in a lot or batch that can be considered satisfactory as a process average.

acceptance sampling Statistical quality control technique used in deciding to accept or reject a shipment of input or output.

active data gathering A method for gathering data that involves approaching respondents to get information.

actively solicited customer feedback Proactive methods for obtaining customer feedback, such as calling customers on the telephone or inviting customers to participate in focus groups.

activity network diagram Also known as a PERT diagram, an activity network diagram is a tool used in controlling projects.

aesthetics A dimension of quality that refers to subjective sensory characteristics such as taste, sound, look, and smell.

affinity diagram A tool that is used to help groups identify the common themes that are associated with a particular problem.

alignment Term that refers to optimal coordination among disparate departments and divisions within a firm.

analyze phase Six-Sigma phase where the collected data are analyzed.

andon A Japanese term that refers to the warning lights on an assembly line that light up when a defect occurs. When the lights go on, the assembly line is usually stopped until the problem is diagnosed and corrected.

annuity relationship This occurs when a business receives many repeat purchases from a customer. The income is received steadily over time from a single customer.

appraisal costs Expenses associated with the direct costs of measuring quality.

assurance A dimension of service quality that refers to the knowledge and courtesy of employees and their ability to inspire trust and confidence.

attribute A binomial state of being.

attrition The practice of not hiring new employees to replace older employees who either quit or retire.

award audits Site visits relating to award programs.

balanced scorecard A tool for monitoring both financial and operational metrics in one document.

Baldrige-lite Term used to depict states' quality award programs using the same criteria as the Malcolm Baldrige National Quality Award but with a simplified process or application.

Baldrige-qualified Term used by firms that have been granted a site visit by the judges in the Malcolm Baldrige National Quality Award competition.

basic events Term used in fault tree analysis. Basic events are initiating faults that do not require events below them to show how they occurred. The symbol used for a basic event is a circle.

basic prototype Nonworking mock-up of a product that can be reviewed by customers prior to acceptance.

basic seven (B7) tools of quality These are the fundamental methods for gathering and analyzing quality-related data. They are: fishbone diagrams, histograms, Pareto analysis, flowcharts, scatter plots, run charts, and control charts.

bathtub-shaped hazard function Reliability model that shows that products are more likely to fail either very early in their useful life or very late in their useful life.

benchmark An organization that is recognized for its exemplary operational performance in one or more areas and is willing to allow others to view its operations and tour its facilities.

benchmarking The process of finding a company that is superior in a particular area, studying what it does, and gathering ideas for improving one's own operation in that area.

best of the best Term used to refer to outstanding world-class benchmark firms.

best in class Term used to refer to firms or organizations that are viewed as the best in an industry on some meaningful criterion.

black belt A designation given to someone who has completed intensive quality training and has demonstrated results from one or more major projects.

business case A mechanism used in Six Sigma and reengineering to outline a basis for improvement.

c chart A chart used to monitor the number of defects in a production process.

capability Likelihood a product will meet specification.

catchball Term used to describe the iterative nature of the Hoshin planning process.

categorizing The act of placing strengths and weaknesses into categories in generic internal assessment.

cause and effect (or fishbone or Ishikawa) diagram A diagram designed to help workers focus on the causes of a problem rather than the symptoms.

certification audits Audits relating to registration (e.g., ISO 9000:2000 audits).

chain of customers A philosophy that espouses the idea that each worker's "customer" is the next worker in the chain of people that produce a finished product or service.

champion Sponsor of a Six-Sigma project.

change In the context of quality management, this means to move from one state of operation to another state of operation.

check sheets Data-gathering tools that can be used in forming histograms. The check sheets can be either tabular or schematic.

churn reduction A process for reducing customer defections.

clickstream The path customers use in navigating Web sites.

compensate (1) To pay or remunerate for some work; (2) To make up for some lack of ability or acuity.

complaint-recovery process Process associated with resolving complaints.

complementary products Products that use similar technologies and can coexist in a family of products.

component reliability The propensity for a part to fail over a given time.

computer-aided design (CAD) A system for digitally developing product designs.

computer-aided inspection (CAI) A system for performing inspection through the use of technology. For example, some systems use infrared to detect defects.

computer-aided testing (CAT) Technology for conducting tests or examinations.

computer-based training A form of training that uses specialized software, known as courseware, to address specific topics.

concept design The process of determining which technologies and processes will be used to produce a product.

concurrent engineering The simultaneous performance of product design and process design. Typically, concurrent engineering involves the formation of cross-functional teams. This allows engineers and managers of different disciplines to work together simultaneously in developing product and process designs.

conformance A dimension of quality that refers to the extent to which a product lies within an allowable range of deviation from its specification.

consultant audits Inspections that are performed by consultants to determine how an organization should be changed for improvement.

Consumer Product Safety Commission (CPSC) An independent federal regulatory agency that helps keep American families safe by reducing the risk of injury or death from consumer products.

consumer's risk The risk of receiving a shipment of poor quality product and believing that it is good quality.

contact personnel The people at the "front lines" who interact with the public in a service setting.

contingency theory A theory that presupposes that there is no theory or method for operating a business that can be applied in all instances.

contract review Contract review involves the steps associated with contracting with suppliers. These steps involve acceptance of the contract or order, the tender of a contract, and review of the contract.

contrition Forgiveness for error or mistake.

control charts Tools for monitoring process variation.

control factors Variables in a Taguchi experiment that are under the control of the operator. These can include things such as temperature or type of ingredient.

control phase Six-Sigma phase where improved process performance is monitored.

control process A process involving gathering process data, analyzing process data, and using this information to make adjustments to the process.

conversion process Aligning the inputs of a process together to form a product or service.

core processes Self-identified processes that are central to the organization and its customers.

criticality A term that refers to how often a failure will occur, how easy it is to diagnose, and whether it can be fixed.

cross-functional teams Teams with members from differing departments and vocations.

cross training Training an employee to do several different jobs.

customer Anyone who is the receiver of the goods or services that are produced.

customer benefits package (CBP) The package of tangibles and intangibles that make up a service.

customer contact A characteristic of services that notes that customers tend to be more involved in the production of services than they are in manufactured goods.

customer coproduction The participation of a customer in the delivery of a service product. For example, in many restaurants it is not uncommon for customers to fill their own drinks.

customer defections The number of customers who do not repeat with a particular firm.

customer-driven quality Term that refers to a proactive approach to satisfying customer needs.

customer expectations (1) What customers expect from a service provider; (2) A part of the SERVQUAL questionnaire.

customer future needs projection Predicting the future needs of customers and designing products that satisfy those needs.

customer perceptions (1) How customers view products or services; (2) The second part of the SERVQUAL survey.

customer rationalization The process of reaching an agreement between marketing and operations as to which customers add the greatest advantage and profits over time.

customer-related ratios Ratios that include customer satisfaction, customer dissatisfaction, and comparisons of customer satisfaction relative to competitors.

customer-relationship management A view of the customer that asserts that the customer is a valued asset that should be managed.

customer resource management systems (CRM) Computerized systems for managing customer-related information.

customer retention The percentage of customers who return to a service provider or continue to purchase a manufactured product.

customer service surveys Instruments that consist of a series of items (or questions) that are designed to elicit customer perceptions.

dashboards Tools for easily tracking metrics.

deduction An approach to theory development based on modeling.

defects per million opportunities (DPMO) Six-Sigma measure of the goodness of a product.

defects per unit (DFU) An overall average of number of defects occurring in a particular product.

Deming prize A Japanese quality award for individuals and groups that have contributed to the field of quality control.

design control A set of steps focused on managing the design of a product.

design for disassembly A method for developing products so that they can easily be taken apart.

design for maintainability A concept that states that products should be designed in a way that makes them easy for consumers to maintain.

design for manufacture (DFM) The principle of designing products so that they are cost effective and easy to make.

design for remanufacture A method for developing products so that the parts can be used in other products. Associated with green manufacturing.

design for reuse Designing products so they can be used in later generations of products.

design for Six Sigma A process for designing products that results in robust designs.

design of experiments (DOE) An approach to product design that involves identifying and testing alternative inputs to the production of a product to identify the best mix of inputs.

design phase Involves Six-Sigma project identification and selection.

design review The process of checking designs for accuracy.

development plan A plan that identifies the skills that will be required for a particular employee to move up in an organization.

distance learning Training that is conducted in one location and is observed in a distant location through telecommunications technology.

DMADV process Design for Six-Sigma process with the define, measure, analyze, design, and verify stages.

DMAIC process Six-Sigma improvement process with define, measure, analyze, improve, and control.

downstream Processes that are closer to the end customer.

dual sourcing Using only a few suppliers for a single component.

durability A dimension of quality that refers to a product's ability to withstand stress or trauma.

electronic data interchange (EDI) Using computers to share data between customers and suppliers.

empathy A dimension of service quality that refers to the amount of caring and individualized attention exhibited by the service firm.

empowerment A management initiative designed to move decision making to the lowest level in the organization.

end user The ultimate user of a product or service.

engineering analysis The process of applying engineering concepts to the design of a product, including tests such as heat transfer analysis, stress analysis, or analysis of the dynamic behavior of the system being designed.

enterprise capabilities Capabilities that make firms unique and attractive to customers.

enterprise resource planning (ERP) system A system that integrates financial, planning, and control systems into a single architecture. Examples include the SAP R/3 system and Oracle.

ethical attributes Attributes having to do with the honesty and goodness of people in a firm.

evaluation Assessment of how relevant resources and capabilities are to generic strategies in generic internal assessment.

experiential training techniques Training that is hands-on and provides the recipients of training the opportunity to experience in some manner the concepts that are being taught.

exporter A firm that sells its product in another country.

extended value stream mapping Flowcharts used to map an entire supply chain.

external customers The ultimate consumers of the goods that an organization produces.

external events A term used in fault-tree analysis. An external event is an event that is normally expected to occur and thus is not considered a fault when it occurs by itself.

external failure costs These are monetary losses associated with product failures after the customer has possession of the product. These may include warranty or field repair costs.

external services Services that are provided by companies other than yours.

external validation Using benchmarking as a way to ensure that a firm's current practices are comparable to those being used by benchmark firms.

facilitation Helping a team or individual achieve a goal. Often used in meetings or with teams to help the teams achieve their objectives.

facilitator The person who performs facilitation. This person may be trained in group dynamics, teamwork, and meeting management methods.

failure costs Two sets of costs—internal failure costs and external failure costs. Internal failure costs include those costs that are associated with failure during production, whereas external failure costs are associated with product failure after the production process.

failure mode, effect, and criticality analysis (FMECA) FMECA is an extensive but simple method for identifying ways in which an engineered system could fail. The primary goal of FMECA is to develop priorities for corrective action based on estimated risk.

failure modes and effects analysis (FMEA) Method for systematically considering each component of a system by identifying, analyzing, and documenting the possible failure modes within a system and the effects of each failure on the system.

fault-tree analysis An analytical tool that graphically renders the combination of faults that lead to the failure of a system.

features A dimension of quality that refers to those attributes of a product that supplement the item's basic performance.

final product definition The process of articulating the final drawings and specifications for a product.

financial benchmarking A type of benchmarking that typically involves using CD-ROM databases such as Lexis/Nexis or Compact Disclosure to gather information about competing firms to perform financial analyses and compare results.

financial ratios Numerical ratios of firm performance such as return on equity, return on assets, and earnings per share.

five S's A process for inducing discipline in an organization.

5w2h Who, what, when, where, why, how, and how much.

flowcharts A pictorial representation of the progression of a particular process over time.

focus group A group of people who are brought together and are asked to share their opinions about a particular product or service.

forming The first stage of team development, where the team is formed and the objectives for the team are set.

force-field analysis A tool for evaluating forces for or against change.

full-Baldrige approach Term used to depict states' quality award programs using the same criteria as the Malcolm Baldrige National Quality Award.

functional benchmarking A type of benchmarking that involves the sharing of information among firms that are interested in the same functional issues.

gap The difference between desired levels of performance and actual levels of performance.

gap analysis A term associated with the SERVQUAL survey instrument, gap analysis is a technique designed to assess the gap that can exist between a service that is offered and customer expectations.

gauge R&R A process for determining if measurements from gauges are repeatable and reproducible. Used during the measurement stage of Six-Sigma projects.

geometric modeling A technique used to develop a computer-based mathematical description of a part.

globalization An approach to international markets that requires a firm to make fundamental changes in the nature of its business by establishing production and marketing facilities in foreign markets.

green belt Someone who has completed green-belt Six-Sigma training.

green manufacturing A method for manufacturing that minimizes waste and pollution. These goals are often achieved through product and process design.

group decision support system A computer system that allows users to anonymously input comments in a focus group type of setting.

group technology A component of CAD that allows for the cataloging and standardization of parts and components for complex products.

hard costs Costs that actually reduce company expenditures. Usually associated with cost-reduction efforts.

hard data Measurements data such as height, weight, volume, or speed that can be measured on a continuous scale.

hardware mock-ups Physical representations of hardware that show designers, managers, and users how an eventual system will work.

heterogeneous A characteristic of services that means that for many companies, no two services are exactly the same. For example, an advertising company would not develop the same advertising campaign for two different clients.

hidden factory A term introduced by Wickham Skinner that refers to firm activities that have no effect on the customer.

histogram A representation of data in a bar chart format.

horizontal deployment A term that denotes that all of the departments of a firm are involved in the firm's quality efforts.

Hoshin planning process A policy deployment approach to strategic planning originated by Japanese firms.

house of quality Another name for quality function deployment.

human resource measures Ratios that are used to measure the effectiveness of a firm's human resource practices.

ideal quality A reference point identified by Taguchi for determining the quality level of a product or service.

improve phase Six-Sigma phase where improvements to products and processes are implemented.

inbound logistics Associated with the movement of purchased products.

individual needs assessment A method for determining training needs at the worker level prior to developing and implementing training programs. Often associated with company literacy programs.

induction An approach to theory development based on observation and description. Although the process of induction is useful, it is subject to observer bias and misperception.

initiator firm The firm that is interested in benchmarking and initiates contact with benchmark firms.

in-process inspection The practice of inspecting work, by the workers themselves, at each stage of the production process.

intangible A characteristic of services that means that services (unlike manufactured goods) cannot be inventoried or carried in stock over a long period of time.

interference checking A feasibility test for product designs to make sure that wires, cabling, and tubing in products such as airplanes don't conflict with each other.

internal assessment The act of searching for strengths and areas for improvement in quality deployment.

internal customers Individuals within the organization who receive the work that other individuals within the same organization do.

internal failure costs Losses that occur while the product is in possession of the producer. These include rework and scrap costs.

internal services Services that are provided by internal company personnel. For example, data processing personnel are often considered providers of internal services.

internal validation Method of studying the quality system to find gaps in quality deployment.

international sourcing Purchasing from foreign suppliers.

interrelationship digraph A tool designed to help identify the causal relationships between the issues affecting a particular problem.

investigation Ability to find sources of competitive advantage in generic internal assessment.

involuntary services A classification for services that are not sought by customers. These include hospitals, prisons, and the Internal Revenue Service.

ISO 9000:2000 The updated registration standard from the International Organization for Standardization.

ISO/TS 16949 A standard for evaluating and improving automotive suppliers.

job analysis The process of collecting detailed information about a particular job. This information includes tasks, skills, abilities, and knowledge requirements that relate to certain jobs.

just-in-time (JIT) (1) A method for optimizing processes that involves continual reduction of waste; (2) The Toyota Motor Company production system; (3) An umbrella term that encompasses several Japanese management techniques.

just-in-time (JIT) purchasing An approach to purchasing that requires long-term agreements with few suppliers.

key business factors (KBF) Those measures or indicators that are significantly related to the business success of a particular firm.

knowledge-growth systems A compensation system that increases an employee's pay as he or she establishes competencies at different levels relating to job knowledge in a single job classification.

knowledge management The process of maintaining and using company information.

knowledge work Jobs that consist primarily of working with information.

law of diminishing marginal returns A law that stipulates that there is a point at which investment in quality improvement will become uneconomical.

leader behavior A view of leadership stating that leadership potential is related to the behaviors an individual exhibits.

leader skills A view of leadership stating that leadership potential is related to the skills possessed by an individual.

leadership The process by which a leader influences a group to move toward the attainment of a group of superordinate goals.

leading The power relationship between two or more individuals where the power is distributed unevenly.

lean production An approach for reducing waste in production processes using just-in-time concepts.

learning curve effect A theoretical concept that suggests that the more you do something, the better you become at doing it.

licensing A method of reaching international markets that does not require the establishment of international supply chains or marketing arms.

life testing A facet of reliability engineering that concerns itself with determining whether a product will fail under controlled conditions during a specified life.

line-stop authority The approval authority to stop a production line whenever a problem is detected.

loss to society According to Taguchi, this occurs every time a dimension in a product varies from its target dimension. This is associated with Taguchi's "ideal quality."

lot tolerance percent defective (LTPD) The maximum level of percent defective acceptable in production lots.

Malcolm Baldrige National Quality Award (MBNQA) A U.S. national quality award sponsored by the U.S. Department of Commerce and private industry. The award is named after former Secretary of Commerce Malcolm Baldrige.

malpractice The result of mistakes made by a professional service provider.

management by fact A core value of the Baldrige award that focuses on data-based decision making.

manufacturing-based Dimensions of quality that are production related.

manufacturing system design The process of designing a manufacturing system.

market share data A comparative measure that determines relative positions of firms in the marketplace.

master black belt An experienced black belt who is used as an organizational resource, trainer, and specialist.

matrix diagram A brainstorming tool that can be used in a group to show the relationships between ideas or issues.

mean time between failures (MTBF) The overall average time between product breakdowns. Usually expressed in terms of operating hours.

mean time to repair (MTTR) The average time it takes for a product to be repaired.

measurement system analysis (MSA) A set of tools for determining the accuracy of measurements.

measure phase Six-Sigma phase for collecting data.

meeting management A term that refers to the effective management of meeting in an organization.

median chart Control chart for monitoring variation in process central tendency when it is difficult to compute averages.

moment of truth In a service context, the phrase "moment of truth" refers to the point in a service experience at which the customer expects something to happen.

mourning The final stage of the team life cycle, where team members regret the ending of the project and the breaking up of the team.

***MR* chart** A chart for plotting variables when samples are not possible.

multilevel approach Term used to depict state quality award programs that include two levels: a top level based on the full-Baldrige criteria and a second level based on the Baldrige-lite approach.

multiple-skills systems A method for developing employees so that they can perform more than a single task.

multiuser CAD systems Computer aided design systems that are networked so that multiple designers can work on a single design simultaneously.

natural work groups A term used to describe teams that are organized according to a common product, customer, or service.

new seven (N7) tools Managerial tools that are used in quality improvement.

norming The third stage of team development, where the team becomes a cohesive unit, and interdependence, trust, and cooperation are built.

nonrandom variation Controllable variation.

***np* chart** A chart used to monitor the number of items defective for a fixed sample size.

off-line experimentation A method for determining the best configurations of processes. Usually uses a design of experiments (DOE) format such as the Taguchi method or Plackett-Burman experiments.

on-the-job training Training that an employee receives at work during the normal workday.

operating characteristic (OC) curve An assessment of the probability of accepting a shipment, given the existing level of quality of the shipment.

operating results Measures that are important to monitoring and tracking the effectiveness of a company's operations.

operational auditing Modern auditing practices that focus on operational efficiencies.

ordinal data Ranked information.

organizational design The process of defining the best structure to meet company objectives.

organizational learning The sum of the changes in knowledge among the employees of a firm.

orthogonal arrays Experimental design tools that ensure independence between iterations of an experiment.

outbound logistics Associated with the movement of products to the customer.

over-the-wall syndrome Difficulties that arise when different types of engineers work in totally different departments in the same firm.

p chart A chart used to monitor proportion defective.
paper prototypes A series of drawings that are developed by the designer on CAD systems and are reviewed by decision makers prior to acceptance.
parallel processing in focused teams Performing work simultaneously rather than sequentially.
parameter design Designing control factors such as product specifications and measurements for optimal product function.
Pareto analysis An economic concept identified by Joseph Juran that argues that the majority of quality problems are caused by relatively few causes. This economic concept is called Pareto's law or the 80/20 rule. Juran dichotomized the population of causes of quality problems as the vital few and the trivial many.
Pareto chart Chart used to identify and prioritize problems to be solved.
Pareto's law (the 80/20 rule) 80 percent of the problems are a result of 20 percent of the causes.
parking lot A term used in meetings that refers to a flipchart or whiteboard where topics that are off-the-subject are "parked" with the agreement that these topics will be candidates for the agenda in a future meeting.
partnering An approach to selling in foreign markets that involves the collaborative effort of two organizations.
passive data gathering This occurs when the customer initiates the data gathering for a firm such as filling out a customer complaint card or sending an e-mail. The firm provides the mechanism for feedback, the customer must initiate the use of the mechanism.
passively solicited customer feedback A method of soliciting customer feedback that is left to the customer to initiate, such as filling out a restaurant complaint card or calling a toll-free complaint line.
payback period The amount of time required to recoup an investment. Usually associated with projects where costs are saved.
pay-for-learning programs Programs that involve compensating employees for their knowledge and skills rather than singularly for the specific jobs they perform.
perceived quality A dimension of quality identified by David Garvin that refers to a subjective assessment of a product's quality based on criteria defined by the observer.
performance A dimension of quality that refers to the efficiency in which a product performs its intended purpose.
performance attributes Attributes having to do with the functioning of a product such as horsepower, signal-to-noise ratio, or decibel output.
performance benchmarking A type of benchmarking that allows initiator firms to compare themselves against benchmark firms on performance issues such as cost structures, various types of productivity performance, speed of concept to market, and quality measures.
performing The fourth stage of team development, where a mutually supportive, steady state is achieved.
physical environment The geographic area that is in the proximity of an organization.
plan-do-check-act (PDCA) cycle A process for improvement pioneered by W. E. Deming.
presidential audits Annual audits where the president leads the quality audit.
prevention costs Costs associated with preventing defects and imperfections from occurring.
preventive maintenance Maintaining scheduled upkeep and improvement to equipment so equipment can actually improve with age.
prioritization grid A tool used to make decisions based on multiple criteria.
process benchmarking A type of benchmarking that focuses on the observation of business processes including process flows, operating systems, process technologies, and the operation of target firms or departments.
process charts Tools for monitoring process stability.
process decision program chart A tool that is used to help brainstorm possible contingencies or problems associated with the implementation of some program or improvement.
process improvement teams Teams that are involved in identifying opportunities for improving select processes in a firm.
producer's risk The risk associated with rejecting a lot of material that has acceptable quality.
product A tangible good that is produced for a customer.
product-based The context of Garvin's quality dimensions.
product benchmarking A type of benchmarking that firms employ when designing new products or upgrades to current products.
product data management A method for gathering and evaluating product-related data.
product design and evaluation Activities that include the definition of the product architecture and the design, production, and testing of a system (including its subassemblies) for production.
product design engineering A form of engineering that involves activities associated with concept development, needs specification, final specification, and final design of a product.
product idea generation The process of generating product ideas from external and internal sources.

product liability The risk a manufacturer assumes when there is a chance that a consumer could be injured by the manufacturer's product.

product marketing and distribution preparation The process of developing the marketing-related activities associated with a product or service.

product manufacture, delivery, and use Stages of the supply chain.

product traceability The ability to trace a component part of a product back to its original manufacturer.

productivity ratios Ratios that are used in measuring the extent to which a firm effectively uses its resources.

profound organizational learning Quality-based learning that occurs as people discover the causes of errors, defects, and poor customer service in a firm.

project charter A document showing the purposes, participants, goals, and authorizations for a project.

project risk assessment A method for determining the propensity for a Six-Sigma project to achieve desired results.

prototyping An iterative approach to design in which a series of mock-ups or models are developed until the customer and the designer come to agreement as to the final design.

Pugh matrix A method of concept selection used to identify conflicting requirements and to prioritize design tradeoff.

QS 9000 A supplier development program developed by a Daimler Chrysler/Ford/General Motors supplier requirement task force. The purpose of QS 9000 is to provide a common standard and a set of procedures for the suppliers of the three companies.

quality assurance Those activities associated with assuring the quality of a product or service.

quality at the source A method of process control whereby each worker is responsible for his or her own work and performs needed inspections at each stage of the process.

quality circles Brainstorming sessions involving employees of a firm whose goal is improving processes and process capability.

quality control The process relating to gathering process data and analyzing the data to determine whether the process exhibits nonrandom variation.

quality dimensions Aspects of quality that help to better define what quality is. These include perceived quality, conformance, reliability, durability, and so on.

quality function deployment (QFD) QFD involves developing a matrix that includes customer preferences and product attributes. A QFD matrix allows a firm to quantitatively analyze the relationship between customer needs and design attributes.

quality improvement system The result of the interactions between the various components that defines the quality policy in a firm.

quality loss function (QLF) A function that determines economic penalties that the customer incurs as a result of purchasing a nonconforming product.

quality management The management processes that overarch and tie together quality control and quality assurance activities.

quality maturity analysis (QMA) A study in which a firm's level of maturity relating to quality practices is assessed.

quality measures Ratios that are used to measure a firm's performance in the area of quality management.

***R* chart** A variables chart that monitors the dispersion of a process.

random variation Variation that is uncontrollable.

reactive customer-driven quality (RCDQ) A state that is characterized by a supplier "reacting" to the quality expectations of a customer rather than proactively anticipating customer needs and expectations.

readiness Used in a leadership context, the term refers to the extent to which a follower has the ability and willingness to accomplish a specific task.

ready–fire–aim A method that focuses on getting new technology to market and then determining how to sell the products.

recall procedures Steps for taking defective products from market. For example, Tylenol and Firestone Wilderness AT tires used these procedures to recall their products.

redundancy A technique for avoiding failure by putting backup systems in place that can take over if a primary system fails. For example, many redundant systems are used on the space shuttle to protect the crew if a primary system fails.

reengineering (1) A method for making rapid, radical changes to a company's organization and processes; (2) Taking apart a competitor's products to see how they are designed and then designing similar products.

reinventing government Clinton administration effort to reduce government waste.

relationship management A method for developing long-term associations with customers.

reliability Propensity for failure of a product or component.

replications Number of runs of an experiment.

responsiveness A dimension of service quality that refers to the willingness of the service provider to be helpful and prompt in providing service.

reverse engineering The process of dismantling a competitor's products to understand the strengths and weaknesses of the designs.

robust design Designing such that an increase in variability will not result in defective products.

RUMBA Realistic, understandable, measurable, believable, actionable.

sample A part representing a whole.

sampling plan A determination of how data are to be gathered and evaluated.

scatter diagram A scatter plot used to examine the relationships between variables.

***s* chart** Standard deviation chart for monitoring changes in process variation.

seiketsu A term that refers to standardization.

selri A term that refers to organizing or throwing away things you don't use.

seiso A term that suggests that a highly productive workplace should be clean.

seiton A term that refers to neatness in the workplace.

selection The process of evaluating and choosing the best qualified candidate for a particular job.

self-directed work teams Work teams that have a considerable degree of autonomy.

self-direction A term that refers to providing autonomy to employees (or other recipients of training) in terms of facilitating their own training needs.

sensory attributes Attributes having to do with our physical senses such as fragrance, taste, or feel.

sequential or departmental approach to design An approach to design that requires product designers, marketers, process designers, and production managers to work through organizational lines of authority to perform work.

service A mix of intangibles and tangibles that are delivered to the customer.

services blueprinting A chart that depicts service processes and potential fail points in a process.

service reliability A dimension of service quality that refers to the ability of the service provider to perform the promised service dependably and accurately.

service transaction analysis (STA) A process for understanding how a firm interacts with customers. For customer service improvement.

serviceability A dimension of quality that refers to a product's ease of repair.

SERVQUAL A survey instrument designed to assess service quality along five specific dimensions consisting of tangibles, reliability, responsiveness, assurance, and empathy.

shitsuke A term that refers to the discipline required to maintain the changes that have been made in a workplace.

signal factors Factors in a Taguchi experiment that are not under control of the operator. Examples include small variations in ambient temperature and variability in material dimensions.

single sourcing Using only one supplier for a single component.

situational leadership model A model of leadership proposed by Hersey and Blanchard that clarifies the interrelation between employee preparedness and effectiveness in leadership.

Six Sigma An approach to process and product design improvement that emphasizes rapid results and payoffs.

societal environment The portion of a firm's environment pertaining to cultural factors such as language, business customs, customer preferences, and patterns of communication.

soft costs Savings that result in increased organizational slack but not actual cost savings.

soft data Data that cannot be measured or specifically quantified, such as survey data that asks respondents to provide their "opinion" about something.

sole-source filters External validation measures of quality programs such as the Baldrige criteria and ISO 9000.

spider charts Charts used for tracking *n* metrics in a two-dimensional space.

stability The likelihood a process will be random.

statistical process control (SPC) A technique that is concerned with monitoring process capability and process stability.

statistical thinking Deming's concept relating to data-based decision making.

storming The second stage of team development, in which the team begins to get to know each other but agreements have not been made to facilitate smooth interaction among team members.

strategic benchmarking A type of benchmarking that involves observing how others compete. This type of benchmarking typically involves target firms that have been identified as "world class."

strategic partnership An association between two firms by which they agree to work together to achieve a strategic goal. This is often associated with long-term supplier-customer relationships.

strategy (1) The art of planning military operations; (2) What a firm does; (3) A firm's long-term plan for attaining objectives.

stretch target A challenging goal or objective requiring significant effort to achieve.

structural attributes Attributes having to do with physical characteristics of a product such as power steering or red paint.

structural measures Measures that include objectives, policies, and procedures that are followed by a firm.

superordinate goals Goals that transcend individual needs to reflect group objectives.

supplier audit The auditing portion of supplier development programs.

supplier certification or qualification programs Programs designed to certify suppliers as acceptable for a particular customer.

supplier development The process of improving supplier performance.

supplier development programs Training and development programs provided by firms to their suppliers.

supplier evaluation A tool used by many firms to differentiate and discriminate among suppliers. Supplier evaluations often involve report cards where potential suppliers are rated based on different criteria such as quality, technical capability, or ability to meet schedule demands.

supplier filters Hurdles suppliers must pass in order to be considered by a potential customer.

supplier qualification The process of grading suppliers. Usually associated with single sourcing.

supplier partnering A term used to characterize the relationship between suppliers and customers when a high degree of linkages and interdependencies exist.

supply chain A network of facilities that procures raw materials, transforms them into intermediate subassemblies and final products, and then delivers the products to customers through a distribution system.

surveying Generating a list of strengths and weaknesses in a firm in generic internal assessment.

system reliability The probability that components in a system will perform their intended function over a specified period of time.

systems view A management viewpoint that focuses on the interactions between the various components (i.e., people, policies, machines, processes, and products) that combine to produce a product or service. The systems view focuses management on the system as the cause of quality problems.

Taguchi method An approach to quality management developed by Genichi Taguchi in 1980. The Taguchi method provides: (1) a basis for determining the functional relationship between controllable product or service design factors and the outcomes of a process, (2) a method for adjusting the mean of a process by optimizing controllable variables, and (3) a procedure for examining the relationship between random noise in the process and product or service variability.

tangibles A dimension of service quality that refers to the physical appearance of the service facility, the equipment, the personnel, and the communications material.

target firm The firm that is being studied or benchmarked against.

task environment The portion of a firm's environment pertaining to structural issues, such as the skill levels of employees, remuneration policies, technology, and the nature of government agencies.

task needs assessment The process of assessing the skills that are needed within a firm.

team A group of individuals working to achieve a goal with activities requiring close coordination.

team building A term that describes the process of identifying roles for team members and helping the team members succeed in their roles.

teamware Computer software that is used in making group decisions.

technology feasibility statement A feasibility statement used in the design process to assess a variety of issues such as necessary parameters for performance, manufacturing imperatives, limitations in the physics of materials, and conditions for quality testing the product.

technology selection for product development The process of selecting materials and technologies that provide the best performance for the customer at an acceptable cost.

temporal attributes Attributes relating to time such as on-time performance and meeting delivery schedules.

360-degree evaluation A method for evaluating performance with input from supervisors, peers, and employees.

three spheres of quality Quality management, assurance, and control.

three T's The task, treatment, and tangibles in service design.

tiger teams Teams with a specific defined goal and a short time frame to attain the goal.

tolerance design The act of determining the amount of allowable variability around parameters.

total quality human resources management (TQHRM) An approach to human resources management that involves many of the concepts of quality management. The primary purpose of this approach is to provide employees a supportive and empowered work environment.

training needs analysis The process of identifying organizational needs in terms of capabilities, task needs assessment in terms of skill sets that are needed within the firm, and individual needs analysis to determine how employee skills fit with company needs.

training needs assessment A process for gathering organizational data relative to finding areas where training is most needed.

training program design A term that describes the process of tailoring a course or set of courses to meet the needs of a company.

trait dimension A view of leadership that states that leadership potential is related to the "traits" of an individual, such as height.

transcendent A definition of quality that states that quality is something we all recognize but we cannot verbally define.

tree diagram A tool used to identify the steps needed to address a particular problem.

TS 16949 ISO standard for supplier development and management. Replacing QS 9000 as an automotive industry standard.

***U* chart** A chart used to monitor the number of defects in sequential production lots.

undeveloped events A term used in fault-tree analysis. Undeveloped events are faults that do not have a significant consequence or are not expanded because there is not sufficient information available.

unified theory for services management A set of propositions relative to managing services.

upstream Processes that are closer to raw materials.

user-based A definition of service or product quality that is customer centered.

value-added A customer-based perspective on quality that is used by services, manufacturing, and public sector organizations. The concept of value-added involves a subjective assessment of the efficacy of every step in the process for the customer.

value-based A definition of quality relating to the social benefit from a product or service.

value chain A tool, developed by Michael Porter, that decomposes a firm into its core activities.

value chain activities Porter's chain of activities, including inbound logistics, production, and outbound logistics.

value stream mapping Flowcharts used in process improvement.

value system A network of value chains.

variable A measurement.

variety The range of product and service choices offered to customers.

vertical deployment A term denoting that all of the levels of the management of a firm are involved in the firm's quality efforts.

virtual teams Teams that do not physically meet but are linked together through intranets and the Internet.

voice of the customer A term that refers to the wants, opinions, perceptions, and desires of the customer.

whack-a-mole A novel term that describes the process of solving a problem only to have another problem surface.

working prototype A functioning mock-up or model of a product.

***X* chart** A chart used to monitor the mean of a process for population values.

$\bar{x}$ chart A chart that monitors the mean of a process for variables.

***XY* matrix** A matrix used to demonstrate and quantify the strength of relationships between dependent and independent variables.

$Y = f(x)$ A formula used in Six Sigma to illustrate the relationships between dependent and independent variables.

yellow belt One who has completed yellow-belt Six-Sigma training.

Index

D

N

O

P

Q

R

S

T

U

V

W

X

Y

Z

LICENSE AGREEMENT AND LIMITED WARRANTY

READ THIS LICENSE CAREFULLY BEFORE USING THIS PACKAGE. BY USING THIS PACKAGE, YOU ARE AGREEING TO THE TERMS AND CONDITIONS OF THIS LICENSE. IF YOU DO NOT AGREE, DO NOT USE THE PACKAGE. PROMPTLY RETURN THE UNUSED PACKAGE AND ALL ACCOMPANYING ITEMS TO THE PLACE YOU OBTAINED IT. ***THESE TERMS APPLY TO ALL LICENSED SOFTWARE ON THE DISK EXCEPT THAT THE TERMS FOR USE OF ANY SHAREWARE OR FREEWARE ON THE DISKETTES ARE AS SET FORTH IN THE ELECTRONIC LICENSE LOCATED ON THE DISK***:

1. GRANT OF LICENSE and OWNERSHIP: The enclosed computer programs and data ("Software") are licensed, not sold, to you by Prentice-Hall, Inc. ("We" or the "Company") in consideration of your purchase or adoption of the accompanying Company textbooks and/or other materials, and your agreement to these terms. We reserve any rights not granted to you. You own only the disk(s) but we and/or licensors own the Software itself. This license allows individuals who have purchased the accompanying Company textbook to use and display their copy of the Software on a single computer (i.e., with a single CPU) at a single location for academic use only, so long as you comply with the terms of this Agreement. You may make one copy for back up, or transfer your copy to another CPU, provided that the Software is usable on only one computer. This license allows instructors at educational institutions [only] who have adopted the accompanying Company textbook to install, use and display the enclosed copy of the Software on individual computers in the computer lab designated for use by any students of a course requiring the accompanying Company textbook and only for as long as such textbook is a required text for such course, at a single campus or branch or geographic location of an educational institution, for academic use only, so long as you comply with the terms of this Agreement.

2. RESTRICTIONS: You may *not* transfer or distribute the Software or documentation to anyone else. You may *not* copy the documentation or the Software. You may *not* reverse engineer, disassemble, decompile, modify, adapt, translate, or create derivative works based on the Software or the Documentation. You may be held legally responsible for any copying or copyright infringement which is caused by your failure to abide by the terms of these restrictions.

3. TERMINATION: This license is effective until terminated. This license will terminate automatically without notice from the Company if you fail to comply with any provisions or limitations of this license. Upon termination, you shall destroy the Documentation and all copies of the Software. All provisions of this Agreement as to limitation and disclaimer of warranties, limitation of liability, remedies or damages, and our ownership rights shall survive termination.

4. LIMITED WARRANTY AND DISCLAIMER OF WARRANTY: Company warrants that for a period of 60 days from the date you purchase this Software (or purchase or adopt the accompanying textbook), the Software, when properly installed and used in accordance with the Documentation, will operate in substantial conformity with the description of the Software set forth in the Documentation, and that for a period of 30 days the disk(s) on which the Software is delivered shall be free from defects in materials and workmanship under normal use. The Company does *not* warrant that the Software will meet your requirements or that the operation of the Software will be uninterrupted or error-free. Your only remedy and the Company's only obligation under these limited warranties is, at the Company's option, return of the disk for a refund of any amounts paid for it by you or replacement of the disk. THIS LIMITED WARRANTY IS THE ONLY WARRANTY PROVIDED BY THE COMPANY AND ITS LICENSORS, AND THE COMPANY AND ITS LICENSORS DISCLAIM ALL OTHER WARRANTIES, EXPRESS OR IMPLIED, INCLUDING WITHOUT LIMITATION, THE IMPLIED WARRANTIES OF MERCHANTABILITY AND FITNESS FOR A PARTICULAR PURPOSE. THE COMPANY DOES NOT WARRANT, GUARANTEE OR MAKE ANY REPRESENTATION REGARDING THE ACCURACY, RELIABILITY, CURRENTNESS, USE, OR RESULTS OF USE, OF THE SOFTWARE.

5. LIMITATION OF REMEDIES AND DAMAGES: IN NO EVENT, SHALL THE COMPANY OR ITS EMPLOYEES, AGENTS, LICENSORS, OR CONTRACTORS BE LIABLE FOR ANY INCIDENTAL, INDIRECT, SPECIAL, OR CONSEQUENTIAL DAMAGES ARISING OUT OF OR IN CONNECTION WITH THIS LICENSE OR THE SOFTWARE, INCLUDING FOR LOSS OF USE, LOSS OF DATA, LOSS OF INCOME OR PROFIT, OR OTHER LOSSES, SUSTAINED AS A RESULT OF INJURY TO ANY PERSON, OR LOSS OF OR DAMAGE TO PROPERTY, OR CLAIMS OF THIRD PARTIES, EVEN IF THE COMPANY OR AN AUTHORIZED REPRESENTATIVE OF THE COMPANY HAS BEEN ADVISED OF THE POSSIBILITY OF SUCH DAMAGES. IN NO EVENT SHALL THE LIABILITY OF THE COMPANY FOR DAMAGES WITH RESPECT TO THE SOFTWARE EXCEED THE AMOUNTS ACTUALLY PAID BY YOU, IF ANY, FOR THE SOFTWARE OR THE ACCOMPANYING TEXTBOOK. SOME JURISDICTIONS DO NOT ALLOW THE LIMITATION OF LIABILITY IN CERTAIN CIRCUMSTANCES, THE ABOVE LIMITATIONS MAY NOT ALWAYS APPLY.

6. GENERAL: THIS AGREEMENT SHALL BE CONSTRUED IN ACCORDANCE WITH THE LAWS OF THE UNITED STATES OF AMERICA AND THE STATE OF NEW YORK, APPLICABLE TO CONTRACTS MADE IN NEW YORK, AND SHALL BENEFIT THE COMPANY, ITS AFFILIATES AND ASSIGNEES. This Agreement is the complete and exclusive statement of the agreement between you and the Company and supersedes all proposals, prior agreements, oral or written, and any other communications between you and the company or any of its representatives relating to the subject matter. If you are a U.S. Government user, this Software is licensed with "restricted rights" as set forth in subparagraphs (a)–(d) of the Commercial Computer-Restricted Rights clause at FAR 52.227-19 or in subparagraphs (c)(1)(ii) of the Rights in Technical Data and Computer Software clause at DFARS 252.227-7013, and similar clauses, as applicable.

Should you have any questions concerning this agreement or if you wish to contact the Company for any reason, please contact in writing:

Director, Media Production
Pearson Education
1 Lake Street
Upper Saddle River, NJ 07458